# Lecture Notes in Computer Science  11949

More information about this series at http://www.springer.com/series/7407

Yingshu Li · Mihaela Cardei ·
Yan Huang (Eds.)

# Combinatorial Optimization and Applications

13th International Conference, COCOA 2019
Xiamen, China, December 13–15, 2019
Proceedings

Springer

*Editors*
Yingshu Li
Georgia State University
Atlanta, GA, USA

Mihaela Cardei
Florida Atlantic University
Boca Raton, FL, USA

Yan Huang
Kennesaw State University
Marietta, GA, USA

ISSN 0302-9743     ISSN 1611-3349 (electronic)
Lecture Notes in Computer Science
ISBN 978-3-030-36411-3     ISBN 978-3-030-36412-0 (eBook)
https://doi.org/10.1007/978-3-030-36412-0

LNCS Sublibrary: SL1 – Theoretical Computer Science and General Issues

This Springer imprint is published by the registered company Springer Nature Switzerland AG
The registered company address is: Gewerbestrasse 11, 6330 Cham, Switzerland

# Preface

This volume contains the papers presented at COCOA 2019: the 13th Annual International Conference on Combinatorial Optimization and Applications held during December 13–15, 2019, in Xiamen, China. The conference was motivated by the recent advances in the areas of combinatorial optimization and its applications. COCOA was designed to be a forum for the researchers working in the area of theoretical computer science, combinatorics, and corresponding applications to discuss and express their views on current trends, challenges, and state-of-the-art solutions related to various theoretical issues as well as real-world problems.

The technical program of the conference included 49 contributed papers selected by the Program Committee from a number of 108 full submissions received in response to the call for papers. All the papers were peer-reviewed by at least three Program Committee members or external reviewers. The topics cover most aspects of theoretical computer science and combinatorics related to computing, including combinatorial optimization, geometric optimization, complexity and data structures, graph theory, etc.

We would like to thank the Program Committee members and external reviewers for volunteering their time and effort to review and discuss the conference papers. We would like to extend a special thanks to the steering and general chairs of the conference for their leadership, and to the finance, publication, publicity, and local organization chairs for their hard work in making COCOA 2019 a successful event. Last but not least, we would like to thank all the authors for contributing and presenting their work at the conference.

October 2019

Yingshu Li
Mihaela Cardei
Yan Huang

# Organization

## Program Committee

| | |
|---|---|
| Ran Bi | Dalian University of Technology, China |
| Zhipeng Cai | Georgia State University, USA |
| Mihaela Cardei | Florida Atlantic University, USA |
| Vincent Chau | Shenzhen Institute of Advanced Technology, China |
| Yong Chen | Hangzhou Dianzi University, China |
| Ovidiu Daescu | The University of Texas at Dallas, USA |
| Thomas Erlebach | University of Leicester, UK |
| Neng Fan | University of Arizona, USA |
| Meng Han | Kennesaw State University, USA |
| Yan Huang | Kennesaw State University, USA |
| Michael Khachay | Krasovsky Institute of Mathematics and Mechanics, Russia |
| Donghyun Kim | Kennesaw State University, USA |
| Joong-Lyul Lee | University of North Carolina at Pembroke, USA |
| Xianyue Li | Lanzhou University, China |
| Yingshu Li | Georgia State University, Atlanta, USA |
| Xianmin Liu | Harbin Institute of Technology, USA |
| Hengzhao Ma | Harbin Institute of Technology, USA |
| Dongjing Miao | Harbin Institute of Technology, USA |
| Viet Hung Nguyen | LIP6 - Sorbonne Université, France |
| Erfang Shan | Shanghai University, China |
| Pavel Skums | Georgia State University, USA |
| Xiang Song | Massachusetts Institute of Technology, USA |
| Guangmo Tong | University of Delaware, USA |
| Weitian Tong | Eastern Michigan University, USA |
| Jinbao Wang | Harbin Institute of Technology, China |
| Wei Wang | Xi'an Jiaotong University, China |
| Yingjie Wang | Yantai University, China |
| Yishui Wang | Shenzhen Institutes of Advanced Technology, Chinese Academy of Sciences, China |
| Yicheng Xu | Shenzhen Institutes of Advanced Technology, Chinese Academy of Sciences, China |
| Boting Yang | University of Regina, Canada |
| Yong Zhang | Shenzhen Institutes of Advanced Technology, Chinese Academy of Sciences, China |
| Zhao Zhang | Zhejiang Normal University, China |
| Xu Zheng | University of Science and Technology of China, China |
| Martin Ziegler | KAIST, South Korea |

# Contents

# Exact Algorithms for the Bounded Repetition Longest Common Subsequence Problem

Yuichi Asahiro[1], Jesper Jansson[2], Guohui Lin[3], Eiji Miyano[4(✉)],
Hirotaka Ono[5], and Tadatoshi Utashima[4]

[1] Kyushu Sangyo University, Fukuoka, Japan
asahiro@is.kyusan-u.ac.jp
[2] The Hong Kong Polytechnic University, Kowloon, Hong Kong
jesper.jansson@polyu.edu.hk
[3] University of Alberta, Edmonton, Canada
guohui@ualberta.ca
[4] Kyushu Institute of Technology, Iizuka, Japan
miyano@ces.kyutech.ac.jp, q236010t@mail.kyutech.jp
[5] Nagoya University, Nagoya, Japan
ono@nagoya-u.jp

**Abstract.** In this paper, we study exact, exponential-time algorithms for a variant of the classic LONGEST COMMON SUBSEQUENCE problem called the $r$-REPETITION LONGEST COMMON SUBSEQUENCE problem (or $r$-RLCS, for short): Given two sequences $X$ and $Y$ over an alphabet $S$, find a longest common subsequence of $X$ and $Y$ such that each symbol appears at most $r$ times in the obtained subsequence. Without loss of generality, we will assume that $|X| \leq |Y|$ from here on. The special case of 1-RLCS, also known as the REPETITION-FREE LONGEST COMMON SUBSEQUENCE problem (RFLCS), has been studied previously; e.g., in [1], Adi *et al.* presented an (exponential-time) integer linear programming-based exact algorithm for 1-RLCS. However, they did not analyze its time complexity, and to the best of our knowledge, there are no previous results on the running times of any exact algorithms for this problem. In this paper, we first propose a simple algorithm for 1-RLCS based on the strategy used in [1] and show explicitly that its running time is bounded by $O(1.44225^{|X|}|X||Y|)$. Next, we provide a DP-based algorithm for $r$-RLCS and prove that its running time is $O((r + 1)^{|X|/(r+1)}|X||Y|)$ for any $r \geq 1$. In particular, our new algorithm runs in $O(1.41422^{|X|}|X||Y|)$ time for 1-RLCS, which is faster than the previous one.

## 1 Introduction

An *alphabet* $S$ is a finite set of *symbols*. Let $X$ be a sequence over the alphabet $S$ and $|X|$ be the length of the sequence $X$. For example, $X = \langle x_1, x_2, \cdots, x_n \rangle$ is a sequence of length $n$, where $x_i \in S$ for $1 \leq i \leq n$, i.e., $|X| = n$. For a sequence $X = \langle x_1, x_2, \cdots, x_n \rangle$, another sequence $Z = \langle z_1, z_2, \cdots, z_c \rangle$ is a *subsequence* of

© Springer Nature Switzerland AG 2019
Y. Li et al. (Eds.): COCOA 2019, LNCS 11949, pp. 1–12, 2019.
https://doi.org/10.1007/978-3-030-36412-0_1

$X$ if there exists a strictly increasing sequence $\langle i_1, i_2, \cdots, i_c \rangle$ of indices of $X$ such that for all $j = 1, 2, \cdots, c$, we have $x_{i_j} = z_j$. Then, we say that a sequence $Z$ is a *common subsequence* of $X$ and $Y$ if $Z$ is a subsequence of both $X$ and $Y$. Given two sequences $X$ and $Y$ as input, the goal of the LONGEST COMMON SUBSEQUENCE problem (LCS) is to find a *longest* common subsequence of $X$ and $Y$.

The problem LCS is clearly a fundamental problem and has a long history [4, 11,14,24]. The comparison of sequences via a longest common subsequence has been applied in several contexts where we want to find the maximum number of symbols that appear in the same order in two sequences. The problem LCS is considered as one of the important computational primitive, and thus has a variety of applications such as bioinformatics [3,18,20], data compression [23], spelling correction [19,24], and file comparison [2].

The problem LCS of two sequences has been deeply investigated, and polynomial time algorithms are well-known [14,15,20,21,24]. It is possible to generalize LCS to a set of three or more sequences; the goal is to compute a longest common subsequence of all input sequences. This LCS of multiple sequences is NP-hard even on binary alphabet [17] and it is not approximable within factor $O(n^{1-\varepsilon})$ on arbitrary alphabet for sequences of length $n$ and any constant $\varepsilon > 0$ [18]. Furthermore, some researchers introduced a constraint on the number of symbol occurrences in the solution. Bonizzoni, Della Vedova, Dondi, Fertin, Rizzi and Vialette considered the EXEMPLAR LONGEST COMMON SUBSEQUENCE problem (ELCS) [9,22]. In ELCS, an alphabet $S$ of symbols is divided into the mandatory alphabet $S_m$ and the optional alphabet $S_o$, and ELCS restricts the number of symbol occurrences in $S_m$ and $S_o$ in the obtained solution. The problem ELCS is APX-hard even for instances of two sequences [9]. In [10], Bonizzoni, Della Vedova, Dondi and Pirola proposed the following DOUBLY-CONSTRAINED LONGEST COMMON SUBSEQUENCE problem (DCLCS): Let a sequence constraint $C$ be a set of sequences over an alphabet $S$ and let an occurrence constraint $C_{occ}$ be a function $C_{occ} : S \to \mathbb{N}$, assigning an upper bound on the number of occurrences of each symbol in $S$. Given two sequences $X$ and $Y$ over an alphabet $S$, a sequence constraint $C$ and an occurrence constraint $C_{occ}$, the goal of DCLCS is to find a longest common subsequence $Z$ of $X$ and $Y$, so that each sequence in $C$ is a subsequence of $Z$ and $Z$ contains at most $C_{occ}(s)$ occurrences of each symbol $s \in S$. Bonizzoni *et al.* showed that DCLCS is NP-hard over an alphabet of three symbols [10]. Adi, Braga, Fernandes, Ferreira, Martinez, Sagot, Stefanes, Tjandraatmadja, and Wakabayashi introduced the REPETITION-FREE LONGEST COMMON SUBSEQUENCE problem (RFLCS) [1]: Given two sequences $X$ and $Y$ over an alphabet $S$, the goal of RFLCS is to find a longest common subsequence of $X$ and $Y$, where each symbol appears at most once in the obtained subsequence. In [1], Adi *et al.* proved that RFLCS is APX-hard even if each symbol appears at most twice in each of the given sequences.

In this paper we study exact, exponential-time algorithms for RFLCS and its variant, called the $r$-REPETITION LONGEST COMMON SUBSEQUENCE problem ($r$-RLCS for short): Given two sequences $X$ and $Y$ over an alphabet $S$, the goal of $r$-RLCS is to find a longest common subsequence of $X$ and $Y$, where each symbol appears at most $r$ times in the obtained subsequence. Without loss of generality, we will assume that $|X| \leq |Y|$ from here on. The special case 1-RLCS is identical to RFLCS. Also, it is easy to see that $r$-RLCS is a special case of DCLCS when a sequence constraint $C = \emptyset$ and an occurrence constraint $C_{occ}(s) = r$ for every $s \in S$. In [1], Adi $et$ $al.$ presented an (exponential-time) integer linear programming-based exact algorithm for 1-RLCS. However, they did not analyze its time complexity, and to the best of our knowledge, there are no previous results on the running times of any exact algorithms for this problem. In this paper, we first propose a simple algorithm for 1-RLCS based on the strategy used in [1] and show explicitly that its running time is bounded by $O(1.44225^{|X|}|X||Y|)$. Next, we provide a DP-based algorithm for $r$-RLCS and prove that its running time is $O((r + 1)^{|X|/(r+1)}|X||Y|)$ for any $r \geq 1$. In particular, our new algorithm runs in $O(1.41422^{|X|}|X||Y|)$ time for 1-RLCS, which is faster than the previous one.

**Related Work.** Although this paper focuses on the exact exponential algorithms, we here make a brief survey on previous results for RFLCS, from the viewpoints of heuristic, approximation and parameterized algorithms. In [1], Adi $et$ $al.$ introduced first heuristic algorithms for RFLCS. After that, several meta-heuristic algorithms for RFLCS were proposed in [6,7,12]. A detailed comparison of those metaheuristic algorithms was given in [8]. As for the approximability of RFLCS, Adi $et$ $al.$ showed [1] that RFLCS admits an $occ_{max}$-approximation algorithm, where $occ_{max}$ is the maximum number of occurrences of each symbol in one of the two input sequences. In [5], Blin, Bonizzoni, Dondi, and Sikora presented a randomized fixed-parameter algorithm for RFLCS parameterized by the size of the solution.

## 2   Warm-Up Algorithms

For a while, we focus on RFLCS, i.e., 1-RLCS. One can see that the following brute-force exact algorithm for RFLCS can work clearly in $O(2^n \cdot n \cdot m)$ time for two sequences $X$ and $Y$ where $|X| = n$, $|Y| = m$, and $|X| \leq |Y|$: (i) We first create all the subsequences of $X$, say, $X_1$ through $X_{2^n}$[1]. Then, (ii) we obtain a longest common subsequence of $X_i$ and $Y$ for each $i$ ($1 \leq i \leq 2^n$) by using an $O(|X_i| \cdot m)$-time algorithm for LCS [20,21,24]. Finally, (iii) we find a repetition-free longest subsequence among those $2^n$ common subsequences obtained in (ii) and output it. The running time of the brute-force algorithm is $O(2^n \cdot n \cdot m)$.

In [1], Adi $et$ $al.$ presented the following quite simple algorithm for RFLCS: Let $S$ be an alphabet of symbols. Suppose that each symbol in $S_X \subseteq S$ appears in $X$ fewer times than $Y$, and $S_X = \{s_1, s_2, \cdots, s_{|S_X|}\}$. Also, let $S_Y = S \setminus S_X$ and $S_Y = \{s_{|S_X|+1}, s_{|S_X|+2}, \cdots, s_{|S|}\}$. (i) The algorithm creates all the subsequences, say, $X_1$ through $X_{N_X}$, from the input sequence $X$ such that all the symbols in

---

[1] Here, $X_i$ denotes the $i$th subsequence in $2^n$ subsequences in any order; on the other hand, in Sect. 3, $X_i$ will be defined to be the $i$th prefix of $X$.

$S_X$ occur *exactly once*, but all the occurrences of symbols in $S_Y$ are kept in $X_i$ for every $1 \leq i \leq N_X$. Also, the algorithm creates all the subsequences, say, $Y_1$ through $Y_{N_Y}$, from the input sequence $Y$ such that all the symbols in $S_Y$ occur *exactly once*, but all the occurrences of symbols in $S_X$ are kept in $Y_j$ for every $1 \leq j \leq N_Y$. Then, (ii) we obtain a longest common subsequence of $X_i$ and $Y_j$ for every pair of $i$ and $j$ ($1 \leq i \leq N_X$ and $1 \leq j \leq N_Y$) by using an $O(|X_i| \cdot |Y_j|)$-time algorithm for the original LCS. Finally, (iii) we find the longest subsequence among $N_X \cdot N_Y$ common subsequences obtained in (ii), which must be repetition-free, and output it. Clearly, the running time of this method is $O(N_X \cdot N_Y \cdot n \cdot m)$. In [1], Adi *et al.* only claimed that if the number of symbols which appear twice or more in $X$ and $Y$ is bounded above by some constant, say, $c$, then the running time is $O(m^c \cdot n \cdot m)$, i.e., RFLCS is solvable in polynomial time. However, no upper bound on $N_X \cdot N_Y$ was given in [1].

## 2.1    Repetition-Free LCS

In this section we consider an algorithm called ALG, based on the same strategy as one in [1] for RFLCS: (i) We first create all the subsequences, say, $X_1$ through $X_N$, from the input sequence $X$ such that every symbol appears *exactly once* in $X_i$ for $1 \leq i \leq N$. Then, (ii) we obtain a longest common subsequence of $X_i$ and $Y$ for each $i$ ($1 \leq i \leq N$). Finally, (iii) we find a repetition-free longest subsequence among $N$ common subsequences obtained in (ii) and output it. Therefore, the running time of ALG is $O(N \cdot n \cdot m)$. It is important to note that ALG is completely the same algorithm as one proposed in [1] if $S_X = S$ and thus $S_Y = \emptyset$.

A quite simple argument gives us the first upper bound on $N$:

**Theorem 1.** *The running time of* ALG *is* $O(1.44668^n \cdot n \cdot m)$ *for* RFLCS *on two sequences* $X$ *and* $Y$ *where* $|X| = n$, $|Y| = m$, *and* $|X| \leq |Y|$.

*Proof.* Suppose that $X$ has $k$ symbols, $s_1, s_2, \cdots, s_k$, and $s_i$ occurs $\mathsf{occ}_i$ times in $X$ for each integer $i$, $1 \leq i \leq k$. Since the number $N$ of subsequences in $X$ created in (i) of ALG is bounded by the number of combinations of $k$ symbols, the following is satisfied:

$$N \leq \prod_{i=1}^{k} \mathsf{occ}_i.$$

From the inequality of arithmetic and geometric means, we have:

$$N \leq \left( (\sum_{i=1}^{k} \mathsf{occ}_i)/k \right)^k \leq (n/k)^k.$$

Here, by setting $p \overset{\text{def}}{=} n/k \in \mathbb{R}^+$, we have:

$$N \leq (p)^{n/p} = (p^{1/p})^n.$$

Note that the value of $p^{1/p}$ becomes the maximum when $p = e$, where $e$ denotes the Euler's number. Therefore, $N$ is bounded above by $e^{n/e} < 1.44668^n$. Therefore, the running time of ALG is $O(1.44668^n \cdot n \cdot m)$.          □

By using more refined estimation, we can show a smaller upper bound on $N$:

**Theorem 2.** *The running time of* ALG *is* $O(1.44225^n \cdot n \cdot m)$ *for RFLCS on two sequences* $X$ *and* $Y$ *where* $|X| = n$, $|Y| = m$, *and* $|X| \le |Y|$.

*Proof.* Again suppose that $X$ has $k$ symbols, $s_1$, $s_2$, $\cdots$, $s_k$, and $s_i$ occurs $occ_i$ times in $X$ for each integer $i$, $1 \le i \le k$. Let $\max_{1 \le i \le k} \{occ_i\} = occ_{max}$. Now let $S_i = \{x_j | occ_j = i\}$ for $1 \le i \le occ_{max}$. That is, $S_i$ is a set of symbols which appear exactly $i$ times in $X$. Let $n_i = i \times |S_i|$. Since each symbol in $S_i$ appears $i$ times in $X$, the following equality holds:

$$\sum_{i=1}^{occ_{max}} n_i = n. \tag{1}$$

In the following, we show a smaller upper bound on $N$. From the fact that $n_i = i \times |S_i|$, one sees that the following equality holds:

$$N \le \prod_{i=1}^{k} occ_i = \prod_{i=1}^{occ_{max}} i^{n_i/i}. \tag{2}$$

Here, from the inequality of arithmetic and geometric means, the following is obtained:

$$\left( \prod_{i=1}^{occ_{max}} \left( i^{1/i} \right)^{n_i} \right)^{1/\sum_{i=1}^{occ_{max}} n_i} \le \frac{\sum_{i=1}^{occ_{max}} \left( i^{1/i} \right) \cdot n_i}{\sum_{i=1}^{occ_{max}} n_i}. \tag{3}$$

From the Eqs. (1), (2), and (3), we get:

$$N \le \left( \frac{\sum_{i=1}^{occ_{max}} \left( i^{1/i} \right) \cdot n_i}{n} \right)^n. \tag{4}$$

Now, it is important to note that $i \in \mathbb{N}$, i.e., $i$ is a positive integer while $p = n/k$ was a positive real in the proof of the previous theorem. Therefore, by a simple calculation, one can verify that the following is true:

$$\max_{i \in \mathbb{N}} \left\{ i^{1/i} \right\} = 3^{1/3}.$$

Hence, we can bound the number $N$ of all the possible repetition-free common subsequences as follows:

$$N \le \left( \frac{\sum_{i=1}^{occ_{max}} 3^{1/3} \cdot n_i}{n} \right)^n = \left( \frac{3^{1/3} \cdot \sum_{i=1}^{occ_{max}} n_i}{n} \right)^n = \left( 3^{1/3} \right)^n < 1.44225^n.$$

As a result, the running time of our algorithm is $O(1.44225^n \cdot n \cdot m)$. This completes the proof.          □

**Corollary 1.** *There is an $O(\text{occ}^{n/\text{occ}} \cdot n \cdot m)$-time algorithm to solve RFLCS for two sequences $X$ and $Y$ where $|X| = n$, $|Y| = m$, and $|X| \leq |Y|$ if all the occurrences of symbols in $X$ are exactly occ.*

*Proof.* By the assumption, $\text{occ} \times |S_{\text{occ}}| = n$ and thus $|S_{\text{occ}'}| = 0$ for $\text{occ} \neq \text{occ}'$. From the inequality (4), one can easily obtain the following:

$$N \leq \left( \text{occ}^{1/\text{occ}} \right)^n.$$

$\square$

For example, if each symbol appears exactly twice (five and six times, resp.) in the shorter sequence $X$, then the running time of ALG is $O\left(1.41422^n \cdot n \cdot m\right)$ ($O\left(1.37973^n \cdot n \cdot m\right)$ and $O\left(1.34801^n \cdot n \cdot m\right)$, resp.).

## 2.2   $r$-Repetition LCS, $r \geq 2$

In this section we consider exact exponential algorithms for $r$-RLCS. First, by a straightforward extension of the algorithm for RFLCS, we can design the following algorithm, say, $\text{ALG}_r$ for $r$-RLCS: First, (i) we create all the subsequences, say, $X_1$ through $X_N$, from the input sequence $X$ such that a symbol, say, $s$, appears *exactly* $r$ times in $X_i$ for $1 \leq i \leq N$ if $X$ has more than $k$ $s$'s; otherwise, all the occurrences of $s$ in $X$ are included in $X_i$. Then, (ii) we obtain a longest common subsequence of $X_i$ and $Y$ for each $i$ $(1 \leq i \leq N)$. Finally, (iii) we find a longest subsequence among $N$ common subsequences obtained in (ii), which has at most $r$ occurrences of every symbol, and output it.

Again, suppose that $X$ has $k$ symbols, $s_1$, $s_2$, $\cdots$, $s_k$, and $s_i$ occurs $\text{occ}_i$ times in $X$ for each integer $i$, $1 \leq i \leq k$, and $\max_{1 \leq i \leq k}\{\text{occ}_i\} = \text{occ}_{max}$. Let $S_i = \{s_j | \text{occ}_j = i\}$ for $1 \leq i \leq \text{occ}_{max}$ and $n_i = i \times |S_i|$. Then, we estimate an upper bound on $N$ for each $r$:

**Theorem 3.** *For $r$-RLCS on two sequences $X$ and $Y$ where $|X| = n$, $|Y| = m$, and $|X| \leq |Y|$, the running time of $\text{ALG}_r$ is as follows:*

$$O\left( \left( \max_{i \in \mathbb{N}}\left\{ \left( \frac{i - \frac{r-1}{2}}{(r!)^{1/r}} \right)^{r/i} \right\} \right)^n \times n \cdot m \right).$$

*Proof.* First, the total number $N$ of sequences created in (i) of $\text{ALG}_r$ can be obtained as follows:

$$N = \prod_{i=1}^{k} \binom{\text{occ}_i}{r} = \prod_{i=r+1}^{\text{occ}_{max}} \binom{i}{r}^{n_i/i}.$$

From the inequality of arithmetic and geometric means, we can obtain the following inequality:

$$(i(i-1)(i-2)\cdots(i-r+1))^{1/r} \leq \frac{(2i - r + 1)r/2}{r} = i - \frac{r-1}{2}.$$

Therefore, $N$ is bounded:

$$\prod_{i=r+1}^{occ_{max}} \binom{i}{r}^{n_i/i} \le \prod_{i=r+1}^{occ_{max}} \left( \frac{(i - \frac{r-1}{2})^r}{r!} \right)^{n_i/i}$$

$$= \prod_{i=r+1}^{occ_{max}} \left( \left( \frac{i - \frac{r-1}{2}}{(r!)^{1/r}} \right)^{r/i} \right)^{n_i}$$

$$\le \left( \max_{i \in \mathbb{N}} \left\{ \left( \frac{i - \frac{r-1}{2}}{(r!)^{1/r}} \right)^{r/i} \right\} \right)^n.$$

This completes the proof.                                                      □

We have obtained the specific values of $\max_{i \in \mathbb{N}} \left\{ \left( \frac{i - \frac{r-1}{2}}{(r!)^{1/r}} \right)^{r/i} \right\}$, say, $N(r)$, and $i$ for $r$-RLCS by its empirical implementation. Table 1 shows $N(r)$ and $i$ for each $r = 2, 3, \cdots, 10$.

**Table 1.** $N(r)$ and $i$ for each $r$

| $r$ | 2 | 3 | 4 | 5 | 6 | 7 | 8 | 9 | 10 |
|---|---|---|---|---|---|---|---|---|---|
| $N(r)$ | 1.5884 | 1.66852 | 1.72013 | 1.75684 | 1.78453 | 1.80630 | 1.82394 | 1.83856 | 1.85091 |
| $i$ | 5 | 7 | 9 | 11 | 13 | 15 | 17 | 19 | 21 |

## 3  DP-Based Algorithms

In this section we design a DP-based algorithm, say, $DP_1$, for RFLCS in Sect. 3.1 and then $DP_r$ for $r$-RLCS in Sect. 3.2.

First we briefly review a dynamic programming paradigm for the original LCS. For more details, e.g., see [13]. Given a sequence $X = \langle x_1, x_2, \cdots, x_n \rangle$, we define the $i$th *prefix* of $X$, for $i = 0, 1, \cdots, n$, as $X_i = \langle x_1, x_2, \cdots, x_i \rangle$. $X_n = X$ and $X_0$ is the empty sequence. Let $X = \langle x_1, x_2, \cdots, x_n \rangle$ and $Y = \langle y_1, y_2, \cdots, y_m \rangle$ be sequences and let $Z = \langle z_1, z_2, \cdots, z_h \rangle$ be any longest common subsequence of $X$ and $Y$. It is well known that LCS has the following optimal-substructure property: (1) If $x_n = y_m$, then $z_h = x_n = y_n$ and $Z_{h-1}$ is a longest common subsequence of $X_{n-1}$ and $Y_{m-1}$. (2) If $x_n \ne y_m$, then (a) $z_h \ne x_n$ implies that $Z$ is a longest common subsequence of $X_{n-1}$ and $Y$; (b) $z_h \ne y_m$, then $Z$ is a longest common subsequence of $X$ and $Y_{m-1}$.

We define $L(i, j)$ to be the length of a longest common subsequence of $X_i$ and $Y_j$. Then, the above optimal substructure of LCS gives the following recursive formula:

$$L(i,j) = \begin{cases} 0 & \text{if } i = 0 \text{ or } j = 0, \\ L(i-1, j-1) + 1 & \text{if } i, j > 0 \text{ and } x_i = y_j, \\ \max\{L(i, j-1), L(i-1, j)\} & \text{if } i, j > 0 \text{ and } x_i \neq y_j. \end{cases}$$

The DP algorithm for the original LCS computes each value of $L(i, j)$ and stores it into a two-dimensional table $L$ of size $n \times m$ in row-major order.

In the case of $r$-RLCS, we have to count the number of occurrences of every symbol in the prefix of $Z$. In the following we show a modified recursive formula and a DP-based algorithm for $r$-RLCS.

## 3.1   Repetition-Free LCS

Suppose that $X$ has $k$ symbols $s_1$, $s_2$, $\cdots$, $s_k$ and $s_i$ occurs $occ_i$ times in $X$ for each integer $i$, $1 \leq i \leq k$. A trivial implementation of a dynamic programming approach might be to use the DP-based algorithm for LCS for multiple sequences: We first generate all the permutations of $k$ symbols, i.e., $k!$ repetition-free sequences of $k$ symbols, say, $X_1$ through $X_{k!}$ and then obtain a longest common subsequence of $X_i$, $X$, and $Y$ for each $i$ $(1 \leq i \leq k!)$ by using an $O(|X_i| \cdot n \cdot m)$-time DP-based algorithm solving LCS for multiple (three) sequences proposed in [16]. Therefore, the total running time is $O(k! \cdot k \cdot n \cdot m)$, which is polynomial if $k$ is constant.

In the following we design a faster DP-based algorithm, named $\mathsf{DP}_1$. Let $S_{\geq 2} = \{s_j \mid occ_j \geq 2\}$. Now suppose that $|S_{\geq 2}| = \ell$ and, without loss of generality, $S_{\geq 2} = \{s_1, s_2, \cdots, s_\ell\}$. Then, we prepare a 0-1 "constraint" vector of length $\ell$, say, $\boldsymbol{v} = (v_1, v_2, \cdots, v_\ell) \in \{0, 1\}^\ell$, where the $p$th component $v_p$ corresponds to the $p$th symbol $s_p$ for $1 \leq p \leq \ell$. Roughly speaking, $v_p = 1$ means that if $x_i = y_j = s_p$ and $s_p$ has not appeared yet in the temporally obtained common subsequence, then $s_p$ is allowed to be attended to the current solution; on the other hand, $v_p = 0$ means that $s_p$ is not allowed to be appended to the current solution even if $x_i = y_j = s_p$.

For the 0-1 constraint vector $\boldsymbol{v} = (v_1, v_2, \cdots, v_p, \cdots, v_\ell)$, we define a new vector $\boldsymbol{v}|_{p=0} = (v_1', v_2', \cdots, v_p', \cdots, v_\ell')$ where $v_i' = v_i$ for $i \neq p$ but $v_p' = 0$. Note that if $v_p = 0$ in $\boldsymbol{v}$, then $\boldsymbol{v}|_{p=0} = \boldsymbol{v}$. Let $\boldsymbol{0}$ ($\boldsymbol{1}$, resp.) be a $\ell$-dimensional 0-vector (1-vector, resp.), i.e., the length of $\boldsymbol{0}$ ($\boldsymbol{1}$, resp.) is $\ell$ and all $\ell$ components are 0 (1, resp.).

Similarly to the above, we define $L(i, j, \boldsymbol{v})$ to be the length of a repetition-free longest common subsequence of $X_i$ and $Y_j$, under the constraint vector $\boldsymbol{v}$. Our algorithm for RFLCS computes each value of $L(i, j, \boldsymbol{v})$ and stores it into a three-dimensional table $L$ of size $n \times m \times 2^\ell$.

**Theorem 4 (Optimal substructure of RFLCS).** *Let* $X = \langle x_1, x_2, \cdots, x_n \rangle$ *and* $Y = \langle y_1, y_2, \cdots, y_m \rangle$ *be sequences and let* $Z = \langle z_1, z_2, \cdots, z_h \rangle$ *be any longest common subsequence of* $X$ *and* $Y$. *Let* $S_{\geq 2} = \{s_1, s_2, \cdots, s_\ell\}$ *be a set of* $\ell$ *symbols such that each* $s_i$ *occurs at least twice in* $X$. *The followings are satisfied:*

(1) If $x_n = y_m = s_q$ and $s_q \notin S_{\geq 2}$, then $z_h = x_n = y_m$ and $Z_{h-1}$ is a repetition-free longest common subsequence of $X_{n-1}$ and $Y_{m-1}$.

(2) If $x_n = y_m = s_q$, $s_q \in S_{\geq 2}$ and $v_q = 1$, then
   (a) $z_h = x_n = y_m$ implies that $Z_{h-1}$ is a repetition-free longest common subsequence of $X_{n-1}$ and $Y_{m-1}$ such that $s_q$ does not appear in $Z_{h-1}$;
   (b) $z_h \neq x_n = y_m$ implies that $Z$ is a repetition-free longest common subsequence of $X_{n-1}$ and $Y_{m-1}$.

(3) If $x_n = y_m = s_q$, $s_q \in S_{\geq 2}$ and $v_q = 0$, then $z_h \neq x_n = y_m$ and $Z$ is a repetition-free longest common subsequence of $X_{n-1}$ and $Y_{m-1}$.

(4) If $x_n \neq y_m$, then
   (a) $z_h \neq x_n$ implies that $Z$ is a repetition-free longest common subsequence of $X_{n-1}$ and $Y$;
   (b) $z_h \neq y_m$ implies that $Z$ is a repetition-free longest common subsequence of $X$ and $Y_{m-1}$.

*Proof.*(1) If $z_h \neq x_n$, then by appending $x_n = y_m = s_q$ to $Z$, we can obtain a repetition-free common subsequence of $X$ and $Y$ of length $h + 1$ since $Z$ does not have a symbol $s_q$ from the supposition $s_q \notin S_{\geq 2}$. This contradicts the assumption that $Z$ is a repetition-free longest common subsequence. Therefore, $z_h = x_n = y_m$ holds. What we have to do is to prove that the prefix $Z_{h-1}$ is a repetition-free common subsequence of $X_{n-1}$ and $Y_{m-1}$ with length $h - 1$. Suppose for contradiction that there exists a repetition-free longest common subsequence $Z'$ of $X_{n-1}$ and $Y_{m-1}$ with length greater $h - 1$. Then, by appending $x_n = y_m = s_q$, we obtain a repetition-free common subsequence of $X$ and $Y$ whose length is greater than $h$, which is a contradiction.

(2) (a) If $z_h = x_n = y_m$, then $Z_{h-1}$ is a repetition-free common subsequence of $X_{n-1}$ and $Y_{m-1}$ such that $s_q$ does not appear in $Z_{h-1}$. If there is a repetition-free common subsequence $Z'$ of $X_{n-1}$ and $Y_{m-1}$ such that $s_q$ does not appear in $Z'$ with length greater than $h - 1$, then by appending $x_n = y_m = s_q$ to $Z'$, we can obtain a repetition-free common subsequence of $X$ and $Y$ whose length is greater than $h$, which is a contradiction. (b) If $z_h \neq x_n = y_m$, then a repetition-free common subsequence of $X_{n-1}$ and $Y_{m-1}$ must include $s_q$ and be the longest one.

(3) If $v_q = 0$, then $s_p$ is not allowed to be included into $Z$. Thus, if $z_h \neq x_n = y_m$, then $Z$ is a repetition-free common subsequence of $X_{n-1}$ and $Y_{m-1}$ and it must be the longest one.

(4) (a) ((b), resp.) If $z_h \neq x_n$ ($z_h \neq y_m$, resp.), then $Z$ is a repetition-free common subsequence of $X_{n-1}$ and $Y$ ($X$ and $Y_{m-1}$, resp.). If there is a repetition-free common subsequence $Z'$ of $X_{n-1}$ and $Y$ ($X$ and $Y_{m-1}$, resp.) with length greater than $h$, then $Z'$ would also be a repetition-free common subsequence of $X$ and $Y$, contradicting the assumption that $Z$ is a repetition-free longest common subsequence of $X$ and $Y$.

$\square$

Then, we can obtain the following recursive formula:

$$
L(i,j,\boldsymbol{v})
$$

$$
= \begin{cases}
0 & \text{if } i = 0,\ j = 0,\ \text{or } \boldsymbol{v} = \boldsymbol{0}, \\
L(i-1, j-1, \boldsymbol{v}) + 1 \\
\quad \text{if } i,j > 0,\ x_i = y_j = s_q,\ \text{and } s_q \notin S_{\geq 2} \\
\max\{L(i-1, j-1, \boldsymbol{v}|_{p=0}) + 1, L(i-1, j-1, \boldsymbol{v})\} \\
\quad \text{if } i,j > 0,\ x_i = y_j = s_p,\ s_p \in S_{\geq 2},\ \text{and } v_p = 1, \\
L(i-1, j-1, \boldsymbol{v}) \\
\quad \text{if } i,j > 0,\ x_i = y_j = s_p,\ s_p \in S_{\geq 2},\ \text{and } v_p = 0, \\
\max\{L(i-1, j, \boldsymbol{v}), L(i, j-1, \boldsymbol{v})\} \\
\quad \text{otherwise.}
\end{cases}
$$

Here is an outline of our algorithm $\mathrm{DP_1}$, which computes each value of $L(i,j,\boldsymbol{v})$ and stores it into a three-dimensional table $L$ of size $n \times m \times 2^\ell$: Initially, we set $L(i,j,\boldsymbol{v}) = 0$ and $pre(i,j,\boldsymbol{v}) = \mathtt{null}$ for every $i$, $j$, and $\boldsymbol{v}$. Then, the algorithm $\mathrm{DP_1}$ fills entries from $L(1,1,\boldsymbol{0})$ to $L(1,1,\boldsymbol{1})$, then from $L(1,2,\boldsymbol{0})$ to $L(1,2,\boldsymbol{1})$, next from $L(1,3,\boldsymbol{0})$ to $L(1,3,\boldsymbol{1})$, and so on. After filling all the entries in the first "two-dimensional plane" $L(1,j,\boldsymbol{v})$, the algorithm fills all the entries in the second two-dimensional plane $L(2,j,\boldsymbol{v})$, and so on. Finally, $\mathrm{DP_1}$ fills all the entries in the $n$th plane. The algorithm $\mathrm{DP_1}$ also maintains a three dimensional table $pre$ of size $n \times m \times 2^\ell$ to help us construct an optimal repetition-free longest subsequence. The entry $pre(i,j,\boldsymbol{v})$ points to the table entry corresponding to the optimal subproblem solution chosen when computing $L(i,j,\boldsymbol{v})$. Further details could be appeared in the full version of this paper.

We bound the running time of $\mathrm{DP_1}$:

**Theorem 5.** *The running time of* $\mathrm{DP_1}$ *is* $O(2^\ell \cdot n \cdot m)$ *for RFLCS on two sequences* $X$ *and* $Y$ *where* $|X| = n$, $|Y| = m$, $|X| \leq |Y|$, *and* $|S_{\geq 2}| = \ell$.

*Proof.* The algorithm $\mathrm{DP_1}$ for RFLCS computes each value of $L(i,j,\boldsymbol{v})$ and stores it into the three-dimensional table $L$ of size $n \times m \times 2^\ell$. Clearly, each table entry takes $O(1)$ time to compute. As a result, the running time of $\mathrm{DP_1}$ is bounded above by $O(2^\ell \cdot n \cdot m)$. $\qquad\square$

**Corollary 2.** *The running time of* $\mathrm{DP_1}$ *is* $O(1.41422^n \cdot n \cdot m)$ *for RFLCS on two sequences* $X$ *and* $Y$ *where* $|X| = n$, $|Y| = m$, *and* $|X| \leq |Y|$.

*Proof.* Recall that the number $|S_{\geq 2}|$ of symbols which appear at least twice in $X$ is defined to be $\ell$. This implies that $\ell \leq \frac{n}{2}$. Therefore, $2^\ell \leq 2^{n/2} < 1.41422^n$ is satisfied. $\qquad\square$

## 3.2  $r$-Repetition LCS, $r \geq 2$

The similar strategies of $\mathrm{DP_1}$ can work well for $r$-RLCS; we can design a DP-based algorithm, named $\mathrm{DP}_r$, for $r$-RLCS. The running time is as follows:

**Theorem 6.** *The running time of $DP_r$ is $O((r+1)^\ell \cdot n \cdot m)$ for r-RLCS on two sequences $X$ and $Y$ where $|X| = n$, $|Y| = m$, $|X| \leq |Y|$, and $|S_{\geq 2}| = \ell$.*

*Proof.* It is enough to prepare a three-dimensional table $L$ of size $n \times m \times (r+1)^\ell$ and each table entry takes $O(1)$ time to compute.                                                □

**Corollary 3.** *The running time of $DP_r$ is $O((r+1)^{n/(r+1)} \cdot n \cdot m)$ for r-RLCS on two sequences $X$ and $Y$ where $|X| = n$, $|Y| = m$, and $|X| \leq |Y|$.*

*Proof.* Clearly $\ell \leq \frac{n}{r+1}$, i.e., $(r+1)^\ell \leq (r+1)^{n/(r+1)}$ holds.                   □

For example, by the above corollary, the running time of our algorithm is $O(1.44225^n \cdot n \cdot m)$ for 2-RLCS, $O(1.41422^n \cdot n \cdot m)$ for 3-RLCS, $O(1.37973^n \cdot n \cdot m)$ for 4-RLCS and so on.

## 4   Conclusion

We studied a new variant of the LONGEST COMMON SUBSEQUENCE problem, called $r$-REPETITION LONGEST COMMON SUBSEQUENCE problem ($r$-RLCS). For $r = 1$, 1-RLCS is known as the REPETITION-FREE LONGEST COMMON SUB-SEQUENCE problem. We first showed that for 1-RLCS there is a simple exact algorithm whose running time is bounded above by $O(1.44225^{|X|}|X||Y|)$. Then, for $r$-RLCS ($r \geq 1$), we designed a DP-based exact algorithm whose running time is $O((r+1)^{|X|/(r+1)}|X||Y|)$. This implies that we can solve 1-RLCS in $O(1.41422^{|X|}|X||Y|)$ time. A promising direction for future research is to design faster exact (exponential-time) algorithms for $r$-RLCS. Also, it would be important to design approximation algorithms for $r$-RLCS.

**Acknowledgments.** This work was partially supported by PolyU Fund 1-ZE8L, the Natural Sciences and Engineering Research Council of Canada, JST CREST JPMJR1402, and Grants-in-Aid for Scientific Research of Japan (KAKENHI) Grant Numbers JP17K00016, JP17K00024, JP17K19960 and JP17H01698.

## References

1. Adi, S.S., et al.: Repetition-free longest common subsequence. Disc. Appl. Math. **158**, 1315–1324 (2010)
2. Aho, A., Hopcroft, J., Ullman, J.: Data Structures and Algorithms. Addison-Wesley, Boston (1983)
3. Altschul, S.F., Gish, W., Miller, W., Myers, E.W., Lipman, D.J.: Basic local alignment search tool. J. Mol. Biol. **215**(3), 403–410 (1990)
4. Bergroth, L., Hakonen, H., Raita, T.: A survey of longest common subsequence algorithms. In: Proceedings of SPIRE, pp. 39–48 (2000)
5. Blin, G., Bonizzoni, P., Dondi, R., Sikora, F.: On the parameterized complexity of the repetition free longest common subsequence problem. Info. Proc. Lett. **112**(7), 272–276 (2012)
6. Blum, C., Blesa, M.J., Calvo, B.: Beam-ACO for the repetition-free longest common subsequence problem. Proc. EA **2013**, 79–90 (2014)

7. Blum, C., Blesa, M.J.: Construct, merge, solve and adapt: application to the repetition-free longest common subsequence problem. In: Chicano, F., Hu, B., García-Sánchez, P. (eds.) EvoCOP 2016. LNCS, vol. 9595, pp. 46–57. Springer, Cham (2016). https://doi.org/10.1007/978-3-319-30698-8_4
8. Blum, C., Blesa, M.J.: A comprehensive comparison of metaheuristics for the repetition-free longest common subsequence problem. J. Heuristics 24(3), 551–579 (2018)
9. Bonizzoni, P., Della Vedova, G., Dondi, R., Fertin, G., Rizzi, R., Vialette, S.: Exemplar longest common subsequence. IEEE/ACM Trans. Comput. Biol. Bioinf. 4(4), 535–543 (2007)
10. Bonizzoni, P., Della Vedova, G., Dondi, R., Pirola, Y.: Variants of constrained longest common subsequence. Inf. Proc. Lett. 110(20), 877–881 (2010)
11. Bulteau, L., Hüffner, F., Komusiewicz, C., Niedermeier, R.: Multivariate algorithmics for NP-hard string problems. The Algorithmics Column by Gerhard J Woeginger. Bulletin of EATCS, no. 114 (2014)
12. Castelli, M., Beretta, S., Vanneschi, L.: A hybrid genetic algorithm for the repetition free longest common subsequence problem. Oper. Res. Lett. 41(6), 644–649 (2013)
13. Cormen, T.H., Leiserson, C.E., Rivest, R.L., Stein, C.: Introduction to Algorithms, 3rd edn. The MIT Press, Cambridge (2009)
14. Hirschberg, D.S.: Algorithms for the longest common subsequence problem. J. ACM 24(4), 664–675 (1977)
15. Hirschberg, D.S.: A linear space algorithm for computing maximal common subsequences. Comm. ACM 18(6), 341–343 (1975)
16. Itoga, S.Y.: The string merging problem. BIT 21(1), 20–30 (1981)
17. Maier, D.: The complexity of some problems on subsequences and supersequences. J. ACM 25(2), 322–336 (1978)
18. Jiang, T., Li, M.: On the approximation of shortest common supersequences and longest common subsequences. SIAM J. Comput. 24(5), 1122–1139 (1995)
19. Morgan, H.L.: Spelling correction in systems programs. Comm. ACM 13(2), 90–94 (1970)
20. Needleman, S.B., Wunsch, C.D.: A general method applicable to the search for similarities in the amino acid sequence of two proteins. J. Mol. Biol. 48(3), 443–453 (1970)
21. Sankoff, D.: Matching sequences under deletion/insertion constraints. Proc. Nat. Acad. Sci. U.S.A. 69(1), 4–6 (1972)
22. Sankoff, D.: Genome rearrangement with gene families. Bioinformatics 15(11), 909–917 (1999)
23. Storer, J.A.: Data compression: methods and theory. Computer Science Press (1988)
24. Wagner, R.A., Fischer, M.J.: The string-to-string correction problem. J. ACM 21(1), 168–173 (1974)

# Improved Bounds for Two Query Adaptive Bitprobe Schemes Storing Five Elements

Mirza Galib Anwarul Husain Baig and Deepanjan Kesh[(⊠)]

Indian Institute of Technology Guwahati, Guwahati 781039, Assam, India
{mirza.baig,deepkesh}@iitg.ac.in

**Abstract.** In this paper, we study two-bitprobe adaptive schemes storing five elements. For these class of schemes, the best known lower bound is $\Omega(m^{1/2})$ due to Alon and Feige [1]. Recently, it was proved by Kesh [9] that two-bitprobe adaptive schemes storing three elements will take at least $\Omega(m^{2/3})$ space, which also puts a lower bound on schemes storing five elements. In this work, we have improved the lower bound to $\Omega(m^{3/4})$. We also present a scheme for the same that takes $\mathcal{O}(m^{5/6})$ space. This improves upon the $\mathcal{O}(m^{18/19})$-scheme due to Garg [7] and the $\mathcal{O}(m^{10/11})$-scheme due to Baig *et al.* [5].

**Keywords:** Data structure · Set membership problem · Bitprobe model · Adaptive scheme

## 1 Introduction

The *static membership problem* involves the study and construction of such data structures which can store an arbitrary subset $S$ of size at most $n$ from the universe $\mathcal{U} = \{1, 2, 3, \ldots, m\}$ such that membership queries of the form "Is $x$ in $S$?" can be answered correctly and efficiently. A special category of the static membership problem is the *bitprobe model* in which we evaluate our solutions w.r.t. the following resources – the size of the data structure, $s$, required to store the subset $S$, and the number of bits, $t$, of the data structure read to answer membership queries. It is the second of these resources that lends the name to this model.

In this model, the design of data structures and query algorithms are known as *schemes*. For a given universe $\mathcal{U}$ and a subset $S$, the algorithm to set the bits of our data structure to store the subset is called the *storage scheme*, whereas the algorithm to answer membership queries is called the *query scheme*. Schemes are divided into two categories depending on the nature of our query scheme. Upon a membership query for an element, if the decision to probe a particular bit depends upon the answers received in the previous bitprobes of this query, then such schemes are known as *adaptive schemes*. If the locations of the bitprobes are fixed for a given element of $\mathcal{U}$, then such schemes are called *non-adaptive schemes*.

Y. Li et al. (Eds.): COCOA 2019, LNCS 11949, pp. 13–25, 2019.
https://doi.org/10.1007/978-3-030-36412-0_2

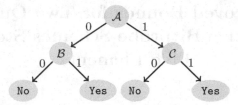

**Fig. 1.** The decision tree of an element.

For any particular setting of $n, m, s$, and $t$, the corresponding scheme is referred to in the literature as a $(n, m, s, t)$-scheme [6,11,12]. Radhakrishnan *et al.* [12] also introduced the convenient notations $s_A(n, m, t)$ and $s_N(n, m, t)$ to denote the space required by an adaptive or a non-adaptive scheme, respectively.

## 1.1   Two-Bitprobe Adaptive Schemes

In this paper, we consider those schemes that use two bitprobes ($t = 2$) to answer membership queries. Data structures in such schemes consist of three tables, namely $\mathcal{A}, \mathcal{B}$, and $\mathcal{C}$. The first bitprobe is always made in table $\mathcal{A}$; the location of the bit being probed, of course, depends on the element which is being queried. The second bitprobe is made in table $\mathcal{B}$ or in table $\mathcal{C}$, depending on whether 0 was returned in the first bitprobe or 1 was returned. The final answer of the query scheme is Yes if 1 is returned by the second bitprobe, otherwise it is No. The data structure and the query scheme can be succinctly denoted diagrammatically by what is known as the *decision tree* of the scheme (Fig. 1).

## 1.2   Our Contribution

In this paper, we study schemes when the number of allowed bitprobes is two ($t = 2$) and the subset size is at most five ($n = 5$). Some progress has been made for subsets of smaller sizes.

When the subset size is odd, the problem is well understood. More particularly, when $n = 1$, there exists a scheme that takes $\mathcal{O}(m^{1/2})$ amount of space, and it has been shown that it matches with the lower bound $\Omega(m^{1/2})$ [1,6,10]. When $n = 3$, Baig and Kesh [2] have shown that there exists a $\mathcal{O}(m^{2/3})$-scheme, and Kesh [9] has proven that it matches with the lower bound $\Omega(m^{2/3})$.

For even sized subsets, tight bounds are yet to be proven. For $n = 2$, Radhakrishnan *et al.* [11] have proposed a scheme that takes $\mathcal{O}(m^{2/3})$ space, and for $n = 4$, Baig *et al.* [4] have presented a $\mathcal{O}(m^{5/6})$-scheme, but it is as of yet unknown whether these bounds are tight.

For subsets of size five ($n = 5$), the best known lower bound was due to Alon and Feige [1] which is $\Omega(m^{1/2})$. The $\Omega(m^{2/3})$ lower bound for $n = 3$ also puts an improved bound for the $n = 5$ case. Our first result improves the bound to $\Omega(m^{3/4})$.

**Result 1 (Theorem 8).** $s_A(5, m, 2) = \Omega(m^{3/4})$.

We also propose an improved scheme for the problem. The best known upper bound was due to Garg [7] which was $\mathcal{O}(m^{18/19})$, which was improved by Baig et al. [5] to $\mathcal{O}(m^{10/11})$. In this paper, we improve the bound to $\mathcal{O}(m^{5/6})$.

**Result 2 (Theorem 9).** $s_A(5, m, 2) = \mathcal{O}(m^{5/6})$.

One thing to note is that the space for the scheme storing five elements now matches the space for the scheme storing four elements. Moreover, the two results stated above combined together significantly reduces the gap between the upper and lower bounds for the problem under consideration.

## 2 Lower Bound

In this section, we present our proof for the lower bound of $s_A(5, m, 2)$. So, the size of our subset $S$ that we want to store in our data structure is at most five.

In table $A$ of our data structure, multiple elements must necessarily map to the same bit to keep the table size to $o(m)$. The set of elements that map to the same bit in this table is referred to in the literature as a *block* (Radhakrishnan *et al.* [11]). We refer by $A(e)$ to the block to which the element $e$ belongs. Elements mapping to the same bit in tables $B$ and $C$ will be referred to as just *sets*. That set of table $B$ to which the element $e$ belongs will be denoted by $B(e)$. $C(e)$ is similarly defined.

Storing a member $e$ of our subset $S$ in table $B$ is an informal way to state the following – the bit corresponding to $A(e)$ is set to 0, and $B(e)$ is set to 1. So, upon query for the element $e$, we will get a 0 in our first bitprobe, query table $B$ at location $B(e)$ to get 1, and finally answer Yes. Similarly, storing an element $f$ which is not in $S$ in table $C$ would entail assigning 1 to $A(f)$ and 0 to $C(f)$.

To start with, we make the following simplifying assumptions about any scheme for the aforementioned problem.

1. All the tables of our datastructure have the same size, namely $s$, and hence the size our data structure is $3 \times s$.
2. If two elements belong to the same block in table $A$, they do not belong to the same sets in either of tables $B$ or $C$.

In the conclusion of this section, we will show that these assumptions do not affect the space asymptotically, but rather by constant factors.

### 2.1 Universe of an Element

We now define the notion of the *universe* of an element. This is similar to the definition of the universe of a set in Kesh [9].

**Definition 1.** *The universe of an element $e$ w.r.t. to table $\mathcal{B}$, denoted by $\mathcal{U}_\mathcal{B}(e)$, is defined as follows.*

$$\mathcal{U}_\mathcal{B}(e) = \bigcup_{f \in \mathcal{B}(e) \setminus \{e\}} \mathcal{A}(f) \setminus \{f\}.$$

*Similarly, the universe of an element $e$ w.r.t. to table $\mathcal{C}$, denoted by $\mathcal{U}_\mathcal{C}(e)$, is defined as follows.*

$$\mathcal{U}_\mathcal{C}(e) = \bigcup_{f \in \mathcal{C}(e) \setminus \{e\}} \mathcal{A}(f) \setminus \{f\}.$$

A simple property of the universe of an element, which will be useful later, is the following.

**Observation 1.** *1. $\mathcal{A}(e) \cap \mathcal{U}_\mathcal{B}(e) = \phi$ and $\mathcal{B}(e) \cap \mathcal{U}_\mathcal{B}(e) = \phi$.*
*2. $\mathcal{A}(e) \cap \mathcal{U}_\mathcal{C}(e) = \phi$ and $\mathcal{C}(e) \cap \mathcal{U}_\mathcal{C}(e) = \phi$.*

*Proof.* Due to Assumption 2, $e$ is the only element of the block $\mathcal{A}(e)$ in set $\mathcal{B}(e)$. So, it follows from the definition that no element of $\mathcal{A}(e)$ is part of $\mathcal{U}_\mathcal{B}(e)$. No element of $\mathcal{B}(e)$ is part of $\mathcal{U}_\mathcal{B}(e)$ as we specifically preclude those elements from $\mathcal{U}_\mathcal{B}(e)$. The other scenarios can be similarly argued. □

We make the following simple observations about the universe of an element to help illustrate the constraints any storage scheme must satisfy to correctly store a subset $\mathcal{S}$.

**Observation 2.** *If $\mathcal{B}(e) \cap \mathcal{S} = \{e\}$, and we want to store $e$ in table $\mathcal{B}$, then all the elements of $\mathcal{U}_\mathcal{B}(e)$ must be stored in table $\mathcal{C}$.*

*Proof.* As $e$ has been stored in table $\mathcal{B}$, the bit corresponding to the set $\mathcal{B}(e)$ must be set to 1. Consider an element $f$, different from $e$, in $\mathcal{B}(e)$. According to Assumption 2, $e$ and $f$ cannot belong to the same block. As $f \notin \mathcal{S}$, so $f$ cannot be stored in table $\mathcal{B}$; if we do so, the query for $f$ will look at the bit $\mathcal{B}(e)$ and incorrectly return 1. So, the element $f$ and its block $\mathcal{A}(f)$ must be stored in table $\mathcal{C}$.

The above argument applies to any arbitrary element of $\mathcal{B}(e) \setminus \{e\}$. So, according to Definition 1, all of the elements of $\mathcal{U}_\mathcal{B}(e)$ must be stored in table $\mathcal{C}$. □

We make the same observation, without proof, in the context of table $\mathcal{C}$.

**Observation 3.** *If $\mathcal{C}(e) \cap \mathcal{S} = \{e\}$, and we want to store $e$ in table $\mathcal{C}$, then all the elements of $\mathcal{U}_\mathcal{C}(e)$ must be stored in table $\mathcal{B}$.*

Next, we define, what could be referred to as, a higher-order universe of an element, built on top of the universe of the element.

**Definition 2.** *The 2-universe of an element $e$ w.r.t. table $\mathcal{B}$, denoted by $\mathcal{U}_\mathcal{B}^2(e)$, is defined as follows.*

$$\mathcal{U}_\mathcal{B}^2(e) = \bigcup_{f \in \mathcal{U}_\mathcal{B}(e)} \mathcal{C}(f) \setminus \{f\}.$$

*Similarly, the 2-universe of an element e w.r.t. table $C$, denoted by $\mathcal{U}_C^2(e)$, is defined as follows.*

$$\mathcal{U}_C^2(e) = \bigcup_{f \in \mathcal{U}_C(e)} \mathcal{B}(f) \backslash \{f\}.$$

The following observations provide more constraints for our storage schemes.

**Observation 4.** *Consider an element e such that $\mathcal{B}(e) \cap S = \{e\}$, and suppose we want to store e in table $\mathcal{B}$. If f is a member of $\mathcal{U}_\mathcal{B}(e)$ such that $C(f) \cap S = \{f\}$, then all the other members of $C(f)$ must be stored in table $\mathcal{B}$.*

*Proof.* If $e$, a member of $S$, is stored in table $\mathcal{B}$, then Observation 2 tells us that all members of $\mathcal{U}_\mathcal{B}(e)$ must be stored in table $C$. As $f \in \mathcal{U}_\mathcal{B}(e) \cap S$, so $f$ must be stored in table $C$, and consequently, the bit corresponding to $C(f)$ must be set to 1. As the other members of $C(f)$ do not belong to $S$, they cannot be stored in table $C$, and hence must be stored in table $\mathcal{B}$.  □

**Observation 5.** *Consider an element e such that $\mathcal{B}(e) \cap S = \{e\}$, and suppose we want to store e in table $\mathcal{B}$. If f is a member $\mathcal{U}_\mathcal{B}(e)$ such that $C(f) \cap S = \{x\}$, where $x \neq f$, then x must be stored in table $\mathcal{B}$.*

*Proof.* As $e$, a member of $S$, is stored in $\mathcal{B}$, Observation 2 tells us that all the members of $\mathcal{U}_\mathcal{B}(e)$, and $f$ in particular, must be stored in table $C$. As $f \notin S$, so the bit corresponding to $C(f)$ has to be 0. As $x \in C(f)$ belongs to $S$, then storing $x$ in table $C$ would imply that $C(f)$ must be set to 1, which is absurd. So, $x$ must be stored in table $\mathcal{B}$.  □

We next state the same observations in the context of table $C$.

**Observation 6.** *Consider an element e such that $C(e) \cap S = \{e\}$, and suppose we want to store e in table $C$. If f is a member of $\mathcal{U}_C(e)$ such that $\mathcal{B}(f) \cap S = \{f\}$, then all the other members of $\mathcal{B}(f)$ must be stored in table $C$.*

**Observation 7.** *Consider an element e such that $C(e) \cap S = \{e\}$, and suppose we want to store e in table $C$. If f is a member $\mathcal{U}_C(e)$ such that $\mathcal{B}(f) \cap S = \{x\}$, where $x \neq f$, then x must be stored in table $C$.*

## 2.2 Bad Elements

We now define the notion of *good* and *bad* elements. These notions are motivated by the notions of *large* and *bounded sets* from Kesh [9].

**Definition 3.** *e is a bad element w.r.t. table $\mathcal{B}$ if one of the following holds.*

1. *Two elements of $\mathcal{U}_\mathcal{B}(e)$ share a set in table $C$.*
2. *The size of $\mathcal{U}_\mathcal{B}^2(e)$ is greater than 2s, i.e. $|\mathcal{U}_\mathcal{B}^2(e)| > 2 \cdot s$.*

*Otherwise, it is said to be good. Bad and good elements w.r.t. to table $C$ are similarly defined.*

The next claims state the consequences of an element being bad due to any of the above properties getting satisfied.

*Claim.* If two elements of $\mathcal{U}_{\mathcal{B}}(e)$ share a set in table $\mathcal{C}$, then $\exists$ a subset $\mathcal{S}$ that contains $e$ and has size two such that to store $\mathcal{S}$, $e$ cannot be stored in $\mathcal{B}$.

*Proof.* Suppose the elements $x, y \in \mathcal{U}_{\mathcal{B}}(e)$ share the set $Y$ in table $\mathcal{C}$. We would prove that to store the subset $\mathcal{S} = \{e, x\}$, $e$ cannot be stored in table $\mathcal{B}$.

Let us say that $e$ indeed can be stored in table $\mathcal{B}$. According to Observation 2, all elements of $\mathcal{U}_{\mathcal{B}}(e)$, including $x$ and $y$, must be stored in table $\mathcal{C}$. As $x \in \mathcal{S}$, the bit corresponding to the set $Y$ must be set to 1. As $y \notin \mathcal{S}$, the bit corresponding to $Y$ must be set to 0. We thus arrive at a contradiction. So, $e$ cannot be stored in table $\mathcal{B}$.                                                           □

*Claim.* If the size of $\mathcal{U}_{\mathcal{B}}^2(e)$ is greater than $2s$, then $\exists$ a subset $\mathcal{S}$ that contains $e$ and has size at most three such that to store $\mathcal{S}$, $e$ cannot be stored in table $\mathcal{B}$.

*Proof.* Consider the set of those elements $f$ of $\mathcal{U}_{\mathcal{B}}^2(e)$ such that it is the only member of its block to belong to $\mathcal{U}_{\mathcal{B}}^2(e)$. As there are a total of $s$ blocks, there could be at most $s$ such elements. Removing those elements from $\mathcal{U}_{\mathcal{B}}^2(e)$ still leaves us with more than $s$ elements in $\mathcal{U}_{\mathcal{B}}^2(e)$. These remaining elements have the property that there is at least one other element from its block that is present in $\mathcal{U}_{\mathcal{B}}^2(e)$. Let this set be denoted by $Z$.

As the size of $Z$ is larger than the size of table $\mathcal{B}$, there must exist at least two elements $x, y \in Z$ that share a set $X$ in table $\mathcal{B}$. According to Definition 2, this implies that there exists elements $z, z' \in \mathcal{U}_{\mathcal{B}}(e)$ such that $x \in \mathcal{C}(z) \setminus \{z\}$ and $y \in \mathcal{C}(z') \setminus \{z'\}$. It might very well be that $z = z'$.

If $x \in \mathcal{A}(e)$, as $e$ has been stored in table $\mathcal{B}$, so all the elements of $\mathcal{A}(e)$, including $x$, must have been stored in table $\mathcal{B}$. Consider the subset $\mathcal{S} = \{e, x, z'\}$. As $x \in \mathcal{S}$, the bit corresponding to set $X$ must be set to 1. As $e$ is stored in table $\mathcal{B}$, Observation 2 tells us that $z' \in \mathcal{U}_{\mathcal{B}}(e)$ must be stored in table $\mathcal{C}$. As $\mathcal{C}(z') \cap \mathcal{S} = \{z'\}$, Observation 4 tells us that $y$ must be stored in table $\mathcal{B}$. So, the bit corresponding to set $X$ must be set to 0, which is absurd. So, to store $\mathcal{S}$, $e$ cannot be stored in table $\mathcal{B}$. This argument holds even if $x = e$. Similar is the case if $y \in \mathcal{A}(e)$.

If $x \in \mathcal{B}(e) \setminus \{e\}$, and as $e$ has been stored in table $\mathcal{B}$, Observation 2 tells us that $x$ must be stored in table $\mathcal{C}$. Consider the subset $\mathcal{S} = \{e, z\}$. Observation 4 tells us that as $z \in \mathcal{U}_{\mathcal{B}}(e)$ is in $\mathcal{S}$, $x \in \mathcal{C}(z)$ cannot be stored in table $\mathcal{C}$, which is absurd. So, to store $\mathcal{S}$, $e$ cannot be stored in table $\mathcal{B}$. We can similarly argue the case $y \in \mathcal{B}(e)$.

We now consider the case when $x, y \notin \mathcal{A}(e)$ and $\notin \mathcal{B}(e)$. If $\mathcal{S}$ contains $e$ and $x$, and we store $e$ in table $\mathcal{B}$, Observation 2 tells us that $z \in \mathcal{U}_{\mathcal{B}}(e)$ must be stored in table $\mathcal{C}$, and as $x \in \mathcal{C}(z)$, Observation 5 tells us that $x$ must be stored in table $\mathcal{B}$. As $x \in \mathcal{S}$, hence the bit corresponding to set of $x$ in table $\mathcal{B}$, which is $X$, must be set to 1.

If $z \neq z'$, we include $z'$ in $\mathcal{S}$, and according to Observation 4, $y \in \mathcal{C}(z')$ must be stored in table $\mathcal{B}$. As $y \in X$ is not in $\mathcal{S}$, $X$ must be set to 0, and we arrive at a contradiction for the subset $\mathcal{S} = \{e, x, z'\}$.

It could also be the case that $z = z'$. As $y \in Z$, there exists an element $y' \in \mathcal{A}(y) \cap \mathcal{U}_{\mathcal{B}}^2(e)$. Let $y' \in C(z'')$, where $z'' \in \mathcal{U}_{\mathcal{B}}(e)$. In this scenario, we consider storing the subset $S = \{e, x, z''\}$. As, $z'' \in S \cap \mathcal{U}_{\mathcal{B}}(e)$, and $y' \notin S$, Observation 4 implies that $y'$, and hence the whole of block $\mathcal{A}(y')$, including $y$, must be stored in table $\mathcal{B}$. As $y \notin S$, the set of $y$ in table $\mathcal{B}$, which is $X$, must be set to 0, and we again arrive at a contradiction.

So, we conclude that $e$ in either of the cases cannot be stored in table $\mathcal{B}$.  $\square$

The two claims above imply the following – if an element $e$ is bad w.r.t. table $\mathcal{B}$ (Definition 3) due to Property 1, or if this property does not hold but Property 2 does, then there exists a subset, say $S_1$, of size at most three containing $e$ such that to store $S_1$, $e$ cannot be stored in table $\mathcal{B}$. The claims above also hold w.r.t. table $C$. So, we can claim that if $e$ is bad w.r.t. to table $C$, then there exists a subset $S_2$ containing $e$ of size at most three such that to store $S_2$, $e$ cannot be stored in table $C$.

Consider the set $S = S_1 \cup S_2$. As $e$ is common in both the subsets, size of $S$ is at most five. If $e$ is bad w.r.t. to table $\mathcal{B}$ and table $C$, then to store subset $S$, we cannot store $e$ in either of the tables, which is absurd. We summarise the discussion in the following lemma.

**Lemma 1.** *If an element $e$ is bad w.r.t. $\mathcal{B}$, then it must be good w.r.t $C$.*

## 2.3   Good Schemes

Based on the above lemma, we can partition our universe $\mathcal{U}$ into two sets $\mathcal{U}_1$ and $\mathcal{U}_2$ – one that contains all the good elements w.r.t. to table $\mathcal{B}$, and one that contains the bad elements. We now partition each block and each set of the three tables of our datastructure into two parts, one containing elements from $\mathcal{U}_1$, and one containing the elements from $\mathcal{U}_2$. For elements of $\mathcal{U}_1$, only those blocks and sets that contain elements of $\mathcal{U}_1$ will be affected; similarly for the elements of $\mathcal{U}_2$.

In effect, we have two independent schemes, one for $\mathcal{U}_1$ and one for $\mathcal{U}_2$. In the scheme for $\mathcal{U}_1$, all the elements in table $\mathcal{B}$ are good. In the scheme for $\mathcal{U}_2$, all the elements in table $\mathcal{B}$ are bad, and consequently, Lemma 1 tells us that all the elements of table $C$ are good. In the scheme for $\mathcal{U}_2$, we now relabel the table $\mathcal{B}$ to $C$ and relabel the table $C$ to $\mathcal{B}$. To make the new scheme for $\mathcal{U}_2$ work, we now have to store 0 in the blocks of table $\mathcal{A}$ for $\mathcal{U}_2$ when earlier we were storing 1, and have to store 1 when earlier we were storing 0.

This change gives us a new scheme with two important properties – the size of the datastructure has doubled from the earlier scheme, and all the elements in table $\mathcal{B}$ are now good.

**Lemma 2.** *Given a $(5, m, s, 2)$-scheme, we can come up with a $(5, m, 2 \times s, 2)$-scheme such that all the elements of $\mathcal{U}$ are good w.r.t. to table $\mathcal{B}$ in the new scheme.*

## 2.4   Space Complexity

Consider a $(5, m, 3 \times s, 2)$-scheme all of whose elements are good w.r.t. table $\mathcal{B}$. The table sizes then are each equal to $s$. According to Lemma 2, the 2-universe of each element w.r.t. to $\mathcal{B}$ will be at most $2s$. So, the sum total of all the 2-universe sizes of all the elements is upper bounded by $m \times 2s$.

We now consider how much each set of table $\mathcal{C}$ contribute to the total. From Definition 2, we have the following –

$$\sum_{e \in \mathcal{U}} \mid \mathcal{U}_{\mathcal{B}}^2(e) \mid = \sum_{e \in \mathcal{U}} \mid \bigcup_{f \in \mathcal{U}_{\mathcal{B}}(e)} \mathcal{C}(f) \setminus \{f\} \mid = \sum_{e \in \mathcal{U}} \left( \sum_{f \in \mathcal{U}_{\mathcal{B}}(e)} \mid \mathcal{C}(f) \setminus \{f\} \mid \right).$$

As all the elements are good, and hence for every element $e$, no two elements of $\mathcal{U}_{\mathcal{B}}(e)$ share a set in table $\mathcal{C}$, we can thus convert the union in Definition 2 to summation.

Resolving the above equation, the details of which can be found in the extended version of the paper [3], we have

$$\sum_{e \in \mathcal{U}} \mid \mathcal{U}_{\mathcal{B}}^2(e) \mid \geq c \cdot \frac{m^4}{s^3},$$

for some constant $c$. The proof show that the minimum value is achieved when all the blocks and the sets in the three tables are of the same size, i.e. $m/s$. This combined with the upperbound for total sum of the sizes of all 2-universes gives us

$$c \cdot \frac{m^4}{s^3} \leq m \times 2s.$$

Resolving the equation gives us

$$s = \Omega(m^{3/4}).$$

This bound applies to good schemes that respect the two assumptions declared at the beginning of this section.

Suppose we have an arbitrary adaptive $(5, m, s, 2)$-scheme. If we want to make all the tables in this scheme of the same size, we can add extra bits which will make the size of the data structure at most $3 \cdot s$. So, we get a $(5, m, 3 \times s, 2)$-scheme that respect Assumption 1.

In Kesh [9], sets which have multiple elements from the same block were referred to as *dirty sets*. *Clean sets* were those which contain elements from distinct blocks. It was shown in Sect. 3 that any scheme with dirty sets can be converted into a scheme with only clean sets by using twice amount of space. Though the final claim was made in context of $n = 3$, but the proof applies to any $n$. So, we can now have a $(5, m, 6 \times s, 2)$-scheme that respects both of our assumptions.

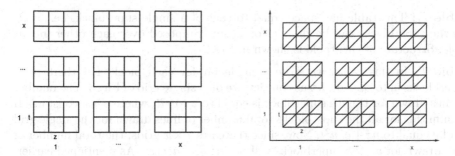

**Fig. 2.** Figure showing structure of a superblock

**Fig. 3.** Lines drawn in the first superblock

Such a scheme can be converted into a scheme with only good elements in table $\mathcal{B}$ by using twice the amount of space as before. We now have a $(5, m, 12 \times s, 2)$-scheme, where the table sizes are all $4s$, and we have shown that $4s = \Omega(m^{3/4})$.

We summarise our discussion in the following theorem on the lower bound for two-adaptive bitprobes schemes storing five elements.

**Theorem 8.** $s_A(5, m, 2) = \Omega(m^{3/4})$.

## 3   Our Scheme

In this section, we will present a scheme which stores an arbitrary subset of size at most five from a universe of size $m$, and answers the membership queries in two adaptive bitprobes. This scheme improves upon the $\mathcal{O}(m^{10/11})$-scheme by the authors [5], and is fundamentally different from that scheme in the way that here block sizes are nonuniform, and any two blocks in table $\mathcal{C}$ share at most one bit. As per the convention of that scheme, we will use the label $T$ to refer to the table $\mathcal{A}$, $T_0$ to refer to the table $\mathcal{B}$, and $T_1$ to refer to the table $\mathcal{C}$.

**Superblock:** In this scheme, we use the idea in Kesh [8] of mapping the elements of the universe on a square grid. Furthermore, we have used the idea of Radhakrishnan *et al.* [11] to divide the universe into blocks and superblocks. Our scheme divides the universe of size $m$ into superblock of size $x^2zt$. Each superblock is made up of rectangular grids of size $t \times z$, and there are $x^2$ of them as shown in Fig. 2. Further, each integral point on a grid represents a unique element.

**Block:** For the 1st superblock we draw lines with slope 1 as shown in Fig. 3. Each line drawn represents a block. From Fig. 3, we can see that some blocks are of equal size and some are of different size. We do this for all the superblocks, and hence partitioning the universe into blocks. For the $i$th superblock we draw lines with slope $1/i$.

**Table $T_1$:** This table has space equal to that of a single superblock, i.e., $x^2zt$. All the superblocks can be thought of as superimposed over each other in this table. Structure of this table is shown in Fig. 2.

**Table $T$:** In this table, we store a single bit for each block. Let there be $n$ superblocks in total. Now let us concentrate on a single grid of Fig. 3. The number of lines drawn for the $i$th superblock is equal to $z + c \cdot it$, where $c$ is a constant. If we sum this for all the superblocks total number of lines drawn for the single grid will be equal to $nz + c \cdot n^2t$. Now, since there are $x \times x$ grids, the total number of lines drawn for all the superblocks will be $(nz + c \cdot n^2t)x^2$. As mentioned earlier, each line represents a block, and for each block, we have one bit of space in table $T$. So the size of this table $T$ is $(nz + c \cdot n^2t)x^2$ bits.

**Table $T_0$:** In addition to lines drawn in superblocks to divide them into blocks, we also draw dotted lines in all the superblocks, as shown in Fig. 4. For the $i$th superblock we draw dotted lines with slope $1/i$. Further, we store a block of size $t$ in table $T_0$ for each dotted lines drawn. Now, we can see that for a specific superblock there could be many blocks belonging to that superblock which lies on the same dotted line. All the blocks which lie completely on the same dotted line query the same block in table $T_0$ kept for the dotted line.

Now let us talk about the space taken by table $T_0$. Using the idea shown in Fig. 4 to draw the dotted lines, if we sum the total number of dotted lines drawn for all the superblocks which pass through x-axis, we will get $nzx$. Further, if we sum the total number of dotted lines drawn for all the superblocks from the y-axis, we get it to be less than or equal to $c_1 \cdot n^2t \times x$, where $c_1$ is a positive constant. If we sum the total number of dotted lines drawn for all the superblocks from x and y-axis, we get $nzx + c_1 \cdot n^2t \times x$. Since we store a block of size $t$ for each dotted line drawn, total space for table $T_0$ is $(nzx + c_1 \cdot n^2t \times x) \times t$.

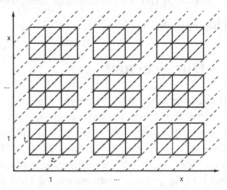

**Fig. 4.** Dotted lines drawn for the first superblock

**Size of Data Structure:** Summing up the space taken by all the tables we get the following equation:

$$s(x, z, t) = x^2zt + (nz + c \cdot n^2t)x^2 + (nzx + c_1 \cdot n^2t \times x) \times t \qquad (1)$$

As mentioned earlier size of each superblock is $x^2zt$, so the total number of superblocks are $n = m/(x^2zt)$. Substituting this in the above equation, we get the following:

$$s(x, z, t) = c_1 \cdot \frac{m^2}{x^3z^2} + c \cdot \frac{m^2}{x^2z^2t} + \frac{m}{x} + \frac{m}{t} + x^2zt \qquad (2)$$

Choosing $x = t = m^{1/6}$ and $z = m^{2/6}$, we get the space taken by our data structure to be $\mathcal{O}(m^{5/6})$.

## 3.1   Query Scheme

Our query scheme has three tables $T, T_0$ and $T_1$. Given a query element, we first find out the blocks to which it belongs. Further, we query the bit stored for this block in table $T$. If the bit returned is zero, we make the next query to table $T_0$ otherwise to table $T_1$. We say that query element is part of the set given to be stored if and only if last bit returned is one.

## 3.2   Storage Scheme

Our storage scheme sets the bits of tables $T, T_0$ and $T_1$ to store an arbitrary subset of size at most five in such a way that membership queries can be answered correctly. Storage scheme sets the bit of data structure depending upon the distribution of elements in various superblocks. Distribution of elements into various superblocks leads to various cases of the storage scheme. While generating various cases we consider an arbitrary subset $S = \{n_1, n_2, n_3, n_4, n_5\}$ of size five given to be stored. Each block is either sent to Table $T_0$ or $T_1$, and we store its bit-vector there. While sending blocks to either $T_0$ or $T_1$, we make sure that no two blocks sharing a bit have conflicting bit common in either of the tables, the correctness of the scheme relies on this fact. Keeping in mind the space constraints, we have discussed a few cases in this section, and for the sake of completeness the rest of the cases which can be handled in a similar fashion are mentioned in the extended version of the paper [3]. Most of the cases generated and assignment made are similar to those generated in the previous paper on the problem by Baig *et al.* [5] to store an arbitrary subset of size at most five.

**Case 1:** All the elements belonging to $S$ belongs to the same superblock. In this case, we send all the blocks having elements from the set given to be stored to table $T_1$. All the empty blocks, i.e., blocks which do not have any elements from $S$ are sent to table $T_0$.

**Case 2:** Four elements $S_1 = \{n_1, n_2, n_3, n_4\}$ lies in one superblock and one $S_2 = \{n_5\}$ in other. In this case, we send the block having element $n_5$ to table $T_1$ and rest all the blocks belonging to superblock which contains this element to table $T_0$. All the blocks which are having conflicting bit common with the block having element $n_5$ are sent to table $T_0$. Remaining all the blocks of superblocks which contains elements from $S_1$ are sent to table $T_1$. Furthermore, rest all the empty blocks of all the superblocks are sent to table $T_0$.

**Case 3:** All the elements $n_1, n_2, n_3, n_4$ and $n_5$ lies in the different superblocks. In this case, we send all the blocks having elements to table $T_0$ and all the empty blocks to table $T_1$.

We conclude this section with the following theorem:

**Theorem 9.** *There is a fully explicit two adaptive bitprobe scheme, which stores an arbitrary subset of size at most five, and uses $\mathcal{O}(m^{5/6})$ space.*

## 4    Conclusion

In this paper, we have studied those schemes that store subsets of size at most five and answer membership queries using two adaptive bitprobes. Our first result improves upon the known lower bounds for the problem by generalising the notion of universe of sets in Kesh [9] to what may be referred to as *second order universe*. We hope that suitably defining still higher order universes will help address the lower bounds for subsets whose sizes are larger than five. Though the lower bound of $\Omega(m^{3/4})$ is an improvement, we believe that it is not tight.

We have also presented an improved scheme for the problem. It refines the approach taken by Baig *et al.* [5] and alleviates the need for blocks that overlap completely to save space. This approach helps us achieve an upper bound of $\mathcal{O}(m^{5/6})$ which is a marked improvement over existing schemes.

## References

1. Alon, N., Feige, U.: On the power of two, three and four probes. In: Proceedings of the Twentieth Annual ACM-SIAM Symposium on Discrete Algorithms, SODA 2009, New York, NY, USA, 4–6 January 2009, pp. 346–354 (2009)
2. Baig, M.G.A.H., Kesh, D.: Two new schemes in the bitprobe model. In: Rahman, M.S., Sung, W.-K., Uehara, R. (eds.) WALCOM 2018. LNCS, vol. 10755, pp. 68–79. Springer, Cham (2018). https://doi.org/10.1007/978-3-319-75172-6_7
3. Baig, M.G.A.H., Kesh, D.: Improved bounds for two query adaptive bitprobe schemes storing five elements. CoRR abs/1910.03651 (2019), http://arxiv.org/abs/1910.03651
4. Baig, M.G.A.H., Kesh, D., Sodani, C.: An improved scheme in the two query adaptive bitprobe model. In: Colbourn, C.J., Grossi, R., Pisanti, N. (eds.) IWOCA 2019. LNCS, vol. 11638, pp. 22–34. Springer, Cham (2019). https://doi.org/10.1007/978-3-030-25005-8_3
5. Baig, M.G.A.H., Kesh, D., Sodani, C.: A two query adaptive bitprobe scheme storing five elements. In: Das, G.K., Mandal, P.S., Mukhopadhyaya, K., Nakano, S. (eds.) WALCOM 2019. LNCS, vol. 11355, pp. 317–328. Springer, Cham (2019). https://doi.org/10.1007/978-3-030-10564-8_25
6. Buhrman, H., Miltersen, P.B., Radhakrishnan, J., Venkatesh, S.: Are bitvectors optimal? In: Proceedings of the Thirty-Second Annual ACM Symposium on Theory of Computing, Portland, OR, USA, 21–23 May 2000, pp. 449–458 (2000)
7. Garg, M.: The bit-probe complexity of set membership. Ph.D. thesis, School of Technology and Computer Science, Tata Institute of Fundamental Research, Homi Bhabha Road, Navy Nagar, Colaba, Mumbai 400005, India (2016)
8. Kesh, D.: On adaptive bitprobe schemes for storing two elements. In: Gao, X., Du, H., Han, M. (eds.) COCOA 2017. LNCS, vol. 10627, pp. 471–479. Springer, Cham (2017). https://doi.org/10.1007/978-3-319-71150-8_39

9. Kesh, D.: Space complexity of two adaptive bitprobe schemes storing three elements. In: 38th IARCS Annual Conference on Foundations of Software Technology and Theoretical Computer Science, FSTTCS 2018, 11–13 December 2018, Ahmedabad, India, pp. 12:1–12:12 (2018)
10. Lewenstein, M., Munro, J.I., Nicholson, P.K., Raman, V.: Improved explicit data structures in the bitprobe model. In: Schulz, A.S., Wagner, D. (eds.) ESA 2014. LNCS, vol. 8737, pp. 630–641. Springer, Heidelberg (2014). https://doi.org/10.1007/978-3-662-44777-2_52
11. Radhakrishnan, J., Raman, V., Srinivasa Rao, S.: Explicit deterministic constructions for membership in the bitprobe model. In: auf der Heide, F.M. (ed.) ESA 2001. LNCS, vol. 2161, pp. 290–299. Springer, Heidelberg (2001). https://doi.org/10.1007/3-540-44676-1_24
12. Radhakrishnan, J., Shah, S., Shannigrahi, S.: Data structures for storing small sets in the bitprobe model. In: de Berg, M., Meyer, U. (eds.) ESA 2010. LNCS, vol. 6347, pp. 159–170. Springer, Heidelberg (2010). https://doi.org/10.1007/978-3-642-15781-3_14

# Critical Rows of Almost-Factorable Matrices

Peter Ballen[(✉)]

University of Pennsylvania, Philadelphia, PA, USA
pballen@seas.upenn.edu

**Abstract.** An almost-factorable matrix is a matrix that has a k-rank nonnegative factorization after deleting a single row. We call such a row a critical row. In graph optimization, a critical edge is an edge which can be deleted to reduce a graph measure, such as the size of the minimum vertex cover. We prove a reduction between critical rows in matrices and critical edges in graphs. Additionally, we describe and experimentally test an algorithm to identify the critical row of a matrix.

**Keywords:** Matrix factorization · Matrix optimization · Critical graph

## 1 Introduction

Nonnegative matrix factorization (NMF) takes a $n \times m$ matrix $\mathbf{V}$ and factors it into a $n \times k$ nonnegative matrix $\mathbf{W}$ and a $k \times m$ nonnegative matrix $\mathbf{W}$, with $\mathbf{V} \approx \mathbf{WH}$ and $k \ll n, m$. NMF has been applied in numerous domains, including image and hyperspectral analysis, biomedicine, and clustering. NMF also has theoretical connections to the minimum bipartite clique problem and problems in communication theory and computation geometry [6].

Finding the optimal NMF is NP-hard, but several algorithms work well in practice. In this work, we consider almost-factorable matrices. A matrix $\mathbf{V}$ is almost-factorable if deleting a single row from $\mathbf{V}$ either makes $\mathbf{V}$ perfectly factorable or substantially increases the quality of the factorization of $\mathbf{V}$. We call this row a 'critical row', and our goal is to find this row.

In graph optimization with measure $\alpha(G)$, an edge or vertex is called $\alpha$-critical if deleting that edge or vertex reduces $\alpha$. Examples of such graph measure include the size of the smallest vertex cover and the smallest number of colors needed to color a graph. Greedy algorithms have often proven effective at solving these optimization problems.

Our first main result is to extend the idea of minimum-vertex-cover critical edges in a graph to nonnegative-rank critical rows in a matrix. We prove a theoretical connection between the two, which allows us to prove that finding critical rows is NP-hard. This also provides a novel proof that finding the optimal NMF is NP-hard. The hardness of NMF was already proven by Vavasis [15], but our proof actually constructs a hard-to-factor matrix, which may be useful for other matrix hardness problems.

© Springer Nature Switzerland AG 2019
Y. Li et al. (Eds.): COCOA 2019, LNCS 11949, pp. 26–38, 2019.
https://doi.org/10.1007/978-3-030-36412-0_3

Our second main result is to develop an algorithm that finds the critical row of a matrix. No polynomial time algorithm can be theoretically guaranteed to find the critical row, but our algorithms perform well and generate a 3–10% improvement in the quality of the factorization on test datasets.

*Notation:* To avoid confusion, we use italics font $(\mathcal{V}, \mathcal{E})$ when discussing graphs and use bold font $(\mathbf{V}, \mathbf{W}, \mathbf{H})$ when discussing matrices. Additionally, $n$ and $m$ will always refer to the size of a matrix - we will use $|\mathcal{V}|$ and $|\mathcal{E}|$ when we want to refer to the number of vertices or edges in a graph. We always assume our graphs are simple and connected.

# 2    Background Work

## 2.1    Graph Background

The concept of $\alpha$-critical graphs was first proposed by Erdős and Gallai [5]. It has been studied when $\alpha$ is the number of colors needed to color a graph, the number of cliques needed to cover a graph, and other graph measures.

**Definition 1.** *Let $G = (\mathcal{V}, \mathcal{E})$ be a graph and let $\alpha(G)$ be a measure. An edge $e \in \mathcal{E}$ is $\alpha$-critical if the subgraph $G' = (\mathcal{V}, \mathcal{E} - e)$ has $\alpha(G') < \alpha(G)$. A graph is $\alpha$-critical if every edge is $\alpha$-critical.*

In this paper, we will take $\alpha(G)$ to be the size of the minimum vertex cover (MVC), which has been previously studied by Jakoby et al. [7].

**Definition 2.** *Let $G = (\mathcal{V}, \mathcal{E})$. A set $\mathcal{C} \subseteq \mathcal{V}$ is a vertex cover if for every edge $(v_i, v_j) \in \mathcal{E}$, either $v_i \in \mathcal{C}$ or $v_j \in \mathcal{C}$ or both. $\mathcal{C}$ is a minimum vertex cover of $G$ if no vertex cover with strictly fewer vertices exists.*

Finding the MVC of a graph is one of Karps' 21 NP-hard problems [9] and has been widely studied. Clarkson [3] proposes a simple greedy algorithm which iteratively adds the vertex adjacent to the most uncovered edges to the cover. Balaji [1] propose a more complex greedy algorithm which iteratively adds the vertex with the largest 'support' to the cover, where the support of a vertex is the total number of uncovered edges adjacent to neighbors of that vertex. These greedy algorithms - and alternative greedy approaches - are not theoretically guaranteed to provide good candidate covers. However, in Khan and Khans' [10] benchmark testing on the DIMASC benchmarks, greedy algorithms found the MVC about 50% of the time and have an average case approximation ratio of about 1.01. Additionally, these greedy algorithms naturally provide candidate critical edges (the last edges to be covered by the greedy algorithm).

Finally, we will need the following lemma about MVC-critical graphs.

**Lemma 1 (Jakoby et al. [7]).** *If $G = (\mathcal{V}, \mathcal{E})$ is a connected MVC-critical graph, then for every vertex $v_i \in \mathcal{V}$, there exists a minimum vertex cover $\mathcal{C}$ with $v_i \notin \mathcal{C}$.*

*Proof.* Suppose the MVC of $G$ has $k$ elements. Let $(v_i, v_j)$ be an edge in $G$ and define $G' = (\mathcal{V}, \mathcal{E} - (v_i, v_j))$. If $G$ is a MVC-critical graph, then $G'$ must contain a MVC $\mathcal{C}'$ with strictly less than $k$ vertices. Additionally, if either $v_i \in \mathcal{C}'$ or $v_j \in \mathcal{C}'$, then $\mathcal{C}'$ would be a MVC of $G$. This would contradict our assumption that the MVC of $G$ has $k$ elements. Therefore, neither $v_i$ nor $v_j$ is in $\mathcal{C}'$. Define $\mathcal{C} = \mathcal{C}' + v_j$; note that $v_i \notin \mathcal{C}$. $\mathcal{C}$ must be a MVC of $G$, since $v_j$ covers $(v_i, v_j)$ and $\mathcal{C}'$ covers every other edge.

## 2.2   Matrix Background

The linear rank of a matrix is the dimension of the subspace spanned by its rows or columns. Computing the linear rank can be done in polynomial time using singular value decomposition (SVD). In contrast, the nonnegative rank of a matrix is the dimension of the space when both the matrix and the coefficients are required to be nonnegative.

**Definition 3.** *Let* $\mathbf{V}$ *be a nonnegative* $n \times m$ *matrix. The nonnegative rank (NNR) of* $\mathbf{V}$ *is the smallest integer* $k$ *such that there exist a nonnegative* $n \times k$ *matrix* $\mathbf{W}$ *and a nonnegative* $k \times m$ *matrix* $\mathbf{H}$ *such that* $\mathbf{V} = \mathbf{WH}$.

Vavasis [15] proved that computing the NNR of a matrix is NP-hard. His reduction is circuitous, requiring a reduction from NNR to a linear algebra problem about singular matrices, to a geometric problem about simplices, and finally to 3SAT. Solving NMF can theoretically be done in $O((nm)^{k^2})$ using an algorithm developed by Morita [13], but this algorithm is not practical to implement due to exorbitant memory requirements, even on $4 \times 4$ matrices.

Instead, many iterative algorithms have been proposed to find the nonnegative matrix factorization. These algorithms do not guarantee an optimal solution, but work well in practice and run in polynomial time. One of the most widely used algorithms are the Lee and Seung [11] multiplicative update rules, which alternate between updating $\mathbf{W}$ and updating $\mathbf{H}$ according to the rules below.

$$\mathbf{W} \leftarrow \mathbf{W} \frac{\mathbf{V}\mathbf{H}^T}{\mathbf{W}\mathbf{H}\mathbf{H}^T} \qquad \mathbf{H} \leftarrow \mathbf{H} \frac{\mathbf{W}^T\mathbf{V}}{\mathbf{W}^T\mathbf{W}\mathbf{H}} \tag{1}$$

Another popular algorithm is coordinate descent, which chooses a single index in $\mathbf{W}$ or $\mathbf{H}$ and finds the optimal value for this index, keeping all other indices constant. This procedure is then repeated on every index in the two matrices.

Nonnegative factorization can be viewed as a vector problem. A vector $\mathbf{v}$ is a nonnegative linear combination of some set of nonnegative vectors $\mathbf{h}_1 \ldots \mathbf{h}_k$ if the exist nonnegative constants $c_1 \ldots c_k \geq 0$ such that $\mathbf{v} = \sum c_j \mathbf{h}_j$.

**Lemma 2.** *Let* $\mathbf{V}$ *be a nonnegative* $n \times m$ *matrix.* $\mathbf{V}$ *has* $k$*-dimensional NNR iff there exist* $k$ $m$*-dimensional vectors* $\mathbf{h}_1 \ldots \mathbf{h}_k$ *such that each row of* $\mathbf{V}$ *can be written as a nonnegative linear combination of* $\mathbf{h}_1 \ldots \mathbf{h}_k$. *We say the vectors* $\mathbf{h}_1 \ldots \mathbf{h}_k$ *generate the rows of* $\mathbf{V}$.

# 3   Critical Matrices

We begin by formally defining an $\alpha$-critical row in a matrix.

**Definition 4.** *Let* **V** *be a nonnegative* $n \times m$ *matrix* **V** *and let* $\alpha(\mathbf{V})$ *be a matrix measure. A set of indices* $\Psi$ *is an* $\alpha$-critical set *if there exists a nonnegative* $n \times m$ *matrix* **V**′ *such that* **V** *and* **V**′ *agree on all elements not in* $\Psi$ *and* $\alpha(\mathbf{V}') < \alpha(\mathbf{V})$. *A row is an* $\alpha$-critical row *if the indices in that row form a critical set. A matrix is* $\alpha$-critical *if every row is an* $\alpha$-critical row.

Note that setting every element of a critical row to zero is equivalent to deleting that row. However, this definition also enables discussion of critical sets that are not rows - we prove one such result in Corollary 2.

**Theorem 1.** *For any connected graph* $G = (\mathcal{V}, \mathcal{E})$, *there exists a nonnegative matrix* **V** *such that* $G$ *has a MVC of size* $s$ *iff* **V** *has a NNR of size* $k = s + |\mathcal{V}| + 4|\mathcal{E}|$. *Additionally, if* $e \in \mathcal{E}$ *is a MVC-critical edge in* $G$, *then the corresponding rows in* **V** *are NNR-critical rows.*

We break this section into three parts. We first define the reduction from graphs to matrices (Eq. 2). Lemma 3 proves one direction of the reduction and Lemma 4 proves the other direction. This reduction takes inspiration from the work of L.B. Thomas [14] on the gap between rank and nonnegative rank, who use the $5 \times 4$ matrix that appears in the bottom-right corner of Eq. 2. It also takes inspiration from the work of Miettinen et al. [12] and Jiang et al. [8], who use similar ideas on binary rank and finite state automata respectively. However, because of the existence of the critical row, and because **H** is real-valued and not a subset of a finite collection of sets, these proofs do not apply and new analysis is needed.

*Reduction:* Let $G = (\mathcal{V}, \mathcal{E})$ be a connected graph. We construct a nonnegative matrix **V** with $|\mathcal{V}| + 5|\mathcal{E}|$ rows and $2|\mathcal{V}| + 4|\mathcal{E}|$ columns. We begin by defining the columns. For each vertex $v_i \in \mathcal{G}$, we create two corresponding columns labeled $x_i$ and $y_i$. For each edge $(v_i, v_j) \in \mathcal{E}$, we create four corresponding columns labeled $a_{ij}, b_{ij}, c_{ij}$, and $d_{ij}$.

We next define the rows and entries on **V**. For each vertex $v_i \in \mathcal{V}$, we define a single row $r_i$. In each $r_i$ row, set the corresponding $x_i$ and $y_i$ to be 1 and set all other entries in this row to 0. This creates the first $|\mathcal{V}|$ rows. Then, for each edge $(v_i, v_j) \in \mathcal{E}$, we define five rows as follows. All values not explicitly set to 1 are set to 0.

- The first row of **V** for $(v_i, v_j)$ sets $x_i, a_{ij}$, and $b_{ij}$ to 1.
- The second row of **V** for $(v_i, v_j)$ sets $y_i, c_{ij}$, and $d_{ij}$ to 1.
- The third row of **V** for $(v_i, v_j)$ sets $x_j, b_{ij}$, and $c_{ij}$ to 1.
- The fourth row of **V** for $(v_i, v_j)$ sets $y_j, a_{ij}$, and $d_{ij}$ to 1.
- The fifth row of **V** for $(v_i, v_j)$ sets $a_{ij}, b_{ij}, c_{ij}, d_{ij}$ to 1.

This construction is visually depicted in Eq. 2, which depicts $v_i, v_j \in \mathcal{V}$ with edge $(v_i, v_j) \in \mathcal{E}$.

$$
\begin{array}{c}
x_i \; y_i \; x_j \; y_j \; \ldots \; a_{ij} \; b_{ij} \; c_{ij} \; d_{ij} \\
\mathbf{V} = \left(
\begin{array}{cccccccc}
\vdots & & & & & \vdots & & \\
1 & 1 & 0 & 0 & \ldots & 0 & 0 & 0 & 0 \\
0 & 0 & 1 & 1 & \ldots & 0 & 0 & 0 & 0 \\
\vdots & & & & & \vdots & & \\
1 & 0 & 0 & 0 & \ldots & 1 & 1 & 0 & 0 \\
0 & 1 & 0 & 0 & \ldots & 0 & 0 & 1 & 1 \\
0 & 0 & 1 & 0 & \ldots & 0 & 1 & 1 & 0 \\
0 & 0 & 0 & 1 & \ldots & 1 & 0 & 0 & 1 \\
0 & 0 & 0 & 0 & \ldots & 1 & 1 & 1 & 1 \\
\vdots & & & & & \vdots & &
\end{array}
\right)
\begin{array}{l}
\\
\text{row for } v_i \\
\text{row for } v_j \\
\\
\text{row1 for } (v_i, v_j) \\
\text{row2 for } (v_i, v_j) \\
\text{row3 for } (v_i, v_j) \\
\text{row4 for } (v_i, v_j) \\
\text{row5 for } (v_i, v_j)
\end{array}
\end{array}
\qquad (2)
$$

**Lemma 3.** *If $G = (\mathcal{V}, \mathcal{E})$ has a minimum vertex cover $\mathcal{C}$ of size $s$, then $\mathbf{V}$ as defined in Eq. 2 has a nonnegative rank of $k = s + |\mathcal{V}| + 4|\mathcal{E}|$. Additionally, if $e \in \mathcal{E}$ is a MVC-critical edge, then all five rows of $\mathbf{V}$ corresponding to that edge are NNR-critical rows.*

*Proof.* If $v_i \in \mathcal{C}$, add one row to $\mathbf{H}$ with $x_i = 1$ and all other entries set to zero, then add a second row to $\mathbf{H}$ with $y_i = 1$ and all other entries set to zero. If $v_i \notin \mathcal{C}$, add one row to $\mathbf{H}$ with $x_i = y_i = 1$ and all other entries set to zero. This process defines the first $s + |\mathcal{V}|$ rows of $\mathbf{H}$.

Next, for each edge $(v_i, v_j) \in \mathcal{G}$, either $v_i$ or $v_j$ must be in the vertex cover. Suppose that $v_i$ is in the cover. Then we will add four rows to $\mathbf{H}$, defined as follows. All values not explicitly set to 1 are set to 0.

- The first row of $\mathbf{H}$ for $(v_i, v_j)$ sets $a_{ij}$, and $b_{ij}$ to 1.
- The second row of $\mathbf{H}$ for $(v_i, v_j)$ sets $c_{ij}$, and $d_{ij}$ to 1.
- The third row of $\mathbf{H}$ for $(v_i, v_j)$ sets $x_j$, $b_{ij}$, and $c_{ij}$ to 1.
- The fourth row of $\mathbf{H}$ for $(v_i, v_j)$ sets $y_j$, $a_{ij}$, and $d_{ij}$ to 1.

This defines the remaining $4|\mathcal{E}|$ rows of $\mathbf{H}$. If $v_j$ was in the cover, we would instead modify these four rows so that $x_i = 1$ in the first row, $y_i = 1$ in the second row, $x_j = 0$ in the third row, and $y_j = 0$ in the fourth row. This construction is visually depicted in Eq. 3.

$$x_i\ x_j\ y_i\ y_j\ \ldots\ a_{ij}\ b_{ij}\ c_{ij}\ d_{ij}$$

$$\mathbf{H} = \begin{pmatrix} & \vdots & & & & & \vdots & & \\ 1 & 0 & 0 & 0 & \ldots & 0 & 0 & 0 & 0 \\ 0 & 1 & 0 & 0 & \ldots & 0 & 0 & 0 & 0 \\ 0 & 0 & 1 & 1 & \ldots & 0 & 0 & 0 & 0 \\ & \vdots & & & & & \vdots & & \\ 0 & 0 & 0 & 0 & \ldots & 1 & 1 & 0 & 0 \\ 0 & 0 & 0 & 0 & \ldots & 0 & 0 & 1 & 1 \\ 0 & 0 & 1 & 0 & \ldots & 0 & 1 & 1 & 0 \\ 0 & 0 & 0 & 1 & \ldots & 1 & 0 & 0 & 1 \\ & \vdots & & & & & \vdots & & \end{pmatrix} \begin{matrix} \\ \text{row1 for } v_i \in \mathcal{C} \\ \text{row2 for } v_i \in \mathcal{C} \\ \text{row1 for } v_j \notin \mathcal{C} \\ \\ \text{row1 for } (v_i, v_j) \\ \text{row2 for } (v_i, v_j) \\ \text{row3 for } (v_i, v_j) \\ \text{row4 for } (v_i, v_j) \\ \\ \end{matrix} \qquad (3)$$

We now observe that every row of $\mathbf{V}$ can be written as a nonnegative linear combination of the rows in $\mathbf{H}$. Every row in $\mathbf{V}$ corresponding to vertex $v_i$ can be written as the sum of the one or two rows in $\mathbf{H}$ corresponding to vertex $v_i$. Additionally, for each edge $(v_i, v_j)$ with $v_i$ in the vertex cover.

1. Row1 (resp. row2) for $(v_i, v_j)$ in $\mathbf{V}$ is the sum of the row1 (resp. row2) for $v_i$ in $\mathbf{H}$ and row1 (resp. row2) for $(v_i, v_j)$ in $\mathbf{H}$.
2. Row3 (resp. row4) for $(v_i, v_j)$ in $\mathbf{V}$ is equal to row3 (resp. row4) for $(v_i, v_j)$ in $\mathbf{H}$.
3. Row5 for $(v_i, v_j)$ in $\mathbf{V}$ is the sum of the row1 and row2 for $(v_i, v_j)$ in $\mathbf{H}$.

Therefore, every row of $\mathbf{V}$ can be written as a nonnegative linear combination of the rows of $\mathbf{H}$, and $\mathbf{H}$ has $k = s + |\mathcal{V}| + 4|\mathcal{E}|$ rows.

For the second part of the lemma, assume $e \in \mathcal{E}$ is a MVC-critical edge of $G$. Then there exists a cover $\mathcal{C}'$ with $s - 1$ vertices that covers every edge except $e$. To construct $\mathbf{H}'$, construct $(s - 1) + |\mathcal{V}| + 4|\mathcal{E}| - 4$ rows associated with each vertex in the cover and each edge other than $e$. Every row of $\mathbf{V}$ - other than the five rows of $\mathbf{V}$ corresponding to $e$ - can be written as a nonnegative combination of the rows in $\mathbf{H}$. Choose one of these rows as the critical row, then insert the remaining four rows into $\mathbf{H}'$. Now $\mathbf{H}'$ has $(s - 1) + |\mathcal{V}| + 4|\mathcal{E}|$ rows, and every row of $\mathbf{V}$ except the critical row can be written as a nonnegative combination of the rows in $\mathbf{H}'$.

**Lemma 4.** *If* $\mathbf{V}$ *as defined in Eq. 2 has a nonnegative rank of* $k = s + |\mathcal{V}| + 4|\mathcal{E}|$, *then the corresponding* $G = (\mathcal{V}, \mathcal{E})$ *has a minimum vertex cover* $\mathcal{C}$ *of size* $s$. *Additionally, if a row in* $\mathbf{V}$ *that corresponds to* $e \in \mathcal{E}$ *is a NNR-critical row, then* $e$ *is a MVC-critical edge in* $G$.

*Proof.* Let $\mathbf{H}$ be a set of $k$ rows such that every row of $\mathbf{V}$ can be written as a nonnegative linear combination of the elements in $\mathbf{V}$. We make the following two observations.

*Observation 1:* For the row in $\mathbf{V}$ corresponding to a $v_i$, either

- The row can be written as a scalar multiple of a row in $\mathbf{H}$ with $x_i = y_i \neq 0$ and all other entries are 0. –or–
- The row can be written as a nonnegative linear combination of two rows in $\mathbf{H}$: one row with $x_i > 0$ and all other entries equal zero and one row with $y_i > 0$ and all other entries equal to zero. In this case, $v_i$ is called a saturated vertex.

*Observation 2:* For the five rows in $\mathbf{V}$ corresponding to edge $(v_i, v_j)$, these five rows cannot be written as the nonnegative linear combination of three or fewer rows, and that any row with that has a nonzero value in a column other than $x_i, y_i, x_j, y_j, a_{ij}, b_{ij}, c_{ij}, d_{ij}$ is useless to represent these rows. Thus, either:

- $v_i$ (resp. $v_j$) is saturated, and these five rows in $\mathbf{V}$ can be written as the linear combination of the rows in $\mathbf{H}$ associated with $v_i$ (resp. $v_j$), plus four additional rows.
- These five rows can be written as the linear combination of five or more rows in $\mathbf{H}$. In this case, edge $(v_i, v_j)$ is called a saturated edge.

In both cases, the rows in $\mathbf{H}$ uses to represent the rows that are not associated with a vertex cannot be used to represent other rows. Thus, $k$ is equal to $|\mathcal{V}| + 4|\mathcal{E}|$ plus the number of saturated vertices, plus the number of rows in excess of four required to represent the saturated edges. This immediately implies that the number of saturated vertices plus the number of saturated edges is less than or equal to $k - |\mathcal{V}| - 4|\mathcal{E}|$. Furthermore, for every edge $(v_i, v_j)$, either one of the vertices is saturated or the edge itself is saturated.

Define $\mathcal{C}$ to be the set of saturated vertices. Then, for each saturated edge, add either endpoint of the edge to $\mathcal{C}$. $\mathcal{C}$ is now a vertex cover with at most $k - |\mathcal{V}| - 4|\mathcal{E}| = s$ elements, proving the first part of the lemma.

For the second part of the lemma, assume one of the rows in $\mathbf{V}$ corresponding to edge $(v_i, v_j)$ is a critical row. Let $k' = k - 1$. Observe that the four remaining rows in $\mathbf{V}$ corresponding to the edge $(v_i, v_j)$ still cannot be written as the nonnegative linear combination of three or fewer rows. Then $k'$ is equal to $|\mathcal{V}| + 4|\mathcal{E}|$, plus the number of saturated vertices, plus the number of rows in excess of four required to represent the saturated edges. However it is possible that neither $v_i$ nor $v_j$ nor edge $(v_i, v_j)$ is saturated. Construct $\mathcal{C}'$ identically as before. $\mathcal{C}'$ is a set with at most $k' - |\mathcal{V}| - 4|\mathcal{E}| = s - 1$ elements that covers every edge except $(v_i, v_j)$, proving the second part of the lemma.

## 3.1   Additional Results

Although the focus of this paper is detecting critical rows in a matrix, we can also use Theorem 1 to prove some related results about critical matrices and critical sets. The first result gives a way to construct nonnegative matrices where each row is a critical row: take a MVC-critical graph (such as the graphs generated by Jacoby et al. [7]) and apply the reduction to those graphs. Any matrix with

NNR $n$ (such as the identity matrix) is a trivial example of such a matrix, but this reduction a way to construct more interesting instances. The second result proves that finding critical sets remains NP-hard even when given certain types of side-channel information. The third result is the observation that the above proofs also hold when considering boolean rank instead of nonnegative rank.

**Corollary 1.** *G is a MVC-critical graph iff every row of* **V** *is a NNR-critical row.*

*Proof.* Lemmas 3 and 4 handle most of this result. However, we still need to argue that if $G$ is a MVC-critical graph, then every row in **V** that corresponds to a vertex is a critical row (Lemma 3 only proves that rows in **V** that correspond to edges are critical). Let row $r$ be the row in **V** that corresponds to vertex $v_i$. Since $G$ is MVC-critical, by Lemma 1 there exists a MVC $\mathcal{C}$ of size $s$ that does not contain $v_i$. Let **H** be the matrix with $s + (|\mathcal{V}| - 1) + 4|\mathcal{E}|$ rows defined by Eq. 3 and $\mathcal{C}$, excluding the vertex row in **H** associated with $v_i$. Every row of **V** other than $r$ can be written as a nonnegative linear combination of the rows of **H**, so $r$ is a critical row.

**Corollary 2.** *Let* **V** *be a nonnegative matrix, and for each column $j$, let $q_j \geq 0$ be an integer. It is NP-hard to determine whether* **V** *has a critical set that contains exactly $q_j$ elements from column $j$.*

*Proof.* Define **V**$'$ as follows. Take **V** as defined in Eq. 2 and define a new column called the checksum column. For each row in **V** corresponding to a vertex, set the checksum to 1. For each row in **V** corresponding to an edge, set the checksum to 0. Define $q_j = 2s$ for the checksum column and $q_j = 0$ for the other columns.

We prove $G$ has a MVC of size $s$ iff **V**$'$ has such a critical set. For the forward direction, define **H**$'$ by taking **H** as defined in Eq. 3 and add the checksum column. If $v_i \in \mathcal{C}$, the checksum of row1 and row2 for $v_i$ is 0 (these are the elements of the critical set). If $v_j \notin \mathcal{C}$, set the checksum of row1 for $v_j$ to be 1 (these rows are never used to generate the other rows of **V**$'$, so they can be safely modified). The other direction of the reduction follows from Lemma 4: the checksums do not modify Observation 1 and 2, or the subsequent argument.

**Corollary 3.** *Theorem 1, Lemma 3 and 4 apply when $\alpha(\mathbf{V})$ is the boolean rank of a zero-one matrix instead of the nonnegative rank.*

The boolean rank of **V** is the rank where **V**, **W**, **H** are zero-one matrices and standard arithmetic is replaced with boolean arithmetic ($1 =$ true, $0 =$ false).

# 4    Detecting Critical Rows

Section 3 proved that finding a NNR-critical row corresponds to finding a MVC-critical vertex in a graph. However, real world datasets are subject to noise and perfect factorizations are rare. We modify Definition 4 to handle this limitation. Definition 4 with $\alpha$ as the NNR is equivalent to Definition 5 with $\epsilon = 0$.

**Definition 5.** *Let* **V** *be a nonnegative* $n \times m$ *matrix* **V**, *let* $0 \leq \epsilon \leq 1$, *and let* $\beta_k(\mathbf{V})$ *be a graph measure. A set of indices* $\Psi$ *is a* $\beta_{k,\epsilon}$-*critical set if* $\Psi$ *if there exists a* $n \times m$ *matrix* **V**′ *such that* **V** *and* **V**′ *agree on all elements not in* $\Psi$ *and* $\beta_k(\mathbf{V}') < \epsilon * \beta_k(\mathbf{V})$.

We take $\beta_k$ to be the error of the optimal $k$-rank NMF $\beta_k(\mathbf{V}) = \sum (\mathbf{V}_{ij} - (\mathbf{WH})_{ij})^2$. We now give two simple algorithms to identify a $\beta$-critical row.

**Algorithm 1:** Run a standard nonnegative matrix factorization algorithm assuming no critical row exists, then return the row with the highest error as the candidate critical row.

Recall that the Clarkson MVC greedy algorithm [3] iteratively takes the vertex adjacent to the largest number of uncovered vertices and adds it to the cover. If the 'error' of a vertex is defined as the number of adjacent uncovered edges, this is equivalent to a greedy algorithm that reduces the error by the largest amount. We now take this idea and extend it to matrices, where $topr(a, \text{ERR})$ is the indices of the rows in ERR with the greatest error.

---

**Algorithm 2.** Critical Row Identification

---

1: Initialize $SUSPECT = \{\}$
2: **for** $t = 1, 2, 3, \ldots$ **do**
3:     $\widetilde{\mathbf{V}} = \mathbf{V}$ after deleting all rows in $SUSPECT$ from **V**
4:     $\widetilde{\mathbf{W}} = \mathbf{W}$ after deleting all rows in $SUSPECT$ from **W**
5:     $\mathbf{H} = \mathbf{H}\dfrac{\widetilde{\mathbf{W}}^T\widetilde{\mathbf{V}}}{\widetilde{\mathbf{W}}^T\widetilde{\mathbf{W}}\mathbf{H}}$          ▷ Suspect rows do not affect the **H** update
6:     $\mathbf{W} = \mathbf{W}\dfrac{\mathbf{VH}^T}{\mathbf{WHH}^T}$          ▷ Suspect rows are still updated in **W**
7:     **if** $t \bmod 25 = 0$ **then**
8:         Compute $\text{ERR}[i] = \sum_{j=1}^m (\mathbf{V}_{ij} - (\mathbf{WH})_{ij})^2$
9:         Set $SUSPECT = topr(.01 * n, \text{ERR})$
10:     **end if**
11: **end for**
12: Compute $ERR_i = \sum_{j=1}^m (\mathbf{V}_{ij} - (\mathbf{WH})_{ij})^2$
13: Return $topr(1, ERR_i)$

---

For the first 25 iterations, run nonnegative matrix factorization as normal. Then compute the total error of each row (ERR) and mark the 1% rows with the largest error as suspect. For iterations 26–50, ignore suspect rows when updating **H**, but still allow suspect rows to be updated in **W**. This corresponds to isolating the suspect rows; suspect rows can't corrupt the factorization by altering values in **H**, but they still get updated in **W** and the error associated with those rows will still decrease.

Observe that while the algorithm ultimately returns a single candidate row, a benefit of this approach over the naive approach is that the algorithm can consider multiple suspect rows to isolate. Thus, the algorithm could also be used to find a critical set that contains multiple rows.

## 5    Experiments

*Datasets:* The first dataset we consider is synthetic. To generate the matrices, we set $\mathbf{W} = \text{RAND}(100, 5)$ and $\mathbf{H} = \text{RAND}(5, 1000)$, where $\text{RAND}(n, m)$ is a $n \times m$ matrix with each entry drawn from the uniform $[0, 1]$ distribution. Values in $\mathbf{V}$ range from 0 to 5, with a mean of 1.25 and a stdev of 0.25.

The second dataset we consider is generated from a flattened hyperspectral satellite image [4], which has been used before in NMF experiments. The original data is a $145 \times 145$ image with 224 values associated with each pixel (a three dimensional matrix). We collapse the matrix into a $224 \times 21025$ matrix, where each row corresponds to a specific spectral band and each column corresponds to a pixel in the image. Values in $\mathbf{V}$ range from 0 to 10000, with a mean of 2652 and a stdev of 1592.3. The goal of finding the critical row is equivalent to finding the spectral band that is returning incorrect values.

The third dataset we consider is Medulloblastoma data [2] (a type of brain tumor), which has been used before in NMF experiments. $\mathbf{V}$ is a $34 \times 5893$ matrix, where each row corresponds to an individual and each column corresponds to a gene. Values in $\mathbf{V}$ range from 20 to 1600 with a mean of 372 and a standard deviation of 972. The goal of finding the critical row is equivalent to finding the individual whose gene scan is abnormal.

In all three experiments, we randomly permute the generated matrix. We then choose a random row from $\mathbf{V}$ to be the critical row. We add independent $Normal(0, \sigma)$ contamination to each entry in the critical row, varying $\sigma$. All negative values are set to zero. As $\sigma$ increases, the critical row drifts further and further away from the underlying factorization, thus making it easier to identify.

*Algorithmic Comparisons:* We implement Algorithm 2 in Python. To implement Algorithm 1, we use scikit-learn's implementation of NMF as the naive critical row algorithm and return the row with the largest squared error. We ran both multiplicative updates and coordinate descent as the underlying algorithm and the results were statistically identical. Both algorithms use the same random initializations of $\mathbf{W}$ and $\mathbf{H}$.

*Error Metrics:* For each value of $\sigma$, we run the experiment 50 times. For each trial, we compute ERR, the total squared error in each row. We sort this list in descending order and compute the index of the critical row in this list. For example, an index of 3 means that the critical row had the third highest error

Table 1. Summary of experimental datasets

| Dataset | $n$ | $m$ | $k$ | mean($\mathbf{V}$) | stdev($\mathbf{V}$) | $(n-1)/n$ |
|---|---|---|---|---|---|---|
| Synthetic | 100 | 1000 | 5 | 1.25 | 0.50 | 0.990 |
| Hyperspectral | 224 | 21025 | 20 | 2652 | 1592 | 0.995 |
| Medulloblastoma | 34 | 5893 | 2 | 372 | 972 | 0.970 |

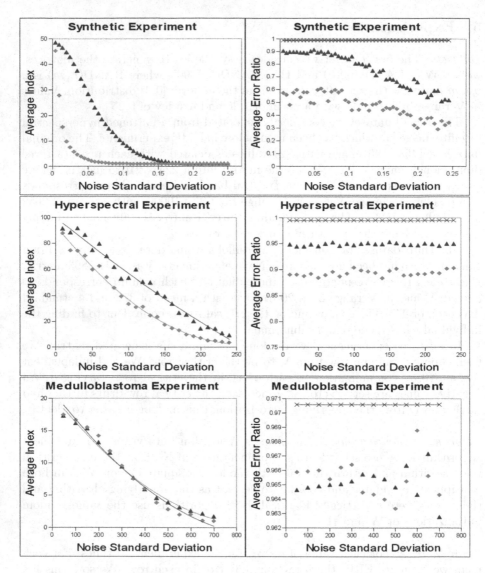

**Fig. 1.** Experimental results. Algorithm 1 is blue triangles, Algorithm 2 is pink dots, Baseline $(n-1)/n$ is green crosses. x-axis denotes noise stdev - larger values indicate the critical row is further from the underlying representation. y-axis denotes average index and average $\beta$-ratio. (Color figure online)

among all rows, and that the critical row would have been the algorithm's third choice. The optimal index is 1, and lower numbers indicate better performance.

We also compute the error ratio. To compute the original error, we run NMF on the input matrix to get $\mathbf{W}, \mathbf{H}$ and set $\beta_k(\mathbf{V}) = \sum (\mathbf{V} - \mathbf{W}\mathbf{H})^2/n$. To compute the new error, we compute the squared error between $\mathbf{V}'$ and $\mathbf{W}', \mathbf{H}'$ returned by the algorithm, excluding the critical row. We then set $\beta_k(\mathbf{V}') = \sum(\mathbf{V}' - \mathbf{W}'\mathbf{H}')^2/(n-1)$. This averaging procedure ensures that $\mathbf{V}'$ does indeed have a better factorization than $\mathbf{V}$, and the improvement is not a result of simply having less rows in the matrix. The error ratio is equal to $\beta_k(\mathbf{V}')/\beta_k(\mathbf{V})$, which corresponds to $\epsilon$ from Definition 5. We also include a baseline of $(n-1)/n$ (Fig. 1).

Both algorithms successfully identify the critical row, as long as the critical row is sufficiently distant from the true data. On synthetic data, removing the critical row can improve the factorization (as measured by $\beta$) by over 50%. On real data that lacks a perfect factorization, removing the critical row can still improve the factorization quality by about 3–10% (Table 1).

## 6   Conclusion

We define critical rows of a matrix, and theoretically connect them to the critical edges of a graph. We also prove results about critical sets and critical matrices. In addition to the theoretical results, we give an algorithm to find the critical rows. One direction for future work is to consider alternative matrix measures $\alpha(\mathbf{V})$, instead of focusing on matrix factorization. An second direction is to consider critical sets with more complex structure, instead of just critical rows.

## References

1. Balaji, S., Swaminathan, V., Kannan, K.: Optimization of unweighted minimum vertex cover. World Acad. Sci. Eng. Technol. **43**, 716–729 (2010)
2. Brunet, J.P., Tamayo, P., Golub, T.R., Mesirov, J.P.: Metagenes and molecular pattern discovery using matrix factorization. Proc. Nat. Acad. Sci. **101**(12), 4164–4169 (2004)
3. Clarkson, K.L.: A modification of the greedy algorithm for vertex cover. Inf. Process. Lett. **16**(1), 23–25 (1983)
4. Computational Intelligence Group, U.o.B.C. http://www.ehu.eus/ccwintco/index.php/hyperspectral_remote_sensing_Scenes
5. Erdos, P., Gallai, T.: On the minimal number of vertices of a graph representing the edges of a graph (1961)
6. Gillis, N.: The why and how of nonnegative matrix factorization. Regul. Optim. Kernels Support. Vector Mach. **12**(257), 257–291 (2014)
7. Jakoby, A., Goswami, N.K., List, E., Lucks, S.: Critical graphs for minimum vertex cover. CoRR abs/1705.04111 (2017)
8. Jiang, T., Ravikumar, B.: Minimal NFA problems are hard. SIAM J. Comput. **22**(6), 1117–1141 (1993)
9. Karp, R.M.: Reducibility among combinatorial problems. In: Miller, R.E., Thatcher, J.W., Bohlinger, J.D. (eds.) Complexity of computer computations. The IBM Research Symposia Series, pp. 85–103. Springer, Boston (1972). https://doi.org/10.1007/978-1-4684-2001-2_9
10. Khan, I., Khan, S.: Experimental comparison of five approximation algorithms for minimum vertex cover. Int. J. u-and e-Serv. **7**(6), 69–84 (2014)
11. Lee, D.D., Seung, H.S.: Algorithms for non-negative matrix factorization. In: Advances in Neural Information Processing Systems, pp. 556–562 (2001)

12. Miettinen, P., Mielikäinen, T., Gionis, A., Das, G., Mannila, H.: The discrete basis problem. In: Fürnkranz, J., Scheffer, T., Spiliopoulou, M. (eds.) PKDD 2006. LNCS (LNAI), vol. 4213, pp. 335–346. Springer, Heidelberg (2006). https://doi.org/10.1007/11871637_33
13. Moitra, A.: An almost optimal algorithm for computing nonnegative rank. SIAM J. Comput. **45**(1), 156–173 (2016)
14. Thomas, L.: Rank factorization of nonnegative matrices. SIAM Rev. **16**(3), 393–394 (1974). https://doi.org/10.1137/1016064
15. Vavasis, S.A.: On the complexity of nonnegative matrix factorization. SIAM J. Optim. **20**(3), 1364–1377 (2009)

# Minimum-Width Drawings
## of Phylogenetic Trees

Juan Jose Besa$^{(\boxtimes)}$ ⓘ, Michael T. Goodrich ⓘ, Timothy Johnson,
and Martha C. Osegueda

University of California, Irvine, USA
{jjbesavi,goodrich,tujohnso,mosegued}@uci.edu

**Abstract.** We show that finding a minimum-width orthogonal upward
drawing of a phylogenetic tree is NP-hard for binary trees with unconstrained combinatorial order and provide a linear-time algorithm for
ordered trees. We also study several heuristic algorithms for the unconstrained case and show their effectiveness through experimentation.

**Keywords:** Phylogenetic · Tree drawing · Orthogonal · Upward

## 1 Introduction

A *phylogenetic tree* is a rooted tree that represents evolutionary relationships
among a group of organisms. The branching represents how species are believed
to have evolved from common ancestors and the vertical height of an edge represents the genetic difference or time estimates for when the species last diverged.
In this paper, we study the algorithmic complexity of producing minimum-width
drawings of phylogenetic trees, in particular we describe it as finding the minimum width drawing of upward orthogonal trees of fixed edge-length. Edge-length
must equal vertical distance because edges can only extend vertically when the
tree is both orthogonal and upward; to allow for more than one outgoing edge,
vertices must extend horizontally.[1] See Fig. 1.

Given a phylogenetic tree, $T$, with a length, $L_e$, defined for each edge $e \in
T$, the *min-width phylogenetic tree* drawing problem is to produce an upward
planar orthogonal drawing of $T$ that satisfies each edge-length constraint (so the
drawn vertical length of each edge $e$ is $L_e$)[2] and minimizes the width of the
drawing of $T$.

The motivation for this problem is to optimize the area of the drawing of a
phylogenetic tree, since the height of the drawing is fixed by the sum of lengths
of the edges on a longest root-to-leaf path. From an algorithmic complexity
perspective this problem is trivial in clock trees for non-extinct species, since all
root-to-leaf paths are of the same length and all leaves must therefore be drawn

---

[1] Alternatively, an upward node-link tree with 1-bend edges and fixed node height.

[2] W.l.o.g., each node is embedded below its parent; other orientations, such as drawing
nodes above their parents, are equivalent to this one via rotation.

© Springer Nature Switzerland AG 2019
Y. Li et al. (Eds.): COCOA 2019, LNCS 11949, pp. 39–55, 2019.
https://doi.org/10.1007/978-3-030-36412-0_4

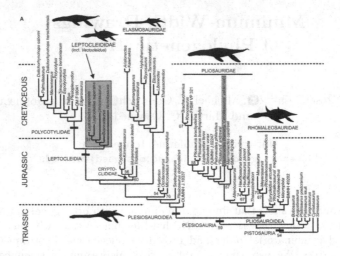

**Fig. 1.** An orthogonal upward drawing of a phylogenetic tree, from [6].

on the same level. Thus, we are interested in the general case, as shown in Fig. 1, where width improvement is possible by allowing subtrees of extinct species to lay above the branches of surviving species.

In this paper, we show that min-width phylogenetic tree is NP-hard if the order of the leaves can be chosen (i.e. unconstrained) and provide a linear-time algorithm for trees with a fixed leaf ordering. Also, we describe several heuristic algorithms for the former and show their effectiveness by experimentation.

*Related Work.* There is considerable prior work on methods for producing combinatorial representations of phylogenetic trees, that is, to determine the branching structures and edge lengths for such trees, e.g., see [17, 18, 22, 28, 30]. For this paper, we assume that a phylogenetic tree is given as input.

Existing software systems produce orthogonal drawings of phylogenetic trees, e.g., see [7, 10, 19, 23, 24, 31], but we are not familiar with any previous work on characterizing the algorithmic complexity of the minimum-width orthogonal phylogenetic tree drawing problem. Bachmaier *et al.* [2] present linear-time algorithms producing other types of drawings of ordered phylogenetic trees, including radial and circular drawings, neither of which are orthogonal drawings.

Several researchers have studied area optimization problems for planar upward tree drawing without edge-length constraints, e.g., see [1, 11–15, 20, 26, 27]. In terms of hardness, Biedl and Mondal [5] show it is NP-hard to decide whether a tree has a strictly upward straight-line drawing in a given $W \times H$ grid. Bhatt and Cosmadakis [4] show that it is NP-complete to decide whether a degree-4 tree has a straight-line (non-upward) orthogonal grid drawing where every edge has length 1. Gregori [16] extends their result to binary trees and Brunner and Matzeder [9] extend their result to ternary trees. In addition, Brandes and Pampel [8] show that several order-constrained orthogonal graph

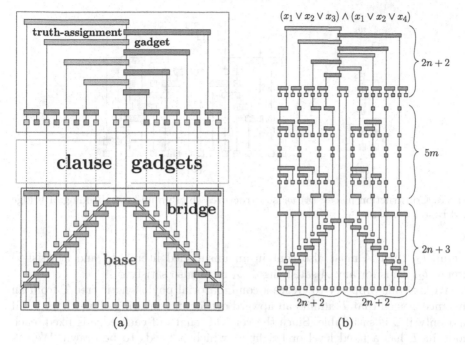

**Fig. 2.** Overview of Theorem 1 (a) General structure. (b) Drawing of reduction corresponding to a satisfying assignment, $\mathcal{A} = \{true, true, false, false\}$

drawing problems are NP-hard, and Bannister and Eppstein [3] show that compacting certain nonplanar orthogonal drawings is difficult to approximate in polynomial time unless $P = NP$. Previous hardness proofs do not apply to the min-width phylogenetic tree drawing problem, however.

## 2    Min-Width Phylogenetic Tree Drawing is NP-hard

**Theorem 1.** *Computing the minimum width required for an upward planar orthogonal drawing of a binary tree with fixed vertical edge lengths is NP-hard.*

We prove this via a reduction from NAE-3SAT, a variant of 3SAT in which an assignment is considered satisfying when each clause's boolean values are not all the same. An instance, $\phi$, of NAE-3SAT is defined by $n$ variables $X = \{x_1, ..., x_n\}$ and $m$ clauses $C = \{c_1, ..., c_m\}$. Each clause consists of exactly 3 literals, where a *literal* is a negated or non-negated variable. Given a truth assignment, a literal is considered *satisfied* if the literal evaluates to true. A truth assignment $\mathcal{A}$ satisfies $\phi$ when each clause contains only one or two satisfied literals. Each clause must therefore also have either one or two unsatisfied literals.

Given a truth assignment, $\mathcal{A}$, we define $\overline{\mathcal{A}}$ to be the truth assignment where each assigned truth value is the negation of the truth value assigned in $\mathcal{A}$. If $\mathcal{A}$ satisfies $\phi$ each clause must contain a satisfied literal, $l_1$, and an unsatisfied

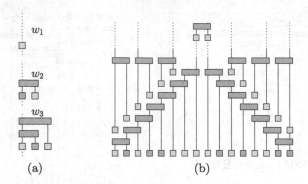

**Fig. 3.** Construction pieces: (a) $w_k$ substructures (b) Pyramidal embedding of bridge and base for $n = 3$

literal, $l_2$, then $\overline{A}$ must also contain an unsatisfied literal, $l_1$, and a satisfied literal, $l_2$. Thus for any $A$ satisfying $\phi$, $\overline{A}$ must also satisfy $\phi$.

We use this property to create a combinatorial phylogenetic tree $T$ from an instance $\phi$ such that $T$ admits an upward orthogonal drawing of width $4n + 4$ if and only if $\phi$ is satisfiable. Since the vertical length of each edge is fixed, each node in $T$ has a fixed level or height at which it needs to be drawn. We say that two nodes are *horizontally aligned* when they lie at the same level. Our reduction follows the general structure shown in Fig. 2a. The top part of the tree forms a *truth-assignment gadget*, where each variable has two branches, one corresponding to a true value (shown in green) and the other corresponding to a false value (shown in red). The truth-assignment gadget's combinatorial order therefore defines truth assignments for $A$ and $\overline{A}$, on the left and right, respectively. Figure 2a illustrates all true and all false truth assignments (respectively) whereas Fig. 2b illustrates a satisfying assignment for a two-clause formula. The middle part comprises a sequence of *clause gadgets*, with 3 rows for each clause gadget, plus rows of *alignment nodes*, separating the clause gadgets. The bottom parts of the tree comprise *bridge* and *base* gadgets, which ensure that the pair of branches corresponding to the true and false values for each variable must be on opposite sides (see Lemma 2).

Let $P(k)$ be a path of $k$ nodes with unit edge-lengths, where $p_1$ is the root of the path and $p_i \in P, 2 \le i \le k$ is the i-th node along the path. A node $p_i$ may have two children, the node $p_{i+1}$ (if it exists) and some other new node (or two if $i = k$). We define the substructures $w_1, w_2$, and $w_3$ (Fig. 3a), consisting of the minimal tree with 1, 2, and 3 leaves at the same height.

We proceed from top to bottom to describe the structure of $T$. The tree, $T$, is rooted on the first node of a path $P(2n)$, called the *variable path*. We first add a $w_2$ component to $p_{2n}$ a unit-length away as its first child, which we use to root the base. We then add a $w_2$ component as the second child of each node $p_i, 1 \le i \le 2n$ so the $2n$ new $w_2$ components lay in horizontal alignment with the one rooting the base. These $2n$ $w_2$ components connect each node in the variable path to a branch. Of the $2n$ branches, there are two for each of the $n$ variables in $\phi$. We label the branch originating from $p_i$ with the truth assignment setting $x_{\lfloor (i+1)/2 \rfloor}$ to 'true' if $i$ is odd and 'false' if $i$ is even. These branches are now called *assignment branches*.

Each assignment branch consists of $m$ clause components connected in a path by unit-length edges, where the $j$-th component (for $1 \leq j \leq m$) belongs to the clause gadget of clause $c_j$. All clause components belong to the same clause have the same height, so each clause occupies a distinct horizontal band of space in the tree. Each clause component consists of one of three possible $w_k$ components and two surrounding alignment nodes, one above the $w_k$ component and one below. Each $w_k$ component has height 3 and for clause components corresponding to the same clause the leaves of all $w_k$ are horizontally aligned.

Recall that each variable has two assignment branches associated to it, one corresponding to setting the variable to true and one to false. A clause component is defined by both the clause, $c_j$, and the branch's labeled truth assignment, $x_i = \{true, false\}$. If $x_i$ does not participate in clause $c_j$ then its clause component has a $w_2$ substructure. If $x_i$ appears as a literal in $c_j$ and evaluating it with the assigned truth value satisfies the literal then the clause component has a $w_3$ substructure. However, if the literal is unsatisfied then it contains a $w_1$ substructure. For example, in Fig. 2b clause $c_1 = (x_1 \vee x_2 \vee x_3)$, the branch labeled $x_1 = true$ contains $w_3$, whereas $x_1 = false$ contains $w_1$ and both $x_4$ assignment branches contain $w_2$.

Top alignment nodes have incoming edges connecting from the bottom alignment nodes of the previous clause. These two consecutive rows of alignment nodes enable the clause gadgets along an assignment branch to be shifted left or right to efficiently align into the available space within that gadget (see Appendix A.1).

After the last clause component in each branch (labeled with $x_i$) we attach a node one unit away from the last shifter and give it two children, one $2n - 2i + 1$ units away and the other $2n - 2i + 2$. These nodes together form the bridge gadget.

To build the base, we set a path $P(2n + 2)$ as a child $5m + 2$ units away for each of the two leaves in the remaining copy of $w_2$ attached to $p_{2n}$. For each non-leaf node in the path we just connected, we set their remaining child to be a single node horizontally aligned with the leaf.

**Lemma 1.** *At minimal width, the base and bridge can only assume a pyramidal embedding. We define a pyramidal embedding as the embedding of the base in which nodes closer to the root lie closer to the center and nodes further from the root approach the outer sides, as shown in Fig. 3b.*

*Proof.* This proof is based on the fact that a pyramidal embedding fully occupies every cell, and thus occupies the least area and minimum width. Furthermore, the pyramidal embedding is the only minimum width embedding for these gadgets. We present the complete proof in Appendix A.2.

**Lemma 2.** *The embedding of the truth-assignment gadget defines assignments $\mathcal{A}$ and $\overline{\mathcal{A}}$.*

*Proof.* As a consequence of Lemma 1, the two edges coming into the base must be centered. We refer to these two edges as 'the split'. The base takes up width $2n + 1$ on either side of these two edges, and each branch needs width two, so at most $n$ can fit on either side.

Furthermore, in order for the bridge to assume the pyramidal embedding, each leaf on the left side must have a corresponding leaf at the same level on the right side. Note that the height of the leaves in the bridge gadget depends on which variable branch they are attached to, so that leaves on the same level correspond to the two assignments of the same variable. Therefore, there is one assignment branch for each variable on each side, so the assignments labeling the branches on one side must all be of different variables and thus describe a truth assignment. The branches on the opposite side have the opposite assignment for each variable. Let the truth assignment on the left side be $\mathcal{A}$, and the one on the right side be $\overline{\mathcal{A}}$.                    ∎

**Lemma 3.** *Given truth assignments $\mathcal{A}$ and $\overline{\mathcal{A}}$, a clause in $\phi$ is satisfied if and only if the width used by the clause gadget is at most $2n + 1$ on both sides of the split.*

*Proof.* The maximum number of horizontally aligned nodes on the left side of the clause gadget is equal to the sum of the width of the $w_k$'s, $k \in \{1, 2, 3\}$, in each assignment branch for assignment $\mathcal{A}$. Recall that within a clause gadget each assignment branch has an embedded copy of either $w_1$, $w_2$ or $w_3$. We define function $p_{i,j}$, which describes the width of an embedded copy $w_k$, and $S_j(\mathcal{A})$, which describes the sum of all the embedded element's widths, as follows:

$$S_j(\mathcal{A}) = \sum_{i=1}^{n} p_{i,j}(\mathcal{A}), \text{ where } p_{i,j}(\mathcal{A}) = \begin{cases} 3 & \text{if } A \text{ does satisfy } x_i\text{'s literal in } c_j \\ 1 & \text{if } A \text{ doesn't satisfy } x_i\text{'s literal in } c_j \\ 2 & \text{if } c_j \text{ has no literal of } x_i \end{cases}$$

By definition, a clause can only be satisfied if one or two of the clause's literals are satisfied and the remaining one or two literals must be unsatisfied. W.l.o.g. we can assume that each clause consists of two or three literals from distinct variables.[3] For clauses with literals from three distinct variables, any clause $c_j$ satisfied by a truth assignment $\mathcal{A}$ must have only one or two satisfied literals. If $\mathcal{A}$ satisfies only one literal in clause, $c_j$, then $S_j(\mathcal{A})$ evaluates to $3+1+1+2(n-3) = 2n - 1$, since only 3 of the $n$ variables participate in a clause. If it satisfies two literals, it evaluates to $3 + 3 + 1 + 2(n - 3) = 2n + 1$ instead. If $\mathcal{A}$ satisfies $c_j$, then $\overline{\mathcal{A}}$ also satisfies $c_j$ implying $S_j(\mathcal{A})$ and $S_j(\overline{\mathcal{A}})$ are both at most $2n + 1$. On the other hand, if $\mathcal{A}$ doesn't satisfy $c_j$, then it either satisfies all three literals, and $S_j(\mathcal{A})$ evaluates to $3 + 3 + 3 + 2(n - 3) = 2n + 3$, or $\overline{\mathcal{A}}$ satisfies all three

---

[3] Degenerate cases to consider include cases when a variable contributes multiple literals to a single clause. We can safely ignore cases when all three identical literals are present (which is not satisfiable) and when positive and negated literals of the same variable are present (since the clause is always satisfied). When a literal is repeated exactly twice, we handle it as a clause of only the two distinct literals.

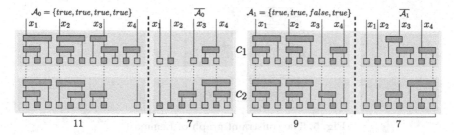

**Fig. 4.** Original and satisfied clause gadgets for $\phi = (x_1 \vee x_2 \vee x_3) \wedge (x_1 \vee x_2 \vee \overline{x_4})$

literals and $S_j(\overline{A}) = 2n + 3$. Therefore if $A$ doesn't satisfy $c_j$, $S_j(A)$ or $S_j(\overline{A})$ will exceed $2n + 1$.

For clauses with literals from two distinct variables, any $A$ can only satisfy $c_j$ with one satisfied literal and one unsatisfied literal. $S_j(A)$ and $S_j(\overline{A})$ both evaluate to $3 + 1 + 2(n - 2) = 2n$, both remaining strictly less than $2n + 1$. However if $A$ doesn't satisfy $c_j$ then either $A$ satisfies both literals, and $S_j(A) = 3 + 3 + 2(n - 2) = 2n + 2$, or $\overline{A}$ satisfies both literals. Therefore if $A$ doesn't satisfy $c_j$, $S_j(A)$ or $S_j(\overline{A})$ will exceed $2n + 1$.

We have now proved that $S_j(A)$ and $S_j(\overline{A})$ are both at most $2n + 1$ if and only if $A$ and $\overline{A}$ both satisfy $c_j$. As long as the clause gadget is able to assume a dense embedding, the width necessary on opposite sides should be exactly equal to $S_j(A)$ and $S_j(\overline{A})$. The alignment nodes in each clause gadget are sufficient to guarantee a dense embedding is possible, which we prove in Appendix A.1. ∎

Wrapping up the main proof, if any clause is not satisfied the clause gadget will exceed the allowable space of $2n+1$ and increase the width to at least $4n+5$. Therefore only a satisfying assignment $A$ would retain a width of $4n+4$, proving that if a satisfying assignment for $\phi$ exists then there exists an embedding of $T$ with width $4n + 4$ (Fig. 4).

On the other hand, if a tree $T$ has a drawing of width $4n + 4$ then every clause was satisfied (following Lemma 3), and thus $A$ must describe a satisfying assignment for $\phi$. This proves that T can be embedded with width $4n + 4$ if and only if $\phi$ is satisfiable.

Furthermore, our reduction features a multi-linear number of nodes: the variable gadget has $8n + 3$ nodes, the clause gadgets exactly $5mn$ nodes, the bridge gadget $6n$ and the base gadget $8n + 6$ totaling exactly $22n + 5mn + 12$ nodes. This completes our proof of Theorem 1.

## 3   Linearity for Fixed-Order Phylogenetic Trees

**Theorem 2.** *A minimum width upward orthogonal drawing of a fixed-order $n$-node phylogenetic tree can be computed in $O(n)$ time.*

We provide an algorithm that computes a minimum width drawing. The key idea is to construct a directed acyclic graph (DAG) of the positional constraints

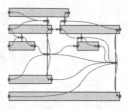

**Fig. 5.** The constraint graph of Lemma 4

between nodes and edges. The DAG can then be processed efficiently to determine a positioning of each node and edge that ensures the minimum width. Let $S$ be a set of non-intersecting orthogonal objects (e.g., rectangles and segments) in the plane. Two objects $s$ and $s'$ are horizontally visible if there exists a horizontal segment that intersects $s$ and $s'$ but no other object of $S$. Since the height of each object of our drawing is fully determined by the edge lengths of the tree, determining which objects are horizontally visible is essential to construct a minimum width drawing. For a fixed order combinatorial phylogenetic tree $T = (V, E)$, the *Constraint Graph* $D = (U, A)$ of $T$ is a directed graph with a vertex for each left and right side (of the rectangle representing the node in the drawing) of each node of $T$ and one for each edge of $T$. (See Fig. 5.) An arc $e = uv \in A$ if the objects corresponding to $u$ and $v$ are horizontally visible and $u$ precedes $v$ as determined by the fixed order.

**Lemma 4.** *The constraint graph $D$ of the fixed order $n$-node phylogenetic tree $T = (V, E)$ is a DAG with $3n - 1$ vertices and $O(n)$ edges, where $n = |V|$.*

*Proof.* The key is that $D$ must be planar, for the full proof see Appendix A.3.

The algorithm has two main steps. First, we construct the constraint graph $D$, and then we process the constraint graph to find a minimum-width drawing. As we have mentioned, the vertices of $D$ can be constructed directly from the vertices and edges of $T$. We now show that the arcs in $D$ can be created using a single pre-order (node, then children left to right) traversal of the tree, while growing a frontier indicating the rightmost object seen at each height. We maintain the frontier efficiently as an ordered list of height ranges. Whenever we update the frontier we have found a new rightmost object. If we are not extending the frontier (i.e. adding to the end of the list), then we have covered/partially covered some object. The two objects must be horizontally visible so we add a new directed arc from the left object to the right.

For a linear algorithm, we must avoid searching in the frontier for the position of each object. The key observation is that, while processing a node $v$ of $T$, the edges and nodes of the subtree rooted at $v$ only affect the frontier below $v$. In other words, the position of $v$ in the frontier doesn't change while processing its subtree. When a child is completely processed we can find the next sibling's position in the frontier by looking at the position of their parent.

Once the constraint graph is constructed, it must be processed to find the positions at which to draw each object. We process the vertices of $D$ in topological order. Vertices that have no incoming arcs, which are the sources of the constraint graph, must be the left side of vertices of $T$ and can be positioned at x-coordinate 0. At each remaining vertex, we check its incoming arcs and assign it the leftmost position that is to the right of every vertex in its in-neighborhood. Because the arcs represent the necessary order at each height and the sources of the DAG are positioned as far left as possible, a simple inductive argument proves that the resulting drawing has minimum width.

Traversing the tree to construct our DAG requires us to update the frontier once for every arc, source, and sink of the DAG. Each update takes constant time, so by Lemma 4 determining the arcs of the constraint DAG takes a total of $O(n)$ time. The time taken for the processing step includes the topological sort and the time to check each incoming arc at each vertex. Both of these are bounded by the number of arcs, so by Lemma 4 processing the DAG also takes $O(n)$ time. In conclusion both steps take $O(n)$ time so the algorithm takes $O(n)$ in total. This completes the proof of Theorem 2.

## 4 Heuristics and Experiments

Let $T$ be a combinatorial phylogenetic tree. Once the order of the children of each vertex is determined, we can use Theorem 2 to find a minimum width drawing that respects the edge lengths. Consequently, a heuristic only needs to define the ordering of the children in each vertex. We define the *flip* of a tree (or subtree) rooted at $v$ as the operation of reversing the order of the children of $v$ and every descendant of $v$. Flipping a tree corresponds to flipping its drawing and does not affect its minimum width.

The *greedy* heuristic proceeds bottom up from the leaves to the tree's root. For each vertex $v$ with children $c_0, \ldots, c_k$ this heuristic assumes that the order of the subtrees rooted at its children are fixed and finds the way to arrange its children to minimize width. To do so it considers every possible permutation and combination of flipped children. In general, for a degree $d$ vertex, the greedy heuristic checks $O(d!3^{d-1})$ possible orderings, bounded degree trees therefore take $O(1)$ time per vertex. Because the algorithm calculates the minimum width drawing using the $O(n)$ algorithm from Theorem 2, and runs it $O(1)$ times per vertex, the total running time of the heuristic is $O(n^2)$ for bounded degree trees.

**Theorem 3.** *The greedy heuristic has an approximation ratio of at least $\Omega(\sqrt{n})$, even for binary phylogenetic trees.*

*Proof.* Recall the structures for $w_k$ as described in Fig. 3a and consider equivalent structures for larger values of $k$ where all $k$ leaves lie in horizontal alignment. Using this definition of $w_k$, Fig. 6 shows a tree structure where a minimum width embedding of the subtree in yellow makes it impossible for the entire drawing to admit minimum width. For a minimum width subtree, the greedy heuristic must choose the smallest width possible $(k+2)$ thus placing the long edges on opposite

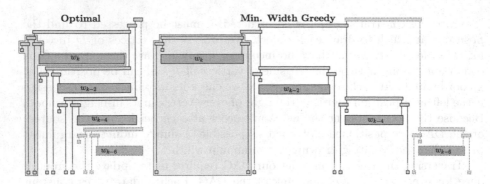

**Fig. 6.** Tree structures causing worst case performance for the greedy heuristic.

sides. In an optimal ordering the subtree's embedding would need to be a unit wider $(k + 3)$ and place both of long edges adjacent to each other, making the space below $w_k$ available for other subtrees. Label each subtree with the size of the $w_k$ structure inside it, and consider the tree structure with $k/2 - 1$ subtrees $w_k, w_{k-2}...w_2$. This structure will have an optimal width of $(k + 3) + (k/2 - 2)$ where the first term accounts for the top-most subtree for $w_k$ (in yellow) and the second from the number of edges connecting to remaining subtrees underneath. The greedy heuristic must instead place each subtree, enclosed by the long edges, side-by-side forcing most leaves into distinct columns. Only one pair of leaves per subtree share their column, therefore the width is equal to the number of leaves $(n + 1)/2$ minus the $k/2 - 1$ overlapping leaves (where $n$ is the total number of nodes). In total, there are $n = \sum_{i=0}^{k/2-1}(7 + 2(k - 2i) + 1) - 1 = k^2/2 + 5k - 1$ nodes, from which we find $k$. We find that $k \approx \sqrt{2n}$, and therefore the approximation ratio achieved by the greedy heuristic for this tree is $\frac{(n+1)/2 - k/2 - 1}{3k/2+1} \approx \frac{n - \sqrt{2n}}{3\sqrt{2n}} = \Theta(\sqrt{n})$, which proves that greedy can have an approximation at least as bad as our tree, thus proving the ratio is at least $\Omega(\sqrt{n})$. ∎

Similar to the greedy heuristic, the *minimum area heuristic* proceeds bottom up from leaves to root and finds the best way of arranging its children assuming their sub-trees have a fixed order. While the greedy heuristic minimizes the area of the tree's bounding rectangle, the minimum area heuristic minimizes the area of the orthogonal y-monotone bounding polygon at the expense of a potential larger total width. The running time is the same as the greedy heuristic and the approximation ratio is also at least $\Omega(\sqrt{n})$ (see Theorem 4 in Appendix A.4).

The *hill climbing* algorithm is a standard black-box optimization approach. Beginning from an initial configuration, it repeatedly tests small changes and keeps them if they do not hurt the quality of the solution. The quality of the solution is exactly equal to the width of the resulting drawing of the tree, and each change tested corresponds to reordering one node's children.

Our *simulated annealing* algorithm is another black-box optimization approach [21], with the same procedure as hill climbing. The main difference is that changes hurting the solution's quality are kept with probability inversely related

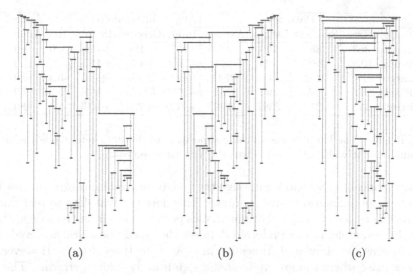

(a)                    (b)                    (c)

**Fig. 7.** Drawings of a tree with 93 nodes (a) Input order, width = 37 (b) Greedy order, width = 33 (c) Simulated annealing order, width = 26

to both the difference in quality and the number of steps taken so far. Once the number of steps is large enough, simulated annealing mimics the behavior of the hill climbing algorithm. Compared to hill climbing, simulated annealing has the advantage of not being trapped in a local minimum when a poor starting point was chosen.

*Experiments.* We evaluate five data sets of real phylogenetic trees obtained from the online phylogenetic tree database TreeBase [25]. The size and compositions of the datasets can be seen in Fig. 8. Each tree in TreeBase originates from a scientific publication, which unfortunately means there are too many for us to list on this paper; we instead provide a complete list of the studies associated with phylogenetic trees used in the data sets and complete experiment source code at `github.com/UC-Irvine-Theory/MinWidthPhylogeneticTrees` (along with some interesting drawings).

Each dataset is read using Dendropy [29], an open source Python library for phylogenetic computing. Each tree is read with an induced order from the source file, which we will serve as the initial configuration. The datasets are filtered to contain only trees with existing edge-lengths and maximum degree 3. Edge-lengths are normalized into discrete values that preserve nodes' original vertical ordering. For trees with few missing edge-lengths we assume missing edges are of unit length. These normalized datasets are used to evaluate the heuristics and produce easily comparable drawings.

*Results.* The first thing that stands out from our results is that all of our proposed approaches improve on the original input order. A typical example is shown in Fig. 7, where the input width is improved by the greedy heuristic and

| | Data Set | | | | Avg. Width Difference from Anneal. | | | | |
|---|---|---|---|---|---|---|---|---|---|
| Name | #Trees | Smallest | Largest | Average | Input | Greedy | MinArea | Hill | Anneal. |
| Small | 363 | 85 | 100 | 92 | +26% | +4% | +13% | ±0 | ±0 |
| Medium | 1026 | 188 | 399 | 271 | +26% | +4% | +12% | -1% | ±0 |
| Large | 28 | 2151 | 3305 | 2541 | +40% | +5% | +9% | -6% | ±0 |
| Plant | 80 | 195 | 3305 | 754 | +57% | +11% | +19% | -3% | ±0 |
| Preferred | 175 | 21 | 2387 | 192 | +21% | +5% | +10% | -1% | ±0 |

**Fig. 8.** Results. The left side shows the composition of the data sets, while the right side compares the width obtained versus the simulated annealing.

simulated annealing. Secondly, although the greedy and minimum area heuristics have a bad approximation ratio guarantee, this does not translate to real world trees. As can be seen in Fig. 8, both heuristics performed well for trees regardless of size. For example, in the Preferred data set the greedy heuristic achieved the same width as the Simulated Annealing in 50% of the trees (Fig. 9). However, a few cases exist where the greedy heuristic significantly under-performs. This is notable for two reasons: the first is that the greedy heuristic produces a drawing 60% wider than hill climbing, and the second is that the greedy heuristic is outperformed by the minimum area heuristic.

Finally, it is clear that black-box approaches are useful to find small-width drawings as they rarely produce drawings wider than those from the heuristics. However the width decrease achieved by the black box algorithms comes at a cost in running time, since in our implementation they took around 40 times longer to converge on average.

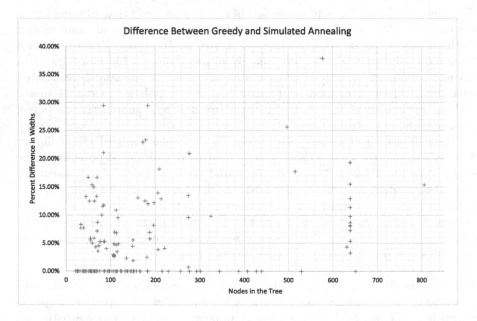

**Fig. 9.** Percent width difference between greedy and simulated annealing depending on tree size for Preferred dataset.

# A    Additional Proofs

## A.1    Alignment Nodes

Alignment nodes only need to be able to fully realign satisfied clauses, therefore within the three structures at least one must be of width three and at least one of width one. Therefore if we consider the periodicity of the column at which the edge drops, the maximum possible phase difference in these periods is shown in Fig. 10. Considering this order as an extreme case is sufficient because after using both satisfied literal structures (after $x_3$ in $c_1$) the only remaining widths could be two (which maintains the phase difference) or one (which reduces the phase difference). The same argument can be made after using both unsatisfied literal structures after $x_2$.

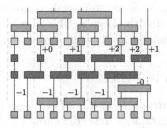

**Fig. 10.** Set of consecutive clauses, $c_1 = x_2 \vee x_3 \vee x_6$ and $c_2 = x_1 \vee \overline{x_2} \vee x_6$ requiring the largest realignment, with satisfying assignment $x = \{false, true, true, ...\}$ (Color figure online)

Without alignment nodes it would be impossible to connect $x_2$ and $x_3$ to the next clause, but adding the two row of alignment nodes (shown in blue) between clauses enable them to remain connected. This allows each clause to remain tight and assume width of at most $2n + 1$ whenever they are satisfied, regardless of the previous clause.

## A.2    Pyramidal Structure

Recall **Lemma** 1: *At minimal width, the base and bridge can only assume a pyramidal embedding. We define a pyramidal embedding as the embedding of the base in which nodes closer to the root lie closer to the center and nodes further from the root approach the outer sides, as shown in Fig. 3b.*

*Proof.* We define the filled area of a drawing as the sum of the space used by each node (equal to its width), and the space used by each edge (equal to its length). The length of the edges is fixed, but we can change the filled area by changing the layout to minimize the width of the non-leaf nodes. When the base is in the pyramidal embedding, it fills the least possible area, since every non-leaf node must have width equal to its number of children. We now show that this is the only configuration with minimal area.

We begin from the parents of the base gadget, which belong to a copy of $w_2$ (Fig. 3a). The two incoming edges from this structure must be next to one another, with no gap between them. The two nodes at the top level of the base each then have both a leaf and a large subtree attached.

The width of each of these top-level nodes must be two, and the only way to achieve this is that each node's leaf must lie on the inside and its subtree on the outside. Similarly, for each subsequent node along the path, the same argument shows that its leaf must lie on the inside. This proves by induction that the base needs to be in a pyramidal embedding. The bridge then must fit against the base. The bridge nodes with the lowest leaves must be on the outside, with the next lowest leaves next to them, and so on by induction back to the center. This shows that the pyramidal embedding is the unique embedding that minimizes the filled area. Since the levels containing the base and bridge nodes are completely packed with no gaps, this also implies that the pyramidal embedding is the unique embedding that minimizes the width of these levels. ■

### A.3   Constraint Graph is a DAG

Recall **Lemma** 4: *The Constraint Graph D of the fixed order n-node phylogenetic tree $T = (V, E)$ is a DAG with $3n - 1$ vertices and $O(n)$ edges, where $n = |V|$.*

*Proof.* Our objects are the left and right sides of each vertex in $T$, and the edges in $T$. This gives us two vertices in $D$ for each vertex in $T$, and one for each edge. Since $T$ is a tree, it must have $n-1$ edges, so $D$ has $3n-1$ vertices. If two objects are horizontally visible, then there is a segment between them that crosses only those two objects. We will use these segments to build a planar embedding of $D$, which will imply that $D$ has $O(n)$ edges.

Let the collection of segments connecting our objects be $S$. We first construct a larger planar graph $D'$, in which the vertices are the endpoints of $S$. The edges of $D'$ include all of the segments in $S$. We also add edges connecting each vertex to the vertices immediately above and below it that represent the same object. By the definition of horizontally visible, each segment in $S$ can only intersect two objects, so none of the segments in $S$ can intersect. The additional edges also cannot intersect, since they are ordered by the height of the vertices. Therefore, $D'$ is planar.

We then contract all of the edges that connect two vertices in $D'$ corresponding to the same object. This produces the DAG $D$. Since $D$ is a contraction of a planar graph, it must also be planar.                                    ■

### A.4   Approximation Guarantee for Minimum Area Heuristic

We first describe the running time of the heuristic. For each ordering of the children, the minimum width drawing is calculated using the algorithm from Theorem 2 and the bounding polygon is calculated by traversing the tree once

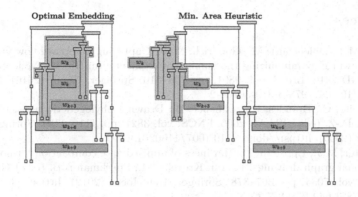

**Fig. 11.** Tree structures causing worst case performance for minimum area heuristic.

to find the extreme-most branches, running in $O(n)$. We repeat this $O(1)$ times per vertex for a total running time of $O(n^2)$ for bounded degree trees.

**Theorem 4.** *The minimum area heuristic has an approximation ratio of at least* $\Omega(\sqrt{n})$.

*Proof.* Recall, the structures for $w_k$ as defined in Theorem 3 and further constrain it to be a complete binary tree with all its $k$ leaves in horizontal alignment. Recall the subtrees used in Theorem 3 and note the subtrees used in this tree instead increase the size of their $w_k$ by 3 each time (with the exception of the first two which have the same $w_k$). Furthermore each subtrees nodes end immediately before the first node in the subtree two subtrees away, the latter subtree also has a node aligned with the former's $w_k$ leaves. The leaves in $w_k$ are horizontally aligned with the top node in the next subtree.

Using these definitions Fig. 11 demonstrates a tree structure where a minimum area embedding of the two subtrees in yellow makes it impossible for the entire drawing to admit minimum width and minimum area. The heuristic achieves the right embedding for the subtree but fails to choose the right embedding for the two sibling subtrees. Although the optimal's embedding uses a larger area for the combination of both siblings with $w_k$, it occupies an almost rectangular space resulting in a really small area (and width) increase when adding the next subtree. Define each subtree by the size of the $w_k$ structure inside it, consider the tree with $2k/3$ subtrees $w_k, w_{k+3}...w_{3k}$. This structure will have an optimal width of $3k + 6$. The minimum area heuristic would instead have two subtrees with their $w_k$ on opposite sides and the $w_{k+3}$ overlapping the bottom-most $w_k$ and every next pair of subtrees overlapping in the same way. The total width achieved by the minimum area heuristic would therefore be $\sum_{i=0}^{2k/6}(k + 6i + 5) = 2k^2/3 + 11k/3 + 5$. The total number of nodes is $n = \sum_{i=0}^{2k/3}(7 + 2(k - 2i) + 1) + 6 + k = 4k^2/9 + 7k + 14$, which we can use to find $k$ in terms of $n$. We find that $k \approx \sqrt{n/2}$, and therefore the approximation ratio achieved by the greedy heuristic for this tree is $\frac{2k^2/3 + 11k/3 + 5}{3k + 6} \approx \frac{2k}{9} = \Omega(\sqrt{n})$. ∎

# References

1. Alam, M.J., Dillencourt, M., Goodrich, M.T.: Capturing Lombardi flow in orthogonal drawings by minimizing the number of segments. In: Hu, Y., Nöllenburg, M. (eds.) GD 2016. LNCS, vol. 9801, pp. 608–610. Springer, Cham (2016). https://doi.org/10.1007/978-3-319-50106-2

2. Bachmaier, C., Brandes, U., Schlieper, B.: Drawing phylogenetic trees. In: Deng, X., Du, D.-Z. (eds.) ISAAC 2005. LNCS, vol. 3827, pp. 1110–1121. Springer, Heidelberg (2005). https://doi.org/10.1007/11602613_110

3. Bannister, M.J., Eppstein, D.: Hardness of approximate compaction for nonplanar orthogonal graph drawings. In: van Kreveld, M., Speckmann, B. (eds.) GD 2011. LNCS, vol. 7034, pp. 367–378. Springer, Heidelberg (2012). https://doi.org/10.1007/978-3-642-25878-7_35

4. Bhatt, S.N., Cosmadakis, S.S.: The complexity of minimizing wire lengths in VLSI layouts. Inf. Process. Lett. **25**(4), 263–267 (1987)

5. Biedl, T., Mondal, D.: On upward drawings of trees on a given grid. In: Frati, F., Ma, K.-L. (eds.) GD 2017. LNCS, vol. 10692, pp. 318–325. Springer, Cham (2018). https://doi.org/10.1007/978-3-319-73915-1_25

6. Benson, R.B., Ketchum, H., Naish, D., Turner, L.E.: A new leptocleidid (sauropterygia, plesiosauria) from the vectis formation (early barremian-early aptian; early cretaceous) of the isle of wight and the evolution of leptocleididae, a controversial clade. J. Syst. Palaeontol. **11**, 233–250 (2013)

7. Boc, A., Diallo, A.B., Makarenkov, V.: T-REX: a web server for inferring, validating and visualizing phylogenetic trees and networks. Nucleic Acids Res. **40**(W1), W573–W579 (2012)

8. Brandes, U., Pampel, B.: Orthogonal-ordering constraints are tough. J. Graph Algorithms Appl. **17**(1), 1–10 (2013)

9. Brunner, W., Matzeder, M.: Drawing ordered $(k-1)$–ary trees on $k$–grids. In: Brandes, U., Cornelsen, S. (eds.) GD 2010. LNCS, vol. 6502, pp. 105–116. Springer, Heidelberg (2011). https://doi.org/10.1007/978-3-642-18469-7_10

10. Carrizo, S.F.: Phylogenetic trees: an information visualisation perspective. In: Proceedings of the 2nd Conference on Asia-Pacific Bioinformatics, pp. 315–320 (2004)

11. Chan, T.M.: Tree drawings revisited. Discret. Comput. Geom. 1–22 (2018)

12. Chan, T.M., Goodrich, M.T., Kosaraju, S.R., Tamassia, R.: Optimizing area and aspect ratio in straight-line orthogonal tree drawings. Comput. Geom. **23**(2), 153–162 (2002)

13. Di Battista, G., Didimo, W., Patrignani, M., Pizzonia, M.: Orthogonal and quasi-upward drawings with vertices of prescribed size. In: Kratochvíyl, J. (ed.) GD 1999. LNCS, vol. 1731, pp. 297–310. Springer, Heidelberg (1999). https://doi.org/10.1007/3-540-46648-7_31

14. Frati, F.: Straight-line orthogonal drawings of binary and ternary trees. In: Hong, S.-H., Nishizeki, T., Quan, W. (eds.) GD 2007. LNCS, vol. 4875, pp. 76–87. Springer, Heidelberg (2008). https://doi.org/10.1007/978-3-540-77537-9_11

15. Garg, A., Goodrich, M.T., Tamassia, R.: Planar upward tree drawings with optimal area. Int. J. Comput. Geom. Appl. **06**(03), 333–356 (1996)

16. Gregori, A.: Unit-length embedding of binary trees on a square grid. Inf. Process. Lett. **31**(4), 167–173 (1989)

17. Gusfield, D.: Efficient algorithms for inferring evolutionary trees. Networks **21**(1), 19–28 (1991)

18. Huelsenbeck, J.P., Ronquist, F., Nielsen, R., Bollback, J.P.: Bayesian inference of phylogeny and its impact on evolutionary biology. Science **294**(5550), 2310–2314 (2001)
19. Huson, D.H., Scornavacca, C.: Dendroscope 3: an interactive tool for rooted phylogenetic trees and networks. Syst. Biol. **61**(6), 1061–1067 (2012)
20. Kim, S.K.: Simple algorithms for orthogonal upward drawings of binary and ternary trees. In: Canadian Conference on Computational Geometry (CCCG), pp. 115–120 (1995)
21. Metropolis, N., Rosenbluth, A.W., Rosenbluth, M.N., Teller, A.H., Teller, E.: Equation of state calculations by fast computing machines. J. Chem. Phys. **21**(6), 1087–1092 (1953)
22. Miller, M.A., Pfeiffer, W., Schwartz, T.: Creating the CIPRES science gateway for inference of large phylogenetic trees. In: Gateway Computing Environments Workshop (GCE), pp. 1–8, November 2010
23. Müller, J., Müller, K.: TREEGRAPH: automated drawing of complex tree figures using an extensible tree description format. Mol. Ecol. Notes **4**(4), 786–788 (2004)
24. Page, R.D.: Visualizing phylogenetic trees using treeview. Curr. Protoc. Bioinform. (1), 6.2.1–6.2.15 (2003)
25. Piel, W.H., Chan, L., Dominus, M.J., Ruan, J., Vos, R.A., Tannen, V.: Treebase v. 2: a database of phylogenetic knowledge. e-BioSphere (2009)
26. Rusu, A., Fabian, A.: A straight-line order-preserving binary tree drawing algorithm with linear area and arbitrary aspect ratio. Comput. Geom. **48**(3), 268–294 (2015)
27. Shin, C.S., Kim, S.K., Chwa, K.Y.: Area-efficient algorithms for straight-line tree drawings. Comput. Geom. **15**(4), 175–202 (2000)
28. Stamatakis, A., Meier, H., Ludwig, T.: RAxML-III: a fast program for maximum likelihood-based inference of large phylogenetic trees. Bioinformatics **21**(4), 456–463 (2004)
29. Sukumaran, J., Holder, M.T.: Dendropy: a python library for phylogenetic computing. Bioinformatics **26**(12), 1569–1571 (2010)
30. Warnow, T.: Tree compatibility and inferring evolutionary history. J. Algorithms **16**(3), 388–407 (1994)
31. Zainon, W.N.W., Calder, P.: Visualising phylogenetic trees. In: Proceedings of 7th Australasian User Interface Conference, pp. 145–152 (2006)

# Balanced Connected Subgraph Problem in Geometric Intersection Graphs

Sujoy Bhore[1], Satyabrata Jana[2(✉)], Supantha Pandit[3(✉)], and Sasanka Roy[2]

[1] Algorithms and Complexity Group, TU Wien, Vienna, Austria
sujoy.bhore@gmail.com
[2] Indian Statistical Institute, Kolkata, India
satyamtma@gmail.com, sasanka.ro@gmail.com
[3] Dhirubhai Ambani Institute of Information and Communication Technology,
Gandhinagar, Gujarat, India
pantha.pandit@gmail.com

**Abstract.** We study the BALANCED CONNECTED SUBGRAPH (shortly, **BCS**) problem on geometric intersection graphs such as interval, circular-arc, permutation, unit-disk, outer-string graphs, etc. Given a *vertex-colored* graph $G = (V, E)$, where each vertex in $V$ is colored with either "*red*" or "*blue*", the BCS problem seeks a maximum cardinality induced connected subgraph $H$ of $G$ such that $H$ is *color-balanced*, i.e., $H$ contains an equal number of red and blue vertices. We study the computational complexity landscape of the BCS problem while considering geometric intersection graphs. On one hand, we prove that the BCS problem is NP-hard on the unit disk, outer-string, complete grid, and unit square graphs. On the other hand, we design polynomial-time algorithms for the BCS problem on interval, circular-arc and permutation graphs. In particular, we give algorithms for the STEINER TREE problem on both interval and circular-arc graphs, and those algorithms are used as subroutines for solving the BCS problem on the same classes of graphs. Finally, we present a FPT algorithm for the BCS problem on general graphs.

**Keywords:** Balanced connected subgraph · Interval graphs · Permutation graphs · Circular-arc graphs · Unit-disk graphs · Outer-string graphs · NP-hard · Color-balanced · Fixed parameter tractable

## 1 Introduction

The *intersection graph* of a collection of sets is a graph where each vertex of the graph represents a set and there is an edge between two vertices if their corresponding sets intersect. Any graph can be represented as an intersection graph over some sets. Geometric intersection graph families consist of intersection graphs for which the underlying collection of sets are some geometric

S. Bhore—The author is supported by the Austrian Science Fund (FWF) grant P 31119.

Y. Li et al. (Eds.): COCOA 2019, LNCS 11949, pp. 56–68, 2019.
https://doi.org/10.1007/978-3-030-36412-0_5

objects. Some of the important graph classes in this family are interval graphs (intervals on real line), circular-arc graphs (arcs on a circle), permutation graphs (line segments with endpoints lying on two parallel lines), unit-disk graphs (unit disks in the Euclidean plane), unit-square graphs (unit squs in the Euclidean plane), outer-string graphs (curves lying inside a disk, with one endpoint on the boundary of the disk), etc. In the past several decades, geometric intersection graphs became very popular and extensively studied due to their interesting theoretical properties and applicability.

In this paper, we consider an interesting problem on general vertex-colored graphs called the BALANCED CONNECTED SUBGRAPH (shortly, BCS) problem. A subgraph $H = (V', E')$ of $G$ is called *color-balanced* if it contains an equal number of red and blue vertices.

---

**BALANCED CONNECTED SUBGRAPH (BCS) Problem**
**Input:** A graph $G = (V, E)$, with node set $V = V_R \cup V_B$ partitioned into red nodes $(V_R)$ and blue nodes $(V_B)$.
**Output:** Maximum-sized color-balanced induced connected subgraph.

---

### 1.1 Previous Work

The BCS problem has been studied on various graph families such as trees, planar graphs, bipartite graphs, chordal graphs, split graphs, etc [2]. Most of the findings suggest that the problem is NP-hard for general graph classes, and it is possible to design polynomial time algorithms for the restricted classes with involved approaches. In [2], we have pointed out a connection between the BCS problem and GRAPH MOTIF problem (see, e.g., [6,7,11]). In the graph motif problem, we are given the input as a graph $G = (V, E)$, a color function $col : V \rightarrow C$ on the vertices, and a multiset $M$, called motif, of colors of $C$; the objective is to find a subset $V' \subseteq V$ such that the induced subgraph on $V'$ is connected and $col(V') = M$. Note that, if $C = \{red, blue\}$ and the motif has the same number of red and blues then, the solution of the graph motif problem gives a balanced connected subgraph. Indeed, a solution to the graph motif problem provides one balanced connected subgraph, with an impact of a polynomial factor in the running time. However, it does not guarantee the maximum size balanced connected subgraph. Nonetheless, the NP-hardness result for the BCS problem on any particular graph class implies the NP-hardness result for the graph motif problem on the same class. Graph motif problem has wide range of applications in bioinformatics [5], DNA physical mapping [8], perfect phylogeny [4], metabolic network analysis [12], protein–protein interaction networks and phylogenetic analysis [3]. This problem was introduced in the context of detecting patterns that occur in the interaction networks between chemical compounds and/or reactions [12].

## 1.2   Our Results

We present a collection of results on the BCS problem on geometric intersection graphs, that advances the study of this problem on diverse graph families.

➡ On the hardness side, in Sect. 2, we show that the BCS problem is NP-hard on unit-disk graphs, outer-string graphs, complete grid graphs, and unit square graphs.

➡ On the algorithmic side, in Sect. 3, we design polynomial-time algorithms for interval graphs ($\mathcal{O}(n^4 \log n)$ time), circular-arc graphs ($\mathcal{O}(n^6 \log n)$ time) and permutation graphs ($\mathcal{O}(n^6)$ time, and the result is described in the full version[1]). Moreover, we give an algorithm for the STEINER TREE problem on the interval graphs, that is used as a subroutine in the algorithm of the BCS problem for intervals graphs. Finally, we show that the BCS problem is fixed-parameter tractable for general graphs ($2^{\mathcal{O}(k)} n^2 \log n$) while parameterized by the number of vertices in a balanced connected subgraph.

## 2   Hardness Results

### 2.1   Unit-Disk Graphs

We show that the BCS problem is NP hard for the unit-disk graphs. We give a reduction from the RECTILINEAR STEINER TREE *(RST)* problem [9]. In this problem, we are given a set $P$ of integer coordinate points in the plane and an integer $L$. The goal is to find a Steiner tree $T$ (if one exists) of length at most $L$.

During the reduction, we first generate a geometric intersection graph from an instance $X(P, L)$ of the *RST* problem. The vertices of this graph are having integer coordinates and each edge is of unit length. Next, we show that this graph is a unit-disk graph.

**Reduction:** Suppose we have an instance $X(P, L)$ of the *RST* problem. For any point $p \in P$, let $p(x)$ and $p(y)$ denote the $x$- and $y$-coordinates of $p$, respectively. Let $p_t$, $p_b$, $p_l$, and $p_r$ be the topmost (largest $y$-coordinate), bottom-most (smallest $y$-coordinate), leftmost (smallest $x$-coordinate), and rightmost (largest $x$-coordinate) points in $P$. We now take a unit integer rectangular grid graph $D$ on the plane such that the coordinates of the lower-left grid vertex is $(p_l(x), p_b(y))$ and upper-right grid vertex is $(p_r(x), p_t(y))$. Now we associate each point $p$ in $P$ with a grid vertex $d_p$ in $D$ having the same $x$- and $y$-coordinates of $p$. Now we assign colors to the points in $D$. The vertices in $D$ that are corresponding to the points in $P$ are colored with red and the remaining grid vertices in $D$ are colored with blue. We add some additional vertices to $D$ as follows:

Observe that if there exists a Steiner tree $T$ of length $L + 1 = |P|$ then $T$ does not include any blue vertex in $D$. Further, if there exists a Steiner tree $T$ of length $L + 1 = 2|P|$ then $T$ contains equal number of red and blue vertices in $D$. Based on this observation we consider two cases to add some additional vertices (not necessarily forming a grid structure) to $D$.

---

[1]   Some results are described in the full version, because of page limitations here.

**Case 1.** $[L + 1 \geq 2|P|]$: In this case the number of blue vertices in a Steiner tree $T$ (if exists) is more than or equals to red vertices in $D$. We consider a path $\delta$ of $(L - 2|P| + 1)$ red vertices starting and ending with vertices $r_1$ and $r_{L-2|P|+1}$, respectively. The coordinates of $r_i$ is $(p_l(x) - i, p_l(y))$, for $1 \leq i \leq L - 2|P| + 1$. We connect this path with $D$ using an edge between the vertices $r_1$ and $p_l$. See Fig. 1(a) for an illustration of this construction. Let the resulting graph be $G_1 = D \cup \delta$.

**Case 2.** $[L + 1 < 2|P|]$: In this case the number of red vertices in a Steiner tree $T$ (if exists) is more than the number of blue vertices in $D$. We consider a path $\delta$ of $(2|P| - L)$ blue vertices starting and ending with vertices $b_1$ and $b_{2|P|-L}$, respectively. The coordinates of $b_i$ is $(p_l(x) - i, p_l(y))$, for $1 \leq i \leq 2|P| - L$. We connect this path with $D$ using an edge between the vertices $b_1$ and $p_l$. We add one more red vertex $r'$ whose coordinates are $(p_{2|P|-L}(x) - 1, p_l(y))$ and connect it with $b_{2|P|-L}$ using an edge. See Fig. 1(b) for an illustration of this construction. Let the resulting graph be $G_2 = D \cup \delta \cup \{r'\}$

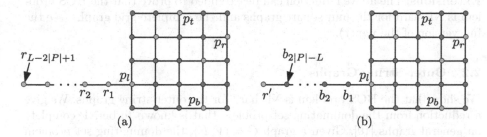

**Fig. 1.** (a) Construction of the instance $G_1$. (b) Construction of the instance $G_2$. (Color figure online)

This completes the construction. Clearly, the construction (either $G_1$ or $G_2$) can be done in polynomial time. Now we prove the following lemma.

**Lemma 1.** *The instance $X$ of the RST problem has a solution $T$ if and only if*

- **For Case 1:** *the instance $G_1$ has a balanced connected subgraph $H$ with $2(L - |P| + 1)$ vertices.*
- **For Case 2:** *the instance $G_2$ has a balanced connected subgraph $T$ with $2(|P| + 1)$ vertices.*

*Proof.* We prove this lemma for Case 1. The proof of Case 2 is similar.

**For Case 1:** Assume that $X$ has a Steiner tree $T$ of length $L$, where $L+1 \geq 2|P|$. Let $U$ be the set of those vertices in $G_1$ corresponds to the vertices in $T$. Clearly, $U$ contains $|P|$ red vertices and $L - |P| + 1$ blue vertices. Since $L + 1 \geq 2|P|$, to make $U$ balanced it needs $L - |P| + 1 - |P|$ more red vertices. So we can add the path $\delta$ of $L - 2|P| + 1$ red vertices to $U$. Therefore, $U \cup \delta$ becomes connected and balanced (contains $L - |P| + 1$ vertices in each color).

On the other hand, assume that there is a balanced connected subgraph $H$ in $G$ with $(L - |P| + 1)$ vertices of each color. We can observe that $H$ is a tree and no blue vertex in $H$ is a leaf vertex. The number of red vertices in $G_1$ is exactly $(L - |P| + 1)$. So the $H$ must pick all the $(L - |P| + 1)$ blue vertices that connect the vertices in $G_1$ corresponding to $P$. We take the set $A$ of all the grid vertices corresponding to the vertices in $H$ except the vertices $\{r_i; 1 \leq i \leq (L - 2|P| + 1)\}$. We output the Steiner tree $T$ that contains the vertex set $A$ and edge set $E_A$ connecting the vertices of $A$ according to the edges in $H$. As $|A| = 2(L - |P| + 1) - (L - 2|P| + 1) = L + 1$, so we output a Steiner tree of length $L$.                                               □

We now show that either $G_1$ or $G_2$ is a unit-disk graph. Let us consider the graph $G_1$. For each vertex $v$ in $G_1$ we take a unit disk whose radius is $\frac{1}{2}$ and center is on the vertex $v$. Therefore from the Lemma 1, we conclude that,

**Theorem 1.** *The* BCS *problem is NP-hard for unit-disk graphs.*

**Extensions:** The above reduction can be extended to prove that the BCS problem is NP-hard for the unit square graphs and the complete grid graphs (see the full version of the paper).

## 2.2  Outer-String Graphs

We show that the BCS problem is NP-hard for the outer-string graphs. We give a reduction from the dominating set problem that is known to be NP complete on general graphs [10]. Given a graph $G = (V, E)$, the dominating set problem asks whether there exists a set $U \subseteq V$ such that $|U| \leq k$ and $N[U] = V$, where $N[U]$ denotes neighbours of $U$ in $G$ including $U$ itself.

During the reduction, we first generate a geometric intersection graph $H = (R \cup B, E')$ from an instance $X(G, k)$ of the dominating set problem on general graph. Next, we show that $H$ is an outer-string graph.

**Reduction:** Let $G = (V, E)$ be graph with vertex set $V = \{v_1, v_2, \ldots, v_n\}$. For each vertex $v_i \in V$ we add a red vertex $v_i$ and a blue vertex $v_i'$ in $H$. For each edge $(v_i, v_j) \in E$, we add two edges $(v_i, v_j'), (v_i', v_j)$ in $E'$. Take a path of $k$ red vertices starting at $r_1$ and ending at $r_k$. Also take a path of $n$ blue vertices starting at $b_1$ and ending at $b_n$. Add the edges $(b_n, r_k), (b_1, v_1)$ into $E'$. We add edges between all pair of vertices in $\{v_1', v_2', \ldots v_n'\}$. Our construction ends with adding $n$ edges $(v_i, v_i')$ into $E'$ for $1 \leq i \leq n$. This completes the construction. See Fig. 2 for an illustration of this construction that can be made in polynomial time.

**Lemma 2.** *The instance $X$ has a dominating set of size $k$ if and only if $H$ has a balanced connected subgraph $T$ with $2(n + k)$ vertices.*

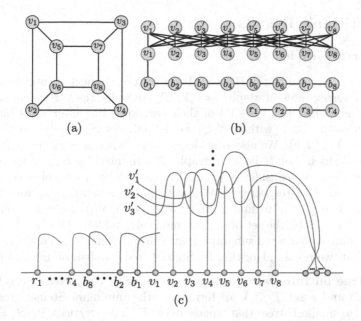

**Fig. 2.** (a) A graph $G$. (b) Construction of $H$ from $G$ with $k = 4$. For the clarity of the figure we omit the edges between each pair of vertices $v_i'$ and $v_j'$, for $i \neq j$. (c) Intersection model of $H$. (Color figure online)

*Proof.* Assume that $G$ has a dominating set $U$ of size $k$. Now we take the subgraph $T$ of $H$ induced by $\{v_i' : v_i \in U\}$ along with the vertices $\{r_i : 1 \leq i \leq k\} \cup \{b_j : 1 \leq j \leq n\} \cup \{v_j : 1 \leq j \leq n\}$ in $H$. Now clearly $H$ is connected and balanced with $2(n + k)$ vertices.

On the other hand, assume that there is a balanced connected tree $T$ in $H$ with $(n + k)$ vertices of each color. The number of red vertices in $H$ is exactly $(n + k)$. So the solution must pick the blue vertices $\{b_i : 1 \leq i \leq n\}$ that connect $v_1$ with $r_1$. As $T$ has exactly $(n + k)$ blue vertices then it $T$ should pick exactly $k$ vertices from the set $\{v_i' : 1 \leq i \leq n\}$. The set of vertices in $V$ corresponding to these $k$ vertices gives us a dominating set of size $k$ in $G$.           $\square$

We now verify that $H$ is an outer-string graph. For an illustration see Fig. 2(c). We draw a horizontal line $y = 0$. For each vertex $v_i \in H$, draw the line segment $L_i = \overline{(i, 0)(i, 1)}$. For each vertex $v_j' \in H$, we draw a curve $C_j$, having one endpoint on the line $y = 0$, in such a way that for each edge $(v_j', v_i) \in H$, $C_j$ bend above $L_i$ (the line segment corresponding to $v_i$) and intersects the lines $L_i$ from top. Also all the curves $C_j$'s intersect each other. Now we add the curves corresponding to $\{r_i : 1 \leq i \leq k\} \cup \{b_j : 1 \leq j \leq n\}$ having one endpoint on the line $y = 0$ with satisfying the adjacency. Finally, using Lemma 2, we conclude:

**Theorem 2.** *The* BCS *problem is NP-hard for the outer-string graphs.*

# 3   Algorithmic Results

## 3.1   Interval Graphs

In this section, we study the BCS problem on the connected interval graphs. Given an $n$ vertex interval graph $G = (V, E)$, we order the vertices of $G$ based on the left endpoints $\{l_v \colon \forall v \in V\}$ of their corresponding intervals. Consider a pair of vertices $u, v \in V$ with $l_u \leq l_v$, we define a set $S_{u,v} = \{w \colon w \in V, l_u \leq l_w < r_w \leq r_v\} \cup \{u, v\}$. We also consider the case when $u = v$, and define $S_{u,u}$ in a similar fashion. Let $H$ be a subgraph of $G$ induced by $S_{u,v}$ (resp. $S_{u,u}$) in $G$. For any $u, v \in V$, consider $S_{u,v}$ and let $r$ and $b$ be the number of red and blue vertices in $S_{u,v}$, respectively. Without loss of generality, we may assume that $b \leq r$. The goal is to find a BCS, with cardinality $2b$ in $H$, containing $u, v \in V(H)$. Let $B$ be the set of all blue intervals and $T = B \cup \{u, v\}$.

We compute a connected subgraph that contains $T$ and some extra red intervals. For that, we use an algorithm for Steiner tree problem on interval graphs.

**Steiner Tree on Interval Graphs:** Given a simple connected interval graph $G = (V, E)$ and a set $T \subseteq V$ of terminals, the minimum Steiner tree problem seeks a smallest tree that spans over $T$. The vertices in $S = V \backslash T$ are denoted as Steiner vertices. First we describe a greedy algorithm called $Select\_Steiners(G = (V, E), T)$ that computes a minimum Steiner tree $T \cup D$ on $G$.

We first partition $T$ into $m$ ($m \in [n]$) components $\{C_1, \ldots, C_m\}$ sorted from left to right based on the right endpoints of the components (note that, the union of the intervals in each component is an interval on the real line). Let $I_{C_i}$ be the rightmost interval of the $i$-th component $C_i$. We consider the first component $C_1$ and the neighborhood set $N(I_{C_1})$ of $I_{C_1}$. Let $I_j$ be the rightmost interval in $N(I_{C_1})$. By rightmost, we mean that the interval having the rightmost endpoint. We add $I_j$ in $D$. Now, we recompute the components based on $T \cup D$. Note, $C_1 \cup I_j$ is now contained in one component. We repeat this produce until $T \cup D$ becomes a single component. Finally we return $T \cup D$ as a solution.

**Correctness:** It is easy to verify that the graph induced by $T \cup D$ is connected. Now we prove the optimality of the Algorithm $Select\_Steiners(G = (V, E), T)$ by induction. The base case is that we have to connect the first two components $C_1$ and $C_2$. We choose the rightmost interval from the neighborhood of $I_{C_1}$. Note, that we have to connect $C_1$ and therefore it is inevitable that we have to pick an interval from $N(I_{C_1})$. We choose the rightmost interval (say, $(I_\ell)$). Now, if this choice already connects $C_2$, we are done. Otherwise, $C_1 = C_1 \cup I_\ell$. Now, let us assume that we have obtained an optimal solution until step $i$. At step $i + 1$, we have to connect the first two components. By applying the same argument as the base case, we choose the rightmost interval from the first component and proceed. It is easy to verify that this algorithm runs in $\mathcal{O}(n^2)$ time.

Now, we go back to the BCS problem. Recall that, for any pair of intervals $u, v \in V$, our objective is to compute a BCS with vertex set $T$ of cardinality $2b$ (if exists), where $b$ and $r$ is the number of red and blue intervals in $S_{u,v}$,

**Algorithm 1.** $BCS\_Interval(H)$

1: $T \leftarrow B \cup \{u, v\}$
2: $D = Select\_Steiners(H, T)$
3: $r' \leftarrow$ number of red vertices in $D \cup \{u, v\}$
4: $b' \leftarrow$ number of blue vertices in $T$
5: **if** $r' > b'$ **then**
6:    Return $\phi$
7: **if** $r' = b'$ **then**
8:    Return $G[D \cup T]$
9: **if** $r' < b'$ **then**
10:    Return $G[D \cup T \cup X]$
     where $X \subset V(H)$ is the set of $(b' - r')$ red vertices with $X \cap (D \cup T) = \phi$.

respectively, and $b \leq r$. Let $H$ be the subgraph of $G$ induced by $S_{u,v}$. Let $R$ and $B$ denote the set of red and blue intervals in $S_{u,v}$, respectively. We describe this process in Algorithm 1. We repeat this procedure for every pair of intervals and report the solution set with the maximum number of intervals.

**Correctness:** We prove that our algorithm yields an optimum solution. Let $G'$ be such a solution. Let $u$ and $v$ be the intervals with leftmost endpoint and rightmost endpoint of $G'$, respectively. Now we show that $V(G') = \min\{2r, 2b\}$, where $r$ and $b$ be the number of red and blue color vertices is $S_{u,v}$, respectively. Let us assume $V(G') \neq \min\{2r, 2b\}$. Then there exists at least one blue interval $z$ and one red interval $z'$ that belong to $S_{u,v} \setminus V(G')$. However, we know that $S_{u,v}$ induces an intersection graph of intervals corresponding to the vertices $\{w : w \in V, l_u \leq l_w < r_w \leq r_v\} \cup \{u, v\}$, and $G'$ contain both $u$ and $v$. So, $N[z] \cap G' \neq \phi$, $N[z'] \cap G' \neq \phi$. Therefore $V(G') \cup \{z, z'\}$ induces a balanced connected subgraph in $G$. It contradicts our assumption and hence the proof.

**Time Complexity:** Here we use the algorithm $Select\_Steiners(G = (V, E), T)$ as a subroutine. Using range tree the set $S_{u,v}$ can be obtained in $\mathcal{O}(\log n)$ time. Hence total running time is $\mathcal{O}(n^4 \log n)$.

**Theorem 3.** *Let $G$ be an interval graph whose $n$ vertices are colored either red or blue. Then BCS problem on $G$ can be solved in $\mathcal{O}(n^4 \log n)$ time.*

## 3.2 Circular-Arc Graphs

We study the BCS problem on circular-arc graphs. We are given a bicolored the circular-arc graph $G = (V_R \cup V_B, E)$, where the set $V_R$ and $V_B$ contains a set of red and blue arcs, respectively. With out loss of generality we assume that the given input arcs fully cover the circle. Otherwise it becomes an interval graph and we use the algorithm of the interval graph to get an optimal solution.

Let us assume that $H$ be a resulting maximum balanced connected subgraph of $G$, and let $S$ denote the set of vertices in $H$. Since $H$ is a connected subgraph of $G$, $H$ covers the circle either partially or entirely. We propose an algorithm that computes a maximum size balanced connected subgraph $H$ of $G$ in polynomial time. Without loss of generality we assume that $V_B \leq V_R$. For any arc $u \in V$, let $l(u)$ and $r(u)$ denote the two endpoints of $u$ in the clockwise order of the endpoints of the arc. To design the algorithm, we shall concentrate on the following two cases – **Case A** and **Case B**. In Case A, we check all possible instances while the the the output set does not cover the circle fully. Then, in Case B, we handle all those instances while the output covers the entire circle. Later, we prove that the optimum solution lies in one of these instances. The objective is to reach the optimum solution by exploiting these instances exhaustively.

**Case A: $H$ covers the circle partially:** In this case, there must be a clique of arcs $K$ ($|K| \geq 1$) that is not present in the optimal solution. We consider any pair of arcs $u, v \in V$ such that $r(u) \prec l(v)$ in the clockwise order of the endpoints of the arcs, and consider the vertex set $S_{u,v} = \{w : w \in V, l_v \leq l_w < r_w \leq r_u\} \cup \{u, v\}$. Then, we use the Algorithm 1 to compute maximum BCS on $G[S_{u,v}]$. This process is repeated for each such pair of arcs, and report the BCS with maximum number of arcs.

**Case B: $H$ covers the circle entirely:** In this case, $S$ must contains $2|V_B|$ number of arcs and in fact that is the maximum number of arcs $S$ can opt. In order to compute such a set $S$, first we add the vertices in $V_B$ to $S$, then consider the vertices in $V_B$ as a set $T$ of terminal arcs and we need to find a minimum number of red arcs $D \in V_R$ to span over $T$. We further assume two cases.

**B.1. [$T \cup D$ covers the circle partially]** This case is similar to Case A without some extra red arcs that would be added afterwards to ensure that $S$ contains $2|V_B|$ arcs. Similar to Case A, we again try all possible subsets obtained by pair of vertices $u, v$ with $r(u) \prec l(v)$ and $S_{u,v}$ contains all blue vertices and we find optimal Steiner tree by using Algorithm $Select\_Steiners(G = (V, E), T)$. Then, we add $(|V_R| - |D|)$ (where $D$ is the set of Steiner arcs) red arcs from $V_R$ to $S$.

**B.2. [$T \cup D$ covers the circle entirely]** First, we obtain a set $\mathcal{C}$ of $m$ (for some $m \in [n]$) components from $T$. We may see each component $C \in \mathcal{C}$ as an arc and the neighborhood set $N(C)$ as the union of the neighborhoods of the arcs contained in $C$. Observe that, for any component $C \in \mathcal{C}$, $D$ contains either one arc from $N(C)$ that covers $C$, or two arcs from $N(C)$ where none of them individually covers $C$. Let us consider one component $C \in \mathcal{C}$. Let $l(C)$ and $r(C)$ be the left and right end points of $C$, respectively. If $|N(C) \cap D| = 1$, we consider each arc from $N(C)$ separately that contains $C$. For each such arc $I(C) \in N(C)$, we do the following three step operations –(1) include $I(C)$ in $D$, (2) remove $N(C)$ from the graph, (3) include two blue arcs $(l(I(C)), r(C))$ and $(r(C), r(I(C)))$ in the vertex set of the graph. Now, $T = T \cup \{[l(I(C)), r(C)), (r(C), r(I(C)))]\}$. We give this processed graph that is an interval graph, as an input of the Steiner tree (Algorithm $Select\_Steiners(G = (V, E), T)$) and look for a tree with at most $(|D| - 1)$ Steiner red arcs. Else, when $|N(C) \cap D| = 2$, we take the arcs $C_\ell$ and $C_r$ from $N(C)$ with leftmost and rightmost endpoints, respectively, in $D$. We do the same three step operations –(1) include $C_\ell$ and $C_r$ in $D$, (2) remove $N(C)$ from the graph, (3) include two blue arcs $(l(C_\ell), l(C)))$,

$(r(C), r(C_r)))$. Now, $T = T \cup \{[l(C_\ell), l(C))), (r(C), r(C_r))]\}$. We give this processed graph that is an interval graph, as an input of the Steiner tree (Algorithm *Select_Steiners*$(G = (V, E), T)$) and look for a tree with at most $(|D| - 2)$ Steiner red arcs. This completes the procedure.

**Correctness:** We prove that our algorithm yields an optimum solution. The proof of correctness follows from the way we have designed the algorithm. The algorithm is divided into two cases. For case A, the primary objective is to construct the instances from a circular-arc graph to some interval graph. Thereafter, we can solve it optimally. Now, Case B is further divided into two sub-cases. Here we know that all blue vertices present in optimum solution. Therefore, our goal is to employ the Steiner tree algorithm with terminal set $T = V_B$. Note, for B 1 we again try all possible subsets obtained by pair of vertices $u, v$ where $r(u) \prec l(v)$ and $S_{u,v}$ contains all blue vertices. Note, $G[S_{u,v}]$ is an interval graph and $V_B$ is the set of terminals (since we assumed, w.l.o.g, $V_B \leq V_R$). Therefore we directly apply the Steiner tree procedure and obtain the optimum subset for each such pair. Indeed, this process reports the maximum BCS. In the case B 2 we modify the input graph in three step operations. Moreover, we update the expected output size to make it coherent to the modified input instance. This process is done for one arbitrary component of $T$ (of size $\geq 1$), which gives an interval graph. Clearly, the choice of this component makes no impact on the size of BCS. Thereafter, we follow the Steiner tree algorithm on this graph. Moreover, the algorithm exploits all possible cases and reduce the graph into interval graph without affecting the size of the BCS. Thereby, putting everything together, we conclude the proof.

**Time Complexity:** For case A, we try all pairs of arcs that holds certain condition and consider the subset (note, such subset can be computed in $\mathcal{O}(\log n)$ time given the clockwise order of the vertex set and a range tree where the arcs are stored). For each such subset we use the algorithm for interval graph to compute the maximum BCS. This whole process takes $\mathcal{O}(n^6 \log n)$ time. The complexity of Case A dominates complexity of Case B and we get the total running time $\mathcal{O}(n^6 \log n)$.

**Theorem 4.** *Given an $n$ vertex circular-arc graph $G$ whose vertices are colored either red or blue, the BCS problem on $G$ can be solved in $\mathcal{O}(n^6 \log n)$ time.*

### 3.3   FPT Algorithm

In this section, we show that the BCS problem is fixed-parameter tractable for general graphs while parameterized by the solution size. Let $G = (V, E)$ be a simple connected graph, and let $k$ be a given parameter. A family $\mathcal{F}$ of functions from $V$ to $\{1, 2, \ldots, k\}$ is called a *perfect hash family* for $V$ if the following condition holds. For any subset $U \subseteq V$ with $|U| = k$, there is a function $f \in \mathcal{F}$ which is injective on $U$, i.e., $f|_U$ is one-to-one. For any graph of $n$ vertices and a positive integer $k$, it is possible to obtain a perfect hash family of size $2^{\mathcal{O}(k)} \log n$ in $2^{\mathcal{O}(k)} n \log n$ time; see [1]. Now, the $k$-BCS problem can be defined as follows.

Given a bicolored (red and blue) graph $G = (V, E)$, and a parameter $k$, decide if $G$ contains a BCS of size $k$.

We employ two methods to solve the $k$-BCS problem: (i) *color coding technique*, and (ii) *batch procedure*. Our approach is motivated by the approach of Fellows et al. [7], where they have used these techniques to provide a FPT-algorithm for the graph motif problem. Suppose $H$ is a solution to the $k$-BCS problem and $\mathcal{F}$ is a perfect hash family for $V$. This ensures us that at least one function of $\mathcal{F}$ assigns vertices of $H$ with $k$ distinct labels. Therefore, if we iterate through all functions of $\mathcal{F}$ and find the subsets of $V$ of size $k$ that are distinctly labeled by our hashing, we are guaranteed to find $H$. Now, we try to solve the following problem: Given a hash function $f \colon V \to \{1, 2, \ldots, k\}$ from perfect hash family $\mathcal{F}$, decide if there is a subset $U \subseteq V$ with $|U| = k$ such that $G[U]$ is a balanced connected subgraph of $G$ and $f|_U$ is one-to-one.

First, we create a table, denoted by $M$. For a label $L \subseteq \{1, 2, \ldots, k\}$ and a color-pair $(b, r)$ of non-negative integers where $b + r = |L|$, we put $M[v \ ; \ L, \ (b, r)] = 1$ if and only if there exists a subset $U \subseteq V$ of vertices such that the conditions holds: (i) $v \in U$, (ii) $f|_U = L$, (iii) $G[U]$ is connected, and (iv) $U$ consisting exactly $b$ blue vertices and $r$ red vertices.

Notice that, the total number of entries of this table is $\mathcal{O}(2^k k n)$. If we can fill all the entries of the table $M$, then we can just look at the entries $M[v \ ; \ \{1, 2, \ldots, k\}, \ (\frac{k}{2}, \frac{k}{2})]$, $\forall v \in V$, and if any of them is one then we can claim that the $k$-BCS problem has a solution. Now we use the batch procedure to compute $M[v \ ; \ L, \ (b, r)]$ for each subset $L \subseteq \{1, 2, \ldots k\}$, and for each pair $(b, r)$ of non-negative integers such that $(b + r) = |L|$. Now, we explain the batch procedure. Without loss of generality we assume that $L = \{1, 2, \ldots, t\}$, $f(v) = t$, and the color of $v$ is red.

**Batch Procedure $(v, L, (b, r))$:**
(1) **Initialize:** Construct the set $\mathcal{S}$ of pairs $(L', (b', r'))$, where $b' + r' = |L'|$ such that $L' \subseteq \{1, 2, \ldots, t - 1\}$, $b' \leq b$, $r' \leq r - 1$ and $M[u \ ; \ L', \ (b', r')] = 1$ for some neighbour $u$ of $v$.
(2) **Update:** If there exists two pairs $\{(L_1, (b_1, r_1)), (L_2, (b_2, r_2))\} \in \mathcal{S}$ such that $L_1 \cap L_2 = \phi$ and $(b_1, r_1) + (b_2, r_2) \leq (b, r - 1)$, then add $(L_1 \cup L_2, (b_1 + b_2, r_1 + r_2))$ into $\mathcal{S}$. Repeat this step until unable to add any more.
(3) **Decide:** Set $M[v \ ; \ L, \ (b, r)] = 1$ if $(\{1, 2, \ldots, t - 1\}, (b, r - 1)) \in \mathcal{S}$, else 0.

**Lemma 3.** *The batch procedure correctly computes $M[v \ ; \ L, \ (b, r)]$ for all $v$, $L \subseteq \{1, 2, \ldots, k\}$ with $b + r = |L|$.*

*Proof.* Without loss of generality we assume that $L = \{1, 2, \ldots, t\}$, $f(v) = t$ and color of $v$ is red. We have to show that $M[v \ ; \ L, \ (b, r)] = 1 \Leftrightarrow$ there exists a connected subgraph, containing $v$, and having exactly $b$ blue vertices and $r$ red vertices. Firstly, we assume that $M[v \ ; \ L, \ (b, r)] = 1$. So, $(\{1, 2, \ldots, t - 1\}, (b, r - 1)) \in \mathcal{S}$. So, there must exist some neighbours $\{v'_1, v'_2, \ldots, v'_l\}$ of $v$ such that $M[v'_1; \ L_1, \ (b_1, r_1)] = M[v'_2; \ L_2, \ (b_2, r_2)] = \cdots = M[v'_l; \ L_l, \ (b_l, r_l)] = 1$ with $\bigcup_{i=1}^{l} L_i = L \setminus \{t\}$, $\sum_{i=1}^{l} b_i = b$, $\sum_{i=1}^{l} r_i = r - 1$ where $L_1, L_2, \ldots, L_l$ are pairwise disjoint. Thus, there exists a connected subgraph containing $\{v, v'_1, \ldots, v'_l\}$ having exactly $b$ blue vertices and $r$ red vertices. Other direction of the proof follows from the same idea.

**Lemma 4.** *Given a hash function* $f \colon V \to \{1, 2, \ldots, k\}$, *batch procedure fill all the entries of table* $M$ *in* $\mathcal{O}(2^{4k} k^3 n^2)$ *time.*

*Proof.* The initialization depends on the number of the search process in the entries correspond to the neighbour of $v$. Now, this number is bounded by the size of $M$. The first step takes $\mathcal{O}(2^k kn)$ time. Now the size of $\mathcal{S}$ can be at most $2^k k$. Each update takes $\mathcal{O}(2^{2k} k^2)$ time. So step 2 takes $\mathcal{O}(2^{3k} k^3)$ time. Now the value of $M[v \, ; \, L, \, (b, r)]$ can be decided in $\mathcal{O}(2^k k)$ time. As the number of entries in $M$ is $\mathcal{O}(2^k kn)$, so the total running time is $\mathcal{O}(2^{4k} k^3 n^2)$.    $\square$

The algorithm for the $k$-BCS problem is following:

**1.** Construct a perfect hash family $\mathcal{F}$ of $2^{\mathcal{O}(k)} \log n$ functions in $2^{\mathcal{O}(k)} n \log n$ time.

**2.** For each function, $f \in \mathcal{F}$ build the table $M$ using batch procedure. For each function $f \in \mathcal{F}$ it takes $\mathcal{O}(2^{4k} k^3 n^2)$ time.

**3.** As each $f \in \mathcal{F}$ is perfect, by an exhaustive search through all function in $\mathcal{F}$ our algorithm correctly decide whether there exists a balanced connected subgraph of $k$ vertices. We output yes, if and only if there is a vertex $v$ and $f \in \mathcal{F}$ for which the corresponding table $M$ contains the entry one in $M[v \, ; \, \{1, 2, \ldots, k\}, \, (\frac{k}{2}, \frac{k}{2})]$.

**Theorem 5.** *The $k$-BCS problem can be solved in time* $2^{\mathcal{O}(k)} n^2 \log n$ *time.*

**Acknowledgement.** We thank Joseph S. B. Mitchell for his useful suggestions.

# References

1. Alon, N., Yuster, R., Zwick, U.: Color-coding. J. ACM **42**(4), 844–856 (1995)
2. Bhore, S., Chakraborty, S., Jana, S., Mitchell, J.S.B., Pandit, S., Roy, S.: The balanced connected subgraph problem. In: Pal, S.P., Vijayakumar, A. (eds.) CALDAM 2019. LNCS, vol. 11394, pp. 201–215. Springer, Cham (2019). https://doi.org/10.1007/978-3-030-11509-8_17
3. Bodlaender, H.L., Fellows, M.R., Langston, M.A., Ragan, M.A., Rosamond, F.A., Weyer, M.: Quadratic kernelization for convex recoloring of trees. In: Lin, G. (ed.) COCOON 2007. LNCS, vol. 4598, pp. 86–96. Springer, Heidelberg (2007). https://doi.org/10.1007/978-3-540-73545-8_11
4. Bodlaender, H.L., Fellows, M.R., Warnow, T.J.: Two strikes against perfect phylogeny. In: Kuich, W. (ed.) ICALP 1992. LNCS, vol. 623, pp. 273–283. Springer, Heidelberg (1992). https://doi.org/10.1007/3-540-55719-9_80
5. Bodlaender, H.L., de Fluiter, B.: Intervalizing $k$-colored graphs. In: Fülöp, Z., Gécseg, F. (eds.) ICALP 1995. LNCS, vol. 944, pp. 87–98. Springer, Heidelberg (1995). https://doi.org/10.1007/3-540-60084-1_65
6. Bonnet, É., Sikora, F.: The graph motif problem parameterized by the structure of the input graph. Discret. Appl. Math. **231**, 78–94 (2017)

7. Fellows, M.R., Fertin, G., Hermelin, D., Vialette, S.: Upper and lower bounds for finding connected motifs in vertex-colored graphs. J. Comput. Syst. Sci. **77**(4), 799–811 (2011)
8. Fellows, M.R., Hallett, M.T., Wareham, H.T.: DNA physical mapping: three ways difficult. In: Lengauer, T. (ed.) ESA 1993. LNCS, vol. 726, pp. 157–168. Springer, Heidelberg (1993). https://doi.org/10.1007/3-540-57273-2_52
9. Garey, M.R., Johnson, D.S.: The rectilinear steiner tree problem is NP-complete. SIAM J. Appl. Math. **32**(4), 826–834 (1977)
10. Kikuno, T., Yoshida, N., Kakuda, Y.: The NP-completeness of the dominating set problem in cubic planer graphs. IEICI Trans. (1976–1990) **63**(6), 443–444 (1980)
11. Lacroix, V., Fernandes, C.G., Sagot, M.: Motif search in graphs: application to metabolic networks. IEEE/ACM Trans. Comput. Biol. Bioinf. **3**(4), 360–368 (2006)
12. Lacroix, V., Fernandes, C.G., Sagot, M.F.: Motif search in graphs: application to metabolic networks. IEEE/ACM Trans. Comput. Biol. Bioinform. (TCBB) **3**(4), 360–368 (2006)

# Approximating Bounded Job Start Scheduling with Application in Royal Mail Deliveries Under Uncertainty

Jeremy T. Bradley[1], Dimitrios Letsios[2(✉)], Ruth Misener[2], and Natasha Page[2]

[1] GBI/Data Science Group, Royal Mail, London, UK
jeremy.bradley@royalmail.com
[2] Department of Computing, Imperial College London, London, UK
{d.letsios,r.misener,natasha.page17}@imperial.ac.uk

**Abstract.** Motivated by mail delivery scheduling problems arising in Royal Mail, we study a generalization of the fundamental makespan scheduling problem $P||C_{\max}$ which we call the *Bounded Job Start Scheduling Problem*. Given a set of jobs, each one specified by an integer processing time $p_j$, that have to be executed non-preemptively by a set of $m$ parallel identical machines, the objective is to compute a minimum makespan schedule subject to an upper bound $g \leq m$ on the number of jobs that may simultaneously begin per unit of time. We show that Longest Processing Time First (LPT) algorithm is tightly 2-approximate. After proving that the problem is strongly $\mathcal{NP}$-hard even when $g = 1$, we elaborate on improving the 2-approximation ratio for this case. We distinguish the classes of long and short instances satisfying $p_j \geq m$ and $p_j < m$, respectively, for each job $j$. We show that LPT is 5/3-approximate for the former and optimal for the latter. Then, we explore the idea of scheduling long jobs in parallel with short jobs to obtain solutions with tightly satisfied packing and bounded job start constraints. For a broad family of instances excluding degenerate instances with many very long jobs and instances with few machines, we derive a 1.985-approximation ratio. For general instances, we require machine augmentation to obtain better than 2-approximate schedules. Finally, we exploit machine augmentation and a state-of-the-art lexicographic optimization method for $P||C_{\max}$ under uncertainty to propose a two-stage robust optimization approach for bounded job start scheduling under uncertainty attaining good trade-offs in terms of makespan and number of used machines. We substantiate this approach numerically using Royal Mail data.

**Keywords:** Bounded job start scheduling · Approximation algorithms · Robust scheduling · Mail deliveries

## 1 Introduction

Royal Mail provides mail collection and delivery services for all United Kingdom (UK) addresses. With a small van fleet of 45,000 vehicles and 90,000 drivers

© Springer Nature Switzerland AG 2019
Y. Li et al. (Eds.): COCOA 2019, LNCS 11949, pp. 69–81, 2019.
https://doi.org/10.1007/978-3-030-36412-0_6

delivering to 27 million locations in UK, efficient resource allocation is essential
to guarantee the business viability. The backbone of the Royal Mail distribution
network is a three-layer hierarchical network with 6 regional distribution centers
serving 38 mail centers. Each mail center receives, processes and distributes mail
for a large geographically-defined area via 1,250 delivery offices, each serving
disjoint sets of neighboring post codes. Mail is collected in mail centers, sorted
by region and forwarded to an appropriate onward mail center, making use of
the regional distribution centers for cross-docking purposes. From the onward
mail center it is transferred to the final delivery office destination. This process
has to be completed within 12 to 16 h for 1st class post and 24 to 36 h for 2nd
class post depending on when the initial collection takes place.

In a delivery office, post is sorted, divided into routes and delivered to
addresses using the a combination of small fleet vans and walked trolleys. Allocation of delivery itineraries to vans is critical. Each delivery office has an exit
gate for vans upper bounding the number of vehicles departing per unit of time.
Thus, we deal with the problem of scheduling a set $\mathcal{J}$ of jobs (delivery itineraries)
each one associated with an integer processing time $p_j$, on $m$ parallel identical
machines (vehicles), s.t. the makespan, i.e. the last job completion time, is minimized. Parameter $g$ upper bounds the number of jobs that may simultaneously
begin per unit of time. Each job has to be executed non-preemptively, i.e. by a
single machine in a continuous time interval without interruptions. We call this
problem the *Bounded Job Start Scheduling Problem (BJSP)*.

BJSP is strongly related to the fundamental makespan scheduling problem
$P||C_{\max}$ [5]. BJSP generalizes $P||C_{\max}$ as the problems become equivalent when
$g = m$. Furthermore, $P||C_{\max}$ is the BJSP relaxation obtained by dropping the
BJSP constraint. Note that the $P||C_{\max}$ optimal solution is a factor $\Omega(m)$ from
the BJSP one, in the worst case. For example, take an arbitrary $P||C_{\max}$ instance
and construct a BJSP one with $g = 1$ by adding a large number of unit jobs.
The BJSP optimal schedule requires time intervals during which $m-1$ machines
are idle at each time while the $P||C_{\max}$ optimal schedule is perfectly balanced
and all machines are busy until the last job completes. On the positive side, we
may convert any $\rho$-approximation algorithm for $P||C_{\max}$ into $2\rho$-approximation
algorithm for BJSP using naive bounds. Given that $P||C_{\max}$ admits a PTAS, we
obtain an $O(n^{1/\epsilon} \cdot poly(n))$-time $(2+\epsilon)$-approximation algorithm for BJSP. Here,
a main goal is to obtain tighter performance guarantees. Similarly to $P||C_{\max}$,
provably good BJSP solutions must attain low imbalance $\max_i\{T-T_i\}$, where $T$
and $T_i$ are the makespan and completion time of machine $i$, respectively. Because
of the BJSP constraint, feasible schedules may require idle machine time before
all jobs have begun. So, BJSP exhibits the additional difficulty of effectively
bounding the total idle period $\sum_{t\le r}(m - |\mathcal{A}_t|)$, where $r$ and $\mathcal{A}_t$ are the last job
start time and set of jobs executed during $[t, t+1)$, respectively.

BJSP relaxes the scheduling problem with forbidden sets, i.e. non-overlapping
constraints, where subsets of jobs cannot run in parallel [10]. For the latter problem, better than 2-approximation algorithms are ruled out, unless $\mathcal{P} = \mathcal{NP}$
[10]. Even when there is a strict order between jobs in the same forbidden

set, the scheduling with forbidden sets problem is equivalent to the precedence-constrained scheduling problem $P|prec|C_{\max}$ and cannot be approximated by a factor lower than $(2 - \epsilon)$, assuming a variant of the unique games conjecture [12]. Also, BJSP relaxes the scheduling with forbidden job start times problem, where no job may begin at certain time points, which does not admit constant-factor approximation algorithms [2,3,8,9]. Despite the commonalities with the afore-mentioned literature, to the authors' knowledge, there is a lack of approximation algorithms for scheduling problems with bounded job starts.

*Contributions and Paper Organization.* Section 2 formally defines BJSP, proves the problem's $\mathcal{NP}$-hardness, and derives a $O(\log m)$ integrality gap for a natural integer programming formulation. Section 3 investigates *Longest Processing Time First (LPT)* algorithm, i.e. the probably simplest option for BJSP, and derives a tight 2-approximation ratio. The remainder of the paper elaborates on improving this ratio for the special case $g = 1$. Section 2 shows that BJSP remains strongly $\mathcal{NP}$-hard even in this case. Note that several arguments in our analysis can be extended to the arbitrary $g$ case. Focusing on $g = 1$ allows to avoid many floors, ceilings, and simplifies our presentation. Furthermore, any Royal Mail instance can be converted to this case using small discretization.

Section 4 distinguishes between long and short instances. An instance $\langle m, \mathcal{J} \rangle$ is *long* if $p_j \geq m$ for each $j \in \mathcal{J}$ and *short* if $p_j < m$ for all $j \in \mathcal{J}$. This distinction is motivated by the observation that idle time occurs mainly because of (i) simultaneous job completions in the former case, and (ii) limited allowable parallel job executions in the latter case. Section 4 proves that LPT is 5/3-approximate for long instances and optimal for short instances. A key ingredient for establishing the ratio in the case of long instances is a concave relaxation for bounding the idle machine time. Section 4 also obtains an improved approximation ratio for long instances when the maximum job processing time is relatively small using the *Shortest Processing Time First (SPT)* algorithm.

Greedy scheduling policies, e.g. LPT and SPT, that sequence long jobs first and short jobs next, or vice versa, cannot achieve an approximation ratio better than 2. Section 5 proposes the *Long-Short Mixing (LSM)* algorithm that devotes a certain number of machines to long jobs and uses all remaining machines for short jobs. By executing the two types of jobs in parallel, LSM achieves a 1.985-approximation ratio for a broad family of instances. For degenerate instances with many very long jobs or with few machines, we require constant-factor machine augmentation, i.e. $fm$ machines where $f > 1$ is constant, to achieve a strictly lower than 2 approximation ratio.

Because Royal Mail delivery scheduling is subject to uncertainty, the full paper version exploits machine augmentation and a state-of-the-art lexicographic optimization method for $P||C_{\max}$ under uncertainty [6,11] to construct a two-stage robust optimization approach [1,4,7] for the BJSP under uncertainty. We substantiate the proposed approach empirically using Royal Mail data.

Section 6 concludes with a collection of intriguing future directions. Due to space constraints, omitted proofs and parts of the paper are deferred to the full version.

## 2  Problem Definition and Preliminary Results

An instance $I = \langle m, \mathcal{J} \rangle$ of the *Bounded Job Start Scheduling Problem (BJSP)* is specified by a set $\mathcal{M} = \{1, \ldots, m\}$ of parallel identical machines, and a set $\mathcal{J} = \{1, \ldots, n\}$ of jobs. A machine may execute at most one job per unit of time. Job $j \in \mathcal{J}$ is associated with an integer processing time $p_j$. Time is partitioned into a set $D = \{1, \ldots, \tau\}$ of discrete time slots. Time slot $t \in T$ corresponds to time interval $[t - 1, t)$. Each job should be executed non-preemptively, i.e. in a continuous time interval without interruptions, by a single machine. This interval should consist of an integer number of time slots. The remainder of the manuscript assumes time intervals $[s, t] = \{s, s + 1, \ldots, t\}$ of time slots. *BJSP parameter g* imposes an upper bound on the number of jobs that may begin per unit of time. The goal is to assign each job $j \in \mathcal{J}$ to a machine and decide its starting time so that this BJSP constraint is not violated and the makespan, i.e. the time at which the last job completes, is minimized. Consider a feasible schedule $\mathcal{S}$ with makespan $T$. Denote the start time of job $j$ by $s_j$. Job $j \in \mathcal{J}$ must be entirely executed during the interval $[s_j, C_j)$, where $C_j = s_j + p_j$ is the completion time of $j$. So, $T = \max_{j \in \mathcal{J}} \{C_j\}$. We say that job $j$ is *alive* at time slot $t$ if $t \in [s_j, C_j)$. Let $\mathcal{A}_t = \{j : t \in [s_j, C_j)\}$ and $\mathcal{B}_t = \{j : s_j = t - 1\}$ be the set of alive and beginning jobs during time slot $t$, respectively. Schedule $\mathcal{S}$ is feasible only if $|\mathcal{A}_t| \leq m$ and $|\mathcal{B}_t| \leq g$, for all $t$.

BJSP is strongly $\mathcal{NP}$-hard because it becomes equivalent with $P||C_{\max}$ in the special case where $g = \min\{m, n\}$. Theorem 1 shows that BJSP is strongly $\mathcal{NP}$-hard also when $g = 1$, through a reduction from 3-Partition.

**Theorem 1.** *BJSP is strongly $\mathcal{NP}$-hard in the special case $g = 1$.*

Theorem 2 shows that a natural integer programming formulation has non-constant integrality gap. Thus, stronger linear programming (LP) relaxations are required for obtaining constant-factor approximation LP rounding algorithms.

**Theorem 2.** *A natural integer programming formulation using binary variables indicating the start time of each job has integrality gap $\Omega(\log m)$.*

## 3  LPT Algorithm

Longest Processing Time first algorithm (LPT) schedules the jobs on a fixed number $m$ of machines w.r.t. the order $p_1 \geq \ldots \geq p_n$. Recall that $|\mathcal{A}_t|$ and $|\mathcal{B}_t|$ is the number of alive and beginning jobs, respectively, at time slot $t \in D$. We say that time slot $t \in D$ is *available* if $|\mathcal{A}_t| < m$ and $|\mathcal{B}_t| < g$. LPT schedules the jobs greedily w.r.t. their sorted order. Each job $j$ is scheduled in the earliest available time slot, i.e. at $s_j = \min\{t : |\mathcal{A}_t| < m, |\mathcal{B}_t| < g, t \in D\}$. Theorem 3 proves a tight approximation ratio of 2 for LPT.

**Theorem 3.** *LPT is tightly 2-approximate for minimizing makespan.*

*Proof.* Denote by $\mathcal{S}$ and $\mathcal{S}^*$ the LPT and a minimum makespan schedule, respectively. Let $\ell$ be the job completing last in $\mathcal{S}$, i.e. $T = s_\ell + p_\ell$. For each time slot $t \leq s_\ell$, either $|\mathcal{A}_t| = m$, or $|\mathcal{A}_t| < m$. Since $\ell$ is scheduled at the earliest available time slot, for each $t \leq s_\ell$ s.t. $|\mathcal{A}_t| < m$, we have that $|\mathcal{B}_t| = g$. Let $\lambda$ be the total length of time s.t. $|\mathcal{A}_t| < m$ in $\mathcal{S}$. Because of the BJSP constraint, exactly $g$ jobs begin per unit of time, which implies that $\lambda \leq \lceil \frac{\ell}{g} \rceil$. Therefore, schedule $\mathcal{S}$ has makespan

$$T = s_\ell + p_\ell \leq \frac{1}{m} \sum_{j \neq \ell} p_j + \lambda + p_\ell \leq \frac{1}{m} \sum_{j=1}^{n} p_j + \left( \left\lceil \frac{\ell}{g} \right\rceil + p_\ell \right)$$

Denote by $s_j^*$ the starting time of job $j$ in $\mathcal{S}^*$ and let $\pi_1, \ldots, \pi_n$ the job indices ordered in non-decreasing schedule $\mathcal{S}^*$ starting times, i.e. $s_{\pi_1}^* \leq \cdots \leq s_{\pi_n}^*$. Because of the BJSP constraint, $s_{\pi_j}^* \geq \lceil j/g \rceil$. In addition, there exists $j' \in [j, n]$ s.t. $p_{\pi_{j'}} \geq p_j$. Thus, $\max_{j'=j}^{n} \{ s_{\pi_{j'}}^* + p_{\pi_{j'}} \} \geq \lceil j/g \rceil + p_j$, for $j = 1, \ldots, n$. Then,

$$T^* \geq \max \left\{ \frac{1}{m} \sum_{j=1}^{n} p_j, \max_{j=1}^{n} \left\{ \left\lceil \frac{j}{g} \right\rceil + p_j \right\} \right\}$$

We conclude that $T \leq 2T^*$.

For the tightness of the analysis, consider an instance $I = \langle m, \mathcal{J} \rangle$ with $m(m-1)$ long jobs of processing time $p$, where $p = \omega(m)$ and $m = \omega(1)$, $m(p-m)$ unit jobs, and BJSP parameter $g = 1$. LPT schedules the long jobs into $m-1$ batches, each one with exactly $m$ jobs. All jobs of a batch are executed in parallel for their greatest part. In particular, the $i$-th job of the $k$-th batch is executed by machine $i$ starting at time slot $(k-1)p + i$. All unit jobs are executed sequentially by machine 1 starting at $(m-1)p+1$. Observe that $\mathcal{S}$ is feasible and has makespan $T = (m-1)p + m(p-m) = (2m-1)p - m^2$. The optimal solution $\mathcal{S}^*$ schedules all jobs in $m$ batches. The $k$-the batch contains $(m-1)$ long jobs and $(p-m+1)$ unit jobs. Specifically, the $i$-th long job is executed by machine $i$ beginning at $(k-1)p+i$, while all short jobs are executed consecutively by machine $m$ starting at $(k-1)p+m$ and completing at $kp$. Schedule $\mathcal{S}^*$ is feasible and has makespan $T^* = mp$. Because $\frac{m}{p} \to 0$ and $\frac{1}{m} \to 0$, i.e. both approach zero, $T \to 2T^*$.    □

## 4    Long and Short Instances

Next, we consider two natural classes of BJSP instances, in the case where $g = 1$, for which LPT achieves an approximation ratio better than 2. Instance $\langle m, \mathcal{J} \rangle$ is (i) *long* if $p_j \geq m$ for each $j \in \mathcal{J}$, and (ii) *short* if $p_j < m$ for every $j \in \mathcal{J}$. This section proves that LPT is 5/3-approximate for long instances and optimal for short instances. Finally, when the longest job is relatively small compared to the total load, we show a better than 5/3-approximation ratio for SPT in the case of long instances.

Consider a feasible schedule $\mathcal{S}$ and let $r = \max_{j \in \mathcal{J}} \{s_j\}$ be the last job start time. We say that $\mathcal{S}$ is a *compact schedule* if it holds that either (i) $|\mathcal{A}_t| = m$, or (ii) $|\mathcal{B}_t| = 1$, for each $t \in [1, r]$. Lemma 1 shows the existence of an optimal compact schedule and derives a lower bound on the optimal makespan.

**Lemma 1.** *For each instance $I = \langle m, \mathcal{J} \rangle$, there exists a feasible compact schedule $\mathcal{S}^*$ which is optimal. Let $\mathcal{J}^L = \{j : p_j \geq m, j \in \mathcal{J}\}$. If $|\mathcal{J}^L| \geq m$, then $\mathcal{S}^*$ has makespan $T^* \geq \frac{m-1}{2} + \frac{1}{m} \sum_{j=1}^{n} p_j$.*

Next, we analyze LPT in the case of long instances. Similarly to the Lemma 1 proof, we may show that LPT produces a compact schedule $\mathcal{S}$. So, we may partition the interval $[1, r]$ into a sequence $P_1, \ldots, P_k$ of maximal periods satisfying the following invariant: for each $q \in \{1, \ldots, k\}$, either (i) $|\mathcal{A}_t| < m$ for each $t \in P_q$, or (ii) $|\mathcal{A}_t| = m$ for each $t \in P_q$. That is, there is no pair of time slots $s, t \in P_q$ such that $|\mathcal{A}_s| < m$ and $|\mathcal{A}_t| = m$. We say that $P_q$ is a *slack period* if $P_q$ satisfies (i). Otherwise, we refer to $P_q$ as a *full* period. For a given period $P_q$ of length $\lambda_q$, denote by $\Lambda_q = \sum_{t \in P_q} (m - |\mathcal{A}_t|)$ the idle machine time. Note that $\Lambda_q = 0$, for each full period $P_q$. Lemma 2 upper bounds the total idle machine time of slack periods in the LPT schedule $\mathcal{S}$, except the very last period $P_k$. In the case where $P_k$ is slack, the length $\lambda_k$ of $P_k$ is upper bounded by Lemma 3.

**Lemma 2.** *Let $k' = k - 1$. Consider a long instance $I = \langle m, \mathcal{J} \rangle$, with $|\mathcal{J}| \geq m$, and the LPT schedule $\mathcal{S}$. It holds that (i) $\lambda_q \leq m - 1$, and (ii) $\Lambda_q \leq \frac{\lambda_q(\lambda_q - 1)}{2}$ for each slack period $P_q$, where $q \in \{1, \ldots, k'\}$. Furthermore, (iii) $\sum_{q=1}^{k'} \Lambda_q \leq \frac{nm}{2}$.*

*Proof.* For part (i), let $P_q = [s, t]$ be a slack time period in $\mathcal{S}$ and assume for contradiction that $\lambda_q \geq m$, i.e. $t \geq s + m - 1$. Given that $p_j \geq m$ for each $j \in \mathcal{J}$, we have that $\{j : s_j \in [s, s + m - 1], j \in \mathcal{J}\} \subseteq \mathcal{A}_{s+m-1}$. That is, all jobs starting during $[s, s + m - 1]$ are alive at time $s + m - 1$. Since $P_q$ is a slack period, it holds that $a_u < m$, for each $u \in [s, s + m - 1]$. Because $\mathcal{S}$ is compact, it must be the case that $b_u = 1$, i.e. exactly $g = 1$ jobs begin, at each $u \in [s, s + m - 1]$. These observations imply that $|\mathcal{A}_{s+m-1}| \geq m$, which contradicts the fact that $P_q$ is a maximal slack period.

For part (ii), we consider the partitioning $\mathcal{A}_u = \mathcal{A}_u^- + \mathcal{A}_u^+$ for each time slot $u \in P_q = [s, t]$, where $\mathcal{A}_u^-$ and $\mathcal{A}_u^+$ is the set of alive jobs at time $u$ completing inside $P_q$, i.e. $C_j \in [s, t]$, and after $P_q$, i.e. $C_j > t$, respectively. Since $\lambda_q \leq m - 1$, every job $j$ beginning during $P_q$, i.e. $s_j \in [s, t]$, must complete after $P_q$, i.e. $C_j > t$. We modify schedule $\mathcal{S}$ by removing every occurrence of a job $j$ executed inside $P_q$ such that $C_j \in P_q$. Clearly, in the modified schedule $\mathcal{S}'$ the idle time $\Lambda_q'$ during $P_q$ may only have increased, i.e. $\Lambda_q \leq \Lambda_q'$. Furthermore, no job $j$ with $s_j \in P_q$ is removed. Because $|\mathcal{B}_u| = 1$ for each $u \in P_q$, we have that $|\mathcal{A}_u| = |\mathcal{A}_{u+1}| - 1$ for $u = s, \ldots, t$. Furthermore, $|\mathcal{A}_{t+1}| = m$. So,

$$\Lambda_q' = \sum_{u=s}^{t} (m - |\mathcal{A}_u|) = \sum_{u=s}^{t} [m - (t - s + 1)] = \sum_{u=1}^{\lambda_q - 1} u = \frac{\lambda_q(\lambda_q - 1)}{2}.$$

Now, we proceed with part (iii). Consider a slack period $P_q$, for $q \in \{1, \ldots, k'\}$. By part (i), $\lambda_q \leq m - 1$. Because of the BJSP constraint, at most

$g = 1$ jobs begin at each $t \in P_q$ and, thus, $\sum_{q=1}^{k'} \lambda_q \leq n - 1$. So, by part (ii), the Eq. (1) concave program upper bounds $\sum_{q=1}^{k'} \Lambda_q$.

$$\max_{\lambda_q} \quad \sum_{q=1}^{k'} \frac{\lambda_q(\lambda_q - 1)}{2} \tag{1a}$$

$$1 \leq \lambda_q \leq m \qquad\qquad q \in \{1, \ldots, k'\} \tag{1b}$$

$$\sum_{q=1}^{k'} \lambda_q \leq n \tag{1c}$$

Assume without loss of generality that $n/m$ is integer. If $k' \leq n/m$, by setting $\lambda_q = m$, for $q \in \{1, \ldots, k'\}$, we obtain that $\sum_{q=1}^{k'} \lambda_q(\lambda_q-1)/2 \leq k'm(m-1)/2 \leq nm/2$. If $k' > n/m$, we argue that the solution $\lambda_q = n/m$, for $q \in \{1, \ldots, n/m\}$, and $\lambda_q = 0$, otherwise, is optimal for the concave program (1). In particular, for any solution $\lambda$ with $0 < \lambda_q, \lambda_{q'} < m$ such that $q \neq q'$, we may construct a modified solution with higher objective value by applying Jensen's inequality $f(\lambda) + f(\lambda') \leq f(\lambda + \lambda')$ for any $\lambda, \lambda' \in [0, \infty)$, with respect to the single variable, convex function $f(\lambda) = \lambda(\lambda-1)/2$. If $\lambda_q + \lambda_{q'} \leq m$, we may set $\tilde{\lambda}_q = \lambda_q + \lambda_{q'}$ and $\tilde{\lambda}_{q'} = 0$. Otherwise, $m < \lambda_q + \lambda_{q'} \leq 2m$ and we set $\tilde{\lambda}_q = m$ and $\tilde{\lambda}_{q'} = \lambda_q + \lambda_{q'} - m$. In both cases, $\lambda_{q''} = \tilde{\lambda}_{q''}$, for each $q'' \in \{1, \ldots, k'\} \setminus \{q, q'\}$. Clearly, $\tilde{\lambda}$ attains higher objective value than $\lambda$, for concave program (1).

**Lemma 3.** *Suppose that the last period $P_k$ is slack and let $\mathcal{J}_k$ be the set of jobs beginning during $P_k$. Then, it holds that $\lambda_k \leq \frac{1}{m} \sum_{j \in \mathcal{J}_k} p_j$.*

**Theorem 4.** *LPT is 5/3-approximate for long instances.*

*Proof.* Denote the LPT and optimal schedules by $\mathcal{S}$ and $\mathcal{S}^*$, respectively. Let $\ell \in \mathcal{J}$ be a job completing last in $\mathcal{S}$, i.e. $T = s_\ell + p_\ell$. Recall that LPT sorts the jobs s.t. $p_1 \geq \ldots \geq p_n$. W.l.o.g. we may assume that $\ell = \arg\min_{j \in \mathcal{J}}\{p_j\}$. Indeed, we may discard every job $j > \ell$ and bound the algorithm's performance w.r.t. instance $\tilde{I} = \langle m, \mathcal{J} \setminus \{j : j > \ell, j \in \mathcal{J}\}\rangle$. Let $\tilde{\mathcal{S}}$ and $\tilde{\mathcal{S}}^*$ be the LPT and an optimal schedule attaining makespan $\tilde{T}$ and $\tilde{T}^*$, respectively, for instance $\tilde{I}$. Showing that $\tilde{T} \leq (5/3)\tilde{T}^*$ is sufficient for our purposes because $T = \tilde{T}$ and $\tilde{T}^* \leq T^*$. We distinguish two cases based on whether $p_n > T^*/3$, or $p_n \leq T^*/3$.

In the former case, we claim that $T \leq (3/2)T^*$. Initially, observe that $n \leq 2m$. Otherwise, there would be a machine $i \in \mathcal{M}$ executing at least $|\mathcal{S}_i^*| \geq 3$ jobs, say $j, j', j'' \in \mathcal{J}$, in $\mathcal{S}^*$. This machine would have last job completion time $T_i^* \geq p_j + p_{j'} + p_{j''} > T^*$, which is a contradiction. If $n \leq m$, LPT has clearly makespan $T = T^*$. So, consider that $n > m$. Then, some machine executes at least two jobs in $\mathcal{S}^*$, i.e. $p_n \leq T^*/2$. To prove our claim, it suffices to show that $s_n \leq T^*$. Let $c = \max_{1 \leq j \leq m}\{C_j\}$ be the time at which the last among the $m$ biggest jobs completes. If $s_n \leq c$, then we have that $s_n \leq \max_{1 \leq j \leq m}\{j + p_j\} \leq T^*$. Otherwise, let $\lambda = s_n - c$. Because $n \leq 2m$, it must be the case that $\lambda \leq m$. Furthermore, $|\mathcal{A}_t| < m$ and, thus, $|\mathcal{B}_t| = 1$, for each $t \in [c + 1, s_n - 1]$. That is, exactly one job begins per unit of time during $[c+1, s_n]$. Due to the LPT ordering, these are exactly the jobs $\{n - \lambda, \ldots, n\}$. Since $\lambda \leq m$ and $p_j \geq m$, at least $m - k$ units of time of job $n - k$ are executed from time $s_n$ and onwards, for $k \in \{1, \ldots, \lambda\}$. Thus, the total processing load which executed not earlier than $s_n$ is $\mu \geq \sum_{k=1}^{\lambda}(m-k)$. On the other hand, at most $m - k$ machines are idle at time slot $c + k$, for $k \in \{1, \ldots, \lambda\}$. So, the total idle machine time during $[m+1, s_n - 1]$ is $\Lambda \leq \sum_{k=1}^{\lambda-1}(m-k)$. We conclude that $\mu \geq \Lambda$ which implies that $s_n \leq \frac{m(m-1)}{2} + \frac{1}{m}\sum_{j \in \mathcal{J}} p_j$. By Lemma 1, our claim follows.

Now, consider the case $p_n \leq T^*/3$. Then,

$$T = s_n + p_n = \frac{1}{m}\left(\sum_{t=1}^{s_n} |\mathcal{A}_t| + \sum_{t=1}^{s_n}(m - |\mathcal{A}_t|)\right) + p_n$$

$$\leq \frac{1}{m}\left(\sum_{i=1}^{n} p_i + \sum_{q=1}^{\ell} \Lambda_q\right) + p_n \leq \frac{1}{m}\sum_{i=1}^{n} p_i + \frac{n}{2} + p_n \leq \frac{5}{3}T^*.$$

The above inequalities hold by (i) the fact that job $n$ completes last, (ii) the definition of alive jobs, (iii) a simple packing argument with job processing times and machine idle time, (iv) Lemmas 1–3, and (v) the bound $T^* \geq \max\{\frac{1}{m}\sum_{j \in \mathcal{J}} p_j, n + p_n, 3p_n\}$, respectively.  □

We complement Theorem 4 with a long instance $I = \langle m, \mathcal{J}\rangle$ example for which LPT is 3/2-approximate and leave closing the gap between the two as an open question. Instance $I$ contains $m + 1$ jobs, where $p_j = 2m - j$, for $j \in \{1, \ldots, m\}$, and $p_{m+1} = m$. In the LPT schedule $\mathcal{S}$, job $j$ is executed at time $s_j = j$, for $j \in \{1, \ldots, m\}$, and $s_{m+1} = 2m - 1$. Hence, $T = 3m - 1$. An optimal schedule $\mathcal{S}^*$ assigns job $j$ to machine $j + 1$ at time $s_j = j + 1$, for $j \in \{1, \ldots, m - 1\}$. Moreover, jobs $m$ and $m + 1$ are assigned to machine 1 beginning at times $s_m = 1$ and $s_{m+1} = m$, respectively. Clearly, $T^* = 2m$.

Theorem 5 uses a simple argument to show that LPT is optimal for short instances.

**Theorem 5.** *LPT is optimal for short instances.*

Finally, we investigate the performance of *Long Job Shortest Processing Time First (LSPT)*. LSPT creates an ordering of the jobs in which (i) each long job precedes every short job, (ii) long jobs are sorted according to *Shortest Processing Time First (SPT)*, and (iii) short jobs are sorted similarly to LPT. LSPT schedules the jobs greedily, in the same vein with LPT, by using this new order of jobs. For long instances, when the largest processing time $p_{\max}$ is relatively small compared to the average machine load, Theorem 6 shows that LSPT achieves an approximation ratio better than the 5/3. From a worst-case analysis viewpoint, the main difference between LSPT and LPT is that the former requires significantly lower idle machine time until the last job start, but at the price of much higher difference between the last job completion times in different machines.

**Theorem 6.** *LSPT is 2-approximate for minimizing makespan. For long instances, SPT is $(1+\min\{1, 1/\alpha\})$-approximate, where $\alpha = (\frac{1}{m}\sum_{j \in \mathcal{J}} p_j)/p_{\max}$.*

# 5  Parallelizing Long and Short Jobs

This section proposes *Long Short Job Mixing (LSM)* algorithm which is 1.985-approximate for a broad family of instances, e.g. with at most $\lceil 5m/6 \rceil$ jobs of processing time (i) $p_j > (1-\epsilon)(\frac{1}{m}\sum_{j'} p_{j'})$, or (ii) $p_j > (1-\epsilon)\max_{j'}\{j' + p_{j'}\}$) assuming non-increasing $p_j$'s, for small constant $\epsilon > 0$. For degenerate instances with more than $\lceil 5m/6 \rceil$ jobs having $p_j > T^*/2$, where $T^*$ is the optimal objective value, LSM requires bounded machine augmentation to achieve an approximation ratio lower than 2. Note that there can be at most $m$ such jobs. For simplicity, we also assume that $m = \omega(1)$, but the approximation can be adapted for smaller values of $m$. However, we require that $m \geq 7$.

*Algorithm Description.* LSM attempts to construct a feasible schedule in which long jobs are executed in parallel with short jobs. To this end, LSM uses $m^L < m$ machines for executing long jobs. The remaining $m^S = m - m^L$ machines are reserved for performing only short jobs. Carefully selecting $m^L$ allows to obtain a good trade-off in terms of (i) delaying long jobs, and (ii) achieving many short job starts at time slots where many long jobs are executed in parallel. Here, we set $m^L = \lceil 5m/6 \rceil$. Before formally presenting LSM, we slightly modify the notions of long and short jobs. In particular, we set $\mathcal{J}^L = \{j : p_j \geq m^L, j \in \mathcal{J}\}$ and $\mathcal{J}^S = \{j : p_j < m^L, j \in \mathcal{J}\}$, respectively. Both $\mathcal{J}^L$ and $\mathcal{J}^S$ are sorted in non-increasing order of processing times. LSM schedules jobs greedily by traversing time slots in increasing order. Let $\mathcal{A}_t^L$ be the set of alive long jobs at time slot $t \in D$. For $t = 1, \ldots, \tau$, LSM proceeds as follows: (i) if $|\mathcal{A}_t^L| < m^L$ and $|\mathcal{J}^L| > 0$, then the next long job begins at $t$, (ii) else if $|\mathcal{J}^S| > 0$ and $m^L \leq |\mathcal{A}_t| < m$, LSM schedules the next short job to start at $t$, (iii) otherwise, LSM considers the next time slot. From a complementary viewpoint, the set $\mathcal{M}$ of machines is partitioned into $\mathcal{M}^L = \{i : i \leq m^L, i \in \mathcal{M}\}$ and $\mathcal{M}^S = \{i > m^L, i \in \mathcal{M}\}$. LSM prioritizes long jobs on machines $\mathcal{M}^L$ and assigns only short jobs to machines $\mathcal{M}^S$. A job may undergo processing on machine $i \in \mathcal{M}^S$ only if all machines in $\mathcal{M}^L$ are busy. Theorem 7 analyzes LSM.

**Theorem 7.** *LSM is 1.985-approximate (i) for instances with no more than $\lceil 5m/6 \rceil$ jobs s.t. $p_j > (1-\epsilon)\max\{\frac{1}{m}\sum_{j'} p_{j'}, \max_{j'}\{j' + p_{j'}\}\}$ for sufficiently large $\epsilon > 0$, and (ii) for general instances with 1.2-machine augmentation.*

*Proof.* Let $\mathcal{S}$ be the LSM schedule and $\ell = \arg\max\{C_j : j \in \mathcal{J}\}$ the job completing last. That is, $\mathcal{S}$ has makespan $T = C_\ell$. For notational convenience, given a subset $P \subseteq D$ of time slots, we denote by $\lambda(P) = |P|$ and $\mu(P) = \sum_{t \in P} |\mathcal{A}_t|$ the number of time slots and executed processing load, respectively, during $P$. Furthermore, let $n^L = |\mathcal{J}^L|$ and $n^S = |\mathcal{J}^S|$ be the number of long and short jobs, respectively. We distinguish two cases: (i) $\ell \in \mathcal{J}^S$, or (ii) $\ell \in \mathcal{J}^L$.

*Case $\ell \in \mathcal{J}^S$.* We partition time slots $\{1, \ldots, T\}$ into five subsets. Let $r^L = \max_{j \in \mathcal{J}^L}\{s_j\}$ and $r^S = \max_{j \in \mathcal{J}^S}\{s_j\}$ be the maximum long and short job start time, respectively, in $\mathcal{S}$. Since $\ell \in \mathcal{J}^S$, it holds that $r^L < r^S$. For each time slot $t \in [1, r^L]$ in $\mathcal{S}$, either $|\mathcal{A}_t^L| = m^L$ long jobs simultaneously run at $t$, or not. In the latter case, it must be the case that $t = s_j$ for some $j \in \mathcal{J}^L$. On the other hand, for each time slot $t \in [r^L + 1, s_\ell]$, either $|\mathcal{A}_t| = m$, or $t = s_j$ for some $j \in \mathcal{J}^S$. Finally, $[s_\ell, T(\mathcal{S})]$ is exactly the interval during which job $\ell$ is executed. If $F^L = \{t : |\mathcal{A}_t^L| = m^L\}$, $B^L = \{t : |\mathcal{A}_t^L| < m^L, t = s_j, j \in \mathcal{J}^L\}$, $F^S = \{t : t > r^L, |\mathcal{A}_t| = m\}$, $B^S = \{t : t > r^L, |\mathcal{A}_t| < m, t = s_j, j \in \mathcal{J}^S\}$, then

$$T \leq \lambda(F^L) + \lambda(B^L) + \lambda(F^S) + \lambda(B^S) + p_\ell. \qquad (2)$$

Next, we upper bound a linear combination of $\lambda(F^L)$, $\lambda(B^S)$, and $\lambda(F^S)$ taking into account the fact that certain short jobs begin during a subset $\hat{B}^S \subseteq F^L \cup F^S$ of time slots. By definition, $\lambda(B^S) \leq n^S - \lambda(\hat{B}^S)$. We claim that $\lambda(\hat{B}^S) \geq (m^S/m^L)(\lambda(F^L) + \lambda(F^S))$. For this, consider the time slots $F^L \cup F^S$ as a continuous time period by disregarding intermediate $B^L$ and $B^S$ time slots. Partition this $F^L \cup F^S$ time period into subperiods of equal length $m^L$. Note that no long job begins during $F^L \cup F^S$ and the machines in $\mathcal{M}^S$ may only execute small jobs in $\mathcal{S}$. Because of the greedy nature of LSM and the fact that $p_j < m^L$ for $j \in \mathcal{J}^S$, there are at least $m^S$ short job starts in each subperiod. Hence, our claim is true and we obtain that $\lambda(B^S) \leq n^S - (m^S/m^L)(\lambda(F^L) + \lambda(F^S))$, or

$$m^S \lambda(F^L) + m^S \lambda(F^S) + m^L \lambda(B^S) \leq m^L n^S. \qquad (3)$$

Subsequently, we upper bound a linear combination of $\lambda(F^L)$, $\lambda(B^L)$, and $\lambda(F^S)$ using a simple packing argument. The part of the LSM schedule for long jobs is exactly the LPT schedule for a long instance with $n^L$ jobs and $m^L$ machines. If $|\mathcal{A}_{r^L}^L| < m$, we make the convention that $\mu(B^L)$ does not contain any load of jobs beginning in the maximal slack period completed at time $r^L$. Observe that $\mu(F^L) = m^L \lambda(F^L)$ and $\mu(F^S) = m\lambda(F^S)$. Additionally, by Lemma 2, we get that $\mu(B^L) \geq m^L \lambda(B^L)/2$, except possibly the very last slack period. Then, by Lemma 3, $\mu(F^L) + \mu(B^L) + \mu(F^S) \leq \sum_{j \in J} p_j$. Hence, we obtain that

$$m^L \lambda(F^L) + \frac{1}{2} m^L \lambda(B^L) + m\lambda(F^S) \leq \sum_{j \in \mathcal{J}} p_j. \qquad (4)$$

By summing expressions (3) and (4),

$$(m^L + m^S)\lambda(F^L) + \frac{1}{2} m^L \lambda(B^L) + (m + m^S)\lambda(F^S) + m^L \lambda(B^S) \leq \sum_{j \in \mathcal{J}} p_j + m^L n^S.$$

Because $m = m^L + m^S$, if we divide by $m$, the last expression gives

$$\lambda(F^L) + \frac{1}{2}\left(\frac{m^L}{m}\right)\lambda(B^L) + \lambda(F^S) + \left(\frac{m^L}{m}\right)\lambda(B^S) \leq \frac{1}{m}\sum_{j \in \mathcal{J}} p_j + \left(\frac{m^L}{m}\right)n^S \qquad (5)$$

Now, we distinguish two subcases based on whether $\lambda(F^S) + \lambda(F^L) \geq 5n^S/6$ or not. Obviously, $\lambda(B^L) \leq n^L$. In the former subcase, inequality (3) gives that $\lambda(B^S) \leq (1 - \frac{5m^S}{6m^L})n^S$. Hence, using inequality (5), expression (2) becomes

$$T \leq \frac{1}{m} \sum_{j \in \mathcal{J}} p_j + (1 - \frac{m^L}{2m})n^L + \left[(\frac{m^L}{m}) + (1 - \frac{m^L}{m})(1 - \frac{5m^S}{6m^L})\right] n^S + p_\ell$$

For $m^L = \lceil 5m/6 \rceil$, we have that (i) $5/6 \leq m^L/m \leq 5/6 + 1/m$, and (ii) $m^S/m^L \geq \frac{1/6 - 1/m}{5/6 + 1/m}$. Given that $m = \omega(1)$,

$$T \leq \frac{1}{m} \sum_{j \in \mathcal{J}} p_j + (1 - \frac{5}{12})n^L + \left[\frac{5}{6} + (1 - \frac{5}{6})(1 - \frac{1}{5})\right] n^S + p_\ell$$

Note that an optimal solution $\mathcal{S}^*$ has makespan

$$T^* \geq \max \left\{ \frac{1}{m} \sum_{j \in \mathcal{J}} p_j, n^L + n^S + p_\ell \right\}$$

Because the instance has both long and short jobs and $\ell \in \mathcal{J}^S$, it holds that $p_\ell \geq T^*/2$. Therefore, $T \leq (1 + \frac{29}{30} + (\frac{29}{30})\frac{1}{2}) \leq 1.985T^*$. Now, consider the opposite subcase where $\lambda(F^L) + \lambda(F^S) \leq 5n^S/6$. Given that $\lambda(B^L) \leq n^L$ and $\lambda(B^S) \leq n^S$, expression (2) becomes $T \leq \frac{11}{6}(n^S + n^L + p_\ell) \leq 1.835 \cdot T^*$.

*Case $\ell \in \mathcal{J}^L$.* Recall that $\mathcal{A}_t^L$ and $\mathcal{B}_t^L$ are the sets of long jobs which are alive and begin, respectively, at time slot $t$. Furthermore, $r^L = \max\{s_j : j \in \mathcal{J}^L\}$ is the last long job starting time. Because LSM greedily uses $m^L$ machines for long jobs, either $|\mathcal{A}_t^L| = m^L$, or $|\mathcal{B}_t^L| = 1$, for each $t \in [1, r^L]$. So, we may partition time slots $\{1, \ldots, r^L\}$ into $F^L = \{t : |\mathcal{A}_t^L| = m\}$ and $B^L = \{t : |\mathcal{A}_t^L| < m, |\mathcal{B}_t^L| = 1\}$ and obtain $T \leq \lambda(F^L) + \lambda(B^L) + p_\ell$. Because $m^L$ long jobs are executed at each time slot $t \in F^L$,

$$\lambda(F^L) \leq \frac{1}{m^L} \left[ \sum_{j \in \mathcal{J}^L} p_j - \mu(B^L) \right].$$

Then, Lemma 2 implies that $\mu(B^L) \geq n^L m^L/2$. Furthermore, $\lambda(B^L) \leq n^L$. Therefore, by considering Lemma 3, we obtain

$$T \leq \frac{m}{m^L} \left( \frac{1}{m} \sum_{j \in \mathcal{J}} p_j \right) + \frac{1}{2}(n^L + p_\ell) + \frac{1}{2}p_\ell$$

In the case $p_\ell \leq T^*/2$, since $T^* \geq n^L + p_\ell$, we obtain an approximation ratio of $(\frac{m}{m^L} + \frac{3}{4}) \leq 1.95$, when $m^L = \lceil 5m/6 \rceil$, given that $m = \omega(1)$.

Next, consider the case $p_\ell > T^*/2$. Let $\mathcal{J}^V = \{j : p_j > T^*/2\}$ be the set of very long jobs and $n^V = |\mathcal{J}^V|$. Clearly, $n^V \leq m$. By using resource augmentation, i.e. allowing LSM to use $m' = \lceil 6m/5 \rceil$ machines, we guarantee that LSM assigns at most one job $j \in \mathcal{J}^V$ in each machine. The theorem follows.

*Remark.* If $\lceil 5m/6 \rceil < |\mathcal{J}^V| \leq m$, LSM is not better than 2-approximate. This can be illustrated with a simple instance consisting of only $\mathcal{J}^V$ jobs. Assigning two such jobs on the same machine is pathological. Thus, better than 2-approximate schedules require assigning all jobs in $\mathcal{J}^V$ to different machines.

## 6  Conclusion

We conclude with a collection of future directions. Because BJSP relaxes scheduling problems with non-overlapping constraints, which do not admit better than 2-approximation algorithms, the existence of such an algorithm without resource augmentation for BJSP is an intriguing open question. A positive answer combining LSM with a new algorithm for instances with many very long jobs is possible. Analyzing the price of robustness of the proposed robust scheduling approach may reveal new insights for effectively solving BJSP under uncertainty. Also, machine augmentation enables more efficient BJSP solving. Integrating multiple delivery offices in a unified setting and performing vehicle sharing on a daily basis enables a practical implementation of machine augmentation when mail delivery demand fluctuates.

## References

1. Bertsimas, D., Sim, M.: The price of robustness. Oper. Res. **52**(1), 35–53 (2004)
2. Billaut, J.-C., Sourd, F.: Single machine scheduling with forbidden start times. 4OR **7**(1), 37–50 (2009)
3. Gabay, M., Rapine, C., Brauner, N.: High-multiplicity scheduling on one machine with forbidden start and completion times. J. Sched. **19**(5), 609–616 (2016)
4. Goerigk, M., Schöbel, A.: Algorithm engineering in robust optimization. In: Kliemann, L., Sanders, P. (eds.) Algorithm Engineering. LNCS, vol. 9220, pp. 245–279. Springer, Cham (2016). https://doi.org/10.1007/978-3-319-49487-6_8
5. Graham, R.L.: Bounds on multiprocessing timing anomalies. SIAM J. Appl. Math. **17**(2), 416–429 (1969)
6. Letsios, D., Ruth, M.: Exact lexicographic scheduling and approximate rescheduling. arXiv 1805.03437 (2018)
7. Liebchen, C., Lübbecke, M., Möhring, R., Stiller, S.: The concept of recoverable robustness, linear programming recovery, and railway applications. In: Ahuja, R.K., Möhring, R.H., Zaroliagis, C.D. (eds.) Robust and Online Large-Scale Optimization. LNCS, vol. 5868, pp. 1–27. Springer, Heidelberg (2009). https://doi.org/10.1007/978-3-642-05465-5_1
8. Mnich, M., Bevern, R.V.: Parameterized complexity of machine scheduling: 15 open problems. Comput. Oper. Res. **100**, 254–261 (2018)
9. Rapine, C., Brauner, N.: A polynomial time algorithm for makespan minimization on one machine with forbidden start and completion times. Discrete Optim. **10**(4), 241–250 (2013)

10. Schäffter, M.W.: Scheduling with forbidden sets. Discrete Appl. Math. **72**(1–2), 155–166 (1997)
11. Skutella, M., Verschae, J.: Robust polynomial-time approximation schemes for parallel machine scheduling with job arrivals and departures. Math. Oper. Res. **41**(3), 991–1021 (2016)
12. Svensson, O.: Hardness of precedence constrained scheduling on identical machines. SIAM J. Comput. **40**(5), 1258–1274 (2011)

# Contact Representations of Directed Planar Graphs in 2D and 3D

Chun-Hsiang Chan and Hsu-Chun Yen[✉]

Department of Electrical Engineering,
National Taiwan University, Taipei, Taiwan
kennyhchan@gmail.com, hcyen@ntu.edu.tw

**Abstract.** In applications such as VLSI floorplanning and cartogram design, vertices of a graph are represented by geometric objects and edges are captured by contacts between those objects, which are examples of a drawing style called *contact graph representations*. We study the feasibility of using line segments, triangles and tetrahedra to realize point-side contact representations for a number of graph classes including oriented versions of outerplanar graphs, 2-trees and 3-trees. Our main results show that every orientation of a maximal outerplanar graph of out-degree at most two, a 2-tree of out-degree at most two, and a planar 3-tree of out-degree at most four enjoy point-side contact representations using line segments, triangles, and tetrahedra, respectively. Unlike undirected graphs for which a fairly large amount of results can be found in the literature in the study of contact representations, directed graphs remain largely unexplored, and our study advances this line of research a step further.

## 1 Introduction

In *contact representations* of graphs, vertices are drawn as interior-disjoint geometric objects with edges corresponding to contacts between those objects. The Koebe's circle packing theorem [9], showing that every planar graph can be drawn as touching circles, is a classical example of a contact graph representation. With potential applications in various practical areas, the study of contact graph representations from an algorithmic viewpoint has received increasing attention in computational geometry in recent years. From the theoretical aspect, contact graph representations also pose many interesting and challenging mathematical questions, partly due to a wide diversity of the underlying contact styles (point vs. side contact, for instance) and object shapes (circle, triangle, ..., etc), as well as whether the drawings are in 2-dimensional or 3-dimensional space.

As triangles represent the simplest form of convex polygons, much work along the line of research in contact representations of graphs has focused on representing vertices by triangles, and edges by point- or side-contacts of triangles.

H.-C. Yen—Research supported in part by Ministry of Science and Technology, Taiwan, under grant MOST 106-2221-E-002-036-MY3.

© Springer Nature Switzerland AG 2019
Y. Li et al. (Eds.): COCOA 2019, LNCS 11949, pp. 82–93, 2019.
https://doi.org/10.1007/978-3-030-36412-0_7

The reader is referred to [1,6–8], for instance. Somewhat surprisingly, with the exception of a recent work [4], all the previous work on contact representations of graphs dealt with undirected graphs only. In [4], a new notion of *point-side contacts* was proposed, in which each vertex is drawn as a triangle and if a point of a triangle associated with a vertex $v$ touches a side of a triangle associated with vertex $u$, then $(v, u)$ is a directed edge in the graph. The main result of [4] showed that the class of outerplanar graphs of out-degree at most 3 always admit point-side triangle contact representations.

In this paper, we continue and extend the study of contact representations of directed graphs in two aspects. First, we consider objects degenerated and extended from triangles, namely, *line segments*, and *tetrahedra*. Notice that tetrahedrons are geometric objects in 3D. For more about contact representations of undirected graphs in 3D, the reader is referred to, e.g., [2,3]. Let the drawing styles psSLCR, psTriCR, and psTetraCR denote *straight-line* (more precisely, *line segment*), *triangle*, and *tetrahedron* contact representations, respectively. Second, unlike the work of [4] which dealt only with outerplanar graphs, we also look at 2-trees and 3-trees. Throughout the rest of our discussion, the central problem is to ask whether every oriented graph (i.e., directed graph without multiple edges) in a given graph class admits a contact graph representation respecting a given drawing style? Our main results are summarized in Table 1.

**Table 1.** Feasibility results, where $\mathcal{O}^d_{\mathcal{G}_{mo}}$ (resp., $\mathcal{O}^d_{\mathcal{G}_{2t}}$ and $\mathcal{O}^d_{\mathcal{G}_{p3t}}$) denotes the class of maximal outerplanar graphs (resp., 2-trees and planar 3-trees) of out-degree at most $d$. Those un-annotated $\bigcirc$ (resp., $\times$) follow immediately from results concerning larger (resp., smaller) graph classes. The "?" stands for the case for which the answer remains open.

|  | $\mathcal{O}^2_{\mathcal{G}_{mo}}$ | $\mathcal{O}^2_{\mathcal{G}_{2t}}$ | $\mathcal{O}^3_{\mathcal{G}_{2t}}$ | $\mathcal{O}^3_{\mathcal{G}_{p3t}}$ | $\mathcal{O}^4_{\mathcal{G}_{p3t}}$ |
|---|---|---|---|---|---|
| psSLCRs | $\bigcirc$ [Thm 3] | $\times$ [Thm 5] | $\times$ | $\times$ | $\times$ |
| psTriCRs | $\bigcirc$ | $\bigcirc$ [Thm 2] | ? | $\times$ [Thm 5] | $\times$ |
| psTetraCRs | $\bigcirc$ | $\bigcirc$ | $\bigcirc$ | $\bigcirc$ | $\bigcirc$ [Thm 4] |

## 2    Preliminaries

A graph is *planar* iff it can be drawn in the Euclidean plane without crossings. A *plane graph* is a planar graph with a fixed combinatorial embedding and a designated outer face. An *outerplanar graph* is a graph for which there exists a planar embedding with all vertices of the graph belonging to the outer face. We say that a planar graph (resp. outerplanar graph) is maximal if no more edges can be added while preserving planarity (resp., outerplanarity). A *k-tree* is an undirected graph formed by starting with a $(k + 1)$-vertex complete graph and then recursively adding vertices such that each added vertex $v$ forms a

$(k + 1)$-clique with its $k$ neighbors. Maximal outerplanar graphs are a sub-class of 2-trees, and planar 3-trees are a sub-class of maximal planar graphs. The following graph classes will be dealt with throughout this paper:

- $\mathcal{G}_o$ (resp., $\mathcal{G}_{mo}$): the class of outerplanar (resp., maximal outerplanar) graphs,
- $\mathcal{G}_{2t}$: the class of 2-trees,
- $\mathcal{G}_{p3t}$: the class of planar 3-trees.

An *orientation* of an undirected graph is an assignment of a direction to each edge, turning such an undirected graph into a directed graph. We use $\mathcal{O}_{\mathcal{G}}^d$ to denote the set of orientations with out-degree at most $d$ of an undirected graph class $\mathcal{G}$. For example, $\mathcal{O}_{\mathcal{G}_{2t}}^3$ is a set including the orientations with out-degree at most 3 of 2-trees. Basically, oriented graphs are directed graphs without multiple edges between vertices.

In the Euclidean plane or space, let $X$ denote a set of geometric objects (i.e., triangles, straight-lines, etc.) with the so-called *points* and *sides* associated with each object. A *point-side $X$ contact representation* of a planar digraph $G = (V, E)$ is a drawing meeting the following conditions:

1. each vertex $v \in V$ is drawn as a $X$-shaped geometric object, and
2. each directed edge $e = (v, u) \in E$ corresponds to a contact between a side of the object associated with $u$ and a point of the object associated with $v$.

In this paper, we focus on objects including line segments, triangles (in 2D), and tetrahedra (in 3D), for which points and sides associated with such objects are easy to identify For tetrahedrons, points refer to the four vertex corners and sides are associated with the four triangular faces. A contact refers to a vertex corner touching a triangular face. We let *psTriCR* (resp., *psSLCR* and *psTetraCR*) denote the style of point-side triangle (resp., line segment and tetrahedron) contact representations.

A *contact system of pseudo-segments* (or a *contact system*, for short) is a set of non-crossing Jordan arcs where any two of them intersect in at most one point, and each intersecting point is internal to at most one arc. If a contact system is *stretchable*, then there exists a homeomorphism transforming the contact system into a drawing where each arc is a straight line. Stretchable contact systems were characterized in [5] based on the notion of *extremal points*. A point $p$ is an extremal point of a contact system $S$ if the following three conditions are satisfied:

1. $p$ is an endpoint of a pseudo-segment in $S$,
2. $p$ is not interior to any pseudo-segment in $S$, and
3. $p$ is incident to the unbounded region of $S$.

The following result characterizes a necessary and sufficient condition for a contact system to be stretchable, based on extremal points. Throughout the rest of this paper, we let $ext(S)$ denote the number of extremal points of a contact system $S$.

**Theorem 1.** *([5]). A contact system $S$ of pseudo-segments is stretchable iff each of its subsystems $S'$ of cardinality greater than 1 has at least 3 extremal points (i.e., $ext(S') \geq 3$).*

# 3 PsTriCRs for Directed 2-Trees

In this section, we show one of our main results that every oriented graph in $\mathcal{O}^2_{\mathcal{G}_{2t}}$ admits a *psTriCR*. As a comparison, it was shown in [4] that graphs in $\mathcal{O}^3_{\mathcal{G}_o}$ (outerplanar digraphs of out-degree at most 3) admit *psTriCRs*. Here we demonstrate a nontrivial class of non-outerplanar graphs that also have *psTriCRs*. Interestingly, there are graphs in $\mathcal{O}^2_{\mathcal{G}_{2t}}$ that do not have *psSLCRs*, as will be shown later.

To proceed further, we require some definitions. As a 2-tree $G = (V, E)$ can be constructed iteratively by adding a vertex in each step to form a 3-clique, such a procedure yields a *construction ordering* $\pi = v_1, v_2, ..., v_n$ of $V$ such that $(v_1, v_2) \in E$ and for all $3 \leq i \leq n$, with respect to the subgraph induced by $v_1, v_2, ..., v_{i-1}, v_i$ is connected to exactly two vertices $v_j, v_k$, where $1 \leq j < k \leq i - 1$ and $(v_j, v_k) \in E$. Let $G_i, 1 \leq i \leq n$, be the subgraph induced by $v_1, v_2, ..., v_{i-1}, v_i$. We define the *parent clique* of a vertex $v_i$ to be the unique clique $(v_i, p, q)$ contained in the subgraph $G_i$, i.e., the 3-clique formed when $v_i$ is added. The edge $(p, q)$ is called the *parent edge* with respect to its *child vertex* $v_i$. Notice that an edge $(p, q)$ might have more than one child vertex, and we denote the set of child vertices of $(p, q)$ as $C_{(p,q)}$. Note that the parent-child relation is with respect to the underlying construction ordering $\pi$.

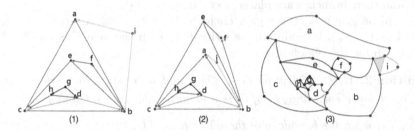

**Fig. 1.** (1): a layer embedding. $a, e, d$ are the child vertices of edge $(c, b)$. The branch $Br_{a(c,b)}$ is the subgraph induced by vertices $c, b, a, i$, excluding the edge $(c, b)$; (2): not a layer embedding, since $a$ is in the internal region of $e$, but $N_{(c,b)}(a) > N_{(c,b)}(e)$. (3): A layer pseudo-segment system corresponding to the layer embedding graph in (1).

With respect to an edge $(p, q)$, we define a subgraph called a *branch* as follows.

**Definition 1.** *A branch* $Br_{v(p,q)} = (V', E')$ *of an edge* $(p, q)$ *and a vertex* $v \in C_{(p,q)}$ *in digraph* $G = (V, E) \in \mathcal{O}^2_{\mathcal{G}_{2t}}$ *is a subgraph of* $G$ *constructed iteratively in the following way:*

*(1) Initially,* $V' = \{v, p, q\}$,
*(2) Repeat the following*
    *Add* $u$ $(\notin C_{(p,q)})$ *to* $V'$, *if* $\exists k, l \in V'$ *such that* $(u, k, l)$ *forms a 3-clique in* $G$
*(3)* $E' = \{(i,j) | i, j \in V', (i,j) \in E, (i,j) \neq (p,q)\}$.

For convenience, we write $V'$ (resp., $E'$) in the above definition as $V(Br_{v(p,q)})$ (resp., $E(Br_{v(p,q)})$).

Let $v \in C_{(p,q)}$, we define $N_{(p,q)}(v)$ to be the number of edges pointing from $p$ or $q$ to $V(Br_{v(p,q)}) \setminus \{p,q\}$. We order vertices in $C_{(p,q)}$ as $\Delta_{(p,q)} = v_{i_1}, ..., v_{i_k}$ such that $N_{(p,q)}(v_{i_j}) \leq N_{(p,q)}(v_{i_{j+1}}), 1 \leq j \leq k-1$, i.e., in ascending order in their $N_{(p,q)}$ values.

Consider Fig. 1(1). Suppose $c,b$ are the first two vertices in the ordering $\pi$. The set of child vertices $C_{(c,b)} = \{a,e,d\}$. The edge $(c,b)$ (resp., clique $(x,c,b)$) is the parent edge (resp., clique) of vertex $x$, where $x \in \{a,e,d\}$. In addition $Br_{a(c,b)}$ is the branch induced by the vertices $a,b,c,i$ (excluding the edge $(c,b)$). Note that $N_{(c,b)}(a)$ is equal to 2 (witnessed by edges $(c,a)$ and $(b,a)$). It is easy to see that $N_{(c,b)}(e)=1$ (witnessed by edge $(b,f)$) and $N_{(c,b)}(d)=0$. Hence, $\Delta_{(c,b)} = d, e, a$.

Given a graph $G = (V,E) \in \mathcal{O}^2_{\mathcal{G}_{2t}}$, there might be several ways to embed $G$ in the plane. In what follows, we define a so-called *layer embedding* of $G$ with respect to a construction ordering $\pi$, on which the feasibility of a *psTriCR* is based. Intuitively, given $\pi = v_1, v_2 ...$, a layer embedding is to start with edge $(v_1, v_2)$ as the parent edge and place it on the outer boundary of the layout. Then place each of the branches of $(v_1, v_2)$ layer by layer inside out respecting the order of $\Delta_{(v_1,v_2)}$. The layout of a branch $Br_{v(v_1,v_2)}$, where $v$ is a child vertex of $(v_1, v_2)$ is done recursively by treating $(v, v_1)$ and $(v, v_2)$ as the new parent edges, and their branches are placed away from $(v_1, v_2)$.

In a plane graph $\tilde{G}$, the region enclosed by a cycle of vertices $v_i, ..., v_k$ is denoted as $F_{v_i,...,v_k}$. In addition, we write $F \subset F'$ if the face $F$ is totally (i.e., properly) inside the region of $F'$.

**Definition 2.** *Given a digraph* $G \in \mathcal{O}^2_{\mathcal{G}_{2t}}$ *and a construction ordering* $\pi = v_1, ..., v_n$, *a layer embedding* $\tilde{G}$ *of* $G$ *is a planar embedding satisfying:*

*(1)* $(v_1, v_2)$ *is on the boundary of the outer face of* $\tilde{G}$,
*(2) For every edge* $(p,q) \in \tilde{G}$ *and* $v \in C_{(p,q)}$, *if* $w \in C_{(v,p)} \cup C_{(v,q)}$, $w$ *cannot lie in the face enclosed by the 3-clique* $(v,p,q)$,
*(3) For every edge* $(p,q) \in \tilde{G}$, *if the ordering* $\Delta_{(p,q)} = v_{i_1}, ..., v_{i_k}$, *the embedding must satisfy* $F_{(v_{i_j},p,q)} \subset F_{(v_{i_{j+1}},p,q)}, 1 \leq j \leq k-1$.

Consider Fig. 1, in which (1) is a layer embedding; (2), however, violates Condition (3) of the above definition, for $\Delta_{(c,b)} = d, e, a$.

Note that a layer embedding naturally induces an ordering (which may not be unique) that respects the structure of the layers, and is also a construction ordering. For instance, in Fig. 1, if we start with edge $(c,b)$ (the first two vertices of $\pi$), the inner-most 3-clique is $(d,c,b)$. Then $(h,c,d)$ forms a 3-clique in the next layer. Repeating the above, we yield a sequence $\pi' = c, b, d, h, g, e, f, a, i$ which is a construction ordering that respects the layer relation of the embedding. In our subsequent discussion, such a sequence $\pi'$ is called a *layer ordering*. It is not hard to see that from $\pi'$ the layer embedding is easily obtained.

The following algorithm LO can be used for obtaining a layer ordering $\pi'$ from a digraph $G \in \mathcal{O}_{\mathcal{G}_{2t}}^2$ and a construction ordering $\pi$.

```
Algorithm LO(p, q)
if Δ(p,q) is null, exit;
    else (suppose Δ(p,q) = vi₁, ..., viₘ)
    for j = 1...m
        output vij
        LO(vij, p)
        LO(vij, q)
```

It is easy to see that $v_1, v_2, LO(v_1, v_2)$, where $v_1$ and $v_2$ are the first two vertices of $\pi$, yields a layer ordering.

Our next step is to construct a specific system $S$ of pseudo-segments from a layer embedding of a graph $G \in \mathcal{O}_{\mathcal{G}_{2t}}^2$, and then show $S$ to be stretchable. Furthermore, stretching $S$ yields a $psTriCR$ of $G$.

We follow a definition in [4], where the notion of a so-called *unit of pseudo-segments* (or simply *unit*, for short) was defined. We use $s = \{a, a'\}$ to denote the pseudo-segment $s$ with endpoints $a$ and $a'$. A *unit* $U$ is a set of three pseudo-segments $s_1 = \{a_1, a_2\}$, $s_2 = \{a_2, a_3\}$, and $s_3 = \{a_3, a_1\}$, for some $a_1, a_2$, and $a_3$, and the closed region $\mathcal{F}_{a_1, a_2, a_3}$ does not contain any other segments. We write $P_U$ as the set of the three endpoints associated with a unit $U$. It is easy to see that stretching a unit yields a triangle. An endpoint $x$ in a unit $U$ is a *free point* if $x$ is not a touching point between $U$ and a segment of another unit. A Trinity Contact System of pseudo-segments (for short, TCS) $S$ is a set of pseudo-segments which can be partitioned into units such that for each pair of units $U_1, U_2$ in $S$,

1. $U_1 \cap U_2 = \emptyset$ and $P_{U_1} \cap P_{U_2} = \emptyset$.
2. $U_1$ and $U_2$ intersect in at most one point in $S$, and each intersecting point is internal to at most one segment.

The following is easy to show:

**Proposition 1.** *Every planar embedding $\tilde{G}$ of a digraph $G$ induces a TCS $S(\tilde{G})$, and vice versa.*

In the above proposition, the TCS $S(\tilde{G})$ is said to respect the embedding $\tilde{G}$, and is called a *layer pseudo-segment system* associated with $\tilde{G}$. See Fig. 1(3), which is a layer TCS corresponding to the layer embedding in Fig. 1(1).

Given three units $U_a, U_b, U_c$, we use $\mathcal{F}_{a,b,c}$ to denote the closed region $\mathcal{F}' \setminus (\mathcal{F}_a \cup \mathcal{F}_b \cup \mathcal{F}_c)$, where $\mathcal{F}'$ is the maximum region enclosed by the segments of $U_a, U_b, U_c$, and $\mathcal{F}_{\{a, b, c\}}$ is the region whose boundary vertices are the contact points of the three units. In Fig. 1(3), for instance, $\mathcal{F}_{a,b,i}$ is the region colored in yellow.

The key idea behind the result in this section is to find a layer embedding of a directed 2-tree, such that the corresponding pseudo-segment system is stretchable. To give the reader a better feeling for why requesting a layer embedding is crucial, consider Fig. 2(1) and (3), in which two embeddings of the same diagraph are shown. Figure 2(1) (a layer embedding) admits a $psTriCR$ as shown in Fig. 2(2). Figure 2(3) (not a layer embedding) does not admit a $psTriCR$ as the corresponding TCS is not stretchable. To see this, consider the subsystem consisting of the two units of $c$ and $b$, plus segment $r$ associated with $e$. Clearly the subsystem only has two extremal points. As a result, the system is not stretchable.

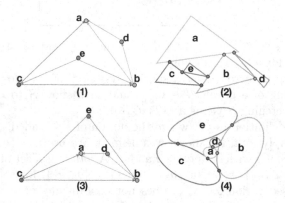

**Fig. 2.** Two embeddings of a 2-tree.

**Lemma 1.** *Given a directed graph $G \in \mathcal{O}^2_{\mathcal{G}_{2t}}$ and a construction ordering $\pi$, there exists a layer embedding $\tilde{G}$ whose corresponding layer contact system $S(\tilde{G})$ is stretchable.*

*Proof.* Let $\pi'$ be a layer ordering which can be constructed from $\pi$ based on our earlier discussion. The proof of our lemma is by induction on $\pi' = v_1, v_2, ..., v_n$. Let the corresponding sequence of layer embedding subgraphs (resp., layer systems of pseudo-segments) be $\tilde{G}_1 \subset \tilde{G}_2 \subset ... \subset \tilde{G}_n = \tilde{G}$ (resp., $S_1 \subset S_2 \subset ... \subset S_n = S(\tilde{G})$).

• The stretchability of the base case (i.e., $S_2$) is trivial.

• Assume that $S_{i-1}$, $\forall i, 2 < i \leq n$, is stretchable. Now consider $S_i$ corresponding to $\tilde{G}_i$. Let $U_{v_i}$ be the unit of vertex $v_i$ and $U_p$, $U_q$ be the units of vertices $p, q$ to which $v_i$ is connected, forming a 3-clique in $\tilde{G}_i$.

To use Lemma 1 to yield the stretchability of $S_i$, one has to show that for every subsystem $S \subseteq S_i$ of two or more pseudo-segments, $ext(S) \geq 3$. To simplify our argument, we may further restrict the contents of $S$ without loss of generality, as the following discussion shows.

Let $\mathcal{F}_U$ be the region enclosed by unit $U$ ($\mathcal{F}_U$ is only bounded by the three segments in $U$), and the bounded region $\mathcal{F}_{v_i,p,q} = \mathcal{M}_{v_i,p,q} \setminus (\mathcal{F}_{v_i} \cup \mathcal{F}_p \cup \mathcal{F}_q)$, where $\mathcal{M}_{v_i,p,q}$ is the maximal bounded region enclosed by segments in $U_{v_i}$, $U_p$ and $U_q$. For instance, in Fig. 3(1) $\mathcal{M}_{v_i,p,q}$ and $\mathcal{F}_{v_i,p,q}$ are the regions enclosed by boundary points $a, f, h, g, i, d, e, c$, and $f, g, c, b$, respectively. Let $S_{\mathcal{F}_{v_i,p,q}}$ be the set of segments enclosed by the bounded region $\mathcal{F}_{v_i,p,q}$.

First notice that the stretchability of $S_i' = S_i \setminus S_{\mathcal{F}_{v_i,p,q}}$ is sufficient to guarantee the stretchability of $S_i$, as no extremal points can be contributed from segments in $F_{v_i,p,q}$. That is, when we consider some subsystem of $S_i$ that has the closed region $\mathcal{F}_{v_i,p,q}$, it is sufficient to consider the subsystem belonging to $S_i \setminus S_{\mathcal{F}_{v_i,p,q}}$. Also notice that among the three segments in $U_{v_i}$, only one segment (say $r$) touches $S_i' \cap S_{i-1}$ at two points (e.g., $f$ and $g$ in Fig. 3(1)). Let the other two segments of $U_{v_i}$ be $r'$ and $r''$. Again, without loss of generality, we only need to consider subsystem $S \cap \{r', r''\} = \emptyset$, and $S$ contains the closed region $\mathcal{F}_{v_i,p,q}$, where $S \subseteq S_i'$.

Now consider two cases:

1. (The vertex $v_i$ points to both $p$ and $q$): Without loss of generality, we further assume that $(p, q) \in E$, i.e., the edge direction is from $p$ to $q$; the other case is symmetric. It is not hard to see the following:
   (1) Each of $P_{U_p}$ and $P_{U_q}$ contains at least one (free) endpoint along the outer boundary of $S_i$, as $G$ is of out-degree at most 2,
   (2) If some endpoint in $P_{U_p} \cup P_{U_q}$ is along the boundary of $\mathcal{F}_{v_i,p,q}$ (corresponding to an outgoing edge from $p$ or $q$), then one of $P_{U_p}$ and $P_{U_q}$ must contain an additional endpoint other than the two stated in (1) above.
   Otherwise, the layout contradicts the definition of layer pseudo-segments.
   To understand the above better, we consider Fig. 3(1). $a$ and $d$ are the two endpoints stated in Condition (1) above. As the end point $b \in P_{U_p}$ is along the boundary of $\mathcal{F}_{v_i,p,q}$, $i$ is an endpoint of $P_{U_q}$ guaranteed by Condition (2) above. Clearly, any subsystem $S$ enclosing $F_{v_i,p,q}$ must contain segments $\{a, b\}, \{b, c\}$ of $U_p$, and $\{i, e\}$ of $U_q$, with $a$ and $i$ as its two extremal points. If $S$ also contains segment $\{i, d\}$ or $\{d, e\}$ of $U_q$, $d$ will be another extremal point, implying that $ext(S) \geq 3$. Now consider the case when neither segment $\{i, d\}$ nor $\{d, e\}$ is in $S$ (see Fig. 3(2)). Let $S'$ be the subsystem depicted in Fig. 3(3). Clearly $S' \subseteq S_{i-1}$, which is stretchable. Hence, in addition to $a$ and $i$, the subsystem $S'$ contains at least one more extremal point which is also an extremal point in $S$. Hence, $ext(S) \geq 3$.
2. (Either $p$ or $q$ points to $v_i$): In this case, we know that both $P_{U_p}$ and $P_{U_q}$ must contain at least one external point along the outer boundary, and at least one endpoint of $r$ will be an extremal point of $S$. Hence, $ext(S) \geq 3$. See Fig. 3(4)–(5). □

According to the above induction and the fact that stretching a unit yields a triangle, we have the following:

**Theorem 2.** *Every digraph $G \in \mathcal{O}_{\mathcal{G}_{2t}}^2$ has a psTriCR.*

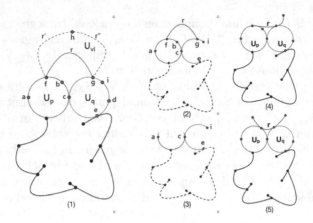

**Fig. 3.** (1), (2) and (3): A contact system in which $v_i$ points to both $p$ and $q$. (4) and (5): A contact system in which $p$ or $q$ points to $v_i$.

## 4    PsSLCRs for Maximal Outerplanar Digraphs

In this section, we show that graphs in $\mathcal{O}^2_{\mathcal{G}_{mo}}$ admit *psSLCRs*. To derive this result, we employ a technique used in [4] which shows a linear-time algorithm for constructing *psTriCRs* for digraphs in $\mathcal{O}^3_{\mathcal{G}_{mo}}$.

As each of the simple cycles in a maximal outerplanar graph $G = (V, E)$ is a triangle, our procedure takes advantage of the existence of an ordering $\pi = v_1, v_2, ..., v_n$ of $V$ such that $(v_1, v_2) \in E$ and for all $3 \le i \le n$, with respect to the subgraph induced by $v_1, v_2, ..., v_{i-1}$, $v_i$ is connected to exactly two vertices $v_j, v_k, 1 \le j < k \le i - 1$ and $(v_j, v_k) \in E$.

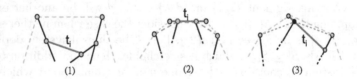

**Fig. 4.** The straight-line $t_i$ corresponds to a $v_i$ with two outgoing edges.

**Theorem 3.** *Given a plane digraph* $G = (V, E)$ *in* $\mathcal{O}^2_{\mathcal{G}_{mo}}$, *there is an algorithm which can construct a psSLCR of* $G$ *in linear time.*

*Proof.* (Sketch) Given any planar embedding of $G = (V, E)$ in $\mathcal{O}^2_{\mathcal{G}_{mo}}$, let $\pi = v_1, v_2, ..., v_n$ be an ordering of $V$ mentioned in the beginning of this section. Using a similar strategy as reported in [4], we construct a *psSLCR* $T$ of $G$ in a greedy fashion in the order specified by $\pi$. Let $t_i$ denote the straight-line corresponding to $v_i$ in $T$, $G_i$ denote the subgraph induced by $v_1, v_2, ..., v_i$, and $T_i$ denote the

$psSLCR$ with respect to subgraph $G_i$. As we shall see later, in each step $i$ of the algorithm, the set of free points in $psSLCR$ $T_i$ always forms a convex polygon.

(**Initial step**) We start with a $psSLCR$ corresponding to a subgraph induced by $v_1$ and $v_2$. If $(v_1, v_2) \in E$ (resp. $(v_2, v_1) \in E$), then an endpoint of $t_1$(resp. $t_2$) touches the side of $t_2$(resp. $t_1$); The convexity of $T_2$ is trivial.

(**Iteration steps**) From $i = 3$ to $|V|$, we add a line segment $t_i$ to $T_{i-1}$ and keep the $psSLCR$ $T_i$ convex. Since $v_i$ is connected to exactly two neighboring vertices among $v_1, ..., v_{i-1}$, say $v_j$ and $v_{j+1}$ for $1 \leq j < i - 1$. There are three cases to be considered, each of which shows how the line segment associated with $t_i$ is placed. The figures are for illustrating purpose, as there are other possibilities which are similar.

(a) ($v_i$ has two out-going edges): See Fig. 4(1).

(b) ($v_i$ has two incoming edges): See Fig. 4(2).

(c) ($v_i$ has one incoming edge and one outgoing edge): See Fig. 4(3).

Finally, the run-time is linear in the number of vertexes.                □

## 5   PsTetraCRs for Directed Planar 3-Trees

We now consider $\mathcal{O}^4_{\mathcal{G}_{p3t}}$. As planar 3-trees can be constructed by a sequence of vertices $v_1, \ldots, v_n$ such that $v_1, v_2, v_3$ form a triangle, and $v_i$, $4 \leq i \leq n$, subdivides a triangular face in the subgraph induced by $v_1, \ldots, v_{i-1}$.

A vertex $p$ of a tetrahedron in a psTetraCR of a planar 3-tree is said to be *external* if $p$ is located on the outer boundary of the psTetraCR, and $p$ is not a contact point with other tetrahedron; otherwise, $p$ is called *internal*.

**Theorem 4.** *Given a plane graph* $G \in \mathcal{O}^4_{\mathcal{G}_{p3t}}$, *there is an algorithm which can construct a psTetraCR of* $G$ *in linear time.*

*Proof.* (Sketch) Given any planar embedding of $G \in \mathcal{O}^4_{\mathcal{G}_{p3t}}$, let $\pi = v_1, v_2, ..., v_n$ be an ordering of $V$, such that $(v_1, v_2, v_3)$ is a 3-clique and for all $4 \leq i \leq n$, with respect to the subgraph induced by $v_1, v_2, ..., v_{i-1}$, $v_i$ is connected with three vertices $v_j, v_k, v_l, 1 \leq j < k < l \leq i - 1$ and forms a 4-clique.

We construct a psTetraCR $T$ of $G$ in a greedy fashion in the order specified by $\pi$. Let $t_i$ denote the tetrahedron corresponding to $v_i$ in $T$, $G_i$ denote the subgraph induced by $v_1, v_2, ..., v_i$, and $T_i$ denote the psTetraCR with respect to subgraph $G_i$. As we shall see later, in each step $i$ of the algorithm, the free vertices (i.e., those not serving as touching points) of psTetraCR $T_i$ always form a convex polyhedron (called a convex body).

(**Initial step**) We start with a psTetraCR corresponding to a subgraph induced by $v_1$, $v_2$ and $v_3$. For all pairs of vertices $(v_p, v_q), p, q \in \{1, 2, 3\}$, if $(v_p, v_q) \in E$, then a corner of $t_p$ touches an face of $t_q$. The convexity of $T_3$ is trivial.

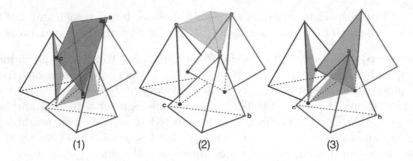

(1)                    (2)                    (3)

**Fig. 5.** (1): An example of adding $t_i$ to $T_{i-1}$ (2): The region for placing a top point; (3): The regions (faces) for placing internal points.

**(Iteration steps)** From $i = 4$ to $|V|$, we add a tetrahedron $t_i$ to $T_{i-1}$ and keep the psTetraCR $T_i$ a convex body. A *bottom face* is formed by the three points $p_{b_1}, p_{b_2}, p_{b_3}$ of a tetrahedron. A *top point* $p_t$ is a point that does not belong to the bottom face in the tetrahedron. Since $v_i$ is connected to exactly three neighboring vertices among $v_1, ..., v_{i-1}$, we first deal with the placement of the three points $p_{b_1}, p_{b_2}, p_{b_3}$ in the bottom face, and then show that there exist a region to place the top point $p_t$. For instance, in Fig. 5(1), the points $a, b, c$ form the bottom face and the top point is $p_t$.

a. (Bottom face) Let $v_j, v_k,$ and $v_l$ be the three neighboring vertices that $v_i$ is connected with among $v_1, ..., v_{i-1}$. Suppose $p_j, p_k,$ and $p_l$ are the corresponding top points of $t_j, t_k,$ and $t_l$, respectively. Since $T_{i-1}$ is a convex body, we know that the plane formed by $p_j, p_k,$ and $p_l$ does not intersect any other tetrahedron in $T_{i-1}$. We now consider how internal or external points associated with the bottom face of $t_i$ are added.

1. (Internal points) Given a directed edge $(v_i, v_j)$, a point $p_b$ of $t_i$ must touch a face of $t_j$. Consider Fig. 5(3). Suppose $t_j$ corresponds to the tetrahedron colored in green. Then $p_b$ can be placed in the region of the green triangle, formed by the top point of $t_j$ and the two connected points between $t_j$ and $t_k$, as well as $t_j$ and $t_l$.

2. (External points) Given a directed edge $(v_k, v_i)$, the top point of $t_k$ must touch the bottom face of $t_i$. Thus, an external point $p_b$ belonging to the bottom face of $t_i$ will be added in the following way. Consider Fig. 5 (1) in which two other bottom points $b, c$ of $t_i$ are internal. The point $a'$ of $t_j$ touches the bottom face of $t_i$, and the point $p_b$ $(= a)$ is placed by extending the triangle $\triangle(b, c, a')$ in the $a'$ direction, as shown in Fig. 5(1). It is not hard to see the feasibility of finding a $p_b$ so that $T_i$ remains convex.

b. (Top point) Finally, we are going to place the top point $p_{t_i}$ of $t_i$ to $T_{i-1}$. Since, $T_{i-1}$ is a convex body, we can place the $p_{t_i}$ in the purple region as seen in Fig. 5(2), while maintaining $T_i$ as a convex body.    □

# 6  Graphs Without PsSLCRs or PsTriCRs

Somewhat surprisingly, the availability of psSLCRs (resp., psTriCRs) is no longer guaranteed for a slightly larger graph class. See Fig. 6.

**Theorem 5.** *(1) There is a digraph in $\mathcal{O}^2_{\mathcal{G}_{2t}}$ which does not admit a psSLCR. (2) There is a digraph in $\mathcal{O}^3_{\mathcal{G}_{p3t}}$ which does not admit a psTriCR.*

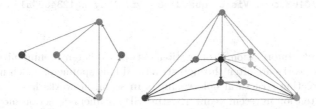

**Fig. 6.** (Left): An oriented 2-tree without a *psSLCR*; (Right): An oriented planar 3-tree without a *psTriCR*.

# References

1. Aerts, N., Felsner, S.: Straight line triangle representations. Discrete Comput. Geom. **57**(2), 257–280 (2017)
2. Alam, J., Evans, W., Kobourov, S., Pupyrev, S., Toeniskoetter, J., Ueckerdt, T.: Contact representations of graphs in 3D. In: Dehne, F., Sack, J.-R., Stege, U. (eds.) WADS 2015. LNCS, vol. 9214, pp. 14–27. Springer, Cham (2015). https://doi.org/10.1007/978-3-319-21840-3_2
3. Alam, J., Kobourov, S., Liotta, G., Pupyrev, S., Veeramoni, S.: 3D proportional contact representations of graphs. In: 5th International Conference on Information, Intelligence, Systems and Applications, IISA, Chania, Crete, Greece, pp. 27–32 (2014)
4. Chan, C.-H., Yen, H.-C.: On contact representations of directed planar graphs. In: Wang, L., Zhu, D. (eds.) COCOON 2018. LNCS, vol. 10976, pp. 218–229. Springer, Cham (2018). https://doi.org/10.1007/978-3-319-94776-1_19
5. de Fraysseix, H., de Mendez, P.O.: Barycentric systems and stretchability. Discrete Appl. Math. **155**, 1079–1095 (2007)
6. Fowler, J.J.: Strongly-connected outerplanar graphs with proper touching triangle representations. In: Wismath, S., Wolff, A. (eds.) GD 2013. LNCS, vol. 8242, pp. 155–160. Springer, Cham (2013). https://doi.org/10.1007/978-3-319-03841-4_14
7. Gansner, E.R., Hu, Y., Kobourov, S.G.: On touching triangle graphs. In: Brandes, U., Cornelsen, S. (eds.) GD 2010. LNCS, vol. 6502, pp. 250–261. Springer, Heidelberg (2011). https://doi.org/10.1007/978-3-642-18469-7_23
8. Kobourov, S.G., Mondal, D., Nishat, R.I.: Touching triangle representations for 3-connected planar graphs. In: Didimo, W., Patrignani, M. (eds.) GD 2012. LNCS, vol. 7704, pp. 199–210. Springer, Heidelberg (2013). https://doi.org/10.1007/978-3-642-36763-2_18
9. Koebe, P.: Kontaktprobleme der konformen Abbil-dung. Ber. Verh. Sachs. Akademie der Wissenschaften Leipzig, Math.-Phys. Klasse **88**, 141–164 (1936)

# Identifying Structural Hole Spanners in Social Networks via Graph Embedding

Sijie Chen, Ziwei Quan, and Yong Liu$^{(\boxtimes)}$

School of Computer Science and Technology, Heilongjiang University, Harbin, China
chensjier@163.com, Vision_quan@163.com, liuyong123456@hlju.edu.cn

**Abstract.** Information Propagation between different communities in social networks are often through structural hole spanners, which makes that detecting structural hole spanners in social networks has received great attention in recent years. Existing SH spanners detection methods usually rely on graph theory knowledge. However, these methods have obvious drawbacks of poor performance and expensive computation for largescale networks. In this work, we propose a novel solution to identify SH spanner based on graph embedding learning. Considering the special topological nature of SH spanner in the network, we design a deep embedding learning method for detecting SH spanner. Extensive experimental results on real world networks demonstrate that our proposed method outperforms several state-of-the-art methods in the SH spanner identification task in terms of several metrics.

**Keywords:** Structural hole · Social network · Graph embedding

## 1 Introduction

Structural hole (abbreviated as SH) has important value in social network analysis, which possess non-redundant information between different objects. From another perspective, SH spanners not only control the communication of information between different communities, but also hold more information benefits. Structural hole theory has been used in different application scenarios. For example, mining anomalous nodes in the brain network is helpful for diagnosis of brain dysfunction [1]. In the spread of rumors and privacy protection, effective isolation of SH spanners can prevent rumors from spreading to other communities [2,3] and sensitive information from being leaked in social networks to protect data privacy [4–6]. In commercial development and market competition, SH spanners not only obtain innovative resources and competitive opportunities [7], but maximize influence as well [8].

Due to the wide application of the SH spanners in practice, many methods for identifying SH spanners have been proposed. For example, [9] proposes two SH spanners mining algorithms, HIS and MaxD, based on information flow in the case of given communities. The authors assume that the removal of SH spanners can maximize the minimal cut between communities. However, the

© Springer Nature Switzerland AG 2019
Y. Li et al. (Eds.): COCOA 2019, LNCS 11949, pp. 94–106, 2019.
https://doi.org/10.1007/978-3-030-36412-0_8

division of community boundaries is subject to certain uncertainties. In addition, the accuracy of both methods depends on the quality of given communities. [10] uses the number of shortest paths through a node is used to describe the indicator of SH spanners. [11] believes that the removal of SH spanners will increase the shortest path distance between two nodes, and if a group of SH spanners in the network is removed, the average distance of the network will increase. Recently, [12] proposes a method called HAM for measuring the harmony and difference between nodes and their neighbors to identify SH spanners. The above methods are all based on network topology to detect SH spanners. However, there is a strong coupling relationship between nodes in real networks, which will lead to an increase of computational complexity and the difficulty of processing and analyzing large-scale networks.

In this paper, we propose a novel method based on network representation to identify SH spanners in networks. This motivation comes from the success of network representation learning in dealing with social recommendation [13], community detection [14] and link prediction [15] in recent years. Most of the existing graph embedding tasks are aimed at mapping nodes into a latent low-dimensional space that can reconstruct the original graph structure and solve subsequent application problems. Although these works have good performance in solving node classification, link prediction and other problems, the special properties of SH spanners are not taken into account when learning latent representaion for nodes, which makes them unsuitable to solve the problem of detecting SH spanners. Specifically, SH spanners in the social network are responsible for the transmission of information between communities. Compared with ordinary nodes in the same community, SH spanners are more susceptible to the influence of other communities, which means that the embedding representation of ordinary nodes will be different from that of structural hole nodes. Therefore, if we can take SH spanners into account when embedding a graph, the resulting embedding representation will help us find SH spanners.

Inspired by the first-order and second-order similarity [16], we propose a deep structural hole embedding (DSHE) model for detecting SH spanners based on the first-order harmony and second-order similarity. The first-order harmony focuses on the local structure in the network. With the help of some harmonic function, the nodes in the same community maintain a harmonious state as much as possible, thus better capturing the local network structure. As shown in Fig. 1, the relationship between the red node $v_1$ and its blue neighbor nodes in $C_1$ is more harmonious than that between $v_1$ and $v_2$. Specifically, the relationship among users within the same community in the network can be in a state of mutual harmony because of the common interests and hobbies. On the contrary, there are certain differences among users in different communities. The first-order harmony reflects the difference between the SH spanners and its neighbors in different communities. The second-order similarity represents the similarity of the vertex neighborhood structure, and also compensates for the network sparsity to obtain the global network structure. Therefore, the network structure

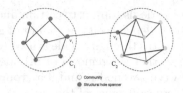

**Fig. 1.** Structural hole spanners between two communities

can be well characterized by first-order harmony and second-order similarity. Our contributions are summarized as follows:

1. Considering the characteristics of SH spanners, we propose a method for detecting SH spanners based on graph embedding. To the best of our knowledge, this is the first attempt to use network representation learning to solve the problem of SH spanners detection in real networks.
2. In the novel graph embedding learning, we combine the first-order harmony and second-order similarity for SH spanners detection. The experimental results show that DSHE method can capture more accurate SH spanners than other detection methods.

## 2   Related Work

Our work is related to both Graph embedding techniques and SH spanners detection. We briefly discuss both of them.

### 2.1   Graph Embedding

Our work is related to classical methods of graph embedding. LINE [17] designed two loss functions to capture local and global network structures. It adopts the network embedding method of shallow model and achieves success. Recently, SDNE adopts the deep model, which designs an explicit objective function to preserve the local and global structure of the network by maintaining the similarity of the first and second order. The above method is used to preserve the first-order similarity of local structure, which refers to the similarity between two connected nodes. By extension, a node has similarities with all its neighbors. However, due to the existence of different communities, the expression of such similarity is deficient. Because even if two nodes in different communities have edges, there is no guarantee that they are similar to each other, such as structural hole nodes. Therefore, DSHE replaces the similarity between nodes with harmony, which not only takes into account the similarity between neighbor nodes in the same community but also the difference between neighbor nodes in different communities, so that the local network structure can be retained more accurately.

## 2.2   Structural Hole Spanners Detection

The structural hole is a classical sociological theory proposed by Burt. In recent years, some methods have been proposed for identifying SH spanners. [11] believes that the removal of SH spanners will increase the distance of the shortest path between two nodes, and If a group of SH spanners in the network is removed, the average distance of the network will increase. Most recently, [12] only given network topology, the model of joint community and SH spanner detection is proposed based on the harmonic equation by the differences in neighboring nodes of different communities and the community indicators of each node, then, get top-k SH spanners by sorting indicators. the topology of the directly utilized in large-scale complex networks has high expensive computation, and node topology is intricacy. Because the network representation learning uses the low-dimensional vector data form to represent the network structure, the capture node structure features are better. Therefore, we propose DSHE for structural hole node design, which can detect SH spanners more effectively.

# 3   Problem Definition

Throughout this paper, we assume that there is a social network represented as an undirected graph $G = (V, E)$, where $V = \{v_1, \ldots, v_n\}$ is set of nodes and $E \subset V \times V$ is the set of edges. Each edge $e_{ij} = (v_i, v_j)$ is associated with a weight $a_{ij} \geq 0$. For $v_i$ and $v_j$ not linked by an edge, $a_{ij} = 0$. Otherwise, for unweighted graph $a_{ij} = 1$ and for weighted graph $a_{ij} > 0$. In particular, assume that the nodes of the network can be grouped into $l$ communities $C = \{C_1, \cdots, C_l\}$ , with $V = C_1 \cup \cdots \cup C_l$ and $C_i \cap C_j = \emptyset$ for every pair $i$, $j$ with $i \neq j$.

**Definition 1. (First-order Harmony)** The first-order harmony represents the harmony of the network structure between a node and all its neighbors. For nodes of the same neighborhood, the higher the number of the same community nodes are, the higher the extent of first-order harmony will be, and vice versa.

The first-order harmony explains that the nodes in the same neighborhood in the real network can maintain a harmonious relationship because of the similarities between them. For example, in social media, users who follow each other in the same community have common points of interest, and sentences in the same context have the similar semantics in the semantic analysis of NLP. However, although many nodes have similarities but no link, it is not enough to preserve the network structure only by the first-order harmony. Then, we introduce the second-order similarity to get the global network structure.

**Definition 2. (Second-order Similarity)** The second-order similarity refers to the similarity between a pair of nodes in their neighborhood network structure. Let $R_u = \{m_{u,1} \ldots, m_{u,|v|}\}$ to denote the first-order harmony of $u$ with all its neighbors, then the second-order proximity between $u$ and $v$ is dependent on the resemblance between $R_u$ and $R_v$. If there is no connection between $u$ and $v$ neighborhood, second-order proximity between $u$ and $v$ is 0.

**Definition 3. (Graph Embedding)** Given a social network $G = (V, E)$, graph embedding aim to use a low-dimensional vector represent the structural information of the nodes in the graph. i.e. learning a mapping function $f : v_i \rightarrow y_i \in \mathbb{R}^d$, where $d \ll |V|$.

## 4 Solution

### 4.1 Model Framework

In this paper, we present a semi-supervised graph embedding model to identify SH spanners, and its framework is shown in Fig. 2. For detecting SH spanners with high precision, we must capture highly non-linear network structures. Therefore, a deep architecture composed of multiple non-linear mapping functions is proposed to map the input data into a highly non-linear latent space for capturing network structures.

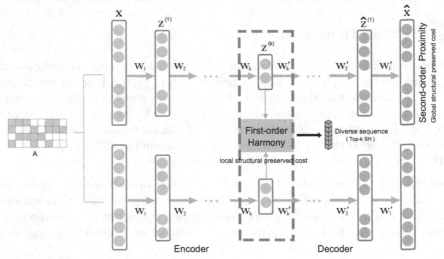

**Fig. 2.** The framework of the deep embedding for structural hole spanners detection.

We design an unsupervised learning component to exploit second-order similarity, which relies on the reconstruction of the neighborhood structure of each node. Similarly, we design components for supervised learning to achieve the first-order harmony, which can be used to refine representations in a latent space. At the same time, since the SH spanner plays the role of crossing the community boundary, compare to the harmony of neighbors from the same community, the harmony of neighbors from different communities will be lower. By jointly optimizing components in the deep model, which can detect SH spanners more accurately while preserving highly non-linear network structures.

**Table 1.** Notations and terms in this paper.

| Symbol | Description |
|---|---|
| $K$ | Number of layers |
| $X = \{a_i, ..., a_n\}$ | Adjacency matrix for the network |
| $X = \{x_i\}_{i=1}^n, \hat{X} = \{\hat{x}_i\}_{i=1}^n$ | Input data and reconstructed data |
| $Z^{(k)} = \{z_i^{(k)}\}_{i=1}^n$ | k-th layer hidden representations |
| $W^{(k)}, \hat{W}^{(k)}$ | k-th layer weight matrix |
| $b^{(k)}, \hat{b}^{(k)}$ | k-th layer biases |
| $\theta = \{W^{(k)}, \hat{W}^{(k)}, b^{(k)}, \hat{b}^{(k)}\}$ | Overall parameters |

## 4.2 Loss Function

We define some of terms and notations in Table 1, which will be used below. We first describe how unsupervised components survey second-order similarities to preserve the global network structure. Second-order similarity can be modeled by simulating the neighborhood of each node. Given a network $G = (V, E)$ we can obtain its adjacency matrix $A$, which contains $n$ instances $a_1, ..., a_n$. $a_i$ describes the neighborhood structure of node $v_i$, while $A$ describes the neighborhood Structural information of all nodes.

We use second-order similarity by extending the traditional autoencoder [18], which consists of an encoder and a decoder. The encoder consists of multiple layers of nonlinear functions that map the input data to the representation space. The decoder consists of multiple layers of nonlinear functions that map the representation from the representation space to the reconstruction space. Below we give input $x_i = a_i$, the hidden representation of each layer in the encoder is shown below:

$$z_i^{(1)} = \sigma(W^{(1)}x_i + b^{(1)})$$
$$z_i^{(k)} = \sigma(W^{(1)}z_i^{(k-1)} + b^{(k)}), k = 2, ..., K \quad (1)$$

After $z_i^{(k)}$ is obtained, the output reconstruction $\hat{x}_i$ is calculated inversely by the decoder. The autoencoder is to minimize the error between the input and the reconstruction. The loss function is shown as follows:

$$L = \sum_{i=1}^n \|\hat{x}_i - x_i\|_2^2 \quad (2)$$

Although minimizing the loss of the reconstruction cannot completely preserve the similarity of samples, the reconstruction can basically obtain the data manifold and preserve the similarity between samples. This reconstruction process enables nodes with similar neighborhood structures to have similar representations in the latent space. However, due to the sparsity of the network, the number of non-zero elements in adjacency matrix $A$ is far less than that of zero elements. If we use $A$ as the input directly, it will result in more zero elements in the reconstruction. In order to solve this problem, we can impose more penalty to the reconstruction error of the non-zero elements than that of zero elements. The improved objective function is as follows:

$$L_{2nd} = \sum_{i=1}^{n} \|(\hat{x}_i - x_i) \odot b\|_2^2 = \left\| (\hat{X} - X) \odot B \right\|_F^2 \tag{3}$$

In the objective function Eq. 3, $m_i = \{b_{i,j}\}_{j=1}^n$. We can set when $a_{i,j} = 0$, $a_{i,j} = 1$, or $a_{i,j} > 1$. Where $\odot$ is Hadamad product. In this way, the similarity of node neighborhood can be mapped to the representation space through the autoencoder, and the global network structure can be preserved. Next we use the supervisory component to investigate first-order harmony. The loss function is defined as follows:

$$L_{1st} = \sum_{i=1}^{n} \left\| z_i^{(K)} - \frac{1}{d_i} \sum_{(v_i, v_j) \in E} z_j^{(K)} \right\|_2^2 = \sum_{i=1}^{n} \left\| z_i - \frac{1}{d_i} \sum_{(v_i, v_j) \in E} z_j \right\|_2^2 \tag{4}$$

The objective function Eq. 4 stems from the idea that SH spanners can cross community boundaries. In graph theory, the community is seen as a group of closely connected nodes. Intuitively, the structure of the community portrays the neighborhood relationship between nodes, that is, nodes with community tags such as similar interests and hobbies gather together. In a graph, the relationship between a node and its neighbor nodes in the same community should be in a harmonious state, that is, the difference between the value of $z_i$ and the averaged value of its neighbors should be minimized.

However, when a node is connected to nodes in other communities, this neighbor relationship is diversity. In particular, the more neighbors of a node belong to other communities, the large value of the Eq. 4 will be, so we get the SH spanners by selecting the maximum value of top-k $\left\| Z_i - 1 \big/ d_i \sum_{(v_i, v_j) \in E} Z_j \right\|$, $d_i$ is the degree of node $v_i$. In addition, the better the $z_i$ representation of hidden space is, the higher the accuracy of identifying SH spanners will be the ultimate objective function for model is:

$$
\begin{aligned}
L_{mix} &= L_{2nd} + \alpha L_{1st} + \nu L_{reg} \\
&= \left\| (\hat{X} - X) \odot B \right\|_F^2 + \alpha \sum_{i=1}^{n} \left\| z_i - \frac{1}{d_i} \sum_{(v_i, v_j) \in E} z_j \right\|_2^2 + \nu L_{reg}
\end{aligned}
\tag{5}
$$

$$L_{reg} = \frac{1}{2} \sum_{K=1}^{K} \left( \left\| \hat{W}^{(K)} \right\|_F^2 + \left\| W^{(K)} \right\|_F^2 \right)$$

### 4.3   Optimization

To optimize our model, our goal is to minimize the loss function. The main thing is to calculate the partial derivative of $\partial L_{mix}/\partial \hat{W}^{(K)}$ and $\partial L_{mix}/\partial W^{(K)}$. The specific mathematical derivation process is as follows:

$$\frac{\partial L_{mix}}{\partial \hat{W}^{(k)}} = \frac{\partial L_{2nd}}{\partial \hat{W}^{(k)}} + \nu \frac{\partial L_{reg}}{\partial \hat{W}^{(k)}}$$
$$\frac{\partial L_{mix}}{\partial W^{(k)}} = \frac{\partial L_{2nd}}{\partial W^{(k)}} + \alpha \frac{\partial L_{1st}}{\partial W^{(k)}} + \nu \frac{\partial L_{reg}}{\partial W^{(k)}}, k = 1, ..., K \tag{6}$$

$\partial L_{2nd}/\partial \hat{W}^{(K)}$ can be rephrased as follows:

$$\frac{\partial L_{2nd}}{\partial \hat{W}^{(K)}} = \frac{\partial L_{2nd}}{\partial \hat{X}} \cdot \frac{\partial \hat{X}}{\partial \hat{W}^{(K)}} \tag{7}$$

According to Eq. 3, for the first term we have:

$$\frac{\partial L_{2nd}}{\partial \hat{X}} = 2(\hat{X} - X)B \tag{8}$$

According to $\hat{X} = \sigma(\hat{Y}^{(K-1)}\hat{W}^{(K)} + \hat{b}^{(K)})$, we can calculate $\partial L_{2nd}/\partial \hat{W}^{(K)}$. Finally, and is obtained based on back-propagation iteration. The can be converted to the following form

$$L_{1st} = \sum_{i=1}^{n} \left\| z_i - \frac{1}{d_i} \sum_{(v_i,v_j)\in E} z_j \right\|_F^2 = \left\| Z - D^{-1}AZ \right\|_2^2 \tag{9}$$

Then, we calculate the $\partial L_{1st}/\partial W^{(K)}$:

$$\frac{\partial L_{1st}}{\partial W^{(K)}} = \frac{\partial L_{1st}}{\partial Z} \cdot \frac{\partial Z}{\partial W^{(K)}} \tag{10}$$

where $D \in \mathbb{R}^{n \times n}$ is a diagonal matrix. We can calculate $Z = \sigma(Z^{(K-1)}W^{(K)} + b^{(K)})$ easily by for the first term of $\partial L_{1st}/\partial Z$ we have:

$$\frac{\partial L_{1st}}{\partial Z} = 2(I - D^{-1}A) \cdot B \tag{11}$$

So, we can finish the calculation of partial derivative of $L_{1st}$ by back-propagation. According to initializing the parameters, the model can be optimized by the stochastic gradient descent method. Because the model has high nonlinearity, there are many local optima in the parameter space. In order to find the optimal place for the parameter space, we first use the Deep Belief Network (DBE) to pretrain the parameters as parameter initialization [19]. The complete algorithm is given in Algorithm 1.

---

**Algorithm 1:** Graph Embedding for SH Spanners Detection

---

Input: $G = (V, E)$, the adjacency matrix $A$, the parameters $\alpha$ and $\nu$

Output: the top-k SH spanners and the network representation $Z$

1: Initialize parameters $\theta = \{\theta^{(1)}, \ldots, \theta^{(K)}\}$ through DBE

2: $X = A$

3: While not converge do

4:    Computer $\hat{X}$ and $Z = Z^K$ according to Eq.1 based on $X, \theta$

5:    $L_{mix}(X; \theta) = \left\|(\hat{X} - X)B\right\|_F^2 + 2\alpha \left\|Z - D^{-1}AZ\right\|_F^2 + \nu L_{reg}$

6:    Apply back-propagation through the entire network to updated $\theta$

7: End While

8: each node according $\left\|Z - D^{-1}AZ\right\|$ in descending order and select the top-k ranked ones as SH spanners

---

## 5   Experiments

In this section, we evaluate the performance of the proposed SH spanners detection algorithm on three real-world datasets, and compare DSHE with several baseline algorithms.

### 5.1   Datasets

**Karate Club**[1]. A social network for a university karate Club that depicts the relationships of its members. Because of the conflict between the coach and the manager, the club was divided into two teams.

**DBLP**[2]. An academic paper citation network built upon the DBLP repository. In the coauthorship network, two authors have published at least one article together. Publication venue serves as a ground-truth community. The authors published articles in the same journals or conferences belong to the same community. four sub-datasets were collected from the original datasets.

**Coauthor**[3]. A network of authors collected by ArnetMinerl. We created a subnetwork containing papers published at 28 major computer science conferences. These conferences cover six research areas, each of which has its meeting. Each conference organizing committee has program committee (PC) members serving multiple domains. We see PC members serving different domains as SH spanners. For experimental evaluation, we extracted 4 subdatasets from them and found a total of 2429 PC members, among which 375 PC members as ground-truth SH spanners.

The detailed statistics of the datasets can be summarized in the Table 2.

---

[1] https://networkdata.ics.edu/data.php?id=105.

[2] https://www.aminer.cn/billboard/aminernetwork.

[3] https://www.aminer.cn/structural-hole.

**Table 2.** Statistics of the datasets.

| Datasets | #Nodes | #Edge | #SH spanners | #Communities |
|---|---|---|---|---|
| Karate club | 34 | 78 | 12 | 2 |
| DBLP | 4236 | 13261 | 253 | 15 |
| Coauthor | 5783 | 24735 | 375 | |

## 5.2 Baseline Algorithms

We will use the following five method seat baselines, each of which represents a different method of identifying SH spanners.

**MaxD.** [9] looks for top-k nodes, which are SH spanners if their removal significantly reduces the minimal cut of the community. This is a greedy algorithm strategy, in each round, the removal of its selected vertices will result in the maximum reduction of the minimum cut.

**HIS.** [9] suppose the community has given, assigns each vertex $v$ a score that simulates the likelihood of $v$ as a SH spanner across the given subset of communities, then select the highest top-k score as the SH spanners.

**2-Step.** [10] sets a score for each node in the network that represents the number of unconnected node pairs in the neighbor nodes of this node. Then select the node with the highest top-k score as the SH spanners.

**AP_BICC.** [11] selects the top-k SH spanners based on articulation points (AP) (that is, nodes of a graph that connect two or more otherwise unconnected parts of the graph and bounded inverse closeness centrality, such that after removing these nodes the increase of the mean distance of the network will be maximized.

**HAM.** [12] is a recently proposed SH spanner detection method, which selects the nodes that have neighbors belonging to more different communities as the SH spanners.

## 5.3 Evaluation Metrics

For quantitatively evaluate the proposed method, we used the following performance indicators:

**Accuracy.** Given the ground-truth SH spanners, we obtain the set of SH spanners obtained by the detection algorithms and then calculate the accuracy of the detection algorithms for measuring the proportion of the effective SH spanners in the set. Gset represents ground-truth SH spanners sets, and $V_{SH}$ represents the set of SH spanners obtained by the detection algorithm. The accuracy is defined as follows:

$$ACC = \frac{|G_{set} \cap V_{SH}|}{|G_{set}|}$$

**Structural Hole Influence Factor.** Since there are no standard metrics to evaluate the performance of SH spanners. Based on the information diffusion process in the social network, we designed a performance evaluation scheme for the characteristics of SH spanners. Considering SH spanners are usually responsible for the transfer of information between communities. When SH spanners are used as seed nodes, information can be transmitted more efficiently between communities. However, the number of affected nodes is related to the size of the community in the network. Furthermore, community center nodes can also influence other communities through neighboring nodes. Therefore, an evaluation criterion for SH spanners detection is designed, and the model is defined as follows:

$$SHIF = \frac{\sum_{i \in V_{SH}} C_i / d_i + \sum_{i \in V_{SH}} IC(v_i) / |V|}{|V_{SH}|}$$

In SHIF, $s$ is the SH spanner seed, $C_i$ is the number of node belonging to different communities. $|V|$ is the number of node in networks. We also borrowed the classical model of information dissemination model: Independent Cascade (IC) model [20] to calculate the impact of SH spanners.

### 5.4 Experiment Results

In order to evaluate the performance of DSHE in SH spanners detection, we compute accuracy and SHIF as evaluation metrics to conduct comparative experiments for different methods in different datasets.

First, on the Coauthor dataset, we respectively used the benchmark algorithm to identify SH spanners. The accuracy of each algorithm is reported in Table 3. It can be seen from the results that DSHE's accuracy in SH spanners detection is higher than other benchmark methods on the same ground-truth SH spanners.

Next, we will perform SH spanners detection on two different datasets, and Table 4 lists the SHIF for each method. By comparison, DSHE performs better than other methods. This indirectly indicates that SH spanners play an important role in information transmission. In fact, most methods of SH spanners detection by graph theory tend to be more likely to become SH spanners with the higher degree of nodes. However, these nodes are often the central nodes of the community. In addition, in network topological space, we can only get network structure information. However, in the embedded latent space, not only the structural information but also the network evolution information can be expressed. It is easier to capture the characteristics of nodes in low-dimensional space, which explains why our method is superior to other methods.

**Table 3.** Evaluation: ACC

| Methed | Coauthor |
|--------|----------|
| 2-step | 0.753 |
| HIS | 0.687 |
| MaxD | 0.754 |
| AP-BICC | 0.765 |
| HAM | 0.735 |
| DSHE | 0.864 |

**Table 4.** Evaluation: SHIF

| Method | Karate club | DBLP |
|--------|-------------|------|
| 2-step | 0.258 | 0.204 |
| HIS | 0.257 | 0.351 |
| MaxD | 0.232 | 0.286 |
| AP-BICC | 0.284 | 0.348 |
| HAM | 0.416 | 0.485 |
| DSHE | 0.435 | 0.524 |

# 6    Conclusion

In this paper, we propose a novel method for detecting SH spanners based on network representation learning. Specially, considering the special position of SH spanners in the network topology, we exploit the first-order harmony and second-order similarity to design a deep graph embedding model for SH spanners identification. Extensive experiments on real datasets demonstrate good improvement of our method compared with state-of-the-art methods. In the future work, we desire to identify SH spanners on dynamic networks.

**Acknowledgments.** This work is funded by the National Natural Science Foundation of China under numbers 61972135 and 61602159, Natural Science Foundation of Heilongjiang Province under number F201430, Innovation Talents Project of Science and Technology Bureau of Harbin under numbers 2017RAQXJ094 and 2017RAQXJ131, Fundamental Research Funds of Universities in Heilongjiang Province, Special Fund of Heilongjiang University under numbers HDJCCX-201608, KJCX201815, and KJCX201816.

# References

1. Wang, S.: Structural deep brain network mining. In: SIGKDD International Conference. ACM (2017)
2. Lin, Y.G., Cai, Z.P., Wang, X.M., Hao, F.: Incentive mechanisms for crowdblocking rumors in mobile social networks. IEEE Trans. Veh. Technol. **68**, 9220–9232 (2019)
3. He, Z.B., Cai, Z.P., Yu, J.G., Wang, X.M., Sun, Y.C., Li, Y.S.: Cost-efficient strategies for restraining rumor spreading in mobile social networks. IEEE Trans. Veh. Technol. **66**(3), 2789–2800 (2017)
4. Cai, Z.P., He, Z.B., Guan, X., Li, Y.S.: Collective data-sanitization for preventing sensitive information inference attacks in social networks. IEEE Trans. Dependable Secure Comput. **15**(4), 577–590 (2018)
5. He, Z.B., Cai, Z.P., Yu, J.G.: Latent-data privacy preserving with customized data utility for social network data. IEEE Trans. Veh. Technol. **67**(1), 665–673 (2018)
6. Zheng, X., Cai, Z.P., Yu, J.G., Wang, C.K., Li, Y.S.: Follow but no track: privacy preserved profile publishing in cyber-physical social systems. IEEE Internet Things **4**(6), 1868–1878 (2017)

7. Swedberg, R.: Structural holes: the social structure of competition. Soc. Sci. Electron. Publishing **42**(22), 7060–6 (1995)
8. Li, J., Cai, Z.P., Yan, M.Y., Li, Y.S.: Using crowdsourced data in location-based social networks to explore influence maximization. In: IEEE International Conference on Computer Communications (INFOCOM 2016) (2016)
9. Lou, T., Tang, J.: Mining structural hole spanners through information diffusion in social networks. In: Proceedings of the 22nd International Conference on World Wide Web, pp. 825–836. ACM (2013)
10. Tang, J., Lou, T., Kleinberg, J.: Inferring social ties across heterogenous networks. In: Proceedings of the Fifth ACM International Conference on Web Search and Data Mining, pp. 743–752. ACM (2012)
11. Rezvani, M., Liang, W., Xu, W., et al.: Identifying top-k structural hole spanners in large-scale social networks. In: Proceedings of the 24th ACM International on Conference on Information and Knowledge Management, pp. 263–272. ACM (2015)
12. He, L., Lu, C.T., Ma, J., et al.: Joint community and structural hole spanner detection via harmonic modularity. In: Proceedings of the 22nd ACM SIGKDD International Conference on Knowledge Discovery and Data Mining, pp. 875–884. ACM (2016)
13. Fan, W., Ma, Y., Li, Q., et al.: Graph neural networks for social recommendation. In: The World Wide Web Conference, pp. 417–426. ACM (2019)
14. Li, Y., Sha, C., Huang, X., et al.: Community detection in attributed graphs: an embedding approach. In: Thirty-Second AAAI Conference on Artificial Intelligence (2018)
15. Chen, H.: PME: projected metric embedding on heterogeneous networks for link prediction. In: Proceedings of the 24th ACM SIGKDD International Conference on Knowledge Discovery & Data Mining. ACM (2018)
16. Wang, D., Cui, P., Zhu, W.: Structural deep network embedding. In: Proceedings of the 22nd ACM SIGKDD International Conference on Knowledge Discovery and Data Mining, pp. 1225–1234. ACM (2016)
17. Tang, J., Qu, M., Wang, M., et al.: Line: large-scale information network embedding. In: Proceedings of the 24th International Conference on World Wide Web, pp. 1067–1077 (2015)
18. Hinton, G.E., Salakhutdinov, R.R.: Reducing the dimensionality of data with neural networks. Science **313**(5786), 504–507 (2006)
19. Hinton, G.E., Osindero, S., Teh, Y.W.: A fast learning algorithm for deep belief nets. Neural Comput. **18**(7), 1527–1554 (2006)
20. Zachary, W.W.: An information flow model for conflict and fission in small groups. J. Anthropol. Res. **33**(4), 452–473 (1977)

# The Price of Anarchy for the Load Balancing Game with a Randomizing Scheduler

Xujin Chen[1,2], Xiaodong Hu[1,2], and Xiaoying Wu[1,2(✉)]

[1] Academy of Mathematics and Systems Science, Chinese Academy of Sciences, Beijing 100190, China
{xchen,xdhu,xywu}@amss.ac.cn
[2] School of Mathematical Sciences, University of Chinese Academy of Sciences, Beijing 100049, China

**Abstract.** We study the price of anarchy (PoA) for the load balancing game with a randomizing scheduler. Given a system of facilities and players, each facility is associated with a linear cost function, while all players are in an ordering randomly taken by a scheduler, from the set of permutations of all players. Each player chooses exactly one of these facilities to fulfill his task, which incurs to him a cost depending on not only the facility he chooses but also his position in the ordering. In competing for the facility usage, each player tries to optimize his own objective determined by a certain decision-making principle. On the other hand, for balancing the load, it is desirable to minimize the maximum cost of all players. Concerning this system goal of load balance, we estimate the PoAs of this class of games, provided all players follow one of the four decision-making principles, namely the bottom-out, win-or-go-home, average-case-analysis, and minimax-regret principles.

**Keywords:** Load balancing game · Price of anarchy · Decision-making principle · Nash equilibrium

## 1 Introduction

Our study is based on the load balancing games which constitute a special class of the bottleneck congestion games. In a (classical) load balancing game, given a set $E$ of $m$ facilities, $n$ weighted players compete for facility selections, where each (noncooperative) player chooses exactly one facility from $E$ and imposes his weight on the facility. Each facility is associated with a cost function. The cost experienced by a player who chooses facility $e \in E$ depends on $e$'s cost function and the total weight imposed on $e$. Each player tries to choose a facility to fulfill

Research supported in part by NNSF of China under Grant No. 11531014, MOST of China under Grant 2018AAA010100 for Major Project on New Generation of Artificial Intelligence, and CAS under Grant ZDBS-LY-7008.

Y. Li et al. (Eds.): COCOA 2019, LNCS 11949, pp. 107–118, 2019.
https://doi.org/10.1007/978-3-030-36412-0_9

his task, aiming to minimize the cost he experiences, while the system designer often cares about the minimization of the social cost, *i.e.*, the maximum cost any player experiences.

As a central solution concept of rational behaviors, Nash equilibrium (NE) is a state where no player has the incentive to deviate unilaterally. To assess the equilibrium inefficiency, we use the standard measure, the *price of anarchy* (PoA), *i.e.*, the ratio between the social cost of a worst NE and that of an optimal solution.

Following Georgios *et al.* [10], we study the load balancing game where a randomizing scheduler randomly takes one of the permutations of the players, and uses it as the player ordering across all facilities. The cost of a player depends on not only the facility he chooses but also his position in the ordering. Roughly, the more players ahead of a player in the ordering, the higher the cost he would potentially experience. This model captures many real-world applications. It can be considered a refinement of tiered systems where the players with higher grades are given priorities for early services. It is worth noting that though the cost depends on both the strategies (choices) and ordering of players, the social cost only depends on the choices of players rather than the ordering. Once all players have made their choices, regardless of what the ordering is, the social cost is determined – it is the maximum among the costs of last players on the corresponding facilities, which only depends on the set of players choosing the facilities.

In load balancing games with a randomizing schedular, players may prefer to optimize some other objectives instead of minimizing their random costs described above. For instance, suppose that the ticket-buying channels for a popular show are about to open. People could buy tickets from different channels such as telephone booking, online booking, or queueing at the ticket offices. On one hand, the organizer hopes for balanced loads so that no channel would crash due to congestion, and tickets would be sold out as soon as possible. On the other hand, people are classified into different grades according to previous credit, consumption levels or some random factors. The higher the grade one has, the earlier the service he gets. Everyone wants to be the first one (among all people in all channels) to be served so that he could choose from as many available seats as possible. Therefore, the purpose of each person is to maximize the probability of being the first one. This provides an illustrative example for the *win-or-go-home* principle. Given any *instance* of the load balancing game (consisting of players and facilities along with their cost functions), different principles, in general, define different NE sets. In this way, the principles would affect the values of PoAs.

*Our Results.* In this paper, we focus on a class of symmetric loading balancing games with a randomizing scheduler, where the players are unit-weighted. Each facility $e_j \in E$ is associated with a linear cost function $f_j(x) = a_j x + b_j$, where $a_j > 0$, $b_j \geq 0$ and $x$ denotes the number of players who choose $e_j$. For convenience, we abbreviate this model to LBGRs.

We study the PoA of LBGRs (which is the worst among the PoAs of all instances) with all players following one of the four decision-making principles. Under the *bottom-out* principle, each player tries to choose a facility that minimizes the probability of being the loser, *i.e.*, the one who bears the largest cost. In contrast the conservative criterion, the *win-or-go-home* principle represents an opposite extreme attitude: maximizing the probability of being the winner, *i.e.*, the one who experiences the smallest cost. Two more moderate principles concern with average case analysis and minimax regret, respectively. When following the *average-case-analysis* principle, each player minimizes his average cost by assuming that he is the middle one in the list of players who choose the same facility as him (no matter which ordering the scheduler selects). When following the *minimax-regret* principle, each player aims at minimizing his maximum regret over all possible orderings, where his regret over an ordering is the difference between his current cost and the lowest one he could potentially ensure by choosing another facility under the ordering. Our main results are summarized in Table 1, where $\delta \geq 0$ equals to the nonnegative difference between the smallest two values (which might be equal) among all summations $a_j + b_j$, $j \in \{1, \ldots, m\}$. In addition, we will show that the PoA under the bottom-out principle is at most 4 if $b_1 = \cdots = b_m$.

**Table 1.** The PoAs of LBGRs with different decision making principles

| Decision-making principle | Upper bound of PoA | The limit of PoA as $n \to \infty$ and $m = o(n)$ |
|---|---|---|
| Bottom-out | $2 + \frac{\max_{j=1}^{m} a_j}{\min_{j=1}^{m} a_j}$ | 1 |
| Win-or-go-home | Unbounded | Unbounded |
| Average-case analysis | 3 | 1 |
| Minmax-regret | $2 + (\sum_{j=1}^{m} \frac{\delta}{a_j})/n$ | 1 |

These results might be useful for identifying good principles in LBGRs. When players choose facilities conservatively according to the bottom-out principle, the efficiency of all NEs is satisfactory. When players follow the win-or-go-home principle, all striving for the first place, some NEs might suffer from inefficiency far beyond a tolerable extent, which implies that excessive competition might not be a good idea as far as load balancing is concerned. The average-case analysis principle guarantees high NE efficiency, which indicates that an attitude that is neither conservative nor radical might be conducive to load balancing. If players adopt the minimax-regret principle, then the NE outcomes enjoy high efficiency when $\delta$ is small or the number of players far outweighs the number of facilities.

*Related Work.* Koutsoupias and Papadimitriou [6] initiated the study of PoAs of load balancing games. They provided a lower bound $\Omega(\ln m / \ln \ln m)$ of the PoA for mixed NEs in the game with identical facilities. Later, Czumaj and Vöcking

[4] gave a tight bound $\Theta(\log m/\log\log\log m)$ for mixed NEs in linear load balancing games. Vöcking [11] proved upper bound $2 - 2/(m+1)$ of the PoA for pure NEs in load balancing games on identical facilities. He also showed that the counterpart for uniformly related facilities is $\Theta(\log m/\log\log m)$. Interestingly, as shown independently by Awerbuch et al. [3] and Gairing et al. [5], this is also the PoA for identical facilities and restricted assignments.

Uncertain delays often appear in congestion games and motivate the study on the various settings where players hold different attitudes towards risk. Nikolova and Stir-Moses [7–9] studied routing games with players trying to minimize the sum of the expectation and a multiple of the standard deviation of delay. They discussed the equilibrium existences under exogenous and endogenous variability of delays. Later, Georgios et al. [10] proved that the PoA is unbounded in linear congestion games where players choose strategies to optimize the same objective as that in [8]. The model by Georgios et al. is the most related one to our work. They studied linear congestion games with a randomizing scheduler, for which they obtained the exact values or ranges of the PoAs when players follow various principles: 5/3 for the minimum-expected-cost and average-case-analysis principles, 2 for the minimax-cost principle, $[4/3, 3]$ for the minimax-regret principle, and unbounded PoAs for the win-or-go-home and second-moment-method principles. The difference between their model and ours mainly lies in different definitions of social costs. The social cost in their model is the sum over all players' costs, while our model concerns the bottleneck: the maximum cost among all players. Moreover, among the four principles considered in our model, the bottom-out principle is brand new – it was not mentioned in their paper [10].

Furthermore, Angelidakis et al. [1] studied the atomic congestion game on a parallel-link network with risk-averse players. They introduced two variants of atomic congestion games, one with stochastic players (each of which imposes a load on his chosen edge independently with a given probability), and the other with stochastic edges (whose cost functions are random). They proved that the game admits an efficiently computable pure NE if the players have either the same risk attitude or the same participation probability. In the case where the edge cost functions are linear, they showed that the PoA is $\Theta(n)$, where $n$ is the number of stochastic players. The PoA may be unbounded in the case of stochastic edges. Ashlagi et al. [2] studied the congestion game on a parallel-link network with an unknown number of risk-neutral players whose cost is defined as the expected cost of the chosen facility. They proved the existence and uniqueness of a symmetric safety-level equilibrium for this game.

*Organization.* The rest of this paper is organized as follows. In Sect. 2, we give a formal descriptions of our LBGRS model and decision-making principles to be studied. In Sect. 3, we present analysis on the PoAs of LBGRS coupled with bottom-out, win-or-go-home, average-case-analysis or minmax-regret principle. In Sect. 4, we conclude with remarks on future research.

## 2 Model

An instance of LBGRS is specified by a tuple $\mathcal{I} = (N, E, (f_j)_{e_j \in E})$, where $N = \{p_1, \ldots, p_n\}$ is the set of $n$ players and $E = \{e_1, \ldots, e_m\}$ is the set of $m$ facilities. Each facility $e_j \in E$ is associated with a nonnegative linear function $f_j(x) = a_j x + b_j$ with $a_j > 0$ and $b_j \geq 0$. Each player $p_i \in N$ has $E$ as his strategy set. The schedular selects a random ordering $r$ of players in a way that each player in $N$ is assigned to each slot of the ordering with equal probability $1/n$. We use $r \sim u$ to denote the fact that $r$ is a random ordering of players which is chosen uniformly from all the permutations over $N$.

Given a strategy profile $s = (s_1, \ldots, s_n) \in E^n$, meaning that each player $p_i$ $(i = 1, \ldots, n)$ has chosen a facility $s_i \in E$, for every $e_j \in E$, let $N_j(s)$ be the set of players who choose facility $e_j$ under $s$. All players in $N_j(s)$ are ranked from $1$ to $n_j(s) = |N_j(s)|$ on $e_j$ with the same relative order as in $r$. For each player $p_i \in N_j(s)$, his position in the ranking is a random variable, which is denoted by $r_{ij}(s)$. The (random) cost of player $p_i$ who chooses $s_i = e_j$ is $C_i(s, r) = f_j(r_{ij}(s))$, which is only affected by the number of players in front of $p_i$ under the ranking. In contrast, the cost of facility $e_j$ is deterministic $f_j(n_j(s))$. As aforementioned, the social cost $\max_{i=1}^{n} C_i(s, r) = \max_{e_j \in E, n_j(s) \neq 0} f_j(n_j(s))$ is independent of the ordering $r$. Given this fact, we define the *social cost* of strategy profile $s$ as

$$C(s) = \max_{e_j \in E, n_j(s) \neq 0} f_j(n_j(s)).$$

*Decision-Making Principles.* Given an instance $\mathcal{I}$ of LBGRS, players make their decisions according to the bottom-out (resp. win-or-go-home, average-case-analysis, minimax-regret) principle, for which we mark the instance as $\mathcal{I}_b$ (resp. $\mathcal{I}_w, \mathcal{I}_a, \mathcal{I}_m$), and each player $p_i \in N$ wishes to minimize his *loss* $\ell_i$ (or equivalently, maximize his *gain* $-\ell_i$). Let $s = (s_1, \ldots, s_n)$ be a given strategy profile.

– In $\mathcal{I}_b$, the loss of $p_i$ under $s$ is his probability of losing

$$\ell_i(s) := \mathbf{Pr}_{r \sim u}[C_i(s, r) \geq \max_{h=1}^{n} C_h(s, r)].$$

– In $\mathcal{I}_w$, the gain of $p_i$ under $s$ is his probability of winning

$$-\ell_i(s) := \mathbf{Pr}_{r \sim u}[C_i(s, r) \leq \min_{h=1}^{n} C_h(s, r)].$$

– In $\mathcal{I}_a$, the loss of $p_i$, who chooses facility $e_j$ under $s$, is his average cost

$$\ell_i(s) := f_j(\frac{n_j(s) + 1}{2}).$$

– In $\mathcal{I}_m$, the loss of $p_i$ under $s$ is his maximum regret

$$\ell_i(s) := \max_{r \in \mathcal{P}} \left( C_i(s, r) - \min_{s' \in E^n, s'_i \neq s_i} C_i(s', r) \right),$$

where $\mathcal{P}$ denotes the set of permutations over $N$.

*Price of Anarchy.* Based on the individual losses (gains) defined above, a NE of the LBGRS instance $\mathcal{I}$ under consideration refers to a strategy profile $s$ in which every player $p_i \in N$ has minimized his loss given the strategies of others, *i.e.,* $\ell_i(s) \le \ell_i(s_1, \ldots, s_{i-1}, s'_i, s_{i+1}, \ldots, s_n)$ holds for all $s'_i \in E$. Let $\mathbb{NE}$ denotes the set of NEs in $\mathcal{I}$. The PoA of $\mathcal{I}$ is the worst-case ratio between the social cost of a NE and the minimum social cost:

$$PoA(\mathcal{I}) = \frac{\max_{s \in \mathbb{NE}} C(s)}{\min_{s \in \mathcal{S}} C(s)}.$$

We conclude this section with a lower bound on the minimum social cost of $\mathcal{I}$, which involves two important parameters to be used throughout our analysis on PoAs for LBGRS:

$$\alpha = \sum_{j=1}^{m} 1/a_j, \ \beta = \sum_{j=1}^{m} b_j/a_j. \tag{2.1}$$

**Lemma 1.** *The minimum social cost of $\mathcal{I}$ is at least $(n + \beta)/\alpha$.*

*Proof.* The minimum social cost of $\mathcal{I}$ is the optimal objective value of the following minimum makespan problem:

$$
\begin{aligned}
\min \ & M \\
\text{s.t} \ & \textstyle\sum_{j=1}^{m} x_{ij} = 1, & & i = 1, 2, \ldots, n; \\
& a_j \left( \textstyle\sum_{i \in N} x_{ij} \right) + b_j \le M, & & j = 1, 2, \ldots, m; \\
& x_{ij} \in \{0, 1\}, & & i = 1, 2, \ldots, n, \ j = 1, 2, \ldots, m.
\end{aligned}
$$

The optimal objective value is lower bounded by that of the LP relaxation, *i.e.,* $(n + \beta)/\alpha$, which can be easily computed. □

# 3    PoAs for Load Balancing with a Randomizing Schedular

Throughout this section, we use $\mathcal{I}$ to denote an arbitrary instance $(N, E, (f_j)_{e_j \in E})$ of LBGRS, and set

$$a_{\min} = \min_{j=1}^{m} a_j, a_{\max} = \max_{j=1}^{m} a_j, \text{ and } b_{\min} = \min_{j=1}^{m} b_j, b_{\max} = \max_{j=1}^{m} b_j.$$

Given any strategy profile $s$ of $\mathcal{I}$, in a mild abuse of notions, we set $f_j(s) := f_j(n_j(s))$. We analyze the PoA of $\mathcal{I}$ under the bottom-up principle in Sect. 3.1, win-or-go-home principle in Sect. 3.2, average-case-analysis principe in Sect. 3.3, and minimax-regret principle in Sect. 3.4.

## 3.1    The Bottom-Out Principle

In $\mathcal{I}_b$, following the bottom-out principle, each player $p_i \in N$ tries to choose a facility so that, under the resulting strategy profile $s$, his probability of *losing* $\mathbf{Pr}_{r \sim u}[C_i(s, r) \ge C_j(s, r)$ for all $p_j \in N]$ is minimized. Most of our efforts are devoted to the proof of the following theorem.

**Theorem 1.** *The PoA of $\mathcal{I}_b$ is at most $2 + \frac{a_{\max}}{a_{\min}}$. Furthermore, it converges to 1 as $n$ tends to infinity, provided that $m = o(n)$.*

We first prove some properties enjoyed by the NEs of $\mathcal{I}_b$. The first one is that the social cost of any NE could only be attained at a unique facility.

**Lemma 2.** *If $s$ is a NE of $\mathcal{I}_b$, then $|\{e_j \in E \mid f_j(n_j(s)) = c(S)\}| = 1$.*

*Proof.* Suppose on the contrary that $C(s) = f_l(s) = f_k(s)$ with $1 \le k \ne l \le m$ and $n_l = n_l(s) \ge n_k(s) = n_k > 0$. Since $n_k$ and $n_l$ are integers, we have $1/n_k > 1/(n_l + 1)$, which implies that a player choosing $e_k$ could reduce his probability $1/n_k$ of being the last one by unilaterally deviating to $e_l$. This is a contradiction to the assumption that $s$ is a NE of $\mathcal{I}$. $\square$

**Lemma 3.** *If $s$ is a NE of $\mathcal{I}_b$ with $C(s) = f_t(s)$ and $n_j = n_j(s)$ for every $e_j \in E$, then the following hold for every $e_j \in E \setminus \{e_t\}$:*

*(i) $a_t n_t + b_t > a_j n_j + b_j$;*
*(ii) $a_t(n_t - 1) + b_t \le a_j(n_j + 1) + b_j$;*
*(iii) $a_l n_l + b_l \le a_j(n_j + 1) + b_j$ for every $e_l \in E \setminus \{e_t, e_j\}$;*
*(iv) $n_t \ge n_j + 1$.*

*Proof.* Since $e_t$ is the facility that attains the social cost $C(s)$, the cost $a_j n_j + b_j$ of any other facility $e_j \in E \setminus \{e_t\}$ is no more than $a_t n_t + b_t$. The strict inequality in (i) follows instantly from Lemma 2.

Since $s$ is a NE of $\mathcal{I}$, no one has incentive to deviate. For the players choosing $e_j \in E \setminus \{e_t\}$, their probability of losing is 0, so they would not deviate. For the players choosing $e_t$, they could not reduce their probability of losing by deviating to any other facility $e_j$. It follows that after a player changing his choice from $e_t$ to $e_j$, the cost of $e_j$ becomes the largest among all facilities, which gives (ii) and (iii).

Furthermore, when a player on $e_t$ deviates to $e_j$, his probability of losing changes from $1/n_t$ to $1/(n_j + 1)$ which cannot be smaller. This proves (iv). $\square$

**Lemma 4.** *If $s$ is a NE of $\mathcal{I}_b$, then $C(s) \le \frac{n + m - 2}{\alpha} + a_{\max} + b_{\max}$, where $\alpha$ is as defined in (2.1).*

*Proof.* Let $e_t$ be the unique facility which, as stated in Lemma 2, attains the social cost $C(s) = f_t(s)$. Lemma 3 (ii) and (v) imply that $\frac{a_t}{a_j}(n_t(s) - 1) + \frac{b_t - b_j}{a_j} - 1 \le n_j(s) \le n_t(s) - 1$ holds for every $e_j \in E \setminus \{e_t\}$. Note that $\sum_{k=1}^{m} n_k(s) = n$. We obtain the following inequalities by summing over $j \in \{1, 2, \ldots, m\} \setminus \{t\}$,

$$\alpha \cdot a_t(n_t(s) - 1) - m + 2 + \alpha \cdot b_t - \beta \le n \le m \cdot n_t(s) - m + 1. \quad (3.1)$$

Since $b_j \ge 0$ for any $j \in \{1, 2, \ldots, m\}$, we have

$$C(s) = a_t \cdot n_t(s) + b_t \le (n + m - 2 + \alpha a_t + \beta)/\alpha \quad (3.2)$$
$$\le (n + m - 2)/\alpha + a_{\max} + b_{\max},$$

as desired. $\square$

**Lemma 5.** $PoA(\mathcal{I}_b) \leq 1 + \frac{1}{n}(m-1)(1 + \frac{a_{\max}}{a_{\min}})$.

*Proof.* Combining Lemmas 1 and 4, we derive $PoA(\mathcal{I}_b) \leq 1 + \frac{m-2+\alpha \cdot a_{\max}}{n+\beta}$. Moreover, note that $\alpha \cdot a_{\max} \leq 1 + (m-1)\frac{a_{\max}}{a_{\min}}$, from which we deduce that $PoA(\mathcal{I}_b) \leq 1 + \frac{m-1}{n+\beta}\left(1 + \frac{a_{\max}}{a_{\min}}\right)$. Recalling 2.1, the parameter $\beta$ is nonnegative, which implies the result.     □

We present a matching example, which shows the tightness of the PoA upper bound provided in Lemma 5.

*Example 1.* Let $\mathcal{I}_b$ be the instance of LBGRS with an even number $n \geq 2$ of players and $m = 2$ facilities $e_1, e_2$, which have identical cost functions $f_1(x) = f_2(x) = x$. Lemma 5 gives $PoA(\mathcal{I}_b) \leq 1 + 2/n$.

Clearly the minimum social cost is $n/2$, which is realized by a half of players choosing $e_1$ and the other half choosing $e_2$. On the other hand, $\mathcal{I}_b$ does admit a NE whose social cost is $(1 + 2/n) \cdot n/2$. Indeed, let $s$ denote the strategy profile where $n/2 - 1$ players choose $e_1$ and $n/2 + 1$ players choose $e_2$. It is clear that under the profile no player on $e_1$ has an incentive to deviate. Moreover, for any player on $e_2$, his probability of losing under $s$ is $1/(n/2 + 1)$, while the probability would increase to $2/n$ if he deviates to $e_1$. Hence, $s$ is a NE with social cost $n/2 + 1$, implying $PoA(\mathcal{I}_b) \geq 1 + 2/n$.     □

**Lemma 6.** *If $m \geq n$, then there is a set $E'$ of $n$ facilities in $E$ such that the LBGRS instance $\mathcal{I}' = (N, E', (f_j)_{e_j \in E'})$ has $PoA(\mathcal{I}'_b) \geq PoA(\mathcal{I}_b)$.*

*Proof.* Let $s$ denote a NE of $\mathcal{I}_b$ whose social cost $C(s)$ is the largest among all NEs of $\mathcal{I}_b$. First, we claim that under $s$, at least two players choose the facility, say $e_t$, whose cost $f_t(s)$ attains the social cost $C(s)$. If not, the single player on $e_k$ could get a smaller probability of being the one with the highest cost by just deviating to another facility which has been chosen by some other player(s).

Next, we prove the property that if $n_\xi(s) = 0$ for some $\xi \in \{1, \dots, m\}$, then $n_j(s) = 0$ for all $j$ with $a_j + b_j > a_\xi + b_\xi$. Considering $a_\xi + b_\xi < a_t + b_t$, one of the two or more players on $e_t$ could get a 0 probability of losing by deviating to $e_\xi$. So $a_\xi + b_\xi \geq a_t + b_t$. We give the remainder of the proof by contradiction. Assume on the contrary that there is a facility $e_j$ with $a_j + b_j > a_\xi + b_\xi$ and $n_j(s) \neq 0$. Then a player on $e_k$ could get a 0 probability of losing by deviating to $e_\xi$, since $n_j(s) \neq 0$ guarantees that the cost of the player after deviating is smaller than that of player(s) on $e_j$. This leads to a contradiction of the assumption that $s$ is a NE. Thus we have proved that there is a set $E'$ of $n$ facilities such that under $s$ no player chooses a facility outside $E'$ and

$$\max_{e_j \in E'} (a_j + b_j) \leq \min_{e_j \in E - E'} (a_j + b_j). \tag{3.3}$$

Clearly, $s$ is a NE of $\mathcal{I}'_b$, where $\mathcal{I}' = (N, E', (f_j)_{e_j \in E'})$, and the social cost of $s$ in $\mathcal{I}'_b$ is the same as that in $\mathcal{I}_b$.

To prove the lemma, it suffices to prove that there is an optimal strategy profile $s^*$ of $\mathcal{I}_b$ (meaning that its social cost is minimum) such that $s_i^* \in E'$

for all $i \in \{1, \ldots, n\}$. Take $s^*$ to be an optimal strategy profile of $\mathcal{I}_b$ such that $\{s_1^*, \ldots, s_n^*\} \cap E'$ is as large as possible. If $\{s_1^*, \ldots, s_n^*\} \subseteq E'$, we are done. Otherwise, there exists $s_i^* = e_k \notin E'$, which implies $n_h(s^*) = 0$ for some $e_h \in E'$. Since $a_h + b_h \le a_k + b_k$ by (3.3), it is easy to see that $C(s_1^*, \ldots, s_{i-1}^*, e_h, s_{i+1}^*, \ldots, s_n^*) \le C(s)$, which implies a contradiction to the maximality of $\{s_1^*, \ldots, s_n^*\} \cap E'$.     □

*Proof (of Theorem 1).* The results follow immediately from Lemmas 5 and 6, where Lemma 6 particularly allows us to focus on the case of $m \le n$ in upper bounding PoA for $\mathcal{I}_b$.     □

Moreover, we can obtain an upper bound on $PoA(\mathcal{I}_b)$, which depends on $\alpha(b_{\max} - b_{\min})$ instead of $a_{\max}/a_{\min}$, as stated in the Theorem 2 below. To prove the result, we need one more lemma.

**Lemma 7.** *If positive number $\mu$ satisfies $\mu > \frac{m}{n-1} \cdot (m - 1 + \alpha(b_{\max} - b_{\min}))$, then for any NE $s$ of $\mathcal{I}_b$, the social cost $C(s)$ can only be attained at a unique facilities $e_t$, for which $a_t \le (m + \mu)/\alpha$ holds.*

*Proof.* By Lemma 2, suppose that $C(s) = f_t(s)$ is attained at a unique facility $e_t$. Recalling (3.1), we have

$$n_t(s) \le \frac{m - 1 + \beta - \alpha \cdot b_t}{\alpha \cdot a_t - m} + 1.$$

If $a_t > (m + \mu)/\alpha$, then $n_t(s) \le \frac{1}{\mu}(m - 1 + \beta - \alpha \cdot b_t)$, and

$$n \le m \cdot n_t(s) - m + 1 \le \frac{m(m - 1 + \beta - \alpha b_t)}{\mu} + 1.$$

Since $\beta - \alpha b_t \le \alpha(b_{\max} - b_{\min})$, it follows that $n \le \frac{m}{\mu}(m - 1 + \alpha(b_{\max} - b_{\min})) + 1$, a contradiction to the hypothesis of the lemma.     □

**Theorem 2.** *If positive number $\mu$ satisfies $\mu > \frac{\min\{m,n\}}{n-1}(\min\{m, n\} - 1 + \alpha(b_{\max} - b_{\min}))$, then $PoA(\mathcal{I}_b) \le 1 + \frac{2\min\{m,n\} + \mu - 2}{n}$.*

*Proof.* By Lemma 6, we may assume that $m \le n$, which implies that $\min\{m, n\} = m$ and $\mu > \frac{m}{n-1} \cdot (m - 1 + \alpha(b_{\max} - b_{\min}))$. Let $s$ be a NE of $\mathcal{I}_b$ whose social cost $C(s)$ is the largest one among all NEs of $\mathcal{I}_b$. It follows from Lemma 7 that $C(s) = f_t(s)$ for a unique $t \in \{1, \ldots, m\}$, and $a_t \le (m + \mu)/\alpha$. Recalling (3.2) in the proof of Lemma 4, we derive

$$C(s) \le \frac{n + m - 2 + \alpha \cdot a_t + \beta}{\alpha} \le \frac{n + m + m + \mu - 2 + \beta}{\alpha}.$$

Recalling the lower bound $(n + \beta)/\alpha$ on the minimum social cost of $\mathcal{I}_b$ (Lemma 1), we have

$$PoA(\mathcal{I}_b) \le \frac{C(s)}{(n + \beta)/\alpha} \le 1 + \frac{2m + \mu - 2}{n + \beta} \le 1 + \frac{2\min\{m, n\} + \mu - 2}{n},$$

which proves the theorem.     □

The PoA upper bound in Theorem 2 shows that no matter what $a_{\max}/a_{\min}$ is (it may even depend on $n$), as long as $n$ is not small in comparison with $\alpha(b_{\max} - b_{\min})$, the PoA can be guaranteed to be bounded above by a constant. Particularly, we have the following immediate corollary.

**Corollary 1.** *If $b_{\max} = b_{\min}$, then $PoA(\mathcal{I}_b) \leq 4$.*    □

### 3.2    The Win-or-Go-Home Principle

This subsection shows an unbounded PoA for the win-or-go-home principle, under which each player $p_i \in N$ tries to maximize his probability of winning $\mathbf{Pr}_{r \sim u}[C_i(s,r) \leq C_j(s,r)$ for all $p_j \in N]$ under the resulting strategy profile $s$.

*Example 2.* Let $\mathcal{I}_w$ be an instance of LBGRS in which $m = n$ and the cost functions are $f_1(x) = x$ and $f_j(x) = 2x$ for $j = 2, \ldots, m$.

The minimum social cost equals 2, which is attained when each player occupies exclusively the whole facility he chooses. Consider the strategy profile where all players choose $e_1$, and therefore, everyone has a winning probability $1/n$. This is a NE because the cost of a deviating player would be at least 2, and he would never have a chance to be the winner. The social cost of the NE equals $n$, which implies an unbounded $PoA(\mathcal{I}_w)$ at least $n/2$.

### 3.3    The Average-Case-Analysis Principle

In this subsection, each player $p_i \in N$ of $\mathcal{I}_b$ follows the average-case-analysis principle to choose a facility $e_j \in E$ such that his average cost $f_j(\frac{n_j(s)+1}{2})$ under the resulting strategy profile $s$ is minimized.

**Lemma 8.** *If $s$ is a NE of $\mathcal{I}_a$, then $C(s) \leq \frac{n+2m+2\beta}{\alpha}$.*

*Proof.* Let $e_t$ denote a facility that attains the social cost of $s$. Since $s$ is a NE, $f_t(\frac{n_t(s)+1}{2}) \leq f_j(\frac{n_j(s)}{2}+1)$ for all $e_j \in E - \{e_t\}$. Thus $(a_t \cdot n_t(s) + 2b_t + a_t - 2b_j - 2a_j)/a_j \leq n_j(s)$ for all $j \in \{1, \ldots, m\}$. Summing these $m$ inequalities, we have $\alpha(a_t \cdot n_t(s) + 2b_t + a_t) - 2\beta \leq n + 2m$, giving $C(s) = a_t \cdot n_t(s) + b_t \leq (n + 2m + 2\beta)/\alpha$ as desired.    □

**Lemma 9.** *If $m \geq n$, then there is a set $E'$ of $n$ facilities in $E$ such that the LBGRS instance $\mathcal{I}' = (N, E', (f_j)_{e_j \in E'})$ has $PoA(\mathcal{I}'_a) \geq PoA(\mathcal{I}_a)$.*

*Proof.* Let $s$ denote a NE of $\mathcal{I}_a$ with $C(s)$ being the largest among all NEs of $\mathcal{I}_a$. First we verify that if $n_\xi(s) = 0$ for some $\xi \in \{1, \ldots, m\}$, then $n_j(s) = 0$ for all $j$ with $a_j + b_j > a_\xi + b_\xi$. Suppose otherwise, one of players on $e_j$ could reduce his average cost from $(a_j \cdot n_j(s) + a_j)/2 + b_j \geq a_j + b_j$ to smaller $a_\xi + b_\xi$ by deviating to $e_\xi$, a contradiction to the assumption that $s$ is a NE. The remaining proof is a verbatim copy from the counterpart in the proof of Lemma 6.    □

**Theorem 3.** *The PoA of $\mathcal{I}_a$ is at most 3. Furthermore, it converges to 1 when $n$ tends to infinity, provided that $m = o(n)$.*

*Proof.* Let $s$ denote a NE of $\mathcal{I}_a$ whose social cost $C(s)$ is the largest among all NEs of $\mathcal{I}_m$. Combing Lemmas 8 and 9, we deduce that $C(s) \le \frac{n+2\min\{m,n\}+2\beta}{\alpha}$. By Lemma 1, we have $PoA(\mathcal{I}_a) \le 1 + \frac{2\min\{m,n\}+\beta}{n+\beta}$, which implies the results. $\Box$

### 3.4 The Minimax-Regret Principle

In this subsection, we study $\mathcal{I}_m$ where each player $p_i \in N$ tries to choose a facility so that his maximum regret under the resulting strategy profile $s$, *i.e.*, $\ell_i(s) = \max_{r \in \mathcal{P}}(C_i(s,r) - \min_{s' \in E^n, s'_i \neq s_i} C_i(s',r))$, is minimized. For notational convenience, we assume in this subsection that $a_1 + b_1 \le \cdots \le a_m + b_m$ and set $\delta := a_2 + b_2 - (a_1 + b_1) \ge 0$.

Given any strategy profile $s$, suppose that player $p_i$ chooses facility $s_i = e_j$. For any fixed ordering $r \in \mathcal{P}$, it is clear that $\min_{s' \in E^n, s'_i \neq s_i} C_i(s',r) = \min_{e_h \in E - \{e_j\}}(a_h + b_h)$ equals to $a_2 + b_2$ if $j = 1$ and $a_1 + b_1$ otherwise. On the other hand, the worst case (among all possible ordering) for player $p_i$ is that he is the last one on $e_j$, and experiences a cost $f_j(s) = a_j \cdot n_j(s) + b_j$. Accordingly, the maximum regret of player $p_i$ with $s_i = e_j$ is $\ell_i(s) = a_j \cdot n_j(s) + c_j$, where $c_1 = b_1 - (a_2 + b_2)$ and $c_j = b_j - (a_1 + b_1)$ for all $2 \le j \le m$.

**Lemma 10.** *If $s$ is a NE of $\mathcal{I}_m$, then $C(s) \le \frac{n+m-1+\beta}{\alpha} + \delta$.*

*Proof.* Suppose that $C(s) = f_t(s)$. Since $s$ is a NE, $a_t \cdot n_t(s) + c_t \le a_j \cdot (n_j(s) + 1) + c_j$ holds for all $j \in \{1, \ldots, m\} \setminus \{t\}$. Adding these $m-1$ inequalities $(a_t \cdot n_t(s) + c_t - c_j)/a_j \le n_j(s) + 1, j \in \{1, 2, \ldots, m\} \setminus \{t\}$, we obtain $\alpha \cdot a_t \cdot n_t(s) + \alpha \cdot c_t - (\sum_{j=1}^m c_j/a_j) \le n + m - 1$, giving $a_t \cdot n_t(s) + c_t \le (n + m - 1 + (\sum_{j=1}^m c_j/a_j))/\alpha$. It follows from $c_t \ge b_t - \max\{a_1 + b_1, a_2 + b_2\} = b_t - (a_2 + b_2)$ that $C(s) = a_t \cdot n_t(s) + b_t$ is upper bounded by

$$\frac{n + m - 1 + (\sum_{j=1}^m c_j/a_j)}{\alpha} + a_2 + b_2 \le \frac{n + m - 1 + \beta}{\alpha} + \delta,$$

proving the lemma. $\Box$

**Lemma 11.** *If $m \ge n$, then there is a set $E'$ of $n$ facilities in $E$ such that the LBGRS instance $\mathcal{I}' = (N, E', (f_j)_{e_j \in E'})$ has $PoA(\mathcal{I}'_m) \ge PoA(\mathcal{I}_m)$.*

*Proof.* Let $s$ denote a NE of $\mathcal{I}_a$ with $C(s)$ being the largest among all NEs of $\mathcal{I}_m$. We verify that if $n_\xi(s) = 0$ for some $\xi \in \{1, \ldots, m\}$, then $n_j(s) = 0$ for all $j$ with $a_j + c_j > a_\xi + c_\xi$. Suppose otherwise, one of players on $e_j$ could reduce his maximum regret from $a_j \cdot n_j(s) + c_j \ge a_j + c_j$ to smaller $a_\xi + c_\xi$ by deviating to $e_\xi$, a contradiction to the assumption that $s$ is a NE. The remaining proof is an almost verbatim copy from the counterpart in the proof of Lemma 6 (with $c_j, c_h, c_k$ in place of $b_j, b_h, b_k$ over there). $\Box$

Combining Lemmas 1, 10 and 11, we obtain the following result.

**Theorem 4.** $PoA(\mathcal{I}_m) \le 1 + \frac{\min\{m,n\}-1+\alpha\delta}{n+\beta} < 2 + \frac{\alpha\delta}{n}$. *In particular, the PoA of $\mathcal{I}_m$ converges to 1 when $n$ tends to infinity, provided $m = o(n)$.* $\Box$

# 4    Conclusion

In this paper, we have studied the load balancing game with a randomizing scheduler, and provided upper bounds on the PoAs when players follow the bottom-out, average-case-analysis, minmax-regret, or win-or-go-home principle. The first three principles are shown to be able to ensure good (in some sense) load balance at NEs; in particular, NEs are approximately optimal when the number of players is large enough. In contrast, the win-or-go-home principle does not necessarily work well as far as load balancing is concerned. Regarding future research, tighter upper bounds for the PoAs studied in the paper, more general network topologies than the current parallel network, and more decision-making principles against more complex uncertainty deserve good research efforts.

# References

1. Angelidakis, H., Fotakis, D., Lianeas, T.: Stochastic congestion games with risk-averse players. In: Vöcking, B. (ed.) SAGT 2013. LNCS, vol. 8146, pp. 86–97. Springer, Heidelberg (2013). https://doi.org/10.1007/978-3-642-41392-6_8
2. Ashlagi, I., Monderer, D., Tennenholtz, M.: Resource selection games with unknown number of players, vol. 2006, pp. 819–825, May 2006
3. Awerbuch, B., Azar, Y., Richter, Y., Tsur, D.: Tradeoffs in worst-case equilibria. In: Solis-Oba, R., Jansen, K. (eds.) WAOA 2003. LNCS, vol. 2909, pp. 41–52. Springer, Heidelberg (2004). https://doi.org/10.1007/978-3-540-24592-6_4
4. Czumaj, A., Vöcking, B.: Tight bounds for worst-case equilibria. In: Thirteenth ACM-SIAM Symposium on Discrete Algorithms (2002)
5. Gairing, M., Mavronicolas, M., Monien, B.: The price of anarchy for restricted parallel links. Parallel Process. Lett. **16**(01), 117–131 (2008)
6. Koutsoupias, E., Papadimitriou, C.: Worst-case equilibria. In: Meinel, C., Tison, S. (eds.) STACS 1999. LNCS, vol. 1563, pp. 404–413. Springer, Heidelberg (1999). https://doi.org/10.1007/3-540-49116-3_38
7. Nikolova, E., Stier-Moses, N.E.: A mean-risk model for the traffic assignment problem with stochastic travel times. Soc. Sci. Electron. Publishing **62**(2), 366–382 (2013)
8. Nikolova, E., Stier-Moses, N.E.: Stochastic selfish routing. In: Persiano, G. (ed.) SAGT 2011. LNCS, vol. 6982, pp. 314–325. Springer, Heidelberg (2011). https://doi.org/10.1007/978-3-642-24829-0_28
9. Nikolova, E., Stier-Moses, N.E.: The burden of risk aversion in mean-risk selfish routing (2014)
10. Piliouras, G., Nikolova, E., Shamma, J.S.: Risk sensitivity of price of anarchy under uncertainty. In: Fourteenth ACM Conference on Electronic Commerce (2013)
11. Vöcking, B.: Selfish load balancing. Algorithmic Game Theor. **20**, 517–542 (2007)

# A Randomized Approximation Algorithm for Metric Triangle Packing

Yong Chen[1], Zhi-Zhong Chen[2(✉)], Guohui Lin[3(✉)], Lusheng Wang[4], and An Zhang[1]

[1] Department of Mathematics, Hangzhou Dianzi University, Hangzhou, China
{chenyong,anzhang}@hdu.edu.cn
[2] Division of Information System Design, Tokyo Denki University, Saitama, Japan
zzchen@mail.dendai.ac.jp
[3] Department of Computing Science, University of Alberta, Edmonton, Canada
guohui@ualberta.ca
[4] Department of Computer Science, City University of Hong Kong, Kowloon, Hong Kong SAR
cswangl@cityu.edu.hk

**Abstract.** Given an edge-weighted complete graph $G$ on $3n$ vertices, the maximum-weight triangle packing problem (MWTP for short) asks for a collection of $n$ vertex-disjoint triangles in $G$ such that the total weight of edges in these $n$ triangles is maximized. Although MWTP has been extensively studied in the literature, it is surprising that prior to this work, no nontrivial approximation algorithm had been designed and analyzed for its metric case (denoted by MMWTP), where the edge weights in the input graph satisfy the triangle inequality. In this paper, we design the first nontrivial polynomial-time approximation algorithm for MMWTP. Our algorithm is randomized and achieves an expected approximation ratio of $0.66745 - \epsilon$ for any constant $\epsilon > 0$.

**Keywords:** Triangle packing · Metric · Approximation algorithm · Randomized algorithm · Maximum cycle cover

## 1 Introduction

An instance of the *maximum-weight triangle packing* problem (MWTP for short) is an edge-weighted complete graph $G$ on $3n$ vertices, where $n$ is a positive integer. Given $G$, the objective of MWTP is to compute $n$ vertex-disjoint triangles such that the total weight of edges in these $n$ triangles is maximized.

The unweighted (or edge uniformly weighted) variant, denoted MTP for short, is to compute the maximum number of vertex-disjoint triangles in the input graph, which is edge unweighted and is not complete.

In their classic book, Garey and Johnson [8] show that MTP is NP-hard. Kann [14] and van Rooij *et al.* [16] show that MTP is APX-hard even restricted on graphs of maximum degree 4. Chlebik and Chlebikova [5] show that unless

© Springer Nature Switzerland AG 2019
Y. Li et al. (Eds.): COCOA 2019, LNCS 11949, pp. 119–129, 2019.
https://doi.org/10.1007/978-3-030-36412-0_10

P = NP, no polynomial-time approximation algorithm for MTP can achieve an approximation ratio of 0.9929. Moreover, Guruswami *et al.* [9] show that MTP remains NP-hard even restricted on chordal, planar, line or total graphs.

MTP can be easily cast as a special case of the *unweighted* 3-*set packing* problem (U3SP for short). Recall that an instance of U3SP is a family $\mathcal{F}$ of sets each of size 3 and the objective is to compute a sub-family $\mathcal{F}' \subset \mathcal{F}$ of the maximum number of disjoint sets. Hurkens and Schrijver [13] (also see Hall-dorsson [10]) present a nontrivial polynomial-time approximation algorithm for U3SP which achieves an approximation ratio of $\frac{2}{3} - \epsilon$ for any constant $\epsilon > 0$. This ratio has been improved to $\frac{3}{4} - \epsilon$ [6,7]. Manic and Wakabayashi [15] present a polynomial-time approximation algorithm for the special case of MTP on graphs of maximum degree 4; their algorithm achieves an approximation ratio of 0.833.

Analogously, MWTP can be cast as a special case of the *weighted* 3-*set packing* problem (W3SP for short). Two different algorithms both based on local search have been designed for W3SP [1,2] and they happen to achieve the same approximation ratio of $\frac{1}{2} - \epsilon$ for any constant $\epsilon > 0$. For MWTP specifically, Hassin and Rubinstein [11,12] present a better randomized approximation algorithm with an expected approximation ratio of $\frac{43}{83} - \epsilon$ for any constant $\epsilon > 0$. This ratio has been improved to roughly 0.523 by Chen *et al.* [3,4] and Zuylen [17].

This paper focuses on a common special case of MWTP, namely, the *metric* MWTP problem (MMWTP for short), where the edge weights in the input graph satisfy the triangle inequality. One can almost trivially design a polynomial-time approximation algorithm for MMWTP to achieve an approximation ratio of $\frac{2}{3}$; but surprisingly, prior to this work, no nontrivial approximation algorithm had been designed and analyzed. In this paper, we design the first nontrivial polynomial-time approximation algorithm for MMWTP. Our algorithm is randomized and achieves an expected ratio of $0.66745 - \epsilon$ for any constant $\epsilon > 0$. At the high level, given an instance graph $G$, our algorithm starts by computing the maximum-weight cycle cover $\mathcal{C}$ in $G$ and then uses $\mathcal{C}$ to construct three triangle packings $T_1$, $T_2$, and $T_3$, among which the heaviest one is the output solution. The computation of $T_1$ and $T_2$ is deterministic but that of $T_3$ is randomized.

The details of the algorithm are presented in the next section. We conclude the paper in the last Sect. 3, with some remarks.

## 2    The Randomized Approximation Algorithm

Hereafter, let $G$ be a given instance of the problem, and we fix an optimal triangle packing $B$ of $G$ for the following argument. Note that there are $3n$ vertices in the input graph $G$.

The algorithm starts by computing the maximum weight cycle cover $\mathcal{C}$ of $G$ in polynomial time. Obviously, $w(\mathcal{C}) \geq w(B)$, since $B$ is also a cycle cover. Let $\epsilon$ be any constant such that $0 < \epsilon < 1$. A cycle $C$ in $\mathcal{C}$ is *short* if its length is at most $\lceil \frac{1}{\epsilon} \rceil$; otherwise, it is *long*. It is easy to transform each long cycle $C$ in $\mathcal{C}$ into two or more short cycles whose total weight is at least $(1 - \epsilon) \cdot w(C)$. So, we hereafter assume that we have modified the long cycles in $\mathcal{C}$ in this way. Then, $\mathcal{C}$ is a collection of short cycles and $w(\mathcal{C}) \geq (1 - \epsilon) \cdot w(B)$.

We will compute three triangle packings $T_1$, $T_2$, $T_3$ in $G$. The computation of $T_1$ and $T_2$ will be deterministic but that of $T_3$ will be randomized. Our goal is to prove that for a constant $\rho$ with $0 < \rho < 1$, $\max\{w(T_1), w(T_2), \mathcal{E}[w(T_3)]\} \geq \left(\frac{2}{3} + \rho\right) \cdot w(B)$, where $\mathcal{E}[X]$ denotes the expected value of a random variable $X$.

## 2.1  Computing $T_1$

We first compute the maximum weight matching $M_1$ of size $n$ (*i.e.*, $n$ edges) in $G$. We then construct an auxiliary complete bipartite graph $H_1$ as follows. One part of $V(H_1)$, denoted as $V \backslash V(M_1)$, consists of the vertices of $G$ that are not endpoints of $M_1$; the vertices of the other part of $V(H_1)$, denoted as $M_1$, one-to-one correspond to the edges in $M_1$. For each edge $\{x, e = \{u, v\}\}$ in the bipartite graph $H_1$, where $x \in V \backslash V(M_1)$ and $e \in M_1$, its weight is set to $w(u, x) + w(v, x)$. Next, we compute the maximum weight matching $M_1'$ in $H_1$ and transform it into a triangle packing $T_1$ with $w(T_1) = w(M_1) + w(M_1')$.

To compare $w(T_1)$ against $w(B)$, we fix a constant $\delta$ with $0 \leq \delta < 1$ and classify the triangles in $B$ into two types as follows. A triangle $t$ in $B$ is *balanced* if the minimum weight of an edge in $t$ is at least $1 - \delta$ times the maximum weight of an edge in $t$; otherwise, it is *unbalanced*.

**Lemma 1.** *Let $B_{\bar{5}}$ be the set of unbalanced triangles in $B$, and $\gamma = \frac{w(B_{\bar{5}})}{w(B)}$. Then,*
$$w(T_1) \geq \left(\frac{2}{3} + \frac{2\gamma\delta}{9 - 3\delta}\right) \cdot w(B).$$

*Proof.* For each $t$ in $B$, let $a_t$ (respectively, $b_t$) be the maximum (respectively, minimum) weight of an edge in $t$. Further let $a = \sum_{t \in B} a_t$ and $b = \sum_{t \in B} b_t$. If $t \in B_{\bar{5}}$, then $b_t < (1 - \delta)a_t$ and in turn $(3 - \delta)a_t > w(t)$. Thus, $\sum_{t \in B_{\bar{5}}} a_t \geq \frac{1}{3-\delta} w(B_{\bar{5}}) \geq \frac{\gamma}{3-\delta} w(B)$. Hence, $w(B) \leq 2a + b \leq 3a - \delta \sum_{t \in B_{\bar{5}}} a_t \leq 3a - \frac{\delta\gamma}{3-\delta} w(B)$ and in turn $a \geq \left(\frac{1}{3} + \frac{\delta\gamma}{9-3\delta}\right) w(B)$. Now, since $w(T_1) \geq 2a$, we finally have
$$w(T_1) \geq \left(\frac{2}{3} + \frac{2\gamma\delta}{9-3\delta}\right) \cdot w(B). \qquad \square$$

## 2.2  Computing $T_2$

Several definitions are in order. A *partial-triangle packing* in a graph is a subgraph $P$ of the graph such that each connected component of $P$ is a vertex, edge, or triangle. A connected component $C$ of $P$ is a *vertex-component* (respectively, *edge-component* or *triangle-component*) of the graph if $C$ is a vertex (respectively, edge or triangle). The *augmented weight* of $P$, denoted by $\hat{w}(P)$, is $\sum_t w(t) + 2\sum_e w(e)$, where $t$ (respectively, $e$) ranges over all triangle-components (respectively, edge-components) of $P$. Intuitively speaking, if $P$ has at least as many vertex-components as edge-components, then we can trivially augment $P$ into a triangle packing $P'$ (by adding more edges) so that $w(P')$ is no less than the augmented weight of $P$.

We classify the triangles $t$ in $B$ into three types as follows.

– $t$ is *completely internal* if all its vertices fall on the same cycle in $\mathcal{C}$.
– $t$ is *partially internal* if exactly two of its vertices fall on the same cycle in $\mathcal{C}$.
– $t$ is *external* if no two of its vertices fall on the same cycle in $\mathcal{C}$.

An edge $e$ of $B$ is *external* if the endpoints of $e$ fall on different cycles in $\mathcal{C}$; otherwise, $e$ is *internal*. In particular, an internal edge $e$ of $B$ is *completely* (respectively, *partially*) *internal* if $e$ appears in a completely (respectively, partially) internal triangle in $B$. A vertex $v$ of $G$ is *external* if it is incident to no internal edges of $B$. Let $B_{\bar{e}}$ be the partial-triangle packing in $G$ obtained from $B$ by deleting all external edges.

Now, we are ready to explain how to construct $T_2$ so that $w(T_2) \geq \hat{w}(B_{\bar{e}})$. Let $C_1, \ldots, C_\ell$ be the cycles in $\mathcal{C}$, and $V_1, \ldots, V_\ell$ be their vertex sets. For each $i \in \{1, \ldots, \ell\}$, let $n_i = |V_i|$, $p_i$ be the number of partially internal edges $e$ in $B$ such that both endpoints of $e$ appear in $C_i$, $q_i$ be the number of external vertices in $C_i$, and $E_i$ be the set of edges $\{u, v\}$ in $G$ with $\{u, v\} \subseteq V_i$. Obviously, $n_i - 2p_i - q_i$ is a multiple of 3. For each $i \in \{1, \ldots, \ell\}$, let $\tilde{n}_i = \sum_{h=1}^{i} n_h$, $\tilde{p}_i = \sum_{h=1}^{i} p_h$, and $\tilde{q}_i = \sum_{h=1}^{i} q_h$.

Although we do not know $p_i$ and $q_i$, we easily see that $0 \leq q_i \leq n_i$ and $0 \leq p_i \leq \lfloor \frac{n_i - q_i}{2} \rfloor$. So, for every $j \in \{0, 1, \ldots, n_i\}$ and every $k \in \{0, 1, \ldots, \lfloor \frac{n_i - j}{2} \rfloor\}$, we compute the maximum-weight (under $\hat{w}$) partial-triangle packing $P_i(j, k)$ in the subgraph of $G$ induced by $V_i$ such that $P_i(j, k)$ has exactly $j$ vertex-components and exactly $k$ edge-components. Since $|V_i|$ is bounded by a constant (namely, $\lceil \frac{1}{\epsilon} \rceil$) from above, the computation of $P_i(j, k)$ takes $O(1)$ time.

Although we do not know $\tilde{p}_i$ and $\tilde{q}_i$, we easily see that $0 \leq \tilde{q}_i \leq \tilde{n}_i$ and $0 \leq \tilde{p}_i \leq \lfloor \frac{\tilde{n}_i - \tilde{q}_i}{2} \rfloor$. For every $j \in \{0, 1, \ldots, \tilde{n}_i\}$ and every $k \in \{0, 1, \ldots, \lfloor \frac{\tilde{n}_i - j}{2} \rfloor\}$, we want to compute the maximum-weight (under $\hat{w}$) partial-triangle packing $\tilde{P}_i(j, k)$ in the graph $(\bigcup_{h=1}^{i} V_h, \bigcup_{h=1}^{i} E_h)$ such that $\tilde{P}_i(j, k)$ has exactly $j$ vertex-components and exactly $k$ edge-components. This can be done by dynamic programming in $O(n^3)$ time as follows. Clearly, $\tilde{P}_1(j, k) = P_1(j, k)$ for every $j \in \{0, 1, \ldots, \tilde{n}_1\}$ and every $k \in \{0, 1, \ldots, \lfloor \frac{\tilde{n}_1 - j}{2} \rfloor\}$. Suppose that $1 \leq i < \ell$ and we have computed $\tilde{P}_i(j, k)$ for every $j \in \{0, 1, \ldots, \tilde{n}_i\}$ and every $k \in \{0, 1, \ldots, \lfloor \frac{\tilde{n}_i - j}{2} \rfloor\}$. For every $j \in \{0, 1, \ldots, \tilde{n}_{i+1}\}$ and every $k \in \{0, 1, \ldots, \lfloor \frac{\tilde{n}_{i+1} - j}{2} \rfloor\}$, we can compute $\tilde{P}_{i+1}(j, k)$ by finding a pair $(j', k')$ such that $j' \in \{0, 1, \ldots, n_{i+1}\}$, $k' \in \{0, 1, \ldots, \lfloor \frac{n_{i+1} - j'}{2} \rfloor\}$, and $\hat{w}(P_{i+1}(j', k')) + \hat{w}(\tilde{P}_i(j - j', k - k'))$ is maximized. Obviously, $\tilde{P}_{i+1}(j, k) = P_{i+1}(j', k') \cup \tilde{P}_i(j - j', k - k')$.

Finally, we have $\tilde{P}_\ell(j, k)$ for every $j \in \{0, 1, \ldots, 3n\}$ and every $k \in \{0, 1, \ldots, \lfloor \frac{3n - j}{2} \rfloor\}$. We now find a pair $(j', k')$ such that $j' \in \{0, 1, \ldots, 3n\}$, $k' \in \{0, 1, \ldots, \lfloor \frac{3n - j'}{2} \rfloor\}$, $k' \leq j'$, and $\hat{w}(\tilde{P}_\ell(j', k'))$ is maximized. Obviously, $\hat{w}(\tilde{P}_\ell(j', k')) \geq \hat{w}(B_{\bar{e}})$. Moreover, we can easily transform $\tilde{P}_\ell(j', k')$ into a triangle packing $T_2$ of $G$ with $w(T_2) \geq \hat{w}(\tilde{P}_\ell(j', k'))$ as follows.

1. Arbitrarily select $k'$ vertex-components of $\tilde{P}_\ell(j', k')$ and connect them to the edge-components of $\tilde{P}_\ell(j', k')$ so that $k'$ vertex-disjoint triangles are formed.
2. Arbitrarily connect the remaining $(j' - k')$ vertex-components of $\tilde{P}_\ell(j', k')$ into $\frac{j' - k'}{3}$ vertex-disjoint triangles.

In summary, we have shown the following lemma:

**Lemma 2.** *We can construct a triangle packing $T_2$ of $G$ with $w(T_2) \geq \hat{w}(B_{\bar{e}})$ in $O(n^3)$ time.*

## 2.3  Computing a Random Matching in $\mathcal{C}$

We compute a random matching $M$ in $\mathcal{C}$ as follows.

1. Initialize two sets $L = \emptyset$ and $M = \emptyset$.
2. For each even cycle $C_i$ in $\mathcal{C}$, perform the following three steps:
   (a) Partition $E(C_i)$ into two matchings $M_{i,1}$ and $M_{i,2}$.
   (b) Select a $j_i \in \{1, 2\}$ uniformly at random.
   (c) Add the edges in $M_{i,j_i}$ to $L$.
3. For each odd cycle $C_i$ in $\mathcal{C}$, perform the following five steps:
   (a) Select an edge $e_i \in E(C_i)$ uniformly at random.
   (b) Partition $E(C_i) \setminus \{e_i\}$ into two matchings $M_{i,1}$ and $M_{i,2}$.
   (c) Select a $j_i \in \{1, 2\}$ uniformly at random.
   (d) Select an edge $e_i' \in M_{i,j_i}$ uniformly at random and add $e_i'$ to $M$.
   (e) Add the edges in $M_{i,j_i} \setminus \{e_i'\}$ to $L$.
4. Select two thirds of edges from $L$ uniformly at random and add them to $M$.

**Lemma 3.** *Let $c_o$ be the number of odd cycles in $\mathcal{C}$. Then, immediately before Step 4, $|L| = \frac{3}{2} \cdot (n - c_o)$.*

*Proof.* Immediately before Step 4, $2|L| = 3n - 3c_o$ and hence $|L| = \frac{3}{2} \cdot (n - c_o)$. $\square$

**Lemma 4.** $|M| = n$.

*Proof.* Immediately before Step 4, $|M| = c_o$. So, by Lemma 3, $|M| = c_o + (n - c_o) = n$ after Step 4. $\square$

**Lemma 5.** *For every vertex $v$ of $G$, $\Pr[v \notin V(M)] = \frac{1}{3}$.*

*Proof.* First consider the case where $v$ appears in an even cycle in $\mathcal{C}$. In this case, $v \in V(M)$ immediately before Step 4. So, after Step 4, $\Pr[v \notin V(M)] = \frac{1}{3}$.

Next consider the case where $v$ appears in an odd cycle $C_i$ in $\mathcal{C}$. There are two subcases, depending on whether or not $v$ is an endpoint of the edge $e_i$ selected in Step 3a. If $v$ is incident to $e_i$, then $\Pr[v \notin V(M_{i,j_i})] = \frac{1}{2}$ and $\Pr[v \in V(M_{i,j_i}) \wedge v \notin V(e_i')] = \frac{1}{2} \cdot \left(1 - \frac{2}{n_i-1}\right)$. Hence, $\Pr[v \notin V(M) \mid v \in V(e_i)] = \frac{1}{2} + \frac{1}{2} \cdot \left(1 - \frac{2}{n_i-1}\right) \cdot \frac{1}{3} = \frac{2n_i-3}{3(n_i-1)}$. On the other hand, if $v$ is not an endpoint of $e_i$, then $\Pr[v \in V(M_{i,j_i})] = 1$ and $\Pr[v \in V(M_{i,j_i}) \wedge v \notin V(e_i')] = 1 \cdot \left(1 - \frac{2}{n_i-1}\right) = \frac{n_i-3}{n_i-1}$. Thus, $\Pr[v \notin V(M) \mid v \notin V(e_i)] = \frac{n_i-3}{n_i-1} \cdot \frac{1}{3} = \frac{n_i-3}{3(n_i-1)}$. Therefore, $\Pr[v \notin V(M)] = \frac{2}{n_i} \cdot \frac{2n_i-3}{3(n_i-1)} + \left(1 - \frac{2}{n_i}\right) \cdot \frac{n_i-3}{3(n_i-1)} = \frac{1}{3}$. $\square$

**Lemma 6.** *For every edge $e$ of $\mathcal{C}$, $\Pr[e \in M] = \frac{1}{3}$.*

*Proof.* First consider the case where $e$ appears in an even cycle in $\mathcal{C}$. In this case, $\Pr[e \in M] = \frac{1}{2} \cdot \frac{2}{3} = \frac{1}{3}$.

Next consider the case where $e$ appears in an odd cycle $C_i$ in $\mathcal{C}$. There are two subcases, depending on whether or not $e$ is the edge $e_i$ selected in Step 3a. If $e = e_i$, then $\Pr[e \notin M] = 1$. Hence, $\Pr[e \notin M \mid e = e_i] = 1$. On the other hand, if $e \neq e_i$, then $\Pr[e \notin M_{i,j_i}] = \frac{1}{2}$ and $\Pr[e \neq e_i' \mid e \in M_{i,j_i}] = 1 - \frac{2}{n_i - 1} = \frac{n_i - 3}{n_i - 1}$. Thus, $\Pr[e \notin M \mid e \neq e_i] = \frac{1}{2} \cdot 1 + \frac{1}{2} \cdot \frac{n_i - 3}{n_i - 1} \cdot \frac{1}{3} = \frac{2n_i - 3}{3(n_i - 1)}$. Therefore, $\Pr[e \notin M] = \frac{1}{n_i} \cdot 1 + \left(1 - \frac{1}{n_i}\right) \cdot \frac{2n_i - 3}{3(n_i - 1)} = \frac{2}{3}$.    $\square$

**Lemma 7.** *For every vertex $v$ of $G$ and every edge $e$ of $\mathcal{C}$ such that $v$ and $e$ appear in different cycles in $\mathcal{C}$, $\Pr[e \in M \wedge v \notin V(M)] \geq \frac{1}{9}$.*

*Proof.* Suppose that $v$ and $e$ appear in $C_{i'}$ and $C_{i''}$, respectively. We distinguish four cases as follows.

*Case 1: Both $n_{i'}$ and $n_{i''}$ are even.* In this case, $\Pr[v \in V(M_{i',j_{i'}})] = 1$ and $\Pr[e \in M_{i'',j_{i''}}] = \frac{1}{2}$. So, $\Pr[v \in V(M_{i',j_{i'}}) \wedge e \in M_{i'',j_{i''}}] = \frac{1}{2}$. Moreover, by Lemma 3, $\Pr[e \in M \wedge v \notin V(M) \mid v \in V(M_{i',j_{i'}}) \wedge e \in M_{i'',j_{i''}}] = \frac{\binom{|L|-2}{\frac{2}{3}|L|-1}}{\binom{|L|}{\frac{2}{3}|L|}} = \frac{(n-c_o)\cdot\frac{1}{2}(n-c_o)}{\frac{3}{2}(n-c_o)\cdot\left(\frac{3}{2}(n-c_o)-1\right)} \geq \frac{2}{9}$. Thus, $\Pr[e \in M \wedge v \notin V(M)] \geq \frac{2}{9} \cdot \frac{1}{2} = \frac{1}{9}$.

*Case 2: $n_{i'}$ is even but $n_{i''}$ is odd.* In this case, $\Pr[v \in V(M_{i',j_{i'}})] = 1$ and $\Pr[e \in M_{i'',j_{i''}}] = \frac{1}{2} \cdot \frac{n_{i''}-1}{n_{i''}} = \frac{n_{i''}-1}{2n_{i''}}$. Moreover, $\Pr[e = e_{i''}' \mid e \in M_{i'',j_{i''}}] = \frac{2}{n_{i''}-1}$, $\Pr[e = e_{i''}'] = \frac{1}{n_{i''}}$, and $\Pr[e \in M_{i'',j_{i''}} \backslash \{e_{i''}'\}] = \frac{n_{i''}-1}{2n_{i''}} \cdot \left(1 - \frac{2}{n_{i''}-1}\right) = \frac{n_{i''}-3}{2n_{i''}}$. Furthermore, $\Pr[v \notin V(M) \mid e = e_{i''}'] = \frac{1}{3}$ by Lemma 5, and $\Pr[v \notin V(M) \wedge e \in M \mid e \in M_{i'',j_{i''}} \backslash \{e_{i''}'\}] = \frac{\binom{|L|-2}{\frac{2}{3}|L|-1}}{\binom{|L|}{\frac{2}{3}|L|}} = \frac{(n-c_o)\cdot\frac{1}{2}(n-c_o)}{\frac{3}{2}(n-c_o)\cdot\left(\frac{3}{2}(n-c_o)-1\right)} \geq \frac{2}{9}$. Thus, $\Pr[e \in M \wedge v \notin V(M)] \geq \frac{1}{3} \cdot \frac{1}{n_{i''}} + \frac{2}{9} \cdot \frac{n_{i''}-3}{2n_{i''}} = \frac{1}{9}$.

*Case 3: $n_{i'}$ is odd but $n_{i''}$ is even.* In this case, $\Pr[v \in V(M_{i',j_{i'}})] = \frac{2}{n_{i'}} \cdot \frac{1}{2} + \left(1 - \frac{2}{n_{i'}}\right) \cdot 1 = \frac{n_{i'}-1}{n_{i'}}$ and $\Pr[e \in M_{i'',j_{i''}}] = \frac{1}{2}$. So, $\Pr[v \in V(M_{i',j_{i'}}) \wedge e \in M_{i'',j_{i''}}] = \frac{n_{i'}-1}{2n_{i'}}$ and $\Pr[v \notin V(M_{i',j_{i'}}) \wedge e \in M_{i'',j_{i''}}] = \frac{1}{2n_{i'}}$. Moreover, by Lemma 3, $\Pr[e \in M \wedge v \notin V(M) \mid v \in V(M_{i',j_{i'}}) \wedge e \in M_{i'',j_{i''}}] = \frac{\binom{|L|-2}{\frac{2}{3}|L|-1}}{\binom{|L|}{\frac{2}{3}|L|}} = \frac{(n-c_o)\cdot\frac{1}{2}(n-c_o)}{\frac{3}{2}(n-c_o)\cdot\left(\frac{3}{2}(n-c_o)-1\right)} \geq \frac{2}{9}$ and $\Pr[e \in M \wedge v \notin V(M) \mid v \notin V(M_{i',j_{i'}}) \wedge e \in M_{i'',j_{i''}}] = \frac{2}{3}$. Thus, $\Pr[e \in M \wedge v \notin V(M)] \geq \frac{n_{i'}-1}{2n_{i'}} \cdot \frac{2}{9} + \frac{1}{2n_{i'}} \cdot \frac{2}{3} \geq \frac{1}{9}$.

*Case 4: Both $n_{i'}$ and $n_{i''}$ are odd.* In this case, $\Pr[v \in V(M_{i',j_{i'}})] = \frac{n_{i'}-1}{n_{i'}}$ and $\Pr[e \in M_{i'',j_{i''}}] = \frac{1}{2} \cdot \frac{n_{i''}-1}{n_{i''}} = \frac{n_{i''}-1}{2n_{i''}}$. Moreover, $\Pr[e = e_{i''}' \mid e \in M_{i'',j_{i''}}] =$

$\frac{2}{n_{i''}-1}$, $\Pr[e = e'_{i''}] = \frac{1}{n_{i''}}$, and $\Pr[e \in M_{i'',j_{i''}} \setminus \{e'_{i''}\}] = \frac{n_{i''}-1}{2n_{i''}} \cdot \left(1 - \frac{2}{n_{i''}-1}\right) = \frac{n_{i''}-3}{2n_{i''}}$. So, $\Pr[v \notin V(M_{i',j_{i'}}) \wedge e = e'_{i'}] = \frac{1}{n_{i'}n_{i''}}$, $\Pr[v \notin V(M_{i',j_{i'}}) \wedge e \in M_{i'',j_{i''}} \setminus \{e'_{i''}\}] = \frac{n_{i''}-3}{2n_{i'}n_{i''}}$, $\Pr[v \in V(M_{i',j_{i'}}) \wedge e = e'_{i'}] = \frac{n_{i'}-1}{n_{i'}n_{i''}}$, $\Pr[v \in V(M_{i',j_{i'}}) \wedge e \in M_{i'',j_{i''}} \setminus \{e'_{i''}\}] = \frac{(n_{i'}-1)(n_{i''}-3)}{2n_{i'}n_{i''}}$. Obviously, $\Pr[e \in M \wedge v \notin V(M) \mid v \in V(M_{i',j_{i'}}) \wedge e = e'_{i''}] = \frac{1}{3}$, $\Pr[e \in M \wedge v \notin V(M) \mid v \notin V(M_{i',j_{i'}}) \wedge e = e'_{i''}] = 1$, and $\Pr[e \in M \wedge v \notin V(M) \mid v \notin V(M_{i',j_{i'}}) \wedge e \in M_{i'',j_{i''}} \setminus \{e'_{i''}\}] = \frac{2}{3}$. Furthermore, by Lemma 3, $\Pr[e \in M \wedge v \notin V(M) \mid v \in V(M_{i',j_{i'}}) \wedge e \in M_{i'',j_{i''}} \setminus \{e'_{i''}\}] = \frac{\binom{\frac{2}{3}|L|-2}{\frac{1}{3}|L|-1}}{\binom{|L|}{\frac{2}{3}|L|}} = \frac{(n-c_o) \cdot \frac{1}{2}(n-c_o)}{\frac{3}{2}(n-c_o) \cdot \left(\frac{3}{2}(n-c_o)-1\right)} \geq \frac{2}{9}$. Thus, $\Pr[e \in M \wedge v \notin V(M)] \geq \frac{1}{3} \cdot \frac{n_{i'}-1}{n_{i'}n_{i''}} + 1 \cdot \frac{1}{n_{i'}n_{i''}} + \frac{2}{3} \cdot \frac{n_{i''}-3}{2n_{i'}n_{i''}} + \frac{2}{9} \cdot \frac{(n_{i'}-1)(n_{i''}-3)}{2n_{i'}n_{i''}} \geq \frac{1}{9}$. □

## 2.4   Computing $T_3$

Fix a constant $\tau$ with $0 < \tau < 1$. A *good triplet* is a triplet $(x, y; z)$, where $\{x, y\}$ is an edge of some cycle $C_i$ in $\mathcal{C}$ and $z$ is a vertex of some other cycle $C_j$ in $\mathcal{C}$ with $i \neq j$ such that $w(x, y) \leq (1 - \tau) \cdot (w(x, z) + w(y, z))$.

To compute $T_3$, we initialize $T_3 = \emptyset$ and proceed as follows.

1. Construct an auxiliary edge-weighted and edge-labeled multi-digraph $H_3$ as follows. The vertex set of $H_3$ is $V(G)$. For each good triplet $(x, y; z)$, $H_3$ contains the two arcs $(z, x)$ and $(z, y)$, each of the two arcs has a weight of $w(x, z) + w(y, z)$ in $H_3$, the label of $(z, x)$ is $y$, and the label of $(z, y)$ is $x$.
2. Compute the maximum-weight matching $M_3$ in $H_3$ (by ignoring the direction of each arc).
3. Compute a random matching $M$ in $\mathcal{C}$ as in Sect. 2.3.
4. Let $N_3$ be the set of all arcs $(z, x) \in M_3$ such that $z \notin V(M)$ and $\{x, y\} \in M$, where $y$ is the label of $(z, x)$. (*Comment:* Since both $M$ and $N_3$ are matchings, no two arcs in $N_3$ can share a label. Moreover, the endpoints of each edge in $M$ can be the heads of at most two arcs in $N_3$.)
5. Initialize $N'_3 = N_3$. For every two arcs $(z, x)$ and $(z', y)$ in $N'_3$ such that $\{x, y\} \in M$, select one of $(z, x)$ and $(z', y)$ uniformly at random and delete it from $N'_3$.
6. For each $(z, x) \in N'_3$, let $T_3$ include the triangle $t$ with $V(t) = \{x, y, z\}$, where $y$ is the label of $(z, x)$. (*Comment:* By Step 5 and the comment on Step 4, the triangles included in $T_3$ in this step are vertex-disjoint.)
7. Let $M'$ be the set of edges $(x, y)$ in $M$ such that neither $x$ nor $y$ is the head or the label of an arc in $N'_3$. Further let $Z$ be the set of vertices $z$ in $G$ such that $z \notin V(M)$ and $z$ is not the tail of an edge in $N'_3$. (*Comment:* Since $|M| = n$ by Lemma 4, the comment on Step 6 implies $|Z| = |M'|$.)
8. Select an arbitrary one-to-one correspondence between the edges in $M'$ and the vertices in $Z$. For each $z \in Z$ and its corresponding edge $(x, y)$ in $M'$, let $T_3$ include the triangle $t$ with $V(t) = \{x, y, z\}$.

We classify external balanced triangles in $B$ into two types as follows. An external balanced triangle $t$ in $B$ is of *Type 1* if for each vertex $v$ of $t$, the weight of each edge incident to $v$ in $\mathcal{C}$ is at least $\frac{1}{2}(1-\frac{1}{2}\delta)(1-\tau)w(t)$; otherwise, $t$ is of *Type 2*.

Similarly, we classify partially internal balanced triangles in $B$ into two types as follows. A partially internal balanced triangle $t$ in $B$ is of *Type 1* if the weight of each edge incident to the external vertex of $t$ in $\mathcal{C}$ is at least $\frac{1}{2}(1-\frac{1}{2}\delta)(1-\tau)w(t)$; otherwise, $t$ is of *Type 2*.

**Lemma 8.** *Let $B_1^e$ be the set of Type-1 external balanced triangles in $B$. Further let $B_1^p$ be the set of Type-1 partially internal balanced triangles in $B$. Then, $w(T_1) \geq \frac{2}{3}w(B) + \frac{2-3\delta-6\tau+3\delta\tau}{54}w(B_1^e) + \frac{2-3\delta-6\tau+3\delta\tau}{162}w(B_1^p)$.*

*Proof.* For the analysis, we use the triangles in $B_1^e \cup B_1^p$ to construct a random matching $N$ in $\mathcal{C}$ as follows.

1. Initialize $N' = \emptyset$. For each triangle $t$ in $B$, select one edge $e_t$ of $t$ uniformly at random and add it to $N'$.
2. For each triangle $t$ in $B_1^e$, choose one neighbor $v_t'$ of $v_t$ in $\mathcal{C}$ uniformly at random, where $v_t$ is the vertex of $t$ not incident to $e_t$.
3. For each triangle $t$ in $B_1^p$ such that $e_t$ is internal, choose one neighbor $v_t'$ of $v_t$ in $\mathcal{C}$ uniformly at random, where $v_t$ is the external vertex of $t$.
4. Initialize $X = \emptyset$. For each $t \in B_1^e \cup B_1^p$, if $v_t' \notin V(N')$, then add $(v_t, v_t')$ to $X$.
5. Let $D$ be the digraph with vertex set $V(G)\backslash V(N')$ and arc set $X$. Partition $X$ into three matchings $X_1, X_2, X_3$ in $D$. (*Comment:* Each connected component of the underlying undirected graph of $D$ is either a cycle of $\mathcal{C}$ or a graph of maximum degree at most 3 whose simplified version is a path. Therefore, the partition in this step can be done.)
6. Select a set $Y$ among $X_1, X_2, X_3$ uniformly at random.
7. Initialize $N = \{e_t \mid t \in B\backslash(B_1^e \cup B_1^p)\}$. For each $t \in B_1^e$, if $(v_t, v_t') \notin Y$, then add $e_t$ to $N$; otherwise add $\{v_t, v_t'\}$ to $N$. Similarly, for each $t \in B_1^p$, if $e_t$ is external or $(v_t, v_t') \notin Y$, then add $e_t$ to $N$; otherwise add $\{v_t, v_t'\}$ to $N$.

For each triangle $t \in B_1^e$, let $E_t$ be the set of edges $e$ in $\mathcal{C}$ such that $e$ is incident to a vertex of $t$. Similarly, for each triangle $t \in B_1^p$, let $E_t$ be the set of edges $e$ in $\mathcal{C}$ such that $e$ is incident to the external vertex of $t$. Consider a $t \in B_1^e$ and an $e = \{x, y\} \in E_t$ with $x \in V(t)$. Obviously, $\Pr[x = v_t] = \frac{1}{3}$ and $\Pr[y = v_t' \mid x = v_t] = \frac{1}{2}$; hence $\Pr[\{v_t, v_t'\} = e] = \frac{1}{6}$. Moreover, $\Pr[v_t' \notin V(N')] = \frac{1}{3}$ and in turn $\Pr[\{v_t, v_t'\} = e \wedge v_t' \notin V(N')] = \frac{1}{18}$. Furthermore, $\Pr[e \in N \mid \{v_t, v_t'\} = e \wedge v_t' \notin V(N')] = \frac{1}{3}$. So, $\Pr[e \in N] = \frac{1}{3} \cdot \frac{1}{18} = \frac{1}{54}$. Now, if $t \in B_1^e$, then $|E_t| = 6$ and in turn $\Pr[e_t \notin N] = 6 \cdot \frac{1}{54} = \frac{1}{9}$. On the other hand, if $t \in B_1^p$, then $|E_t| = 2$ and in turn $\Pr[e_t \notin N] = 2 \cdot \frac{1}{54} = \frac{1}{27}$.

By the discussions in the last paragraph, $\mathcal{E}[w(N)] \geq \frac{1}{3} \sum_{t \in B\backslash(B_1^e \cup B_1^p)} w(t) + \frac{8}{9} \cdot \frac{1}{3} \sum_{t \in B_1^e} w(t) + \frac{1}{9} \cdot \frac{1}{2}(1-\frac{1}{2}\delta)(1-\tau) \sum_{t \in B_1^e} w(t) + \frac{26}{27} \cdot \frac{1}{3} \sum_{t \in B_1^p} w(t) + \frac{1}{27} \cdot \frac{1}{2}(1-\frac{1}{2}\delta)(1-\tau) \sum_{t \in B_1^p} w(t) = \frac{1}{3}w(B) + \frac{2-3\delta-6\tau+3\delta\tau}{108}w(B_1^e) + \frac{2-3\delta-6\tau+3\delta\tau}{324}w(B_1^p)$. So, $w(T_1) \geq 2 \cdot \mathcal{E}[w(N)] \geq \frac{2}{3}w(B) + \frac{2-3\delta-6\tau+3\delta\tau}{54}w(B_1^e) + \frac{2-3\delta-6\tau+3\delta\tau}{162}w(B_1^p)$.    $\square$

**Lemma 9.** *Let $B_2^e$ be the set of Type-2 external balanced triangles in $B$ to $w(B)$. Further let $B_2^p$ be the set of Type-2 partially internal balanced triangles in $B$. Then, $\mathcal{E}[w(T_3)] \geq \frac{2(1-\epsilon)}{3}w(B) + \frac{(1-\delta)\tau}{54-18\delta} \cdot w(B_2^e) + \frac{(1-\delta)\tau}{54-18\delta} \cdot w(B_2^p)$.*

*Proof.* For a set $F$ of edges in $H_3$, let $\tilde{w}(F)$ denote the total weight of edges of $F$ in $H_3$. Further let $W_2$ be the total weight of triangles in $B_2^e \cup B_2^p$.

Consider an arbitrary $t \in B_2^e \cup B_2^p$. Since $t$ is of Type 2, $t$ has a vertex $v_t$ such that some neighbor $v_t'$ of $v_t$ in $\mathcal{C}$ satisfies $w(v_t, v_t') < \frac{1}{2}(1 - \frac{1}{2}\delta)(1 - \tau)w(t)$. Let $z_t$ and $z_t'$ be the vertices in $V(t)\backslash\{v_t\}$. By the triangle inequality, $w(z_t, v_t') \geq \frac{1}{2}w(z_t, z_t')$ or $w(z_t', v_t') \geq \frac{1}{2}w(z_t, z_t')$. Without loss of generality, we may assume that $w(z_t, v_t') \geq \frac{1}{2}w(z_t, z_t')$. We claim that $(v_t, v_t'; z_t)$ is a good triplet. To see this, first recall that $(1 - \delta)w(z_t', v_t) \leq w(z_t, v_t)$ because $t$ is balanced. So, $(1 - \frac{1}{2}\delta)w(z_t', v_t) \leq (1 + \frac{1}{2}\delta)w(z_t, v_t) + \frac{1}{2}\delta w(z_t, z_t')$ by the triangle inequality. Thus, $(1 - \frac{1}{2}\delta)(w(z_t, v_t) + w(z_t, z_t') + w(z_t', v_t)) \leq 2w(z_t, v_t) + w(z_t, z_t') \leq 2w(z_t, v_t) + 2w(z_t, v_t')$. Hence, $\frac{1}{2}(1 - \frac{1}{2}\delta)w(t) \leq w(z_t, v_t) + w(z_t, v_t')$. Therefore, $w(v_t, v_t') < \frac{1}{2}(1 - \frac{1}{2}\delta)(1 - \tau)w(t) \leq (1 - \tau)(w(z_t, v_t) + w(z_t, v_t'))$. Consequently, the claim holds.

By the claim in the last paragraph, the set $X$ of all $\{z_t, v_t\}$ with $t \in B_2^e \cup B_2^p$ is a matching in $H_3$. Moreover, $\tilde{w}(M_3) \geq \tilde{w}(X) = \sum_{t \in B_2^e \cup B_2^p} w(z_t, v_t) \geq \frac{1-\delta}{3-\delta} \sum_{t \in B_2^e \cup B_2^p} w(t) = \frac{1-\delta}{3-\delta}W_2$, where the second inequality holds because $t$ is balanced and in turn $w(z_t, v_t) \geq \frac{1-\delta}{3-\delta}w(t)$. Now, by Lemma 7, $\mathcal{E}[\tilde{w}(N_3)] \geq \frac{1}{9}\tilde{w}(M_3) \geq \frac{1-\delta}{27-9\delta}W_2$ and in turn $\mathcal{E}[\tilde{w}(N_3')] \geq \frac{1-\delta}{54-18\delta}W_2$. Obviously, $w(T_3) \geq 2w(M) + \tau \cdot \tilde{w}(N_3')$ by the triangle inequality. Therefore, by Lemma 6, $\mathcal{E}[w(T_3)] \geq \frac{2}{3} \cdot w(\mathcal{C}) + \frac{(1-\delta)\tau}{54-18\delta}W_2 \geq \frac{2(1-\epsilon)}{3} \cdot w(B) + \frac{(1-\delta)\tau}{54-18\delta}W_2$.  $\qquad\square$

## 2.5   Analyzing the Approximation Ratio

Let $B^i$ be the set of completely internal balanced triangles in $B$. For convenience, let $\alpha_1 = \frac{w(B^i)}{w(B)}$, $\alpha_2 = \frac{w(B_1^e)}{w(B)}$, $\alpha_3 = \frac{w(B_2^e)}{w(B)}$, $\alpha_4 = \frac{w(B_1^p)}{w(B)}$, and $\alpha_5 = \frac{w(B_2^p)}{w(B)}$. Then, $\gamma + \alpha_1 + \alpha_2 + \alpha_3 + \alpha_4 + \alpha_5 = 1$.

We choose $\delta = 0.08$ and $\tau = 0.22$. Then, by Lemmas 1, 2, 8, and 9, we have the following inequalities:

$$\frac{w(T_1)}{w(B)} \geq \frac{2}{3} + \frac{4}{219}\gamma \tag{1}$$

$$\frac{w(T_2)}{w(B)} \geq \alpha_1 + \frac{2}{3}\alpha_4 + \frac{2}{3}\alpha_5 \tag{2}$$

$$\frac{w(T_1)}{w(B)} \geq \frac{2}{3} + \frac{0.2464}{27}\alpha_2 + \frac{0.2464}{81}\alpha_4 \tag{3}$$

$$\frac{\mathcal{E}[w(T_3)]}{w(B)} \geq \frac{2(1-\epsilon)}{3} + \frac{2.53}{657}\alpha_3 + \frac{2.53}{657}\alpha_5. \tag{4}$$

Suppose that we multiply both sides of Inequalities (1), (2), (3), and (4) by 0.1288, 0.00235, 0.2578, and 0.611, respectively. Then, one can easily verify that the summation of the left-hand sides of the resulting inequalities is

$$0.1288 \cdot \frac{w(T_1)}{w(B)} + 0.00235 \cdot \frac{w(T_2)}{w(B)} + 0.2578 \cdot \frac{w(T_1)}{w(B)} + 0.611 \cdot \frac{\mathcal{E}[w(T_3)]}{w(B)},$$

while the summation of the right-hand sides is at least

$$\frac{1.9952}{3} - \frac{1.222}{3}\epsilon + 0.00235(\gamma + \alpha_1 + \alpha_2 + \alpha_3 + \alpha_4 + \alpha_5).$$

Now, using $\gamma + \alpha_1 + \alpha_2 + \alpha_3 + \alpha_4 + \alpha_5 = 1$, we finally have

$$(0.3866 + 0.00235 + 0.611) \cdot \max\left\{\frac{w(T_1)}{w(B)}, \frac{w(T_2)}{w(B)}, \frac{\mathcal{E}[w(T_3)]}{w(B)}\right\} \geq \frac{2.00225}{3} - \frac{1.222}{3}\epsilon.$$

That is,

$$\max\left\{w(T_1), w(T_2), \mathcal{E}[w(T_3)]\right\} \geq (0.66745 - 0.41\epsilon) \cdot w(B).$$

In summary, we have proven the following theorem, stating that the MMWTP problem admits a better approximation algorithm than the trivial $\frac{2}{3}$-approximation.

**Theorem 1.** *For any constant $0 < \epsilon < 0.00078$, the expected approximation ratio achieved by our randomized approximation algorithm is at least $0.66745 - \epsilon$.*

## 3   Conclusions

We studied the maximum-weight triangle packing problem on an edge-weighted complete graph $G$, in which the edge weights satisfy the triangle inequality. Although the non-metric variant has been extensively studied in the literature, it is surprising that prior to our work, no nontrivial approximation algorithm had been designed and analyzed for this common metric case. We designed the first nontrivial polynomial-time approximation algorithm for MMWTP, which is randomized and achieves an expected approximation ratio of $0.66745 - \epsilon$ for any positive constant $\epsilon < 0.00078$. This improves the almost trivial deterministic $\frac{2}{3}$-approximation.

Perhaps more dexterous tuning of the parameters inside our algorithm could lead to certain better worst-case performance ratio, but we doubt it will be significantly better. New ideas are needed for the next major improvement.

**Acknowledgements.** YC and AZ are supported by the NSFC Grants 11971139, 11771114 and 11571252; and supported by the CSC Grants 201508330054 and 201908330090, respectively. ZZC is supported by in part by the Grant-in-Aid for Scientific Research of the Ministry of Education, Science, Sports and Culture of Japan, under Grant No. 18K11183. GL is supported by the NSERC Canada. LW is supported by a grant for Hong Kong Special Administrative Region, China (CityU 11210119).

# References

1. Arkin, E.M., Hassin, R.: On local search for weighted packing problems. Math. Oper. Res. **23**, 640–648 (1998)
2. Berman, P.: A d/2 approximation for maximum weight independent set in d-claw free graphs. SWAT 2000. LNCS, vol. 1851, pp. 214–219. Springer, Heidelberg (2000). https://doi.org/10.1007/3-540-44985-X_19
3. Chen, Z.-Z., Tanahashi, R., Wang, L.: An improved randomized approximation algorithm for maximum triangle packing. Discrete Appl. Math. **157**, 1640–1646 (2009)
4. Chen, Z.-Z., Tanahashi, R., Wang, L.: Erratum to "an improved randomized approximation algorithm for maximum triangle packing". Discrete Appl. Math. **158**, 1045–1047 (2010)
5. Chlebík, M., Chlebíková, J.: Approximation hardness for small occurrence instances of NP-hard problems. In: Petreschi, R., Persiano, G., Silvestri, R. (eds.) CIAC 2003. LNCS, vol. 2653, pp. 152–164. Springer, Heidelberg (2003). https://doi.org/10.1007/3-540-44849-7_21
6. Cygan, M.: Improved approximation for 3-dimensional matching via bounded pathwidth local search. In: Proceedings of FOCS 2013, pp. 509–518 (2013)
7. Fürer, M., Yu, H.: Approximating the k-set packing problem by local improvements. In: Fouilhoux, P., Gouveia, L.E.N., Mahjoub, A.R., Paschos, V.T. (eds.) ISCO 2014. LNCS, vol. 8596, pp. 408–420. Springer, Cham (2014). https://doi.org/10.1007/978-3-319-09174-7_35
8. Garey, M.R., Johnson, D.S.: Computers and Intractability: A Guide to the Theory of NP-Completeness. W. H. Freeman and Company, San Francisco (1979)
9. Guruswami, V., Rangan, C.P., Chang, M.S., Chang, G.J., Wong, C.K.: The vertex-disjoint triangles problem. In: Hromkovič, J., Sýkora, O. (eds.) WG 1998. LNCS, vol. 1517, pp. 26–37. Springer, Heidelberg (1998). https://doi.org/10.1007/10692760_3
10. Halldórsson, M.M.: Approximating discrete collections via local improvement. In: ACM-SIAM Proceedings of the Sixth Annual Symposium on Discrete Algorithms (SODA 1995), pp. 160–169 (1995)
11. Hassin, R., Rubinstein, S.: An approximation algorithm for maximum triangle packing. Discrete Appl. Math. **154**, 971–979 (2006)
12. Hassin, R., Rubinstein, S.: Erratum to "an approximation algorithm for maximum triangle packing". Discrete Appl. Math. **154**, 2620 (2006)
13. Hurkens, C.A.J., Schrijver, A.: On the size of systems of sets every t of which have an SDR, with an application to the worst-case ratio of heuristics for packing problems. SIAM J. Discrete Math. **2**, 68–72 (1989)
14. Kann, V.: Maximum bounded 3-dimensional matching is MAX SNP-complete. Inform. Process. Lett. **37**, 27–35 (1991)
15. Manic, G., Wakabayashi, Y.: Packing triangles in low degree graphs and indifference graphs. Discrete Math. **308**, 1455–1471 (2008)
16. van Rooij, J.M.M., van Kooten Niekerk, M.E., Bodlaender, H.L.: Partition into triangles on bounded degree graphs. Theory Comput. Syst. **52**, 687–718 (2013)
17. van Zuylen, A.: Deterministic approximation algorithms for the maximum traveling salesman and maximum triangle packing problems. Discrete Appl. Math. **161**, 2142–2157 (2013)

# Approximation Algorithms for Maximally Balanced Connected Graph Partition

Yong Chen[1], Zhi-Zhong Chen[2], Guohui Lin[3]([✉]), Yao Xu[4], and An Zhang[1]

[1] Department of Mathematics, Hangzhou Dianzi University, Hangzhou, China
{chenyong,anzhang}@hdu.edu.cn
[2] Division of Information System Design, Tokyo Denki University, Saitama, Japan
zzchen@mail.dendai.ac.jp
[3] Department of Computing Science, University of Alberta, Edmonton, Canada
guohui@ualberta.ca
[4] Department of Computer Science, Kettering University, Flint, MI, USA
yxu@kettering.edu

**Abstract.** Given a simple connected graph $G = (V, E)$, we seek to partition the vertex set $V$ into $k$ non-empty parts such that the subgraph induced by each part is connected, and the partition is maximally balanced in the way that the maximum cardinality of these $k$ parts is minimized. We refer this problem to as *min-max balanced connected graph partition* into $k$ parts and denote it as $k$-BGP. The general vertex-weighted version of this problem on trees has been studied since about four decades ago, which admits a linear time exact algorithm; the vertex-weighted 2-BGP and 3-BGP admit a 5/4-approximation and a 3/2-approximation, respectively; but no approximability result exists for $k$-BGP when $k \geq 4$, except a trivial $k$-approximation. In this paper, we present another 3/2-approximation for our cardinality 3-BGP and then extend it to become a $k/2$-approximation for $k$-BGP, for any constant $k \geq 3$. Furthermore, for 4-BGP, we propose an improved 24/13-approximation. To these purposes, we have designed several local improvement operations, which could be useful for related graph partition problems.

**Keywords:** Graph partition · Induced subgraph · Connected component · Local improvement · Approximation algorithm

## 1 Introduction

We study the following graph partition problem: given a connected graph $G = (V, E)$, we want to partition the vertex set $V$ into $k$ non-empty parts denoted as $V_1, V_2, \ldots, V_k$ such that the subgraph $G[V_i]$ induced by each part $V_i$ is connected, and the cardinalities (or called sizes) of these $k$ parts, $|V_1|, |V_2|, \ldots, |V_k|$, are maximally balanced in the way that the maximum cardinality is minimized. We call this problem as *min-max Balanced connected Graph k-Partition* and denote it as $k$-BGP for short. $k$-BGP and several closely related problems with

© Springer Nature Switzerland AG 2019
Y. Li et al. (Eds.): COCOA 2019, LNCS 11949, pp. 130–141, 2019.
https://doi.org/10.1007/978-3-030-36412-0_11

various applications (in image processing, clustering, computational topology, information and library processing, to name a few) have been investigated in the literature.

Dyer and Frieze [6] proved the NP-hardness for $k$-BGP on bipartite graphs, for any fixed $k \geq 2$. When the objective is to maximize the minimum cardinality, denoted as MAX-MIN $k$-BGP, Chlebíková [5] proved its NP-hardness on bipartite graphs (again), and that for any $\epsilon > 0$ it is NP-hard to approximate the maximum within an absolute error guarantee of $|V|^{1-\epsilon}$. Chataigner et al. [2] proved further the strong NP-hardness for MAX-MIN $k$-BGP on $k$-connected graphs, for any fixed $k \geq 2$, and that unless P = NP, there is no $(1+\epsilon)$-approximation algorithm for MAX-MIN 2-BGP problem, where $\epsilon \leq 1/|V|^2$; and they showed that when $k$ is part of the input, the problem, denoted as MAX-MIN BGP, cannot be approximated within 6/5 unless P = NP.

When the vertices are non-negatively weighted, the weight of a part is the total weight of the vertices inside, and the objective of vertex-weighted $k$-BGP (vertex-weighted MAX-MIN $k$-BGP, respectively) becomes to minimize the maximum (maximize the minimum, respectively) weight of the $k$ parts. The vertex weighted $k$-BGP problem is also called the *minimum spanning $k$-forest* problem in the literature. Given a vertex-weighted connected graph $G = (V, E)$, a *spanning $k$-forest* is a collection of $k$ trees $T_1, T_2, \ldots, T_k$, such that each tree is a subgraph of $G$ and every vertex of $V$ appears in exactly one tree. The weight of the spanning $k$-forest $\{T_1, T_2, \ldots, T_k\}$ is defined as the maximum weight of the $k$ trees, and the weight of the tree $T_i$ is measured as the total weight of the vertices in $T_i$. The objective of this problem is to find a *minimum* weight spanning $k$-forest of $G$. The equivalence between these two problems is seen by the fact that a spanning tree is trivial to compute for a connected graph. The minimum spanning $k$-forest problem is defined on general graphs, but was studied only on trees in the literature [1,7,8,11], which admits an $O(|V|)$-time exact algorithm.

Not too many positive results from approximation algorithms perspective exist in the literature. Chlebíková [5] gave a tight 4/3-approximation algorithm for the vertex-weighted MAX-MIN 2-BGP problem; Chataigner et al. [2] proposed a 2-approximation algorithm for vertex-weighted MAX-MIN 3-BGP on 3-connected graphs, and a 2-approximation algorithm for vertex-weighted MAX-MIN 4-BGP on 4-connected graphs. Approximation algorithms for the vertex-weighted $k$-BGP problem on some *special classes* of graphs can be found in [13–15]. Recently, on general vertex-weighted graphs, Chen et al. [3] showed that the algorithm by Chlebíková [5] is also a 5/4-approximation algorithm for the vertex-weighted 2-BGP problem; and they presented a 3/2-approximation algorithm for the vertex-weighted 3-BGP problem and a 5/3-approximation algorithm for the vertex-weighted MAX-MIN 3-BGP problem.

Motivated by an expensive computation performed by the computational topology software RIVET [9], Madkour et al. [10] introduced the edge-weighted variant of the $k$-BGP problem, denoted as $k$-EBGP. Given an edge non-negatively weighted connected graph $G = (V, E)$, the weight of a tree subgraph $T$ of $G$ is measured as the total weight of the edges in $T$, and the weight of a spanning $k$-forest $\{T_1, T_2, \ldots, T_k\}$ is defined as the maximum weight among the $k$ trees. The $k$-EBGP problem is to find a *minimum* weight spanning $k$-forest of $G$, and it can be re-stated as asking for a partition of the vertex set $V$ into $k$ non-empty parts $V_1, V_2, \ldots, V_k$ such that for each part $V_i$ the induced subgraph $G[V_i]$ is connected and its weight is measured as the weight of the minimum spanning tree of $G[V_i]$, with the objective to minimize the maximum weight of the $k$ parts. Madkour et al. [10] showed that the $k$-EBGP problem is NP-hard on general graphs for any fixed $k \geq 2$, and proposed two $k$-approximation algorithms. Vaishali et al. [12] presented an $O(k|V|^3)$-time exact algorithm when the input graph is a tree, and proved that the problem remains NP-hard on edge uniformly weighted (or unweighted) graphs. It follows that our $k$-BGP problem is NP-hard (again), for any fixed $k \geq 2$. However, the two $k$-approximation algorithms for $k$-EBGP do not trivially work for our $k$-BGP problem.

This paper focuses on designing approximation algorithms for the vertex uniformly weighted (or unweighted) $k$-BGP problem for a fixed $k \geq 4$, *i.e.*, to minimize the maximum cardinality of the $k$ parts in a partition. One can probably easily see a trivial $k$-approximation algorithm, since the maximum cardinality is always at least one $k$-th of the order of the input graph. We remark that the $3/2$-approximation algorithm for the vertex-weighted 3-BGP problem by Chen et al. [3] could not be extended trivially for $k$-BGP for $k \geq 4$. After some preliminaries introduced in Sect. 2, we present in Sect. 3 another $3/2$-approximation algorithm for 3-BGP based on two intuitive local improvement operations, and extend it to become a $k/2$-approximation algorithm for $k$-BGP, for any fixed $k \geq 4$. In Sect. 4, we introduce several complex local improvement operations for 4-BGP, and use them to design a $24/13$-approximation algorithm. We conclude the paper in Sect. 5.

## 2    Preliminaries

Recall that the $k$-BGP problem seeks for a partition of the vertex set $V$ of the given connected graph $G = (V, E)$ into $k$ non-empty subsets $V_1, V_2, \ldots, V_k$ such that $G[V_i]$ is connected for every $i = 1, 2, \ldots, k$, and $\max_{1 \leq i \leq k} |V_i|$ is minimized. For convenience, we call $\max_{1 \leq i \leq k} |V_i|$ the *size* of the partition $\{V_1, V_2, \ldots, V_k\}$. In the rest of the paper, when we know these cardinalities, we always assume they are sorted into $0 < |V_1| \leq |V_2| \leq \ldots \leq |V_k|$, and thus the size of the partition is $|V_k|$.

For two partitions $\{V_1, V_2, \ldots, V_k\}$ and $\{V_1', V_2', \ldots, V_k'\}$, if their sizes $|V_k'| < |V_k|$, or if $|V_k'| = |V_k|$ and $|V_{k-1}'| < |V_{k-1}|$, then we say the partition $\{V_1', V_2', \ldots, V_k'\}$ is *better* than the partition $\{V_1, V_2, \ldots, V_k\}$.

For any two disjoint subsets $V_1, V_2 \subset V$, $E(V_1, V_2) \subseteq E$ denotes the edge subset between $V_1$ and $V_2$; if $E(V_1, V_2) \neq \emptyset$, then we say $V_1$ and $V_2$ are *adjacent*. If additionally both $G[V_1]$ and $G[V_2]$ are connected, then we also say $G[V_1]$ and $G[V_2]$ are *adjacent*.[1]

---

[1] Basically, we reserve the word "connected" for a graph and the word "adjacent" for two objects with at least one edge between them.

We note that obtaining an initial feasible partition of $V$ is trivial in $O(|V| + |E|)$ time, as follows: one first constructs a spanning tree $T$ of $G$, then arbitrarily removes $k - 1$ edges from $T$ to produce a forest of $k$ trees $T_1, T_2, \ldots, T_k$, and lastly sets $V_i$ to be the vertex set of $T_i$. The following approximation algorithms all start with a feasible partition and iteratively apply some local improvement operations to improve it. For $k = 3$, there are only two intuitive local improvement operations and the performance analysis is relatively simple; for $k = 4$, we introduce several more local improvement operations and the performance analysis is more involved, though the key ideas in the design and analysis remain intuitive.

Given a connected graph $G = (V, E)$, let $n = |V|$ denote its order. Let OPT denote the size of an optimal $k$-part partition of the vertex set $V$. The following lower bound on OPT is trivial, and thus the $k$-BGP problem admits a trivial $k$-approximation.

**Lemma 1.** *Given a connected graph $G = (V, E)$, OPT $\geq \frac{1}{k}n$.*

## 3   A $k/2$-Approximation for $k$-BGP, for a Fixed $k \geq 3$

We consider first $k = 3$, and let $\{V_1, V_2, V_3\}$ denote an initial feasible tripartition (with $|V_1| \leq |V_2| \leq |V_3|$). Our goal is to reduce the cardinality of $V_3$ to be no larger than $\frac{1}{2}n$. It will then follow from Lemma 1 that the achieved tripartition is within $\frac{3}{2}$ of the optimum.

Recently, Chen et al. [3] presented a 3/2-approximation algorithm for the vertex-weighted 3-BGP problem, by noticing that a feasible tripartition "cuts" into at most two blocks (that is, maximal 2-connected components) in the input graph. It is surely a 3/2-approximation algorithm for our vertex unweighted 3-BGP problem too, but no better analysis can be achieved since the algorithm (re-)assigns weights to the cut vertices. Furthermore, it is noted by the authors that the algorithm cannot be extended trivially for $k$-BGP for $k \geq 4$, for which one has to deal with vertex-weighted graphs having exactly three blocks.

Our new 3/2-approximation algorithm for 3-BGP, denoted as APPROX-3 and detailed in the following, does not deal with blocks, and it can be extended to become a $k/2$-approximation for $k$-BGP for any fixed $k \geq 4$.

Clearly, during the execution of the algorithm APPROX-3, if $|V_3| \leq \frac{1}{2}n$, then we may terminate and return the achieved tripartition; otherwise, we will execute one of the two local improvement operations called *Merge* and *Pull*, defined in the following, whenever applicable.

Since the input graph $G$ is connected, for any feasible tripartition $\{V_1, V_2, V_3\}$, $V_3$ is adjacent to at least one of $V_1$ and $V_2$.

**Definition 1.** *Operation* Merge$(V_1, V_2)$:

- precondition: $|V_3| > \frac{1}{2}n$; $V_1$ *and* $V_2$ *are adjacent;*
- effect: *the operation produces a new tripartition* $\{V_1 \cup V_2, V_{31}, V_{32}\}$, *where* $\{V_{31}, V_{32}\}$ *is an arbitrary feasible bipartition of* $V_3$.

**Lemma 2.** *Given a connected graph $G = (V, E)$ and a tripartition $\{V_1, V_2, V_3\}$ with $|V_3| > \frac{1}{2}n$, the achieved partition by the operation Merge($V_1, V_2$) is feasible and better.*

*Proof.* Note from the precondition of the operation Merge($V_1, V_2$) that the size of the new part $V_1 \cup V_2$ is $|V_1| + |V_2| < \frac{1}{2}n < |V_3|$; the sizes of the other two new parts $V_{31}$ and $V_{32}$ partitioned from $V_3$ are clearly strictly less than $|V_3|$. This proves the lemma.    □

**Definition 2.** *Operation Pull($U \subset V_3, V_i$), where $i \in \{1, 2\}$,*

- *precondition: $|V_3| > \frac{1}{2}n$; both $G[U]$ and $G[V_3 \setminus U]$ are connected, $U$ is adjacent to $V_i$, and $|V_i| + |U| < |V_3|$;*
- *effect: the operation produces a new tripartition $\{V_3 \setminus U, V_i \cup U, V_{3-i}\}$.*

**Lemma 3.** *Given a connected graph $G = (V, E)$ and a tripartition $\{V_1, V_2, V_3\}$ with $|V_3| > \frac{1}{2}n$, the achieved partition by the operation Pull($U \subset V_3, V_i$) is feasible and better.*

**Lemma 4.** *Given a connected graph $G = (V, E)$, when none of the Merge and Pull operations is applicable to the tripartition $\{V_1, V_2, V_3\}$ with $|V_3| > \frac{1}{2}n$,*

*(1) $|V_1| + |V_2| < \frac{1}{2}n$ (and thus $|V_1| < \frac{1}{4}n$); $V_1$ and $V_2$ aren't adjacent (and thus both are adjacent to $V_3$);*
*(2) let $(u, v) \in E(V_3, V_1)$; then $G[V_3 \setminus \{u\}]$ is disconnected; suppose $G[V_{31}^u], G[V_{32}^u], \ldots, G[V_{3\ell}^u]$ are the components in $G[V_3 \setminus \{u\}]$, then for every $i$, $|V_{3i}^u| \le |V_1|$, and $V_{3i}^u$ and $V_1$ aren't adjacent;*
*(3) no vertex of $V_1 \cup V_2$ is adjacent to any vertex of $V_3$ other than $u$.*

*Proof.* See for an illustration in Fig. 1.

From $|V_3| > \frac{1}{2}n$, we know $|V_1| + |V_2| < \frac{1}{2}n$ and thus $|V_1| < \frac{1}{4}n$. Since no Merge operation is possible, $V_1$ and $V_2$ aren't adjacent and consequently they both are adjacent to $V_3$. This proves Item (1).

Item (2) can be proven similarly as Lemma 3. If $G[V_3 \setminus \{u\}]$ were connected, then it would enable the operation Pull($\{u\} \subset V_3, V_1$), assuming non-trivially $n \ge 5$; secondly, if $|V_{3i}^u| > |V_1|$ for some $i$, then it would enable the operation Pull($V_3 \setminus V_{3i}^u \subset V_3, V_1$), since $|V_3 \setminus V_{3i}^u| + |V_1| < |V_3|$; lastly, if $V_{3i}^u$ and $V_1$ were adjacent for some $i$, then it would enable the operation Pull($V_{3i}^u \subset V_3, V_1$), since $|V_{3i}^u| + |V_1| \le 2|V_1| < \frac{1}{2}n < |V_3|$. This proves the item.

For Item (3), the above item (2) says that $u$ is the only vertex to which a vertex of $V_1$ can possibly be adjacent. Recall that $V_2$ and $V_3$ are adjacent; we want to prove that for every $i$, $V_{3i}^u$ and $V_2$ aren't adjacent. Assume $V_2$ is adjacent to $V_{3i}^u$ for some $i$. Then, due to $|V_{3i}^u| \le |V_1|$, we have $|V_{3i}^u| + |V_2| \le |V_1| + |V_2| < \frac{1}{2}n < |V_3|$, suggesting an operation Pull($V_{3i}^u \subset V_3, V_2$) is applicable, a contradiction. That is, $u$ is the only vertex to which a vertex of $V_2$ can possibly be adjacent.

This proves the lemma.    □

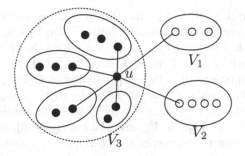

**Fig. 1.** An illustration of the connectivity configuration of the graph $G = (V, E)$, with respect to the tripartition $\{V_1, V_2, V_3\}$ and $|V_3| > \frac{1}{2}n$, on which no Merge or Pull operation is applicable.

From Lemmas 2–4, we can design an algorithm, denoted as APPROX-3, to first compute in $O(|V| + |E|)$ time an initial feasible tripartition of the vertex set $V$ to the 3-BGP problem; we then apply the operations Merge and Pull to iteratively reduce the size of the tripartition, until either this size is no larger than $\frac{1}{2}n$ or none of the two operations is applicable. The final achieved tripartition is returned as the solution. See Fig. 2 for a high-level description of the algorithm APPROX-3. We thus conclude with Theorem 1.

---

The algorithm APPROX-3 for 3-BGP on graph $G = (V, E)$:

**Step 1.** Construct the initial feasible tripartition $\{V_1, V_2, V_3\}$ of $V$;
**Step 2.** while $|V_3| > \frac{1}{2}n$, using Lemma 4,
  if a Merge or a Pull operation is applicable, then update the tripartition;
**Step 3.** return the final tripartition $\{V_1, V_2, V_3\}$.

---

**Fig. 2.** A high-level description of the algorithm APPROX-3 for 3-BGP.

**Theorem 1.** *The algorithm* APPROX-3 *is an* $O(|V||E|)$*-time* $\frac{3}{2}$*-approximation for the* 3-BGP *problem, and the ratio* $\frac{3}{2}$ *is tight for the algorithm.*

*Proof.* Note that in order to apply a Pull operation using Lemma 4, one can execute a graph traversal on $G[V_3 \setminus \{u\}]$ to determine whether it is connected, and if not, to explore all its connected components. Such a graph traversal can be done in $O(|V| + |E|)$ time. A merge operation is also done in $O(|V| + |E|)$ time. The total number of Merge and Pull operations executed in the algorithm is in $O(|V|)$. Therefore, the total running time of the algorithm APPROX-3 is in $O(|V||E|)$.

At termination, if $|V_3| \leq \frac{1}{2}n$, then by Lemma 1 we have $\frac{|V_3|}{\text{OPT}} \leq \frac{3}{2}$.

If $|V_3| > \frac{1}{2}n$, then $|V_1|+|V_2| < \frac{1}{2}n$ and thus $|V_1| < \frac{1}{4}n$, suggesting by Lemma 2 that $V_1$ and $V_2$ aren't adjacent. Therefore, both $V_1$ and $V_2$ are adjacent to $V_3$. By Lemma 4, let $u$ denote the unique vertex of $V_3$ to which the vertices of $V_1 \cup V_2$ can be adjacent. We conclude from Lemma 4 that $G[V_3 \setminus \{u\}]$ is disconnected, there are at least two components in $G[V_3 \setminus \{u\}]$ denoted as $G[V_{31}^u], G[V_{32}^u], \ldots, G[V_{3\ell}^u]$ ($\ell \geq 2$), such that for each $i$, $V_{3i}^u$ is not adjacent to $V_1$ or $V_2$ and $|V_{3i}^u| \leq |V_1|$. That is, $G = (V, E)$ has a very special "star"-like structure, in that these $\ell + 2$ vertex subsets $V_1, V_2, V_{31}^u, V_{32}^u, \ldots, V_{3\ell}^u$ are pairwise non-adjacent to each other, but they all are adjacent to the vertex $u$. Clearly, in an optimal tripartition, the part containing the vertex $u$ has its size at least $|V_3|$, suggesting the optimality of the achieved partition $\{V_1, V_2, V_3\}$.

For the tightness, one can consider a simple path of order 12: $v_1$-$v_2$-$v_3$-$\cdots\cdots$-$v_{11}$-$v_{12}$, on which the algorithm APPROX-3 may terminate at a tripartition of size 6, while an optimal tripartition has size 4. This proves the theorem.     □

**Theorem 2.** *The $k$-BGP problem admits an $O(|V||E|)$-time $\frac{k}{2}$-approximation, for any constant $k \geq 3$.*

*Proof.* Notice that we may apply the algorithm APPROX-3 on the input graph $G = (V, E)$ to obtain a tripartition $\{V_1, V_2, V_3\}$ of the vertex set $V$, with $|V_1| \leq |V_2| \leq |V_3|$.

If $|V_3| \leq \frac{1}{2}n$, then we may continue on to further partition the largest existing part into two smaller parts iteratively, resulting in a $k$-part partition in which the size of the largest part is no larger than $\frac{1}{2}n$ (less than $\frac{1}{2}n$ when $k \geq 4$).

If $|V_3| > \frac{1}{2}n$, then let $u$ be the only vertex of $V_3$ to which the vertices of $V_1 \cup V_2$ can be adjacent; that is, $G[V \setminus \{u\}]$ is disconnected, there are $\ell \geq 4$ connected components in $G[V \setminus \{u\}]$ (see Fig. 1), each is adjacent to $u$ and the largest (which is $G[V_2]$) has size less than $\frac{1}{2}n$ (all the others have sizes less than $\frac{1}{4}n$). When $k \leq \ell$, we can achieve a $k$-part partition by setting the $k - 1$ largest components to be the $k-1$ parts, and all the other components together with $u$ to be the last part. Such a partition has size no greater than $\max\{|V_2|, \text{OPT}\}$, since in an optimal $k$-part partition the part containing the vertex $u$ is no smaller than the last constructed part. When $k > \ell$, we can start with the $\ell$-part partition obtained as above to further partition the largest existing part into two smaller parts iteratively, resulting in a $k$-part partition in which the size of the largest part is less than $\frac{1}{2}n$.

In summary, we either achieve an optimal $k$-part partition or achieve a $k$-part partition in which the size of the largest part is no greater than $\frac{1}{2}n$. Using the lower bound in Lemma 1, this is a $\frac{k}{2}$-approximation.

Running APPROX-3 takes $O(|V||E|)$ time; the subsequent iterative bipartitioning needs only $O(|E|)$ per iteration. Therefore, the total running time is still in $O(|V||E|)$, since $k \leq |V|$.

Lastly, we remark that in the above proof, when $k \geq 4$, if the achieved $k$-part partition is not optimal, then its size is less than $\frac{1}{2}n$. That is, when $k \geq 4$, the ratio $\frac{k}{2}$ is not tight for the algorithm.     □

# 4   A 24/13-approximation for 4-BGP

Theorem 2 states that the 4-BGP problem admits a 2-approximation. In this section, we design a better $\frac{24}{13}$-approximation, which uses three more local improvement operations besides the similarly defined Merge and Pull operations. Basically, these three new operations each finds a subset of the largest two parts, respectively, to merge them into a new part.

Let $\{V_1, V_2, V_3, V_4\}$ denote an initial feasible tetrapartition, with $|V_1| \leq |V_2| \leq |V_3| \leq |V_4|$. Note that these four parts must satisfy some adjacency constraints due to $G$ being connected. We try to reduce the size of $V_4$ to be no larger than $\frac{2}{5}n$, whenever possible; or otherwise we will show that the achieved partition is a $\frac{24}{13}$-approximation. We point out a major difference from 3-BGP, that the two largest parts $V_3$ and $V_4$ in a tetrapartition can both be larger than the desired bound of $\frac{2}{5}n$. Therefore, we need new local improvement operations.

In the following algorithm denoted as APPROX-4, if $|V_4| \leq \frac{2}{5}n$, then we may terminate and return the achieved tetrapartition; it follows from Lemma 1 that the achieved tetrapartition is within $\frac{8}{5}$ of the optimum. Otherwise, the algorithm will execute one of the following local improvement operations whenever applicable.

The first two local improvement operations are similar to the Merge and Pull operations in APPROX-3, except that they now deal with more cases.

**Definition 3.** *Operation* Merge($V_i, V_j$), *for some* $i, j \in \{1, 2, 3\}$:

- precondition: $|V_4| > \frac{2}{5}n$; $V_i$ and $V_j$ are adjacent, and $|V_i| + |V_j| < |V_4|$;
- effect: the operation produces a new tetrapartition $\{V_i \cup V_j, V_{6-i-j}, V_{41}, V_{42}\}$, where $\{V_{41}, V_{42}\}$ is an arbitrary feasible bipartition of $V_4$.

**Definition 4.** *Operation* Pull($U \subset V_j, V_i$), *for some pair* $(i, j) \in \{(1, 3), (1, 4), (2, 3), (2, 4), (3, 4)\}$,

- precondition: $|V_4| > \frac{2}{5}n$ and no Merge operation is applicable; both $G[U]$ and $G[V_j \setminus U]$ are connected, $U$ is adjacent to $V_i$, and $|V_i| + |U| < |V_j|$;
- effect: the operation produces a new tetrapartition $\{V_j \setminus U, V_i \cup U, V_a, V_b\}$, where $a, b \in \{1, 2, 3, 4\} \setminus \{i, j\}$.

**Lemma 5.** (Structure Properties) *Given a connected graph* $G = (V, E)$, *when none of the* Merge *and* Pull *operations is applicable to the tetrapartition* $\{V_1, V_2, V_3, V_4\}$ *with* $|V_4| > \frac{2}{5}n$,

(1) $|V_1| < \frac{1}{5}n$, $|V_2| < \frac{3}{10}n$, $|V_1| + |V_2| < \frac{2}{5}n$, and $V_1$ and $V_2$ aren't adjacent;
(2) if $V_i$ and $V_3$ are adjacent, for some $i \in \{1, 2\}$, then $|V_i| + |V_3| \geq |V_4|$;
(3) if $V_i$ and $V_4$ are adjacent for some $i \in \{1, 2\}$, and there is an edge $(u, v) \in E(V_4, V_i)$, then $G[V_4 \setminus \{u\}]$ is disconnected, every component $G[V_{4\ell}^u]$ in $G[V_4 \setminus \{u\}]$ has its order $|V_{4\ell}^u| \leq |V_i|$, and $V_{4\ell}^u$ and $V_i$ aren't adjacent; furthermore, if $V_{4\ell}^u$ and $V_3$ are adjacent, then $|V_{4\ell}^u \cup V_3| \geq |V_4|$;
(4) if $V_i$ and $V_3$ are adjacent for some $i \in \{1, 2\}$, $|V_i| < \frac{1}{3}|V_4|$, and there is an edge $(v, u) \in E(V_3, V_i)$, then $G[V_3 \setminus \{v\}]$ is disconnected, every component $G[V_{3\ell}^v]$ in $G[V_3 \setminus \{v\}]$ has its order $|V_{3\ell}^v| \leq |V_i|$, and $V_{3\ell}^v$ and $V_i$ aren't adjacent;

(5) if $|V_2| \geq \frac{1}{6}|V_4|$, then the partition $\{V_1, V_2, V_3, V_4\}$ is a $\frac{24}{13}$-approximation; otherwise, we have

$$\begin{cases} |V_2| + |V_3| \geq |V_4|, \\ \quad\quad |V_4| < \frac{1}{2}n, \\ |V_1| \leq |V_2| < \frac{1}{6}|V_4| < \frac{1}{12}n, \\ \quad\quad\quad |V_3| > \frac{1}{3}n; \end{cases} \tag{1}$$

(6) if both $V_1$ and $V_2$ are adjacent to $V_j$ for some $j \in \{3, 4\}$, then the vertices of $V_1 \cup V_2$ can be adjacent to only one vertex of $V_j$.

The readers are referred to our arXiv submission [4] for the detailed proof, and the proofs of the other lemmas and theorems.

**Proposition 1.** *In the following, we distinguish three cases of the tetrapartition* $\{V_1, V_2, V_3, V_4\}$ *with* $|V_4| > \frac{2}{5}n$, *to which none of the* Merge *and* Pull *operations is applicable, and Eq. (1) holds:*

Case 1: *none of* $V_1$ *and* $V_2$ *is adjacent to* $V_3$ *(i.e., both* $V_1$ *and* $V_2$ *are adjacent to* $V_4$ *only and at the vertex* $u \in V_4$ *only; see for an illustration in Fig. 3, to be handled in Theorems 3 and 4);*
Case 2: *none of* $V_1$ *and* $V_2$ *is adjacent to* $V_4$;
Case 3: *one of* $V_1$ *and* $V_2$ *is adjacent to* $V_3$ *and the other is adjacent to* $V_4$.

*The final conclusion is presented as Theorem 5.*

Lemma 5 states several structural properties of the graph $G = (V, E)$ with respect to the tetrapartition, which is yet unknown to be a $\frac{24}{13}$-approximation or not. For Case 1 listed in Proposition 1, Lemma 5 leads to a further conclusion, stated in Theorems 3.

**Theorem 3.** *In Case 1, let* $V_4'$ *denote the union of the vertex sets of all the components of* $G[V_4 \setminus \{u\}]$ *that are adjacent to* $V_3$; *if* $|V_4'| \leq |V_1| + |V_2| + \frac{11}{24}|V_4|$, *then the partition* $\{V_1, V_2, V_3, V_4\}$ *is a* $\frac{24}{13}$-approximation.

We have seen that $G[V_4]$ exhibits a nice star-like configuration (Fig. 1), due to $V_4$ being adjacent to $V_1$ and $V_2$. Since none of $V_1$ and $V_2$ is adjacent to $V_3$ in Case 1, the connectivity configuration of $G[V_3]$ is unclear. We next bipartition $V_3$ as evenly as possible, and let $\{V_{31}, V_{32}\}$ denote the achieved bipartition with $|V_{31}| \leq |V_{32}|$. If $|V_{32}| \leq \frac{2}{3}|V_3|$, and assuming there are multiple components of $G[V_4 \setminus \{u\}]$ adjacent to $V_{3i}$ (for some $i \in \{1, 2\}$) with their total size greater than $|V_1|$, then we find a minimal sub-collection of these components of $G[V_4 \setminus \{u\}]$ adjacent to $V_{3i}$ with their total size exceeding $|V_1|$, denote by $V_4'$ the union of their vertex sets, and subsequently create three new parts $V_4 \cup V_1 \setminus V_4'$, $V_4' \cup V_{3i}$, and $V_{3,3-i}$, while keeping $V_2$ unchanged. One sees that this new tetrapartition is feasible and better, since $|V_4'| + |V_{3i}| \leq 2|V_1| + |V_{3i}| < \frac{1}{3}|V_4| + \frac{2}{3}|V_3| \leq |V_4|$.

In the other case, by Lemma 3, $G[V_3]$ also exhibits a nice star-like configuration centering at some vertex $v$, such that $G[V_3 \setminus \{v\}]$ is disconnected and each component of $G[V_3 \setminus \{v\}]$ has size less than $\frac{1}{3}|V_3|$. See for an illustration in Fig. 3.

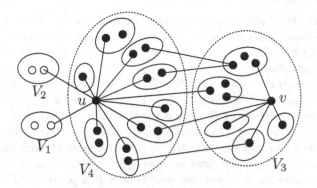

**Fig. 3.** An illustration of the "bi-star"-like configuration of the graph $G = (V, E)$, with respect to the tetrapartition $\{V_1, V_2, V_3, V_4\}$ in Case 1.

The following *Bridge-1* operation aims to find a subset $V_3' \subset V_3$ and a subset $V_4' \subset V_4$ to form a new part larger than $V_1$, possibly cutting off another subset $V_4''$ from $V_4$ and merging it into $V_3$, and merging the old part $V_1$ into $V_4$. This way, a better tetrapartition is achieved. We prove later that when such a bridging operation isn't applicable, each component in the residual graph by deleting the two star centers has size at most $2|V_1| + |V_2|$, and subsequently the tetrapartition can be shown to be a $\frac{12}{7}$-approximation.

**Definition 5.** *Operation* Bridge-1$(V_3, V_4)$:

- precondition: *In Case 1, there are multiple components of $G[V_4 \setminus \{u\}]$ adjacent to $V_3$ with their total size greater than $|V_1| + |V_2| + \frac{11}{24}|V_4|$, and there is a vertex $v \in V_3$ such that $G[V_3 \setminus \{v\}]$ is disconnected and each component has size less than $\frac{1}{3}|V_3|$.*
- effect: *Find a component $G[V_{4x}^u]$ of $G[V_4 \setminus \{u\}]$, if exists, that is adjacent to a component $G[V_{3y}^v]$ of $G[V_3 \setminus \{v\}]$; initialize $V_4'$ to be $V_{4x}^u$ and $V_3'$ to be $V_{3y}^v$; iteratively,*
  - *let $C_3$ denote the collection of the components of $G[V_3 \setminus \{v\}]$ that are adjacent to $V_4'$, excluding $V_3'$;*
    - *if the total size of components in $C_3$ exceeds $2|V_1| - |V_3'|$, then the operation greedily finds a minimal sub-collection of these components of $C_3$ with their total size exceeding $2|V_1| - |V_3'|$, adds their vertex sets to $V_3'$, and proceeds to termination;*
    - *if the total size of components in $C_3$ is less than $2|V_1| - |V_3'|$, then the operation adds the vertex sets of all these components to $V_3'$;*
  - *let $C_4$ denote the collection of the components of $G[V_4 \setminus \{u\}]$ that are adjacent to $V_3'$, excluding $V_4'$;*
    - *if the total size of components in $C_4$ exceeds $|V_1| - |V_4'|$, then the operation greedily finds a minimal sub-collection of these components of $C_4$ with their total size exceeding $|V_1| - |V_4'|$, adds their vertex sets to $V_4'$, and proceeds to termination;*

* *if the total size of components in $C_4$ is less than $|V_1| - |V_4'|$, then the operation adds the vertex sets of all these components to $V_4'$;*
- *if both $C_3$ and $C_4$ are empty, then the operation terminates without updating the partition.*

*At termination, exactly one of $|V_3'| > 2|V_1|$ and $|V_4'| > |V_1|$ holds.*

- *When $|V_3'| > 2|V_1|$, we have $|V_4'| \leq |V_1|$ and $|V_3'| < 2|V_1| + \frac{1}{3}|V_3|$;*
  * *if the collection of the components of $G[V_4 \setminus \{u\}]$ that are adjacent to $V_3'$, excluding $V_4'$, exceeds $|V_1| - |V_4'|$, then the operation greedily finds a minimal sub-collection of these components with their total size exceeding $|V_1| - |V_4'|$, and denotes by $V_4''$ the union of their vertex sets; subsequently, the operation creates three new parts $V_4 \cup V_1 \setminus (V_4' \cup V_4'')$, $(V_4' \cup V_4'') \cup V_3'$, and $V_3 \setminus V_3'$;*
  * *otherwise, the operation greedily finds a minimal sub-collection of the components of $G[V_4 \setminus \{u\}]$ that aren't adjacent to $V_3'$ with their total size exceeding $|V_1| - |V_4'|$, and denotes by $V_4''$ the union of their vertex sets; subsequently, the operation creates three new parts $V_4 \cup V_1 \setminus (V_4' \cup V_4'')$, $V_4' \cup V_3'$, and $(V_3 \setminus V_3') \cup V_4''$.*
- *When $|V_4'| > |V_1|$, we have $|V_4'| \leq 2|V_1|$ and $|V_3'| \leq 2|V_1|$; the operation creates three new parts $V_4 \cup V_1 \setminus V_4'$, $V_4' \cup V_3'$, and $V_3 \setminus V_3'$.*
- *In all the above three cases of updating, the part $V_2$ is kept unchanged.*

**Theorem 4.** *In Case 1, if there are multiple components of $G[V_4 \setminus \{u\}]$ adjacent to $V_3$ with their total size greater than $|V_1| + |V_2| + \frac{11}{24}|V_4|$ and the Bridge-1 operation isn't applicable, then the partition $\{V_1, V_2, V_3, V_4\}$ is a $\frac{12}{7}$-approximation.*

Case 2 and Case 3 can be separately dealt similarly. Combining them, we can design an iterative algorithm which applies one of the Merge and the Pull and the three Bridge operations. And we have the following final conclusion for the 4-BGP problem:

**Theorem 5.** *4-BGP admits an $O(|V|^2|E|)$-time $\frac{24}{13}$-approximation algorithm.*

## 5    Conclusions

We studied the $k$-BGP problem to partition the vertex set of a given simple connected graph $G = (V, E)$ into $k$ parts, such that the subgraph induced by each part is connected and the maximum cardinality of these $k$ parts is minimized. The problem is NP-hard, and approximation algorithms were proposed for only $k = 2, 3$. We focus on $k \geq 4$, and present a $k/2$-approximation algorithm for $k$-BGP, for any fixed $k \geq 3$, and an improved 24/13-approximation for 4-BGP. Along the way, we have designed several intuitive and interesting local improvement operations.

There is no any non-trivial lower bound on the approximation ratio for the $k$-BGP problem, except 6/5 for the problem when $k$ is part of the input. We feel that it could be challenging to design better approximation algorithms for 2-BGP and 3-BGP; but for 4-BGP we believe the parameters in the three

Bridge operations can be adjusted better, though non-trivially, leading to an 8/5-approximation. We leave it open on whether or not $k$-BGP admits an $o(k)$-approximation.

**Acknowledgements.** CY and AZ are supported by the NSFC Grants 11971139, 11771114 and 11571252; and supported by the CSC Grants 201508330054 and 201908330090, respectively. ZZC is supported by in part by the Grant-in-Aid for Scientific Research of the Ministry of Education, Science, Sports and Culture of Japan, under Grant No. 18K11183. GL is supported by the NSERC Canada.

# References

1. Becker, R.I., Schach, S.R., Perl, Y.: A shifting algorithm for min-max tree partitioning. J. ACM **29**, 58–67 (1982)
2. Chataigner, F., Salgado, L.R.B., Wakabayashi, Y.: Approximation and inapproximability results on balanced connected partitions of graphs. Discrete Math. Theor. Comput. Sci. **9**, 177–192 (2007)
3. Chen, G., Chen, Y., Chen, Z.-Z., Lin, G., Liu, T., Zhang, A.: Approximation algorithms for maximally balanced connected graph tripartition problem (2019). Submission under review
4. Chen, Y., Chen, Z.-Z., Lin, G., Xu, Y., Zhang, A.: Approximation algorithms for maximally balanced connected graph partition (2019). arXiv:1910.02470
5. Chlebíková, J.: Approximating the maximally balanced connected partition problem in graphs. Inf. Process. Lett. **60**, 225–230 (1996)
6. Dyer, M.E., Frieze, A.M.: On the complexity of partitioning graphs into connected subgraphs. Discrete Appl. Math. **10**, 139–153 (1985)
7. Frederickson, G.N.: Optimal algorithms for tree partitioning. Proc. SODA **1991**, 168–177 (1991)
8. Frederickson, G.N., Zhou, S.: Optimal parametric search for path and tree partitioning (2017). arXiv:1711.00599
9. Lesnick, M., Wright, M.: Interactive visualization of 2-D persistence modules (2015). arxiv:1512.00180
10. Madkour, A.R., Nadolny, P., Wright, M.: Finding minimal spanning forests in a graph (2017). arXiv:1705.00774
11. Perl, Y., Schach, S.R.: Max-min tree partitioning. J. ACM **28**, 5–15 (1981)
12. Vaishali, S., Atulya, M.S., Purohit, N.: Efficient algorithms for a graph partitioning problem. In: Chen, J., Lu, P. (eds.) FAW 2018. LNCS, vol. 10823, pp. 29–42. Springer, Cham (2018). https://doi.org/10.1007/978-3-319-78455-7_3
13. Wang, L., Zhang, Z., Wu, D., Wu, W., Fan, L.: Max-min weight balanced connected partition. J. Glob. Optim. **57**, 1263–1275 (2013)
14. Wu, B.Y.: A 7/6-approximation algorithm for the max-min connected bipartition problem on grid graphs. In: Akiyama, J., Bo, J., Kano, M., Tan, X. (eds.) CGGA 2010. LNCS, vol. 7033, pp. 188–194. Springer, Heidelberg (2011). https://doi.org/10.1007/978-3-642-24983-9_19
15. Wu, B.Y.: Fully polynomial-time approximation schemes for the max-min connected partition problem on interval graphs. Discrete Math. Algorithms Appl. **4**, 1250005 (2013)

# Edge Exploration of a Graph
# by Mobile Agent

Amit Kumar Dhar[1], Barun Gorain[1(✉)], Kaushik Mondal[2], Shaswati Patra[1], and Rishi Ranjan Singh[1]

[1] Indian Institute of Technology Bhilai, Sejbahar, Raipur, India
{amitkdhar,barun,shaswatip,rishi}@iitbhilai.ac.in
[2] Indian Institute of Information Technology Vadodara, Gandhinagar, India
kaushikmondal85@gmail.com

**Abstract.** In this paper, we study the problem of edge exploration of an $n$ node graph by a mobile agent. The nodes of the graph are unlabeled, and the ports at a node of degree $d$ are arbitrarily numbered $0, \ldots, d-1$. A mobile agent, starting from some node, has to visit all the edges of the graph and stop. The time of the exploration is the number of edges the agent traverses before it stops. The task of exploration can not be performed even for a class of cycles if no additional information, called advice, is provided to the agent a priori. Following the paradigm of algorithms with advice, this priori information is provided to the agent by an Oracle in the form of a binary string. The Oracle knows the graph, but does not have the knowledge of the starting point of the agent. In this paper, we consider the following two problems of edge exploration. The first problem is: "how fast is it possible to explore an $n$ node graph regardless of the size of advice provided to the agent?"

We show a lower bound of $\Omega(n^{\frac{8}{3}})$ on exploration time to answer the above question. Next, we show the existence of an $O(n^3)$ time algorithm with $O(n \log n)$ advice. The second problem then asks the following question: "what is the smallest advice that needs to be provided to the agent in order to achieve time $O(n^3)$?" We show a lower bound $\Omega(n^\delta)$ on size of the advice, for any $\delta < \frac{1}{3}$, to answer the above question.

**Keywords:** Algorithm · Graph · Exploration · Mobile agent · Advice

## 1 Introduction

Exploration of a network by mobile agents is a well studied problem [28] which has various applications like treasure hunt, collecting data from some node in the network or samples from contaminated mines where corridors along with the crossings forms a virtual network. Many real life applications require collection of information from edges of a network as well. In such scenarios, edge explorations are essential to retrieve the required knowledge.

In this paper, we consider the edge exploration problem where a mobile agent, albeit with advice, aims to explore all the edges in a network and stop when

© Springer Nature Switzerland AG 2019
Y. Li et al. (Eds.): COCOA 2019, LNCS 11949, pp. 142–154, 2019.
https://doi.org/10.1007/978-3-030-36412-0_12

done. By *advice* we mean some prior information provided to the agent for the exploration, by an Oracle in the form of a binary string. The length of the string is called the size of advice. We analyze the lower bound on the exploration time with arbitrary size of advice before providing an efficient exploration algorithm.

The network is modeled as a simple connected undirected graph $G = (V, E)$ consisting of $n$ nodes. Nodes are anonymous but all the edges associated to a node of degree $d$ are arbitrarily numbered $0, 1, \cdots, d-1$ at the node. The mobile agent starts from an arbitrary node which we call as the starting node. Before starting the exploration, the agent knows the degree of the starting nodes. When the agent takes the port $i$ at a node $u$ and reaches node $v$, it learns the degree of $v$, and the port of the edge at $v$ through which it reached $v$. The agent does not have the capability to mark any edge or node.

The time of the exploration is the number of edges the agent traverse before it stops. It is evident that some prior information needs to be provided to the agent in order to complete the task of edge exploration. For example, in the class of rings with ports numbered 0,1 in clockwise order at all the nodes, the agent can not learn the size of the ring only by exploring edges if no prior information is provided. Hence, it cannot distinguish between any two oriented rings of different size $k_1, k_2 \geq 3$. Therefore, any exploration algorithm that stops after $t$ steps will fail to explore all the edges a ring with size $t + 2$ or more. In this paper we study the problem of how much knowledge the agent needs to have a priori, in order to explore all the edges of a given graph in given time $t$ by any deterministic algorithm.

Following the paradigm algorithm with advice [5,8], this prior information is provided to the agent by an Oracle. According to the literature [21], there are two kind of Oracles, instance Oracle and map Oracle. The entire instance of the exploration problem, i.e., the port-numbered map of the underlying graph and the starting node of the agent in this map is known by the instance Oracle, where as the map Oracle knows the port-numbered map of the underlying graph but does not know the starting node of the agent. In this work, we consider map Oracle.

Hence to prove possibility of such an exploration, we have to show existence of an exploration algorithm which uses advices of length at most $x$, one for each graph, provided by a map Oracle and explores all the graphs in $\mathscr{G}$ within time $t$ starting from any node. On the other hand, to prove such an exploration is impossible in time $t$ with advice of length $x$, we need to show existence of at least one graph and a starting point, such that no algorithm successfully explores all the edges of this graph within time $t$ with any advice of length at most $x$.

It is natural to investigate the trade-off between exploration time and size of advice for edge exploration. In this paper, we provide two lower bound results, one on exploration time, and another on the size of advice for the edge exploration problem.

## 2   Contribution

Our main result consists of two lower bound results, one on exploration time and the other on size of advice. We prove that it is not possible to complete the task of edge exploration within time $o(n^{\frac{8}{3}})$, regardless of the size of advice. Next, we show the existence of an algorithm which works in time $O(n^3)$ with advice of size $O(n \log n)$. We also show that the minimum size of the advice necessary to explore all the edges of a graph in time $O(n^3)$ is $\Omega(n^\delta)$, for any $\delta < \frac{1}{3}$.

### 2.1   Related Work

Exploration of unknown environments by mobile agents is an extensively studied problem (cf. the survey [28]). We work on a model where the graph is undirected, nodes are anonymous and the mobile agent have some information a priori. Accordingly the mobile agent may traverse in any direction along an edge. The agent either has restricted tank [1] and needs to return to the base for refueling or already attached to the base with a cable of restricted length [9]. Usually in literature, most of the works analyze the time of completing the exploration by measuring the number of edges (counting multiple traversals) the agent traverses. This is considered as the efficiency measure of the algorithms.

Exploring any anonymous graph and to stop when done is impossible due to the anonymities of the nodes. As a solution, agents can have a finite number of *pebbles* [2,3] to drop on nodes which helps recognizing already visited ones or even put a stationary token at the starting node [6,27].

The problem of exploring anonymous graphs without node marking has been studied in several literature [7,16] where the termination condition after exploring all the edges is removed. Hence, in such variation of problems, the number of edge traversal becomes meaningless, instead, finding the minimum memory required for exploration appears to be the key.

For termination after successful exploration, further knowledge about the graph is essential, e.g., an upper bound on its size [6,29]. These information are usually known as advice and the approaches as *algorithms with advice*.

The paradigm of algorithm with advice is also extensively studied for other problems like graph coloring, broadcasting, leader election and topology recognitions where external information is provided to the nodes of a network or a external entity like mobile agent to perform the task efficiently [10–15,17–20,22–24,26].

In [13], the authors studied comparison of advice size in order to solve two information dissemination problems where the number of messages exchanged is linear. In [15], distributed construction of a minimum spanning tree in logarithmic time with constant size advice is discussed. In [10], authors shows that in order to broadcast a message from a source node in a radio network, 2-bits labeling, which also can be view as external advice, to the nodes are sufficient.

The algorithms with advice in the context of online algorithms is studied in [5,8,11]. In [8], online algorithm with advice is considered for a labeled weighted graph. In [25], authors did online exploration assuming upon visiting a node for the first time, the searcher learns all incident edges and their respective traversal costs. In weighted graphs, treasure hunt with advice, which is also a variation of exploration problem, was studied in [25]. Exploration with advice was studied for trees [14] and for general graphs in [21]. In [21] authors have described node exploration of an anonymous graph with advice. Two kind of Oracles are considered in this paper. Map Oracle, that knows the unlabeled graph, but does not know the starting node of the agent, and Instance Oracle, that knows the graph as well as the starting node of the agent. Trade-off between exploration time and size of advice is shown for both type of oracles.

## 3 Lower Bound on Exploration Time

In this section, we give a lower bound on time for edge exploration on a graph. More precisely, we show an exploration time of $\Omega(n^{\frac{8}{3}})$ for exploring all the edges of some graph regardless of the size of advice. To establish the lower bound, we construct an $n$ node graph $\widehat{G}$ such that even if the agent is provided the map of the graph as advice, the time taken by the agent to explore all the edges is $\Omega(n^{\frac{8}{3}})$. The construction of the graph $\widehat{G}$ is given below.

**Construction of $\widehat{G}$:** We use the graphs discussed in [4] as building blocks to construct the graph for the lower bound result. For the sake of completeness, we discuss below the construction of the graphs (discussed in [4]).

Let $m > 0$ be an even positive integer. Let $H$ be a $m$ node regular graph with degree $\frac{m}{2}$. In this case we take $H$ as a complete bipartite graph with the partition $U$ and $V$ of same size. Let $T$ be any spanning tree of $H$ with $E(T)$ being the spanning tree edges. Let $S$ be the set of edges in $H$ which are not in $E(T)$. Let $S = \{e_1, e_2, \cdots, e_s\}$ where $s = \frac{m^2}{4} - m + 1$. Let $X = x_1, x_2, \cdots, x_s$ be a binary string of length $s$, where not all $x_i$'s are zero. A graph $H_X$ is constructed from $H$ using $X$ as follows: take two copies of $H$, say $H_1$ and $H_2$ with the bipartitions $U_1, V_1$ and $U_2, V_2$, respectively. For all $i$, $1 \leq i \leq s$, if $x_i = 1$, then delete the edges $e_i$ from both $H_1$ and $H_2$ and cross two copies of $e_i$ between the corresponding vertices of $H_1$ and $H_2$. More precisely, let $e_i = (u, v)$ be an edge of $H$ with port numbers $p$ at $u$ and port number $q$ at $v$. Let $u_1, v_1$ and $u_2, v_2$ be the nodes corresponding to $u, v$ in $H_1$ and $H_2$, respectively. Delete $(u_1, v_1)$ from $H_1$ and $(u_2, v_2)$ from $H_2$, and connect two edges $(u_1, v_2)$ and $(u_2, v_1)$, (See Figs. 1 and 2). The port numbers of the newly added edges are $p$ at both $u_1, u_2$ and $q$ at both $v_1$ and $v_2$.

According to the result from [4], for each edge $e_i = (u, v) \in S$, there exists some sequence $X_i \in \{0, 1\}^s \setminus \{0\}^s$ such that if any exploration algorithm explores $H$ starting from a node $v_0 \in H$ using a sequence of port numbers $Q$ which traverse the edge $e_i$ less than $s$ times, then at least one of the edges $(u_1, v_2)$ or $(u_2, v_1)$ in $H_{X_i}$ will remains unexplored while exploring $H_{X_i}$ by $Q$. Let $\mathscr{H} = \{H_{X_i} : 1 \leq i \leq s\}$.

We use the above class of graphs $\mathscr{H}$ as building blocks to construct a graph for our lower bound. Our constructed graph will have the property that, if every

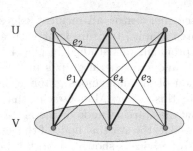

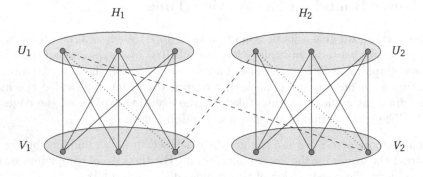

**Fig. 1.** Graph $H$ with spanning tree edges shown in bold and non-tree edges labelled.

**Fig. 2.** The graph $H_X$, for $X = 0100$. The deleted edges are shown with dots and newly added edges with dash.

edge of the graph is not visited at least a fixed number of times, then some edge of the graph remains unexplored.

With the above discussions, we are ready to construct our final graph $\widehat{G}$. The high level idea of the construction is as follows. We will construct $\widehat{G}$ consisting of all the graphs from $\mathscr{H} = \{H_{X_i} : i \in [1, s]\}$ and systematically add some extra edges between every pair of $H_{X_i}$ and $H_{X_j}$.

**Vertices of $\widehat{G}$:** The vertex set $V$ of $\widehat{G}$ consists of all the vertices of all the graphs in $\mathscr{H}$. Hence, $|V| = (\frac{m^2}{4} - m + 1) * 2m = O(m^3)$. For the rest of the construction, we denote the independent sets of $H_{X_i}$ as $H_{X_i}(U_1), H_{X_i}(V_1), H_{X_i}(U_2), H_{X_i}(V_2)$ (as shown in Fig. 3).

**Edges of $\widehat{G}$:** For every $H_{X_i}$ with $i \in [1, s]$, we add the edges of $H_{X_i}$ among the vertices of $H_{X_i}(U_1), H_{X_i}(V_1), H_{X_i}(U_2), H_{X_i}(V_2)$. Let these set of edges be denoted by $E_{i,i}$. For every ordered pair $(i, j)$ such that $1 \le i \ne j \le s$, we add a set of edges to $\widehat{G}$, defined as $E_{i,j}$. For every edge $(u, v)$ with $v \in H_{X_i}(V_k)$ we add the edge $(u, v') \in E_{i,j}$ with $v' \in H_{X_j}(V_k)$ for some $k \in \{1, 2\}$. The edge set $E$ of $\widehat{G}$ is defined as $E = \bigcup_{i,j \in [1,s]} E_{i,j}$. Note that the subgraph of $\widehat{G}$ formed using the edge set $E_{i,j}$ and only the vertex sets $H_{X_i}(U_1), H_{X_i}(U_2), H_{X_j}(V_1), H_{X_j}(V_2)$

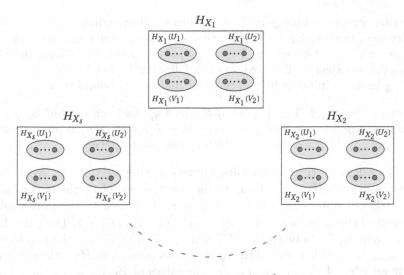

**Fig. 3.** Set of vertices in $\widehat{G}$

is isomorphic to $H_{X_i}$. In figure Fig. 4, we show the edges corresponding to $\bigcup\limits_{j\in[1,s]} E_{1,j}$ that are added to $\widehat{G}$.

**Port Numbers of $\widehat{G}$:** For all ordered pair $(i,j)$, $1 \leq i,j \leq s$, for each edge $e = (u,v) \in H_{X_i}$ with port numbers $(p,q)$, the port number of the corresponding edge in $E_{i,j}$ is $(k\frac{m}{2}+p, k\frac{m}{2}+q)$, where $k = j-i \mod s$.

Let $n$ be the number of nodes of $\widehat{G}$. Then $n = 2sm \geq \frac{m^3}{5}$.

Note that by construction of $H, H_{X_i}$ and $\widehat{G}$ the degree of all the vertices in $\widehat{G}$ is exactly same. Suppose that the agent starts exploration from the node $v_1$ in $H_{X_1}$. Consider the following exploration sequence of the edges by the agent: visit all the ports from $0, 1, \cdots, \frac{m}{2}-1$ attached at $v_1$ one by one, i.e, for every port $i$, visit the edge with port $i$, $0 \leq i \leq \frac{m}{2}-1$ to reach a new vertex; come back to $v_1$ using the last visited edge. The above exploration sequence of edges will visit all the edges of attached with $v_1$ which are corresponding to the edges attached with $v_1$ in $H_{X_i}$. Now, if we change the staring node as the node $v_1$ in $H_{X_2}$, the same exploration sequence visits all the edges of attached with $v_1$ which are corresponding to the edges attached with $v_1$ in $H_{X_2}$. The construction of the graph $\widehat{G}$ guarantees that the agent cannot distinguish between this change in the starting node. In other words, any valid exploration algorithm for a particular graph $\widehat{G}$ gives a sequence of ports to be visited, that will remain same irrespective of the starting point of the algorithm as the agent cannot distinguish between different starting points. Therefore, any such exploration algorithm can be uniquely coded as a sequence of outgoing port numbers and the agent follows the ports according to this sequence in consecutive steps of exploration.

Let $\mathscr{B}$ be an exploration algorithm using which the agent explores all the edges of the graph starting from $v_0 \in H_{X_j}$, for any $j$, $1 \leq j \leq s$. Let $U$ be the exploration sequence of outgoing port numbers corresponding to $\mathscr{B}$. Note that irrespective of the starting node, the sequence of port numbers $U$ must visit all the edges of $\widehat{G}$, i.e., for every $j$, $1 \leq j \leq s$, if the agent starts from the node $v_0 \in H(X_j)$, it explores all the edges of $\widehat{G}$ following $U$. Let $U = q_1, q_2, \cdots, q_w$. Following lemma will be useful to prove our main lower bound result.

**Lemma 1.** *For any $j$, $1 \leq j \leq w$, if the port $q_j$ visits an edge of $E_{x,y}$ when the starting node is $v_0 \in H_{X_i}$, for some $i$, $1 \leq i \leq s$, then $q_j$ visits an edge of $E_{x+t-i \mod s, y+t-i \mod s}$ when the starting node is $v_0 \in H_{X_t}$.*

*Proof.* We will prove this lemma using induction. Suppose that the agent starts from the node $v_0 \in H_{X_i}$. According to the construction of $\widehat{G}$, the port $q_1$ must visit an edge of some $E_{i,y}$, where $1 \leq y \leq s$. According to the port number assignment of the edges of $\widehat{G}$, $(y-i)\frac{m}{2} \leq q_1 \leq (y-i+1)\frac{m}{2} - 1$. The ports from the node $v_0$ in $H(X_t)$ between $(y-i)\frac{m}{2}$ and $(y-i+1)\frac{m}{2} - 1$ are the part of edge set $E_{t,t+y-i \mod s}$. Therefore, if the agent starts from $v_0$ in $H_{X_t}$, then the port $q_1$ visits an edge of $E_{t,t+y-i \mod s}$, i.e., the edges of $E_{t+i-i \mod s, t+y-i \mod s}$. Hence the lemma is true for $j = 1$.

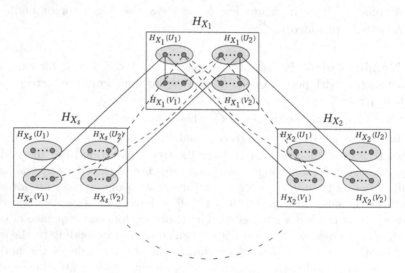

**Fig. 4.** A subset of edges of $\widehat{G}$ ( $\bigcup\limits_{j \in [1,s]} E_{1,j}$)

Suppose that the lemma is true for any integer $j \leq f$. Let $q_f$ be a port between $d\frac{m}{2}$ and $(d+1)\frac{m}{2} - 1$, for some $d$ and $q_f$ visits an edge of $E_{x,y}$ when the starting node is $v_0 \in H_{X_i}$. This implies that before taking the port $q_f$, then agent is in a vertex of $H_{X_x}$ and after taking the port $q_f$, the agent reaches a vertex of $H_{X_z}$, where $z = x + d \mod s$. Therefore, using the induction hypothesis, when the starting node in $v_0 \in H_{X_t}$, before taking the port $q_f$, the agent is in a vertex of $H_{X_{x+t-i} \mod s}$ and after taking the port $q_f$, the agent reaches a vertex of $H_{X_{z'}}$, where $z' = x + t - i + d \mod s$.

Now, lets consider the port $q_{f+1}$ when the agent starts from $v_0 \in H_{X_i}$. Let $d'\frac{m}{2} \le q_{f+1} \le (d'+1)\frac{m}{2} - 1$, for some $d'$. Then $q_{f+1}$ visits an edge of $E_{x',y'}$, where $x' = x + d \mod s$, $y' = x + d + d' \mod s$. Suppose that, when the starting node is $v_0 \in H_{X_t}$, the port $q_{f+1}$ visits an edge of $E_{x'',y''}$. As, in this case, the agent is in a vertex of $H_{X_{z'}}$, where $z' = x + t - i + d \mod s$ and $d'\frac{m}{2} \le q_{f+1} \le (d'+1)\frac{m}{2} - 1$, according to the port assignments of the edges of $\widehat{G}$, $x'' = x+t-i+d \mod s$ and $y'' = x + t - i + d + d' \mod s$, i.e., $x'' = x' + t - i \mod s$ and $y'' = y' + t - i \mod s$. This proves that the lemma is true for $j = f + 1$ and hence the lemma is proved by induction. $\square$

With the above discussion, we are ready to prove our lower bound result.

**Theorem 1.** *Any exploration algorithm using any advice given by a Oracle must take time $\Omega(n^{\frac{8}{3}})$ time to explore all the edges of $\widehat{G}$.*

*Proof.* It is enough to prove the theorem for sufficiently large values of $n$, assuming that the advice given by the Oracle is $\widehat{G}$. Let $\mathscr{B}$ be an exploration algorithm using which the agent explores all the edges of the graph starting from $v_0 \in H_{X_j}$, for any $i$, $1 \le j \le s$. Let $U$ be the exploration sequence of port numbers corresponding to $\mathscr{B}$. Consider the execution of the movement of the agent along the edges of $\widehat{G}$ when the starting node in $v_0 \in H(X_i)$, for some $i$. The sequence of port numbers $U$ can be written as $U = B_1.(p_1).B_2.(p_2).\cdots.B_k.(p_k).B_{k+1}$, where each $B_\ell$ is a sequence of port numbers corresponding to continuous movements of the mobile agent moving according to $\mathscr{B}$ ($B_\ell$ is a sequence involving zero or more port numbers, for each $\ell$) and $p_1, p_2, \cdots, p_k$, are the ports that the agent takes to visit only the edges from $E_{x,y}$. Note that the sequence of port numbers $W = p_1, p_2, \cdots, p_k$ explores all the edges of $E_{x,y}$ in a scattered manner. We can convert $W$ to a continuous sequence of port numbers $W'$ such that $W'$ explores all the edges in $E_{x,y}$ continuously, starting from $v_0 \in H_{X_x}$ as follows. Suppose that $u_1, u_2, \cdots, u_k$ be the vertices from where the agent takes the ports $p_1, p_2, \cdots, p_k$, respectively. Construct $W' = C_1 p_1 C_2 p_2 \cdots C_k p_k$, where $C_1$ is the sequence of port numbers corresponding to a shortest path from $v_0 \in H_{X_x}$ to $u_1$ and $C_\ell$ is the sequence of port numbers corresponding to a shortest path from the node the agent reached after taking the port $p_\ell$ (say, $v$), to $u_{\ell+1}$. Let $W''$ be the sequence of port number which is constructed from $W'$ such that for each $i$, the value of the $i$-th port of $W''$ is assigned as the value of the $i$-th port of $W' \mod \frac{m}{2}$. Then $W''$ is an exploration sequence for $H_{X_x}$. Since degree of each node in the subgraph $H_{X_x}$ induced by the edge set $E_{X,Y}$ is $\frac{m}{2}$, the length of each of the $C_\ell$'s are constant.

Hence $|W''| \in O(k)$. Also, since $W''$ is an exploration sequence of $H_{X_x}$, $W''$ must be an exploration sequence of $H$ as well.

**Claim:** $k \in O(s^2)$.

We prove the above claim by showing that the exploration sequence $W''$ working on the graph $H$ must visit every edge $e_1, \cdots e_s$ at least $s$ times. Suppose that $W''$ visits some edge $e_{\ell'}$ at most $s-1$ times in $H$. Choose the starting node of the agent as the node $v_0$ of $H_{X_t}$ where $\ell' = t + x - i \mod s$.

According to Lemma 1, using the sequence of port numbers of $W''$, the agent visits all the edges of $E_{\ell',\ell''}$, where $\ell'' = t + y - i \mod s$. Thus, $W''$ is an exploration sequence for $H_{X_{\ell'}}$ which visits the edge $e_{\ell'}$ in $H$ at most $s-1$ times. Therefore by the result of [4], at least one of the edge in $H_{X_{\ell'}}$ is not visited by $W''$. This contradicts the fact that $\mathscr{B}$ is an algorithm that visits all the edges of $\widehat{G}$.

Therefore $|W''| \in O(s^2)$ and hence the claim is true.

Similarly, considering all $E_{x,y}$, for $1 \le x, y \le s$, it can be proved that all the edges of $\widehat{G}$ must be visited at least $s$ times. Since, the total number of edges of the graph $\widehat{G}$ is $s^2 \frac{m^2}{4} \ge \frac{m^6}{20}$, therefore, $|U| \ge s\frac{m^6}{20} \ge \frac{m^8}{100} \in \Omega(n^{\frac{8}{3}})$.     □

## 4   Exploration in $O(n^3)$ Time

In this section, we propose upper bound and lower bound results on size of advice in order to explore all the edges of the graph in time $O(n^3)$.

### 4.1   The Algorithm

Here we propose an algorithm using which the agent explores all the edges of an $n$ node graph in time $O(n^3)$ with advice of size $O(n \log n)$.

Let $G$ be any $n$ node graph. The advice provided to the agent is a port numbered spanning tree of $G$. The spanning tree can be coded as a binary string of size $O(n \log n)$ [21]. The agent, after receiving the advice, will decode the spanning tree and explores all the edges of the graph as described below.

**Algorithm**: EdgeExploration

**Step 1:** After receiving the tree $T$ as advice, the agent locally labels all the nodes of the spanning tree with unique labels from $\{1, 2 \cdots, n\}$. Then it computes $n$ eulerian tours $\mathscr{E}_1, \cdots, \mathscr{E}_n$ from the spanning tree, where $\mathscr{E}_i$ is an eulerian tour that starts and ends at the vertex $i$. These tours are basically sequences of outgoing port numbers starting and ending at the same vertex $i$, for different values of $i$.

**Step 2:** For each $1 \le i \le n$, the agent start exploring according to the sequence of port numbers corresponding to $\mathscr{E}_i$ as follows. Let $p_1, p_2, \cdots, p_k$, be the sequence of port numbers corresponding to the tour $\mathscr{E}_i$. The agent visits all the port incident to the starting vertex and come back to it using the reverse ports from the adjacent node. Then it takes port $p_1$ and stores the port number of the other side of the edge in a stack. Next, it visits all the edges incident to the current node same as before, and come back to it using the reverse ports. Then it takes port $p_2$ and stores the port number of the other side of the edge in a stack. The agent continue visiting the edges in this way until it get stuck (this might happen in a case, when the agent is supposed to take the port $p_i$ but the degree of the current node is less than $p_i + 1$) at some node or the tour

$\mathscr{E}_i$ is completed. At this point, it uses the port numbers which are stored in the stack to backtrack to the starting point.

Since the initial position of the agent is one of the nodes $1 \leq i \leq n$, it will succeed visiting all the edges for at least one $\mathscr{E}_i$. For each of such euler tour, the agent visit all the edges at each vertex, hence will take $O(n^2)$ time. Since there are $n$ such euler tour, the time of exploration would be $O(n^3)$. Hence we will have the following theorem.

**Theorem 2.** *Algorithm* EdgeExploration *explores all the edges of the graph $G$ in time $O(n^3)$ with advice of size $O(n \log n)$.*

### 4.2 Lower Bound

In this section, we prove that the size $\Omega(n^\delta)$ of advice is necessary, for any $\delta < \frac{1}{3}$ in order to perform edge exploration in $O(n^3)$ time. To prove this lower bound result, we construct a class of graphs $\mathscr{G}$ for which if the size of the advice given by the Oracle is $o(n^\delta)$, then there exist a graph in $\mathscr{G}$ for which the time of edge exploration is $\omega(n^3)$. The graphs in $\mathscr{G}$ are constructed in similar fashion as the graph $\widehat{G}$ in Sect. 3.

Let $\delta < \frac{1}{3}$ is a positive real constant. Then there exists a real constant $\epsilon$, $0 < \epsilon < \frac{1}{2}$, such that $\delta > \frac{\epsilon}{1+\epsilon}$. Also, for $\epsilon < \frac{1}{2}$, there exists a real constant $c < \frac{1}{2}$ such that $\epsilon < \frac{(1-c)}{2}$.

Let $S = \{e_1, e_2, \cdots, e_s\}$ be the set of non spanning tree edges in $H$, where $H$ is the complete bipartite graph of $m$ (even) nodes and where each node has degree $\frac{m}{2}$. Let $Z$ be any subset of $S$ of size $m^\epsilon$. Construct the graph $G_Z$ as follows. Let $Z = \{e_{i_1}, e_{i_2}, \cdots, e_{i_p}\}$, where $p = m^\epsilon$. Construct $G_Z$ in the same way as we have constructed $\widehat{G}$ (in Sect. 3) by replacing $s$ with $p$ and $S$ with $Z$. In other words, take one copy of each $H(X_{i_j})$, for $1 \leq j \leq p$, and connect additional edges similarly as explained in Sect. 3 to construct sets of edges $E_{a,b}$, where $a, b \in \{i_1, \cdots i_p\}$. Note that each subset $Z$ of $S$ corresponds to a graph $G_Z$. There are $\binom{s}{p}$ different subsets of $S$ and hence there are $\binom{s}{p}$ different graphs like $G_Z$ can be constructed.

Let $\mathscr{G} = \{G_Z | Z \subset S\}$. Then $|\mathscr{G}| = \binom{\frac{m^2}{4} - m + 1}{m^\epsilon} \geq \binom{\frac{m^2}{5}}{m^\epsilon} \geq (\frac{m^{2-\epsilon}}{5})^{m^\epsilon} \geq m^{(2-\frac{c}{2}-\epsilon)m^\epsilon}$, for large values of $m$. Let $n$ be the number of nodes in each graph of $\mathscr{G}$. Then $n = 2m^{1+\epsilon}$. With this class of graphs, we are ready to prove our lower bound result.

**Theorem 3.** *For any $\delta < \frac{1}{3}$, any exploration algorithm using advice of size $o(n^\delta \log n)$ must take $\omega(n^3)$ time on some $n$ node graph of the class $\mathscr{G}$ for arbitrarily large $n$.*

*Proof.* Suppose that there exists an algorithm $\mathscr{A}$, using which the agent explores all the edges of any graph in $\mathscr{G}$ using advice of size at most $\frac{c}{2} m^\epsilon \log m - 1$. There are at most $m^{\frac{c}{2}m^\epsilon}$ many different binary strings possible with length at most $\frac{c}{2} m^\epsilon \log m - 1$. Since $|\mathscr{G}| \geq m^{(2-\epsilon-\frac{c}{2})m^\epsilon}$, by Pigeon hole principle, for at least

$m^{(2-c-\epsilon)m^\epsilon}$ many graphs the agent must receive same advice. Suppose $\mathscr{G}' \subset \mathscr{G}$ be the set of graphs with same advice.

Let $F(\mathscr{G}') = \{\cup\{e_{i_1}, e_{i_2}, \cdots, e_{i_p}\} | \ Z = \{e_{i_1}, e_{i_2}, \cdots, e_{i_p}\}$ and $G_Z \in \mathscr{G}'\}$. Intuitively, $F(\mathscr{G}')$ is the collection of all such edge $e_{i_k}$ of $S$ for which $H_{i_k}$ is used in the construction of at least one graph in $\mathscr{G}'$.

Next, we claim that $|F(\mathscr{G}')| \geq |\mathscr{G}'|^{\frac{1}{m^\epsilon}}$. To prove this claim, suppose otherwise. That is, $|F(\mathscr{G}')| < |\mathscr{G}'|^{\frac{1}{m^\epsilon}}$. Note that each graph in $\mathscr{G}'$ is constructed using $m^\epsilon$ different $H_{X_{i_j}}$. Therefore, at most $\binom{|F(\mathscr{G}')|}{m^\epsilon}$ different graphs are possible in $\mathscr{G}'$. Hence, $|\mathscr{G}'| \leq \binom{|F(\mathscr{G}')|}{m^\epsilon} \leq |F(\mathscr{G}')|^{m^\epsilon} < |\mathscr{G}'|$, which is a contradiction. Therefore, $|F(\mathscr{G}')| \geq m^{(2-c-\epsilon)}$.

Let $G$ be any graph in $\mathscr{G}'$. We consider the execution of the algorithm $\mathscr{A}$ where the starting node of the agent is $v_0$ of some $H_{X_{i_j}}$ in $\mathscr{G}'$.

Let $U$ be the sequence of port numbers corresponding to $\mathscr{A}$. The sequence of port numbers $U$ can be written as $U = B_1.(p_1).B_2.(p_2).\cdots.B_k.(p_k)$, where each $B_i$ is a sequence of port numbers corresponding to continuous movements of the mobile agent moving according to $\mathscr{A}$ ($B_i$ is a sequence involving zero or more port numbers, for each $i$) and $p_1, p_2, \cdots, p_k$, are the ports that the agent takes to visit only the edges from $E_{i_x, i_y}$. Consider the sequence of port numbers $W = p_1, p_2, \cdots, p_k$ (represents a discontinuous movement of the agent) and construct $W' = C_1 p_1 C_2 p_2 \cdots, C_k p_k$, where $C_1$ is the sequence of port numbers corresponding to a shortest path from $v_0 \in H_{i_x}$ to the end vertex of the port $p_1$ and $C_i$ is the sequence of port numbers corresponding to a shortest path from the node the agent reached after taking the port $p_i$ to the end vertex of the port $p_{i+1}$. Let $W''$ be the sequence of port number which is constructed from $W'$ such that for each $i$, the value of the $i$−th port of $W''$ is assigned as the value of the $i$−th port of $W'$ mod $\frac{m}{2}$. We claim that the sequence of port numbers $W''$, when applied on $H$, starting from $v_0$ in $H$, must visit all the edges in $F(\mathscr{G}')$ at least $s$ times. Otherwise, suppose $W''$ visits the edge $e_{i_\ell} \in F(\mathscr{G}')$ at most $s$ times in $H$, starting from $v_0$ in $H$. Note that, since $e_{i_\ell} \in F(\mathscr{G}')$, there exists a graph $G_Z \in \mathscr{G}$ such that the edge $e_{i_\ell} \in Z$ and $Z \subset S$. Consider the exploration of the mobile agent in $G_Z$. Since the agent received same advice for all the graphs in $\mathscr{G}'$ and the graphs are indistinguishable for the agent, it explores all the graphs in $\mathscr{G}'$ using the same sequence of port numbers.

Therefore, in the exploration of $G_Z$, if the agent starts from $v_0$ in $H_{X_{i_\ell}}$, such that $i_\ell = i_x + i_t - i_j$, the sequence of port numbers $W''$ visits all the edges of $E_{i_\ell, i_{\ell'}}$, where $i_\ell = i_x + i_t - i_j$ (using Lemma 1). Since $W''$ visits the edge $e_{i_\ell}$ at most $s - 1$ times, by the property of $E_{i_\ell, i_{\ell'}}$, the agent can not explore at least one edge in $H_{X_{i_\ell}}$, which is a contradiction that the algorithm $\mathscr{A}$ explores all the edges. This proves that $|W''| \geq s|F(\mathscr{G}')|$. Since $m^\epsilon$ copies of every graph $H(X_{i_t})$ is constructed in $G_Z$, for every edge $e_{i_t} \in Z$, by similar arguments we can prove that $|U| \geq s|F(\mathscr{G}')| \cdot m^\epsilon \cdot m^\epsilon \geq \frac{m^2}{5} m^{(2-\epsilon-c)} \cdot m^{2\epsilon} = \frac{m^{4+\epsilon-c}}{5}$. Since $\epsilon < \frac{1-c}{2}$, therefore $|U| \in \omega(m^{3+3\epsilon})$. Also, since $n = 2m^{1+\epsilon}$, therefore, $|U| \in \omega(n^3)$. Note that the size of the advice provided is $o(m^\epsilon \log m)$. Since $n = 2m^{1+\epsilon}$, therefore the size of advice is $o(n^{\frac{\epsilon}{1+\epsilon}} \log n)$, i.e, $o(n^\delta \log n)$. Therefore, with advice of size $o(n^\delta \log n)$, the time of exploration must be $\omega(n^3)$.    □

## 5    Conclusion

The first lower bound results of $\Omega(n^{\frac{2}{3}})$ time for edge exploration and the proposed algorithm of $O(n^3)$ leaves a small gap of $O(n^{\frac{1}{3}})$ on exploration time. On the other hand, the second lower bound result on the size of advice, compared with the proposed algorithm also leaves a gap less than $O(n^{\frac{2}{3}+\epsilon})$, for any $\epsilon > 0$. Closing up these gaps between upper and lower bounds are natural open problems which can be addressed in the future. Another interesting problem is to study the edge exploration problem where the advice is provided by an instance Oracle.

## References

1. Awerbuch, B., Betke, M., Rivest, R.L., Singh, M.: Piecemeal graph exploration by a mobile robot. Inf. Comput. **152**, 155–172 (1999)
2. Bender, M.A., Fernández, A., Ron, D., Sahai, A., Vadhan, S.P.: The power of a pebble: exploring and mapping directed graphs. Inf. Comput. **176**, 1–21 (2002)
3. Bender, M.A., Slonim, D.: The power of team exploration: two robots can learn unlabeled directed graphs. In: Proceedings of 35th Annual Symposium on Foundations of Computer Science (FOCS 1994), pp. 75–85 (1994)
4. Borodin, A., Ruzzo, W., Tompa, M.: Lower bounds on the length of universal traversal sequences. J. Comput. Syst. Sci. **45**, 180–203 (1992)
5. Boyar, J., Favrholdt, L.M., Kudahl, C., Larsen, K.S., Mikkelsen, J.W.: Online algorithms with advice: a survey. ACM Comput. Surv. **50**(2), 19:1–19:34 (2017)
6. Chalopin, J., Das, S., Kosowski, A.: Constructing a map of an anonymous graph: applications of universal sequences. In: Proceedings of 14th International Conference on Principles of Distributed Systems (OPODIS 2010), pp. 119–134 (2010)
7. Diks, K., Fraigniaud, P., Kranakis, E., Pelc, A.: Tree exploration with little memory. J. Algorithms **51**, 38–63 (2004)
8. Dobrev, S., Kralovic, R., Markou, E.: Online graph exploration with advice. In: Proceedings of 19th International Colloquium on Structural Information and Communication Complexity (SIROCCO 2012), pp. 267–278 (2012)
9. Duncan, C.A., Kobourov, S.G., Anil Kumar, V.S.: Optimal constrained graph exploration. ACM Trans. Algorithms **2**, 380–402 (2006)
10. Ellen, F., Gorain, B., Miller, A., Pelc, A.: Constant-length labeling schemes for deterministic radio broadcast. In: Proceedings of 31st ACM Symposium on Parallelism in Algorithms and Architectures (SPAA), pp. 171–178 (2019)
11. Emek, Y., Fraigniaud, P., Korman, A., Rosen, A.: Online computation with advice. Theor. Comput. Sci. **412**, 2642–2656 (2011)
12. Fraigniaud, P., Gavoille, C., Ilcinkas, D., Pelc, A.: Distributed computing with advice: information sensitivity of graph coloring. Distrib. Comput. **21**, 395–403 (2009)
13. Fraigniaud, P., Ilcinkas, D., Pelc, A.: Communication algorithms with advice. J. Comput. Syst. Sci. **76**, 222–232 (2010). Oracle size: a new measure of difficulty for communication problems. In: Proceedings of 25th Annual ACM Symposium on Principles of Distributed Computing (PODC 2006), pp. 179–187

14. Fraigniaud, P., Ilcinkas, D., Pelc, A.: Tree exploration with advice. Inf. Comput. **206**, 1276–1287 (2008)
15. Fraigniaud, P., Korman, A., Lebhar, E.: Local MST computation with short advice. Theory Comput. Syst. **47**, 920–933 (2010)
16. Fraigniaud, P., Ilcinkas, D.: Directed graphs exploration with little memory. In: Proceedings of 21st Symposium on Theoretical Aspects of Computer Science (STACS 2004), pp. 246–257 (2004)
17. Fusco, E., Pelc, A.: Trade-offs between the size of advice and broadcasting time in trees. Algorithmica **60**, 719–734 (2011)
18. Fusco, E., Pelc, A., Petreschi, R.: Topology recognition with advice. Inf. Comput. **247**, 254–265 (2016)
19. Gavoille, C., Peleg, D., Pérennes, S., Raz, R.: Distance labeling in graphs. J. Algorithms **53**, 85–112 (2004)
20. Glacet, C., Miller, A., Pelc, A.: Time vs. information tradeoffs for leader election in anonymous trees. In: Proceedings of 27th Annual ACM-SIAM Symposium on Discrete Algorithms (SODA 2016), pp. 600–609 (2016)
21. Gorain, B., Pelc, A.: Deterministic graph exploration with advice. ACM Trans. Algorithms **15**, 8:1–8:17 (2018)
22. Gorain, B., Pelc, A.: Short labeling schemes for topology recognition in wireless tree networks. In: Proceedings of 24th International Colloquium Structural Information and Communication Complexity (SIROCCO), pp. 37–52 (2017)
23. Ilcinkas, D., Kowalski, D., Pelc, A.: Fast radio broadcasting with advice. Theor. Comput. Sci. **411**, 1544–1557 (2012)
24. Korman, A., Kutten, S., Peleg, D.: Proof labeling schemes. Distrib. Comput. **22**, 215–233 (2010)
25. Megow, N., Mehlhorn, K., Schweitzer, P.: Online graph exploration: new results on old and new algorithms. Theor. Comput. Sci. **463**, 62–72 (2012)
26. Nisse, N., Soguet, D.: Graph searching with advice. Theoret. Comput. Sci. **410**, 1307–1318 (2009)
27. Pelc, A., Tiane, A.: Efficient grid exploration with a stationary token. Int. J. Found. Comput. Sci. **25**, 247–262 (2014)
28. Rao, N.S.V., Kareti, S., Shi, W., Iyengar, S.S.: Robot navigation in unknown terrains: introductory survey of non-heuristic algorithms. Technical Report ORNL/TM-12410, Oak Ridge National Laboratory, July 1993
29. Reingold, O.: Undirected connectivity in log-space. J. ACM **55**, 17 (2008)

# Fast Diameter Computation Within
# Split Graphs

Guillaume Ducoffe[1,2]([✉]), Michel Habib[3,4], and Laurent Viennot[3,5]

[1] Faculty of Mathematics and Computer Science,
University of Bucharest, Bucharest, Romania
guillaume.ducoffe@ici.ro
[2] National Institute for Research and Development in Informatics,
Bucharest, Romania
[3] Paris University, Paris, France
[4] IRIF, CNRS, Paris, France
[5] Inria, Paris, France

**Abstract.** *When can we compute the diameter of a graph in quasi linear time?* We address this question for the class of *split graphs*, that we observe to be the hardest instances for deciding whether the diameter is at most two. We stress that although the diameter of a non-complete split graph can only be either 2 or 3, under the Strong Exponential-Time Hypothesis (SETH) we cannot compute the diameter of a split graph in less than quadratic time. Therefore it is worth to study the complexity of diameter computation on *subclasses* of split graphs, in order to better understand the complexity border. Specifically, we consider the split graphs with bounded *clique-interval number* and their complements, with the former being a natural variation of the concept of interval number for split graphs that we introduce in this paper. We first discuss the relations between the clique-interval number and other graph invariants and then almost completely settle the complexity of diameter computation on these subclasses of split graphs:
- For the $k$-clique-interval split graphs, we can compute their diameter in truly subquadratic time if $k = \mathcal{O}(1)$, and even in quasi linear time if $k = o(\log n)$ and in addition a corresponding ordering is given. However, under SETH this cannot be done in truly subquadratic time for any $k = \omega(\log n)$.
- For the *complements* of $k$-clique-interval split graphs, we can compute their diameter in truly subquadratic time if $k = \mathcal{O}(1)$, and even in time $\mathcal{O}(km)$ if a corresponding ordering is given. Again this latter result is optimal under SETH up to polylogarithmic factors.

G. Ducoffe—This work was supported by a grant of Romanian Ministry of Research and Innovation CCCDI-UEFISCDI. Project no. 17PCCDI/2018.
M. Habib—Supported by Inria Gang project-team, and ANR project DISTANCIA (ANR-17-CE40-0015).
L. Viennot—Supported by Irif laboratory from CNRS and Paris University, and ANR project Multimod (ANR-17-CE22-0016).

Y. Li et al. (Eds.): COCOA 2019, LNCS 11949, pp. 155–167, 2019.
https://doi.org/10.1007/978-3-030-36412-0_13

Our findings raise the question whether a $k$-clique interval ordering can always be computed in quasi linear time. We prove that it is the case for $k = 1$ and for some subclasses such as bounded-treewidth split graphs, threshold graphs and comparability split graphs. Finally, we prove that some important subclasses of split graphs – including the ones mentioned above – have a bounded clique-interval number. A research report version is deposited on HAL repository with number hal-02307397.

## 1    Introduction

Computing the diameter of a graph (maximum number of edges on a shortest path) is a fundamental problem with countless applications in computer science and beyond. Unfortunately, the textbook algorithm for computing the diameter of an $n$-vertex $m$-edge graph takes $\mathcal{O}(nm)$-time. This quadratic running-time is too prohibitive for large graphs with millions of nodes. As already noticed in [15, 16], an algorithm breaking this quadratic barrier for general graphs is unlikely to exist since it would lead to more efficient algorithms for some disjoint set problems and, as proved in [27], the latter would falsify the Strong Exponential-Time Hypothesis (SETH). This raises the question of *when we can compute the diameter faster than $\mathcal{O}(nm)$*. By restricting ourselves to more structured graph classes, here we hope in obtaining a finer-grained dichotomy between "easy" and "hard" instances for diameter computations – with the former being quasi linear-time solvable and the latter being impossible to solve in subquadratic time under some complexity assumptions.

Specifically, we focus in this work on the class of split graphs, *i.e.*, the graphs that can be bipartitioned into a clique and a stable set. – For any undefined graph terminology, see [5]. – This is one of the most basic idealized models of core/periphery structure in complex networks [7]. We stress that every split graph has diameter at most three. In particular, computing the diameter of a non-complete split graph boils down to decide whether this is either two or three. Nevertheless, under the Strong Exponential-Time Hypothesis (SETH) the textbook algorithm is optimal even for split graphs [6]. We observe that the split graphs are in some sense the *hardest* instances for deciding whether the diameter is at most two. For that, let us bipartition a split graph $G$ into a maximal clique $K$ and a stable set $S$; it takes linear time [22]. The sparse representation of $G$ is defined as $(K, \{N_G[v] \mid v \in S\})^1$. – Note that all our algorithms in this paper run in time linear in the size of this above representation. –

**Observation 1.** *Deciding whether an $n$-vertex graph $G$ has diameter two can be reduced in linear time to deciding whether a $2n$-vertex split graph $G'$ (given by its sparse representation) has diameter two.*

Due to space restrictions, many proofs are omitted (including that of the above observation); they can be found in the research report version of the paper.

---

[1] This sparse representation is also called the neighbourhood set system of the stable set of $G$, see Sect. 2.

We here address the fine-grained complexity of diameter computation on *subclasses* of split graphs. By the above Observation 1, our results can be applied to the study of the diameter-two problem on general graphs.

*Related work.* Exact and approximate distance computations for chordal graphs: a far-reaching generalization of split graphs, have been a common research topic over the last few decades [10,11,17,18]. To the best of our knowledge, this is the first study on the complexity of diameter computation on split graphs. However, there exist linear-time algorithms for computing the diameter on some other subclasses of chordal graphs such as: interval graphs [26], or strongly chordal graphs [9]. These results imply the existence of linear-time algorithms for diameter computations on interval split graphs and strongly chordal split graphs, among other subclasses.

Beyond chordal graphs, the complexity of diameter computation has been considered for many graph classes, *e.g.*, see [16] and the papers cited therein. In particular, the diameter of general graphs with *treewidth* $o(\log n)$ can be computed in quasi linear time [1], whereas under SETH we cannot compute the diameter of split graphs with clique-number (and so, treewidth) $\omega(\log n)$ in subquadratic time [6]. We stress that our two first examples of "easy" subclasses, namely: interval split graphs and strongly chordal split graphs have unbounded treewidth. Our work unifies almost all known tractable cases for diameter computation on split graphs – and offers some new such cases – through a new increasing hierarchy of subclasses.

Relatedly, we proved in a companion paper [20] that on the proper minor-closed graph classes and the bounded-diameter graphs of constant distance VC-dimension (not necessarily split) we can compute the diameter in time $\mathcal{O}(mn^{1-\varepsilon}) = o(mn)$, for some small $\varepsilon > 0$. We consider in this work some subclasses of split graphs of constant distance VC-dimension. However, the time bounds obtained in [20] are barely subquadratic. For instance, although the graphs of constant treewidth fit in our framework, our techniques in [20] do not suffice for computing their diameter in quasi linear time. In fact, neither are they sufficient to explain why we can compute the diameter in subquadratic time on graphs of superconstant treewidth $o(\log n)$. Unlike [20] this article is a new step toward characterizing the graph classes for which we can compute the diameter in quasi *linear time*.

*Our Results.* We introduce a new invariant for split graphs, that we call the *clique-interval number*. Formally, for any $k \geq 1$, a split graph is *k-clique-interval* if the vertices in the clique can be totally ordered in such a way that the neighbours of any vertex in the stable set form at most $k$ intervals. The clique-interval number of a split graph is the minimum $k$ such that it is $k$-clique-interval. Although this definition is quite similar to the one of the interval number [25], we show in Sect. 2 that being $k$-clique-interval does not imply being $k$-interval, and vice-versa[2]. In fact, as we also proved in Sect. 2, the clique-interval number

---

[2] We observe that the *general graphs* with bounded interval number are sometimes called "split interval" [2], that may create some confusion.

of a split graph is more closely related to the *VC-dimension* and the *stabbing number* of the neighbourhood set system of its stable set. Nevertheless, a weak relationship with the interval number can also be derived in this way.

We then study in Sect. 3 what the complexity of computing the diameter is on split graphs parameterized by the clique-interval number. It follows from our results in Sect. 2 and those in [20] that on every subclass of *constant* clique-interval number, there exists a subquadratic-time algorithm for diameter computation. Our work completes this general result as it provides an almost complete characterization of the quasi linear-time solvable instances. As a warm-up, we observe in Sect. 3.1 that on *clique-interval* split graphs (*a.k.a.*, 1-clique-interval), deciding whether the diameter is two is equivalent to testing for a universal vertex. We give a direct proof of this result and another one based on the inclusion of clique-interval split graphs in the subclass of *strongly chordal* split graphs. On the way, we prove more generally – and perhaps surprisingly – that for the intersection of split graphs with many interesting graph classes from the literature, having diameter at most two is equivalent to having a universal vertex! Then, we address the more general case of $k$-clique-interval split graphs, for $k \geq 2$.

- Our first main contribution is the following almost dichotomy result (Theorem 2). For every $n$-vertex $k$-clique-interval split graph, we can compute its diameter in quasi linear-time if $k = o(\log n)$ and a corresponding total ordering of its clique is given. This result follows from an all new application of a generic framework based on $k$-*range trees* [3], that was already used for diameter computations on some special cases [1,19] but with a quite different approach than ours. Furthermore, the logarithmic upper bound on $k$ is somewhat tight. Indeed, we also prove that under SETH we cannot compute in subquadratic time the diameter of $k$-clique-interval split graphs for $k = \omega(\log n)$.
  We note that this above result is quite similar to the one obtained in [1] for treewidth. Indeed, we observe that every split graph of treewidth $k$ is $k$-clique-interval (and even $\lceil \frac{k}{2} \rceil$-clique-interval, see Sect. 2). We use this easy observation so as to prove our conditional time complexity lower-bound.
- Then, we focus on the *complements* of $k$-clique-interval split graphs—that are an interesting subclass in their own right since they generalize, *e.g.*, interval split graphs. For the latter we get a more straightforward algorithm for diameter computation, with a better dependency in $k$. Indeed, we prove that we can compute the diameter of such graphs in time $\mathcal{O}(km)$ if a corresponding ordering is given. This result is conditionally optimal up to polylogarithmic factors because $k = \mathcal{O}(n)$ and, under SETH, we cannot compute the diameter of split graphs with $m = \tilde{\mathcal{O}}(n)$ edges in subquadratic time [1].

It follows from these two above results that having at hands a $k$-clique-interval ordering for a split graph or its complement can help to significantly improve the time complexity for computing its diameter. We so ask whether such orderings can be computed in quasi linear time, for some small values of $k$. In Sect. 4 we prove that it is indeed the case for bounded-treewidth split graphs (trivially) and some other dense subclasses of bounded clique-interval number such as comparability split graphs. Finally our main result in this section is that the clique-interval split graphs can be recognized in linear time.

Overall, we believe that our study of $k$-clique-interval split graphs is a promising framework in order to prove, somewhat automatically, new quasi linear-time solvable special cases for diameter computations on split graphs and beyond.

## 2    Clique-Interval Numbers and Other Graph Parameters

We start by relating the clique-interval number of split graphs with better-studied invariants from Graph theory and Computational geometry.

*Treewidth.* First we observe that if $G = (S \cup K, E)$ is a split graph then, for *any* total order over $K$ and any $v \in S$, $N_G(v)$ is the union of at most $|K|$ intervals. In fact we can improve this rough upper-bound, as follows:

**Lemma 1.** *Every split graph with clique-number (and so, treewidth) at most $k$ is $\lceil \frac{k}{2} \rceil$-clique-interval.*

Conversely, since any complete graph is 1-clique-interval, the clique-interval number cannot be bounded by any function of the treewidth.

*VC-Dimension and Stabbing Number.* For a set system $(X, R)$ (a.k.a., range space or hypergraph), we say that a subset $Y \subseteq X$ is *shattered* if $\{Y \cap r \mid r \in R\}$ is the power-set of $Y$. The *VC-dimension* of a finite set system is the largest size of a shattered subset. We now prove an intriguing connection between the clique-interval number of a split graph and the VC-dimension of a related set system.

**Proposition 1.** *For any split graph $G = (K \cup S, E)$, let $\mathcal{S} = \{N_G(u) \mid u \in S\}$. If $G$ is $k$-clique-interval then $(K, \mathcal{S})$ has VC-dimension at most $2k$.*

It turns our that a week converse of Proposition 1 also holds. We need a bit of terminology from [14]. A *spanning path* for $(X, R)$ is a total ordering $x_1, x_2, \ldots, x_{|X|}$ of $X$. Its *stabbing number* is the maximum number of consecutive pairs $x_i, x_{i+1}$ that a set $r \in R$ can *stab*, i.e., for which $|r \cap \{x_i, x_{i+1}\}| = 1$. Finally, the stabbing number of $(X, R)$ is the minimum stabbing number over its spanning paths.

**Observation 2.** *For any split graph $G = (K \cup S, E)$, let $\mathcal{S} = \{N_G(u) \mid u \in S\}$. The clique-interval number of $G$ is up to one the stabbing number of $(K, \mathcal{S})$.*

The main result of [14] is that every range system of VC-dimension at most $k$ has a stabbing number in $\tilde{O}(f(k) \cdot n^{1 - \frac{1}{f(k)}})$, for some exponential function $f$. We so obtain:

**Corollary 1.** *For any split graph $G = (K \cup S, E)$, let $\mathcal{S} = \{N_G(u) \mid u \in S\}$. If $(K, \mathcal{S})$ has VC-dimension at most $k$ then $G$ is $\tilde{O}(f(k) \cdot n^{1 - \frac{1}{f(k)}})$-clique-interval, for some exponential function $f$.*

*Interval Number.* Finally, we relate the clique-interval number of split graphs with their interval number. A graph $G = (V, E)$ is called $k$-interval if we can map every $v \in V$ to the union of at most $k$ closed interval on the real line, denoted by $I(v)$, in such a way that $uv \in E \iff I(u) \cap I(v) \neq \emptyset$. In particular, 1-interval graphs are exactly the interval graphs.

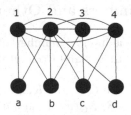

**Fig. 1.** An interval split graph that is not clique-interval.

We observe that there are $k$-interval split graphs that are *not* $k$-clique-interval, already for $k = 1$. For instance, consider the interval split graph of Fig. 1 (note that $\{a, 1, 2\}, \{b, 1, 2, 3\}, \{1, 2, 3, 4\}, \{c, 2, 3, 4\}, \{d, 2, 4\}$ is a linear ordering of its maximal cliques). Suppose by contradiction it is clique-interval, and let us consider a corresponding total ordering of $K = \{1, 2, 3, 4\}$. The intersection $N(b) \cap N(c) = \{2, 3\}$ should be an interval. But then, one of the pairs $\{1, 2\}$ or $\{4, 2\}$ is not consecutive, and so one of $N(a)$ or $N(d)$ is not an interval. Therefore, the graph of Fig. 1 is not clique-interval. Conversely, there are clique-interval split graphs which are not interval graphs. For instance, this is the case of thin spiders, *i.e.*, split graphs such that the edges between the maximal clique and the stable set induce a perfect matching.

Nevertheless we prove a weak connection between the clique-interval number and the interval number of a split graph, by using the VC-dimension. Specifically, Bousquet et al. proved that the neighbourhood set system of any interval graph has VC-dimension at most two [8]. We generalize their result.

**Proposition 2.** *The neighbourhood set system of any $k$-interval graph has VC-dimension at most $(4 + o(1))k \log k$.*

**Corollary 2.** *If $G$ is a $k$-interval split graph then, $G$ is also $\tilde{O}(f(k \log k) \cdot n^{1 - \frac{1}{f(k \log k)}})$-clique-interval, for some exponential function $f$.*

## 3  Diameter Computation in Quasi Linear Time

We now address the time complexity of diameter computation on $k$-clique-interval split graphs. Our first result in this section follows from the relations proved in Sect. 2 with the VC-dimension.

**Theorem 1** ([20]). *For every $d > 0$, there exists a constant $\varepsilon_d \in (0;1)$ such that in time $\tilde{O}(mn^{1-\varepsilon_d})$ we can decide whether a graph whose neighbourhood set system has VC-dimension at most $d$ has diameter two.*

**Corollary 3.** *For every constant $k$, there exists a constant $\eta_k \in (0;1)$ such that in time $\tilde{O}(mn^{1-\eta_k})$ we can decide whether a $k$-clique-interval split graph has diameter two.*

We observe that Corollary 3 also holds for the *complements* of $k$-clique-interval split graphs. In the remainder of this section we focus on the existence of *quasi linear-time* algorithms. It starts with a small digression on clique-interval split graphs.

## 3.1   The Case $k = 1$ and Beyond

**Proposition 3.** *A clique-interval split graph has diameter at most two if and only if it has a universal vertex. In particular, we can compute the diameter of clique-interval split graphs in linear time.*

Our proof of Proposition 3 is based on the Helly property of the intervals on a line. In the remainder of this part we give an alternative proof of Proposition 3 that also applies to larger subclasses of split graphs. It starts with the following inclusion lemma.

**Lemma 2.** *Every clique-interval split graph is strongly chordal. The minimal obstructions to strong chordality are an infinite family of split graphs, sometimes called sun graphs.*

We stress that since every $n$-sun graph is 2-clique-interval, this above Lemma 2 does not hold for $k$-clique-interval graphs, for any $k \geq 2$. Now, a *maximum neighbour* of $v$ is any $u \in V$ such that $\bigcup_{w \in N_G[v]} N_G[w] \subseteq N_G[u]$ (see [9]). We prove that if at least one vertex in a graph has a maximum neighbour then deciding whether the diameter is at most two becomes a trivial task. In particular, this is always the case for strongly chordal graphs [21].

**Lemma 3.** *For every $G = (V, E)$ and $u, v \in V$ such that $u$ is a maximum neighbour of $v$, we have $diam(G) \leq 2$ if and only if $u$ is a universal vertex.*

Proposition 3 now follows from the combination of Lemmas 2 and 3. Furthermore it turns out that many well-structured graph classes ensure the existence of a vertex with a maximum neighbour such as: graphs with a pendant vertex, threshold graphs [24] and interval graphs [12], or even more generally dually chordal graphs [9]. We conclude that:

**Corollary 4.** *We can compute the diameter of split graphs with minimum degree one and dually chordal split graphs in linear time.*

*In particular, we can compute the diameter of interval split graphs and strongly chordal split graphs in linear time.*

We observe that we can easily modify the hardness reduction from [6] in order to show that, under SETH, we cannot compute in subquadratic time the diameter of split graphs with minimum degree two. Indeed, this aforementioned reduction outputs a split graph $G = (K \cup S, E)$ such that: $S$ is partitioned in two disjoint subsets $A$ and $B$; and a diametral pair must have one end in each subset. Let $a, b, v_0 \notin V$ be fresh new vertices, and let $G' = (K' \cup S', E')$ be such that: $K' = K \cup \{a, b\}$, $S' = S \cup \{v_0\}$ and:

$$E' = E \cup \{au \mid u \in A\} \cup \{bw \mid w \in B\} \cup \{av, bv \mid v \in K \cup \{v_0\}\} \cup \{ab\}.$$

By construction, $diam(G') \leq 2$ if and only if $diam(G) \leq 2$. Therefore, Corollary 4 is optimal for the parameter minimum degree.

## 3.2   The General Case

We are now ready to prove the first main result of this paper.

**Theorem 2.** *If $G = (K \cup S, E)$ is an $n$-vertex $k$-clique-interval split graph and a corresponding total order over $K$ is given then, we can compute the diameter of $G$ in time $\mathcal{O}(m + k^2 2^{\mathcal{O}(k)} n^{1+o(1)})$. This is quasi linear-time if $k = o(\log n)$.*

*Conversely, under SETH we cannot compute the diameter of $n$-vertex split graphs with clique-interval number $\omega(\log n)$ in subquadratic time.*

In order to prove Theorem 2, we will use a special instance of $k$-*range tree*. Such a data-structure stores a static set of $k$-dimensional points and it supports the following operation:

- (*Range Query*) Given $k$ intervals $[l_i; u_i]$, $1 \leq i \leq k$, compute the number of stored points $p$ such that, for every $1 \leq i \leq k$, we have $l_i \leq p_i \leq u_i$.[3]

**Proposition 4 ([13]).** *For any $k \geq 2$ and any $n$-set of $k$-dimensional points, we can construct a $k$-range tree in time $\mathcal{O}\big(k\binom{k+1+\lceil \log n \rceil}{k+1}\big)n\big) = \mathcal{O}(k2^{\mathcal{O}(k)}n^{1+o(1)})$ and answer any range query in time $\mathcal{O}\big(2^k\binom{k+1+\lceil \log n \rceil}{k+1}\big)\big) = \mathcal{O}(2^{\mathcal{O}(k)}n^{o(1)})$.*

The use of range queries for diameter computation dates back from [1] (see also [13,19] for some further applications). Roughly, if in a graph $G$ we can find a separator $S$ of size at most $k$ that disconnects a diametral pair of $G$, then the idea is to compute a "distance profile" for every vertex $v \notin S$ w.r.t. $S$, and to see this profile as a $k$-dimensional point. We can compute a diametral pair by constructing a $k$-range tree for these points and computing $\mathcal{O}(kn)$ range queries. Our present approach is different from the one in [1] as we define our multi-dimensional points based on some interval representation of the graph rather than on distance profiles, and we use a different type of range query than in [1].

*Proof of Theorem 2.* Let $G = (K \cup S, E)$ be a $k$-clique-interval split graph, and a corresponding total order over $K$. By the hypothesis for every $v \in S$, we have

---

[3] We refer to [13] for a more general presentation of range trees.

$N_G(v) = \bigcup_{i=1}^{k}[l_i(v); u_i(v)]$ is the union of $k$ intervals such that $l_1(v) \leq u_1(v) \leq l_2(v) \leq u_2(v) \leq \ldots \leq l_k(v) \leq u_k(v)$. Furthermore since the ordering over $K$ is given, the $2k$ endpoints that delimit these intervals can be computed in time $\mathcal{O}(|N_G(v)|)$, simply by scanning the neighbours of vertex $v$. We so map every vertex $v \in S$ to the $2k$-dimensional point $p(v) = (l_1(v), u_1(v), \ldots, l_k(v), u_k(v))$, that takes total time $\mathcal{O}(m)$. Then, we construct a $2k$-range tree for the points $p(v), v \in S$, that takes time $\mathcal{O}(k2^{\mathcal{O}(k)}n^{1+o(1)})$ by Proposition 4.

For every $v \in S$, we are left with computing the number of vertices in $S$ at distance two from $v$. Indeed, $G$ has diameter at most two if and only if this number is $|S| - 1$ for every $v \in S$. More specifically, for every $1 \leq i, j \leq k$ we want to compute the number of vertices $w \in S$ such that:

1. $[l_i(v); u_i(v)] \cap [l_j(w); u_j(w)] \neq \emptyset$;
2. while for every $1 \leq i' < i$, $[l_{i'}(v); u_{i'}(v)] \cap [l_j(w); u_j(w)] = \emptyset$;
3. and for every $1 \leq i' \leq i$ and $1 \leq j' < j$, $[l_{i'}(v); u_{i'}(v)] \cap [l_{j'}(w); u_{j'}(w)] = \emptyset$.

The first and second constraints are equivalent to one of the following three *disjoint* possibilities:

- $u_{i-1}(v) < l_j(w) \leq l_i(v)$ and $u_i(v) \leq u_j(w)$;
- or $u_{i-1}(v) < l_j(w) \leq l_i(v)$ and $l_i(v) \leq u_j(w) < u_i(v)$;
- or $l_i(v) < l_j(w) \leq u_i(v)$.

The third constraint is equivalent to have $\bigcup_{j'<j}[l_{j'}(w); u_{j'}(w)] \subseteq (-\infty, l_1(v)) \cup (\bigcup_{i'<i}(u_{i'}(v), l_{i'+1}(v)))$. Note that in order to subdivide this constraint into disjoint possibilities, it suffices to indicate: (i) the subset of all intervals among $(-\infty, l_1(v)) \cup (\bigcup_{i'<i}(u_{i'}(v), l_{i'+1}(v)))$ that contain an interval $[l_{j'}(w); u_{j'}(w)]$ for some $j' < j$; and (ii) the set of all indices $j' \in (1; j)$ such that $[l_{j'-1}(w); u_{j'-1}(w)]$ and $[l_{j'}(w); u_{j'}(w)]$ are *not* contained in the same such interval. Overall, that divides the third constraint in at most $2^i2^{j-2} \leq 2^{2k-2}$ disjoint events. For a fixed pair $(i, j)$ we so reduce our computation to at most $3 \cdot 2^{2k-2}$ range queries, that takes time $\mathcal{O}(2^{\mathcal{O}(k)}n^{o(1)})$ by Proposition 4. Since there are $\mathcal{O}(k^2)$ such pairs, the total time in order to compute the number of vertices in $S$ at distance two from $v$ is an $\mathcal{O}(k^2 2^{\mathcal{O}(k)}n^{o(1)})$.

Finally, the hardness result for $k = \omega(\log n)$ follows from the fact that $k$-treewidth split graphs are $k$-clique-interval (Lemma 1) and that under SETH, we cannot compute the diameter of split graphs with treewidth $\omega(\log n)$ in sub-quadratic time [1,6].  □

Before ending this section, we state a faster algorithm for computing the diameter on the *complements* of $k$-clique-interval split graphs (see the research report version for the proof). It is similar in spirit to [20, Lemma 6].

**Theorem 3.** *If $G = (K \cup S, E)$ is the complement of a $k$-clique-interval split graph and a corresponding total order over $S$ is given then, we can compute the diameter of $G$ in time $\mathcal{O}(km)$.*

# 4    Recognition of $k$-Clique-Interval Split Graphs

Our two algorithms in Sect. 3.2 show that in order to compute the diameter of $k$-clique-interval split graphs in quasi linear time, it is sufficient to compute a corresponding total order of their maximal clique. This raises the question whether such $k$-clique-interval orderings can always be computed in quasi linear time. A first positive example was given by Lemma 1. Indeed, for a split graph of treewidth at most $k$, we can pick *any* total order of its maximal clique. We complete this easy result by Sect. 4.1 where we give examples of dense subclasses of split graphs with constant clique-interval number and for which a corresponding order can be computed in linear time. Finally, in Sect. 4.2 we prove a stronger result, namely that we can recognize the clique-interval graphs in linear time.

## 4.1    Examples of Subclasses with Bounded Clique-Interval Number

A *threshold graph* is a split graph $G = (K \cup S, E)$ such that: (i) the neighbourhoods of the vertices in $K$ and (ii) the neighbourhoods of the vertices in $S$ are totally ordered by inclusion. Observe that threshold graphs can be dense and of unbounded treewidth.

**Lemma 4.** *Every threshold graph is clique-interval.*

Note that we can easily derive from the proof of Lemma 4 a linear-time algorithm for computing a clique-interval ordering.

Finally, a *comparability graph* is a graph that admits a transitive orientation.

**Lemma 5.** *For every comparability split graph $G = (K \cup S, E)$, we can compute in linear time a total order over $K$ such that, for every $v \in S$, $N_G(v)$ is the union of a prefix and a suffix of this order.*

*In particular, every comparability split graph is 2-clique-interval.*

We also want to stress that the complements of comparability split graphs, i.e., the cocomparability split graphs are just interval split graphs and we have already considered this case in Sect. 3.1.

*Remark 1.* The ordering given by Lemma 5 has some additional properties that can be used for computing the diameter of comparability split graphs in linear time (see the research report version for details).

## 4.2    Linear-Time Recognition of Clique-Interval Graphs

**Theorem 4.** *Clique-interval split graphs can be recognized in linear time.*

*Proof.* Let $G = (K \cup S, E)$ be a split graph. We define a graph $G^+$ from $G$ by first transforming $K$ into a stable set and then, for every $u \in K$, making a clique of $N_G(u) \cap S$. Furthermore, we claim that $G$ is clique-interval if and only if $G^+$ is interval. To see that, let us call *clique-path* of a graph $G'$ an ordering of its maximal cliques such that, for any vertex $v$ of $G'$, the maximal cliques containing $v$ are contiguous in the ordering; a graph is interval if and only if it admits a clique-path [4]. We can now prove our claim, as follows:

- If $G$ is clique-interval then, any clique-ordering over $K$ for $G$ is a total ordering over the maximal cliques $\{u\} \cup (N_G(u) \cap S)$, $u \in K$, for $G^+$. Furthermore as already observed in the proof of Proposition 3 the vertices in a subset $S' \subseteq S$ are pairwise at distance two in $G$ if and only if they have a common neighbour in $K$. As a result, the maximal cliques of $G^+$ are exactly the sets $\{u\} \cup (N_G(u) \cap S)$, $u \in K$, and so $G^+$ admits a clique-path. This implies that $G^+$ is an interval graph.
- Conversely, if $G^+$ is an interval graph then any clique-path of $G^+$ induces a total ordering over the maximal cliques $\{u\} \cup (N_G(u) \cap S)$, $u \in K$, and so, a clique-ordering over $K$ for $G$.

Unfortunately, computing the graph $G^+$ may take super-linear time. We can overcome this issue as follows. First, given a family $\mathcal{C}$ of subsets over $V$, if there exists a chordal graph $G_\mathcal{C}$ whose maximal cliques are exactly those in $\mathcal{C}$ then, we can compute a Lex-BFS ordering of $G_\mathcal{C}$ in time $\mathcal{O}(\sum_{C \in \mathcal{C}} |C|)$ (Algorithm 10 in [23]). Moreover we can also compute a clique-tree of $G_\mathcal{C}$ in time $\mathcal{O}(\sum_{C \in \mathcal{C}} |C|)$ [29, Sect. 3]. Finally given this Lex-BFS ordering and the corresponding clique-tree of $G_\mathcal{C}$, we can apply Algorithm 9 from [23] in order to decide in time $\mathcal{O}(\sum_{C \in \mathcal{C}} |C|)$ whether $G_\mathcal{C}$ is interval. For solving our initial problem, we can take $\mathcal{C} = \{\{u\} \cup (N_G(u) \cap S) \mid u \in K\}$, and in this situation we have $\sum_{C \in \mathcal{C}} |C| = \mathcal{O}(n + m)$. □

We left open the status of the recognition of $k$-clique-interval split graphs, for $k \geq 2$.

## 5   Open Problems

Although the definitions of $k$-clique-interval and $k$-interval split graphs have some similarities, we observe that computing the diameter of 2-interval split graphs in quasi linear time already looks like a challenging task. Indeed, for a 2-interval split graph $G = (K \cup S, E)$ and $v \in S$, the vertices $u \in S$ at distance two from $v$ are exactly those such that one of their 2 intervals intersects one of the $2|N_G(v)|$ intervals that represent the neighbours of $v$. We cannot use our range query framework in order to avoid overcounting these vertices as this would require up to $2^{\mathcal{O}(|N_G(v)|)}$ range queries. More generally, for every fixed $k > 1$, *can we compute the diameter of $k$-interval graphs in quasi linear time?* We stress that every planar graph is 3-interval [28], and that the complexity of diameter computation on this class of graphs is a longstanding open problem. The case $k = 2$ could thus be an interesting intermediate step.

# References

1. Abboud, A., Williams, V.V., Wang, J.: Approximation and fixed parameter sub-quadratic algorithms for radius and diameter in sparse graphs. In: SODA, pp. 377–391. SIAM (2016)
2. Bar-Yehuda, R., Halldórsson, M., Naor, J., Shachnai, H., Shapira, I.: Scheduling split intervals. SIAM J. Comput. **36**(1), 1–15 (2006)
3. Bentley, J.H.: Decomposable searching problems. Inf. Process. Lett. **8**(5), 244–251 (1979)
4. Blair, J.R., Peyton, B.: An introduction to chordal graphs and clique trees. In: George, A., Gilbert, J.R., Liu, J.W.H. (eds.) Graph Theory and Sparse Matrix Computation. IMA, vol. 56, pp. 1–29. Springer, New York (1993). https://doi.org/10.1007/978-1-4613-8369-7_1
5. Bondy, J.A., Murty, U.S.R.: Graph Theory. Springer, Berlin (2008)
6. Borassi, M., Crescenzi, P., Habib, M.: Into the square: on the complexity of some quadratic-time solvable problems. Electron. Notes Theor. Comput. Sci. **322**, 51–67 (2016)
7. Borgatti, S., Everett, M.: Models of core/periphery structures. Soc. Netw. **21**(4), 375–395 (2000)
8. Bousquet, N., Lagoutte, A., Li, Z., Parreau, A., Thomassé, S.: Identifying codes in hereditary classes of graphs and VC-dimension. SIAM J. Discrete Math. **29**(4), 2047–2064 (2015)
9. Brandstädt, A., Chepoi, V., Dragan, F.: The algorithmic use of hypertree structure and maximum neighbourhood orderings. Discrete Appl. Math. **82**(1–3), 43–77 (1998)
10. Brandstädt, A., Chepoi, V., Dragan, F.: Distance approximating trees for chordal and dually chordal graphs. J. Algorithms **30**(1), 166–184 (1999)
11. Brandstädt, A., Dragan, F., Le, H., Le, V.: Tree spanners on chordal graphs: complexity and algorithms. Theoret. Comput. Sci. **310**(1), 329–354 (2004)
12. Brandstädt, A., Hundt, C., Mancini, F., Wagner, P.: Rooted directed path graphs are leaf powers. Discrete Math. **310**(4), 897–910 (2010)
13. Bringmann, K., Husfeldt, T., Magnusson, M.: Multivariate analysis of orthogonal range searching and graph distances parameterized by treewidth. In: IPEC (2018)
14. Chazelle, B., Welzl, E.: Quasi-optimal range searching in spaces of finite VC-dimension. Discrete Comput. Geom. **4**(5), 467–489 (1989)
15. Chepoi, V., Dragan, F.: Disjoint sets problem (1992)
16. Corneil, D., Dragan, F., Habib, M., Paul, C.: Diameter determination on restricted graph families. Discrete Appl. Math. **113**(2–3), 143–166 (2001)
17. Dourisboure, Y., Gavoille, C.: Improved compact routing scheme for chordal graphs. In: Malkhi, D. (ed.) DISC 2002. LNCS, vol. 2508, pp. 252–264. Springer, Heidelberg (2002). https://doi.org/10.1007/3-540-36108-1_17
18. Dragan, F.: Estimating all pairs shortest paths in restricted graph families: a unified approach. J. Algorithms **57**(1), 1–21 (2005)
19. Ducoffe, G.: A new application of orthogonal range searching for computing giant graph diameters. In: 2nd Symposium on Simplicity in Algorithms (SOSA 2019) (2019)
20. Ducoffe, G., Habib, M., Viennot, L.: Diameter computation on $H$-minor free graphs and graphs of bounded (distance) VC-dimension. Technical report 1907.04385 arXiv (2019)

21. Farber, M.: Characterizations of strongly chordal graphs. Discrete Math. **43**(2–3), 173–189 (1983)
22. Golumbic, M.: Algorithmic Graph Theory and Perfect Graphs, vol. 57. Elsevier, Amsterdam (2004)
23. Habib, M., McConnell, R., Paul, C., Viennot, L.: Lex-BFS and partition refinement, with applications to transitive orientation, interval graph recognition and consecutive ones testing. Theoret. Comput. Sci. **234**(1–2), 59–84 (2000)
24. Heggernes, P., Kratsch, D.: Linear-time certifying recognition algorithms and forbidden induced subgraphs. Nord. J. Comput. **14**(1–2), 87–108 (2007)
25. McGuigan, R.: Presentation at NSF-CBMS Conference at Colby College (1977)
26. Olariu, S.: A simple linear-time algorithm for computing the center of an interval graph. Int. J. Comput. Math. **34**(3–4), 121–128 (1990)
27. Roditty, L., Williams, V.V.: Fast approximation algorithms for the diameter and radius of sparse graphs. In: STOC, pp. 515–524. ACM (2013)
28. Scheinerman, E., West, D.: The interval number of a planar graph: three intervals suffice. J. Comb. Theory, Series B **35**(3), 224–239 (1983)
29. Tarjan, R., Yannakakis, M.: Simple linear-time algorithms to test chordality of graphs, test acyclicity of hypergraphs, and selectively reduce acyclic hypergraphs. SIAM J. Comput. **13**(3), 566–579 (1984)

# Approximate Shortest Paths in Polygons with Violations

Binayak Dutta[1](✉) and Sasanka Roy[2]

[1] Tezpur University, Tezpur, India
binayak66@gmail.com
[2] Indian Statistical Institute, Kolkata, Kolkata, India
sasanka.ro@gmail.com

**Abstract.** We study the problem of computing shortest $k$-violation path problem on polygons. Let $P$ be a simple polygon in $\mathbb{R}^2$ with $n$ vertices and let $s, t$ be a pair of points in $P$. Let $int(P)$ represent the interior of $P$. In other words, $int(P) = P \setminus \Delta(P)$, where $\Delta(P)$ is the boundary of $P$. Let $\tilde{P} = \mathbb{R}^2 \setminus int(P)$ represent the exterior of $P$. For an integer $k \geq 0$, the problem of $k$-violation shortest path in $P$ is the problem of computing the shortest path from $s$ to $t$ in $P$, such that at most $k$ path segments are allowed to be in $\tilde{P}$. The path segments are not allowed to bend in $\tilde{P}$. For any $k$, we present a $(1 + \epsilon)$ factor approximation algorithm for the problem, that runs in $O(n^2\sigma^2 k \log n^2\sigma^2 + n^2\sigma^2 k)$ time. Here $\sigma = O(\log_\delta(\frac{|L|}{r}))$ and $\delta, L, r$ are geometric parameters.

**Keywords:** Steiner points · Approximation · Shortest paths · Violations · Polygons · Triangulation

## 1 Introduction

Let $P$ be a simple polygon in $\mathbb{R}^2$ with $n$ vertices. Let $int(P)$ and $\Delta(P)$ represent the interior and the boundary of the polygon $P$ respectively and $\tilde{P} = \mathbb{R} \setminus int(P)$ represent the exterior of $P$.

Shortest path in a polygon $P$, and its variants, is a very widely studied topic in the field of computational geometry. The shortest path problem in $P$ is the problem of finding the shortest path that connects a pair of points $s, t \in P$. The shortest path consists of a sequence of line segments, called path segments, that do not cross the boundary of $P$. For the sake of clarity, we refer to the edges of a path as path segments and the edges of $P$ as edges. The shortest path changes directions only at the vertices of $P$. The sum total of the lengths of the constituent path segments is the length of the shortest path from $s$ to $t$. We refer the reader to [5, 13, 17] for a detailed survey on the topic. The shortest path from a point $s$ to all the vertices of $P$ can be obtained in $O(n)$ time using shortest path maps [7].

For an integer $k \geq 0$, the $k$-violation shortest path in $P$ is the problem of computing the shortest path between a pair of points $s, t \in P$, such that the

© Springer Nature Switzerland AG 2019
Y. Li et al. (Eds.): COCOA 2019, LNCS 11949, pp. 168–180, 2019.
https://doi.org/10.1007/978-3-030-36412-0_14

path's intersection with $\tilde{P}$ is at most $k$ path segments. It must be noted, that this means we do not allow the path segments to bend in $\tilde{P}$. Maheshwari et al. [14] proposed an algorithm that computes the shortest one violation path for a pair of points $s, t \in P$ in $O(n^3)$ time. The problem that we study in this paper is inspired by the shortest one violation path. In the same paper, two other shortest path problems where some constraints are violated in some restricted way are studied. They study the one-stretch violation path problem and monotone rectilinear path problems with violations. They propose an $O(n \log n \log \log n)$ time algorithm for shortest one-stretch violation path and an $O(n \log n)$ time algorithm for shortest one violation monotone rectilinear path problem.

To compute the shortest $k$-violation we use the discretization technique of Aleksandrov et al. [2]. They used the mentioned technique to solve the problem of computing the shortest path in weighted terrains. The technique consists of discretizing the faces of the terrain by inserting *Steiner points* on the edges of the terrain. Then a weighted graph $G$ is constructed where nodes of $G$ are the Steiner points and the vertices of the terrain. Then the shortest path is computed on $G$ using Dijkstra's shortest path algorithm.

## 1.1  Related Works

Lanthier et al. [12] used Steiner points to discretize the edges of a triangulated polyhedral surface, where faces of the triangulation have weights associated with them, to compute a $(1 + \epsilon)$-factor weighted shortest path on the polyhedral surface. Their algorithm runs in $O(n^3 \log n)$ time. This approach of placing Steiner points on the edges of a triangulated weighted polyhedral surface was used by Aleksandrov et. al [2] to compute a $(1 + \epsilon)$, for a given $\epsilon > 0$, approximate shortest path between a pair of points. Their algorithm runs in $O(mn \log mn + nm^2)$, where $m$ and $n$ are the number of Steiner points added and the number of vertices of the polygon respectively. Roy et al. [20] used this technique of discretization using Steiner points to compute a $O(1 + \epsilon)$ shortest descending path between a pair of points on the surface of a polyhedral terrain. The running time of this algorithm is $O(mn \log mn + nm^2)$. Cheng et al. [4] used a strategy of placing Steiner points on the edges of the triangulation of the polyhedral surface different from the previously mentioned works to compute a $(1 + \epsilon)$-factor shortest path on the planar subdivision. In this problem, the weight of a path segment of a path on a face of the triangulation is the distance between the two points where the path intersects with the face. This distance is determined by a convex distance function, possibly asymmetric, associated with the face. Different faces can have different associated convex functions. The running time of the algorithm does not depend on the geometric parameters like the minimum angle in the subdivision. They achieve this by restricting the placement of the Steiner points only on the edges of the triangulation that are contained in an ellipse whose diameter is $\frac{4}{3}\rho\|s\,t\|$, where $\rho \geq 1$ and $\|s\,t\|$ is the global geodesic shortest path between $s$ and $t$. The time complexity of the algorithm is $O(\frac{\rho^2 \log \rho}{\epsilon^2} n^3 \log \frac{\rho n}{\epsilon})$. Their strategy of discretization cannot be used in computing approximate $k$-violation path directly because, while the global geodesic shortest path is a lower bound for the shortest path in their problem, it is an upper bound for the shortest $k$-violation path, because unless the line segment $\overline{st}$ lies completely within a polygon $P$, there always exists a path with at least one violation whose length is shorter than $\overline{st}$ [14].

Maheshwari et al. [14] studied the problem of computing the shortest path between points $s$ and $t$ in a polygon $P$ such that the path's intersection with $\tilde{P}$ is at most one path segment. This is the most related problem to the problem of shortest $k$ violation path in a polygon. They propose an algorithm that computes the shortest one violation path $\pi_1(s,t)$ in $O(n^3)$ time. They use a structure called shortest path map, due to [9], which is used to compute the shortest path from a source $s$ to all points in a polygon $P$. A shortest path map splits the boundary of a polygon into open intervals called component pieces. Each component piece has an associated vertex of the polygon, which is called a parent vertex. For all points in the component piece, the shortest path from $s$ in $P$ passes through the parent vertex. Then they consider the convex hull of $P$, $CH(P)$. Each edge of the convex hull of $P$ that is disjoint from the edges of $P$ induces a region called a pocket. It must be noted here that any one violation path can pass through only one pocket out of the pockets induced by $CH(P)$. Let us call this pocket $P_i$. $\pi_1(s,t)$ enters $P_i$ through a point $b$ on component piece $I$ and leaves $P_i$ through a point $c$ component piece $J$. So once they fix the points $b$ and $c$ on $I$ and $J$ respectively, they compute the shortest one violation path by minimizing the function obtained by adding the lengths of (i) path from $s$ to $u$, where $u$ is the parent of $I$, (ii) path from $v$ to $t$, where $v$ is the parent of $J$, (iii) line segment $\overline{bc}$, (iv) line segment $\overline{ub}$ and (v) line segment $\overline{cv}$. They do this over all the pockets.

A related topic is computing the shortest path in the plane that contains a set of polygonal obstacles. It is a very widely studied topic in the field of computational geometry [3,6,11,15,16,18,19,22]. Hershberger et al. produced an optimal algorithm for the aforementioned problem that runs in $O(n \log n)$ time [10] using the continuous Dijkstra technique. In a recent paper, Agarwal et al. [1] considered the problem of computing shortest Euclidean path in a plane with removable objects. Each object has an associated cost that must be spent to remove it from the plane. The objective is to minimize the length of the path under the constraint of a cost budget $C > 0$. They proved that the problem is NP-hard and produced a $(1 + \epsilon)$-factor FPTAS for a special case where the objects are convex polygons, that runs in $O(\frac{nh}{\epsilon^2} \log n \log \frac{n}{\epsilon})$, where $h$ is the number of obstacles, $n$ is the number of vertices of the obstacles and $\epsilon \in (0,1)$ is a parameter defined by user. The removal cost is at most $C \cdot (1 + \epsilon)$.

Recently, Hershberger et al. [8] worked on a related problem of computing shortest path between a pair of points on a plane with a set of convex obstacles, where $k$ obstacles from the set of obstacles can be violated. This problem can also be viewed as the problem of computing the shortest path in a plane with a set of $n$ obstacle from which $k \leq n$ can be removed. They developed a structure called *shortest $k$-path map*, $SPM_k$, to compute the shortest $k$-path that violates at most $k$ many of the convex obstacles. Their problem is different than the shortest $k$-violation path in a polygon, as obstacles present in the plane are convex in case of shortest $k$-path. In the case of the shortest $k$-violation path the pockets of the given polygon can be convex as well as non-convex. The other difference is the shortest $k$-path can only bend at the vertices of the obstacles. Unlike shortest $k$-path, shortest $k$-violation path can bend at the edges of the given polygon $P$. The most important difference of all is stated by the following lemma and the succeeding observation.

**Lemma 1.** [8]  *Let $\pi_k(t)$ and $\tilde{\pi}_k(t')$ be two subpaths of shortest $k$-paths whose prefix counts are the same. Then $\pi_k(t)$ and $\tilde{\pi}_k(t')$ do not cross each other. Prefix count is the number of objects violated by a path.*

**Observation 1.**  *A valid configuration can be constructed (Fig. 1)where two subpaths of shortest $k$-violation paths originating from the same source with the same prefix counts intersect each other.*

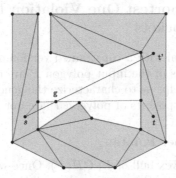

**Fig. 1.** Two subpaths with the same prefix count intersecting at the point $g$.

As it is stated in [8], Lemma 1 is very instrumental in constructing the data structure called *shortest $k$-path map*. Hence the same technique cannot be applied to solve the problem that we study in this paper, as shown in Observation 1.

Because of Observation 1, another nice technique called Bushwhack [23], cannot be applied to our problem. Lemma 1 in [23] states that two optimal paths cannot intersect in the interior of a region.

## 2  Our Contribution

We present an algorithm to compute a $(1 + \epsilon)$-factor approximate shortest $k$-violation path between a pair of points $s$ and $t$ inside a polygon $P$. Our algorithm runs in $O(n^2\sigma^2 k \log n^2\sigma^2 + n^2\sigma^2 k)$ time, where $\sigma = O(\log_\delta(\frac{|L|}{r}))$ and $\delta$, $L$, $r$ are geometric parameters (described in Sect. 4.1). For a special case where the pockets of $P$ are convex and $k = 1$, we present an algorithm that computes a $(1 + \epsilon)$-factor approximate shortest one-violation path in $O(n\sigma \log n\sigma + n\sigma^2)$ time.

## 3    Preliminaries and Notations

Given a simple polygon $P$ of $n$ vertices and two points called the source $s$ and destination $t$ inside the polygon, the objective is to find a $(1 + \epsilon)$-factor shortest $k$-violation path that connects the two points, for any given $\epsilon > 0$. Let $int(P)$ denote the region inside $P$ and $\tilde{P} = \mathbb{R} \setminus int(P)$ be the region outside $P$. A $k$-violation path from $s$ to $t$ $\pi_k(s, t)$ is a path in which at most $k$ path segments lie in $\tilde{P}$. The path segments of $\pi_k(s, t)$ that lie in $\tilde{P}$ are called the violation path segments. The path segments are not allowed to bend in $\tilde{P}$.

**Definition 1.** [14] *A pocket $P_i$ is a polygonal region induced by an edge $e$ of $CH(P)$ that is not an edge of $P$. $P_i$ is bounded by the edge $e$ and a sequence of consecutive edges of $P$. The first and the last edges of the sequence are incident on the two endpoints of $e$.*

## 4    Approximate Shortest One Violation Path for Polygons with Convex Pockets

For simplicity, we first deal with a constrained version of the problem when $(i)$ $k = 1$ and $(ii)$ the pockets of the input polygon $P$ are convex. The solution to the constrained version helps us to characterize the general case when $(i)$ $k$ is a part of the input and $(ii)$ pockets of polygon $P$ are not necessarily convex.

### 4.1    Insertion of Steiner Points

We first compute the convex hull of $P$, $CH(P)$. Once we have the $CH(P)$ and the set of pockets $P_1, P_2, \ldots, P_l$, we triangulate all the pockets $P_i$.

Once we triangulate all the $P_i$, we add Steiner points to the edges of the triangles of the triangulation. We insert the set of Steiner points to discretize the edges bounding the pocket $P_i$ and the faces of triangles of the triangulation.

**Steiner Points:** The technique of placing Steiner points on the edges of the triangles of triangulation is the same as in [2]. For a vertex $v$ of the triangulation, we define the minimum distance from $v$ to the boundary of the union of faces incident on $v$ to be $h_v$. We define $r_v = \epsilon h_v$, for some $\epsilon > 0$. For each vertex $v$ of face $f_i$, let $e_p$ and $e_q$ be the two edges incident on it. We place two Steiner points $p_1$ and $q_1$ on $e_p$ and $e_q$ respectively, at distance $r_v$ from $v$. By definition $|vq_1| = |vp_1| = r_v$. Let $\theta_v$ be the angle between $e_q$ and $e_p$. Now we define:

$$\delta_v = \begin{cases} 1 + \epsilon \cdot \sin \theta_v & \text{for } \theta_v < \frac{\pi}{2} \\ 1 + \epsilon & \text{otherwise} \end{cases}$$

Now, Steiner points $q_1, q_2, \ldots, q_{\mu_q - 1}$ are placed along $e_q$, such that $|vq_j| = r_v \delta_v^{j-1}$, where $\mu_q = \log_{\delta_v}(\frac{|e_q|}{r_v})$. Similarly, $p_1, p_2, \ldots, p_{\mu_p - 1}$ are placed along $e_p$, where $\mu_p = \log_{\delta_v}(\frac{|e_p|}{r_v})$ (see Fig. 2). A set of Steiner points $q_1, q_2, \ldots, q_{\mu_q - 1}$ partition an edge $e_q$ in a set of *intervals* $q_1 q_2, q_2 q_3, \ldots, q_{l-1} q_l$.

Let $dist(a, e)$ be the distance from a point $a$ to an edge $e$.

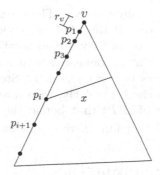

**Fig. 2.** Steiner points.

**Result 1.** [2] $|q_i q_{i+1}| \leq \epsilon \cdot dist(q, e_p)$ and $|p_j p_{j+1}| \leq \epsilon \cdot dist(q, e_q)$ where $0 < i < \mu_q$ and $0 < j < \mu_p$, $q \in q_i q_{i+1}$ and $p \in p_j p_{j+1}$.

In the next step, we create a graph $G_i$ in each pocket $P_i$. Note that this procedure of constructing $G$ and hence the technique discussed in the following section only work when the pockets of $P$ are convex. We consider the Steiner points and the vertices of the triangulation in $P_i$ as the vertices of $G_i$. Two vertices, $u_j$ and $u_l$ of $G_i$ share an arc if and only if they lie on two different edges of the face $f_i$ or they are adjacent points on the same edge of $f_i$. The weight on the arc $(u_j, u_l)$ is the Euclidean distance between $u_j$ and $u_l$. Now we construct the graph $G$. We create shortest path maps [[7], [21]] for $s$ and $t$, called $SPM_s$ and $SPM_t$ respectively. Typically $SPM_s$ is used to compute the shortest paths from $s$ to all points $p \in P$ that stay in $int(P)$. Using $SPM_s$ we connect $s$ to the Steiner points in the boundary of each pocket $P_i$ of $P$. Similarly using $SPM_t$ for $t$, we connect $t$ to the Steiner points in the boundary of each pocket $P_i$.

For the sake of clarity, we call the edges of the triangulation as edges and edges of the Steiner graph $G$ as arcs.

**Lemma 2.** *Total number of Steiner points added is at most $O(n \log_\delta(\frac{|L|}{r}))$, where $|L|$ is the length of the longest edge of the triangulation, $r$ is the minimum among $r_v$ and $\delta$ is the minimum $\delta_v$, over all $v$.*

*Proof.* The number of Steiner points added in a single edge of the triangulation is $O(\log_\delta(\frac{|L|}{r}))$. If there are $n_i$ vertices in $P_i$, then there are $O(n_i)$ edges in the triangulation of $P_i$. Hence in $P_i$ there are $O(n_i \log_\delta(\frac{|L|}{r}))$ Steiner points. So the total number of Steiner points added is $\sum O(n_i \log_\delta(\frac{|L|}{r})) = O(n \log_\delta(\frac{|L|}{r}))$. ■

**Lemma 3.** *$G$ has $O(n \log_\delta(\frac{|L|}{r}))$ vertices and $O(n \log_\delta^2(\frac{|L|}{r}))$ arcs.*

*Proof.* The vertices of $G$ are the Steiner points and the points $s$ and $t$. By Lemma 2, the total number of Steiner points added is at most $O(n \log_\delta(\frac{|L|}{r}))$, hence $G$ has $O(n \log_\delta(\frac{|L|}{r}))$ many vertices. There exists an arc between two Steiner points in $G$ if and only if they belong to two different edges of the same face $f_i$ of the triangulation or they are adjacent points on the same edge of $f_i$. Hence each Steiner point is connected with arcs to $O(\log_\delta(\frac{|L|}{r}))$ many Steiner points. There are arcs between $s$ and the Steiner points on the edges of $\Delta(P)$ that bound the pockets. Similarly, there are arcs between $t$ and the Steiner points on the edges of $\Delta(P)$ that bound the pockets. Since there are $O(n \log_\delta(\frac{|L|}{r}))$ vertices in $G$, therefore there can be at most $O(n \log_\delta^2(\frac{|L|}{r}))$ arcs in $G$.    ∎

For simplicity, hereinafter $\sigma = O(\log_\delta(\frac{|L|}{r}))$.

## 4.2 Approximating Shortest One Violation Path for Polygons with Convex Pockets

Let $\pi_1 = [s, o_1, o_2, \ldots, o_l, t]$ be the optimal one violation path that passes through a sequence edges $e_1, e_2, \ldots, e_l$ of the triangulation in a pocket. In each edge $e_i$ the path $\pi_1$ passes through a point $o_i$ between two Steiner points $a_{\alpha i}$ and $b_{\alpha i}$. We assign the Steiner point closer to $o_i$ among $a_{\alpha i}$ and $b_{\alpha i}$ to $u_{\alpha_i}$. Let $\pi^*(s, t) = [s, u_{\alpha_1}, u_{\alpha_2}, \ldots, u_{\alpha_l}, t]$ be a path in $G$.

**Lemma 4.** *The length of a the path $\pi^*(s, t)$ with path segments through the Steiner points on the edges of the triangular faces of a pocket $P_i$ is at most $(1 + \epsilon) \cdot \pi_1(s, t)$*

*Proof.* $\pi^*(s, t) = |su_{\alpha_1}| + |u_{\alpha_1} u_{\alpha_2}| + |u_{\alpha_2} u_{\alpha_3}| + \ldots + |u_{\alpha_{l-1}} u_{\alpha_l}| + |u_{\alpha_l} t|$
$\leq |so_1| + |o_1 u_{\alpha_1}| + |u_{\alpha_1} o_1| + |o_1 o_2| + |o_2 u_{\alpha_2}| + |u_{\alpha_2} o_2| + |o_2 o_3| + |o_3 u_{\alpha_3}| + |u_{\alpha_3} o_3| + \ldots + |u_{\alpha_{l-1}} u_{\alpha_l}| + |u_{\alpha_l} t|$

$= |so_1| + |o_1 o_2| + |o_2 o_3| + \ldots + |o_{l-1} o_l| + ||o_l t| + 2\{|o_1 u_{\alpha_1}| + |o_2 u_{\alpha_2}| + \ldots + |o_l u_{\alpha_l}|\}$

$\leq |so_1| + |o_1 o_2| + |o_2 o_3| + \ldots + |o_{l-1} o_l| + ||o_l t| + 2\epsilon\{|o_1 o_2| + |o_2 o_3| + \ldots + |o_{l-1} o_l|\}$
(using Result 1)

$\leq (1 + 2\epsilon) \cdot \pi_1(s, t)$    ∎

From the above lemma, we know that $\pi^*(s, t) \leq (1 + 2\epsilon) \cdot \pi_1(s, t)$. Dijkstra's algorithm may produce a different path $\pi^{**}(s, t)$ (Fig. 3a). Dijkstra's algorithm always produces a path of shortest length, therefore $\pi^{**}(s, t) \leq \pi^*(s, t)$. Now, $\pi^{**}(s, t)$ is not a feasible solution to the problem, since we don't allow the one violation path to bend in $\tilde{P}$ and while passing through the edges of the triangulation in a pocket, $\pi^{**}(s, t)$ might have multiple bends. Therefore, once we obtain $\pi^{**}(s, t)$, we can compute the approximate one violation shortest path $\pi_1^*(s, t)$ by replacing the intermediate path segments of $\pi^{**}(s, t)$ between $u_{\alpha_1}$ and $u_{\alpha_l}$ by a line segment $\overline{u_{\alpha_1} u_{\alpha_l}}$, where $u_{\alpha_1}$ and $u_{\alpha_l}$ are the first and last Steiner points between $s$ and $t$ (Fig. 3b). Next we argue that $\pi_1^*(s, t) = [s, u_{\alpha_1}, u_{\alpha_l}, t]$ is a $(1 + \epsilon)$-factor approximate one violation shortest path.

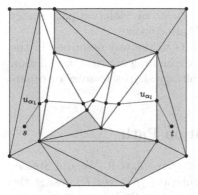 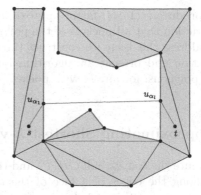

(a) The path $\pi^{**}(s,t)$ returned by Dijk-stra's algorithm

(b) Path $\pi_1^*(s,t)$ produced by joining $u_{\alpha_1}$ and $u_{\alpha_l}$ with a line segment.

**Fig. 3.** Approximate one violation shortest path $\pi_1^*(s,t)$.

**Lemma 5.** $\pi_1^*(s,t) \leq (1 + 2\epsilon) \cdot \pi_1(s,t)$

*Proof.* By Lemma 4, $\pi^{**}(s,t) \leq (1 + 2\epsilon) \cdot \pi_1(s,t)$. Since, to compute $\pi_1^*(s,t)$ we remove all the intermediate path segments between $u_{\alpha_1}$ and $u_{\alpha_l}$, and connect them with a line segment, therefore $\pi_1^*(s,t) \leq \pi^{**}(s,t)$. Thus, $\pi_1^*(s,t) \leq (1 + 2\epsilon) \cdot \pi_1(s,t)$. ∎

**Theorem 1.** *For a polygon $P$ whose pockets are convex, a $(1+\epsilon)$ approximate one violation shortest path between any pair of points $s$ and $t$ can be computed in $O(n\sigma \log n\sigma + n\sigma^2)$, where $\sigma = O(\log_\delta(\frac{|L|}{r}))$.*

*Proof.* The time complexity of our technique is the same as the computation of single source shortest path on $G$ using Dijkstra's algorithm with Fibonacci heap. ∎

### 4.3 Limitations of the Technique

The technique proposed Sects. 4.1 and 4.2 works under the assumption that at least one end of the interval $p_j p_{j+1}$ is visible from at least one end of the interval $q_l q_{l+1}$, where $p_j p_{j+1}$ is the interval through which $\pi_1(s,t)$ enters the pocket and $q_l q_{l+1}$ is the interval through which $\pi_1(s,t)$ leaves the pocket and enters $int(P)$ again. But observe that, if the pockets are not convex, it might not be possible to connect an endpoint of the interval of $p_j p_{j+1}$ to an endpoint of the interval $q_l q_{l+1}$ with a line segment even though the shortest path passes through a point $o_j$ between $p_j p_{j+1}$ and $o_l$ between $q_l q_{l+1}$. Hence the technique discussed in the aforementioned subsections will always produce a $\pi_1^*(s,t)$ if for all $s, t$, if it can be ensured that an endpoint of the interval $p_j p_{j+1}$ is always visible to an endpoint of the interval $q_l q_{l+1}$. As a consequence of this observation, this technique will always produce a $(1 + \epsilon)$-factor approximate shortest one violation path under the restriction that all pockets of the polygon $P$ are convex.

To address this issue we construct the graph $G$ in the way described in the next section and present an algorithm for approximating shortest $k$-violation path in polygons, for any $k$, with pockets of arbitrary shapes (convex or otherwise).

## 5   Approximating Shortest $k$-violation Path

We compute the convex hull $CH(P)$ and then we triangulate it. We add Steiner points along the edges of a face of the triangulation $f_i \in int(P)$, using the strategy as used in Sect. 4.1. In each face $f_i$, each pair of Steiner points $p$ and $q$ on the edges of $f_i$ are connected with an arc if and only if $p$ and $q$ are adjacent Steiner points on the same edge of $f_i$ or $p$ and $q$ are placed on two different edges of $f_i$. The weight of the arc $(pq)$ is the length of the arc.

Now, we remove the edges of the triangulation in each pocket $P_i$. This leaves us with edges of $\Delta(P)$ that bound the $P_i$ and a set of Steiner points on them. We call the piece of an edge between two consecutive Steiner points $p_i p_{i+1}$ an *interval*.

**Definition 2.** *We call a pair of intervals $p_j p_{j+1}$, $q_l q_{l+1}$ weakly visible if and only if there exists at least one pair of points $w \in p_j p_{j+1}$ and $z \in q_l q_{l+1}$, such that $w$ and $z$ are visible to each other.*

If a pair of intervals $p_j p_{j+1}$ and $q_l q_{l+1}$ of a pocket $P_i$ are weakly visible, but either one of the endpoints of $p_j p_{j+1}$ is not visible to either one of the endpoints of $q_l q_{l+1}$, then we find a pair of points $p_\alpha \in p_j p_{j+1}$ and $q_\alpha \in q_l q_{l+1}$ that are visible to each other and add $p_\alpha$ and $q_\alpha$ in the set of Steiner points. We connect $p_\alpha$ and $q_\alpha$ with an arc and add the two points to $V$ and the arc to $E$.

**Definition 3.** *An arc $e \in E$ is called a good arc if and only if $e \subset int(P) \bigcup \Delta(P)$.*

**Definition 4.** *An arc $e \in E$ is called a bad arc if and only if $e \subset \tilde{P}$.*

Once we construct $G$, we add $s$ and $t$ in the set of vertices $V$ and add arcs in $E$ between $s$ (likewise $t$) and the Steiner points on the edges of the face that contains $s$ (likewise $t$).

**Lemma 6.** *$G$ has $O(n^2\sigma^2)$ vertices and edges. $G$ can be constructed in $O(n^2\sigma^2)$ time.*

*Proof.* To construct the *good arcs* of $G$, we connect all pair of Steiner points $p$ and $q$ if and only if $p$ and $q$ are adjacent Steiner points on the same edge of a face $f_i$ in $P$ or $p$ and $q$ are placed on two different edges of $f_i$. The weight of the arc $(pq)$ is the length of the arc.

For the set of bad arcs, in a pocket $P_i$, we check for every pair of Steiner points $p, q$ on the edges of $\Delta(P_i)$, where $\Delta(P_i)$ are the edges of $\Delta(P)$ that bound

$P_i$, if $p$ and $q$ are visible to each other. As mentioned in Sect. 4.3, there may be cases when for a pair of intervals $p_j p_{j+1}$ and $q_l q_{l+1}$, none of the endpoints of $p_j p_{j+1}$ is visible from $q_r q_{r+1}$, but $\pi_k(s,t)$ passes through points $o_j \in p_j p_{j+1}$ and $o_l \in q_r q_{r+1}$. To resolve this issue we check the visibility of every interval $p_j p_{j+1}$ from every other $q_r q_{r+1}$ on the edges of $\Delta(P_i)$. So if at least one endpoint of $p_j p_{j+1}$ is visible to at least one endpoint of $q_r q_{r+1}$, we connect the visible Steiner points with arcs known as *bad arcs*. Otherwise, we check for weak visibility between $p_j p_{j+1}$ and $q_r q_{r+1}$. If there exists at least one pair of points $p_l \in p_j p_{j+1}$ and $q_l \in q_r q_{r+1}$ such that $p_l$ and $q_l$ are visible to each other, we connect $p_l$ and $q_l$ with a bad arc and add $p_l$ and $q_l$ to the set of vertices of $G$. Such newly created Steiner points are called visibility Steiner points. Since there are $O(\sigma)$ Steiner points in each edge of $\Delta(P_i)$, and there exist $n_i$ such edges, if $n_i$ is the number of vertices of $P_i$, there exist $O(n_i \sigma)$ Steiner points. Hence there can be at most $O(n_i \sigma)$ intervals on the edges of $\Delta(P_i)$. Weak visibility polygon of an interval $p_j p_{j+1}$ of $P_i$ can be constructed in $O(n_i \sigma)$ time [5], where we consider the Steiner points to be the vertices of the weak visibility polygon. Now, for every interval $p_j p_{j+1}$ we can construct its weak visibility polygon and check if $p_j p_{j+1}$ is weakly visible from every other interval $q_r q_{r+1}$ in $P_i$. We obtain the pair of visibility Steiner points $p_l$ and $q_l$ that are visible to each other, similarly and connect them with a bad arc. Note that, one of the visibility Steiner points $p_l$ and $q_l$ can possibly be one endpoint of an existing interval. An interval $p_j p_{j+1}$ may be weakly visible to $O(n_i \sigma)$ other intervals on $\Delta(P_i)$ for which we may have to introduce $O(n_i \sigma)$ new pairs of visibility Steiner points and connect them with bad arcs and assign their lengths as their weights. Hence our procedure creates $O(n_i^2 \sigma^2)$ visibility Steiner points in $P_i$. Whenever a pair of visibility Steiner points is added to $G$ a bad arc is also added to $G$. Therefore, there can be $O(n_i^2 \sigma^2)$ bad arcs in $P_i$. Therefore it takes $O(n_i^2 \sigma^2)$ time to add the bad arcs for every pair of intervals in $P_i$. Since $\sum n_i \leq n$, in $O(n^2 \sigma^2)$ time the set of Steiner points and bad arcs can be added to $G$.  ∎

Let $\pi_k^{**}(s,t)$ be the shortest path in $G$ with at most $k$ bad arcs, computed using a modified version of Dijkstra's algorithm proposed in [14] called *Dijkstra's-k-violation algorithm*.

Let the optimal $k$-violation shortest path $\pi_k(s,t)$ pass through a sequence of edges $e_{\alpha_1}, e_{\alpha_2}, \ldots, e_{\alpha_l}$. In each edge $e_{\alpha_j}$, $\pi_k(s,t)$ passes through a point $o_j$ between two Steiner points $a_{\alpha_j}, b_{\alpha_j}$. Therefore, $\pi_k(s,t) = [s, o_1, o_2, \ldots, o_l, t]$.

Now, let us construct a path $\pi_k^*(s,t)$, just as we did in Sect. 4.2. In each edge $e_{\alpha_j}$, we assign to $u_{\alpha_j}$ the Steiner point or visibility Steiner point that is closer to $o_j$ and is an endpoint of a bad arc, among the pair $a_{\alpha_j}, b_{\alpha_j}$. Let $\pi_k^*(s,t) = [s, u_{\alpha_1}, u_{\alpha_2}, \ldots, u_{\alpha_l}, t]$.

**Lemma 7.** *For any edge $o_s o_{s+1}$ of $\pi_k(s,t)$ in a pocket $P_i$, such that $o_s \in p_i p_{i+1}$ and $o_{s+1} \in q_j q_{j+1}$, the length of the line segment connecting one endpoint of $p_i p_{i+1}$ to one endpoint of $q_j q_{j+1}$ is at most $(1 + 2\epsilon) \cdot |o_s o_{s+1}|$.*

*Proof.* Observe that, in each pocket $P_i$, for a pair of intervals $p_i p_{i+1}, q_j q_{j+1}$ that are visible, weakly or totally, one of the following two cases are possible. Either, at least one of the endpoints of the interval $p_i p_{i+1}$ is visible to at least one of the endpoints of the interval $q_j q_{j+1}$ or there exists a point $p_l \in p_i p_{i+1}$ that is visible to $q_l \in q_j q_{j+1}$. $p_l$ and $q_l$ are present in the set $V$ of $G$. Let $\pi_k(s,t)$ pass the boundary of $P_i$ through points $o_s$ and $o_{s+1}$. If $o_s \in p_i p_{i+1}$ and $o_{s+1} \in q_j q_{j+1}$ and $p_i$ is visible to $q_{j+1}$, then by Lemma 4, $|p_i q_{j+1}| \leq (1 + 2\epsilon) \cdot |o_s o_{s+1}|$. If $o_s \in p_i p_l$ and $o_{s+1} \in q_j q_l$, then we have $|p_l q_l| \leq (1 + 2\epsilon) \cdot |o_s o_{s+1}|$ as well, since $p_l \in p_i p_{i+1}$ and $q_l \in q_j q_{j+1}$, hence $|p_l o_s| \leq |p_i o_s|$ and $|q_l o_{s+1}| \leq |q_j o_{s+1}|$. We assign $p_l$ to $u_{\alpha_l}$. ∎

**Lemma 8.** $\pi_k^*(s,t) \leq (1 + 2\epsilon) \cdot \pi_k(s,t)$.

*Proof.* From Lemma 7, the length a bad arc in any pocket $P_i$, connecting the endpoints of the intervals through which $\pi_k(s,t)$ passes, is at most $(1 + 2\epsilon)$ times the length of the path segment of $\pi_k(s,t)$ in $P_i$. Rest of $\pi_k^*(s,t)$ is constructed exactly as we constructed $\pi^*(s,t)$ of Sect. 4.2 and Lemma 4. ∎

**Lemma 9.** Let $\pi_k^{**}$ be the path in $G$ computed by Dijkstra's-k-violation algorithm [14]. Then $\pi_k^{**}(s,t) \leq (1 + 2\epsilon) \cdot \pi_k(s,t)$.

*Proof.* $\pi_k^*(s,t)$ is a path in $G$. Since $\pi_k^{**}$ is the path produced by Dijkstra's-k-violation algorithm, therefore $\pi_k^{**} \leq \pi_k^*$. By Lemma 8, $\pi_k^*(s,t) \leq (1 + 2\epsilon) \cdot \pi_k(s,t)$. Therefore, $\pi_k^{**}(s,t) \leq (1 + 2\epsilon) \cdot \pi_k(s,t)$. ∎

**Theorem 2.** *The $(1 + \epsilon)$-factor approximate k-violation shortest path between any pair of points $s$ and $t$ can be computed in $O(n^2 \sigma^2 k \log n^2 \sigma^2 + n^2 \sigma^2 k)$ on $G$, where $\sigma = O(\log_\delta(\frac{|L|}{r}))$.*

*Proof.* Since $G$ consists at most $O(n^2 \sigma^2)$ vertices and $O(n^2 \sigma^2)$ arcs (by Lemma 6) and $\pi_k^{**}(s,t) \leq (1 + 2\epsilon) \cdot \pi_k(s,t)$ (by Lemma 9), the time complexity is the same as computing single source shortest path with at most $k$ bad edges on a graph with as many vertices and arcs with Dijkstra's-k-violation algorithm [14]. ∎

## 6  Conclusion

In this paper, we studied the shortest $k$-violation path problem in polygons. We produced a $(1 + \epsilon)$-factor algorithm for the problem that runs in $O(n^2 \sigma^2 k \log n^2 \sigma^2 + n^2 \sigma^2 k)$ time, where $\sigma = O(\log_\delta(\frac{|L|}{r}))$. We show that when $k = 1$ and pockets of polygon $P$ are convex, a $(1 + \epsilon)$-factor approximate shortest one violation path can be computed in $O(n\sigma \log n\sigma + n\sigma^2)$ time. A $k$-violation path is allowed to bend only in the interior of a polygon $P$. It will be interesting to see whether an algorithm can be designed to compute a $(1 + \epsilon)$-factor shortest $k$-violation path in sub-quadratic time.

# References

1. Agarwal, P.K., Kumar, N., Sintos, S., Suri, S.: Computing shortest paths in the plane with removable obstacles. In: 16th Scandinavian Symposium and Workshops on Algorithm Theory (SWAT 2018). Schloss Dagstuhl-Leibniz-Zentrum fuer Informatik (2018)
2. Aleksandrov, L., Lanthier, M., Maheshwari, A., Sack, J.-R.: An $\varepsilon$—approximation algorithm for weighted shortest paths on polyhedral surfaces. In: Arnborg, S., Ivansson, L. (eds.) SWAT 1998. LNCS, vol. 1432, pp. 11–22. Springer, Heidelberg (1998). https://doi.org/10.1007/BFb0054351
3. Asano, T., Asano, T., Guibas, L., Hershberger, J., Imai, H.: Visibility-polygon search and Euclidean shortest paths. In: 26th Annual Symposium on Foundations of Computer Science, pp. 155–164. IEEE (1985)
4. Cheng, S.W., Na, H.S., Vigneron, A., Wang, Y.: Approximate shortest paths in anisotropic regions. SIAM J. Comput. **38**(3), 802–824 (2008)
5. Ghosh, S.K.: Visibility Algorithms in the Plane. Cambridge University Press, Cambridge (2007)
6. Ghosh, S.K., Mount, D.M.: An output-sensitive algorithm for computing visibility graphs. SIAM J. Comput. **20**(5), 888–910 (1991)
7. Guibas, L., Hershberger, J., Leven, D., Sharir, M., Tarjan, R.E.: Linear-time algorithms for visibility and shortest path problems inside triangulated simple polygons. Algorithmica **2**(1–4), 209–233 (1987)
8. Hershberger, J., Kumar, N., Suri, S.: Shortest paths in the plane with obstacle violations. In: 25th Annual European Symposium on Algorithms (ESA 2017). Schloss Dagstuhl-Leibniz-Zentrum fuer Informatik (2017)
9. Hershberger, J., Suri, S.: Practical methods for approximating shortest paths on a convex polytope in $\mathcal{R}^3$. Comput. Geom. **10**(1), 31–46 (1998)
10. Hershberger, J., Suri, S.: An optimal algorithm for Euclidean shortest paths in the plane. SIAM J. Comput. **28**(6), 2215–2256 (1999)
11. Kapoor, S., Maheshwari, S.: Efficient algorithms for Euclidean shortest path and visibility problems with polygonal obstacles. In: Proceedings of the Fourth Annual Symposium on Computational Geometry, pp. 172–182. ACM (1988)
12. Lanthier, M., Maheshwari, A., Sack, J.R.: Approximating weighted shortest paths on polyhedral surfaces. In: Proceedings of the Thirteenth Annual Symposium on Computational Geometry, pp. 274–283. ACM (1997)
13. Li, F., Klette, R.: Euclidean shortest paths. In: Li, F., Klette, R. (eds.) Euclidean Shortest Paths, pp. 3–29. Springer, London (2011). https://doi.org/10.1007/978-1-4471-2256-2_1
14. Maheshwari, A., Nandy, S.C., Pattanayak, D., Roy, S., Smid, M.: Geometric path problems with violations. Algorithmica **80**(2), 448–471 (2018)
15. Mitchell, J.S.: A new algorithm for shortest paths among obstacles in the plane. Ann. Math. Artif. Intell. **3**(1), 83–105 (1991)
16. Mitchell, J.S.: Shortest paths among obstacles in the plane. Int. J. Comput. Geom. Appl. **6**(03), 309–332 (1996)
17. Mitchell, J.S.: Geometric shortest paths and network optimization. In: Handbook of Computational Geometry, vol. 334, pp. 633–702 (2000)
18. Overmars, M.H., Welzl, E.: New methods for computing visibility graphs. In: Proceedings of the Fourth Annual Symposium on Computational Geometry, pp. 164–171. ACM (1988)

19. Rohnert, H.: Shortest paths in the plane with convex polygonal obstacles. Inf. Process. Lett. **23**(2), 71–76 (1986)
20. Roy, S., Lodha, S., Das, S., Maheswari, A.: Approximate shortest descent path on a terrain (2007)
21. Snoeyink, J.H.J.: Computing minimum length paths of a given homotopy class. In: Computational Geometry: Theory and Applications. Citeseer (1990)
22. Storer, J.A., Reif, J.H.: Shortest paths in the plane with polygonal obstacles. J. ACM (JACM) **41**(5), 982–1012 (1994)
23. Sun, Z., Reif, J.: BUSHWHACK: an approximation algorithm for minimal paths through pseudo-Euclidean spaces. In: Eades, P., Takaoka, T. (eds.) ISAAC 2001. LNCS, vol. 2223, pp. 160–171. Springer, Heidelberg (2001). https://doi.org/10. 1007/3-540-45678-3_15

# Parametrized Runtimes for Label Tournaments

Stefan Funke[1][(✉)] and Sabine Storandt[2]

[1] Universität Stuttgart, Stuttgart, Germany
funke@fmi.uni-stuttgart.de
[2] Universität Konstanz, Konstanz, Germany

**Abstract.** Given an initial placement of $n$ prioritized labels on a rotatable map, we consider the problem of determining which label subsets shall be displayed in zoomed-out views. This is modelled as a label tournament where the labels are represented as disks growing inversely proportional to a continuously decreasing zoom level. Due to that growth, labels would eventually overlap impairing the readability of the map. Hence whenever two labels touch, the one with lower priority gets eliminated. The goal of the paper is to design efficient algorithms that compute the elimination zoom level of each label. In previous work, it was shown that this can be accomplished within $\mathcal{O}(n^{5/3+\varepsilon})$ time and space. As this is practically infeasible for large $n$, algorithms with a parametrized running time depending not only on $n$ but also on other aspects as the largest disk size or the spread of the disk centers were investigated. This paper contains two results: first, we introduce a new parameter $C$ which denotes the number of different disk sizes in the input. In contrast to previously considered parameters, $C$ is upper bounded by $n$. For the case that disk sizes and priorities coincide, we design an algorithm which runs in time $\mathcal{O}(nC \log^{\mathcal{O}(1)} n)$. Experiments on label sets extracted from OpenStreetMaps demonstrate the applicability of our new approach. As a second result, we present improved running times for a known parametrization of the problem in higher dimensions.

**Keywords:** Map labeling · Parametrization · Nearest neighbor

## 1 Introduction

Map labeling is an extensively studied topic in the realm of geographic information systems as it is crucial for the comprehension of displayed spatial objects. In case the map view is not static but the map can be translated, rotated, and viewed at different zoom levels, it is even more important to choose a labeling that works well across different views. Important aspects are stability and consistency: stability means that the labels remain anchored at the same point on the map and don't move around; consistency means that a label displayed at two different zoom levels has also to be displayed in all the zoom levels in between [3]. In addition, labels should have an appropriate size at each zoom level, should

© Springer Nature Switzerland AG 2019
Y. Li et al. (Eds.): COCOA 2019, LNCS 11949, pp. 181–196, 2019.
https://doi.org/10.1007/978-3-030-36412-0_15

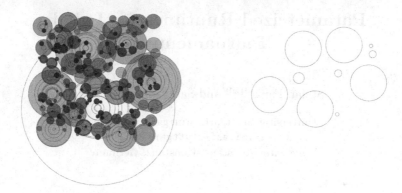

**Fig. 1.** Visualization of the label tournament with 200 labels (left image): Each label is drawn at all sizes at which it induces a collision event with another label, and hence one of the two labels gets eliminated. The darker the label colour, the earlier in time the label was eliminated. The right image depicts the label set to be displayed at a certain point in time according to the elimination times computed with the tournament. Labels in such a set never overlap.

never overlap, and important labels should be given preference. In general, map labeling is an optimization problem, where at any point in time, as many labels as possible should be displayed overlap-free and without violating consistency or stability over time.

It was observed in previous work that label tournaments are a sensible means to achieve all those requirements simultaneously. In a label tournament, we are given a set of label anchor points as well as their initial geometric shape (in this paper we will focus on disk labels anchored at their centers). Zooming is modelled as continuous progression of time; and the label shapes grow linearly over time. If at a certain point in time two labels touch, the less important one (as given by the input) is eliminated. The goal is to compute for each label its elimination time, since then we can simply display for any zoom level (associated with a certain time) the set of labels with a higher elimination time. This then automatically results in an intersection free set of labels; and while zooming in or out consistency can never be impaired. In Fig. 1, a label tournament is visualized along with an induced labeling at one point in time.

In this paper, we present new theoretical and practical results on label tournaments, with a special focus on designing algorithms with parametrized running times that work well on large label sets.

## 1.1 Related Work

The label tournament problem for disks (also called ball tournament problem) was introduced in [7]. There, a simple algorithm with a running time of $\mathcal{O}(n^2 \log n)$ was described, as well as an algorithm with a running time depending on $\Delta$, a parameter which denotes the ratio between the largest and the smallest disk radius in the input. The latter algorithm was considerably simplified and improved in [2], resulting in a running time of $\mathcal{O}(\Delta^2 n(\log n + \Delta^2))$. This algorithm also generalizes to higher dimensions where the labels are represented as balls in $\mathbb{R}^d$. The respective running time was shown to be in $\mathcal{O}(\Delta^d n(\log n + \Delta^d))$. In [1], a multitude of new results on label tournaments was presented. For the generalized problem in $\mathcal{R}^d$, an algorithm with a running time of $\mathcal{O}(d \cdot n^2)$ was described, which works for $d$-dimensional balls as well as boxes. Furthermore, the first subquadratic running times on disks was proven using an advanced query data structure for lower envelopes of algebraic surfaces. In particular, the elimination times of all labels were shown to be computable in time $\mathcal{O}(n^{5/3+\varepsilon})$. Unfortunately, the required space consumption matches that bound which poses a problem when considering large label sets. There were also novel parametrized algorithms provided in [1], with their running time depending on parameter $\Delta$ as well as $\Phi$ (the spread of the label centers). More precisely, one algorithm with a running time of $\mathcal{O}(n \log \Phi \min\{\log \Delta, \log \Phi\})$ and a space consumption of $\mathcal{O}(n \log \Phi)$ was described which leverages quadtrees to efficiently compute which labels will collide next. Using a compressed quadtree data structure, a running time of $\mathcal{O}(n(\log n + \min\{\log \Delta, \log \Phi\}))$ and a space consumption of $\mathcal{O}(n)$ were shown. Those are the currently best results for label tournaments in the plane. They don't generalize to higher dimensions, though. Unfortunately we are not aware of any implementation of the algorithms proposed in [1], probably due to the highly complex data structures that are employed in some of the algorithms. Experimental results are available for the algorithms described in [2]. There, it was shown that for artificial and real-world instances which exhibit a sufficiently small $\Delta$ value, elimination times for millions of labels can be computed within minutes. A related, but different label tournament problem, where instead of eliminating one label, conflicting labels merge into one bigger label was considered and experimentally evaluated in [4] and [5] (with a focus on square labels).

## 1.2 Contribution

Our paper has two main contributions. First, we improve the dependency on parameter $\Delta$ compared to the results in [2] for ball tournaments from $\mathcal{O}(\Delta^d n(\log n + \Delta^d))$ to $\mathcal{O}(n \log \Delta(\log n + \Delta^{d-1}))$ in our new algorithm.

Second, we propose a new parameter $C$ – the number of different disk sizes in the input – and design an efficient label tournament algorithm in case $C$ is small. For our algorithm to work, label priorities have to coincide with the disk size. We argue that this is a natural model for label tournaments, and provide experimental results on real-world labels extracted from OpenStreetMap. The theoretical running time of our algorithm is $\mathcal{O}(nC \log^{\mathcal{O}(1)} n)$. While the running times of the algorithms proposed in [1] only exhibit logarithmic dependency on the parameters $\Delta$ and $\Phi$, respectively, one has to note that those parameters can be arbitrarily large compared to the input size. In contrast, our parameter $C$ is upper bounded by the input size $n$, as there obviously can be no more than $n$ different label sizes in a set of $n$ labels. We demonstrate in the experiments that there are indeed real-world instances with quite large $\Delta$ values for which we can nevertheless compute the elimination times quickly with our new approach.

## 2    Preliminaries

We first provide the formal problem definition of a ball tournament. Then we discuss a simple baseline algorithm, and subsequently review some tools from computational geometry which will be used in our algorithms.

### 2.1    Ball Tournament

The input is a set $B$ of $n$ balls, each ball $b_i = (c_i, r_i, p_i)$ is defined by a center point $c_i \in \mathbb{R}^d$, an associated radius $r_i \in \mathbb{R}^+$, and a priority value $p_i \in \mathbb{R}^+$.

In a *ball tournament*, time $t$ progresses continuously from 0 to $\infty$, and the balls' radii grow linearly with $t$. Hence for any time $t$, if ball $b_i$ was not eliminated yet, its center $c_i$ induces a ball with radius $r_i t$. Accordingly, two 'alive' balls $b_i, b_j$ collide at time $t_{col}(b_i, b_j) = |c_i c_j|/(r_i + r_j)$ and then the one with lower priority gets eliminated. The tournament ends when only one ball remains. This ball $b$ (the one with highest priority) has elimination time $t_{ele}(b) = \infty$. The goal is to compute the (finite) elimination times of all other balls efficiently.

### 2.2    Naive Algorithm and Lower Bound

A straightforward algorithm consists of computing all potential pairwise collision times $t_{col}(b_i, b_j)$ for $i = 1, \ldots, n, j = i+1, \ldots, n$. These $\mathcal{O}(n^2)$ potential *collision events* are then processed in their temporal order. If the two balls involved in the collision are still 'alive', the one with lower priority dies/gets eliminated; otherwise the event is ignored. The running time of this algorithm is dominated by sorting the collision times and hence equals $\mathcal{O}(n^2 \log n)$.

In [1], a lower bound of $\Omega(n \log n)$ for ball tournaments was proven. Therefore, the focus is on the development of algorithms with a (parametrized) running time close to that lower bound.

### 2.3    Nearest Neighbor and Range Queries

The two main questions which will be asked repeatedly in our algorithms are: *What is the ball with its center closest to the center of a ball $b$?* and *Which balls have their center within a certain distance of ball $b$?* More formally, we want to answer nearest neighbor (NN) and range reporting queries (RR).

**Definition 1 (Nearest Neighbor Query).** *For a point set $P \subset \mathbb{R}^d$ and a query point $q$, a nearest neighbor of $q$ is a point $c \in P$ with $|qc| \leq |qc'|, \forall\, c' \in P$.*

**Definition 2 (Range Reporting Query).** *For point set $P \subset \mathbb{R}^d$, query point $q$, and distance $r$, a range reporting query returns the set $S := \{c \in P : |qc| \leq r\}$.*

Both query types are well-studied. For $d = 2$, there is a data structure which supports both query types and that can be constructed in time $\mathcal{O}(n \log^2 n)$. Nearest neighbor queries then take $\mathcal{O}(\log^2 n)$ time and range reporting queries $\mathcal{O}(\log^2 n + k \log n)$ time, where $k = |S|$ is the size of the output of the range reporting query. Furthermore, deletions of points can be handled in expected amortized $\mathcal{O}(\log^6 n)$ time, see [6].

In higher dimension, $d \geq 3$, one has to rely on approximate queries, as no exact DS supporting polylogarithmic query times is known so far. An $\epsilon$-approximate NN (ANN) of $q$ is a point $c \in P$ with $|qc| \leq (1 + \epsilon)|qc'|$, for all $c' \in P$. An $\epsilon$-approximate RR query returns a set $S$ with $S \supseteq \{c \in P : |qc| \leq r\}$ and $S \subseteq \{c \in P : |qc| \leq (1 + \epsilon)r\}$. For arbitrary $d$, a so-called *quadtreap* [8] has the following guarantees: The tree structure is randomized and has height $h = O(\log n)$ with high probability. It supports ANN queries in $O(h + \left(\frac{1}{\epsilon}\right)^{d-1})$, approximate RR queries in $O(h + \left(\frac{1}{\epsilon}\right)^{d-1} + k)$ where $k$ is the output size, and point deletion and insertion in expected time $O(h)$. It is understood that in the following references to running times of quadtreap operations are always in an expected sense or with high probability.

## 2.4   Collision and Update Events

To beat the naive algorithm, we have to avoid creating all possible *collision events* in the first place (as among those $\Theta(n^2)$ potential events only $n - 1$ materialize). Instead we only want to create the ones necessary to find the next collision. Therefore, we create a new so called *update event* which triggers a reexamination of potential upcoming collisions for a certain ball.

Formally, a *collision event* is a triple $(b, b', t_{col}(b, b'))$, consisting of two balls and their collision time. Thereby, the larger of the two balls is denoted by $b$ and is hence always the first entry in the triple. An *update event* is a pair $(b, t)$, consisting of a single ball $b$ and a time $t$ where $b$ needs to be reexamined. We call both events *anchored* at $b$.

An update event $(b, t)$ is created if $b$ certainly does not collide with a smaller ball $b'$ before time $t$. This can be ensured if $b$ – as the larger of the two balls – contains at most half of the segment $cc'$, i.e. $rt < |cc'|/2$. So if we know the NN distance $l$ for $c$, setting $t = l/(2r)$ is a valid update time. As for higher dimensions, we only get the ANN distance $l^* \leq l(1+\epsilon)$, we will use $t = l^*/(2(1+\epsilon)r)$ instead. In [2], $\epsilon = 1$ was used. We will fix a suitable $\Delta$-dependent $\epsilon$ later, which will improve the overall running time.

## 3   An Algorithm with Improved $\Delta$-Dependency

First, we consider ball tournaments in an arbitrary but fixed dimension $d \geq 2$. In this section, we will describe an efficient algorithm with a running time depending on parameter $\Delta$ – the ratio of the largest to the smallest radius in the input.

## 3.1    Algorithm

The algorithm presented in [2] for ball tournaments in arbitrary dimensions uses a priority queue to store collision and update events sorted by their event times.

Our algorithm will be conceptually similar but it will make explicit use of the fact that collisions can always be predicted by the larger of the two collision partners. For ease of exposition, we assume in this section that priorities are unique and no more than two balls collide at the same time. In the following, we describe the basic steps of the algorithm:

1. For all $b \in B$, we create an ANN-based update event and insert it into the priority queue (PQ).
2. As long as the PQ is not empty, we extract the next event. Let $b$ be the anchor of the event, and $t$ the event time. If $b$ is dead, there is nothing to do. Otherwise:
   - If it is an update event, compute the current ANN and the respective new update time $t_{new}$ for $b$. If for $t_{new}$ we have $t_{new} \leq (1 + \epsilon)t$, predict the next collision event for $b$ with a ball $b'$ of smaller or equal radius and insert this event in the PQ. Otherwise, insert the update event $(b, t_{new})$.
   - If it is a collision event and both $b$ and $b'$ are alive, eliminate the one with lower priority and store its elimination time. If $b$ survives, compute an update event for $b$. If the computed update time is before $t$, predict a collision event for $b$ instead.
   - If it is a collision event and $b'$ is dead, compute an update event for $b$. Again, if the update time is before $t$, predict a collision event instead.

**Theorem 1.** *The algorithm computes the correct elimination times.*

*Proof.* We show that for every collision $(b, b', t_{col})$ with $r < r'$, at any time $t \leq t_{col}$, there is either an update or a collision event anchored at $b$ in the PQ with the event time being $\leq t_{col}$, and both $b, b'$ are alive. This enforces that the correct collision event will be inserted latest at time $t_{col}$ and processed accordingly. The conditions are fulfilled after step 1, as update event times underestimate collision times with balls of smaller radius, and for every ball there is such an update event in the PQ. In the loop in step 2, update events are either replaced by new update events or by a collision event. If a collision event anchored at $b$ is inserted in the PQ, the event time is the next possible collision time with a ball of smaller radius. Hence in both cases the event time has to be $\leq t_{col}$. So at the moment the first collision event is extracted from the PQ, it is a real collision. For the smaller ball, there has to be another update or collision event in the PQ anyhow, for the larger ball $b$, in case it does not get eliminated, we insert a new update or collision event. Hence, also after a collision, for all balls still alive there is a respective event in the PQ that does not overestimate the collision time. A collision event where the smaller partner is already dead leads also to the insertion of a new update event or collision for $b$.    □

## 3.2   Implementation Details and Running Time

To have efficient access to balls with smaller or equal radius, we first partition the $n$ balls into $\log_2 \Delta$ classes according to their radii. So class number $i$ contains balls with radii $r \in [2^i r_{min}, 2^{i+1} r_{min}[$ for $i = 0, \cdots, \log \Delta$. Then for each class, we construct a separate quadtreap. In total this takes time $\mathcal{O}(n \log n)$.

**Lemma 1.** *An update event can be computed in time*

$$\mathcal{O}(\log \Delta(\log n + 1/\epsilon^{d-1})).$$

*Proof.* For a ball $b$ we issue an ANN query to the DS of every class with smaller or equal radius, so at most $\log \Delta$ many. We then compute among those candidates the closest to the center of $b$. Each query takes $\mathcal{O}(\log n + 1/\epsilon^{d-1})$ time.     □

Note, that it could be that the returned ANN is actually not a ball smaller than $b$ as desired, but it might have almost twice the radius if it stems from the radius class which $b$ is in. In the following, we first discuss only the case that the NN/ANN has indeed a radius smaller than $b$, and describe the necessary modifications for the other case at the end of the analysis.

Next, we investigate how to predict a collision event induced by processing an update event. We observe that with $l^*$ being the current ANN distance for a ball $b$, the current time is at least $l^*/(2(1+\epsilon)^2 r)$, as $l^*$ is within a factor of $(1+\epsilon)$ of the former ANN $l_0^*$ and this ANN led to the update event $(b, l_0^*/(2(1+\epsilon)r))$.

**Lemma 2.** *For a ball $b = (c, r)$ at time $t \geq l^*/(2(1+\epsilon)^2 r)$, with $l^*$ being the ANN distance, the next collision with a ball $b' = (c', r')$, $r' \leq r$, can be predicted in time $\mathcal{O}(\log \Delta(\log n + 1/\epsilon^{d-1} + \max(\epsilon \Delta^d, \Delta^{d-1})))$ for $\epsilon \in ]0,1]$.*

*Proof.* In case the ANN ball $b_{ANN}$ is *not* the next collision partner for $b$, there either is a ball even closer to $b$ than the ANN or there is a ball further away than the ANN but with a radius proportionally larger.

The NN distance $l$ is $\geq l^*/(1+\epsilon) \geq l^*/(1+\epsilon)^2$. So in the annulus centered at $c$ with $r_1 = l^*/(1+\epsilon)^2$ and $r_2 = l^*$, there can be alternative collision partners. For a ball $b' = (c', r')$ with $l' = |cc'| > l^*$ to be a possible collision partner, the collision time has to be $\leq l^*/r$, as this is an upper bound on $t_{col}(b, b_{ANN})$. So we get $l'/(r + r') \leq l^*/r \Rightarrow l' \leq l^*(1 + r'/r)$. Note that we always have $r'/r \geq 1/\Delta$. For each $x = 1/2^i$, $i = 0, \cdots, \log r$, we issue a RR query $B_{l^*(1+x)}(c)$ to the DS of the class containing radii $rx$. As we only have an approximate RR data structure, the result returns centers up to a distance of $l^*(1 + x)(1 + \epsilon)$. The main question now is how many balls with a radius of $\Omega(rx)$ can fit inside the annulus $r_1 = l^*/(1 + \epsilon)^2$, $r_2 = l^*(1 + x)(1 + \epsilon)$ at time $t = l^*/(2(1 + \epsilon)^2 r)$ (see Fig. 2 for an illustration).

The annulus volume is $V_a = \mathcal{O}(l^{*d}((1 + x)^d(1 + \epsilon)^d - 1/(1 + \epsilon)^{2d}))$. The volume of a ball of radius $rx$ at time $t$ is $V_{b'} = \Omega(l^{*d}x^d/(1 + \epsilon)^{2d})$ (and at least a constant fraction has to be inside the annulus). Hence, $V_a/V_{b'} = \mathcal{O}(((1 + x)^d(1+\epsilon)^{3d} - 1)/x^d)$. Because of the $-1$ in the numerator, after expansion of the product $(1 + x)^d(1 + \epsilon)^{3d}$ all remaining terms are either divisible by $\epsilon$ or $x$. All

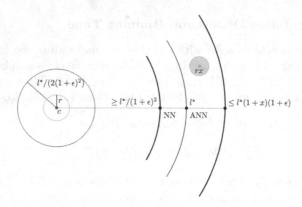

**Fig. 2.** Illustration for the proof of Lemma 2. The annulus for given $x$ is indicated by the two thick arcs.

coefficients are in $\mathcal{O}(2^{4d}) = \mathcal{O}(1)$ for fixed $d$ and can therefore be neglected. As $\epsilon, x \in ]0, 1]$, the largest of the summands after cancellation with the denominator is $\mathcal{O}(\epsilon/x^d)$ in case $\epsilon \geq 1/\Delta$, and $\mathcal{O}(1/x^{d-1})$ otherwise, and we have $\mathcal{O}(d^2) = O(1)$ summands. As $x \geq 1/\Delta$, we get $\mathcal{O}(\max(\epsilon \Delta^d, \Delta^{d-1}))$ as an upper bound on the result size of one approximate RR query. As we issue at most $\log \Delta$ such queries and each of them costs $\mathcal{O}(\log n + 1/\epsilon^{d-1} + \max(\epsilon \Delta^d, \Delta^{d-1}))$, the running time bound follows. □

We observe that the time to predict the next collision is minimized in case $1/\epsilon^{d-1} = \epsilon \Delta^d = \Delta^{d-1}$ which is realized by $\epsilon = 1/\Delta$. Therefore, the time to compute an update as well as a collision is in $\mathcal{O}(\log \Delta(\log n + \Delta^{d-1}))$.

Now that we have running time bounds for the creation of update and collision events, we want to upper bound their total numbers. Let $n_u$ be the number of update events, and $n_c$ the number of collision events inserted in the PQ in the course of the algorithm. We know that there are exactly $n-1$ 'real' collision events, that is, collision events with both balls being still alive at the time of extraction from the PQ. And we know that we initially insert $n$ update events in the PQ. The question now is how many update events lead to new update events and how many collision events in the PQ are affected by a real collision, i.e. one of the collision partners dies before the event time.

**Lemma 3.** *The number of times an update event $(b, t)$ leads to the direct insertion of a new update event $(b, t')$ is in $\mathcal{O}(n)$.*

*Proof.* Creation of a new update event $(b, t')$ is only triggered if $t' > (1 + \epsilon)t$. We observe that this can only happen if the real NN of $b$ was eliminated. Otherwise, if the NN is alive and we get two ANN candidates, their distance difference towards $c$ can never vary more than a factor of $(1 + \epsilon)$. As proven in [2], only a constant number of balls can share the same NN. Hence every elimination of a ball can trigger at most a constant number of new updates. □

**Lemma 4.** *For any ball $b \in B$ and at any time of the algorithm, the number of collision events in the PQ involving $b$ is bounded by a constant.*

*Proof.* The number of collision events with $b$ being the anchor is at most one. So the interesting case is $b = (c, r)$ being the smaller of the two collision partners. We define $b^+ = (c^+, r^+)$ with $r^+ \geq r$ to be the collision partner with the largest center-to-center-distance $k$. In the ball $B_k(c^+)$, there cannot be centers of other balls with a radius $\in [r, r^+]$, as they would be the collision partner for $b^+$ instead of $b$. But there might be balls with a radius $> r^+$ centered in $B_k(c^+)$, as we only looked for the next collision of $b^+$ with a smaller or equal sized ball. So the interesting question is how many balls with a radius $> r^+$ can be centered in $B_k(c^+) \cap B_k(c)$ (see the red marked area in Fig. 3) – as they might be other possible collision partners for $b$.

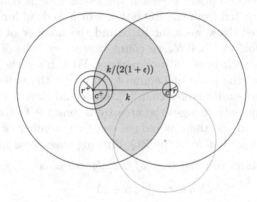

**Fig. 3.** If $(c^+, r^+)$ is the furthest collision partner of $b = (c, r)$ with a distance of $k$, only a constant number of other collision partners of $b$ can be centered in the red area, as they have to have a radius of at least $k/(2(1 + \epsilon))$. Also the red area cuts away a constant fraction of $b$, as does the area induced by the second furthest collision partner outside the red area as illustrated schematically in gray. (Color figure online)

Let $t$ be the time at which the collision event $(b^+, b, t_{col}(b^+, b))$ was inserted in the PQ. We know that $t \geq l^*/(2(1+\epsilon)r^+)$ with $l^*$ being the ANN distance for $b^+$. Furthermore, we know that the distance to a smaller collision partner cannot exceed $2l \leq 2l^*$. So we have $k \leq 2l^*$, and we know that every ball with a radius $\geq r^+$ has to have an induced radius of $\geq l^*/(2(1 + \epsilon)) \geq k/(4(1 + \epsilon)) \geq k/8$. Every ball with a center in $B_k(c^+) \cap B_k(c)$ has at least a constant fraction of its volume inside as well, and the volume of $B_k(c^+) \cap B_k(c)$ is proportional to $k$ as is the volume of the inscribed balls. Hence only a constant number of such balls can be centered there.

So $B_k(c^+)$ cuts away a constant fraction of $B_r(c)$ and we can show that only a constant number of other collision partners can be associated with that volume. We then repeat the argument for the collision partner with next smaller distance to $c$ which was not accounted for, and so on. After constantly many repetitions, the whole volume of $B_r(c)$ has been covered. So in total, only a constant number of larger balls can have $b$ as their current collision partner.                    □

It follows that in total only a linear number of collision events can lead to additional update event insertions. Hence the overall number of update and collision events is $\mathcal{O}(n)$.

In case the ANN ball $b_{ANN}$ of a ball $b$ has a radius $r_{ANN}$ with $2r > r_{ANN} > r$, we need some slight modifications. The respective update event can be inserted as usual. Note also, that Lemma 3 applies unaltered, so the update only triggers another update in case the $NN$ died. But if a collision event has to be predicted, it might happen that the collision partner $b'$ computed by the algorithm described in the proof of Lemma 2 is also a ball of radius larger than $b$. But the collision event $b, b'$ should be anchored at $b'$ and not at $b$. So instead we push a new update $(b, t_{col}(b, b'))$ in the PQ and make sure that collision events are always processed before update events if the event time is equal. So when the update event is extracted from the PQ either $b$ or $b'$ is dead for sure, and a new update gets triggered. Now we want to bound the number of these additional update events. So for each ball $b'$, we count how many balls of at least half its radius can have $b$ as their next collision partner. We can use the very same argumentation here as in the proof of Lemma 4, only that the balls centered in the red area are not necessarily larger than $b^+$, but could have a radius smaller by a factor of 2. As this only changes the area by a constant factor, the remaining calculations stay valid. So there is only a constant number of such additional update events for every ball $b'$ in the PQ, limiting their total number to $\mathcal{O}(n)$.

**Theorem 2.** *The total running time of the algorithm is*

$$\mathcal{O}(n \log \Delta(\log n + \Delta^{d-1})).$$

*Proof.* We need $\mathcal{O}(n \log n)$ to build suitable NN+RR DS for the $\log \Delta$ classes. We then insert a linear number of update and collision events into the PQ (Lemmas 3 and 4) and extract and process them, taking time $\mathcal{O}(n \log n)$ in total. Each of the events can be computed in time $\mathcal{O}(\log \Delta(\log n + \Delta^{d-1}))$ (Lemmas 1 and 2). And we spend time $\mathcal{O}(n \log n)$ on deleting (and reinserting) points in data structures. So the total running time is $\mathcal{O}(n \log \Delta(\log n + \Delta^{d-1}))$.                    □

## 4   Considering Radii Classes

Let $C \leq n$ denote the number of different radii $r_i$ in the input. In this section, we will prove that in case $C$ is small, label tournaments can be performed efficiently if the priorities coincide with the radii.

### 4.1   One Radius Class

If $C = 1$, that is all radii are the same, $\Delta = 1$ follows directly as the largest radius equals the smallest radius. In this setting, the $\Delta$-dependent algorithms exhibit a running time of $\mathcal{O}(n \log n)$ for arbitrary fixed $d$. We note that the lower bound of $\Omega(n \log n)$ also applies to this special input. Hence the above mentioned algorithms have an asymptotically optimal running time.

## 4.2   Larger Balls, Higher Priority

Positive correlation of priority and radius appears naturally in the map labeling domain. The hierarchy of POI, village, town, city, state, country labels, etc. is normally already expressed in the most zoomed-in view. So city labels reserve a larger disk (using a larger font) than town labels, while town labels can be enforced to have (roughly) the same radius. And, if a town and a city label collide, the town label gets eliminated, as a city label is more important. Hence if we have $C$ classes of importance, we can easily translate this into $C$ radii classes where radius and priority coincide – and $C$ is likely to be small in many practical scenarios.

We will now present an algorithm for $d = 2$ which runs in time $\mathcal{O}(Cn \log^{\mathcal{O}(1)} n)$. Note that if there were exact NN and RR data structures for $d \geq 3$ with a running time of $\mathcal{O}(\log^{\mathcal{O}(1)} n + k)$, the same overall running time would apply.

**Algorithm.** The algorithm starts by partitioning the balls into their respective radii classes $R_1, R_2, \cdots, R_C$ with $R_i < R_{i+1}$. With $n_i$ we refer to the number of balls in class $R_i$. Then, for the class $R_C$, we compute the elimination order within. As seen above, this takes time $\mathcal{O}(n_C \log n_C)$.

Thereafter, the suffix of the elimination sequence is already known, as the elimination times for balls in class $R_C$ cannot be influenced by balls with smaller radii. For all other classes $R_i$, there might be collisions with balls in some class $R_j, j \geq i$. We will detect these collision events efficiently in a top-down fashion.

For each $j = C - 1, \ldots, 1$, we consider the balls in $R_j$ in increasing order of their elimination sequence. Let $b$ be the current ball with center $c$ and tentative elimination time $t$. For each class $R_i, i \leq j$ we assume that a RR-data structure is available. We observe that all balls in $R_i$ with their center being further than $(R_j + R_i)t$ away from $c$ can never collide with $b$ before its elimination. So we issue the RR $B_{(R_j+R_i)t}(c)$ to the data structure for class $R_i$ to get all potential collision partners. This takes time $\mathcal{O}(\log^{\mathcal{O}(1)} n + k \log n)$ with $k = |B_{(R_j+R_i)t}(c)|$.

Now all balls in $R_i$ with their center within $(R_j - R_i)t$ of $c$ can only collide with $b$ among all balls in $R_j$ still to be considered, as no such ball can reach them before $b$ does, see Fig. 4 for an illustration. Therefore, we can simply compare the collision time with $b$ with the current collision time assigned to those balls in $R_i$ and update it if necessary.

For balls $q \in R_i$ with their center between $(R_j - R_i)t$ and $(R_j + R_i)t$, $b$ or some other still alive ball $b' \in R_j$ is the collision partner in $R_j$. If $b'$ collides earlier with $q$ than $b$, the distance of the centers of $b'$ and $q$ has to be smaller than $(R_j + R_i)t$. But every such $b'$ is alive at $t$ and hence has a radius of $R_j t$ at the elimination time of $b$. Accordingly, a RR $B_{(R_j+R_i)t}(c(q))$ in the DS for $R_j$ (containing only alive balls at time $t$), will return only a constant number of candidates to check for the real next collision (see again Fig. 4). The data structure for $R_j$ is then maintained by deleting $b$, and all elements in $B_{(R_j+R_i)t}(c(q))$ are removed from the data structure for $R_i$. Note that those deletions are reversed before starting the next round with the next smaller value of $j$.

## 4.3   Analysis

We now want to show that the individual computations for each radii class are not too expensive in case $C$ is small. Hence we analyse the running time of the algorithm described above in dependency of the parameter $C$.

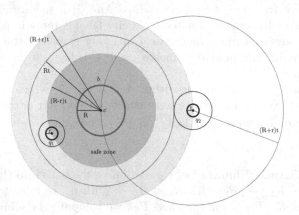

**Fig. 4.** The disk $b$ with radius $R$ gets eliminated at time $t$. Its final size is indicated by the blue cycle. All disks with radius $r < R$ with a center within $(R-r)t$ (green zone) of $c$ never exceed the blue cycle, as shown for $q_1$, and hence cannot collide with any other disk of radius $R$ earlier. In contrast, those with a center in the violet zone exceed the blue cycle before colliding with $b$. The extreme case is illustrated via $q_2$. But any earlier collision partner with radius $R$ for $q_2$ has to have its center within the red circle, while also having a radius of $Rt$ at time $t$ and no two of the candidates are allowed to overlap. Therefore only a constant number of alternatives need to be checked. (Color figure online)

**Lemma 5.** *For a class $R_j$ with known elimination times, all possible collision events with balls in classes $R_i$ for $i = 1, \cdots, j - 1$ can be computed in time $\mathcal{O}((n_j C + n) \log^{\mathcal{O}(1)} n)$.*

*Proof.* The overall running time for collision checks with balls in $R_j$ is composed of the following building blocks:

- $n_j$ range queries per class $R_i, i < j$, taking time $\mathcal{O}(\log^{\mathcal{O}(1)} n + k \log n)$ each, with all $k$ summing up to at most $n$, hence in total $\mathcal{O}(n_j C \log^{\mathcal{O}(1)} n + n \log n)$.
- At most $n$ range queries issued to the data structure of $R_j$ to identify possible collision partners for balls not in the safe zone, demanding time $\mathcal{O}(\log^{\mathcal{O}(1)} n)$ each as $k \in \mathcal{O}(1)$. In total this takes $\mathcal{O}(n \log^{\mathcal{O}(1)} n)$ time.
- Updating the RR-DS for $R_j$, consisting of $n_j$ deletions, hence taking time $\mathcal{O}(n_j \log^{\mathcal{O}(1)} n)$ in total. Similarly updating the RR-DS for all the $R_j$ by performing $O(n)$ deletions (and later reinsertions) takes $\mathcal{O}(n \log^{\mathcal{O}(1)} n)$ in total.

Summing up over the running times of all blocks, we and up with a running time of $\mathcal{O}((n_j C + n) \log^{\mathcal{O}(1)} n)$ for each radii class $j$.     □

The Lemma allows us to bound the running time for the whole algorithm, as shown next.

**Theorem 3.** *The running time to compute the elimination times for all balls in a ball tournament with $C$ radii classes, and priorities being proportional to the radii, is $\mathcal{O}(nC \log^{\mathcal{O}(1)} n)$.*

*Proof.* For each class $R_i$, we need to construct a RR-data structure at most $C$ times, which costs $\mathcal{O}(Cn_i \log^{\mathcal{O}(1)} n)$. Hence summed over all classes, we get construction costs of $\mathcal{O}(Cn \log^{\mathcal{O}(1)} n)$. As the elimination times for $R_C$ are already known, we can invoke Lemma 5 to detect collision events with balls of lower classes. After that, all possible collision times for balls in $R_{C-1}$ with balls in $R_C$ are known. We can now push these $n_{c-1}$ events in a priority queue and then let the algorithm for one radius class, (Sect. 4.1), run using this preinitialized priority queue. This will provide us with the correct collision time for all balls in $R_{C-1}$ in time $\mathcal{O}(n_{c-1} \log n_{c-1})$. Hence, we can now invoke Lemma 5 for class $R_{C-1}$ and so forth. The preinitialized priority queue for class $R_i$ then always contains $n_i$ events, as we only consider the earliest collision event for each ball in $R_i$ with a ball in $R_{j>i}$. So computing the correct elimination times for balls in $R_i$ costs $\mathcal{O}(n_i \log n_i)$ each, and hence $\mathcal{O}(n \log n)$ in total. Therefore, the overall running time can be expressed as $\sum_{j=1}^{C} \mathcal{O}((n_j C + n) \log^{\mathcal{O}(1)} n) = \mathcal{O}(C \log^{\mathcal{O}(1)} n \sum_{j=1}^{C} n_j) + \mathcal{O}(Cn \log^{\mathcal{O}(1)} n) = \mathcal{O}(nC \log^{\mathcal{O}(1)} n)$.     □

According to the theorem, for any $C \in \mathcal{O}(1)$, we end up with a running time of $\mathcal{O}(n \log^{\mathcal{O}(1)} n)$ which matches the lower bound of $\Omega(n \log n)$ up to logarithmic factors. As argued above, in practical applications we expect $C$ to be indeed a small constant. In that case, the resulting running time should be fast enough to manage even huge label sets.

## 5     Experimental Results

To prove feasibility of our algorithm from Sect. 4, we benchmarked a single threaded implementation of this algorithm in C++ (g++ 7.4.0) on an AMD Ryzen 1800X machine with 16 GB of RAM running Ubuntu 18.04. As nearest neighbor and range query data structure we employed the dynamic 2d Delaunay triangulation from CGAL 4.11 allowing for insertions and deletions of points. We compared the algorithm from Sect. 4 with a reimplementation of the algorithm in [2] based on the same nearest neighbor and range query data structure.

We experimented with several synthetically generated data sets as well as real-world data. In the experiments on artificial data, we first fixed the number of classes $C$ and then the radii and number of circles to be generated. Table 1 shows the respective input descriptions and the results. For example, in the row which contains $100 - 10 - 1$ in the first column and $10K - 100K - 1M$

**Table 1.** Benchmarks dependent on the number of radii classes $C$.

| Radii | Instance | C-algorithm | | | | | Δ-algorithm [2] | |
|---|---|---|---|---|---|---|---|---|
| | | NN time(s) | RQ time(s) | RQ max size | RQ avg size | Time (s) | Time (s) | RQ avg size |
| 100 ($C=1$, $\Delta=1$) | 100K | 0.4 | 2.2 | 10 | 4.1 | 3 | 3 | 4.1 |
| | 1M | 5.6 | 177.3 | 10 | 4.1 | 190 | 192 | 4.1 |
| | 10M | 70.5 | 7042.0 | 11 | 4.1 | 7207 | 7213 | 4.1 |
| 60-50-40 ($C=3$, $\Delta=1.5$) | 33K-33K-33K | 0.5 | 2.0 | 37 | 2.1 | 3 | 2 | 4.6 |
| | 333K-333K-333K | 24.3 | 124.1 | 145 | 2.1 | 179 | 149 | 4.6 |
| | 3.33M-3.33M-3.33M | 850.0 | 5541.0 | 86 | 2.1 | 7416 | 5577 | 4.6 |
| 100-10-1 ($C=3$, $\Delta=100$) | 1K-10K-100K | 0.2 | 0.4 | 2206 | 6.2 | 1 | 7 | 109.2 |
| | 10K-100K-1M | 5.1 | 21.4 | 15858 | 6.4 | 38 | 167 | 112.1 |
| | 100K-1M-10M | 444.9 | 963.7 | 92262 | 6.5 | 1611 | 5364 | 113.0 |
| 10000-1000-100-10-1 ($C=5$, $\Delta=10000$) | 10-100-1K-10K-100K | 0.1 | 0.4 | 42086 | 10.4 | 2 | 350 | 7442.3 |
| | 100-1K-10K-100K-1M | 1.9 | 21.3 | 63830 | 10.3 | 42 | 6083 | 11354.8 |
| | 1K-10K-100K-1M-10M | 25.7 | 734.4 | 235622 | 10.7 | 1061 | - | - |
| 10-9-8-7-6-5-4-3-2-1 ($C=10$, $\Delta=10$) | 25-100-400-1.5k-6k-25k 100k-400k-1.6M-6.4M | 269.7 | 2191.0 | 1297700 | 7.1 | 2793 | - | - |
| 50-30-12-6-5-1 ($C=6$, $\Delta=50$) | 195-2k-19k-111k-12M-30M | 492.8 | 16891.2 | 1691000 | 6.2 | 18031 | - | - |

in the second column, we generated $10K$ circles of highest priority and radius 100, $100K$ circles of medium priority and radius 10, and $1M$ circles of radius 1 and low priority. The centers were chosen uniformly at random in $[0, 100000] \times [0, 100000]$, the priorities within each priority class (high, medium, low) were also chosen randomly. The measurements show that our new C-dependent algorithm takes 38 s to compute the elimination order, 5 s of which are spent on nearest neighbor queries and 21 s on range queries. The average result size over all range queries was 6.43, the largest range query result was 15858. Comparing these results to the algorithm from [2], we see that this algorithm takes considerably more time (167 s) mainly due to the more complex (i.e., larger, 113 on average) range query results.

If there is only one class ($C = 1$), however, the two algorithms essentially behave identical. In general, if the circles are all of about the same size (i.e., $\Delta = O(1)$), there is little to no benefit for our new algorithm. On the other hand, if the ratio between largest and smallest radius is high, the algorithm from [2] suffers due to its strong dependence on $\Delta$. For example, in the case with radii $10000, 1000, 100, 10, 1$ and around 11 million circles in total, our new algorithm computed the result within 1060 s, whereas the old algorithm did not complete within a 12 h time frame. Large values of $\Delta$ arise easily in practice if for example icons and text labels are displayed simultaneously. Our algorithm also scales to larger values of $C$, as can be seen from the $10 - 9 - 8 - 7 - 6 - 5 - 4 - 3 - 2 - 1$ instance. We want to emphasize that the computation of an elimination sequence would be part of the preprocessing phase of the whole map rendering pipeline. So running times even of several hours for the planet data set are acceptable.

But again, this instance could not be completed within a reasonable time frame by the old algorithm.

To anticipate the actual instrumentation of our algorithm for visualization of map data, we derived hierarchical data based on OpenStreetMap (OSM). In OSM, there are about 30 million items tagged with keys @amenity, @tourism, @shop, each of which might be drawn as a small icon within a circle of relative size 1. Then there are 1.2 million villages (@place:village), 111 thousand towns (@place:town), 19 thousand cities (@place:city), 2 thousand states (@place:state) and 195 countries, which could be visualized as labels fitting in circles of relative sizes $5, 6, 12, 30, 50$, respectively. The last row in Table 1 reports on the running time with our algorithm. Note that the algorithm from [2] did not complete within 12 h and was hence aborted.

# 6    Conclusions and Future Work

We introduced a new meaningful parameter in the context of label tournaments, namely $C$, the number of different radii in the input. Our novel algorithm, with its running time depending on $C$, was shown to be efficient in theory and practice if $C$ is sufficiently small. Instances which were intractable for prior algorithms with a running time depending on parameter $\Delta$, the ratio of the largest and smallest disk, could be solved with our new approach. In particular, if text *and* icon labels shall be displayed in a map, $\Delta$ is likely to be very large in real-world instances. In that case, our algorithm turns out to be superior.

Future work will focus on closing the gap between the lower bound of $\Omega(n \log n)$ and the currently best known upper bound of $\mathcal{O}(n^{\log 5/3 + \varepsilon})$ for algorithms without any parametrization. For parametrized running times, reduction of the dependencies, e.g., only logarithmic in $C$, as well as lower bound constructions are also of interest. Further research directions include identification of other useful parameters for label tournaments and different ball growth models.

# References

1. Ahn, H.K., et al.: Faster algorithms for growing prioritized disks and rectangles. Comput. Geom. **80**, 23–39 (2019)
2. Bahrdt, D., et al.: Growing balls in $R^d$. In: Algorithm Engineering and Experiments (ALENEX) (2017)
3. Been, K., Daiches, E., Yap, C.: Dynamic map labeling. IEEE Trans. Visual Comput. Graphics **12**(5), 773–780 (2006)
4. Castermans, T., Speckmann, B., Staals, F., Verbeek, K.: Agglomerative clustering of growing squares. In: Bender, M.A., Farach-Colton, M., Mosteiro, M.A. (eds.) LATIN 2018. LNCS, vol. 10807, pp. 260–274. Springer, Cham (2018). https://doi.org/10.1007/978-3-319-77404-6_20
5. Castermans, T., Speckmann, B., Verbeek, K.: A practical algorithm for spatial agglomerative clustering. In: Proceedings of 21st Workshop on Algorithm Engineering and Experiments, ALENEX, pp. 174–185. SIAM (2019)

6. Chan, T.M.: Closest-point problems simplified on the RAM. In: Proceedings of 13th Annual ACM-SIAM Symposium on Discrete Algorithms (SODA), pp. 472–473 (2002)
7. Funke, S., Krumpe, F., Storandt, S.: Crushing disks efficiently. In: Mäkinen, V., Puglisi, S.J., Salmela, L. (eds.) IWOCA 2016. LNCS, vol. 9843, pp. 43–54. Springer, Cham (2016). https://doi.org/10.1007/978-3-319-44543-4_4
8. Mount, D., Park, E.: A dynamic data structure for approximate range searching. In: Proceedings of 26th Annual Symposium on Computational Geometry (SoCG) (2010)

# The $k$-Delivery Traveling Salesman Problem: Revisited

Jinxiang Gan$^{(\boxtimes)}$ and Guochuan Zhang

Zhejiang University, Hangzhou, China
gjx@zju.edu.cn

**Abstract.** The $k$-delivery Travelling Salesman Problem ($k$-dTSP) is one of the most interesting variants of vehicle routing. Given a weighted graph, on which each node either holds an item or requires an item, and a vehicle of capacity of $k$ which can carry at most $k$ items at a time, one is asked to schedule a shortest route for the vehicle so that it starts from a depot and transports all items from the points holding them to the points requiring items, and returns home under the vehicle capacity constraint. It is clearly an NP-hard problem. Among quite a few results the best-known approximation algorithm is due to Charikar, Menuier and Raghavachari (2001). All the previous approaches first construct an approximate TSP tour. It is then expanded to a (much) larger but bounded one by repeatedly moving the vehicle on the tour, so that the pickup/delivery tasks can be completed. It motivates us to consider a simpler approach, that directly applies an optimal algorithm on a circle graph after we derive an approximate TSP tour. It is known that $k$-dTSP on a path graph is polynomially solvable, while the complexity is open on a circle graph. In this paper, we settle this issue by presenting a polynomial algorithm on a circle, and then apply it to $k$-dTSP on general graphs. Although the theoretical bound is yet unclear, experiments show that it outperforms the existing approaches.

**Keywords:** Traveling salesman problem · Pickup and delivery · Circle

## 1 Introduction

The Vehicle Routing Problems (VRP) have been widely studied in computer science and operations research. A survey on these problems can be found in [10]. The Vehicle Routing Problem with Pickups and Deliveries (VRPPD) is one of the significant classes in the vehicle routing problems. VRPPD refers to the problems that goods are transported from pickup to delivery points without exceeding the capacity of the vehicle. There are two versions of VRPPD: the preemptive one and the non-preemptive one. In the preemptive version, the items are droppable, i.e, they can be dropped at temporary locations along the route before being moved to their final destinations. In this paper, we focus on the case that the preemption of loads is not allowed (the non-preemptive version). We formulate the problem, called $k$-dTSP as follows.

© Springer Nature Switzerland AG 2019
Y. Li et al. (Eds.): COCOA 2019, LNCS 11949, pp. 197–209, 2019.
https://doi.org/10.1007/978-3-030-36412-0_16

*The k-Delivery Traveling Salesman Problem (k-dTSP)*

Given an undirected graph $G = (V, E)$, where $V$ is a set of points and $E$ is a set of edges, each edge $e$ has a non-negative cost $c(e)$ satisfying the triangle inequality. The point set $V = V^+ \cup V^-$ is partitioned into a set $V^+$ of positive points and a set $V^-$ of negative points. Each positive point in $V^+$ provides one item, while each negative point in $V^-$ requires one item. In this problem, all items are identical, meaning that any item can be used to satisfy any negative point. A vehicle with capacity $k$ picks up an item from a positive point and delivers it to a negative point. We call that the vehicle serves a point, if it picks up an item at the positive point or delivers an item at the negative point. It can be assumed that $|V^+| = |V^-|$ and $k < |V^+|$. The problem aims to determine a minimum cost tour for the vehicle so that all the items in positive points can be transported to negative points without exceeding the capacity of the vehicle.

*Related Work*

The $k$-delivery TSP is NP-hard as TSP is a special case of the problem. There is a lot of research about $k$-dTSP. For the general problem, Chalasani and Motwani [4] give a 9.5 approximation algorithm. The ratio is improved to 5 by Charikar et al. [5]. It is also currently the best-known approximation ratio for $k$-dTSP in a graph. After that, Hernández-Pérez and Salazar-González [7] introduce a 0-1 integer linear model for this problem and describe a branch-and-cut procedure for finding an optimal solution. Their exact approach is applied to solve instances with 40 points. Furthermore, Hernández-Pérez and Salazar-González [8] also propose a heuristic approach for the problem, which is applied to solve hard instances with up to 500 points.

In addition, there are quite a few studies for $k$-dTSP on special graphs. Wang and Lim [11] study the complexity of $k$-dTSP on a path and on a tree. The authors propose an optimization algorithm in $O(n/\min\{k, n\})$ time for the path case, where $n$ is the number of points in the graph. In the same paper, they reduce the 3-partition problem to $k$-dTSP on a tree. It shows that the problem on tree networks is NP-hard even when the height of the tree is 2. Lim and Wang [9] propose a 2-approximation algorithm for the problem on a tree. After that, Bhattacharya and Hu [3] improve the result of the problem both on a path and on a tree. The authors propose a Come Back rule. With this rule, they propose a linear time algorithm to find an optimal solution for $k$-dTSP on a path and improve the approximation ratio to $\frac{5}{3}$ for $k$-dTSP on a tree.

*Our Results*

- It is known that $k$-dTSP on a path graph is polynomially solvable, while the complexity is open on a circle graph. We settle this issue by presenting a polynomial time algorithm for it. Our approach is to analyze how many times each edge must be traversed and find a feasible solution that traverses each edge exactly these times.
- Based on the circle algorithm, we propose a simple but efficient heuristic approach for $k$-dTSP on a general graph: we directly apply the optimal algorithm on a circle graph after we derive an approximate TSP tour. Although

the theoretical bound is yet unclear, experiments show that it outperforms the best-known approximation algorithm in [5].

We will introduce the outline of our approach and the structure of this paper in next section.

## 2    Outline of the Algorithm

Our work aims at proposing an efficient heuristic algorithm for $k$-dTSP on a general graph. In this section, we give an outline of the algorithm and the motivation of it.

We first review the algorithm for $k$-dTSP on a general graph. The first constant factor approximation algorithm is proposed by Chalasani and Motwani [4]. They find two TSP tours, one contains all positive points and the other contains all negative points. These tours are broken into paths containing $k$ points each. For these paths, find a minimum-weight perfect matching between the paths containing positive points and the paths containing negative points. After that, they get a feasible tour and show that its performance ratio is at most 9.5. Depending on the same two TSP tours and a perfect matching, Anily and Bramel [1] construct feasible tours in two ways (clockwise and counterclockwise) and show the ratio of the smaller one of these two feasible tours is at most 7. The best-known 5-approximation algorithm is proposed by Charikar et al. [5]. They construct only one TSP tour containing all points, break the tour into many paths and combine them by matching positive and negative points in a similar way.

All these algorithms have three steps:

---
*Step* 1.  Find a TSP tour.
*Step* 2.  Cluster points on the TSP tour.
*Step* 3.  Find a perfect matching between positive clusters and negative clusters.

---

The main idea of their algorithms is to expand the TSP cycle to a (much) larger but bounded one by repeatedly moving the vehicle on the cycle, so that the pickup/delivery tasks can be completed. We now give a sufficient and necessary condition for a Hamiltonian Cycle where there is a feasible tour which traverses each edge exactly once.

**Lemma 1.** *Let $a_v \in \{1, -1\}$ indicate $v$ to be a positive point or a negative point. Let $[p, q]$ denote the line segment from $p$ to $q$ along clockwise direction. For a Hamiltonian Cycle $\mathcal{C}$, there is a depot $r$ and the tour, which starts from the depot and traverses each edge $e \in \mathcal{C}$ exactly once along a direction (clockwise or counterclockwise), is feasible for $k$-dTSP iff for any line segment $[p, q] \in \mathcal{C}$, $|\sum_{v \in [p,q]} a_v| \leqslant k$ (see Appendix for the proof).*

Based on Lemma 1, if we can get a Hamiltonian Cycle satisfying some properties, there is a feasible tour naturally. Unfortunately, these properties are too strict to be met. That is why the previous approaches carry out a series of complicated steps after they construct an approximate TSP tour. It motivates us to consider a simpler approach, that directly applies an efficient algorithm on a circle graph after we derive an approximate TSP tour.

---

Step 1.  Find a TSP tour.
Step 2.  Apply the algorithm for a circle to the TSP tour.

---

Now the problem on a general graph is converted into the problem on a circle, and a circle network is similar to a path network. It is known that $k$-dTSP on a path graph is polynomially solvable, while the complexity is open on a circle graph (raised in [2]). For proposing an efficient algorithm for the problem on a circle, we review the result of the problem on a path. Bhattacharya and Hu [3] propose a linear time algorithm for the problem on a path. We will refer to their algorithm as the "Come Back Algorithm", which consists of:

---

Step 1.  Divide the path into some line segments.
Step 2.  Construct a route for each segment, which serves all points in the segment. Each route starts in an end point of the segment and ends in the other.
Step 3.  Combine these routes into a feasible tour.

---

We still want to use these three steps to solve the $k$-delivery problem on a circle. However, for obtaining an optimal algorithm, we need to carefully refine the algorithm in each step:

- In the first step, the line segments have significant properties, which are divided from the path. We will show that these line segments, which have the same properties, exist in a circle and can be found in polynomial time in Sect. 3.1.
- In the second step, the route, constructed by the Come Back Algorithm, in each segment is proven to the optimal, because of the fact that there is only one way to transport an item between two points in a path. In contrast, there are two ways to transport an item between two points in a circle, clockwise and counterclockwise. In Sect. 3.2, we will show that there exists an optimal solution where the route, constructed by the Come Back Algorithm, in each segment is optimal.
- In the third step, there is only one way to combine routes into a feasible tour on a path. In contrast, we will discuss the possibility of combining routes on a circle in Sect. 3.3.

After we deal with the problems from each step which is applied on a circle, we propose an algorithm for $k$-dTSP for a circle and show that it can find an optimal solution in polynomial time in Sect. 4.1. Furthermore, we do some experiments to test the performance of our heuristic algorithm in Sect. 4.2. Finally, we will conclude this paper in Sect. 5.

## 3    The k-dTSP on a Circle

In this section, we will investigate the properties of the optimal solution on a circle.

### 3.1    Zero Segment

Firstly, we will introduce the line segments which are divided from a path in the Come Back Algorithm.

A subset of points is called **balanced** if it is of even cardinality, where half of its points are positive and the remaining are negative. Each edge divides a path into two parts. In the first step of the Come Back Algorithm, if an edge $e$ divides a path into two parts which are balanced, the algorithm will remove it from the path. Hence, after the first step, each line segments is balanced. Bhattacharya and Hu [3] show that there exists an optimal solution where the vehicle will not transport the items from one segment to another. These line segments are called **zero segment**.

There are some properties in each zero segment:

*Property 1.*

- Each zero segment is balanced.
- For the two end points of a zero segment, one is positive and the other is negative.
- If we remove an edge in the segment, remaining two line segments are unbalanced.

We now show that there exist zero segments in an optimal solution of the problem on a circle. All items in positive points have to be transported to negative points in $k$-dTSP. Hence there exists a one-to-one matching between a positive point and a negative point for any feasible solution. The transport route of a positive item to a negative point is called a *service route*. If the item is transported from $u$ to $v$, it is denoted by $u \to v$. The definition of the service route is directional, as shown in Fig. 1:

**Fig. 1.** The arrow line means that $u$ serve $v$

By analyzing the optimal solution of the problem, it is found that the service route of the optimal solution can be divided into some segments. If two service routes have common edges, we call the common part of two service routes is the overlap. If the point $w$ is in the route $u \rightarrow v$, it is called that $w$ is covered by $u \rightarrow v$. For a route $u_1 \rightarrow v_1$, if $u_1$ and $v_1$ are covered by $u_2 \rightarrow v_2$, it is called that $u_1 \rightarrow v_1$ is covered by $u_2 \rightarrow v_2$.

**Lemma 2.** *There exists an optimal solution such that there is no overlap in two service routes which have different directions* (see Appendix for the proof).

By Lemma 2, there exists an optimal solution where there are no overlaps in two service routes which have different directions. We define a starting point and an ending point by service routes, for dividing the optimal solution into several pieces.

**Definition 1.**

- **Starting point:** *A positive point is called a starting point if it is not covered by a service route.*
- **Ending point:** *A negative point is called a ending point if it is not covered by a service route.*

It can be observed that the starting points and the ending points are one-to-one correspondent. We can show that a line segment from the starting point to the ending point is a zero segment.

For the convenience, we define a function $n(e)$ to describe the difference between the number of positive points and the number of negative points in a line segment. Fix a depot $r$ on the circle and determine a direction(clockwise or counterclockwise). The direction determined is called the positive direction. After that, define a function $n(e)$.

$$\forall e \in E, \ n(e) = POS(e) - NEG(e)$$

where $POS(e)$ is the number of positive points between $r$ and edge $e$ (including $r$) along the direction determined, and $NEG(e)$ is the number of negative points between $r$ and edge $e$ (including $r$) along the same direction. Now, we have:

**Theorem 1.** *A line segment from the starting point to the ending point is a zero segment* (see Appendix for the proof).

For any solutions in the $k$-delivery TSP on a path, it is easy to find starting points and ending points because the end point of the path is either the starting point or the ending point. However, there are no end points in a circle, so we need to show that the existence of starting points and ending points in some optimal solutions.

**Lemma 3.** *There exists an optimal solution whose service routes have starting points and ending points* (see Appendix for the proof).

Lemmas 2, 3 and Theorem 1 show that there exists an optimal solution whose service routes can be divided into some zero segments. Now, we define zero segments on a circle.

In the following definition, we break the circle into zero segments by the function $n(e)$:

**Definition 2.** *After fixing a depot $r$ and determining a direction, we compute $n(e)$ for each $e \in E$. Break the circle into some segments by removing all edges $e$ which $n(e) = 0$. Call these edges cut edges. In a segment, it can be observed that $n(e)$ of all edges have the same sign. Each segment is called a **zero segment**.*

*Furthermore, for a segment $P$:*

– *If $\forall e \in P$, $n(e) > 0$, we call $P$ is a **positive direction zero segment**.*
– *If $\forall e \in P$, $n(e) < 0$, we call $P$ is a **negative direction zero segment**.*

It can be observed that a division of a circle is determined after the depot and the direction of $n(e)$ are determined. In addition, there are $n$ points in the circle and two directions (clockwise and counterclockwise). Therefore, all possibilities of divisions can be enumerated in polynomial time. Furthermore, it means that we can find the zero segments which are divided from an optimal solution in polynomial time.

### 3.2 Lower Bounds in a Zero Segment

In this subsection, we will show that the Come Back Algorithm still find an optimal route in each zero segment.

**Definition 3.** *In a zero segment, for each edge $e$, it divides the zero segment into two parts. Let $S^+(e)$ and $S^-(e)$ respectively denote the part which has more positive points and the part which has more negative points. As shown in Fig. 2, A black point represents a positive point and a white point represents a negative point.*

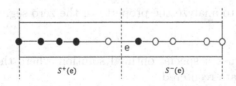

**Fig. 2.** The figure is a zero segment. Black points are positive points and white points are negative points. The edge $e$ divides the segment into two parts: $S^+(e)$ and $S^-(e)$

In a zero segment $P$, for each $e \in P$, the items in $S^+(e)$ must be transported to $S^-(e)$ by passing $e$. Since the vehicle can carry at most $k$ items at a time, each edge $e$ must be traversed at least $\lceil \frac{n(e)}{k} \rceil$ times from $S^+(e)$ to $S^-(e)$. After serving some negative points in $S^-(e)$, the vehicle will return to $S^+(e)$ which still has items. The vehicle must return to $S^+(e)$ from $S^-(e)$ at least $\lceil \frac{n(e)}{k} \rceil - 1$ times. If the zero segment is in a path graph, it can be observed clearly that each edge must be traversed $2 \cdot \lceil \frac{n(e)}{k} \rceil - 1$ times.

However, because of the property of a circle, there are two ways to go back to $S^+(e)$ from $S^-(e)$: through $e$ or not through $e$, so it is hard to get a tight lower bound of the times that each edge must be traversed. Let $\lceil \frac{n(e)}{k} \rceil - 1 - t$ denote the times that the vehicle return to $S^+(e)$ from $S^-(e)$ by passing $e$.

If $t = 0$, it is the case the same as the path case. Bhattacharya and Hu [3] show that *Come-Back Algorithm* can give an optimal route. In the route from the starting point to the ending point, the vehicle serves all points in a zero segment. For an edge $e$ in a zero segment, the route crosses $e$ exactly $2 \cdot \lceil \frac{|n(e)|}{k} \rceil - 1$ times.

If $t > 0$, the edges not in the zero segment $P$ must be traversed at least $t$ times. We give a definition of this case as follows:

**Definition 4.** *For a zero segment $P$, there is an edge $e \in P$ and the vehicle return to $S^+(e)$ from $S^-(e)$ by passing $e$ less than $\lceil \frac{n(e)}{k} \rceil - 1$ times. We call the vehicle **reenter** the zero segment.*

*If the vehicle return to $S^+(e)$ from $S^-(e)$ by passing $e$ exactly $\lceil \frac{n(e)}{k} \rceil - 1 - t$ times $(t > 0)$, it is called that the zero segment is reentered $t$ times.*

If there are no zero segments which are reentered in an optimal solution, we can compute an optimal route for each segment by Come-Back Algorithm. By contrast, we need to pay our attention to the optimal solution where there is a zero segment which is reentered.

After studying, we find that the optimal solution where there is a zero segment which is reentered can be transformed to the optimal solution where there are no zero segment which are reentered.

**Theorem 2.** *There exists an optimal solution whose zero segments are not reentered.*

The proof of the Theorem 2 is divided into the following steps:

(1) Firstly, we need to analyze the property of the zero segment which is reentered.

(2) Then, we construct a special optimal solution where there are some zero segments which are reentered.

(3) Finally, we find a new optimal solution where there is no zero segment which is reentered from transforming the special optimal solution constructed.

We will describe the proof of Theorem 2 in detail (includes Lemma 5–8) in Appendix.

### 3.3    The Possibilities of the Optimal Route

By the Sect. 3.2, we know that there exists an optimal solution whose zero segments are not reentered and we can find an optimal route for each zero segment. In this subsection, we discuss how to combine these routes into an optimal solution.

For serving all points on a circle, there is at most one edge in the circle, which is not traversed by the vehicle. There are two cases of the optimal solution: all the edges have been traversed by the vehicle; only one of the edges is left without being traversed.

**Case 1:** *There is an edge $e_0$ which is not traversed in the optimal solution.*

It is the case of $k$-dTSP on a path. By analyzing in [3,11], each edge on the circle except $e_0$ is traversed at least $2 \cdot \max\{\lceil \frac{|n(e)|}{k} \rceil, 1\}$ times in an optimal solution.

**Case 2:** *All edges on the circle are traversed in the optimal solution.*

**Lemma 4.** *In order to merge optimal routes in each segment into a **Case 2** optimal solution, the edge $e$ in a negative direction zero segment are traversed at least $2 \cdot \lceil \frac{|n(e)|}{k} \rceil + 1$ times (see Appendix for the proof).*

We analyze how many times each edge must be traversed in an optimal solution.

- For the cut edge ($n(e) = 0$), it is traversed at least once.
- For the edge in the positive direction zero segment, it is traversed at least $2 \cdot \lceil \frac{|n(e)|}{k} \rceil - 1$ times for serving all points.
- For the edge in the positive direction zero segment, it is traversed at least $2 \cdot \lceil \frac{|n(e)|}{k} \rceil + 1$ times.

## 4    Algorithm

Based on the properties of an optimal solution, we propose an algorithm for $k$-dTSP on a circle. After that, we do some experiments to test the performance of our heuristic algorithm.

### 4.1    Algorithm for a Circle

The structure of the algorithm of a circle is the same to the algorithm for a path. We now give details of the algorithm for a circle:

**Step 1:** For each point in the circle, we set it as the depot respectively. After depot is fixed, we run the algorithm with clockwise and counterclockwise as the positive direction respectively.

**Step 2:** After fixing a depot $r$ and determining a direction, we compute $n(e)$ for each $e \in E$. Break the circle into some segments by removing all edges $e$, where $n(e) = 0$.

**Step 3:** Run the Come-Back Algorithm in each zero segment.

**Step 4:** The two final tours is as follows: (The solid line represents the route constructed by the Come-Back Algorithm. The dashed line represents that the edges are traversed once. The left and the right figure is the optimal solution in Case 2 and Case 1 respectively)

**Case 1:** The vehicle traverses all edges in the circle along the positive direction. When entering a positive direction zero segment, the vehicle follows the route constructed by the Come Back Algorithm and serves all points in it. When entering a negative direction zero segment, the vehicle goes to the starting point of the segment. Then, the vehicle follows the route constructed by the Come Back Algorithm and serves all points in it along the negative direction. Finally, it returns to the starting point of the segment. When encountering a cut edge, the vehicle traverses it along the positive direction until returning to the depot. As shown in the left.

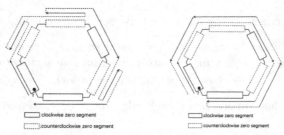

**Case 2:** Let $e_0$ be the edge satisfying $r = v^-(e_0)$ (where $v^-(e)$ is the point following $e$ along the direction determined), the vehicle traverses all edges except the edge $e_0$. Let the vehicle goes to $v^+(e_0)$ (where $v^+(e)$ is the point preceding $e$ along the direction determined) along the positive direction. When entering a positive direction zero segment, the vehicle follows the route constructed by the Come Back Algorithm and serves all points in it. When entering a negative direction zero segment, the vehicle passes it without serving any points in it.

After the vehicle arrives at $v^+(e_0)$, it returns to the depot $r$ along the negative direction. When entering a negative direction zero segment, the vehicle follows the route constructed by the Come Back Algorithm and serves all points in it. When entering a positive direction zero segment, the vehicle passes it without serving any points in it. As shown in the right.

**Step 5:** Return the shortest tour from amongst the $4n$ tours constructed, where $n$ is $|V|$.

In this algorithm, Step 1 and Step 2 are to find the zero segments which are divided from an optimal solution, Step 3 is to compute an optimal route for each zero segment and Step 4 is to combine these routes into an optimal solution. We now analyze the quality of the algorithm for a circle.

**Theorem 3.** *The algorithm finds an optimal solution for k-dTSP on a circle in polynomial time.*

*Proof.* For each point $r$, used as a deport, we run the Come-Back Algorithm whose running time is $O(n)$. Therefore, the running time of the algorithm which we propose is $O(n^2)$.

For serving all points, each edge $e$ in a zero segment is traversed $2 \cdot \lceil \frac{|n(e)|}{k} \rceil - 1$ times by Come-Back Algorithm.

If there is an edge $e_0$ is not traversed in the optimal solution, our algorithm will find the $e_0$ by enumerating the depot and the positive direction. Each edge $e$ in a zero segment is traversed $2 \cdot \lceil \frac{|n(e)|}{k} \rceil - 1$ times. In addition, all edges in a zero segment are traversed once. The cut edges except $e_0$ are traversed twice. It is exactly the lower bound that each edge must be traversed.

If all edges on the circle are traversed in the optimal solution, our algorithm will find the division by enumerating the depot. After depot is fixed, we run the algorithm with clockwise and counterclockwise as the positive direction respectively. Hence, we will find the direction along which all edges are traversed. In our algorithm, the edges in a positive direction zero segment are traversed $2 \cdot \lceil \frac{|n(e)|}{k} \rceil - 1$ times. The edges in a negative direction zero segment are traversed $2 \cdot \lceil \frac{|n(e)|}{k} \rceil + 1$ times. The cut edges are traversed once. From Lemma 4, we know that it is exactly the lower bound that each edge must be traversed.

Therefore, we find an optimal solution in polynomial time.

### 4.2 Heuristic Algorithm and Experiments

In this paper, our work aims to propose an efficient heuristic algorithm for $k$-dTSP on a general graph. The structure of the algorithm has two steps: (1) find a TSP tour for all points, (2) use the algorithm for a circle on the TSP tour. Specifically, the optimal algorithm for $k$-dTSP on a circle is shown in Sect. 4.1.

We now test the performance of our approach and compare with the 5-approximation algorithm, which is currently the best-known [5]. We generate 30 random instances. For each $n \in \{300, 500, 1000\}$, where $n$ is the number of points in the graph, we generate $n$ random pairs in the square $[0, 5000] * [0, 5000]$, each corresponding to a point in the graph. After that, we randomly select half of $n$ points as positive points and the other half as negative points. The cost $c_{ij}$ is computed as the Euclidean distance between the points $i$ and $j$.

To evaluate the quality of the generated feasible solutions, we first compute a TSP tour for the generated graph by Christofides algorithm [6]. Based on the same TSP tour, two algorithms each constructs feasible solutions respectively. We then make a comparison between the two tours.

The experiment result is shown in the Appendix, due to the page limit. In the experiment, the running time of our approach is dominated by the running time of Christofides Algorithm, so the instances scale which we can compute depends on the scale which Christofides Algorithm can compute. In addition, the running time of our approach is less than the 5-approximation algorithm, which contains a matching procedure.

In all generated instances, the result of our approach is no worse than the 5-approximation algorithm in any test examples. The average improvement is about 15% and the best improvement is 36.4%. When the gap between the capacity and the number of stations becomes larger, the more difference between the TSP tour and a feasible solution. Specifically, when the gap is narrowed to $\frac{50}{1000} = \frac{1}{20}$, the two algorithms make almost no difference. From the experiments, our approach works very well. However, the theoretical error bound is yet open.

## 5   Conclusions and Future Work

In this paper, we aim to propose a heuristic algorithm for $k$-dTSP on a general graph. Our main contribution is to design an optimal algorithm for the problem on a circle. Basing on the circle algorithm, we propose an efficient heuristic approach for $k$-dTSP on a general graph: we directly applies the optimal algorithm on a circle graph after we derive an approximate TSP tour. Although the theoretical bound is yet unclear, experiments show that it outperforms the best-known approximation algorithm in [5].

An obvious remaining question is to figure out an approximation bound for our approach. We believe a small ratio is there. It is also interesting to generalize the model to the scenario that each point holds (or requires) more items, that finds a great application in bike rebalancing in public (or sharing) bike systems.

## References

1. Anily, S., Bramel, J.: Approximation agorithms for the capacitated traveling salesman problem with pick-ups and deliveries. Naval Res. Logist. **46**(6), 654–670 (1999)
2. Benchimol, M., et al.: Balancing the stations of a self service "bike hire" system. RAIRO-Oper. Res. **45**(1), 37–61 (2011)
3. Bhattacharya, B., Hu, Y.: K-delivery traveling salesman problem on tree networks. In: 32nd International Conference on Foundations of Software Technology and Theoretical Computer Science, pp. 325–336 (2012)
4. Chalasani, P., Motwani, R.: Approximating capacitated routing and delivery problems. SIAM J. Comput. **28**(6), 2133–2149 (1999)
5. Charikar, M., Khuller, S., Raghavachari, B.: Algorithms for capacitated vehicle routing. SIAM J. Comput. **31**(3), 665–682 (2001)
6. Christofides, N.: Worst-case analysis of a new heuristic for the traveling salesman problem. Technical Report 388, Graduate School of Industrial Administration, Carnegie-Mellon University (1976)
7. Hernández-Pérez, H., Salazar-González, J.: A branch-and-cut algorithm for a traveling salesman problem with pickup and delivery. Discrete Appl. Math. **145**(1), 126–139 (2004)
8. Hernández-Pérez, H., Salazar-González, J.: Heuristics for the one-commodity pickup-and-delivery traveling salesman problem. Transp. Sci. **38**(2), 245–255 (2004)
9. Lim, A., Wang, F., Xu, Z.: The capacitated traveling salesman problem with pickups and deliveries on a tree. In: Deng, X., Du, D.-Z. (eds.) ISAAC 2005. LNCS, vol. 3827, pp. 1061–1070. Springer, Heidelberg (2005). https://doi.org/10.1007/11602613_105

10. Parragh, S.N., Doerner, K.F., Hartl, R.F.: A survey on pickup and delivery problems. Journal für Betriebswirtschaft **58**(2), 81–117 (2008)
11. Wang, F., Lim, A., Xu, Z.: The one-commodity pickup and delivery travelling salesman problem on a path or a tree. Networks **48**(1), 24–35 (2006)

# Algorithmic Pricing for the Partial Assignment

Guichen Gao[1], Li Ning[1(✉)], Hing-Fung Ting[2], Yong Zhang[1], and Yifei Zou[2]

[1] Shenzhen Institutes of Advanced Technology, Chinese Academy of Sciences, Shenzhen, People's Republic of China
{gc.gao,li.ning,zhangyong}@siat.ac.cn
[2] Department of Computer Science, The University of Hong Kong, Hong Kong, People's Republic of China
{hfting,yfzou}@cs.hku.hk

**Abstract.** A seller has an items set $M = \{1, 2, \ldots, m\}$ and the amount of each item is 1. In the buyer set $N = \{1, 2, ..., n\}$, each buyer $i$ has an interested bundle $B_i \subseteq M$, a valuation $v_i$ on $B_i$, and a set of budgets $e_{ij}$ on each item $j$. The seller knows the whole information of the buyers and decides the price for each item so as to maximize the social welfare. Buyers come in an arbitrary order. When a buyer comes, each item can be only sold integrally. In previous works, if a buyer's interested bundle does not exist, he will buy nothing and cannot be satisfied. In this paper, we consider the buyer can be partially satisfied. If some items in $B_i$ are sold out on the arrival of buyer $i$, he might be partially satisfied by buying a subset of $B_i$ without violating the budget condition. We first show that in this new model, achieving the maximum social welfare is NP-hard. Moreover, an optimal assignment oracle can be used to achieve the maximum social welfare. We then analyze two pricing schemes to approximate the optimal social welfare. In the trade-off pricing scheme, an $O(k)$-approximation algorithm can be achieved if considering the subset of the bundle, where $k$ is the maximal cardinality of the assigned bundle among all buyers. In the item-set pricing scheme, when $d = \frac{1}{2}|x_{iB}^*|$, the social welfare can be approximated within the factor 2, moreover, compared with the traditional (non-partial assignment) setting, the social welfare can be increased by at least $\sum_{S_i} \frac{d}{|x_{iB}^*|} \sum_{j \in S_i} e_{ij}'$, where $S_i$ is the buyer subset containing item $j$, $x_{iB}^*$ represents the ith optimal assignment subset's complete set, i.e, if the optimal assignment item set is one buyer bundle $B_i$'s subset, then $x_{iB}^* = B_i$, $x_{iB}$ is the intersection of $B_i$ and the optimal assignment item set $S$, $d$ is the minimum of $|x_{iB}|$ and $|x_{iB}^*| - |x_{iB}|$, and $e_{ij}'$ is the optimal assignment item's budget.

## 1 Introduction

Selling (or pricing) is a fundamental problem in Economics and is well studied in the field of optimization. Consider a situation in the supermarket, according

Y. Li et al. (Eds.): COCOA 2019, LNCS 11949, pp. 210–222, 2019.
https://doi.org/10.1007/978-3-030-36412-0_17

to the knowledge of the statistical data of buyers, the seller decides the price for each item. The price of an item is the same for all buyers, who come to the supermarket in an arbitrary order. To maximize the social welfare, i.e., the total revenue of the seller and the utilities of all buyers, the prices must be decided carefully. The higher price can get more revenue, however, many buyers will not buy it due to the expensive price. On the other hand, if the price is low, buyers with low utility may get this item and result in a low social welfare. In the real case, each buyer may have an interested bundle (or package) of items to buy, which makes the problem more complicated.

Such pricing problem has been studied for many years. Chawla et al. [5] discussed the pricing problem with the unit-demand requirement, they gave a constant approximation pricing scheme assuming that the buyers valuation are drawn from a distribution. Following the above pricing problem, there are a lot of works on the variants and extensions of the pricing model. Cai et al. analyzed the revenue-optimal mechanism by duality method and gave an $O(\log m)$-approximation algorithm when buyers' valuations are subadditive [4]. In the single minded setting, each buyer is only interested a subset (bundle) of all items. Cheung et al. studied such problem under the limited supply assumption [7]. If each buyer can buy more bundles, Zhang et al. gave an $O(\sqrt{k} \cdot \log h \log k)$-competitive algorithm where $k$ is the number of item types and $h$ is the highest price [13]. Recently, the competitive ratio of the above model was improved to $O(\sqrt{k} \cdot (\log h + \log k))$ by Chin et al. [8]. Besides this, Babaioff et al. considered an additive buyer whose valuation is drawn from a given distribution, and gave a constant-approximation algorithm for the bundle pricing [2]. In recent years, online pricing is the main focus of the selling problem. Online fashion is closer to the reality and is more complicated than the offline version. For the online resource allocation problems, Chawla et al. considered two kinds of buyer characteristics to design the pricing mechanisms and they showed tight or nearly tight approximation algorithm to maximize the social welfare [6]. Besides the above methods, Azar et al. implemented the linear programming duality method to analyze the online lower bounds [1]. Furthermore, some works paid more attention on designing pricing mechanisms [3,11]. Bei et al. proposed a framework that the seller only have some partial information, they analyzed several DSIC mechanisms and showed that a constant approximation mechanism can be achieved with the help of SPM [3]. Jin et al. studied a single item selling problem and analyzed the revenue gaps among four mechanisms [11].

What's more, a simple case is that the seller only have one item to sell, he will sell it to the first buyer whose valuation is higher than the designated price. This case has a good property which is similar to *take-it-or-leave* pricing. A nature extension is whether there exists good assignment for $m$ different items to make the arrival buyer selecting his interested bundle which is not sold. There are some literatures to discuss the above extension that some methods and mechanisms were given. Dütting et al. proposed a prophet inequalities method to solve the combinatorial problem and proved that there is a relationship between the posted price mechanism and some smooth mechanisms [9]. In addition, Feldman et al. studied the combinatorial auctions by a Bayesian framework and gave a 2-approximation algorithm for the anonymous posted price mechanisms [10].

In this paper, we consider the case that even if some item in a buyer's interested bundle is sold out, he may partially satisfied by the remaining items. This model makes sense in reality since some item may appear in many buyers' interested bundles and it is often sold out due to the arbitrary arrival order among buyers. If a buyer buys nothing, his utility cannot be realized and the social welfare will experience loss too. Theoretically speaking, if we ignore the impact caused by the subsets of bundles, our model retrogress to the models in previous works. Thus, it is believed that our model is more general than the previous models. Although there are many existing works for the online pricing problem, the social welfare cannot be improved a lot due to the fact that buyers either buys the whole bundle or nothing.

In this paper, we firstly analyze the complexity of the problem and prove that it is NP-hard to maximize the social welfare. With the help of the optimal item assignment *oracle*, the optimal pricing for the partial assignment can be achieved within polynomial time lookups. Moreover, we give an $O(k)$-approximation algorithm for the problem, where $k$ is the maximal cardinality of the assigned bundle among all buyers. Compared with the previous model, we prove that our algorithm can increase the social welfare by at least $\sum_{S_i} \frac{d}{|x_{iB}^*|} \sum_{j \in S_i} e'_{ij}$, where $S_i$ is the buyer subset containing item $j$, $x_{iB}$ is the intersection of $B_i$ and the optimal assignment item set $S$.

## 2  Problem Statement

The seller has $m$ kinds of items which are denoted as the items set $M = \{1, 2, \ldots, m\}$, and each kind of item has only one. Buyers arrive in an online fashion and let $N = \{1, 2, \ldots, n\}$ represents the buyer set. Each buyer $i$ is associated with his interested bundle $B_i \in M$. Let $B = \{B_1, B_2, \ldots, B_n\}$. For each buyer $i$, he has a valuation $v_i$ w.r.t. $B_i$ and budget $e_{ij}$ for each item $j$.

We assume that the buyer can be partially satisfied. Given the price vector $p = \{p_1, \ldots, p_m\}$, if buyer $i$ gets the whole bundle $B_i$, his utility is $v_i - \sum_{j \in B_i} p_j$; while if he gets a subset $S_i \subset B_i$, his utility is $\sum_{j \in S_i} e_{ij} - \sum_{j \in S_i} p_j$. When a buyer $i$ arrives, he selects the item set to maximize his utility. The trivial conditions is $v_i > \sum_{j \in S_i} e_{ij}$ for all $S_i \subset B_i$. In order to design the seller's selling mechanism, we assume that for each item $j$, there is $e_{ij} \leq \frac{1}{|B_i|} v_i$. And by this assumption, we will give a simple proof that when the buyer expected bundle $B_i$ exists, the buyer doesn't buy the bundle's subset.

**Lemma 1.** *When the buyer expected bundle $B_i$ exists, the buyer will not buy the bundle's subset.*

*Proof.* When there exists the buyer $i$'s expected bundle $B_i$, and the buyer $i$ buys it, his utility is $u_B = (v_i - \sum_{j \in B_i} p_j)^+$. If the buyer doesn't buy $B_i$, he buys the subset $S_i$ of $B_i$, then his utility is $u_{S_i} = (\sum_{j \in S_i} e_{ij} - \sum_{j \in S_i} p_j)^+$. W.l.o.g, considering that the $j \in B_i$ has the same value, then the buyer's utility can be represented as:

$$u_B = v_i - |B_i|p_j \geq 0, \quad u_{S_i} = \sum_{j \in S_i} e_{ij} - |S_i|p_j \geq 0$$

Now, we only need to prove that $u_B - u_{S_i} \geq 0$. And we use the assumption $e_{ij} \leq \dfrac{1}{|B_i|}v_i$. Thus, we give the proof as following:

$$u_B - u_{S_i} = v_i - |B_i|p_j - \sum_{j \in S_i} e_{ij} + |S_i|p_j = v_i - \sum_{j \in S_i} e_{ij} - (|B_i| - |S_i|)p_j$$

$$\geq v_i - \sum_{j \in S_i} \frac{1}{|B_i|}v_i - (|B_i| - |S_i|)p_j \geq \frac{|B_i| - |S_i|}{|B_i|}v_i - (|B_i| - |S_i|)p_j$$

$$= \frac{|B_i| - |S_i|}{|B_i|}(v_i - |B_i|p_j)$$

$$\tag{1}$$

- When $p_j$ is given according to the item set is sold as bundle, then there is $v_i - |B_i|p_j \geq 0$, i.e, $u_B \geq u_{S_i}$;
- When $p_j$ is given according to the item set is sold as subset, then there is $\sum_{j \in S_i} e_{ij} - |S_i|p_j \geq 0$. And, it satisfies:

$$\sum_{j \in S_i} e_{ij} - |S_i|p_j \leq \frac{|S_i|}{|B_i|}v_i - |S_i|p_j = \frac{|S_i|}{|B_i|}(v_i - |B_i|p_j)$$

Then, $(v_i - |B_i|p_j) \geq 0$. Thus, $u_B \geq u_{S_i}$.

And when $j \in B_i$ has different value, we can use $\sum_{j \in B_i - S_i} p_j$ to analyze the buyer's utility. And the whole process is consistent with the above.  ∎

The social welfare is the total revenue of the seller plus the total utilities of the buyers, i.e., $\sum_{B_i} v_i + \sum_{j \in S_{i'}} e_{i'j}$, such that each item belongs to at most one $B_i$ or subset $S_{i'}$.

In this problem, the seller has all information of the buyers, however, the difficulty is buyers may come in an arbitrary order. The objective is to maximize the social welfare by deciding the price of each item before the arrival of buyers. Note that when the prices are determined, they cannot be changed and kept the same for all buyers.

## 3   Complexity Analysis

In this part, we prove that the partial bundle selling problem is NP-hard. The proof will be done by the reduction from the Weighted Set Packing problem, which is one of Karp's 21 NP-Hard problems [12].

**The Weighted Set Packing:** Given a finite set $U = \{e_1, e_2, \ldots, e_m\}$ and a collection of subset $X = \{X_1, X_2, \ldots, X_n\}$, where each $X_i \subseteq U$ is associated with a nonnegative weight $w_i$ and $\cup_{i=1}^{n} X_i = U$. The Weighted Set Packing problem is to find the maximum weighted subset collection $X' \subseteq X$, i.e., maximize $\sum_{X_i \in X'} w_i$ such that $X_i \cap X_j = \emptyset$ for all $X_i, X_j \in X'$.

**Example 1.** *Consider a finite set $U = \{a, b, c, d, e, f\}$, and a collection $X$ contains five subsets, such that $X_1 = (a, b, c)$, $X_2 = (a, d)$, $X_3 = (b, c)$, $X_4 = (a, e)$, $X_5 = (e, f)$. Each subset has the unit weight. For the collection $X$, the optimal packing is $X' = \{X_2, X_3, X_5\}$ with the total weight 3.*

**Theorem 1.** *The partial selling problem is NP-hard.*

*Proof.* Consider a weighted set packing problem instance: $U = \{e_1, e_2, \ldots, e_m\}$. A collection of subset $X = \{X_1, X_2, \ldots, X_n\}$ where each $X_i \subseteq U$ is associated with a nonnegative weight $w_i$ and $\cup_{i=1}^{n} X_i = U$. The construction of the Partial Bundle Selling problem is shown as follows.

1. The set of items: $M = U$, i.e., each element in $U$ is an item in $M$;
2. The subset of items: $B_i = X_i$, i.e., each subset $X_i$ is a bundle w.r.t. buyer $i$.
3. The valuation and budget: for buyer $i$, $v_i = w_i$ for each bundle $B_i$ and $e_{ij} = 0$ for any $S_i \subset B_i$.

The above construction can be done in linear time. Next, we show that the weighted set packing get the optimal value $OPT$ if and only if the partial bundle selling problem get the same maximum social welfare $OPT$.

$\Rightarrow$: Suppose the weighted set packing has a solution to achieve $\sum_{i \in X'} w_i = OPT$ such that $X_i \cap X_j = \emptyset$ for all $X_i, X_j \in X'$. The collection of bundles contains $B_i$ if $X_i \in X'$. In this way, we can see that it is a solution of the partial bundle selling problem with the total welfare $OPT$.

$\Leftarrow$: Suppose the weighted partial bundle selling problem has a solution with the social welfare $OPT$. According to the construction, only the whole bundle sold is counted, otherwise, the utility is non-positive if a buyer is assigned with a partial bundle. Similar to the above analysis, a packing solution with the total weight $OPT$ can be achieved.

Therefore, maximizing the total social welfare of the partial bundle selling problem is NP-hard. ∎

## 4    Two Pricing Schemes

Even through the seller knows all the information for buyers, it is still impossible to design a pricing scheme for the seller to maximize the social welfare in every case. Specifically, we consider an example with three items and two buyers $a$, $b$. Buyer $a$ wants to buy random two items with the bundle $v_1 = 2$, and budget price $e_{1j} = 0.8$ for each item $j$. Buyer $b$ wants to buy all the items with bundle $v_2 = 3$, and budget price $e_{2j} = 0.7$ for each item $j$. In the above assumptions, the maximized social welfare is 3. However, no matter how the seller decides on the pricing, there is always a worst case in which the achieved social welfare is not the maximum one. Thus, it is significant to design a pricing scheme to make the social welfare close to the maximized one.

In the following, two pricing schemes are presented, both of which guarantee a non-trivial social welfare compared with the maximized one. And $x^*$ is the optimal assignment on pricing for seller by executing the optimal assignment oracle. In this paper, we use the widely adopted "oracle" as a tool to get the optimal assignment, which will be used in our schemes.

## 4.1   The Trade-Off Pricing Scheme

When analysing the impact of pricing on social welfare, a trade-off phenomenon between revenue and utilities caused by pricing is observed. When the prices of items set by seller are small enough, every buyer can afford the price for his expected bundle. Then, the sum of utilities of the buyers are large while seller get a very small revenue. When the prices get larger, the sum of utilities decrease while revenue increases. In the case that the prices are sufficient large, items will not be bought by buyers. However, the case with a large revenue but small utilities is also not close to the optimal case. Inspired by the trade-off phenomenon, we propose a trade-off pricing(TOP) scheme to set prices for items as:

$$
p_j = \begin{cases} \sum_{j \in S_i} e_{ij}(x^*_{iS_i})/(2|x^*_{iS_i}|) & j \in S_i \\ v_i(x^*_{iB})/(2|x^*_{iB}|) & j \in B_i \end{cases},
$$

to satisfy the seller's revenue and the buyers' utility, where $p_j$ is the price of item $j$, and $B_i$ is the bundle which the buyer i wants to buy. We assume that all buyers information are known by the seller except the arriving order of the buyers. Thus, there is an optimal assignment for seller to sell items. And we have $x^*_{iB}$ to denote the i-th optimal assignment bundle that the seller wants to sell. $S_i \subset B_i$ is the subset of bundle $B_i$, $x^*_{iS_i}$ is the subset of i-th optimal assignment, $v_i(x^*_{iB})$ represents the valuation of i-th optimal assignment bundle, and $\sum_{j \in S_i} e_{ij}(x^*_{iS_i})$ is the budget price of the i-th optimal assignment subset. Also, it deserves to note that every item $j$ only belongs to at most one set.

In the next, we first show that the price set by our trade-off pricing scheme reach a well balance between the revenue of seller and the sum of utilities of buyers.

- This pricing scheme can guarantee the buyers utilities when seller selling the items, which means that buyers can buy what they want and their utilities will not be too bad. Considering the optimal assignment $x^*$, it can implement the optimal bundle valuation $v(x^*_B)$ and the optimal subset budget $e_j(x^*_j)$. Thus the optimal social welfare $(\mathcal{W}_{OPT})$ can be represented as:

$$
\mathcal{W}_{OPT} = v(x^*_B) + e_j(x^*_j)
$$

When $|x_{iB}^*| = |x_{iS_i}^*| = 1$, since the seller sells items by integer, for each sold item, we have:

$$p_j \le \frac{1}{2} v_i\left(x_{iB}^*\right) + \frac{1}{2} \sum_{j' \in S_i} e_{ij'}\left(x_{iS_i}^*\right)$$

Thus, for all sold items, our TOP scheme makes sure that the revenue ($\mathcal{R}$) at most $\frac{1}{2}\left(v\left(x_B^*\right) + e_j\left(x_j^*\right)\right)$. It is clear that the sum of the pricing is lower than the pricing in optimal partial assignment ($\mathcal{W}_{OPT}$). And for each buyer $i$, his utility is equal to $v_i - \sum_{j \in B_i} p_j$ or $\sum_{j \in S_i} e_{ij} - \sum_{j \in S_i} p_j$. And all of the buyers utilities ($\mathcal{U}$) can be represented as:

$$\mathcal{U} = \sum_i \left(\left(v_i - \sum_{j \in B_i} p_j\right) + \left(\sum_{j \in S_i} e_{ij} - \sum_{j \in S_i} p_j\right)\right) = \sum_i \left(v_i + \sum_{j \in S_i} e_{ij}\right) - \sum_{j \in M} p_j$$

$$\ge \sum_i \left(v_i + \sum_{j \in S_i} e_{ij}\right) - \frac{1}{2}\left(v(x_B^*) + e_j(x_j^*)\right) \ge 0$$

Thus, the buyers can buy the items.

- As the same as above, this pricing scheme can guarantee the sellers revenue by making that the buyer with a lower valuation buys nothing. For $\forall B$ and $S_i$, the total pricing of $B$ and $S_i$ in the optimal assignment $x^*$ is at least $\frac{1}{2k}$ or $\frac{1}{2d}$, with $k = \max |x_{iB}^*|$, and $d = \max |x_{iS_i}^*|$. And due to $max|x_{iB}^*| \ge max|x_{iS_i}^*|$, so $k \ge d$. The total pricing can be represented as:

$$\mathcal{R} \ge \frac{1}{2k} \sum_i v_i\left(x_{iB}^*\right) + \frac{1}{2d} \sum_i \left(\sum_{j \in S_i} e_{ij}(x_{iS_i}^*)\right) \ge \frac{1}{2k}\left(v(x_B^*) + e_j(x_j^*)\right)$$

Similar to the above, the buyers utilities can be represented as:

$$\mathcal{U} = \sum_i \left(\left(v_i - \sum_{j \in B_i} p_j\right) + \left(\sum_{j \in S_i} e_{ij} - \sum_{j \in S_i} p_j\right)\right) = \sum_i \left(v_i + \sum_{j \in S_i} e_{ij}\right) - \sum_{j \in M} p_j$$

$$\le \sum_i \left(v_i + \sum_{j \in S_i} e_{ij}\right) - \frac{1}{2k}\left(v(x_B^*) + e_j(x_j^*)\right)$$

That means the buyers utilities can not be too high, then the seller's revenue can be guaranteed.

Thus, the seller has a certain revenue ($\mathcal{R}$) and the buyers' utility ($\mathcal{U}$) is guaranteed.

Then, with the help of trade-off pricing scheme, the seller assign the items as is described by Algorithm 1.

---

**Algorithm 1.** Item Assignment for seller according to TOP scheme

---

1  **Input:** Sets: $B$, $M$, $N$; and all $v_i$, $e_{ij}$ for $i \in [1,n]$, $j \in [1,m]$;
2  **Output:** The Social Welfare ($\mathcal{W}_A$);
3  **Initialize:** $\mathcal{W}_A = 0$;
4  Seller use the *Optimal Assignment Oracle* to know the $\boldsymbol{x}^*$;
5  Seller sets the prices for each item according to TOP scheme.
6  **for** *each buyer $i = 1$ to $n$* **do**
7      **if** $\exists$ *bundle $B_i$ and* $\sum_{j \in B_i} p_j \leq v_i$ **then**
8          $M = M - B_i$;
9          $\mathcal{W}_A = \mathcal{W}_A + v_i$;
10     **else if** $\exists$ *subset $S_i$ and* $\sum_{j \in S_i} p_j \leq \sum_{j \in S_i} e_{ij}$ **then**
11         $M = M - S_i$;
12         $\mathcal{W}_A = \mathcal{W}_A + \sum_{j \in S_i} e_{ij}$;

---

   **return** $\mathcal{W}$;

---

**Lemma 2.** *By applying the trade-off pricing scheme, our algorithm reaches a $O(k)$-approximate social welfare.*

*Proof.* Let set $I \subseteq N$ represent the buyers set in which the demand of the buyers is satisfied. $\boldsymbol{x}$ is the arrival buyer's purchase decision. The analysis on the social welfare ($\mathcal{W}_A$) of our TOP scheme can be divided into two parts: the seller's revenue ($\mathcal{R}_A$) and the buyers' utility ($\mathcal{U}_A$).

(1) Obviously, for the revenue:

$$\mathcal{R}_A \geq \frac{1}{2k} \sum_{i \in I, j \in B_i} v_i\left(x_{iB}^*\right) + \frac{1}{2d} \sum_{i \in I, j \in S_i} \left( \sum_{j \in S_i} e_{ij}\left(x_{iS_i}^*\right) \right)$$

(2) For utility, due to the order of buyers arrival, they may want to buy the items in $x_i^*$, but they don't achieve it. These buyers are represented by $i \in N - I$. Then there are the buyers utilities from the $x_i^*$. Obviously, for each buyer $i \in N - I$, his utility can be represented as: $v_i(x_{iB}^*)$-the pricing of $x_{iB}^*$ or $\sum_{j \in S_i} e_{ij}\left(x_{iS_i}^*\right)$-the pricing of $x_{iS_i}^*$. And according to the trade-off pricing scheme, the pricing of $x_{iB}^* \leq \frac{1}{2}v(x_B^*)$ and the pricing of $x_{iS_i}^* \leq \frac{1}{2}e_j(x_j^*)$. Thus all of the buyers $i \in N - I$, their utilities are at least as:

$$\mathcal{U}_A \geq \frac{1}{2k} \left( \sum_{i \in N-I,\, j \in B_i} v_i\left(x_{iB}^*\right) - \frac{1}{2}v\left(x_B^*\right) \right) + \frac{1}{2d} \left( \sum_{i \in N-I,\, j \in S_i} e_{ij}\left(x_{iS_i}^*\right) - \frac{1}{2}e_j\left(x_j^*\right) \right)$$

Now, we will analyze the social welfare as following:

$$\mathcal{W}_A = \mathcal{R}_A + \mathcal{U}_A$$

$$\geq \frac{1}{2k} \sum_{i \in I, j \in B_i} v_i\left(x_{iB}^*\right) + \frac{1}{2d} \sum_{i \in I, j \in S_i} \left(\sum_{j \in S_i} e_{ij}\left(x_{iS_i}^*\right)\right)$$

$$+ \frac{1}{2k}\left(\sum_{i \in N-I, \ j \in B_i} v_i\left(x_{iB}^*\right) - \frac{1}{2}v\left(x_B^*\right)\right) + \frac{1}{2d}\left(\sum_{i \in N-I, \ j \in S_i} e_{ij}\left(x_{iS_i}^*\right) - \frac{1}{2}e_j\left(x_j^*\right)\right)$$

$$= \frac{1}{4k}v(x_B^*) + \frac{1}{4d}e_j(x_j^*)$$

Thus, $\mathcal{W}_A \geq \frac{1}{4k}v\left(x_B^*\right) + \frac{1}{4d}e_j\left(x_j^*\right)$. Because of $d \leq k$, so the social welfare satisfies $\mathcal{W}_A \geq \frac{1}{4k}\left(v\left(x_B^*\right) + e_j\left(x_j^*\right)\right)$. And the optimal social welfare $(\mathcal{W}_{OPT})$ is $\mathcal{W}_{OPT} = v\left(x_B^*\right) + e_j\left(x_j^*\right)$, so the competition ratio is $\rho = \frac{\mathcal{W}_{OPT}}{\mathcal{W}_{Algo}} \leq 4k$. Based on the above result, considering buyer bundle's subset, our model is $O(k)$-approximation.

## 4.2   The Item-Set Pricing Scheme

Since the seller can know the optimal assignment $x^*$ according to the optimal assignment oracle, the valuation $v_i'$ and $\sum_{j \in S_i} e_{ij}$ of each optimal assignment are also accessible for the seller. For an optimal assignment set, we say it is an item set as $S$. In the item-set pricing(ISP) scheme, the seller directly set a price for each item set to make sure that the items can be sold as many as possible.

Specifically, for an arbitrary item set $S$, it has a pricing according to the optimal assignment oracle $x^*$. And when the item set $S$ is obtained by some buyers bundle, then it prices as $p_B = \frac{1}{2}max\ v_i = \frac{1}{2}v_i'$. Similar to the above, when the item set $S$ is obtained by some buyers' subset, then it prices as $p_{S_i} = max\frac{|x_{iB}|}{|x_{iB}^*|}\sum_{j \in S_i} e_{ij} = \frac{|x_{iB}|}{|x_{iB}^*|}\sum_{j \in S_i} e_{ij}'$. And it need to explain that $x_{iB}^*$ represents the ith optimal assignment subset's complete set, i.e, if the optimal assignment item set is one buyer bundle $B_i$'s subset, then $x_{iB}^* = B_i$. And $x_{iB}$ is the intersection of $B_i$ and the optimal assignment item set $S$. For example: $x_{iB}^* = B_i = \{1,2,3\}$, $S = \{1,3\}$, then $|x_{iB}| = 2$. in addition, $|x_{iB}|$ and $|x_{iB}^*|$ can be known according to the optimal assignment oracle. Thus, we can set each item's price as the following:

$$p_S = \begin{cases} \dfrac{|x_{iB}|}{|x_{iB}^*|}\displaystyle\sum_{j \in S_i} e_{ij}' & S \subset B_i \\[2ex] \dfrac{1}{2}v_i' & S = B_i \end{cases},$$

We use $\mathcal{W}_{A'} = \mathcal{R}_{A'} + \mathcal{U}_{A'}$ to analyze the social welfare based on a single item set. In addition, $\triangle\ \mathcal{W}_1$ and $\triangle\ \mathcal{W}_2$ respectively represent the single item set's increased social welfare in different case.

Then, with the help of item-set pricing scheme, the seller assign the items as is described by Algorithm 2.

---

**Algorithm 2.** Item Assignment for seller according to ISP scheme

---
1 **Input:** Sets: $B$, $M$, $N$; and all $v_i$, $e_{ij}$ for $i \in [1, n]$, $j \in [1, m]$;
2 **Output:** The Social Welfare($\mathcal{W}_{A'}$);
3 **Initialize:** $\mathcal{W}_{A'} = 0$;
4 Seller use the *Optimal Assignment Oracle* to know the $x^*$;
5 Seller sets the prices for each item according to ISP scheme;
6 Seller sells items as an item set.
7 **for** *each buyer* $i = 1$ *to* $n$ **do**
8      **if** $\exists$ *bundle* $B_i$ *and* $p_B \leq v_i$ **then**
9          $M = M - B_i$;
10          $\mathcal{W}_{A'} = \mathcal{W}_{A'} + v_i$;
11      **else if** $\exists$ *subset of* $B_i$ *and* $p_{S_i} \leq \sum_{j \in S_i} e_{ij}$ **then**
12          $M = M - S_i$;
13          $\mathcal{W}_{A'} = \mathcal{W}_{A'} + \sum_{j \in S_i} e_{ij}$;

     **return** $\mathcal{W}$;

---

**Lemma 3.** *By applying the item-set pricing scheme, $d = min\{|x_{iB}|, |x_{iB}^*| - |x_{iB}|\}$ and when $d = \frac{1}{2}|x_{iB}^*|$, our algorithm implements 2-approximate social welfare.*

*Proof.* In the following, we consider the two cases that (1) the item set $S$ is sold as bundle $B_i$, and (2) the item set $S$ is sold as the subset and $S \cap B_i \neq \emptyset$, since the other cases makes no sense for the final welfare. In addition, in case (1), we use $\triangle \mathcal{R}_1$ and $\triangle \mathcal{U}_1$ to respectively represent the revenue and the utility. In case (2), we use $\triangle \mathcal{R}_2$ and $\triangle \mathcal{U}_2$ to resprectively represent the revenue and the utility.

In case (1), let $X = 1$ if the item set $S$ is sold as $B_i$, otherwise $X = 0$. Then:

$$\triangle \mathcal{R}_1 = \frac{1}{2}v_i' * Pr[X = 1] \tag{2}$$

When the first buyer whose valuation is higher than $\frac{1}{2}v_i^*$, the seller will sell the item set $S$ as bundle $B_i$. The revenue of the seller is represented as above. And the other buyers who can buy this item set $S$. Let $\mathcal{E}_i$ denotes that $i$ has a chance to purchase S, then the utilities can be represented as following:

$$\triangle \mathcal{U}_1 = \sum_i (v_i - p_B)^+ * Pr[\mathcal{E}_i] \tag{3}$$

It means when the item set isn't sold, other buyers still have a chance to buy it. Then the lower bound of formula (3) can be represented as:

$$\left(\sum_i (v_i - p_B)^+\right) * Pr[X = 0] \geq \max(v_i - p_B)^+ * Pr[X = 0]$$

$$\geq (\max\ v_i - p_B) * Pr[X = 0] \qquad (4)$$

$$\geq \frac{1}{2}v_i' * Pr[X = 0]$$

Thus for arbitrary item set $S$, it increased social welfare ($\triangle \mathcal{W}_1$) is:

$$\triangle \mathcal{W}_1 = \triangle \mathcal{R}_1 + \triangle \mathcal{U}_1 \geq \frac{1}{2}v_i' * Pr[X = 1] + \frac{1}{2}v_i' * Pr[X = 0] = \frac{1}{2}v_i'$$

In case (2), let $X' = 1$ if the item set $S$ are sold as subset, and $S \cap B_i \neq \emptyset$, otherwise $X' = 0$. Then:

$$\triangle \mathcal{R}_2 = \frac{|x_{iB}|}{|x_{iB}^*|} \sum_{j \in S_i} e_{ij}' * Pr[X' = 1] \qquad (5)$$

In this model, when the buyers can't be satisfied with their bundle, they will consider the subset $S_i$ of their bundles. Then the seller will sell the items to the first arriving buyer whose budget price is higher than $\frac{|x_{iB}|}{|x_{iB}^*|} \sum_{j \in S_i} e_{ij}'$. And other buyers who can buy this subset $S_i$. Similarly, $\mathcal{E}_i'$ denotes that $i$ has a chance to purchase $S_i$, the utilities can be represented as following:

$$\triangle \mathcal{U}_2 = \sum_i \left(\sum_{j \in S_i} e_{ij} - p_{S_i}\right)^+ \times Pr[\mathcal{E}_i'] \qquad (6)$$

It means that when the subset of the item set $S$ isn't sold, other buyers still have a chance to buy it. Considering the lower bound of the formula (6), we have:

$$\left(\sum_i \left(\sum_{j \in S_i} e_{ij} - p_{S_i}\right)^+\right) * Pr[X' = 0] \geq \max(\sum_{j \in S_i} e_{ij} - p_{S_i})^+ * Pr[X' = 0]$$

$$\geq (\max \sum_{j \in S_i} e_{ij} - p_{S_i}) * Pr[X' = 0]$$

$$= \frac{|x_{iB}^*| - |x_{iB}|}{|x_{iB}^*|} \sum_{j \in S_i} e_{ij}' * Pr[X' = 0]$$

$$(7)$$

Thus for arbitrary subset of the item set $i$, it increased social welfare ($\triangle \mathcal{W}_2$) is:

$$\triangle \mathcal{W}_2 = \triangle \mathcal{R}_2 + \triangle \mathcal{U}_2$$

$$\geq \frac{|x_{iB}|}{|x_{iB}^*|} \sum_{j \in S_i} e_{ij}' * Pr[X' = 1] + \frac{|x_{iB}^*| - |x_{iB}|}{|x_{iB}^*|} \sum_{j \in S_i} e_{ij}' * Pr[X' = 0] \qquad (8)$$

Now, we will analyze the relationship between $|x_{iB}|$ and $|x_{iB}^*| - |x_{iB}|$.

- When $|x_{iB}| = |x_{iB}^*| - |x_{iB}|$:

$$\triangle W_2 \geq \frac{|x_{iB}|}{|x_{iB}^*|} \sum_{j \in S_i} e_{ij}' = \frac{1}{2} \sum_{j \in S_i} e_{ij}'$$

In this case, the pricing of subset also implements *2-approximation* as same as the bundle.

- When $|x_{iB}| \neq |x_{iB}^*| - |x_{iB}|$, we define $d = min\{|x_{iB}|, |x_{iB}^*| - |x_{iB}|\}$. Thus the social welfare can be represented as following:

$$\triangle W_2 \geq \frac{d}{|x_{iB}^*|} \sum_{j \in S_i} e_{ij}'$$

It means that the arbitrary subset can implement at least $\frac{d}{|x_{iB}^*|} \sum_{j \in S_i} e_{ij}'$ social welfare. We can see that the $|x_{iB}^*|$, and the $|x_{iB}|$ have an important influence to the subset's profit.

Based on the above analysis, we can get the total social welfare can be represented as following:

$$W_{A'} = \sum_{B_i} \triangle W_1 + \sum_{S_i} \triangle W_2 \geq \frac{1}{2} \sum_{B_i} v_i' + \sum_{S_i} \frac{d}{|x_{iB}^*|} \sum_{j \in S_i} e_{ij}'$$

Comparing with only sold items in bundle, our model which considers the bundle's subset increases the social welfare at least $\sum_{S_i} \frac{d}{|x_{iB}^*|} \sum_{j \in S_i} e_{ij}'$. And when $d = \frac{1}{2} |x_{iB}^*|$, the algorithm implements *2-approximate* social welfare.

# 5 Conclusion

In this work, we presented a new model to depict the online pricing problem, in which the buyers could buy a subset of their bundle items to maximize the social welfare. To the best of our knowledge, this model is the first one with such an assumption. Compared with the various pricing models in previous work, all of which assumed that buyers could only buy the bundle items together or buy nothing, our model is more comprehensive and realistic. Under the new-proposed model, two pricing algorithms were given, with their theoretical performances better than those of the previous works and close to the performance of the optimal solution. Also, it is believed that our algorithms are more suitable for practical applications because of the realistic model. A further study may start to achieve a better bounded guarantee for this problem and its extension on unknown buyer valuation.

**Acknowledgements.** This research is supported by China's NSFC grants (No. 61433012), Hong Kong GRF-17208019, Shenzhen research grant (No. KQJSCX2018033 0170311901, JCYJ 20180305180840138 and GGFW2017073114031767), Shenzhen Discipline Construction Project for Urban Computing and Data Intelligence.

# References

1. Azar, Y., Cohen, I.R., Roytman, A.: Online lower bounds via duality. In: Proceedings of SODA 2017, pp. 1038–1050 (2017)
2. Babaioff, M., Immorlica, N., Lucier, B., Weinberg, S.M.: A simple and approximately optimal mechanism for an additive buyer. In: Proceedings of FOCS 2014, pp. 21–30 (2014)
3. Bei, X., Gravin, N., Lu, P., Tang, Z.G.: Correlation-robust analysis of single item auction. In: Proceedings of SODA 2019, pp. 193–208 (2019)
4. Cai, Y., Zhao, M.: Simple mechanisms for subadditive buyers via duality. In: Proceedings of STOC 2017, pp. 170–183 (2017)
5. Chawla, S., Hartline, J., Kleinberg, R.: Algorithmic pricing via virtual valuations. In: Proceedings of EC 2007, pp. 243–251 (2007)
6. Chawla, S., Miller, J.B., Teng, Y.: Pricing for online resource allocation: intervals and paths. In: Proceedings of SODA 2019, pp. 1962–1981 (2019)
7. Cheung, M., Swamy, C.: Approximation algorithms for single-minded envy-free profit-maximization problems with limited supply. In: Proceedings of FOCS 2008, pp. 35–44 (2008)
8. Chin, F.Y.L., Poon, S.-H., Ting, H.-F., Xu, D., Yu, D., Zhang, Y.: Approximation and competitive algorithms for single-minded selling problem. In: Tang, S., Du, D.-Z., Woodruff, D., Butenko, S. (eds.) AAIM 2018. LNCS, vol. 11343, pp. 98–110. Springer, Cham (2018). https://doi.org/10.1007/978-3-030-04618-7_9
9. Dütting, P., Feldman, M., Kesselheim, T., Lucier, B.: Prophet inequalities made easy: stochastic optimization by pricing non-stochastic inputs. In: Proceedings of FOCS 2017, pp. 540–551 (2017)
10. Feldman, M., Gravin, N., Lucier, B.: Combinatorial auctions via posted prices. In: Proceedings of SODA 2015, pp. 123–135 (2015)
11. Jin, Y., Pinyan, L., Tang, Z.G., Xiao, T.: Tight revenue gaps among simple mechanisms. In: Proceedings of SODA 2019, pp. 209–228 (2019)
12. Karp, R.M.: Reducibility among combinatorial problems. In: Miller, R.E., Thatcher, J.W., Bohlinger, J.D. (eds.) Complexity of Computer Computations. IRSS, pp. 85–103. Springer, Boston (1972). https://doi.org/10.1007/978-1-4684-2001-2_9
13. Zhang, Y., Chin, F.Y., Ting, H.-F.: Online pricing for bundles of multiple items. J. Global Optim. **58**(2), 377–387 (2014)

# Recognizing the Tractability in Big Data Computing

Xiangyu Gao, Jianzhong Li$^{(\boxtimes)}$, Dongjing Miao, and Xianmin Liu

Harbin Institute of Technology, Harbin 150001, Heilongjiang, China
{gaoxy,lijzh,miaodongjing,liuxianmin}@hit.edu.cn

**Abstract.** Due to the limitation on computational power of existing computers, the polynomial time does not works for identifying the tractable problems in big data computing. This paper adopts the sublinear time as the new tractable standard to recognize the tractability in big data computing, and the random-access Turing machine is used as the computational model to characterize the problems that are tractable on big data. First, two pure-tractable classes are first proposed. One is the class PL consisting of the problems that can be solved in polylogarithmic time by a RATM. The another one is the class ST including all the problems that can be solved in sublinear time by a RATM. The structure of the two pure-tractable classes is deeply investigated and they are proved $PL^i \subsetneq PL^{i+1}$ and $PL \subsetneq ST$. Then, two pseudo-tractable classes, PTR and PTE, are proposed. PTR consists of all the problems that can solved by a RATM in sublinear time after a PTIME preprocessing by reducing the size of input dataset. PTE includes all the problems that can solved by a RATM in sublinear time after a PTIME preprocessing by extending the size of input dataset. The relations among the two pseudo-tractable classes and other complexity classes are investigated and they are proved that $PT \subseteq P$, $\Pi'T_Q^0 \subsetneq PTR_Q^0$ and $PT_P = P$.

**Keywords:** Big data computing · Tractability · Sublinear · Complexity theory

## 1  Introduction

Due to the limitation on computational power of existing computers, the challenges brought by big data suggest that the tractability should be re-considered for big data computing. Traditionally, a problem is tractable if there is an algorithm for solving the problem in time bounded by a polynomial (PTIME) in size of the input. In practice, PTIME no longer works for identifying the tractable problems in big data computing.

*Example 1.1.* Sorting is a fundamental operation in computer science, and many efficient algorithms have been developed. Recently, some algorithms are proposed for sorting big data, such as Samplesort and Terasort. However, these algorithms are not powerful enough for big data since their time complexity is still $O(n \log n)$

© Springer Nature Switzerland AG 2019
Y. Li et al. (Eds.): COCOA 2019, LNCS 11949, pp. 223–234, 2019.
https://doi.org/10.1007/978-3-030-36412-0_18

in nature. We performed Samplesort and Terasort algorithms on a dataset with size 1 peta bytes, 1PB for short. The computing platform is a cluster of 33 computation nodes, each of which has 2 Intel Xeon CPUs, interconnected by a 1000 Mbps ethernet. The Samplesort algorithm took more than 35 days, and the Terasort algorithm took more than 90 days.

*Example 1.2.* Even using the fastest solid state drives in current market, whose I/O bandwidth is smaller than 8 GB per second [1], a linear scan of a dataset with size 1 PB, will take 34.7 h. The time of the linear scan is the lower bound of many data processing problems.

Example 1.1 shows that PTIME is no more the good yardstick for tractability in big data computing. Example 1.2 indicates that the linear time, is still unacceptable in big data computing. In addition, many of the problems in database management have been proven to be with high complexity [12–17].

This paper suggests that the sublinear time should be the new tractable standard in big data computing. Besides the problems that can be solved in sublinear time directly, many problems can also be solved in sublinear time by adding a one-time preprocessing. For example, searching a special data in a dataset can be solved in $O(\log n)$ time by sorting the dataset first, a $O(n \log n)$ time preprocessing, where $n$ is the size of the dataset.

To date, some effort has been devoted to providing the new tractability standard on big data. In 2013, Fan et al. made the first attempt to formally characterize query classes that are feasible on big data. They defined a concept of $\Pi$-tractability, *i.e.* a query is $\Pi$-tractable if it can be processed in parallel polylogarithmic time after a PTIME preprocessing [9]. Actually, they gave a new standard of tractability in big data computing, that is, a problem is tractable if it can be solved in parallel polylogarithmic time after a PTIME preprocessing. They showed that several feasible problems on big data conform their definition. They also gave some interesting results. This work is impressive but still need to be improved for the following reasons. (1) Different from the traditional complexity theory, this work only focuses on the problem of boolean query processing. (2) The work only concerns with the problems that can be solved in parallel polylogarithmic time after a PTIME preprocessing. Actually, many problems can be solved in parallel polylogarithmic time without PTIME preprocessing. (3) The work is based on parallel computational models or parallel computing platforms without considering the general computational models. (4) The work takes polylogarithmic time as the only standard for tractability. Polylogarithmic time is a special case of the sublinear time, and it is not sufficient for characterizing the tractability in big data computing.

Similar to the $\Pi$-tractability theory [9], Yang et al. placed a logarithmic-size restriction on the preprocessing result and relaxed the query execution time to PTIME and introduced the corresponding $\Pi'$-tractability [20]. They clarified that a short query is tractable if it can be evaluated in PTIME after a one-time preprocessing with logarithmic-size output. This work just pursued Fan et al.'s methodology, and there is no improvement on Fan et al.'s work [9]. Besides, the logarithmic restriction on the output size of preprocessing is too strict to cover all query classes that are tractable on big data.

In addition, computation model is the fundamental in the theory of computational complexity. Deterministic Turing machine (DTM) is not suitable to characterize sublinear time algorithms since the sequential operating mode makes only the front part of input can be read in sublinear time. For instance, searching in an ordered list is a classical problem that can be solved in logarithmic time. However, if DTM is used to describe the computation procedure of it, the running time comes to polynomial. To describe sublinear time computation accurately and make all of the input can be accessed in sublinear time, random access is very significant.

This paper is to further recognize the tractability in big data computing. A general computational model, random-access Turing machine, is used to characterize the problems that are tractable on big data, not only query problems. The sublinear time, rather than polylogarithmic time, is adopted as the tractability standard for big data computing. Two classes of tractable problems on big data are identified. The first class is called as pure-tractable class, including the problems that can be solved in sublinear time without preprocessing. The second class is called as pseudo-tractable class, consisting of the problems that can be solved in sublinear time after a PTIME preprocessing. The structure of the two classes is investigated, and the relations among the two classes and other existing complexity classes are also studied. The main contributions of this paper are as follows.

(1) To describe sublinear time computation more accurately, the random-access Turing machine (RATM) is formally defined. RATM is used in the whole work of this paper. It is proved that the RATM is equivalent to the deterministic Turing machine in polynomial time. Based on RATM, an efficient universal random-access Turing machine $U$ is devised. The input and output of $U$ are $(x, c(M))$ and $M(x)$ respectively, where $c(M)$ is the encoding of a RATM $M$, $x$ is an input of $M$, and $M(x)$ is the output of $M$ on input $x$. Moreover, if $M$ halts on input $x$ within $T$ steps, then $U$ halts within $cT \log T$ steps, where $c$ is a constant.

(2) Using RATM and taking sublinear time as the tractability standard, the classes of tractable problems in big data computing are defined. First, two pure-tractable complexity classes are defined. One is a polylogarithmic time class PL, which is the set of problems that can be solved by a RATM in polylogarithmic time. The another one is a sublinear time class ST, which is the set of all decision problems that can be solved by a RATM in sublinear time. Then, two pseudo-tractable classes, PTR and PTE, are first defined. PTR is the set of all problems that can be solved in sublinear time after a PTIME preprocessing by reducing the size of input dataset. PTE is the set of all problems that can be solved in sublinear time after a PTIME preprocessing by extending the size of input dataset.

(3) The structure of the pure-tractable classes is investigated deeply. It is first proved that $PL^i \subsetneq PL^{i+1}$, where $PL^i$ is the class of the problems that can

be solved by a RATM in $O(\log^i n)$ time and $n$ is the size of the input. Thus, a polylogarithmic time hierarchy is obtained. It is proved that $\mathrm{PL} = \bigcup_i \mathrm{PL}^i \subsetneq \mathrm{ST}$. This result shows that there is a gap between polylogarithmic time class and linear time class.

(4) The relations among the complexity classes PTR, PTE, $\sqcap'\mathrm{T}_Q^0$ [20] and P is studied. They are proved that $\mathrm{PT} \subseteq \mathrm{P}$ and $\sqcap'\mathrm{T}_Q^0 \subsetneq \mathrm{PTR}_Q^0$. Finally, it is concluded that all problems in P can be made pseudo-tractable.

The remainder of this paper is organized as follows. Section 2 formally defines the complexity model RATM, proves that RATM is equivalent to DTM and there is an efficient URATM, and defines the problem in big data computing. Section 3 defines the pure-tractable classes, and investigates the structure of the pure-tractable classes. Section 4 defines the pseudo-tractable classes, and studies the relations among the complexity classes PTR, PTE, $\sqcap'\mathrm{T}_Q^0$ [20] and P. Finally, Sect. 5 concludes the paper.

## 2   Preliminaries

To define sublinear time complexity classes precisely, a suitable computation model should be chosen since sublinear time algorithms may read only a minuscule fraction of its input and thus random access is very important. The random-access Turing machine is chosen as the foundation of the work in this paper. This section gives the formal definition of the random-access Turing machine. They are proved that the RATM is equivalent to the determinate Turing machine (DTM) in polynomial time and there is a universal random-access Turing machine. Finally, a problem in big data computing is defined.

### 2.1   Random-Access Turing Machine

A random-access Turing machine (RATM) $M$ is a $k$-tape Turing machine with an input tape and an output tape and is additionally equipped with $k$ binary index tapes that are write-only. One of the binary index tapes is for $M$'s read-only input tape and the others for the $M$'s $k-1$ work tapes. Note that $k \geq 2$. The formally definition of RATM is as follows.

**Definition 2.1.** *A RATM $M$ is a 8-tuple $M = (Q, \Sigma, \Gamma, \delta, q_0, B, q_f, q_a)$, where*

$Q$: *The finite set of states.*
$\Sigma$: *The finite set of input symbols.*
$\Gamma$: *The finite set of tape symbols, and $\Sigma \subseteq \Gamma$.*
$\delta$: $Q \times \Gamma^k \to Q \times \Gamma^{k-1} \times \{0, 1, B\}^k \times \{L, S, R\}^{2k}$, *where $k \geq 2$.*
$q_0 \in Q$: *The start state of $M$.*
$B \in \Gamma \setminus \Sigma$: *The blank symbol.*
$q_f \in Q$: *The accepting state.*
$q_a \in Q$: *The random access state. If $M$ enters state $q_a$, $M$ will move the heads of all non-index tapes to the cells described by the respective index tapes automatically.*

Assuming the first tape of a RATM $M$ is the input tape, if $M$ is in state $q \in Q$, $(\sigma_1, \cdots, \sigma_k)$ are the symbols currently being read in the $k$ non-index tapes of $M$, and the related transition function is $\delta(q, (\sigma_1, \cdots, \sigma_k)) = (q', (\sigma'_2, \cdots, \sigma'_k), (a_1, \cdots, a_k), (z_1, \cdots, z_{2k}))$, $M$ will replace $\sigma_i$ with $\sigma'_i$, where $2 \le i \le k$, write $a_j$ $(1 \le j \le k)$ on the corresponding index tape, move heads Left, Right, or Stay in place as given by $(z_1, \cdots, z_{2k})$, and enter the new state $q'$.

The following lemmas state that RATM is equivalent to the deterministic Turing machine (DTM). Due to space limitation, the proofs can be found in the full version of the paper [10].

**Theorem 2.1.** *For a Boolean function $f$ and a time-constructible function [2] $T$,*

(1) *if $f$ is computable by a DTM within time $T(n)$, then it is computable by a RATM within time $T(n)$, and*
(2) *if $f$ is computable by a RATM within time $T(n)$, then it is computable by a DTM within time $T(n)^2 \log T(n)$.*

**Corollary 2.1.** *If a Boolean function $f$ is computable by a RATM within time $o(n)$, then it is computable by a DTM within time $o(n)n \log n$.*

## 2.2    The Universal Random-Access Turing Machine

Just like DTM, RATM can be encoded by a binary string. The code of a RATM $M$ is denoted by $c(M)$. The encoding method of RATM is the same as that of DTM [8]. The encoding of RATM makes it possible to devise a universal random-access Turing machine (URATM) with input $(x, c(M))$ and outputs $M(x)$, where $x$ is the input of a RATM $M$, $c(M)$ is the code of $M$, and $M(x)$ is the output of $M$ on $x$. Before the formal introduction of URATM, we first present two lemmas. Due to space limitation, the detailed proofs can be found in the full version of the paper [10].

**Lemma 2.1.** *For every function $f$, if $f$ is computable in time $T(n)$ by a RATM $M$ using alphabet $\Gamma$, then it is computable in time $c_1 T(n)$ by a RATM $\tilde{M}$ using alphabet $\{0, 1, B\}$.*

**Lemma 2.2.** *For every function $f$, if $f$ is computable in time $T(n)$ by a RATM $M$ using $k$ tapes, then it is computable in time $c_2 T(n) \log T(n)$ by a 5-tape RATM $\tilde{M}$.*

**Theorem 2.2.** *There exists a universal random-access Turing machine $\mathcal{U}$, whose input is $(x, c(M))$ and outputs is $M(x)$, where $x$ is an input of $M$, $c(M)$ is the code of $M$, and $M(x)$ is the output of $M$ on $x$. Moreover, if $M$ halts on input $x$ in $T$ steps then $\mathcal{U}$ halts on input $(x, c(M))$ in $O(cT \log T)$ steps, where $c$ is a constant depending on $M$.*

Theorem 2.2 is an encouraging result which can help us to investigate the structure of sublinear time complexity classes.

## 2.3   Problems in Big Data Computing

To reflect the characteristics in big data computing, a problem in big data computing is defined as follows.

INPUT: big data $D$, and a function $\mathcal{F}$ on $D$.

OUTPUT: $\mathcal{F}(D)$.

Unlike the traditional definition of a problem, the input of a big data computing problem must be big data, where big data usually has size greater than or equal to 1 PB. The problem defined above is often called as big data computing problem. The big data set in the input of a big data computing problem may consists of multiple big data sets. The problems discussed in the rest of paper are big data computing problems, and we will simply call them problems in the rest of the paper.

# 3   Pure-Tractable Classes

In this section, we first give the formal definitions of the pure-tractable classes PL and ST, and then investigate the structure of the pure-tractable classes.

## 3.1   Polylogarithmic-Tractable Class PL

As mentioned in [19], the class DLOGTIME consists of all problems that can be solved by a RATM in $O(\log n)$ time, where $n$ is the length of the input. DLOGTIME was underestimated before [3,4]. However, it is very impotent in big data computing [9] and there are indeed many interesting problems in this class [3]. In this section, we propose the complexity class PL by extending the DLOGTIME to characterize problems that are tractable on big data, and inspired by DLOGTIME and NC hierarchy, we use $\mathrm{PL}^i$ to reinterpret PL as a hierarchy.

**Definition 3.1.** *The class* PL *consists of decision problems that can be solved by a* RATM *in polylogarithmic time.*

**Definition 3.2.** *For each $i \geq 1$, the class $\mathrm{PL}^i$ consists of decision problems that can be solved by a* RATM *in $O(\log^i n)$, where $n$ is the length of the input.*

According to the definition, $\mathrm{PL}^1$ is equivalent to DLOGTIME. It is clear that $\mathrm{PL}^1 \subseteq \mathrm{PL}^2 \subseteq \cdots \subseteq \mathrm{PL}^i \subseteq \cdots \subseteq \mathrm{PL}$, which forms the PL hierarchy. The following Theorem 3.1 shows that $\mathrm{PL}^i \subsetneqq \mathrm{PL}^{i+1}$ for $i \in \mathbb{N}$.

**Lemma 3.1.** [7] *There is a logarithmic time RATM, which takes $x$ as input and generates the output $n$ encoded in binary such that $n = |x|$.*

**Theorem 3.1.** *For any $i \in \mathbb{N}, \mathrm{PL}^i \subsetneqq \mathrm{PL}^{i+1}$.*

*Proof.* We prove this theorem by constructing a RATM $M^*$ such that $L(M^*) \in \mathrm{PL}^{i+1} - \mathrm{PL}^i$.

According to Lemma 3.1, there exists a RATM $M_1$, which takes $x$ as input and outputs the binary form of $n = |x|$ in $c \log n$ time.

Since $n^{i+1}$ is a polynomial time constructible function for any $i \in \mathbb{N}$, there exists a DTM $M_2$ that takes $x$, whose length is $n$, as input and outputs the binary form of $n^{i+1}$ in time $n^{i+1}$.

By combining $M_1$ and $M_2$, we can construct a RATM $M$ that works as follows. Given an input $x$, $M$ first simulates $M_1$ on $x$ and outputs the binary form of $n = |x|$. Then, $M$ simulates $M_2$'s on the binary form of $n = |x|$. The total running time of $M$ is $\log^{i+1} n$.

Now, we are ready to construct $M^*$. On any input $x$, $M^*$ works as follows:

(1) $M^*$ simulates the computation of $\mathcal{U}$ on input $(x, x)$ and the computation of $M$ on input $x$ simultaneously.
(2) Any one of $\mathcal{U}$ and $M$ halts, $M^*$ halts, and the state entered by $M^*$ is determined as follow:
   (a) If $\mathcal{U}$ halts first and enters the accept state, then $M^*$ halts and enters the reject state.
   (b) If $\mathcal{U}$ halts first and enters the reject state, then $M^*$ halts and enters the accept state.
   (c) If $M$ halts first and enters state $q$, then $M^*$ halts and enters $q$.

The running time of $M^*$ is at most $\log^{i+1} n$, so $L(M^*) \in \mathrm{PL}^{i+1}$.

Assume that there is a $\log^i n$ time RATM $N$ such that $L(N) = L(M^*)$. Since $\lim\limits_{n \to \infty} \frac{\log^i n \log \log^i n}{\log^{i+1} n} = 0$, there must be a number $n_0$ such that $\log^i n \log \log^i n < \log^{i+1} n$ for each $n \geq n_0$. Let $x$ be a string representing the machine $N$ whose length is at least $n_0$. Such string exists since a string of a RATM can be added any long string behind the encoding of the RATM. We have $Time_{\mathcal{U}}(x, x) \leq cTime_N(x) \log (Time_N(x)) \leq c \log^i n \log \log^i n \leq \log^{i+1} n$. It means that $\mathcal{U}$ halts before $M$, which is in contradiction with (a) and (b) of $M^*$'s work procedure.

$\square$

## 3.2 Sublinear-Tractable Class ST

To denote all problems can be solve in sublinear time, the complexity class ST is proposed in this subsection. And the relation between ST and PL is investigated. We first give the formal definition of ST.

**Definition 3.3.** *The class* ST *consists of the decision problems that can be solved by a RATM in $o(n)$ time, where $n$ is the size of the input.*

There are indeed many problems that can solved in $o(n)$ time, such as searching in a sorted list, point location in two-dimensional Delaunay triangulations, and checking whether two convex polygons intersect mentioned in [5, 6, 18].

To understand the structure of pure-tractable classes, we study the relation between ST and PL. Theorem 3.2 shows that ST contains PL properly. This result indicates that there is a gap between polylogarithmic time class and linear time class. The detailed proof can be found in the full version of the paper [10].

**Theorem 3.2.** PL $\subsetneq$ ST.

## 4    Pseudo-Tractable Classes

In this section, we study the big data computing problems that can be solved in sublinear time after a PTIME preprocessing. We propose two complexity classes, PTR and PTE, and investigate the relations among PTR, PTE and other complexity classes. For easy to understand, we first review the definition of a problem in big data computing, that is,

INPUT: big data $D$, and a function $\mathcal{F}$ on $D$.
OUTPUT: $\mathcal{F}(D)$.

### 4.1    Pseudo-Tractable Class by Reducing $|D|$

We will use PTR to express the pseudo-tractable class by reducing $|D|$, which is defined as follows.

**Definition 4.1.** *A problem $\mathcal{F}$ is in the complexity class* PTR *if there exists a* PTIME *preprocessing function $\Pi$ such that for big data $D$,*

*(1) $|\Pi(D)| < |D|$ and $\mathcal{F}(\Pi(D)) = \mathcal{F}(D)$.*
*(2) $\mathcal{F}(\Pi(D))$ can be solved by a* RATM *in $o(|D|)$ time.*

Data $D$ is preprocessed by a preprocessing function $\Pi$ in polynomial time to generate another structure $\Pi(D)$. Besides PTIME restriction on $\Pi$, it is required that the size of $\Pi(D)$ is smaller than $D$. This allows many of previous polynomial time algorithms to be used. For example, $\mathcal{F}(\Pi(D))$ can be solved by a quadratic polynomial time algorithm if $|D| = n$ and $|\Pi(D)| = n^{1/3}$, and the time needed for solving $\mathcal{F}(\Pi(D))$ is $O(n^{2/3}) \in o(n)$. To make $\Pi(D)$ less than $D$, $\Pi$ can be data compression, sampling, etc.

The following simple propositions show the time complexity of the problem after preprocessing.

**Proposition 4.1.** *If there is a preprocessing function $\Pi$ and a constant $c > 1$ such that $|\Pi(D)| = |D|^{1/c}$ for any $D$, and there is a algorithm to solve $\mathcal{F}(\Pi(D))$ in time of polynomials of degree $d$, where $d < c$, then $F(\Pi(D))$ can be solved in $o(|D|)$ time, and thus $\mathcal{F}$ is in* PTR.

**Proposition 4.2.** *If there is a preprocessing function $\Pi$ and a constant $c$ such that $|\Pi(D)| = \log^c |D|$ for any $D$, and there is a PTIME algorithm to solve $\mathcal{F}(\Pi(D))$, then $\mathcal{F}(\Pi(D))$ can be solved in $o(|D|)$ time, and thus $\mathcal{F}$ is in* PTR.

**Proposition 4.3.** *If there is a preprocessing function $\Pi$ and a constant $c \in (0, 1)$ such that $|\Pi(D)| = c \log |D|$ for any $D$, and there is a $O(2^{|\Pi(D)|})$ time algorithm to solve $\mathcal{F}(\Pi(D))$, then $\mathcal{F}(\Pi(D))$ can be solved in $o(|D|)$ time, and thus $\mathcal{F}$ is in* PTR.

## 4.2  Pseudo-Tractable Class by Extending $|D|$

Obviously, PTR does not characterize all problems that can be solved in sublinear time after preprocessing. We define another complexity class PTE to denote remaining problems that can be solved in sublinear time after preprocessing.

**Definition 4.2.** *A problem $\mathcal{F}$ is in the complexity class* PTE *if there exists a* PTIME *preprocessing $\Pi$ such that for big data $D$,*

*(1) $|\Pi(D)| \geq |D|$ and $\mathcal{F}(\Pi(D)) = \mathcal{F}(D)$.*
*(2) $\mathcal{F}(\Pi(D))$ can be solved by a* RATM *in $o(|D|)$ time.*

The only difference between PTE and PTR is that the former requires that preprocessing results in a larger dataset compared to original dataset. Intuitively, if a problem $\mathcal{F}$ is in PTE, then $\mathcal{F}$ can be solved in sublinear time by sacrificing space. For example, many queries are solvable in sublinear time by building index before query processing. Many non-trivial data structures are also designed for some problems to accelerate the computation. Extremely, all the problems in P can be solved in polynomial time and stored previously, and can be solved in $O(1)$ time later, that is, P $\subseteq$ PTE.

There are also some propositions show the complexity class of the problem after preprocessing.

**Proposition 4.4.** *If there is a preprocessing function $\Pi$ and a constant $c \geq 1$ such that $|\Pi(D)| = c|D|$ for any $D$, and there is a sublinear time algorithm to solve $\mathcal{F}(\Pi(D))$, then $\mathcal{F}(\Pi(D))$ can be solved in $o(|D|)$ time, and thus $\mathcal{F}$ is in* PTE.

**Proposition 4.5.** *If there is a preprocessing function $\Pi$ and a constant $c \geq 1$ such that $|\Pi(D)| = |D|^c$ for any $D$, and there is a polylogarithmic time algorithm to solve $\mathcal{F}(\Pi(D))$, then $\mathcal{F}(\Pi(D))$ can be solved in $o(|D|)$ time, and thus $\mathcal{F}$ is in* PTE.

## 4.3  Relations Between Pseudo-Tractable Classes and Other Classes

In the rest of the paper, we use PT to express PTR $\cup$ PTE. The proofs of the following theorem and corollary can be found in the full version of the paper [10].

**Theorem 4.1.** PT $\subseteq$ P.

**Corollary 4.1.** *No* NP-*Complete problem is in* PT *if* P $\neq$ NP.

Recalling the definitions of PTR and PTE, PTR and PTE concern all big data computing problems rather than only queries. To investigate the relations among PTR, PTE, and $\sqcap'T_Q^0$ [20], we restrict $\mathcal{F}$ in the definitions of PTR and PTE to boolean query class $\mathcal{Q}$ to define two pseudo-tractable query complexity classes in the following.

Following the conventions in [9], a boolean query class $\mathcal{Q}$ can be encoded as $S = \{\langle D, Q \rangle\} \subseteq \Sigma^* \times \Sigma^*$ such that $\langle D, Q \rangle \in S$ if and only if $Q(D)$ is true, where $Q \in \mathcal{Q}$ is defined on $D$.

Now we define $\mathrm{PTR}_Q^0$ and $\mathrm{PTE}_Q^0$ as follows.

**Definition 4.3.** *A language* $S = \{\langle D, Q \rangle\}$ *is* pseudo-tractable *by reducing* datasets *if there exist a* PTIME *preprocessing function $\Pi$ and a language $S'$ such that for all $D, Q$,*

*(1) $\langle D, Q \rangle \in S$ iff $\langle \Pi(D), Q \rangle \in S'$, $|\Pi(D)| < |D|$ and*
*(2) $S'$ can be solved by a RATM in $o(|D|)$ time.*

A class of queries, $\mathcal{Q}$, is *pseudo-tractable by reducing datasets* if $S$ is pseudo-tractable by reducing datasets, where $S$ is the language for $\mathcal{Q}$.

**Definition 4.4.** *The query complexity class* $\mathrm{PTR}_\mathrm{Q}^0$ *is the set of $\mathcal{Q}$ that is pseudo-tractable by reducing datasets.*

**Definition 4.5.** *A language* $S = \{\langle D, Q \rangle\}$ *is* pseudo-tractable *by extending* datasets *if there exist a* PTIME *preprocessing function $\Pi$ and a language $S'$ of pairs such that for all $D, Q$,*

*(1) $\langle D, Q \rangle \in S$ iff $\langle \Pi(D), Q \rangle \in S'$, $|\Pi(D)| \geq |D|$, and,*
*(2) $S'$ can be solved by a RATM in $o(|D|)$ time.*

A class of queries $\mathcal{Q}$ is *pseudo-tractable by extending datasets* if $S$ is extended pseudo-tractable, where $S$ is the language for $\mathcal{Q}$.

**Definition 4.6.** *The query complexity class* $\mathrm{PTE}_\mathrm{Q}^0$ *is the set of $\mathcal{Q}$ that is pseudo-tractable by extending datasets.*

Obviously, $\mathrm{PTR}_\mathrm{Q}^0 \subseteq \mathrm{PTR}$ and $\mathrm{PTE}_\mathrm{Q}^0 \subseteq \mathrm{PTE}$.

The following theorem shows that $\sqcap'\mathrm{T}_\mathrm{Q}^0$ is strictly smaller than $\mathrm{PTR}_\mathrm{Q}^0$ and also indicates that the logarithmic restriction on output size of preprocessing [20] is too strict. The following theorem is proved by showing that the boolean queries for breadth–depth search, which is not in $\sqcap'\mathrm{T}_\mathrm{Q}^0$ [20], is in $\mathrm{PTR}_\mathrm{Q}^0$. The detailed proof can be found in the full version of the paper [10].

**Theorem 4.2.** $\sqcap'\mathrm{T}_\mathrm{Q}^0 \subsetneq \mathrm{PTR}_\mathrm{Q}^0$.

*Proof.* Let $\mathcal{Q} \in \sqcap'\mathrm{T}_\mathrm{Q}^0$. For any $\langle D, Q \rangle \in S_\mathcal{Q}$, $Q$ satisfies short-query property, *i.e.*, $|Q| \in O(\log |D|)$, there is a PTIME preprocessing function $\Pi$ such that $|\Pi(D)| \in O(\log |D|)$ and a language $S'$ such that $\langle D, Q \rangle \in S$ if and only if $\langle \Pi(D), Q \rangle \in S'$, and $S'$ is in P, that is, the time of computing $Q(D')$ is $O((|D'|)^c) = O(\log^c |D|)$, according to the definition of $\sqcap'\mathrm{T}_\mathrm{Q}^0$. Thus, $\sqcap'\mathrm{T}_\mathrm{Q}^0 \subseteq \mathrm{PTR}_\mathrm{Q}^0$ since $O(\log^c |D|) \subseteq o(|D|)$.

It has been proved that the boolean queries for breadth–depth search, denoted by $\mathcal{Q}_{\mathrm{BDS}}$, is not in $\sqcap'\mathrm{T}_\mathrm{Q}^0$ [20]. We need to prove that $\mathcal{Q}_{\mathrm{BDS}}$ is in $\mathrm{PTR}_\mathrm{Q}^0$.

Given an undirected graph $G = (V, E)$ with $n$ nodes with numbers and $m$ edges and a pair of nodes $u$ and $v$ in $V$, the query is whether $u$ visited before $v$ in the breadth-depth search of $G$. The query is processed as follows.

We first define preprocessing function $\Pi$ that performs breadth-depth search on $G$ [11], and return a array $M$, where the index of $M$ is the node number and if $M[i] < M[j]$ then node $i$ is visited before node $j$.

After the preprocessing, the query can be answered in two following steps: (1) Access $M[u]$ and $M[v]$. (2) If $M[u] < M[v]$ then $u$ is visited before $v$ else $v$ is visited before $u$.

It is obvious that the query can be answered by a RATM in $O(\log n)$ time.

Moreover, the size of $G$ encoded by adjacency list is $O(n^2)$ and the size of $M$ is $O(n \log n)$, that is, $|M| < |G|$.

Therefore, $\mathcal{Q}_{BDS} \in \mathrm{PTR}^0_Q$. Consequently, $\sqcap' \mathrm{T}^0_Q \subsetneq \mathrm{PTR}^0_Q$.    □

Let $\mathrm{PT}^0_Q = \mathrm{PTR}^0_Q \cup \mathrm{PTE}^0_Q$. We extend the definition of $\mathrm{PT}^0_Q$ to the decision problems by using the definition of factorization in [9]. A factorization of a language $L$ is $\Upsilon = (\pi_1, \pi_2, \rho)$, where $\pi_1$, $\pi_2$, and $\rho$ are in NC and satisfy that $\langle \pi_1(x), \pi_2(x) \rangle \in S_{(L,\Upsilon)}$ and $\rho(\pi_1(x), \pi_2(x)) = x$ for all $x \in L$.

**Definition 4.7.** *A decision problem $L$ can be made pseudo-tractable if there exists a factorization $\Upsilon$ of $L$ such that the language $S_{(L,\Upsilon)}$ of pairs for $(L, \Upsilon)$ is pseudo-tractable, i.e, $S_{(L,\Upsilon)} \in \mathrm{PT}^0_Q$.*

In the rest of the paper, we use $\mathrm{PT}_P$ to denote the set of all decision problems that can be made pseudo-tractable.

The following theorem tells us that all problem in P can be made pseudo-tractable. Intuitively, we prove that Bread-Depth Search, which is P-complete under NC-reduction [11], is in $\mathrm{PT}_P$. Detailed proof can be found in the full version of the paper [10].

**Theorem 4.3.** $\mathrm{PT}_P = \mathrm{P}$.

## 5    Conclusion

This work aims to recognize the tractability in big data computing. The RATM is used as the computational model in our work, and an efficient URATM is devised.

Two pure-tractable classes are proposed. One is the class PL consisting of the problems that can be solved in polylogarithmic time by a RATM. The another one is the class ST including all the problems that can be solved in sublinear time by a RATM. The structure of pure-tractable classes is deeply investigated. The polylogarithmic time hierarchy, $\mathrm{PL}^i \subsetneq \mathrm{PL}^{i+1}$, is first proved. Then, is proved that $\mathrm{PL} \subsetneq \mathrm{ST}$.

Two pseudo-tractable classes, PTR and PTE, are also proposed. PTR consisting of all the problems that can solved by a RATM in sublinear time after a PTIME preprocessing by reducing the size of input dataset. PTE consisting of all the problems that can solved by a RATM in sublinear time after a PTIME preprocessing by extending the size of input dataset. The relations among the pseudo-tractable classes and other complexity classes are investigated. They are proved that $\mathrm{PT} \subseteq \mathrm{P}$ and $\sqcap' \mathrm{T}^0_Q \subsetneq \mathrm{PTR}^0_Q$. And for the decision problems, we proved that all decision problems in P can be made pseudo-tractable.

**Acknowledgment.** This work was supported by the National Natural Science Foundation of China under grants 61732003, 61832003, 61972110 and U1811461.

# References

1. https://www.wikiwand.com/en/PCI_Express
2. Arora, S., Barak, B.: Computational Complexity: A Modern Approach. Cambridge University Press, Cambridge (2009)
3. Mix Barrington, D.A., Immerman, N., Straubing, H.: On uniformity within NC1. J. Comput. Syst. Sci. **41**(3), 274–306 (1990)
4. Buss, S.R.: The Boolean formula value problem is in ALOGTIME. In: Nineteenth ACM Symposium on Theory of Computing (1987)
5. Chazelle, B., Liu, D., Magen, A.: Sublinear geometric algorithms. SIAM J. Comput. **35**(3), 627–646 (2005)
6. Czumaj, A., Sohler, C.: Sublinear-time algorithms. In: Property Testing-current Research & Surveys (2010)
7. Dowd, M.: Notes on log space representation (1986)
8. Du, D.-Z., Ko, K.-I.: Theory of Computational Complexity, vol. 58. Wiley, New York (2011)
9. Fan, W., Geerts, F., Neven, F.: Making queries tractable on big data with pre-processing: (through the eyes of complexity theory). Proc. VLDB Endow. **6**(9), 685–696 (2013)
10. Gao, X., Li, J., Miao, D., Liu, X.: Recognizing the tractability in big data computing (2019)
11. Greenlaw, R.: Breadth-depth search is $\mathcal{P}$-complete. Parallel Process. Lett. **3**(03), 209–222 (1993)
12. Liu, X., Cai, Z., Miao, D., Li, J.: Tree size reduction with keeping distinguishability. Theor. Comput. Sci. **749**, 26–35 (2018)
13. Miao, D., Cai, Z., Li, J.: On the complexity of bounded view propagation for conjunctive queries. IEEE Trans. Knowl. Data Eng. **30**(1), 115–127 (2018)
14. Miao, D., Cai, Z., Li, Y.: SEF view deletion under bounded condition. Theor. Comput. Sci. **749**, 17–25 (2018)
15. Miao, D., Cai, Z., Jiguo, Y., Li, Y.: Triangle edge deletion on planar glasses-free RGB-digraphs. Theor. Comput. Sci. **788**, 2–11 (2019)
16. Miao, D., Liu, X., Li, J.: On the complexity of sampling query feedback restricted database repair of functional dependency violations. Theor. Comput. Sci. **609**, 594–605 (2016)
17. Miao, D., Liu, X., Li, Y., Li, J.: Vertex cover in conflict graphs. Theor. Comput. Sci. **774**, 103–112 (2019)
18. Rubinfeld, R.: Sublinear time algorithms. Marta Sanz Solé **34**(2), 1095–1110 (2011)
19. Van Leeuwen, J., Leeuwen, J.: Handbook of Theoretical Computer Science, vol. 1. Elsevier, Amsterdam (1990)
20. Yang, J., Wang, H., Cao, Y.: Tractable queries on big data via preprocessing with logarithmic-size output. Knowl. Inf. Syst. **56**(12), 1–23 (2017)

# A Novel Virtual Traffic Light Algorithm Based on V2V for Single Intersection in Vehicular Networks

Longjiang Guo[1,2,3,4], De Wang[4], Peng Li[1,2,3], Lichen Zhang[1,2,3(✉)],
Meirei Ren[1,2,3(✉)], Hong Liu[2,3], and Ana Wang[2,3]

[1] Key Laboratory of Modern Teaching Technology, Ministry of Education,
Xi'an 710062, China
[2] Engineering Laboratory of Teaching Information Technology of Shaanxi Province,
Xi'an 710119, China
[3] School of Computer Science, Shaanxi Normal University, Xi'an 710119, China
{longjiangguo,zhanglichen,meiruiren}@snnu.edu.cn
[4] School of Computer Science and Technology, Heilongjiang University,
Harbin 150080, China

**Abstract.** For the vehicle passing priority, the existing virtual traffic light algorithm only considered the factor of the arrival time, a group of vehicles won't transfer the wayleave until they all pass completely. This method cannot achieve the maximum road traffic throughput capacity. This paper designs a queuing chain model and proposes a virtual traffic lights algorithm for Single Intersection (for short SVTL) based on the queuing chain model, considering comprehensively three parameters: the waiting time of each lane, the length of queue chain and the number of queues. The SVTL algorithm eliminates the phenomenon of waiting in vain basically, saves the waiting time of the vehicle, and avoids unnecessary parking conditions. The simulation shows that the SVTL shortens the average waiting time when the traffic density is high, compared with the Common Traffic Lights (CTL) and the existing virtual traffic light algorithm DVTL, the average speed is increased by 36.9% and 16.2% respectively.

**Keywords:** Intelligent transport system · Internet of Vehicles ·
Virtual traffic lights · V2V

## 1 Introduction

In recent years, large-scale urban traffic congestion has taken place in more and more cities for a long time, especially in morning and evening rush hours and bad weather period, which also increases rapidly in the second-tier and third-tier city. According to a report from Amap's China Major Urban Traffic Analysis in 2017, the big data analysis of 364 cities and highway networks nationwide shows

L. Guo, D. Wang and P. Li—Contributed equally to this work.

© Springer Nature Switzerland AG 2019
Y. Li et al. (Eds.): COCOA 2019, LNCS 11949, pp. 235–251, 2019.
https://doi.org/10.1007/978-3-030-36412-0_19

that one third of the cities are threatened by congestion during peak hours in China. More than half of city's commuter peaks are in a slow state, and 35% cities are in a slow state during the flat hump period.

There are usually three common ways to alleviate traffic congestion: reducing the number of vehicles, increasing infrastructure capacity (widening roads, increasing road network density, optimizing road structure, etc.) and increasing throughput at intersections. The throughput bottleneck at the intersection has a lot to do with the management of the intersection. Efficient traffic signal control is an effective and economical solution.

Internet of Vehicles (IoV) is an important part of the Intelligent Traffic System (ITS). Communication between vehicles and vehicles or infrastructure (Vehicle to everything, V2X) can significantly improve the safety, manageability and transport efficiency of the transportation system, and reduce energy consumption. In the field of traffic control, the Internet of Vehicles provides a more efficient means for the realization of virtual traffic signal control system solutions. Virtual traffic lights rely on the Internet of Vehicles environment to negotiate an appropriate road-rights distribution method through information exchange among vehicles, which provides a new direction for improving the throughput and efficiency of road junctions.

The contributions of this paper are given as follows.

1. This paper proposed a vehicle queuing chain model in the Internet of Vehicle, this model is closer to the application scenario of urban multi Lane mode. The common three-lane queuing chain model is adopted in this paper.
2. Based on the queuing chain model, a data collection algorithm between vehicles is proposed in this paper. The information queue organized by this algorithm is just like the real vehicle queue on the road, which is convenient for data collection and maintenance.
3. This paper proposes a virtual traffic lights algorithm for Single Intersection, to calculating the priority of traffic with the three parameters which are the waiting time of each lane, the length of queue chain and the number of queues. According to the intersection conflict table, find out the non-conflicting lanes, and let these lanes enter the green time at the same time to increase the road throughput. This algorithm eliminates the phenomenon of waiting in vain basically, save waiting time and avoids unnecessary parking.
4. The simulation shows that the SVTL shortens the average waiting time when the traffic density is high, compared with the Common Traffic Lights (CTL) and the existing virtual traffic light algorithm DVTL, the average speed is increased by 36.9% and 16.2% respectively.

## 2   Related Work

In vehicular networks, there are many interesting research fields. Some researchers who have focused on intersection-based forwarding protocol, privacy vehicular data and capacity analysis for vehicular networks. Xiong et al. have proposed a privacy-preserving approach to protect vehicular camera data [24].

Guan et al. have presented an intersection-based forwarding protocol for vehicular networks [7]. Huang et al. studied the theory based capacity analysis for vehicular networks [9].

There are quantities of researches about traffic signal phase position algorithm, one of which is based on traffic history data, SCOOT [19] is the widely used traffic signal timing system, which detects traffic flow on the upstream road segment and predicts traffic flow to the intersection to make some pre-adjustments. Although the control signal can be sent in real time, it sacrifices part of the accuracy in order to speed up the response of the system and cannot be optimized. Australia's SCATS system [11] constructs a traffic signal task plan based on historical data which collects real-time traffic flow by arranging sensors at the intersection stop line, and then makes plans according to the flow data. But this kind of system is not real-time; it detects the traffic flow and selects one of the best options from the existing solution, and then deploys the solution at the intersection. The method is less accurate and reliable. It is widespread that using wireless sensor nodes to monitor and direct traffic [14]: by establishing sensor network topology to monitor traffic flow and improve driving environment, use the acquired traffic flow design algorithm to avoid congestion, and use particle swarm optimization (PSO) method [10] to develop sensor placement position to optimize sensor coverage. The WSN network is used to obtain the traffic flow of the upstream road segment, and the fuzzy controller is used to estimate the traffic volume of the future intersection, thereby formulating the mission plan of the traffic signal [3].

Some researchers have studied the joint task formulation of multiple traffic junctions in cities. For example, some traffic control algorithms have adopted the particle swarm optimization algorithm [5], genetic algorithm [2], neural network computing model [4,21], machine learning [12,22] and other methods. He et al. combined particle swarm optimization algorithm with neural network model [8], using evolutionary particle swarm optimization algorithm to deal with multiple intersection traffic flow model optimization problems, and using artificial neural network model to learn multi-junction traffic flow control strategy. And in the simulation environment, the experimental results of this method are verified to be better than the single method. Although these methods take the overall traffic situation into account, they require a lot of calculation time and cannot meet the real-time performance, and the data used for calculation are all out-of-date. When the traffic flow fluctuates greatly, the result of historical data can not meet the real-time requirement.

Another solution is based on the Internet of vehicles, in which vehicles can communicate via V2I and V2V to monitor real-time information of vehicles. Some researchers have proposed that the formation of an adaptive traffic control system [20] by using the short-distance communication of vehicles, using a very short path (1–2 hops) between the vehicle and the signal to communicate. Wu et al. established a two-layer pipeline model using the V2I communication of the Internet of Vehicles system at the intersection [23] which is used to accurately detecting vehicle information and assign different weights to vehicles according to the distance from the vehicle to the intersection. The adaptive skip signal control method uses the principle of on-demand distribution to prevent the vehicle from waiting at the idle intersection. Another mode is to use the vehicle terminal to collect traffic data and send them to the traffic signal [25], and the result is calculated by the remote computer, which does not take the delay of communication into account between the traffic signal and the remote computer.

The introduction of Virtual Traffic Lights (VTL) [6] opened the other direction into intelligent traffic lights. In the Internet of Vehicles, virtual traffic lights are used to solve the traffic congestion problem. Instead of any street infrastructure, the direct communication between vehicles is used to negotiate the right-of-way assignment rules, and the driver is informed by the vehicle display device when to pass the intersection.

The Internet of Vehicles for IEEE802.11p still takes a long time to become popular. Some researchers use the Wi-Fi Direct of smartphones to communicate [13,16] to implement the virtual signal system, but the authors only implemented the system in the laboratory, without considering that Wi-Fi Direct could not communicate efficiently while the actual vehicle was running.

Some researchers have proposed V2V-based distributed algorithm named DVTL [1] to implement the basic model of the virtual signal light mentioned in [6]. When the vehicle information is exchanged, the location and the vehicle basic information are transmitted by broadcasting, and the "leader" election information is exchanged by using the unicast method. But DVTL has some deficiencies in the formulation of traffic signals: the only factor that the arrival time at the intersection is taken into account when setting the priority of the traffic, without including the number of queues and other factors; There is no strict regulation on the passage time of each lane, and the right of way will be transferred to the next lane only if the group of vehicles completely pass the intersection in the lane with the right of way; The established lane models are all single-lane models, but in reality, urban roads are mostly two or more lanes.

In this paper, a queuing chain model is proposed. Based on the model, V2V communication is used to count and collect the vehicle information. According to the collected number of vehicles, the queue length and the waiting time, the traffic signal phase position of the intersection are determined by the vehicle self-organization.

## 3    System Model

### 3.1    VTL System Model

Virtual traffic signal system requires vehicles to communicate and coordinate with each other to formulate traffic signal phase position. When the system is running, the system endows the vehicle with different identities to complete different tasks that required by the signal phase position according to different location of the vehicle.

The identity of vehicles in the system are defined as follows:

**Definition 1** *Free node. The identity of the vehicle is a free node when the vehicle passes the intersection but does not drive beyond the threshold.*

The threshold is a logical line. The distance between the threshold and the nearest intersection is usually not more than a third of the distance between two adjacent intersections.

**Definition 2** *Obey node. The vehicle passes the driving threshold, but it is not the nearest car to the next intersection within the lane communication range.*

**Definition 3** *Group aggregation node. The vehicle passes the driving threshold and it is the nearest car to the next intersection within the lane communication range.*

**Definition 4** *The total aggregation node. Before entering the intersection area, all of the group aggregation nodes negotiate and select a vehicle from themselves as the total aggregation node to calculate the traffic signal phase position.*

The VTL module needs to obtain its own information such as location, time and vehicle status, and broadcasts the data to other intersection vehicles through DSRC or LTE-V2X at the intersection. According to the received data, it is processed and handed over to the VTL group node decision module, which is used to elect the group node of each lane. Then, the group nodes share the lane data, and an aggregation node is elected. The vehicle calculates the phase of the signal and broadcasts the result to the intersection vehicle. The vehicles at the intersection pass the intersection in an orderly manner according to the result.

### 3.2 Intersection Conflict Model

We can deduce the conflict situation. These conflicts should be taken into account when we formulate the phase rules of the intersection. Table 1 lists the conflict situation at the intersection, $L$ indicates the left turn, the $F$ is straight, 0 represents the conflict, and the 1 indicates no conflict. $A$, $B$, $C$, and $D$, indicate the north, east, south and west of a crossroad respectively.

### 3.3 Vehicle Queuing Model

There are significant differences in its size and initial acceleration when vehicles are in queues at intersections due to different vehicle models. When arranging traffic lights according to the queuing of vehicles at intersections, it is necessary to consider the case that new queued vehicles are added. Therefore, the traffic signal phase positions formulation should take the impact of the vehicle model and the new queued vehicle into account. Although the actual vehicle models are complicated, there are no obvious differences in the acceleration for same type of vehicle. Therefore, in the experiment, three models were used to simulate different vehicle lengths and accelerations to eliminate the effects caused by vehicle type. In addition, the queuing of the vehicle can be modeled using Markov

**Table 1.** Intersection conflict table

|    | AL | AF | BL | BF | CL | CF | DL | DF |
|----|----|----|----|----|----|----|----|----|
| AL | 1  | 1  | 0  | 0  | 1  | 0  | 0  | 1  |
| AF | 1  | 1  | 1  | 0  | 0  | 1  | 0  | 0  |
| BL | 0  | 1  | 1  | 1  | 0  | 0  | 1  | 0  |
| BF | 0  | 0  | 1  | 1  | 1  | 0  | 0  | 1  |
| CL | 1  | 0  | 0  | 1  | 1  | 1  | 0  | 0  |
| CF | 0  | 1  | 0  | 0  | 1  | 1  | 1  | 0  |
| DL | 0  | 0  | 1  | 0  | 0  | 1  | 1  | 1  |
| DF | 1  | 0  | 0  | 1  | 0  | 0  | 1  | 1  |

queuing theory, which estimates the possibility of joining the vehicle in turn, and develop the traffic light phase.

According to the defined vehicle attributes and the vehicle type distribution rules at the intersection, the time required for vehicles to pass through the intersection can be obtained.

The time required for a car to pass through an intersection can be divided into: driver response delay and vehicle start delay $T_d$, vehicle acceleration time $T_a$ and vehicle uniform speed time $T_v$. The total time the motorcade passes through the intersection is equal to the time when the last car completely crossed the intersection. The time for the last car to pass the intersection is $T_{last} = T_a + T_v + T_d$, vehicle acceleration time $T_a$ and vehicle constant speed time $T_v$ depend on the distance that the last car will travel.

The last vehicle crossing the intersection needs to travel the distance:

$$S_{last} = \frac{1}{2}aT_a{}^2 + vT_v = \sum_i^N L_i + S_{cr} + (N-1) \times L_{gap}(aT_a \le v) \qquad (1)$$

where, $N$ is the number of vehicles queuing at the junction, $S_{cr}$ is the length of the intersection, $L_i$ is the length of the vehicle, $L_{gap}$ is the average length of the vehicle interval, $a$ is the average acceleration of the vehicle, $v$ is the maximum speed allowed by the vehicle, and $aT_a \le v$, when $aT_a < v$, $T_v = 0$. $T_d$ are all driver response delays and vehicle startup delays in the driveway, where $T_{idelay}$ is the $i^{th}$ driver response delay and vehicle startup delay:

$$T_d = \sum_{i=1}^N T_{idelay} \qquad (2)$$

In conclusion, when the last vehicle passes the intersection, the time $T_{last}$ can be obtained, as shown in Eq. 3.

$$T_{last} = \begin{cases} \sqrt{\frac{2S_{last}}{a}} + T_d & aT_a < v \\ \frac{S_{last} - \frac{v^2}{2a}}{v} + \frac{v}{a} + T_d & aT_a = v \end{cases} \qquad (3)$$

By calculating the $T_{last}$ of each lane and obtaining the lane combination of green light according to the conflict Table 1, the total time of vehicles passing through the intersection at the current intersection is minimized.

# 4  Data Collection Algorithm Based on Queuing Chain

## 4.1  Message Definition

Three kinds of data packets are used for vehicle communication in the system: information packet, notification packet and control packet. There are two types of information packets: information packets for a single car and group aggregation information packets. Information packets are used to convey their own information to surrounding vehicles, which is the basis for the aggregation node selection. The notification data packet is used to send a notice to the surrounding vehicles, to inform the aggregation node of the selection result and to inform the vehicle that the vehicle is about to enter the intersection. The function of controlling data packets is to send the phase data of the traffic lights to the waiting vehicles at the intersection.

The information packet contains the following information: Information type, Vehicle ID, Vehicle Length, Distance, Lane Number, Direction and Time Stamp. The type of information indicates which type the packet belongs to. The vehicle ID is the only code that identifies the vehicle identity. The length of the vehicle, the distance from the vehicle to the intersection and the lane number of the vehicle are used to calculate the traffic light phase. The direction indicates the direction that the vehicle will be driving at the intersection.

The group aggregation information packets represent the aggregation information of the waiting vehicles in a lane for determining the total aggregation node and calculate the traffic signal phase position.

The notification packet is sent by the aggregation node to inform other vehicle aggregation nodes identity changes, and the packet has synchronous timestamp function.

The transfer notification packet is used to transfer the queuing chain of the group aggregation node. The packet contains the data information collected from all vehicles.

The control packet contains the allocation of road rights of all vehicles in all lanes of the intersection. The start time, end time and the sequence of signal lights constitute a set of signal lamp phases, and the corresponding signal sequences of each lane are executed in the starting time and ending time.

## 4.2  Queuing Chain Model

In a real traffic environment, the state of the vehicle is constantly changing, and behaviors, such as lane change and overtaking, are frequent. In the multi-lane case, the group aggregation node vehicle must maintain the member information in the group, and the data structure of the linked list is very suitable for the maintenance of the queue chain. The formation of the queuing chain requires two steps: the formation of a queuing chain and the maintenance of a queuing chain.

Queuing chain formation: Before the vehicle travels past the threshold through the intersection, if the vehicle does not receive the aggregation node notification packet sent by the preceding vehicle, the vehicle changes the identity to the group aggregation node, and periodically broadcasts the notification packet to each channel. If the notification packet of the lane group aggregation node is received before the threshold is exceeded, the vehicle becomes a obey node, and the obey node periodically sends its own information to the group aggregation node to maintain the queued chain order.

Queueing chain maintenance:

Case 1: The group aggregation node receives the vehicle information sent by the own lane obey node. If the obey node is in front of the vehicle, then the group aggregation node hands over the group aggregation node identity to the obey node, otherwise the obey node is added to the queuing chain according to the Distance data in the information.

Case 2: The group aggregation node vehicle receives notification information of other vehicles in the same lane with the group aggregation node. If the front group aggregation node does not enter the decision stage of the total aggregation node, the group aggregation node in the rear group transfers the group aggregation node identity to the queue chain order first. If the front group aggregation node has entered the decision stage of the total aggregation node, then it maintains the identity of the group aggregation node.

Case 3: The obey node receives the notification information of the plurality of own lane group aggregation nodes, and selects the first group aggregation node to send the vehicle information.

Case 4: The group aggregation node leaves the queue and hands over the identity to the first car in the rear before leaving.

Case 5: The group aggregation node has not received certain obey node information in the queuing chain for a long time, and deletes the node from the queuing chain.

Case 6: The vehicle becomes a free node through the intersection.

Case 7: After the group aggregation node enters the intersection area, the identity is no longer changed, and new nodes are not accepted to join.

## 4.3   Data Collection Algorithm Based on Queue Chain

The vehicle identity in the virtual traffic light system is divided into the free node, the obey node, the group aggregation node and the total aggregation node. The queuing chain is composed of the group aggregation node as the chain head and the obey node as the chain body. The formation and maintenance of the queuing chain is information-driven, and the formation and maintenance process of the queuing chain is the collection process of vehicle information data. The head of the chain node (group aggregation node) has the right to broadcast in all channels, and other nodes can only communicate with the sink node directly in one-hop mode within the limited channel of the lane in which they belong to. In order to reduce communication, vehicles are allowed to send messages only after they have passed the threshold line. The obey node can only send information within the limited channel, and the group aggregation node can broadcast notification information to the whole network periodically, and control packet broadcasting only in the channel of the queuing chain in the lane.

At the intersection, according to the lane of the vehicle and the direction (left turn, right turn, straight ahead) in which the vehicle is moving, the vehicle determines which queuing chain the vehicle belongs to. According to the queuing chain model, the aggregation node of the queuing chain is the queue head, the group aggregation node of the left lane is the first left-turn vehicle, if the first car runs straight forward, it will be classified as the middle lane and the identity is the middle lane obey node. The group aggregation node of the middle lane is the first vehicle runs straight forward; otherwise it is classified as the obey node of the corresponding lane. The group aggregation node of the right lane is still the first vehicle runs straight forward. If the first vehicle is a right-turn vehicle, it is a free node and not included in the queuing chain.

Since at the intersection which queuing chain the vehicle belongs to is determined by the lane of the vehicle and the direction in which the vehicle is moving. Therefore, the lane change will have an impact on the queue chain of the lane in which it is located and the queue chain to be entered. There are two identities in a queuing chain: the first node of the queue chain (group aggregation node) and the obey node.

When a group aggregation node changes lanes, it will cause changes to the group aggregation node. Therefore, when a lane change occurs, the identity of the aggregation node is handed over to the vehicle in the original lane list that matches the identity of the aggregation node (handed over the entire queue chain). That is, the first left-turning vehicle in the left lane, the first vehicle runs straight forward in the middle lane or in the right lane. Upon completion of identity transfer, it becomes obey node.

The selection of the group aggregation node is the chain head vehicle of each lane queue chain. The selection and maintenance algorithm of the group aggregation node is essentially to maintain the queue chain of each lane. The queue chain satisfies two conditions. Firstly, the vehicle on each chain must pass the intersection in the green light phase time of the lane. Secondly, the group aggregation node on the chain must be the first vehicle on the lane queue chain. Because each lane has a corresponding traffic signal, the first condition can ensure the accuracy of the time of passing the intersection in the lane, that is, to ensure the phase accuracy of the traffic light at the intersection. The second condition ensures that the group aggregation node is the first vehicle to enter the intersection area, and the total aggregation node vehicle is selected from the group aggregation node, so that the communication is as unobstructed as possible to ensure communication quality.

Through the above rules, each lane constitutes a respective lane queuing chain, and the first node of the queuing chain is a group aggregation node, and the information of the queuing chain vehicle is collected in the group aggregation node vehicle. The total aggregation node is elected according to the information of the vehicles in the queue chain between the lanes. The purpose of the total aggregation node is to replace the physical traffic lights at the intersection to direct traffic. Therefore, in a custom traffic light phase cycle, the queue of the total aggregation node should pass through the intersection as late as possible. The system follows the first-come first-pass and long-queue priority principles to dynamically adjust the transit time and traffic order of each lane. The total aggregation node is chosen only refer to two indicators: the arrival time and the queue length of the lane. The time reaching the intersection can be used to ensure fairness. Vehicles arriving at the intersection early have the right to pass through the intersection first. The length of the queue chain represents the degree of congestion in the lane. Allowing longer queues to pass first can appropriately relieve road pressure.

## 5    Virtual Traffic Light Based on V2V for Single Intersection

The phase of traffic lights generated in the virtual intelligent traffic light system is real-time. Every time a group of vehicles pass through the intersection, the corresponding phase of the signal lights will be formulated according to the vehicles in each lane. The signal lights generated by each group are customized, so it can greatly reduce the phenomenon of waiting in vain and improve the throughput of the intersection.

A set of traffic light phases consists of two elements: the order of travel and the travel time. In other word, to formulate a set of traffic light phase is to formulate a combination of a set of travel time and traffic order. The goal is to make all the vehicles at the intersection pass through the intersection as soon as possible. Calculating the transit time firstly requires calculating the traffic distance. According to the calculation formula defined by the vehicle queuing model, the distance that the last vehicle in each lane needs to pass through the intersection is obtained. The distance of the last car in the left lane through the intersection is $S_{Llast}$, according to the joint calculation of the middle lane and the right lane we can obtain that the distance of the last car passing through the intersection is $S_{Mlast}$. The distances obtained are brought into Eq. 3 to determine the time when the vehicles pass through the intersections in each lane. In order to ensure the fairness of each group, the transit time is less than 90 s.

The order of passage for each lane should consider three factors: waiting time, number of queues, and queue length. The large number of queues means that the traffic volume of the lane is large and the traffic pressure is high, so it should be preferred. (The number of queues in the middle lane is the sum of the straight movement vehicles in the middle and right lanes). In order to ensure the fairness of traffic, waiting time should have a higher priority and the longer the waiting time, the higher the priority. The length of the queue and the number of queues affect the time passing through the intersection. According to the traffic conflict list at the intersection defined in the vehicle queuing model, after calculating the transit time of each lane, it needs to make overall consideration, so that the vehicles in multiple lanes can pass through the intersection at the same time without conflict.

The priority of the pass order is determined by the wait time and the number of queues. The weights of the two factors shown in Eq. 4 are divided according to the waiting time gradient. The weight of the waiting time in 0–30 s is 0.4, the weight increases linearly within 30–90 s, and the weight of the waiting time exceeds 90 s is 1. That is, when the vehicle waiting time exceeds 90 s, the number of queues is not referred to when calculating the priority of the traffic order.

$$W_T = \begin{cases} 0.4 & 0 \leq t < 30\,\mathrm{s} \\ 0.01t + 0.1 & 30 \leq t < 90\,\mathrm{s} \\ 1 & t > 90\,\mathrm{s} \end{cases} \tag{4}$$

The waiting time and the number of queues are two kinds of metrics, and the two types of data need to be normalized separately so that the two types of data can be added to generate a priority. Normalize the waiting time, using the minimum-maximum normalization method, the minimum value is 0, and the maximum value is 90 s (After waiting for more than 90 s, regardless of the impact of the number of queues on the priority, the priority may be greater than 1), so the time-based priority is:

$$P_T = \frac{T - T_{MIN}}{T_{MAX} - T_{MIN}} = \frac{T}{90} \tag{5}$$

Then, the data of the number of queues is normalized still using the minimum maximum normalization method. The minimum value is 1 (0 means no waiting for the vehicle, does not participate in the calculation), and the maximum value is the maximum number of vehicles that can pass the intersection at an average of 60 s without congestion. In the simulation experiment, 60 s can pass through 25 vehicles. The priority based on the number of queues is:

$$P_N = \frac{N - N_{MIN}}{N_{MAX} - N_{MIN}} = \frac{N - 1}{24} \tag{6}$$

In summary, the priority of a lane passing through the intersection is a weighted sum of the priority PY based on the waiting time and the priority PN based on the number of queues, as shown in Eq. 7:

$$P = W_T P_T + (1 - W_T) P_N \tag{7}$$

Calculate the priority of each lane according to Eq. 7, and select the lane with the highest priority to pass first (Long queues with the same priority take precedence). Lookup Table 1 finds that the lanes that do not conflict are added to the group of traffic phases by priority. If the time-taken is less than the highest priority lane supplemented by the next non-collision lane, if the time-taken is greater than the highest priority lane, set the time to be the same as the highest priority lane. So reciprocating until all vehicles in the group have completed the transit time, then the next group of vehicle traffic phases are turned on.

In summary, the basic steps for calculating the phase of a traffic light by the SVTL system are:

Step 1: Determine the identity of the total aggregation node vehicle. The group aggregation node vehicles elect the total aggregation node according to their respective queuing chain information. The total aggregation node needs to complete the formulation work of the traffic light.

Step 2: calculating the transit time, using the queue chain information collected by the group aggregation node of each lane to calculate the transit time required by the vehicles in each lane.

Step 3: data normalization, using the two parameters of waiting time and queuing number to determine the pass priority, so the two parameters need to be normalized.

Step 4: Calculate the priority and calculate the priority of each lane according to Eq. 6.

Step 5: Develop the intersection traffic light phase, use the priority calculated in the fourth step, select the first green light phase, and then select the phase that does not conflict with the phase and has the highest priority according to the conflict table and priority. Co-execute the first green light.

Step 6: Distribute the results. The total aggregation node first unicasts the results to the group aggregation node, and then the group aggregation node broadcasts to all the nodes in the group.

## 6    Simulation Results

In this section, we use the OMNeT++ [15], SUMO [18] and VEINS [17] frameworks to run the simulation. The simulation runs the Common Traffic Lights (CTL), the Distributed Virtual Traffic Lights (DVTL) [1] system and the single virtual traffic lights (SVTL) under the same experimental environment. Compare the average parking time and average driving speed of the three algorithms.

### 6.1    Experimental Parameter Setting

Road network environment setting: There are 49 roads in the road network, each road is 300 m long, the road width is 20 m, the road is six-way, the road max speed is 19.5 m/s, the blocking density is 300 pcu/km, the three lanes at one direction intersection are left turn lane, straight lane and right turn or straight lane; a total of 64 intersections, including 36 intersections, 24 T-junctions and 4 right-angle intersections. Set the right turn to be unrestricted, left turn and straight line must be exercised according to the signal indication.

Vehicle Setting: In the simulation process, a total of 9000 vehicles were built to participate in the simulation experiment. 9000 vehicles entered the road network at the same time interval according to the serial number. The vehicle has three models, the lengths are 4 m, 6 m and 12 m respectively, and the acceleration is set to 5 m/s². The parking gap is 1.5 m. In the case where the starting and ending of 9000 vehicles remain unchanged, the tools provided by SUMO are used to generate three sets of routes, and three sets of routes can be used to perform three experiments separately.

Network Settings: use perfect network status, that is, no packet loss, no network conflicts. The vehicle communication radius is 100 m, strong penetration mode. The communication protocol uses the IEEE802.11P protocol. Use the same road network configuration and vehicle settings to run CTL, DVTL and SVTL respectively. Compare the differences between the three systems and set different traffic flows to simulate the difference in performance of traffic lights under different traffic flows.

### 6.2    Vehicle Average Parking Time

Figure 1 shows the average parking time of the vehicle under different traffic flows. In the case of CTL, when the number of vehicles in the road network is between 200 and 1000, the parking time increases steadily, but even in the case of only 200 vehicles in the road network, the average waiting time is still 80 s. DVTL and SVTL have an average parking time of less than 2 s in the case of 200 vehicles in the road network, but as the number of vehicles increases, the average parking time also rises rapidly. This is because DVTL and SVTL basically eliminate the phenomenon of waiting in vain, but as the number of vehicles increases, vehicles waiting to pass at various intersections began to increase, resulting in a faster growth rate of average waiting time. As the density of vehicles increases, the parking time under DVTL grows faster than SVTL. This is because DVTL only considers the order of arrival at the intersection when setting priorities, so that the intersection traffic cannot be optimal.

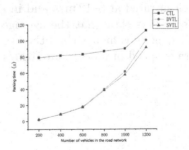

**Fig. 1.** Average parking time

## 6.3   Vehicle Average Driving Speed

Figure 2 shows the average driving speed of the vehicle under the control of three traffic lights under different traffic flows. Similar to the average parking time in the previous section, CTL performed poorly in a sparsely traffic environment, and the average speed was pulled down due to the large number of waiting in vain conditions. DVTL and SVTL basically eliminate the waiting in vain, so that its speed can better reflect the road congestion. From the flow density formula, it can be known that the vehicle density in the road will affect the speed of the vehicle. That is to say, when the traffic volume is large, the factor limiting the average speed of the vehicle is mainly the vehicle density rather than the waiting time of the traffic signal.

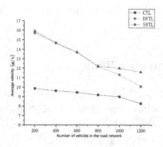

**Fig. 2.** Average driving speed

Figure 3 shows a comparison of average speeds in an average of 1200 vehicles participating in traffic at the same time. The average speed of the vehicle under CTL is 8.4 m/s, and the average speed of the vehicle under DVTL is about 9.9 m/s. The total average speed of the vehicle under the SVTL is approximately 11.5 m/s. Compared with CTL and DVTL, the average speed increased by 36.9% and 16.2%, respectively. In terms of speed distribution, the average speed of vehicles in a CTL environment is concentrated at 4–12 m/s, in the DVTL environment is concentrated at 8–12 m/s and in the SVTL environment is concentrated at 8–16 m/s. It is clear that the average speed that the SVTL can achieve at this vehicle density is higher than the other two algorithms, and the vehicles with an average speed of more than 16 m/s are much higher than the other two algorithms.

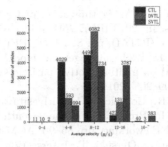

**Fig. 3.** Statistical histogram of average speed in congested state

## 7 Conclusion

This paper defines the queuing chain model and designs the algorithm for generating and maintaining the queuing chain. The establishment and maintenance of the queuing chain can assist the group aggregation node vehicles to collect and maintain the information of the vehicles in the lane, and ensure the accurate and real-time information. Proposed a real-time single virtual traffic lights (SVTL) phase algorithm. Experiments show that compared with the Common Traffic Lights method, this algorithm has better effect no matter how traffic density changes, the average speed of the vehicle is higher, and the parking waiting time is lower. Compared with the DVTL algorithm, when the traffic density is small, the performance of the two algorithms is not much different. When the density of the traffic vehicle is gradually increased, the performance of the DVTL is lower than that of the SVTL. In the subsequent research, multiple intersection factors will be considered at the same time, making the algorithm more suitable for real traffic environment.

**Acknowledgement.** This work is partly supported by the National Natural Science Foundation of China with No. 61877037, the Fundamental Research Funds for the Central Universities of China with No. GK201903090 and GK201801004, and the National Key R&D Program of China with No. 2017YFB1402102.

## References

1. Bazzi, A., Zanella, A., Masini, B.M.: A distributed virtual traffic light algorithm exploiting short range V2V communications. Ad Hoc Netw. **49**, 42–57 (2016)
2. Chin, Y.K., Yong, K., Bolong, N., Yang, S.S., Teo, K.T.K.: Multiple intersections traffic signal timing optimization with genetic algorithm. In: 2011 IEEE International Conference on Control System, Computing and Engineering (ICCSCE), pp. 454–459. IEEE (2011)
3. Collotta, M., Bello, L.L., Pau, G.: A novel approach for dynamic traffic lights management based on wireless sensor networks and multiple fuzzy logic controllers. Expert. Syst. Appl. **42**(13), 5403–5415 (2015)

4. De Oliveira, M.B., Neto, A.D.A.: Optimization of traffic lights timing based on multiple neural networks. In: 2013 IEEE 25th International Conference on Tools with Artificial Intelligence (ICTAI), pp. 825–832. IEEE (2013)
5. El Hatri, C., Boumhidi, J.: Traffic management model for vehicle re-routing and traffic light control based on multi-objective particle swarm optimization. Intell. Decis. Technol. 11(2), 199–208 (2017)
6. Ferreira, M., Fernandes, R., Conceição, H., Viriyasitavat, W., Tonguz, O.K.: Self-organized traffic control. In: Proceedings of the Seventh ACM International Workshop on VehiculAr InterNETworking, pp. 85–90. ACM (2010)
7. Guan, X., Huang, Y., Cai, Z., Ohtsuki, T.: Intersection-based forwarding protocol for vehicular ad hoc networks. Telecommun. Syst. 62(1), 67–76 (2016)
8. He, L., Fu, C., Yang, L., Tong, S., Luo, Q.: Coordinated real-time control algorithm for multi-crossing traffic lights. In: 2014 10th International Conference on Natural Computation (ICNC), pp. 128–133. IEEE (2014)
9. Huang, Y., Chen, M., Cai, Z., Guan, X., Ohtsuki, T., Zhang, Y.: Graph theory based capacity analysis for vehicular ad hoc networks. In: 2015 IEEE Global Communications Conference, (GLOBECOM) 2015, San Diego, CA, USA, 6–10 December 2015, pp. 1–5 (2015)
10. Kulkarni, R.V., Venayagamoorthy, G.K.: Particle swarm optimization in wireless-sensor networks: a brief survey. IEEE Trans. Syst. Man Cybern. Part C (Appl. Rev.) 41(2), 262–267 (2011)
11. Martin, P.T.: Scats, an overview. In: Transportation Research Board Annual Meeting (2001)
12. Mousavi, S.S., Schukat, M., Howley, E.: Traffic light control using deep policy-gradient and value-function-based reinforcement learning. IET Intell. Transp. Syst. 11(7), 417–423 (2017)
13. Nakamurakare, M., Viriyasitavat, W., Tonguz, O.K.: A prototype of virtual traffic lights on android-based smartphones. In: 2013 10th Annual IEEE Communications Society Conference on Sensor, Mesh and Ad Hoc Communications and Networks (SECON), pp. 236–238. IEEE (2013)
14. Nellore, K., Hancke, G.P.: A survey on urban traffic management system using wireless sensor networks. Sensors 16(2), 157 (2016)
15. OMNet++: https://omnetpp.org/
16. Perry, A., Verbin, D., Kiryati, N.: Crossing the road without traffic lights: an android-based safety device. In: Battiato, S., Gallo, G., Schettini, R., Stanco, F. (eds.) ICIAP 2017. LNCS, vol. 10485, pp. 534–544. Springer, Cham (2017). https://doi.org/10.1007/978-3-319-68548-9_49
17. Sommer, C., German, R., Dressler, F.: Bidirectionally coupled network and road traffic simulation for improved IVC analysis. IEEE Trans. Mob. Comput. 10(1), 3–15 (2011)
18. SUMO: http://sumo.sourceforge.net/
19. Unit, Traffic Advisory: The 'SCOOT' urban traffic control system. Traffic Advis. Leafl. 7, 99 (1999)
20. Varga, N., Bokor, L., Takács, A., Kovács, J., Virág, L.: An architecture proposal for V2X communication-centric traffic light controller systems. In: 2017 15th International Conference on ITS Telecommunications (ITST), pp. 1–7. IEEE (2017)
21. Wang, D., Bao, H., Zhang, F.: CTL-DNNet: effective circular traffic light recognition with a deep neural network. Int. J. Pattern Recognit. Artif. Intell. 31(11), 848–854 (2017)

22. Wang, H., Liu, J., Pan, Z., Takashi, K., Shimamoto, S.: Cooperative traffic light controlling based on machine learning and a genetic algorithm. In: 2017 23rd Asia-Pacific Conference on Communications (APCC), pp. 1–6. IEEE (2017)
23. Wu, L.B., Nie, L., Liu, B.Y., Wu, N., Zou, Y.F., Ye, L.Y.: An intelligent traffic signal control method in VANET. Jisuanji Xuebao/Chin. J. Comput. **39**(6), 1105–1119 (2016)
24. Xiong, Z., Li, W., Han, Q., Cai, Z.: Privacy-preserving auto-driving: a gan-based approach to protect vehicular camera data. In: 2019 IEEE International Conference on Data Mining, (ICDM) 2019, Beijing, China, 8–11 November 2019, pp. 1–12. IEEE (2019)
25. Younes, M.B., Boukerche, A.: Intelligent traffic light controlling algorithms using vehicular networks. IEEE Trans. Veh. Technol. **65**(8), 5887–5899 (2016)

# Characterizations for Special Directed Co-graphs

Frank Gurski[✉], Dominique Komander, and Carolin Rehs

Institute of Computer Science, Algorithmics for Hard Problems Group,
Heinrich-Heine-University Düsseldorf, 40225 Düsseldorf, Germany
frank.gurski@hhu.de

**Abstract.** In this paper we consider special subclasses of directed co-graphs and their characterizations. The class of directed co-graphs has been well-studied by now and there are different definitions and applications. But whereas for undirected co-graphs multiple subclasses have been considered and characterized successfully by several definitions, for directed co-graphs very few known subclasses exist by now. Known classes are oriented co-graphs which we obtain omitting the series composition and which have been analyzed by Lawler in the 1970s and their restriction to linear expressions, recently studied by Boeckner. We consider directed and oriented versions of threshold graphs, simple co-graphs, co-simple co-graphs, trivially perfect graphs, co-trivially perfect graphs, weakly quasi threshold graphs and co-weakly quasi threshold graphs. For all these classes we provide characterizations by finite sets of minimal forbidden induced subdigraphs, which lead to first polynomial time recognition algorithms for the corresponding graph classes. Further, we analyze relations between these graph classes.

**Keywords:** Directed co-graphs · Forbidden induced subdigraphs

## 1 Introduction

Undirected co-graphs (short for complement reducible graphs) were developed independently by several authors, see [24,26] for example. Undirected co-graphs are precisely those graphs which can be generated from the single vertex graph by disjoint union and join operations. Co-graphs are interesting from an algorithmic point of view since several hard graph problems can be solved in polynomial time by dynamic programming along the tree structure of the input graph, see [5,8,9]. Furthermore co-graphs are interesting from a graph-theoretic point of view. They are exactly the $P_4$-free graphs (where $P_4$ denotes the path on 4 vertices) [8]. Further, they lead to characterizations for graphs of NLC-width 1 and clique-width 2 by [10,27]. Several subclasses of undirected co-graphs have been studied: trivially perfect graphs [12], threshold graphs [7], weakly quasi threshold graphs [3,25], and simple co-graphs [21], see Table 1.

During the last years classes of directed graphs have received a lot of attention [1], since they are useful in multiple applications of directed networks. The

Y. Li et al. (Eds.): COCOA 2019, LNCS 11949, pp. 252–264, 2019.
https://doi.org/10.1007/978-3-030-36412-0_20

field of directed co-graphs is far from been as well studied as the undirected version, even though it has a similar useful structure. Directed co-graphs can be generated from the single vertex graph by applying disjoint union, order composition and series composition [4]. Directed co-graphs are interesting from an algorithmic point of view since several hard graph problems can be solved in polynomial time by dynamic programming along the tree structure of the input graph, see [2,13,14,16,17]. Furthermore, directed co-graphs are also interesting from a graph-theoretic point of view. Similar as undirected co-graphs by the $P_4$, directed co-graphs can be characterized by excluding eight forbidden induced subdigraphs, see [11] and Table 2. Directed co-graphs are useful to characterize digraphs of directed NLC-width 1 and digraphs of directed clique-width 2 [20]. The only known subclasses of directed co-graphs are oriented co-graphs which can be obtained by omitting the series composition (analyzed by Lawler in the 1970s [23]) and the restriction to linear expressions (recently studied by Boeckner [6]). But there are many more interesting subclasses of directed co-graphs, that were mostly not characterized until now. In order to close this gap in literature, we consider directed versions of threshold graphs, simple co-graphs, trivially perfect graphs and weakly quasi threshold graphs. The subclasses are motivated by the related subclasses of undirected co-graphs given in Table 1. Furthermore, we take a look at the oriented versions of these classes and the related complement classes. All of them have in common that they can be constructed recursively by several different operations. All of these classes are hereditary, just like directed co-graphs, such that they can be characterized by a set of forbidden induced subdigraphs. We will even show a finite number of forbidden induced subdigraphs for the further introduced classes. This is for example very useful to know the existence of polynomial recognition algorithm for these classes. Further, we analyze relations between these digraph classes (see Fig. 1) and how they are related to the undirected classes to give an useful and complete overview of the results.

**Table 1.** Overview on the operations and characterizations by forbidden induced subdigraphs for undirected co-graphs and subclasses $X$. $G$ and $H$ are graphs of the class $X$, $I$ is an edgeless graph and $K$ is a complete graph.

| class $X$ | Notation | operations | Forb($X$) |
|---|---|---|---|
| co-graphs | C | • $G \oplus H$  $G \otimes H$ | $P_4$ |
| trivially perfect graphs | TP | • $G \oplus H$  $G \otimes \bullet$ | $P_4, C_4$ |
| co-trivially perfect graphs | CTP | • $G \oplus \bullet$  $G \otimes H$ | $P_4, 2K_2$ |
| simple co-graphs | SC | • $G \oplus I$  $G \otimes I$ | $P_4, \text{co-}2P_3, 2K_2$ |
| co-simple co-graphs | CSC | • $G \oplus K$  $G \otimes K$ | $P_4, 2P_3, C_4$ |
| weakly quasi threshold graphs | WQT | $I$  $G \oplus G_2$  $G \otimes I$ | $P_4, \text{co-}2P_3$ |
| co-weakly quasi threshold graphs | CWQT | $K$  $G \oplus K$  $G \otimes H$ | $P_4, 2P_3$ |
| threshold graphs | T | • $G \oplus \bullet$  $G \otimes \bullet$ | $P_4, C_4, 2K_2$ |

## 2    Preliminaries

We use the notations of Bang-Jensen and Gutin [1] for graphs and digraphs. For a digraph $G$ let $un(G)$ be the underlying undirected graph and co-$G$ be the edge complement graph. The same notations are also applied to graph classes.

The following notations and results are given in [22, Chapter 2] for undirected graphs but they also can be applied to directed graphs. Classes of (di)graphs which are closed under taking induced sub(di)graphs are called *hereditary*. For some (di)graph class $F$ we define Free($F$) as the set of all (di)graphs $G$ such that no induced sub(di)graph of $G$ is isomorphic to a member of $F$. A class of (di)graphs $X$ is hereditary, if and only if there is a set $F$, such that Free($F$) = $X$. A (di)graph $G$ is a *minimal forbidden induced sub(di)graph* for some hereditary class $X$, if $G$ does not belong to $X$ and every proper induced sub(di)graph of $G$ belongs to $X$. For some hereditary (di)graph class $X$ we define Forb($X$) as the set of all minimal forbidden induced sub(di)graphs for $X$. For every hereditary class of (di)graphs $X$ it holds that $X =$ Free(Forb($X$)). Further, set Forb($X$) is unique and of minimal size.

**Observation 1.** *Let $G$ be a digraph such that $un(G) \in$ Free($X$) for some hereditary class of graphs Free($X$), then for all $X^* \in X$ and all biorientations $b(X^*)$ of $X^*$ it holds that $G \in$ Free($b(X^*)$).*

## 3    Directed Co-graphs and Oriented Co-graphs

We introduce operations in order to recall the definition of directed co-graphs from [4] and introduce several interesting subclasses. Let $G = (V_1, E_1)$ and $H = (V_2, E_2)$ be two vertex-disjoint directed graphs.[1]

- The *disjoint union* of $G$ and $H$, denoted by $G \oplus H$, is the digraph with vertex set $V_1 \cup V_2$ and arc set $E_1 \cup E_2$.
- The *series composition* of $G$ and $H$, denoted by $G \otimes H$, is the digraph with vertex set $V_1 \cup V_2$ and arc set $E_1 \cup E_2 \cup \{(u,v),(v,u) \mid u \in V_1, v \in V_2\}$.
- The *order composition* of $G$ and $H$, denoted by $G \oslash H$, is the digraph with vertex set $V_1 \cup V_2$ and arc set $E_1 \cup E_2 \cup \{(u,v) \mid u \in V_1, v \in V_2\}$.

Directed co-graphs have been defined by Bechet et al. in [4].

**Definition 1 (Directed co-graphs, [4]).** *A directed co-graph is a digraph that can be constructed with the following operations.*

*(i) Every single vertex • is a directed co-graph.*
*(ii) If $G$ and $H$ are vertex disjoint directed co-graphs, then $G \oplus H$, $G \otimes H$, and $G \oslash H$ are directed co-graphs.*

*The class of directed co-graphs is denoted by DC.*

---

[1] We use the same symbols for the disjoint union and join between undirected and directed graphs. We want to emphasize that the meaning becomes clear from the context.

Directed co-graphs can be characterized be forbidden induced subdigraphs [11] and lead to characterizations for graphs of small directed NLC-width [20] and small directed clique-width [10].

**Theorem 2 ($\bigstar^2$).** *For a digraph $G$ the following statements are equivalent.*

1. $G \in DC$.
2. $G \in Free(\{D_1, \ldots, D_8\})$.
3. $G \in Free(\{D_1, \ldots, D_6\})$ and $un(G) \in C$.
4. $G$ has directed NLC-width 1.
5. $G$ has directed clique-width at most 2 and $G \in Free(\{D_2, D_3\})$.

**Table 2.** Eight induced forbidden subdigraphs for directed co-graphs

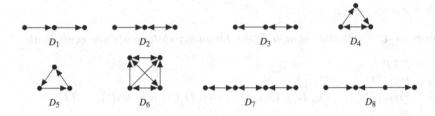

$D_1$          $D_2$          $D_3$          $D_4$

$D_5$          $D_6$          $D_7$          $D_8$

Since $\{D_1, \ldots, D_8\} = co\text{-}\{D_1, \ldots, D_8\}$, it holds that $DC = co\text{-}DC$.

The class of *oriented co-graphs* $OC$ consists of directed co-graphs without bidirectional arcs, by omitting the series composition. This class is interesting when considering oriented coloring of recursively defined oriented graphs [16]. It was already introduced and analyzed by Lawler [23] as the class of transitive series parallel digraphs.

**Theorem 3 ([16]).** *For a digraph $G$ the following statements are equivalent.*

1. $G \in OC$.
2. $G \in Free(\{D_1, D_5, D_8, \overrightarrow{K_2}\})$.
3. $G \in Free(\{D_1, D_5, \overrightarrow{K_2}\})$ and $un(G) \in C$.
4. $G$ has directed NLC-width 1 and $G \in Free(\{\overleftrightarrow{K_2}\})$.
5. $G$ has directed clique-width at most 2 and $G \in Free(\{\overleftrightarrow{K_2}\})$.

# 4 Directed (Oriented) Trivially Perfect Graphs

**Definition 2 (Directed trivially perfect graphs).** *A directed trivially perfect graph is a digraph that can be constructed by the following operations.*

*(i) Every digraph with only one vertex, denoted by $\bullet$ is directed trivially perfect.*

---

2 The proofs of the results marked with a $\bigstar$ are omitted due to space restrictions, see [15].

*(ii)* *If G and H are vertex disjoint directed trivially perfect graphs, then the disjoint union $G \oplus H$ results in a directed trivially perfect graph.*

*(iii)* *If G is a directed trivially perfect graph, then $G \otimes \bullet$, $G \oslash \bullet$ and $\bullet \oslash G$ are directed trivially perfect graphs, too.*

*The class of directed trivially perfect graphs is denoted by DTP.*

Let $G$ be a digraph. Adding an out-dominating vertex to $G$ is equal to $\bullet \oslash G$ and adding an in-dominating vertex is equal to $G \oslash \bullet$.

**Lemma 1** ([19]). *For every digraph G the following properties are equivalent.*

1. *$G$ is a transitive tournament.*
2. *$G \in Free(\{\overrightarrow{C_3}\})$ and $G$ is a tournament.*
3. *$G$ can be constructed from the one-vertex graph by repeatedly adding an out-dominating vertex.*

**Theorem 4.** *For some digraph G the following statements are equivalent.*

1. *$G \in DTP$.*
2. *$G \in Free(\{D_1, \ldots, D_{15}\})$.*
3. *$G \in Free(\{D_1, \ldots, D_6, D_{10}, D_{11}, D_{13}, D_{14}, D_{15}\})$ and $un(G) \in TP$.*

*Proof.* (1. $\Rightarrow$ 2.) The given forbidden digraphs $D_1, \ldots, D_{15}$ (Tables 2 and 3) are not directed trivially perfect graphs and the set of directed trivially perfect graphs is hereditary. (2. $\Rightarrow$ 1.) As $D_1, \ldots, D_8$ are not included in $G$, with Theorem 2 $G$ is a directed co-graph which can be built by the disjoint union, the series and the order composition. Since $G \in Free(\{D_9, D_{10}, D_{11}\})$ for every series composition of two digraphs with at least two vertices, at least one of them must be bidirectional complete. Such a subdigraph can be inserted by a number of feasible operations for DTP. Since $G \in Free(\{D_{12}, D_{13}, D_{14}, D_{15}\})$ holds for every order combination between two digraphs with at least two vertices, at least one of them must be a tournament. Further, we know that by excluding $D_5$ with Lemma 1, at least one of the two digraphs is a transitive tournament. By the same lemma, such a digraph can be defined by a sequence of out-dominating vertices, which are feasible operations for DTP. (2. $\Rightarrow$ 3.) $G \in DTP \Rightarrow un(G) \in TP$. (3. $\Rightarrow$ 2.) By Observation 1 and Forb(TP) = $\{C_4, P_4\}$.                    □

**Table 3.** Forbidden induced subdigraphs

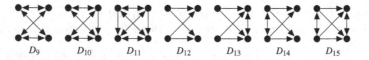

$D_9$        $D_{10}$        $D_{11}$        $D_{12}$        $D_{13}$        $D_{14}$        $D_{15}$

Since DTP $\neq$ co-DTP, we consider the complement class of DTP on its own.

**Definition 3 (Directed co-trivially perfect graphs).** *A directed co-trivially perfect graph is a digraph that can be constructed by the following operations.*

  *(i) Every digraph $G$ with only one vertex, denoted by $\bullet$ is directed co-trivially perfect.*

  *(ii) If $G$ and $H$ are vertex disjoint directed co-trivially perfect graphs, then the disjoint union $G \otimes H$ results in a directed co-trivially perfect graph.*

  *(iii) If $G$ is a directed co-trivially perfect graph, then $G \oplus \bullet$, $G \oslash \bullet$ and $\bullet \oslash G$ are directed co-trivially perfect graphs.*

*The class of directed co-trivially perfect graphs is denoted by DCTP.*

Since we know the characterizations for the complement class DTP, this leads us directly to the following characterization of DCTP by forbidden induced subdigraphs.

**Corollary 1 ($\bigstar$).** *For a digraph $G$ the following statements are equivalent.*

*1. $G \in DCTP$.*
*2. $G \in Free(\{D_1, \ldots, D_8, co\text{-}D_9, co\text{-}D_{10}, co\text{-}D_{11}, D_{12}, \ldots, D_{15}\})$.*
*3. $G \in Free(\{D_1, \ldots, D_6, D_{12}, \ldots, D_{15}\})$ and $un(G) \in CTP$.*

The class of *oriented trivially perfect graphs* OTP consists of directed trivially perfect graphs without bidirectional arcs, by omitting the series composition.

Since OTP is a subclass of DTP the characterizations for the class DTP leads us directly to the following characterization of OTP by forbidden induced subdigraphs.

**Corollary 2 ($\bigstar$).** *For a digraph $G$ the following statements are equivalent.*

*1. $G \in OTP$.*
*2. $G \in Free(\{D_1, D_5, D_8, D_{12}, \overleftrightarrow{K_2}\})$.*
*3. $G \in Free(\{D_1, D_5, \overleftrightarrow{K_2}\})$ and $un(G) \in TP$.*

If we define the complement class of OTP with the operations allowed in directed co-trivially perfect graphs and we omit the series composition, it leads to an equivalent of the class of oriented threshold graphs. It was initially introduced by Boeckner [6], see Sect. 7.

## 5    Directed (Oriented) Weakly Quasi Threshold Graphs

**Definition 4 (Directed weakly quasi threshold graphs).** *A directed weakly quasi threshold graph is a digraph that can be constructed through the following operations.*

  *(i) Every edgeless graph $I$ is a directed weakly quasi threshold graph.*

  *(ii) If $G$ and $H$ are vertex disjoint directed weakly quasi threshold graphs, then $G \oplus H$ is a directed weakly quasi threshold graph.*

**Table 4.** Forbidden induced subdigraphs for DWQT.

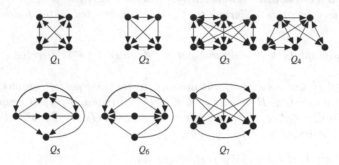

**Table 5.** Digraphs to build the forbidden subdigraphs of DWQT.

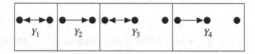

(iii) *If G is a directed weakly quasi threshold graph which is vertex disjoint to an edgeless digraph I, then $G \otimes I$, $G \oslash I$ and $I \oslash G$ are directed weakly quasi threshold graphs.*

*The class of directed weakly quasi threshold graphs is denoted by DWQT.*

**Theorem 5.** *Let G be a digraph, then the following statements are equivalent.*

1. *$G \in DWQT$.*
2. *$G \in Free(\{D_1, \ldots, D_8, Q_1, \ldots, Q_7\})^3$.*
3. *$G \in Free(\{D_1, \ldots, D_6, Q_1, Q_2, Q_4, Q_5, Q_6\})$ and $un(G) \in WQT$.*

*Proof.* (1. $\Rightarrow$ 2.) The given forbidden digraphs $D_1, \ldots, D_8, Q_1, \ldots, Q_7$ (Tables 2 and 4) are not directed weakly quasi threshold graphs and the set of directed weakly quasi threshold graphs is hereditary. (2. $\Rightarrow$ 1.) Let $G$ be a digraph without induced $D_1, \ldots, D_8, Q_1, \ldots, Q_7$. Since there are no induced $D_1, \ldots, D_8$, it holds that $G \in DC$. Thus, $G$ is constructed by the disjoint union, the series and the order composition. $Q_1, Q_3$ and $Q_4$ can only be build by a series composition of two graphs $G_1$ and $G_2$, where $G_1, G_2 \in \{Y_2, Y_3\}$ (cf. Table 5). We note that $Y_2, Y_3$ are contained in every directed co-graph containing more vertices, that is not a sequence of length at least one of series compositions of independent sets. Consequently, there are no bigger forbidden induced subdigraphs that emerged through a series operation, such that the digraphs

$$Q_1 = Y_2 \otimes Y_2, Q_3 = Y_3 \otimes Y_3, Q_4 = Y_2 \otimes Y_3$$

---

[3] Note that $Q_1 = D_{11}$ and $Q_2 = D_{15}$.

characterize exactly the allowed series compositions in DWQT. The digraphs

$$Q_2 = Y_1 \oslash Y_1, Q_5 = Y_1 \oslash Y_4, Q_6 = Y_4 \oslash Y_1, Q_7 = Y_4 \oslash Y_4$$

can only be build by an order composition of two graphs $G_1$ and $G_2$, where $G_1, G_2 \in \{Y_1, Y_4\}$. We note that $Y_1, Y_4$ are contained in every directed co-graph containing more vertices, that is not a sequence of length at least one of order compositions of independent sets. Consequently, there are no forbidden induced subdigraphs containing more vertices that emerged through an order operation, such that the $Q_2, Q_5, Q_6$ and $Q_7$ characterize exactly the allowed order compositions in DWQT. Finally, by excluding these digraphs, we end up in the Definition 4 for DWQT, such that $G \in$ DWQT. (2. $\Rightarrow$ 3.) $G \in$ DWQT $\Rightarrow un(G) \in$ WQT. (3. $\Rightarrow$ 2.) By Observation 1 and Forb(WQT) $= \{P_4, \text{co-}2P_3\}$. $\qquad\square$

Since co-$\{Q_1, \ldots, Q_7\} \neq \{Q_1, \ldots, Q_7\}$ we introduce the complementary class of DWQT. Therefore, we need the definition of *directed cliques*. A directed clique is a bidirectional complete digraph, such that $K = (V, E)$ with $E = \{(u, v) \mid \forall u, v \in V, u \neq v\}$.

**Definition 5 (Directed co-weakly quasi threshold graphs).** *A directed co-weakly quasi threshold graph is a digraph that can be constructed with the following operations.*

(i) *Every directed clique $K$ is a directed co-weakly quasi threshold graph.*
(ii) *If $G$ and $H$ are vertex disjoint directed co-weakly quasi threshold graph, then $G \otimes H$ is a directed co-weakly quasi threshold graph.*
(iii) *If $G$ is a directed co-weakly quasi threshold graph, $K$ is a directed clique and $G$ and $K$ are vertex disjoint, then $G \oslash K$, $K \oslash G$ and $K \oplus G$ are directed co-weakly quasi threshold graphs.*

*The class of directed co-weakly quasi threshold graphs is denoted by DCWQT.*

**Corollary 3 ($\bigstar$).** *For a graph $G$ the following properties are equivalent.*

*1. $G \in DCWQT$*
*2. $G \in Free(\{D_1, \ldots, D_8, \text{co-}Q_1, \ldots, \text{co-}Q_7\})$*

The class of *oriented weakly quasi threshold graphs* OWQT consists of directed weakly quasi threshold graphs without bidirectional arcs, by omitting the series composition.

**Theorem 6 ($\bigstar$).** *For a graph $G$ the following properties are equivalent.*

*1. $G \in OWQT$.*
*2. $G \in Free(\{D_1, D_5, D_8, \overleftrightarrow{K_2}, Q_7\})$.*

In order to define an oriented version of directed co-weakly quasi threshold graphs we omit the series composition and replace the directed clique by a transitive tournament.

**Table 6.** Forbidden induced subdigraphs of OCWQT.

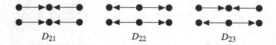

$D_{21}$            $D_{22}$            $D_{23}$

**Definition 6 (Oriented co-weakly quasi threshold graphs).** *An* oriented co-weakly quasi threshold graph *is a digraph that can be constructed through the following operations.*

*(i) Every transitive tournament $T$ is an oriented weakly quasi threshold graph.*

*(ii) If $G$ is an oriented co-weakly quasi threshold graph, $T$ is a transitive tournament and $G$ and $T$ are vertex disjoint, then $G \oplus T$, $T \oslash G$ and $G \oslash T$ are oriented weakly quasi threshold graphs.*

*The class of oriented co-weakly quasi threshold graphs is denoted by OCWQT.*

In OWQT the definitions uses edgeless sets with $n$ vertices, so one might think we can use the $\overleftrightarrow{K_n}$ in the complement class. The reason we chose transitive tournaments instead, is obviously that all biorientations are forbidden in oriented digraphs.

**Theorem 7 ($\star$).** *For a graph $G$ the following properties are equivalent.*

1. *$G \in OCWQT$.*
2. *$G \in Free(\{D_1, D_5, D_8, \overleftrightarrow{K_2}, D_{12}, D_{21}, D_{22}, D_{23}\})$[4] (Table 6).*
3. *$G$ is an oriented co-graph and $G \in Free(\{D_{12}, D_{21}, D_{22}, D_{23}\})$.*

## 6   Directed (Oriented) Simple Co-graphs

**Definition 7 (Directed simple co-graphs).** *A* directed simple co-graph *is a digraph that can be constructed through the following operations.*

*(i) Every single vertex $\bullet$ is a directed simple co-graph.*

*(ii) If $G$ is a directed simple co-graph and $G$ and $I$ are vertex disjoint, then $G \oplus I$, $I \oslash G$, $G \oslash I$ and $G \otimes I$ are directed simple co-graphs.*

*The class of directed simple co-graphs is denoted by DSC.*

**Theorem 8 ($\star$).** *The following statements are equivalent.*

1. *$G \in DSC$.*
2. *$G \in Free(\{D_1, \ldots, D_8, Q_1, \ldots, Q_7, co\text{-}D_9, co\text{-}D_{10}, co\text{-}D_{11}\})$.*
3. *$G \in Free(\{D_1, \ldots, D_8, Q_1, \ldots, Q_7\})$ and $un(G) \in SC$.*

Since DSC $\neq$ co-DSC, we consider the complement class of DSC.

---

[4] $D_{12}$ is equal to co-$Q_2$.

**Definition 8 (Directed co-simple co-graphs).** *An directed co-simple co-graph is a digraph that can be constructed through the following operations.*

*(i) Every single vertex • is a directed co-simple co-graph.*

*(ii) If G is a directed co-simple co-graph and G and K are vertex disjoint, then $G \oplus K$, $K \oslash G$, $G_1 \oslash K$ and $G \otimes K$ are directed co-simple co-graphs.*

*The class of directed co-simple co-graphs is denoted by DCSC.*

**Corollary 4 (★).** *The following statements are equivalent.*

1. $G \in DCSC$.
2. $G \in Free(\{D_1, \ldots, D_8, co\text{-}Q_1, \ldots, co\text{-}Q_7, Q_1, D_9, D_{10}\})$.

The class of *oriented simple co-graphs* OSC consists of directed simple co-graphs without bidirectional arcs, by omitting the series composition.

**Theorem 9 (★).** *The following statements are equivalent.*

1. $G \in OSC$.
2. $G \in Free(\{D_1, D_5, D_8, Q_7, co\text{-}D_{11}, \overleftrightarrow{K_2}\})$.

In order to define an oriented version of directed co-simple co-graphs we omit the series composition and replace the directed clique by a transitive tournament.

**Definition 9 (Oriented co-simple co-graphs).** *An oriented co-simple co-graph is a digraph that can be constructed by the following operations.*

*(i) Every single vertex • is an oriented co-simple co-graph.*

*(ii) If G is an oriented co-simple co-graph, T a transitive tournament and G and T are vertex disjoint, then $G \oplus T$, $T \oslash G$ and $G \oslash T$ are oriented co-simple co-graphs.*

*The class of oriented co-simple co-graphs is denoted by OCSC.*

It is obvious, that the classes OCSC and OCWQT are equal.

# 7  Directed (Oriented) Threshold Graphs

**Definition 10 (Directed threshold graphs).** *A directed threshold graph is a digraph that can be constructed by the following operations.*

*(i) Every single vertex • is a directed threshold graph.*

*(ii) If G is a directed threshold graph, then $G \oplus \bullet$, $G \oslash \bullet$, $\bullet \oslash G$ and $G \otimes \bullet$ are directed threshold graphs.*

*The class of directed threshold graphs is denoted by DT.*

Directed threshold graphs can be characterized be forbidden induced subdigraphs and lead to characterizations for graphs of small directed linear NLC-width, small directed neighbourhood-width, and small directed linear clique-width [18].

**Theorem 10** ([18]). *For a digraph $G$ the following statements are equivalent.*

1. $G \in DT$.
2. $G \in Free(\{D_1, \ldots, D_{15}, co\text{-}D_9, co\text{-}D_{10}, co\text{-}D_{11}\})$.
3. $G \in Free(\{D_1, \ldots, D_6, D_{10}, D_{11}, D_{13}, D_{14}, D_{15}\})$ *and* $un(G) \in T$.
4. $G$ *has directed linear NLC-width* 1.
5. $G$ *has directed neighbourhood-width* 1.
6. $G \in Free(\{D_2, D_3, D_9, D_{10}, D_{12}, D_{13}, D_{14}\})$ *and* $G$ *has directed linear clique-width at most* 2.

By Theorem 10(2.) we observe that DT = co-DT.

The class of oriented threshold graphs was already introduced by Boeckner [6]. We introduce an equivalent definition that is consistent to the other recursively definitions. The class of *oriented threshold graphs* OT consists of directed threshold graphs without bidirectional arcs, by omitting the series composition. By definition it is easy to see that OT = OCTP. Boeckner [6] found equivalent definitions for the class OT, which can be extended as follows.

**Theorem 11 (★).** *For a digraph $G$, the following properties are equivalent.*

1. $G \in OT$.
2. $G \in OCTP$.
3. $G \in Free(\{D_1, D_5, D_8, D_{12}, \overleftrightarrow{K_2}, co\text{-}D_{11}\})$.
4. $G \in Free(\{D_1, D_5, \overleftrightarrow{K_2}\})$ *and* $un(G) \in Free(\{2K_2, C_4, P_4\})$.
5. $G \in Free(\{D_1, D_5, \overleftrightarrow{K_2}\})$ *and* $un(G) \in T$.
6. $G \in Free(\{D_1, D_5, D_{12}, \overleftrightarrow{K_2}\})$ *and* $un(G) \in CTP$.
7. $G \in Free(\{D_1, D_5, D_{12}, co\text{-}D_{11}, \overleftrightarrow{K_2}\})$ *and* $un(G) \in C$.

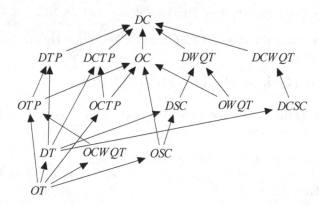

**Fig. 1.** Relations between the subclasses of directed co-graphs. If there is a path from $A$ to $B$, then it holds that $A \subset B$. The classes, that are not connected by a directed path are incomparable.

# 8    Conclusion

We introduced several new digraph classes and gave different characterizations for them. All these classes are subsets of directed co-graphs which have been defined by Bechet et al. in [4] and supersets of oriented co-graphs which were studied by Boeckner [6]. Our digraph classes are motivated by the well known related subclasses of undirected co-graphs, see Table 1. Further, every considered digraph class $X$ generalizes an undirected class $un(X)$, since for every $G \in X$ it holds $un(G) \in un(X)$. Directed co-graphs and their subclasses are algorithmically very interesting, as there are many NP-hard problems on undirected graphs in general, which get polynomial or even linear on special subclasses of directed co-graphs, see [2,13,14,16,17]. By the definitions, we can also give some relations between the different presented graph classes, as shown in Fig. 1.

Our characterizations by finite sets of minimal forbidden induced subdigraphs lead to polynomial time recognition algorithms for the corresponding graph classes. It remains to find more efficient algorithms for this purpose. Up to know, only for directed co-graphs a linear time recognition algorithm is known from [11].

**Acknowledgments.** This work was funded by the Deutsche Forschungsgemeinschaft (DFG, German Research Foundation) – 388221852.

# References

1. Bang-Jensen, J., Gutin, G. (eds.): Classes of Directed Graphs. Springer, Cham (2018). https://doi.org/10.1007/978-3-319-71840-8
2. Bang-Jensen, J., Maddaloni, A.: Arc-disjoint paths in decomposable digraphs. J. Graph Theory **77**, 89–110 (2014)
3. Bapat, R., Lal, A., Pati, S.: Laplacian spectrum of weakly quasi-threshold graphs. Graphs Comb. **24**(4), 273–290 (2008)
4. Bechet, D., de Groote, P., Retoré, C.: A complete axiomatisation for the inclusion of series-parallel partial orders. In: Comon, H. (ed.) RTA 1997. LNCS, vol. 1232, pp. 230–240. Springer, Heidelberg (1997). https://doi.org/10.1007/3-540-62950-5_74
5. Bodlaender, H., Möhring, R.: The pathwidth and treewidth of cographs. SIAM J. Disc. Math. **6**(2), 181–188 (1993)
6. Boeckner, D.: Oriented threshold graphs. Australas. J. Comb. **71**(1), 43–53 (2018)
7. Chvátal, V., Hammer, P.: Aggregation of inequalities in integer programming. Ann. Discrete Math. **1**, 145–162 (1977)
8. Corneil, D., Lerchs, H., Stewart-Burlingham, L.: Complement reducible graphs. Discrete Appl. Math. **3**, 163–174 (1981)
9. Corneil, D., Perl, Y., Stewart, L.: Cographs: recognition, applications, and algorithms. In: Proceedings of 15th Southeastern Conference on Combinatorics, Graph Theory, and Computing (1984)
10. Courcelle, B., Olariu, S.: Upper bounds to the clique width of graphs. Discrete Appl. Math. **101**, 77–114 (2000)
11. Crespelle, C., Paul, C.: Fully dynamic recognition algorithm and certificate for directed cographs. Discrete Appl. Math. **154**(12), 1722–1741 (2006)

12. Golumbic, M.: Trivially perfect graphs. Discrete Math. **24**, 105–107 (1978)
13. Gurski, F.: Dynamic programming algorithms on directed cographs. Stat. Optim. Inf. Comput. **5**, 35–44 (2017)
14. Gurski, F., Komander, D., Rehs, C.: Computing digraph width measures on directed co-graphs. In: Gąsieniec, L.A., Jansson, J., Levcopoulos, C. (eds.) FCT 2019. LNCS, vol. 11651, pp. 292–305. Springer, Cham (2019). https://doi.org/10.1007/978-3-030-25027-0_20
15. Gurski, F., Komander, D., Rehs, C.: On characterizations for subclasses of directed co-graphs. ACM Computing Research Repository (CoRR) abs/1907.00801, 25 pages (2019)
16. Gurski, F., Komander, D., Rehs, C.: Oriented coloring on recursively defined digraphs. Algorithms **12**(4), 87 (2019)
17. Gurski, F., Rehs, C.: Directed path-width and directed tree-width of directed co-graphs. In: Wang, L., Zhu, D. (eds.) COCOON 2018. LNCS, vol. 10976, pp. 255–267. Springer, Cham (2018). https://doi.org/10.1007/978-3-319-94776-1_22
18. Gurski, F., Rehs, C.: Comparing linear width parameters for directed graphs. Theory Comput. Syst. **63**(6), 1358–1387 (2019)
19. Gurski, F., Rehs, C., Rethmann, J.: Characterizations and directed path-width of sequence digraphs. ACM Computing Research Repository (CoRR) abs/1811.02259, 31 pages (2018)
20. Gurski, F., Wanke, E., Yilmaz, E.: Directed NLC-width. Theor. Comput. Sci. **616**, 1–17 (2016)
21. Heggernes, P., Meister, D., Papadopoulos, C.: Graphs of linear clique-width at most 3. Theor. Comput. Sci. **412**(39), 5466–5486 (2011)
22. Kitaev, S., Lozin, V.: Words and Graphs. Springer, Cham (2015). https://doi.org/10.1007/978-3-319-25859-1
23. Lawler, E.: Graphical algorithms and their complexity. Math. Centre Tracts **81**, 3–32 (1976)
24. Lerchs, H.: On cliques and kernels. Technical report, Department of Computer Science, University of Toronto (1971)
25. Nikolopoulos, S., Papadopoulos, C.: A simple linear-time recognition algorithm for weakly quasi-threshold graphs. Graphs Comb. **27**(4), 557–565 (2011)
26. Sumner, P.: Dacey graphs. J. Aust. Soc. **18**, 492–502 (1974)
27. Wanke, E.: k-NLC graphs and polynomial algorithms. Discrete Appl. Math. **54**, 251–266 (1994)

# Scheduling Game with Machine Modification in the Random Setting

Chaoyu He and Zhiyi Tan[(✉)]

School of Mathematics, Zhejiang University,
Hangzhou 310027, People's Republic of China
tanzy@zju.edu.cn

**Abstract.** This paper studies scheduling game with machine modification in the random setting. A set of jobs is to be processed on a set of identical machines. An initial schedule before the modification of machines is given as a prior. Then some machines are removed and some new machines are added. Each job has the right either to stay on its original machine if the machine is not removed, or to move to another machine. If one job changes its machine, it will be behind of all the former jobs scheduled on the target machine. For two jobs moving to the same machine or two jobs staying on the same machine, each one of them has the same probability to be ahead of the other one. The individual cost of each job is its completion time, and the social cost is the makespan. We present properties of the Nash Equilibrium and establish the Price of Anarchy of the game. The bounds are tight for each combination of the number of final machines, the number of added machines and the number of removed machines.

## 1 Introduction

In service industries such as airport security and banking, it is often the case that customers queue in front of multiple parallel service windows. Due to the factors such as the significant difference between the number of customers at different time periods and the need for servers to take turns, the operation department will dynamically adjust the service window. That is close some windows and open some new windows. At this point, not only customers queuing before the closed window must select a new one, but also other customers have an incentive to migrate to a window with a short queue. Naturally, if a customer moves to an existing window instead of a new one, he must be behind of all the former members of the target window. However, for two customers moved to the same window, their relative order in the new queue may depend on some unpredictable factors, for example, the speed of reaction and movement.

Above scenario is suitable for being formulated as a scheduling game with machine modification. A set of jobs is to be processed on a set of identical

C. He—Supported by the National Natural Science Foundation of China (11801505).
Z. Tan—Supported by the National Natural Science Foundation of China (11671356).

Y. Li et al. (Eds.): COCOA 2019, LNCS 11949, pp. 265–276, 2019.
https://doi.org/10.1007/978-3-030-36412-0_21

machines with potential modification. W. l. o. g., we ignore all jobs that have been completed before the modification of machines, and assume that no job is partially processed at the time of modification. An initial schedule before the modification of machines is given as a prior. Then some machines are removed and some new machines are added. Each job has the right either to stay on its original machine if the machine is not removed, or to move to another machine. The choices of all jobs after the modification of machines constitute a new schedule, which is the main research concern.

Although each job can freely select a machine, the order in which jobs are processed on each machine must follow certain rules. If one job changes its machine, it will be behind of all the former jobs scheduled on the target machine. If two jobs move to the same machine, each one of them has the same probability to be ahead of the other one. Such rule is called *Random* policy in the literature [11]. We assume that the relative processing order of any two jobs staying on the same machine also follows Random policy. Given above rules, the completion times of all jobs in any schedule can be determined.

Though the goal of each customer is certainly to reduce his waiting time, the operation department must concern the social interest. With the above background, we take the completion time of a job as its individual cost, and the maximum completion time of all jobs (*makespan*) as the social cost. A schedule is a *Nash Equilibrium* (NE) if no job can reduce its cost by moving to a different machine [12]. Apparently, a NE may not be optimal in terms of makespan due to the lack of central coordination. *Price of Anarchy* (PoA) [12] is used to measure the inefficiency of the NE. The PoA of an instance is defined to be the ratio of the worst NE's makespan to the optimal makespan, and the PoA of the game is the supremum value of the PoA of all instances.

In this paper, we will study the scheduling game with machine modification in the random setting. We show the PoA of the game as a function of the number of final machines, the number of added machines and the number of removed machines. The bounds are tight for each combination of the parameters.

The paper is organized as follows. In Sect. 2, we give a detail description of the game and show our main results. In Sects. 3 and 4, we show the upper and lower bounds of PoA, respectively.

**Related Work:** The basic paradigm of scheduling game was first introduced by Koutsoupias and Papadimitriou [12]. In their original model, the completion time of a job equals to the *load* of the machine it is scheduled on. Here the load of a machine is the total processing time of jobs scheduled on the machine. Thus the completion times of all jobs that are scheduled on the same machine are identical. It was extended later that the completion time of a job is determined by the *coordination mechanism* of the machine it selects [7]. Several coordination mechanisms, deterministic or randomized, were studied and the PoA were obtained [11]. More sophisticated coordination mechanisms in more general machine environment can be found in [1, 2, 4, 8]. However, all these studies assume that the machines are fixed.

There are also some studies on scheduling games which allow the modification of machines. For the scheduling game with migration cost [3], a set of original machines, as well as an initial schedule is given. Machines can either be removed, or be added, but not both. The processing time of a job will increase by a constant if it moves. The individual cost of a job is the load of the machine it is scheduled on, and the social cost is the makespan. The properties of NE were established and the PoA for various situations were obtained. For scheduling game with machine activation costs [5,9,14], the number of machines can be increased from 1 to an arbitrary integer, and an activation cost is needed whenever a machine is activated. The individual cost of a job is the sum of the load of the machine it chooses and its shared activated cost. The social cost is the maximum or total cost of all jobs. The different forms of PoA as functions of one of several parameters were obtained.

Another related model is scheduling games on hierarchical machines [6,10, 13]. Each job and each machine is associated with a hierarchy, and a job can only be processed on a machine whose hierarchy is no more than the hierarchy of the job. Typical coordination mechanisms require that for two jobs scheduled on the same machine, the job with larger hierarchy is processed no earlier than the job with smaller hierarchy, or vice versa, which is similar to the priority of an unmoved job to moved jobs on the original machine of the unmoved job. However, the hierarchy of a job is independent with the machine it is scheduled on, while whether a job is unmoved and moved is depended on the machine it is scheduled on before and after its move.

## 2    Model Description and Main Results

Let $\mathcal{J} = \{J_1, J_2, \ldots, J_n\}$ be the set of jobs. The processing time of $J_j$ is $p_j$, $j = 1, 2, \ldots, n$. Let $\mathcal{M}_0 = \{M_1, M_2, \ldots, M_{m_0}\}$ be the set of original machines. During the modification of machines, $m_-$ machines of $\mathcal{M}_0$ are removed and $m_+$ new machines are added. Denote by $\mathcal{M}_-$ and $\mathcal{M}_+$ the sets of removed machines and new machines, respectively. W. l. o. g., we assume $\mathcal{M}_- = \{M_1, M_2, \ldots, M_{m_-}\}$ and $\mathcal{M}_+ = \{M_{m_0+1}, M_{m_0+2}, \ldots, M_{m_0+m_+}\}$. The set of final machines is

$$\mathcal{M} = \mathcal{M}_0 \setminus \mathcal{M}_- \cup \mathcal{M}_+ = \{M_{m_-+1}, M_{m_-+2}, \ldots, M_{m_0+m_+}\}.$$

The number of final machines in $\mathcal{M}$ is $m = m_0 - m_- + m_+$. To make the problem be different from the classical one, we assume $m_0 - m_- \geq 1$ and $m_+ + m_- \geq 1$.

Let $\sigma_0$ be the original schedule before the modification of machines. The machine a job is scheduled on in $\sigma_0$ is called its *original* machine. After the modification of machines and the possible move of jobs, all jobs fall into two categories, those stay on their original machines, which are called *unmoved* jobs, and those move to a machine different from their original machine, which are called *moved* jobs. For convenience of future discussion, we also classify jobs into *prior* and *regular*. A prior job is an unmoved job that is scheduled on a machine which processes at least one moved job. All moved jobs and those unmoved jobs that are scheduled on a machine which processes only unmoved jobs are regular jobs.

For the jobs scheduled on the same machine, we assume that any moved job would not be completed earlier than any unmoved job. In the random setting, two moved (res. unmoved) jobs scheduled on the same machine are processed in a random order. More specifically, suppose that $J_j$ and $J_l$ are two moved (res. unmoved) jobs scheduled on the same machine. Then the two probabilities that $J_j$ is processed before $J_l$ and $J_l$ is processed before $J_j$ are equal. Thus let $a$ and $b$ be the total processing time of unmoved jobs and moved jobs scheduled on a machine. If $J_j$ is an unmoved or moved job scheduled on that machine, its (expected) completion time equals to $\frac{a-p_j}{2} + p_j = \frac{a+p_j}{2}$ and $a + \frac{b-p_j}{2} + p_j = a + \frac{b+p_j}{2}$, respectively [11].

The main result of the paper is the following theorem.

**Theorem 1.** *For the scheduling game with machine modification in the random setting, the PoA is the following, and the bound is tight.*

(i) *If* $\mathcal{M}_- = \emptyset$,

$$PoA = \begin{cases} \frac{4m}{m+m_+ +3}, & m \leq m_+ + 3, \\ \frac{5m+3m_+}{2(m+m_+ +1)}, & m_+ + 4 \leq m \leq m_+ + 2 + \sqrt{4m_+^2 + 10m_+ + 4}, \\ \frac{3m}{m+m_+ +2}, & m \geq m_+ + 2 + \sqrt{4m_+^2 + 10m_+ + 4}. \end{cases}$$

(ii) *If* $\mathcal{M}_- \neq \emptyset$ *and* $\mathcal{M}_+ \neq \emptyset$,

$$PoA = \begin{cases} \frac{4m}{m+m_+ +2}, & m \leq m_+ + 2, \\ \frac{5m+3m_+ -3}{2(m+m_+)}, & m_+ + 3 \leq m \leq m_+ + 1 + \sqrt{4m_+^2 + 2m_+ - 2}, \\ \frac{3m}{m+m_+ +1}, & m \geq m_+ + 1 + \sqrt{4m_+^2 + 2m_+ - 2}. \end{cases}$$

(iii) *If* $\mathcal{M}_+ = \emptyset$,

$$PoA = \begin{cases} \frac{4m}{m+3}, & m \leq 3, \\ \frac{5m-3}{2m}, & 4 \leq m \leq 5, \\ \frac{3m}{m+2}, & m \geq 6. \end{cases}$$

For any subset $\mathcal{J}_0 \subseteq \mathcal{J}$, let $P(\mathcal{J}_0)$ be the total processing time of jobs in $\mathcal{J}_0$. We simplify $P(\mathcal{J})$ to $P$. For any schedule $\sigma^A$, let $\mathcal{J}_i(\sigma^A)$ denote the set of jobs processed on $M_i$ in $\sigma^A$, and $L_i(\sigma^A) = P(\mathcal{J}_i(\sigma^A))$ be the load of $M_i$ in $\sigma^A$, $i = 1, \ldots, m$. Denote by $C_j(\sigma^A)$ the completion time of $J_j$ in $\sigma^A$. Let $C^A = \max_{M_i \in \mathcal{M}} L_i(\sigma^A) = \max_{j=1,\ldots,n} C_j(\sigma^A)$ be the makespan of $\sigma^A$. For any job $J_j$, let us denote $M_{u_j(\sigma)}$ the machine which it is scheduled on in $\sigma$. More specifically, $M_{u_j(\sigma_0)}$ is the original machine of $J_j$.

## 3    Upper Bounds

Let $\sigma^{NE}$ be an arbitrary NE, and $M_{i'}$ be the machine with the maximum load in $\sigma^{NE}$. Denote by $\mathcal{J}_x$ and $\mathcal{J}_y$ the set of prior and regular jobs which are scheduled on $M_{i'}$, respectively. The job which has the smallest processing time among $\mathcal{J}_y$ is denoted by $J_{j'}$. Let $x = P(\mathcal{J}_x)$ and $y = P(\mathcal{J}_y \setminus \{J_{j'}\})$. In the random setting,

$$C_{j'}(\sigma^{NE}) = x + \frac{y}{2} + p_{j'} \tag{1}$$

and

$$C^{NE} = L_{i'}(\sigma^{NE}) = x + y + p_{j'}. \tag{2}$$

We will present some lemmas which reveal properties of a NE in the random setting.

**Lemma 1.** *Suppose that $M_i$, $i \neq i'$ is an arbitrary machine.*

*(i) If $M_i \in \mathcal{M}_+$, then $L_i(\sigma^{NE}) \geq 2x + y$.*
*(ii) If $M_i \in \mathcal{M}_0 \setminus \mathcal{M}_-$ and $i \neq u_{j'}(\sigma_0)$, then $L_i(\sigma^{NE}) \geq x + \frac{y}{2}$.*
*(iii) If $M_i \in \mathcal{M}_0 \setminus \mathcal{M}_-$ and $i = u_{j'}(\sigma_0)$, then $L_i(\sigma^{NE}) \geq 2x + y$.*

*Proof.* For any given $i$, let $a$ and $b$ be the total processing time of unmoved jobs and moved jobs scheduled on $M_i$ in $\sigma^{NE}$, respectively. Then

$$L_i(\sigma^{NE}) = a + b. \tag{3}$$

Consider the schedule $\sigma'$ obtained from $\sigma^{NE}$ by unilaterally moving $J_{j'}$ from $M_{i'}$ to $M_i$. Since $\sigma^{NE}$ is a NE,

$$C_{j'}(\sigma') \geq C_{j'}(\sigma^{NE}). \tag{4}$$

Note that $J_{j'}$ is an unmoved job in $\sigma'$ iff $u_{j'}(\sigma_0) = i$. Then

$$C_{j'}(\sigma') = \begin{cases} \frac{a}{2} + p_{j'}, & u_{j'}(\sigma_0) = i, \\ a + \frac{b}{2} + p_{j'}, & u_{j'}(\sigma_0) \neq i. \end{cases} \tag{5}$$

(i) If $M_i \in \mathcal{M}_+$, then $a = 0$ and $u_{j'}(\sigma_0) \neq i$. By (5), (4) and (1),

$$\frac{b}{2} + p_{j'} = C_{j'}(\sigma') \geq C_{j'}(\sigma^{NE}) = x + \frac{y}{2} + p_{j'}.$$

Hence, $L_i(\sigma^{NE}) = b \geq 2x + y$ by (3).

(ii) If $M_i \in \mathcal{M}_0 \setminus \mathcal{M}_-$ and $i \neq u_{j'}(\sigma_0)$, then by (5), (4) and (1),

$$a + \frac{b}{2} + p_{j'} = C_{j'}(\sigma') \geq C_{j'}(\sigma^{NE}) = x + \frac{y}{2} + p_{j'}.$$

Hence, $L_i(\sigma^{NE}) = a + b \geq a + \frac{b}{2} \geq x + \frac{y}{2}$ by (3).

(iii) If $M_i \in \mathcal{M}_0 \setminus \mathcal{M}_-$ and $i = u_{j'}(\sigma_0)$, then by (5), (4) and (1),

$$\frac{a}{2} + p_{j'} = C_{j'}(\sigma') \geq C_{j'}(\sigma^{NE}) = x + \frac{y}{2} + p_{j'}.$$

Hence, $L_i(\sigma^{NE}) = a + b \geq a \geq 2x + y$ by (3).     □

**Lemma 2.** (i) If $M_{i'} \in \mathcal{M}_+$ and $M_{u_{j'}(\sigma_0)} \in \mathcal{M}_-$, then $P \geq \frac{m+m_+}{2}y + p_{j'}$.
(ii) If $M_{i'} \in \mathcal{M}_+$ and $M_{u_{j'}(\sigma_0)} \notin \mathcal{M}_-$, then $P \geq \frac{m+m_++1}{2}y + p_{j'}$.
(iii) If $M_{i'} \in \mathcal{M}_0 \setminus \mathcal{M}_-$ and $M_{u_{j'}(\sigma_0)} \in \mathcal{M}_-$, then

$$P \geq (m + m_+)x + \frac{m + m_+ + 1}{2}y + p_{j'}.$$

(iv) If $M_{i'}, M_{u_{j'}(\sigma_0)} \in \mathcal{M}_0 \setminus \mathcal{M}_-$ and $i' \neq u_{j'}(\sigma_0)$, then

$$P \geq (m + m_+ + 1)x + \frac{m + m_+ + 2}{2}y + p_{j'}.$$

(v) If $M_{i'} \in \mathcal{M}_0 \setminus \mathcal{M}_-$ and $i' = u_{j'}(\sigma_0)$, then $P \geq \frac{m+m_++1}{2}y + p_{j'}$.

*Proof.* (i) If $M_{i'} \in \mathcal{M}_+$ and $M_{u_{j'}(\sigma_0)} \in \mathcal{M}_-$, then $x = 0$. By (2) and Lemma 1(i)(ii),

$$P = \sum_{M_i \in \mathcal{M}_0 \setminus \mathcal{M}_-} L_i(\sigma^{NE}) + \sum_{M_i \in \mathcal{M}_+ \setminus \{M_{i'}\}} L_i(\sigma^{NE}) + L_{i'}(\sigma^{NE})$$

$$\geq (m_0 - m_-)(x + \frac{y}{2}) + (m_+ - 1)(2x + y) + (x + y + p_{j'}) = \frac{m + m_+}{2}y + p_{j'}.$$

(ii) If $M_{i'} \in \mathcal{M}_+$ and $M_{u_{j'}(\sigma_0)} \notin \mathcal{M}_-$, then $x = 0$. By (2) and Lemma 1(i)(ii)(iii),

$$P = \sum_{M_i \in \mathcal{M}_0 \setminus \mathcal{M}_- \setminus \{M_{u_{j'}(\sigma_0)}\}} L_i(\sigma^{NE}) + L_{u_{j'}(\sigma_0)}(\sigma^{NE})$$

$$+ \sum_{M_i \in \mathcal{M}_+ \setminus \{M_{i'}\}} L_i(\sigma^{NE}) + L_{i'}(\sigma^{NE})$$

$$\geq (m_0 - m_- - 1)\left(x + \frac{y}{2}\right) + (2x + y) + (m_+ - 1)(2x + y) + (x + y + p_{j'})$$

$$= \frac{m + m_+ + 1}{2}y + p_{j'}.$$

(iii) If $M_{i'} \in \mathcal{M}_0 \setminus \mathcal{M}_-$ and $M_{u_{j'}(\sigma_0)} \in \mathcal{M}_-$, by (2) and Lemma 1(i)(ii),

$$P = \sum_{M_i \in \mathcal{M}_0 \setminus \mathcal{M}_- \setminus \{M_{i'}\}} L_i(\sigma^{NE}) + L_{i'}(\sigma^{NE}) + \sum_{M_i \in \mathcal{M}_+} L_i(\sigma^{NE})$$

$$\geq (m_0 - m_- - 1)\left(x + \frac{y}{2}\right) + (x + y + p_{j'}) + m_+(2x + y)$$

$$= (m + m_+)x + \frac{m + m_+ + 1}{2}y + p_{j'}.$$

(iv) If $M_{i'}, M_{u_{j'}(\sigma_0)} \in \mathcal{M}_0 \backslash \mathcal{M}_-$ and $i' \neq u_{j'}(\sigma_0)$, by (2) and Lemma 1(i)(ii)(iii),

$$P = \sum_{M_i \in \mathcal{M}_0 \backslash \mathcal{M}_- \backslash \{M_{i'}, M_{u_{j'}(\sigma_0)}\}} L_i(\sigma^{NE}) + L_{i'}(\sigma^{NE})$$

$$+ L_{u_{j'}(\sigma_0)}(\sigma^{NE}) + \sum_{M_i \in \mathcal{M}_+} L_i(\sigma^{NE})$$

$$\geq (m_0 - m_- - 2)\left(x + \frac{y}{2}\right) + (x + y + p_{j'}) + (2x + y) + m_+(2x + y)$$

$$= (m + m_+ + 1)x + \frac{m + m_+ + 2}{2}y + p_{j'}.$$

(v) If $M_{i'} \in \mathcal{M}_0 \backslash \mathcal{M}_-$ and $i' = u_{j'}(\sigma_0)$, then $x = 0$. By (2) and Lemma 1(i)(ii),

$$P = \sum_{M_i \in \mathcal{M}_0 \backslash \mathcal{M}_- \backslash \{M_{u_{j'}(\sigma_0)}\}} L_i(\sigma^{NE}) + L_{i'}(\sigma^{NE}) + \sum_{M_i \in \mathcal{M}_+} L_i(\sigma^{NE})$$

$$\geq (m_0 - m_- - 1)\left(x + \frac{y}{2}\right) + (x + y + p_{j'}) + m_+(2x + y)$$

$$= \frac{m + m_+ + 1}{2}y + p_{j'}.$$

$\square$

Let $\sigma^*$ be the optimal schedule. The following lemma presents some lower bounds on the optimum, and the relationship between $p_{j'}$ and $y$. The result is folklore and the proof is straightforward.

**Lemma 3.** *(i) If $|\mathcal{J}_y| = 1$, then $y = 0$ and $C^* \geq p_{j'}$.*
*(ii) If $|\mathcal{J}_y| = 2$, then $C^* \geq y \geq p_{j'}$.*
*(iii) If $|\mathcal{J}_y| \geq 3$, then $p_{j'} \leq \frac{y}{2}$.*
*(iv) $C^* \geq \frac{P}{m}$.*

The rest of the section devotes to the proof of the upper bound of the PoA. Due to the lack of space, only the proof of the situation of $\mathcal{M}_- = \emptyset$ will be given below.

We distinguish two cases according to the set that $M_{i'}$ and $M_{u_{j'}(\sigma_0)}$ belong to. First, assume that $M_{i'} \in \mathcal{M}_+$, or $M_{i'} \in \mathcal{M}_0$ and $i' = u_{j'}(\sigma_0)$. Thus $x = 0$ and $C^{NE} = L_{i'}(\sigma^{NE}) = y + p_{j'}$ by (2). If $|\mathcal{J}_y| = 1$, then

$$C^{NE} = p_{j'} \leq C^* \tag{6}$$

by Lemma 3(i). If $|\mathcal{J}_y| = 2$, then $C^{NE} = y + p_{j'} \leq 2y \leq 2C^*$ and

$$C^{NE} = y + p_{j'} \leq \left(2 - \frac{4}{m + m_+ + 3}\right)y + \frac{4}{m + m_+ + 3}p_{j'}$$

$$= \frac{4}{m + m_+ + 3}\left(\frac{m + m_+ + 1}{2}y + p_{j'}\right) \leq \frac{4}{m + m_+ + 3}P \leq \frac{4m}{m + m_+ + 3}C^*$$

by Lemmas 3(ii)(iv) and 2(ii)(v). Hence,

$$\frac{C^{NE}}{C^*} \le \min\left\{2, \frac{4m}{m + m_+ + 3}\right\}. \tag{7}$$

If $|\mathcal{J}_y| \ge 3$, then by Lemmas 2(ii)(v) and 3(iii)(iv),

$$\frac{C^{NE}}{C^*} \le \frac{C^{NE}}{\frac{P}{m}} \le \frac{m(y + p_{j'})}{\frac{m + m_+ + 1}{2}y + p_{j'}} \le \frac{m(y + \frac{y}{2})}{\frac{m + m_+ + 1}{2}y + \frac{y}{2}} = \frac{3m}{m + m_+ + 2}. \tag{8}$$

Next, assume that $M_{i'} \in \mathcal{M}_0$ and $i' \ne u_{j'}(\sigma_0)$. If $|\mathcal{J}_y| = 1$, then by (2) and Lemmas 2(iv), 3(i)(iv),

$$C^{NE} = x + p_{j'} = \frac{1}{m + m_+ + 1}((m + m_+ + 1)x + p_{j'}) + (1 - \frac{1}{m + m_+ + 1})p_{j'}$$

$$\le \frac{1}{m + m_+ + 1}P + \frac{m + m_+}{m + m_+ + 1}p_{j'} \le \frac{m}{m + m_+ + 1}C^* + \frac{m + m_+}{m + m_+ + 1}C^*$$

$$= \frac{2m + m_+}{m + m_+ + 1}C^*. \tag{9}$$

If $|\mathcal{J}_y| = 2$, then by (2) and Lemmas 2(iv), 3(ii)(iv),

$$C^{NE} = x + y + p_{j'} = \frac{1}{m + m_+ + 1}\left((m + m_+ + 1)x + \frac{m + m_+ + 2}{2}y + p_{j'}\right)$$

$$+ \left(1 - \frac{m + m_+ + 2}{2(m + m_+ + 1)}\right)y + \left(1 - \frac{1}{m + m_+ + 1}\right)p_{j'}$$

$$\le \frac{1}{m + m_+ + 1}P + \frac{m + m_+}{2(m + m_+ + 1)}y + \frac{m + m_+}{m + m_+ + 1}p_{j'}$$

$$\le \frac{m}{m + m_+ + 1}C^* + \frac{m + m_+}{2(m + m_+ + 1)}C^* + \frac{m + m_+}{m + m_+ + 1}C^*$$

$$= \frac{5m + 3m_+}{2(m + m_+ + 1)}C^*.$$

and

$$\frac{C^{NE}}{C^*} \le \frac{C^{NE}}{\frac{P}{m}} \le \frac{m(x + y + p_{j'})}{(m + m_+ + 1)x + \frac{m + m_+ + 2}{2}y + p_{j'}}$$

$$\le \frac{m(x + y + y)}{(m + m_+ + 1)x + \frac{m + m_+ + 2}{2}y + y} = \frac{m(x + 2y)}{(m + m_+ + 1)x + \frac{m + m_+ + 4}{2}y}$$

$$\le \frac{2my}{\frac{m + m_+ + 4}{2}y} = \frac{4m}{m + m_+ + 4}.$$

Hence,

$$\frac{C^{NE}}{C^*} \le \min\left\{\frac{5m + 3m_+}{2(m + m_+ + 1)}, \frac{4m}{m + m_+ + 4}\right\}. \tag{10}$$

If $|\mathcal{J}_y| \geq 3$, then by (2) and Lemmas 2(iv), 3(iii)(iv),

$$
\frac{C^{NE}}{C^*} \leq \frac{C^{NE}}{\frac{P}{m}} \leq \frac{m(x + y + p_{j'})}{(m + m_+ + 1)x + \frac{m + m_+ + 2}{2}y + p_{j'}}
$$

$$
\leq \frac{m\left(x + y + \frac{y}{2}\right)}{(m + m_+ + 1)x + \frac{m + m_+ + 2}{2}y + \frac{y}{2}} = \frac{m\left(x + \frac{3y}{2}\right)}{(m + m_+ + 1)x + \frac{m + m_+ + 3}{2}y}
$$

$$
\leq \frac{\frac{3m}{2}y}{\frac{m + m_+ + 3}{2}y} = \frac{3m}{m + m_+ + 3}. \tag{11}
$$

Note that $\frac{3m}{m + m_+ + 3} \leq \frac{3m}{m + m_+ + 2} \leq \frac{4m}{m + m_+ + 3}$. When $m_+ + 1 \leq m \leq m_+ + 3$, $\frac{2m + m_+}{m + m_+ + 1} \leq \frac{4m}{m + m_+ + 3}$, $\frac{4m}{m + m_+ + 4} \leq \frac{5m + 3m_+}{2(m + m_+ + 1)}$ and $\frac{4m}{m + m_+ + 4} \leq \frac{4m}{m + m_+ + 3} \leq 2$. By (6)–(11), $\frac{C^{NE}}{C^*} \leq \frac{4m}{m + m_+ + 3}$. When $m \geq m_+ + 4$, $\frac{4m}{m + m_+ + 3} \geq 2$ and $\frac{4m}{m + m_+ + 4} \geq \frac{5m + 3m_+}{2(m + m_+ + 1)} \geq 2 \geq \frac{2m + m_+}{m + m_+ + 1}$. By (6)–(11), $\frac{C^{NE}}{C^*} \leq \min\left\{\frac{5m + 3m_+}{2(m + m_+ + 1)}, \frac{3m}{m + m_+ + 2}\right\}$. This completes the proof of the first situation.

## 4    Lower Bounds

In this section, we will present instances to show that the bounds given in Theorem 1 are tight for each combinations of $m_+$, $m_-$ and $m$. Remind that the completion times of jobs in the random setting are less intuitive than in the deterministic setting. We introduce some lemmas which are helpful in verifying a certain schedule is a NE.

**Lemma 4.** *In* $\sigma^A$, $n_1$ *regular jobs with processing time* $p$ *are scheduled on* $M_i$. *The schedule* $\sigma^B$ *is obtained from* $\sigma^A$ *by unilaterally moving a job* $J_j$ *from* $M_i$ *to* $M_k$, $k \neq i$.

  (i) *If* $u_j(\sigma_0) \neq k$ *and* $L_k(\sigma^A) \geq (n_1 - 1)p$, *then* $C_j(\sigma^A) \leq C_j(\sigma^B)$.
  (ii) *If* $u_j(\sigma_0) = k$, $u_l(\sigma_0) = k$ *for any* $J_l \in \mathcal{J}_k(\sigma^A)$ *and* $L_k(\sigma^A) \geq (n_1 - 1)p$, *then* $C_j(\sigma^A) \leq C_j(\sigma^B)$.
  (iii) *If* $u_j(\sigma_0) \neq k$, $u_l(\sigma_0) = k$ *for any* $J_l \in \mathcal{J}_k(\sigma^A)$ *and* $L_k(\sigma^A) \geq \frac{n_1 - 1}{2}p$, *then* $C_j(\sigma^A) \leq C_j(\sigma^B)$.

*Proof.* Clearly, $C_j(\sigma^A) = \frac{n_1 + 1}{2}p$. If $u_j(\sigma_0) \neq k$, then $C_j(\sigma^B) \geq \frac{L_k(\sigma^A)}{2} + p$. If $u_j(\sigma_0) = k$ *and* $u_l(\sigma_0) = k$ *for any* $J_l \in \mathcal{J}_k(\sigma^A)$, then $C_j(\sigma^B) = \frac{L_k(\sigma^A)}{2} + p$. If $u_j(\sigma_0) \neq k$ *and* $u_l(\sigma_0) = k$ *for any* $J_l \in \mathcal{J}_k(\sigma^A)$, then $C_j(\sigma^B) = L_k(\sigma^A) + p$. Thus (i), (ii), (iii) can be obtained by algebraic calculation.    □

**Lemma 5.** *In* $\sigma^A$, $n_1$ *prior jobs with processing time* 1 *and* $n_2$ *regular jobs with processing time* $p$ *are scheduled on* $M_i$. *The schedule* $\sigma^B$ *is obtained from* $\sigma^A$ *by unilaterally moving a job* $J_j$ *from* $M_i$ *to* $M_k$, $k \neq i$.

  (i) *If* $p_j = 1$ *and* $L_k(\sigma^A) \geq n_1 - 1$, *then* $C_j(\sigma^A) \leq C_j(\sigma^B)$.

*(ii)* If $p_j = p$, $L_k(\sigma^A) \geq 2n_1 + (n_2 - 1)p$ and $u_j(\sigma_0) \neq k$, then $C_j(\sigma^A) \leq C_j(\sigma^B)$.

*(iii)* If $p_j = p$, $L_k(\sigma^A) \geq 2n_1 + (n_2 - 1)p$ and $u_j(\sigma_0) = u_l(\sigma_0) = k$ for any $J_l \in \mathcal{J}_k(\sigma^A)$, then $C_j(\sigma^A) \leq C_j(\sigma^B)$.

*(iv)* If $p_j = p$, $L_k(\sigma^A) \geq n_1 + \frac{(n_2-1)p}{2}$, $u_j(\sigma_0) \neq k$ and $u_l(\sigma_0) = k$ for any $J_l \in \mathcal{J}_k(\sigma^A)$, then $C_j(\sigma^A) \leq C_j(\sigma^B)$.

*Proof.* Clearly, if $p_j = 1$, then $C_j(\sigma^A) = \frac{n_1+1}{2}$ and $C_j(\sigma^B) \geq \frac{L_k(\sigma^A)}{2} + 1$. Thus (i) can be obtained by algebraic calculation. If $p_j = p$, then $C_j(\sigma^A) = n_1 + \frac{n_2+1}{2}p$. If $u_j(\sigma_0) \neq k$, then $C_j(\sigma^B) \geq \frac{L_k(\sigma^A)}{2} + p$. If $u_j(\sigma_0) = u_l(\sigma_0) = k$ for any $J_l \in \mathcal{J}_k(\sigma^A)$, then $C_j(\sigma^B) = \frac{L_k(\sigma^A)}{2} + p$. If $u_j(\sigma_0) \neq k$ and $u_l(\sigma_0) = k$ for any $J_l \in \mathcal{J}_k(\sigma^A)$, then $C_j(\sigma^B) \geq L_k(\sigma^A) + p$. Thus (ii), (iii) and (iv) can be obtained by algebraic calculation. □

Based on the discussion of last section, to show the bound is tight, it is sufficient to find an instance, as well as a NE of the instance, such that the ratio between the makespan of the NE and the optimal makespan matches the bound. However, only one of the six instances are presented here due to page limit.

**Instance $I_1$ for $m_- = 0$ and $m_+ + 4 \leq m \leq m_+ + 2 + \sqrt{4m_+^2 + 10m_+ + 4}$**

The job set of $I_1$ consists of $2(m + m_+ + 1)(m - 2)$ *small* jobs $J_1, J_2, \ldots,$ $J_{2(m+m_++1)(m-2)}$ with processing time 1, and 2 *big* jobs $J_{2(m+m_++1)(m-2)+1}$, $J_{2(m+m_++1)(m-2)+2}$ with processing time $2(m + m_+ + 1)$. The original schedule $\sigma_0$ is given by

$$u_j(\sigma_0) = \begin{cases} 1, & j = (m + m_+)(2m - 3) + 1, \ldots, 2(m + m_+ + 1)(m - 2), \\ k, & j = (k - 2)(2m - 3) + 1, \ldots, (k - 1)(2m - 3), \\ & \quad k = 2, \ldots, m - m_+ - 1, \\ m_0, & j = (m - m_+ - 2)(2m - 3) + 1, \ldots, (m + m_+)(2m - 3), \\ & \quad 2(m + m_+ + 1)(m - 2) + 1, 2(m + m_+ + 1)(m - 2) + 2. \end{cases}$$

In $\sigma_0$, $M_1$ processes $2(m + m_+ + 1)(m - 2) - (m - m_+ - 2)(2m - 3) - (m_+ + 1)(4m - 6) = m - m_+ - 4$ small jobs. These jobs are called *Type A*. Each of the $m - m_+ - 2$ machines in $\mathcal{M}_0 \setminus \{M_1, M_{m_0}\}$ processes $2m - 3$ small jobs. These jobs are called *Type B*. The remaining machine $M_{m_0}$ processes two big jobs and $(4m - 6)(m_+ + 1)$ small jobs. Among these small jobs, the first $4m - 6$ jobs are called *Type C*, and the remaining $(4m - 6)m_+$ jobs are called *Type D*.

Consider a new schedule $\sigma_1$ with

$$u_j(\sigma_1) = \begin{cases} 1, & j = (m + m_+)(2m - 3) + 1, \ldots, \\ & \quad 2(m + m_+ + 1)(m - 2), 2(m + m_+ + 1)(m - 2) + 1, \\ & \quad 2(m + m_+ + 1)(m - 2) + 2 \\ k, & j = (k - 2)(2m - 3) + 1, \ldots, (k - 1)(2m - 3), \\ & \quad k = 2, \ldots, m - m_+ - 1, \\ m_0 - 1 + k, & j = (2m - 3)(2k + m - m_+ - 4) + 1, \ldots, \\ & \quad (2m - 3)(2k + m - m_+ - 2), k = 1, 2, \ldots, m_+ + 1. \end{cases}$$

In $\sigma_1$, $M_1$ processes two big jobs and $m - m_+ - 4$ Type A small jobs. Each of the $m - m_+ - 2$ machines in $\mathcal{M}_0 \setminus \{M_1, M_{m_0}\}$ processes $2m - 3$ Type B small jobs.

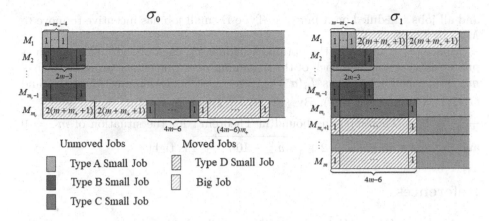

**Fig. 1.** Instance 1

Machine $M_{m_0}$ processes $4m - 6$ Type C small jobs. Each of the $m_+$ machines in $\mathcal{M}_+$ processes $4m - 6$ Type D small jobs. Hence,

$$L_i(\sigma_1) = \begin{cases} 5m + 3m_+, & i = 1, \\ 2m - 3, & i = 2, 3, \ldots, m_0 - 1, \\ 4m - 6, & i = m_0, m_0 + 1, \ldots, m_0 + m_+. \end{cases} \tag{12}$$

Note that all small jobs of Type A, B, C are scheduled on their original machines (Fig. 1). We will prove in the following that $\sigma_1$ is a NE.

Firstly, for any $M_i \in \mathcal{M}_0 \setminus \{M_1, M_{m_0}\}$, $L_i(\sigma_1) = 2m - 3 = (m - m_+ - 4) + \frac{1}{2}2(m + m_+ + 1)$ and $M_i$ is the original machine of all jobs scheduled on it in $\sigma_1$ but is not the original machine of any big job. Thus no big job has incentive to move from $M_1$ to $M_i$ by Lemma 5(iv). Since $L_{m_0}(\sigma_1) = 4m - 6 = 2(m - m_+ - 4) + 2(m + m_+ + 1)$ and $M_{m_0}$ is the original machine of both big jobs and all jobs scheduled on it in $\sigma_1$, no big job has incentive to move from $M_1$ to $M_{m_0}$ by Lemma 5(iii). For any $M_i \in \mathcal{M}_+$, $L_i(\sigma_1) = 4m - 6 = 2(m - m_+ - 4) + 2(m + m_+ + 1)$ and $M_i$ is not the original machine of any big job. Thus no big job has incentive to move to $M_i$ by Lemma 5(ii). Secondly, since $L_i(\sigma_1) \geq m - m_+ - 5$ for any $M_i \in \mathcal{M}$, no Type A small job has incentive to move from its original machine $M_1$ to any other machine by Lemma 5(i). Similarly, since $L_i(\sigma_1) \geq 2m - 4$ for any $M_i \in \mathcal{M}$, no Type B small job has incentive to move from its original machine to any other machine by Lemma 4(i). Thirdly, since $L_i(\sigma_1) \geq 4m - 7$ for any $M_i \in \mathcal{M}_+ \cup \{M_1\}$, no Type C small job or Type D small job whose original machine is not $M_i$ has incentive to move to $M_i$ by Lemma 4(i). For any $M_i \in \mathcal{M}_0 \setminus \{M_1, M_{m_0}\}$, $L_i(\sigma_1) = 2m - 3 \geq \frac{4m - 7}{2}$ and $M_i$ is not the original machine of any Type C or Type D small job but is the original machine of all jobs scheduled on it in $\sigma_1$. Thus no Type C or Type D small job has incentive to move to $M_i$ by Lemma 4(iii). Finally, since $L_{m_0}(\sigma_1) \geq 4m - 7$ and $M_{m_0}$ is the original machine of all Type C small jobs and all jobs scheduled on it in $\sigma_1$, no Type D small job has incentive to move to $M_{m_0}$ by Lemma 4(ii).

Hence, $\sigma_1$ is a NE. In the optimal schedule $\sigma^*$, each of the first 2 machines processes a big job, and each of the remaining $m - 2$ machines processes $2(m + m_+ + 1)$ small jobs. Thus $L_i(\sigma^*) = 2(m + m_+ + 1)$ for any $M_i \in \mathcal{M}$ and $C^*(I_1) = 2(m + m_+ + 1)$. By (12), $C^{NE}(I_1) = 5m + 3m_+$ and $\frac{C^{NE}(I_1)}{C^*(I_1)} = \frac{5m + 3m_+}{2(m + m_+ + 1)}$. Therefore, the bound in Theorem 1 for the situation of $m_- = 0$ and $m_+ + 4 \leq m \leq m_+ + 2 + \sqrt{4m_+^2 + 10m_+ + 4}$ is tight.

# References

1. Abed, F., Huang, C.-C.: Preemptive coordination mechanisms for unrelated machines. In: Epstein, L., Ferragina, P. (eds.) ESA 2012. LNCS, vol. 7501, pp. 12–23. Springer, Heidelberg (2012). https://doi.org/10.1007/978-3-642-33090-2_3
2. Azar, Y., Fleischer, L., Jain, K., Mirrokni, V., Svitkina, Z.: Optimal coordination mechanisms for unrelated machine scheduling. Oper. Res. **63**, 489–500 (2015)
3. Belikovetsky, S., Tamir, T.: Load rebalancing games in dynamic systems with migration costs. Theoret. Comput. Sci. **622**, 16–33 (2016)
4. Caragiannis, I.: Efficient coordination mechanisms for unrelated machine scheduling. Algorithmica **66**(3), 512–540 (2013)
5. Chen, B., Gurel, S.: Efficiency analysis of load balancing games with and without activation costs. J. Sched. **15**(2), 157–164 (2012)
6. Chen, Q., Tan, Z.: Mixed coordination mechanisms for scheduling games on hierarchical machines. Int. Trans. Oper. Res. (2018). https://doi.org/10.1111/itor.12558
7. Christodoulou, G., Koutsoupias, E., Nanavati, A.: Coordination mechanisms. Theoret. Comput. Sci. **410**, 3327–3336 (2009)
8. Cohen, J., Durr, C., Kim, T.N.: Non-clairvoyant scheduling games. Theory Comput. Syst. **49**, 3–23 (2011)
9. Feldman, M., Tamir, T.: Conflicting congestion effects in resource allocation games. Oper. Res. **60**, 529–540 (2012)
10. Guan, L., Li, J.: Coordination mechanism for selfish scheduling under a grade of service provision. Inf. Process. Lett. **113**, 251–254 (2013)
11. Immorlica, N., Li, L., Mirrokni, V.S., Schulz, A.: Coordination mechanisms for selfish scheduling. Theoret. Comput. Sci. **410**, 1589–1598 (2009)
12. Koutsoupias, E., Papadimitriou, C.H.: Worst-case equilibria. Comput. Sci. Rev. **3**, 65–69 (2009)
13. Lee, K., Leung, J.Y.T., Pinedo, M.L.: Coordination mechanisms with hybrid local policies. Discrete Optim. **8**, 513–524 (2011)
14. Lin, L., Xian, X., Yan, Y., He, X., Tan, Z.: Inefficiency of equilibria for scheduling game with machine activation costs. Theoret. Comput. Sci. **607**, 193–207 (2015)

# Tracking Histogram of Attributes over Private Social Data in Data Markets

Zaobo He[1](✉) and Yan Huang[2]

[1] Miami University, Oxford, OH 45056, USA
hez26@miamioh.edu
[2] Kennesaw State University, Kennesaw, GA 30144, USA
yhuang24@kennesaw.edu

**Abstract.** We propose new inference attacks to predict sensitive attributes of individuals in social attribute networks. Our attacks assume powerful adversaries with wide background knowledge leverage innocent individual information which is publicly available to predict unobserved attributes of targeted users. Given the growing availability of individual information online, our results have significant implications for data privacy preservation – the latent correlation between innocent data and sensitive data might be learned unless we sanitize data prior to publishing to protect against inference attacks. Existing inference attacks has high computational complexity by leveraging either the publicly available social friends or individuals' observed social attributes, but not both. To mitigate the gaps, we develop an approach to integrate social structures and attributes into a probabilistic graphical model and predict unobserved attributes with linear complexity. Furthermore, we propose a differentially private approach to track histogram of attributes from social-attribute networks.

**Keywords:** Inference attack · Differential privacy · Probabilistic graphical model · Belief propagation · Histogram

## 1 Introduction

Many kinds of graph data exist, e.g., social networks, wireless sensor networks. Especially, social-attribute networks attach a set of social attributes to each node, such as gender, education, political view, etc. With the help of social network analysis, interesting statistical aggregations can be learned with the increasing availability of social-attribute networks online, which act as critical components for diverse applications. As a fundamental aggregate, histogram of social attributes reveals the distribution information of targeted attributes in networks, which is helpful in diversified applications, such as marketing or disease outbreak prediction. However, directly publishing the attribute histogram may reveal sensitive information of individuals if the network is anonymized [1,2]. With the advancement of machine learning technologies, adversaries can

© Springer Nature Switzerland AG 2019
Y. Li et al. (Eds.): COCOA 2019, LNCS 11949, pp. 277–288, 2019.
https://doi.org/10.1007/978-3-030-36412-0_22

successfully predict sensitive information by feeding available information into advanced algorithms [3]. Researchers have been studying the problem of tracking the attribute histogram from social-attribute networks under Differential Privacy [4,5] which is adopted as "gold standard", allowing unlimited reasoning power and background knowledge of adversaries. An algorithm that satisfies differential privacy guarantees that the output distribution does not change significantly from one dataset to a neighboring dataset. Therefore, this paper first proposes a novel inference attack model for predicting sensitive attributes from an input social attribute network, then this paper proposes a differentially private algorithm for tracking attribute histogram.

Putting differential privacy guarantee into histogram publishing remains a challenging problem. From its proposal, many efforts have been devoted to develop mechanisms for publishing statistical aggregates or sanitized social-attribute graphs. However, it seems to encounter difficulty in practice when trying to apply differential privacy into tracking histogram from high-dimensional social data. As indicated in [6], there are two reasons for the above difficulty: *output scalability* and *signal-to-noise ratio*. Output scalability refers to the dilemma when compared to the size of input dataset, the output is very unwieldy and slow to use. Signal-to-noise ratio refers to the dilemma when noise dominates the original data signal, making the sanitized data next to useless. Therefore, for a social attribute dataset with high-dimensional attributes, how to design a differentially private histogram publishing approach remains a serious challenge.

To establish an inference model for predicting unobserved attributes based on innocent information which is either publicly available in networks, or latent information that might be learned from available information, a model is required to incorporate observed and unobserved attributes, social connections and latent probability dependency among attributes. This problem is challenging since learning the probability dependency among attributes is hard, given high-dimensional attributes. Furthermore, most of existing approaches predict unknown variables by computing marginal probability distributions; however, computation of marginal probability distributions has exponential complexity as it might request to add items with exponential scale together. Therefore, with high-dimensional attributes, how to design an efficient inference attack model also remains a serious challenge.

In this paper, we formulate an inference attack model for predicting unobserved attributes, launched by an adversary with wide background knowledge. The proposed model presents a probabilistic graphical model for inference attacks with improved performance in computational complexity. Specifically, belief propagation over factor graph [7] is adopted, which allows the computation of marginal probability distributions of unknown attributes has linear complexity, instead of exponential complexity through adding items with exponential scale together. To develop the factor graph, we learn the probability dependency among attributes through constructing a *Bayesian network* [8] that provides a model of the probability dependency among the attributes in a network.

We study the problem of releasing attribute histogram under differential privacy guarantee. Given an input social attribute dataset $\mathcal{X}$ for all individuals, the goal is to publish a sanitized histogram of $\mathcal{X}$, $\mathbf{his}^*(\mathcal{X})$, that approximates original histogram $\mathbf{his}(\mathcal{X})$ as accurate as possible while satisfying differential privacy guarantee. In this paper, we propose to first find a synthetic $\mathcal{X}'$ by sampling $\mathcal{X}'$ from the full dimensional distribution of $\mathcal{X}$ leveraging a $(\varepsilon/2)$-differentially private approach. Then the histogram of attributes is returned leveraging a $(\varepsilon/2)$-differentially private approach. As stated previously, however, injecting noise into high-dimensional data might arise two problems, $i.e.$, the output scalability and signal-to-noise ratio. Therefore, we present an approach to seek a group of local functions to approximate the full dimensional distribution of $\mathcal{X}$. Then, after injecting noise into these local functions, we can get the approximate distribution for all attributes. Finally, synthetic attributes $\mathcal{X}$ with differential privacy guarantee can be sampled from the combination of noisy functions.

## 2    Problem Formulation

This section first introduces necessary preliminary knowledge on differential privacy, and belief propagation on factor graph. Then we present the problem definition of inference attack and differentially private histogram publishing.

### 2.1    Differential Privacy

The formal definition of differential privacy is introduced as follows:

**Definition 1. $\varepsilon$-Differential Privacy.** *A randomized algorithm $\mathcal{A}$ satisfies $\varepsilon$-Differential Privacy ($\varepsilon$-DP), if for any two neighboring datasets, $D$ and $D'$ that differ in only one entry and for any possible output $O$ of $\mathcal{A}$, we have:*

$$\Pr[\mathcal{A}(D) = O] \leq e^{\varepsilon} \cdot \Pr[\mathcal{A}(D') = O].$$

There are two widely used approaches to achieve differential privacy, namely, the *Laplace mechanism* [5] and the *exponential mechanism* [9].

### 2.2    Problem Formulation

In this work, we consider undirected social attribute network $\mathcal{G}(\mathcal{V}, \mathcal{E}, \mathcal{X})$. The adversary aims to infer some target attributes of target individuals in the social network. We define $\mathbf{G}$ to be the social graph after removing attributes associated with each node. Note that the attribute IDs are same for all nodes in the network. We use $x_j^i$ to denote the value of attribute $j$ ($j \in \mathcal{X}$) for node $i$ ($i \in \mathcal{V}$). Some $x_j^i$ might be known by the adversity (collected from privacy-unconcerned individuals or learned through inference attacks) but others might be unknown. We denote the set of unknown attributes in $\mathcal{X}$ as $\mathcal{X}_U$ and known attribute as $\mathcal{X}_K$.

A strong adversary is considered in this paper with wide background knowledge. In addition to publicly known attributes and social connections (namely, the social graph), the adversary might learn attribute correlations. For example, $\mathcal{F}(x^i_{j_1}, x^i_{j_2}, x^i_{j_3})$ is the function representing the attribute correlation among attributes $j_1$, $j_2$ and $j_3$, which means the probability distribution of $x^i_{j_3}$ is highly determined in terms of the value of $j_1$ and $j_2$, for any individual $i$. For example, attribute *political view* has close correlation with *gender* and *job position*. $\mathcal{L}(x^{i_1}_j, x^{i_2}_j, x^{i_3}_j)$ is the function representing the social connections among $i_1$, $i_2$ and $i_3$, which means the probability distribution of $x^{i_3}_j$ is highly determined in terms of the value of $j$ of $i_1$ and $i_2$. The adversary carries out an inference attack to predict $x^i_j \in \mathcal{X}_U$, by relying on his background knowledge $\mathcal{F}(x^i_{j_1}, x^i_{j_2}, x^i_{j_3})$, $\mathcal{L}(x^{i_1}_j, x^{i_2}_j, x^{i_3}_j)$ and collected knowledge $\mathcal{X}_K$. Finally, we propose a data sanitization approach based on differential privacy to preserve the individual's privacy in the process of releasing histogram of attributes for the input social attribute network.

## 3    Inference Attacks on Unknown Attributes

We depict the inference attack as computing the marginal probability distribution of unknown attributes $\mathcal{X}_U$, given the known attributes $\mathcal{X}_K$, attribute correlations $\mathcal{F}(x^i_{j_1}, x^i_{j_2}, x^i_{j_3})$ and social connections $\mathcal{L}(x^{i_1}_j, x^{i_2}_j, x^{i_3}_j)$. The marginal probability distribution of an unknown variable $x_i \in \mathcal{X}_U$ is computed as

$$\mathbf{Pr}(x_i | \mathcal{X}_K, \mathcal{X}_U, \mathcal{F}(.), \mathcal{L}(.)) = \sum_{\mathcal{X}_U \setminus x_i} \mathbf{Pr}(\mathcal{X}_U | \mathcal{X}_K, \mathcal{F}(x^i_{j_1}, x^i_{j_2}, x^i_{j_3}), \mathcal{L}(x^{i_1}_j, x^{i_2}_j, x^{i_3}_j))$$

(1)

where $\mathbf{Pr}(\mathcal{X}_U | \mathcal{X}_K, \mathcal{F}(x^i_{j_1}, x^i_{j_2}, x^i_{j_3}), \mathcal{L}(x^{i_1}_j, x^{i_2}_j, x^{i_3}_j))$ is the joint probability distribution of all unknown attributes, conditioned on the accessible information by the adversary.

However, the computational complexity of the above formulation is too high to conduct prediction on big social attribute network with high-dimensional attributes and tens of millions of nodes and connections. Specifically, the summing terms increase with exponential speed with the growth of the number of attributes. Therefore, it is infeasible to predict unknown attributes by naively feeding information into the above formula. We propose to construct a factor graph then execute belief propagation on it, so that the joint probability distribution can be factorized into the product of a group of local functions. Belief propagation [7] is a message-passing algorithm for conducting prediction on unknown variables on probabilistic graphical model (factor graph, Markov random fields, Bayesian networks). Each local function depicts the probability dependency among attributes, by taking a subset of attributes as arguments.

After the message-passing process based on belief propagation over factor graph, the marginal probability distribution can be calculated with linear computational complexity. We construct the factor graph with 3 types of nodes: *attribute variable node*: a known or unknown attribute; *attribute factor node*: the probability dependency among attributes, and *connection factor node*: the probability dependency among attributes derived from social connections. Then, variable and factor nodes are connected in the following way:

- Attribute variable node $x^i_j$ and $x^i_k$ connect to attribute factor node $f^i_l$ if the probability distribution of $x_l$ highly depends on the value of $x_i$ and $x_k$, for any individual $i$.
- Attribute variable node $x^i_j$ and $x^k_j$ connect to connection factor node $g^{j,k}_i$ if $i$ and $k$ are connected by an edge in social graph **G**.

For example, a factor graph consists of three individuals in Fig. 1, and each individual has 3 attributes.

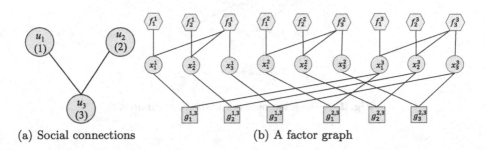

(a) Social connections                          (b) A factor graph

**Fig. 1.** The factor graph representation of a social-attribute network with three nodes $(u_1, u_2, u_3)$ with 3 attributes each node.

After the message-passing process based on belief propagation over factor graph, the joint probability distribution in Eq. 1 can be factorized into the product of a set of local function that take a subset of attributes as arguments:

$$\mathbf{Pr}(\mathcal{X}_U | \mathcal{X}_K, \mathcal{F}(x^i_{j_1}, x^i_{j_2}, x^i_{j_3}), \mathcal{L}(x^{i_1}_j, x^{i_2}_j, x^{i_3}_j)) =$$

$$\frac{1}{Z} \left[ \prod_{i \in V} \prod_{j \in \mathcal{X}} f^i_j(x^i_j, \Theta(x^i_j), \mathcal{F}) \right] \times \left[ \prod_{i \in V} \prod_{j \in \mathcal{X}} g^i_{j,m}(x^i_j, x^i_m, \mathcal{L}) \right], \quad (2)$$

where $Z$ denotes the normalization constant, and $\Theta(x^i_j)$ is the set of attributes that correlate with $j$.

Equation (2) shows that the computation of $f^i_j(x^i_j, \Theta(x^i_j), \mathcal{F}(x^i_{j_1}, x^i_{j_2}, x^i_{j_3}))$ and $g^i_{j,m}(x^i_j, x^i_m, \mathcal{L}(x^{i_1}_j, x^{i_2}_j, x^{i_3}_j))$ are crucial to carry out belief propagation on factor graph through passing messages between variable nodes and factor nodes.

### 3.1 Attribute Correlation $\mathcal{F}$

As shown in Eq. 2, part of local functions describe the attribute correlation, *i.e.*, the distribution of set of attributes conditioned on other attributes. Therefore, the first challenge is to learn attribute correlation $\mathcal{F}(x^i_{j_1}, x^i_{j_2}, \ldots, x^i_{j_k})$, given a group of attributes $j_1$ to $j_k$. We propose to leverage a *Bayesian network* [8] $\mathcal{N}$ which offers a probability model to depict the conditional dependency among the attributes in a network. Bayesian network is widely used in the machine learning community. $\mathcal{N}$ is a directed acyclic graph that represents each variable as a network node and depicts conditional independence between attributes employing directed edges. For example, Fig. 2 shows a Bayesian network over six attributes, namely, *difficulty, intelligence, grade, workless, education* and *income*. Given any two attributes $j, k \in X$, there are three types of possible probability dependency between $j$ and $k$.

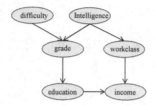

**Fig. 2.** A Bayesian network with six attributes.

- **Direct probability dependence.** There is an edge between $j$ and $k$ (say, from $k$ to $j$).
- **Weak probability dependence.** $j$ and $k$ are connected directly (say, from $k$ to $j$). This means that, given education and workclass of an individual, the relationship between income and difficulty is conditional independence.
- **Strong probability dependence.** $j$ and $k$ are neither connected directly nor indirectly. Then, the relationship of $j$ and $k$ is conditional independence given $j$'s and $k$'s parent sets.

We introduce the notion of *degree* of $\mathcal{N}$ as the maximum number of attributes of an arbitrary parent $\pi$. For instance, the Bayesian network shown in Fig. 2 has degree two since the size of parent set of any node don't exceed two.

We propose to construct a Bayesian network $\mathcal{N}$ with degree $k$ to depict the probability dependence among the attributes $\mathcal{X}$. Ideally, $\mathcal{N}$ can approximate the distribution of $\mathcal{X}$ with high accuracy. Then, a natural question is how to measure the quality of a $\mathcal{N}$. We use the notion of *attribute dependence* from rough set theory [10] to measure the quality of $\mathcal{N}$. Given two variables $x, \pi, x \in \mathcal{X}, \pi \subset \mathcal{X}$, we say that $x$ depends on $\pi$ with degree $k$ ($1 \leq k \leq 1$), denoted by $\pi \rightarrow^k x$, if

$$k = \frac{\left| \bigcup_{Y \in [u]_x} \{u | [u]_\pi \subseteq Y\} \right|}{|\mathcal{V}|}, \tag{3}$$

where $[u]_x$ denotes the nodes whose attributes have the same values in terms of attributes $x$. When $k = 1$, we say that the distribution of $x$ is partially determined by the attribute value of $\pi$, namely, the relationship between $x$ and $\pi$ is direct probability dependence.

We try to find a Bayesian network so that the dependence parameter $k$ is maximized (closed to 1), given any pairs of attribute $j$ and attribute set $\pi$. Therefore, the construction of $\mathcal{N}$ can be depicted as an optimization problem, where the optimization goal is to find a parent set $\pi(j)$ for each attribute $j$ in $\mathcal{X}$ to maximize $\frac{|\bigcup_{Y \in [u]_x} \{u | [u]_\pi \subseteq Y\}|}{|V|}$.

The state of the arts to construct a Bayesian network to approximate attribute distribution accurately are based on the notion of *entropy* from information theory [6,11]. When $k = 1$, [11] shows that linking a next pair of $x$ and $\pi(x)$ greedily based on the maximized mutual information is optimal. However, [12] further proves that the constructing a Bayesian network in [11] is a NP-hard problem when $k > 1$. Therefore, heuristic algorithms are frequently used to solve this problem [13]. By extending the entropy-based approaches proposed in [6], we propose Algorithm 1 to construct a Bayesian network $\mathcal{N}$ with degree $k$.

---

**Algorithm 1:** Greedily construct $\mathcal{N}$

---

**Input:** $\mathcal{X}$, $k$
**Output:** $\mathcal{N}$
1: Initialize $\mathcal{N} = \emptyset$ and $\Theta = \emptyset$
2: Randomly select an attribute $x_1$ from $\mathcal{X}$; add $(x_1, \emptyset)$ to $\mathcal{N}$; add $x_1$ to $\Theta$
3: **for** $j = 2$ to $|\mathcal{X}|$ **do**
4:     Initialize $\Psi = \emptyset$
5:     For each $x \in X \backslash \Theta$, randomly select $k$ attributes (*i.e.*, a $\pi$) from $\Theta$
6:     Add $(x, \pi)$ to $\Psi$
7:     Select a random pair $(x_j, \pi_j)$ from $\Psi$ with a maximum dependence parameter $k$ with $\pi_j \rightarrow^k x_j$
8:     Add $(x_j, \pi_j)$ to $\mathcal{N}$; add $x_j$ to $\Theta$
9: **end for**
10: **return** $\mathcal{N}$

---

To guarantee $\mathcal{N}$ is a directed acyclic graph, Algorithm 1 randomly seek each pair of $(x_j, \pi_j)$ such that $\mathcal{N}$ doesn't contain edge from $x_i$ to $x_j$ for any $j < i$. Finally, each pair of $(x_j, \pi_j)$ indicates the probability distribution of $x_j$ is highly depended on the value of attributes in $\pi_j$.

### 3.2 Social Connection $\mathcal{L}$

In the context of inference attack, the friend-based attacks derive from *homophily* [14], indicating that two linked friends share similar attributes. We propose a voting-based approach to depict the probability distribution of $x_j^i$ in terms of the value of $x_j$ of $i$'s friends.

We adopt a widely used approach [15] to measure the strength of tie given two neighbors $x$ and $y$, the *neighborhood overlap* of an edge $(x, y)$ can be computed as:

$$\omega(x, y) = \frac{Ne(x) \cap Ne(y)}{|Ne(x) - \{y\} \cup (Ne(y) - \{x\})|}, \quad (4)$$

where $Ne(.)$ denotes neighbor set of a node, $Ne(x) \cap Ne(y)$ represents the common neighbors of $x$ and $y$, and $|Ne(x) - \{y\} \cup (Ne(y) - \{x\})|$ represents the neighbors of at least one of $x$ and $y$.

A larger $\omega(x, y)$ means there exists a strong tie between $x$ and $y$. Assume a node $i$ has $n$ neighbors $i_1, i_2, \ldots, i_n$. Then, we can compute the wight $\psi(i, i_k)$ that evaluates the homophily between $i$ and $k$ based on Equation (4), where $1 \le k \le n$.

Therefore, the weight of $i'$ neighbor $i_k$ can be computed as follows:

$$\psi(i, i_k) = \frac{\omega(i, i_k)}{\sum_{j=1}^{n} \omega(i, i_j)}. \tag{5}$$

Then the probability distribution of $x_j^i$ in terms of $x_j^{i_k}$ can be computed as follows:

$$\mathbf{Pr}(x_j^i | x_j^{i_k}) = \frac{\omega(x_j^i, x_j^{i_k})}{\sum_{j=1}^{n} \omega(x_j^i, x_j^{i_k})}. \tag{6}$$

## 3.3   Inference Attack

Belief propagation is a message-passing algorithm that iteratively passes messages between factor nodes and variable nodes, to compute the marginal probability distribution of unknown variables. We represents the messages passing from variable nodes to factor nodes as $\mu$, while the messages passing from attribute factor nodes to variable nodes as $\nu$, and message passing from connection factor nodes to variable nodes as $\lambda$. We use $\mathbb{X}^T = \{x_i^{i^T}, i \in \mathcal{V}, j \in \mathcal{X}\}$ to denote all attribute variables in the iteration $T$. $\mu_{v \to s}^T(x_j^{i^T} = j_l)$ denotes the probability of $x_j^{i^T} = j_l$, at the $T$-th iteration, where $j_l$ is the $l$-th possible value for attribute $j$. Moreover, $\nu_{s \to v}^T(x_j^{i^T} = j_l)$ is the probability of $x_j^{i^T} = j_l$, at the $T$-th iteration, given attribute correlation $\mathcal{F}$. Likewise, $\lambda_{s \to v}^T(x_j^{i^T} = j_l)$ is the probability of $x_j^{i^T} = j_l$, at the $T$-th iteration, given attribute correlation $\mathcal{L}$ between the attributes.

In the process of passing message, a variable node $v$ sends message to one of its neighbors $s$, $\mu_{v \to s}^T(x_j^{i^T})$ by multiplying all messages it receives from its neighbors excluding the factor node $s$, where factor node $s$ might be attribute factor node (denoted as $f$) or connection factor node (denoted as $g$). Take the factor graph in Fig. 1 as an example, the message passing from $x_1^3$ to $f_1^3$, $\mu_{v \to f}^T(x_1^{3^T})$ can be formed as:

$$\mu_{v \to f}^T(x_1^{3^T}) = \frac{1}{Z} \times \prod_{f^* \in (\omega(x_1^3) \sim f_1^3)} \nu_{f^* \to v}^{(T-1)}(x_1^{3^{(T-1)}}) \times \prod_{g^* \in (\varpi(x_1^3))} \lambda_{g^* \to v}^{(T-1)}(x_1^{3^{(T-1)}}), \tag{7}$$

where $\omega(x_1^3) \sim f_1^3)$ is the set of all attribute factor node neighbors of the variable node $x_1^3$, excluding $f_1^3$, . $\varpi(x_1^3)$ is the set of all connection factor node neighbors of $x_1^3$. For example, in Fig. 1, $\omega(x_1^3) \sim f_1^3) = \{f_2^3, f_3^3\}$ and $\varpi(x_1^3) = \{g_1^{1,3}, g_2^{1,3}\}$.

The computation in Eq. (7) above is repeated in the process of passing message to each neighbor of each variable node. When $x_j^i \in \mathcal{X}_U$, the computation in Eq. (7) above are valid; however, when $x_j^i \in \mathcal{X}_K$, assume $x_j^i = \rho$, then $\mu_{v \to s}^T(x_j^{i^T} = \rho) = 1$ and $\mu_{v \to f}^T(x_j^{i^T}) = 0$ for other possible attribute value $x_j^i$.

Factor node $f$ passes messages to its neighbor variable node $v$ by multiplying all of the massages from $f'$ neighbors excluding $v$, and then computing the product of the factor $f$ and the obtained results; finally add the messages from all the neighbors of $f$ excluding $v$ together. We still take $x_1^3$ in Fig. 1 as example, the message passing from attribute factor node $f_1^3$ to the variable node $x_1^3$ at $T$-th iteration can be formed as:

$$\mu_{f \to v}^T(x_1^{3^T}) = \sum_{\{x_2^3, x_3^3\}} f_1^3(x_1^3, x_2^3, x_3^3, \mathcal{F}) \prod_{x^* \in (x_2^3, x_3^3)} \mu_{x^* \to f}^T(x^{*^T}), \qquad (8)$$

Here note that $f_1^3(x_1^3, x_2^3, x_3^3, \mathcal{F}) \propto \Pr(x_1^3 | x_2^3, x_3^3)$ and it can be computed using Table 1. The computation in Eq. (8) above must be repeated for each neighbor of each attribute factor node. Note that $g_1^{2,3}(x_1^2, x_1^3, \mathcal{L}) \propto \mathbf{Pr}(x_1^3 | x_1^2)$ which can be computed as shown in Sects. 3.1 and 3.2.

Initially, each variable node starts sending messages to its neighbor factor nodes. At the first iteration (i.e., $T = 1$), any variable node $x_j^i \in \mathcal{X}_U$, sends message $\mu_{v \to s}^1(x_j^{i^T}) = 1$ for every possible attribute value, since there is currently no message sending for neighbor factor nodes of $x_j^i$. For any variable node $x_j^i \in X_K$ with $x_j^i = \rho$, sends message $\mu_{v \to s}^1(x_j^{i^T} = \rho) = 1$ and $\mu_{v \to s}^1(x_j^{i^T} = \rho') = 0$ for every possible attribute value $\rho'$ except $\rho$. Until the value of all unknown variables are converged after several iterations (or the content of passed massages are converged), the process of iteration can be terminated.

After the message passing process based on the principle of belief propagation on factor graph, the marginal probability distribution of any unknown variable $x_j^i$ can be computed with linear complexity by multiplying all messages passed to $x_j^i$.

## 4    Tracking his($\mathcal{X}$) Under Differential Privacy

This section proposes an approach to release the histogram of attributes $\mathcal{X}$ from a social-attribute network in a $\varepsilon$-differentially private way. We denote the set of attributes of individual $i$ as $\mathcal{X}_i$, $\mathcal{X} = \cup_{1:|\mathcal{V}|} \mathcal{X}_i$. The approach has the following three stages:

1. Develop a $(\varepsilon/2)$-differentially private algorithm to inject Laplace noise into the set of local functions shown in Eq. (2), such that for each $f_j^i(.)$ or $g_{j.m}^i(.)$, we have sanitized version $f_j^i(.)'$ or $g_{j.m}^i(.)'$.
2. Combine the set of sanitized local functions (generated from the first step) to calculate an approximate probability distribution of attributes, and then carry out Bernoulli-sampling to sampling attributes based on the approximate probability distribution to produce a synthetic attribute dataset $\mathcal{X}'$.

3. Develop a $(\varepsilon/2)$-differentially private algorithm to inject Laplace noise into the set of histogram values.

From Eq. (2), we define the amount of attribute factor nodes $f_j^i(.)$ and connection factor nodes $g_{j.m}^i(.)$ are $m$, $m = |\mathcal{V} \times \mathcal{X}|$ and $n$, respectively. For each attribute factor node $f_j^i(.)$, the scale of Laplace noise injected to it is $\frac{12}{m\varepsilon}$, in order for guarantee $(\varepsilon/4)$-differential privacy, since the sensitivity of $f_j^i(.)$ is $\frac{3}{m}$. For the factor node $g_{j.m}^i(.)$, the scale of Laplace noise injected to it is $\frac{8}{m\varepsilon}$, in order for guarantee $(\varepsilon/4)$-differential privacy, since the sensitivity of $g_{j.m}^i(.)$ is $\frac{2}{m}$. Finally, for the returned histogram from the noise $\mathcal{X}$ generated from the operation above, the scale of Laplace noise injected to it is $\frac{4}{m\varepsilon}$, in order for guarantee $(\varepsilon/2)$-differential privacy, since the sensitivity of histogram is $\frac{2}{m}$.

According to the compensability property of differential privacy, our method satisfies $\varepsilon$-differential privacy.

## 5  Related Works

Inference attacks on sensitive information have been investigated extensively in recent years [16–26]. The approach in [27] predicts unknown attributes by estimating the value of unknown attributes first, then repeatedly refine the prediction results from last iteration using a classifier learned from known and estimated attributes (from last iteration) and neighbors' information that is also refined in last iteration. The authors in [14] present a social-behavior-attribute network to predict unknown attributes by combining social friends, individual attributes and behavior records in a unified network, and then the authors propose a vote distribution attack model to assign vote capacity to perform inference attacks. The approach in [28] assumes the adversary can always make optimal decisions in predicting the distribution of unknown attributes. To protect against the inference attack, the authors seek a tradeoff by constructing an optimization problem with the goal of maximizing the estimation error of the adversary. [29] transforms the problem of attribute inference to Bayesian inference on a Bayesian network that is constructed using the social connections between users. [30] incorporates social structures and known and unknown attribute into a Naive Bayes model to perform inference on unknown attributes. For example, to predict a user's income, their approach requires to know the users other known attributes such as title, work class, as well as her social friends and theirs attributes. However, their approach cannot work if user that publish no attributes (unknown or unknown) at all. However, the above works are not applicable to edge cases when there are users who shares no social friends or attributes at all. In our work, in the edge cases above, the proposed attribute inference approach can degrade into either attribute-based or friend-based inference attack naturally.

## 6  Conclusion

This paper studies inference attacks to predict unobserved social attributes based on publicly available information online. Through incorporating social structures, attributes, and learned probability dependency into a probabilistic graphical model, unobserved attributes can be predicted by executing belief propagation on factor graph with linear computational complexity. Existing attribute inference attacks has exponential computational complexity by leveraging either the friends' information or individuals' attribute attributes, but not both. Furthermore, this paper propose a differentially private histogram of attributes publishing approach.

# References

1. Backstrom, L., Dwork, C., Kleinberg, J.: Wherefore art thou r3579x?: anonymized social networks, hidden patterns, and structural steganography. In: Proceedings of the 16th International Conference on World Wide Web, pp. 181–190. ACM (2007)
2. Narayanan, A., Shmatikov, V.: De-anonymizing social networks, arXiv preprint arXiv:0903.3276 (2009)
3. Jia, J., Gong, N.Z.: Attriguard: a practical defense against attribute inference attacks via adversarial machine learning. In: 27th {USENIX} Security Symposium ({USENIX} Security 18), pp. 513–529 (2018)
4. Dwork, C.: Differential privacy. In: Bugliesi, M., Preneel, B., Sassone, V., Wegener, I. (eds.) ICALP 2006. LNCS, vol. 4052, pp. 1–12. Springer, Heidelberg (2006). https://doi.org/10.1007/11787006_1
5. Dwork, C., McSherry, F., Nissim, K., Smith, A.: Calibrating noise to sensitivity in private data analysis. In: Halevi, S., Rabin, T. (eds.) TCC 2006. LNCS, vol. 3876, pp. 265–284. Springer, Heidelberg (2006). https://doi.org/10.1007/11681878_14
6. Zhang, J., Cormode, G., Procopiuc, C.M., Srivastava, D., Xiao, X.: PrivBayes: private data release via Bayesian networks. In: Proceedings of the 2014 ACM SIGMOD International Conference on Management of Data, pp. 1423–1434 (2014)
7. Kschischang, F.R., Frey, B.J., Loeliger, H.-A., et al.: Factor graphs and the sum-product algorithm. IEEE Trans. Inf. Theory $47(2)$, 498–519 (2001)
8. Koller, D., Friedman, N.: Probabilistic Graphical Models: Principles and Techniques. MIT Press, Cambridge (2009)
9. McSherry, F., Talwar, K.: Mechanism design via differential privacy. In: FOCS, vol. 7, pp. 94–103 (2007)
10. Pawlak, Z.: Rough set theory and its applications to data analysis. Cybern. Syst. $29(7)$, 661–688 (1998)
11. Chow, C., Liu, C.: Approximating discrete probability distributions with dependence trees. IEEE Trans. Inf. Theory $14(3)$, 462–467 (1968)
12. Chickering, D.M., Heckerman, D., Meek, C.: Large-sample learning of Bayesian networks is NP-hard. J. Mach. Learn. Res. $5$(Oct), 1287–1330 (2004)
13. Margaritis, D.: Learning Bayesian network model structure from data. Carnegie-Mellon University, Pittsburgh, PA, School of Computer Science, Technical report (2003)
14. Gong, N.Z., Liu, B.: You are who you know and how you behave: attribute inference attacks via users' social friends and behaviors. In: 25th {USENIX} Security Symposium ({USENIX} Security 16), pp. 979–995 (2016)
15. Easley, D., Kleinberg, J.: Strong and weak ties. In: Networks, Crowds, and Markets: Reasoning About a Highly Connected World, pp. 47–84 (2010)

16. Cai, Z., He, Z.: Trading private range counting over big IoT data. In: The 39th IEEE International Conference on Distributed Computing Systems (ICDCS) (2019)
17. He, Z., Li, J.: Modeling SNP-trait associations and realizing privacy-utility tradeoff in genomic data publishing. In: Cai, Z., Skums, P., Li, M. (eds.) ISBRA 2019. LNCS, vol. 11490, pp. 65–72. Springer, Cham (2019). https://doi.org/10.1007/978-3-030-20242-2_6
18. He, Z., Li, Y., Li, J., Li, K., Cai, Q., Liang, Y.: Achieving differential privacy of genomic data releasing via belief propagation. Tsinghua Sci. Technol. **23**(4), 389–395 (2018)
19. He, Z., Yu, J., Li, J., Han, Q., Luo, G., Li, Y.: Inference attacks and controls on genotypes and phenotypes for individual genomic data. IEEE/ACM Trans. Comput. Biol. Bioinform. (2018)
20. He, Z., Li, Y., Wang, J.: Differential privacy preserving genomic data releasing via factor graph. In: Cai, Z., Daescu, O., Li, M. (eds.) ISBRA 2017. LNCS, vol. 10330, pp. 350–355. Springer, Cham (2017). https://doi.org/10.1007/978-3-319-59575-7_33
21. He, Z., Li, Y., Li, J., Yu, J., Gao, H., Wang, J.: Addressing the threats of inference attacks on traits and genotypes from individual genomic data. In: Cai, Z., Daescu, O., Li, M. (eds.) ISBRA 2017. LNCS, vol. 10330, pp. 223–233. Springer, Cham (2017). https://doi.org/10.1007/978-3-319-59575-7_20
22. Cai, Z., Zheng, X., Yu, J.: A differential-private framework for urban traffic flows estimation via taxi companies. IEEE Trans. Ind. Inform. (2019)
23. Cai, Z., Zheng, X.: A private and efficient mechanism for data uploading in smart cyber-physical systems. IEEE Trans. Netw. Sci. Eng. (2018)
24. Huang, Y., Cai, Z., Bourgeois, A.G.: Search locations safely and accurately: a location privacy protection algorithm with accurate service. J. Netw. Comput. Appl. **103**, 146–156 (2018)
25. Liang, Y., Cai, Z., Yu, J., Han, Q., Li, Y.: Deep learning based inference of private information using embedded sensors in smart devices. IEEE Netw. **32**(4), 8–14 (2018)
26. Xiong, Z., Li, W., Han, Q., Cai, Z.: Privacy-preserving auto-driving: a GAN-based approach to protect vehicular camera data. In: 19th IEEE International Conference on Data Mining (2019)
27. Cai, Z., He, Z., Guan, X., Li, Y.: Collective data-sanitization for preventing sensitive information inference attacks in social networks. IEEE Trans. Dependable Secure Comput. **15**(4), 577–590 (2016)
28. He, Z., Cai, Z., Yu, J.: Latent-data privacy preserving with customized data utility for social network data. IEEE Trans. Veh. Technol. **67**(1), 665–673 (2017)
29. He, J., Chu, W.W., Liu, Z.V.: Inferring privacy information from social networks. In: Mehrotra, S., Zeng, D.D., Chen, H., Thuraisingham, B., Wang, F.-Y. (eds.) ISI 2006. LNCS, vol. 3975, pp. 154–165. Springer, Heidelberg (2006). https://doi.org/10.1007/11760146_14
30. Lindamood, J., Heatherly, R., Kantarcioglu, M., Thuraisingham, B.: Inferring private information using social network data. In: Proceedings of the 18th International Conference on World Wide Web, pp. 1145–1146. ACM (2009)

# Feature Selection Based on Graph Structure

Zhiwei Hu[1], Zhaogong Zhang[1]([✉]), Zongchao Huang[1],
Dayuan Zheng[2]([✉]), and Ziliang Zhang[3]

[1] School of Computer Science and Technology, Heilongjiang University,
Harbin, China
zhaogong.zhang@qq.com
[2] School of Data Science and Technology, Heilongjiang University,
Harbin, China
1985022@hlju.edu.cn
[3] Department of Computer Science, College of Engineering,
The University of Alabama, Tuscaloosa, USA

**Abstract.** Feature selection is an important part of data preprocessing. Selecting effective feature subsets can effectively reduce feature redundancy and reduce irrelevant features, and reduce training costs. Based on the theory of feature clusters, this paper proposes a feature selection strategy based on the graph structure. Considering a feature as a node in the graph, using the idea of graph message propagation, integrating the first-order neighbor information of each node, and then selecting the key point of the local maximum score as the selected feature, this can effectively reduce the feature redundancy and reduce features that are not related to the label. Finally, in order to verify the anti-interference of this novel method, the noise dimension was added in the UCI data set, and the comparison test was again performed. The experimental results show that the proposed algorithm can effectively improve the classification accuracy in a specific data set, and the anti-interference is better than other feature selection algorithms.

**Keywords:** Feature cluster · Feature selection · MPNN · Supervised · Anti-interference

## 1  Introduction

In the field of data mining and machine learning research, feature dimension reduction is an important part of feature engineering. An effective feature dimension reduction method can accelerate the learning process of the training model and improve the performance of the model. Some studies have shown that, without losing the accuracy of classification, selecting effective feature subsets can suppress the over-fitting of the model, thus improving the universality of the model. In the popular deep learning field, the process of feature selection is hidden during neural network training. But when the

---

Z. Zhang and D. Zheng—This work was supported by the Natural Science Foundation of Heilongjiang Province (No. F2017024, No. F2017025).

Y. Li et al. (Eds.): COCOA 2019, LNCS 11949, pp. 289–302, 2019.
https://doi.org/10.1007/978-3-030-36412-0_23

amount of data is too large, it is not practical to put the data directly into neural network training. For example, in the analysis of biological genetic information, high-dimensional genetic data, 3 million features, neural network methods can not be directly calculated, at this time must feature selection.

Feature engineering is of great significance in many fields, such as the application of feature engineering in biological information field [1–4]. Feature dimension reduction is an important part of feature engineering. Feature dimension reduction is divided into feature selection and feature extraction. Feature selection selects $M$ features from $N$ features $(N > M)$, and no new features are generated; and feature extraction extracts $M$ features from the original $N$ features, which will generate new features. The most typical feature extraction algorithm is the principal component analysis (PCA) algorithm, which selects the features corresponding to the largest $M$ eigenvalues of the covariance matrix as the basis of the projection space, and finally projects the original features into the new space to generate new features. Feature extraction is only suitable for simple classification problems, and is not suitable for special tasks that finding some features which affects label. For example, in the field of bioinformatics, the goal is to find out some of the factors most relevant to a disease. In this case, feature extraction cannot be used as a method of feature dimension reduction.

According to the process of feature selection, feature selection can be divided into filtering, wrapping and embedding. Feature selection has two major tasks, eliminating redundancy between features and discarding features that are not related to label. The specific details are introduced in the next section.

This paper introduces a new graph-based filtering feature selection algorithm. The feature is regarded as the point of the graph structure. The edge between the point and the point is the correlation between the feature and the feature. The information in each node is the correlation between the feature and the label. And set a threshold and crop some edges with lower correlation to reduce the amount of calculation in the graph. The algorithm uses the idea of Message Passing Neural Nets (MPNN) [5] to integrate the information of each node's first-order neighbors with the designed integration function, and update the information of each node. Finally, we adopt the idea of SIFT (Scale-invariant feature transform) [6] to select the key point and select the local maximum node. The selected nodes are more stable and robust than other nodes, and the features corresponding to such nodes are considered to be selected features.

The rest of the paper is structured as follows. Overview of feature selection will be introduced in Sect. 2. MPNN and mutual information will be discussed in Sect. 3, and the proposed feature selection algorithm will be briefly described in Sect. 4. Finally, the experimental procedure and results of the comparison with the ReliefF and SFS algorithms will be described in Sect. 5. Finally, the conclusion of the work is offered in Sect. 6.

## 2    Overview of Feature Selection Method

The filter feature selection is independent of the subsequent classification (regression) model. Generally, the Pearson correlation index, mutual information, and maximum information coefficient are used to judge the relationship between features. The original filtering algorithm only considers the relationship between features or only considers the relationship between features and labels, so as to sort the features and select the optimal $M$ features. But simply considering a single task of feature selection, the selected feature subset is not optimal. Caruana et al. proposed the SFS (Sequence Forward Search) algorithm [7], the algorithm is based on the idea of greed, and the optimal one is selected each time. The feature subset starts from the empty set, and each time the feature with the best fitness function is added to the feature subset until all the features are traversed. The disadvantage of the SFS algorithm is that it depends on the fitness function and can only be added to the feature subset, and cannot be eliminated. Therefore, there is still a case of feature redundancy. In 1992, Kira proposed the relief algorithm [8], which is only suitable for the feature selection of two-class tasks. A sample is randomly selected, and then $K$ neighbor samples (for example, cosine similarity is used to calculate the similarity between samples) of the same classes as the selected sample are selected, and $K$ neighbor samples of different classes from the selected sample are selected, and determine if the feature makes sense for the classification, and calculate weights based on this. Iterate multiple times, and finally sort according to the weight of each feature to select the appropriate feature. Later, in order to solve the multi-classification problem, Kononenko proposed the ReliefF algorithm [9]. The Relief series of algorithms are simple and efficient, and there are no restrictions on the data types. It belongs to a feature weighting algorithm. Therefore, features with a high correlation with the label will be given higher weights. The limitation of this algorithm is that it cannot effectively eliminate redundant features.

The wrapped feature selection takes the performance of the latter learning model as a reference, for example, the accuracy of the final model is used as a criterion for judging the quality of the feature subset. The main idea of the representative wrap feature selection algorithm is to regard feature selection as $01$ problem, $0$ for no selection, $1$ for selection, and the feature selection problem to find the optimal solution in the solution space. For example, feature selection based on genetic algorithm (GAs) [10], feature selection based on particle swarm optimization algorithm PSO [11, 12] and gray wolf algorithm (GWO) [13] and particle swarm optimization algorithm combined Algorithms [14]. Of course, there are also some feature selection algorithms based on other optimization algorithms [15–18]. Because the wrap feature selection is the most reference to the performance of the model, the resulting solution will reach an approximate optimal solution as the number of iterations increases, with the accompanying calculation being particularly large.

The embedded feature selection is the same as the wrapper feature selection method, which is related to the training model, and the process of feature selection is embedded in the learning process of the model. Common embedded feature selection methods are generally applied to the regression task. The $L_2$ paradigm (Lasso) is embedded in the loss function of the learning model to achieve the compression factor effect. A feature that is not associated with a label, its coefficients are compressed to a small extent, and the feature coefficients associated with the label are amplified. In 2004, Efron proposed the Least Angle Regression Algorithm (LARS) [19], which treats the label $Y$ as a vector and other features as vectors. The algorithm starts with all coefficients being zero, first finds the feature variable $X_1$ most relevant to the label $Y$, and proceeds on the solution path of the selected variable until there is another variable $X_2$, so that the two variables have the same correlation with the current residual. Then repeat this process, LARS guarantees that all the variables of the selected regression model will advance on the solution path, and the correlation coefficient with the current residual is the same. At the end of the article, the author proves that LARS and Lasso regression are equivalent. In addition, there are more traditional feature selection methods Bess (best subset selection) [20] for the feature selection of regression tasks. Consider the coefficient $\beta_i$ of a feature as unknown, and use the loss function $l(\beta_i)$ to perform Taylor expansion, and calculate the difference between the minimum value of the expansion and $l(0)$. Sorting according to the difference, screening a part of the feature, the algorithm also embeds the process of updating the coefficient. Generally, the embedded feature selection algorithm is applied to the regression task, and the embedded feature selection method for the classification task is less.

# 3   Related Work

There are many criteria for the correlation between two variables. The most commonly used are Pearson correlation coefficients, cosine similarity, etc. All of the above criteria measure the value of continuous values, and mutual information has a better performance for discrete attributes. The proposed algorithm is a feature selection for the classification task and needs to measure the correlation between the feature and the label. Therefore, mutual information is used as the correlation criterion in the experiment.

## 3.1   Mutual Information

In 1948, Claude Shannon, the father of information theory, proposed the theory of information entropy. Information entropy represents the average information of multiple possibilities of one thing. The formula of information entropy is defined as follows:

$$H(X) = - \sum_{x \in X} p(x) \log(p(x)) \tag{1}$$

Mutual information is a very useful measure of information in information theory. It can be regarded as the amount of information about another random variable contained in a random variable, or it is the uncertainty of a random variable that is reduced due to the knowledge of another random variable.

**Definition 1:** Let the joint distribution of two random variables $(X; Y)$ be $p(x, y)$, the edge distribution be $p(x), p(y)$, and the mutual information $I(X; Y)$ be the relative entropy between the joint distribution $p(x, y)$ and the edge distribution $p(x)p(y)$, i.e.:

$$I(X; Y) = \sum_{y \in Y} \sum_{x \in X} p(x, y) \log\left(\frac{p(x,y)}{p(x)p(y)}\right) \tag{2}$$

In addition, mutual information can also be expressed by the following formula:

$$I(X;Y) = H(X) + H(Y) - H(X,Y) = H(X) - H(X|Y) \tag{3}$$

## 3.2  MPNN

MPNN is a method designed to extract the features of a topology. We describe MPNNs which operate on undirected graphs $G$ with node features $x_v$ and edge features $e_{vw}$. To enrich the relationship between each node and other nodes, MPNN allows each node to grasp the information from local to global through the spread of the message. The forward propagation of MPNN consists of two steps, a message-passing phase, and a readout phase. The message propagation process is updated by the information transfer function $M_t$ and the node update function $U_t$. The hidden feature of the node after $T$ update is defined as $h_v^T$, and initial state is defined as $h_v^0 = x_v$. Hidden features during messaging are defined according to the following formula:

$$m_v^{t+1} = \sum_{w \in N(v)} M_t\left(h_v^t, h_w^t, e_{vw}\right) \tag{4}$$

$$h_v^{t+1} = U_t\left(h_v^t, m_v^{t+1}\right) \tag{5}$$

$Nv$ represents the first-order neighbor node of $v$. With the times of updates, the information of the node's features is more and more comprehensive, analogous to the convolution of the image. As the convolutional layer increases, the extracted information becomes more and more comprehensive. Finally, the prediction can be made based on the updated node information. The prediction function has the following definitions:

$$\hat{y} = R\left(\{h_v^T | v \in G\}\right) \tag{6}$$

The core idea of MPNN is to integrate the information of the neighbor nodes of a node onto the node. In this paper, the proposed algorithm is modified based on the idea of MPNN. The structure of the feature selection problem is regarded as the graph structure, and the feature is regarded as a node. The weight of the edge in the graph is the correlation between the feature and the label is stored in the node. The proposed algorithm sets the threshold and rejects the edges with small weights, then the graph changes from a complete graph to a normal graph. There are two reasons for this. First, the computational overhead is reduced. Second, the remaining edges are considered valid after the edges with small weights are discarded, because these edges are considered weakly correlated or irrelevant.

## 3.3  Node Information Update

The criterion for a good feature is that the feature is related to the label and is not related to other features. Assuming that mutual information is used as a measure of variable relevance, the greater the mutual information value, the greater the correlation between the variables. The two major tasks of feature selection can be integrated into one formula: the correlation between features and labels minus the redundancy of features and other features. The greater the difference, the better the feature, and the smaller the difference, the more redundant or irrelevant the feature is. Consider this difference as the score of the feature. The process of calculating the difference corresponds to the part of the MPNN node update information. In this paper, the node information is updated according to the following formula:

$$v_i = I(X_i, Y) - \sum_{j \in Ner(i)} I(X_i, X_j) \tag{7}$$

The node updated information is represented by $v_i$, and $Ner(i)$ represents the first-order neighbor node of node $i$.

There are two extreme examples in (a) and (b) in Fig. 1, both of which are first-order neighbor graphs for the feature $X_5$. The $X_5$ in (a) is a redundant feature, because it has a great correlation with other features. After updating the value in the node by formula (7), it will be found that the value of the $X_5$ node is compressed very low; The node $X_5$ in (b) is a feature that has a high correlation with the label and low correlation with other features. Such a feature is an ideal feature. After the update of the formula (4), the value of $X_5$ will be higher than the other node values after the update. In the actual experiment, the edges with small weights are discarded, but the redundancy of related features and the redundancy of unrelated features can be used for reference.

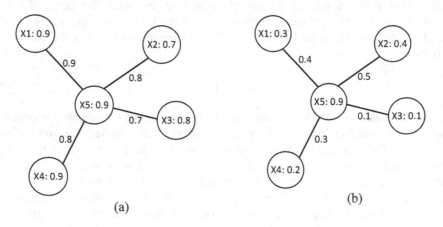

Fig. 1. Partial node diagram

## 4   Proposed Method

In order to solve the problem of feature selection, many researchers solve problems from different angles. The current popular research angle is to regard the problem as an NP-hard problem, and use the optimization algorithm to find the optimal solution in the solution space; Another angle is to treat feature selection as a clustering problem. Based on the hypothesis (1), features with redundancy are considered to be classified into the same cluster. The first one belongs to the wrapper type scheme, which is expensive to calculate, but the effect is good; the second type is filter type, and the calculation consumption is small, but the performance is not as good as the wrapper type, but it can process high-dimensional data.

**Hypothesis (1):** If the similarity between feature $X_1$ and feature $X_2$ is high, then the correlation of $X_1$ with the label approximates the relevance of $X_2$ to the label.

The proposed algorithm is based on the second solution. Many researchers have established different algorithms based on the theory of feature clusters. In [21], the algorithm proposed by Wang uses clustering algorithm to search clusters in different subspaces. According to the idea of KNN, features are added to existing clusters. And the author sets a threshold for the robustness of the cluster, and if the maximum distance between the feature and the cluster is greater than the threshold, a new cluster is established. Besides, in [22], the algorithm proposed by Song Q is also based on feature clusters, but instead of using clustering algorithms, the topology is used to construct the relationship between features and features. Use the Prim algorithm to generate the minimum spanning tree, and then turn it into a forest according to a certain strategy. Each tree in the forest corresponds to a cluster. The same point of the two algorithms is to find the feature cluster and then select the best feature from the feature cluster, but such a method will lose some features and the performance of redundant feature culling of high-dimensional data is not very good. The proposed algorithm does not need to find such a feature cluster. The graph propagated through the MPNNs network will integrate the information of the neighbor nodes. Assuming that the node is redundant, the score of the node is lowered, so that the node (feature) with a high local score is taken as a key node, and such a feature has higher correlation with the label and lower redundancy.

Based on the description of the above information, a novel feature selection algorithm GBFS (graph-based feature selection) is proposed. The algorithm first relies on correlation metrics to calculate the correlation between features and label, and then to filter the correlation between some features. In the experiment, the threshold was selected as the median of all the correlations, and the correlations less than the median were all discarded. The structure of the graph is constructed based on the remaining information. The specific construction process is as follows: The correlation between the feature and the feature is taken as the weight of the edge in the graph, and the correlation between the feature and the label is used as the information of the node. After that, the node information is updated according to the idea of the message propagation network (formula (7)). Only need to spread once, without having to propagate multiple times to get global information, because the graph before the filter edge is a complete graph, and the value less than the median is considered weakly correlated or irrelevant. Then, after screening, for a node, the remaining neighbor nodes are related, and are related to the node in the global scope. Finally, the best node is selected by the principle of local optimality. In theory, the score of a node is greater than that of all its neighbors. Such a node is considered to be the best, but in reality, such conditions are more rigorous, resulting in the sparse features selected. In the experiment, the selection condition is relaxed, and the node whose score is greater than half is considered to be critical. If the dimensions of the dataset are large, you can still limit the selection criteria.

From the description of the above algorithm, the running time of the proposed algorithm is mainly in calculating the correlation between any two features. In this step, the algorithm requires the calculation of $C_n^2$ times of mutual information. Therefore, the time complexity of the correlation between the features of the algorithm is calculated as $O(n^2 * T)$, and the time complexity of calculating the mutual information is assumed to be T. Since the adjacency matrix is used to store the information of the edge, the time complexity of updating the node information is $O(n^2)$, and the time complexity of finding the best node is also $O(n^2)$. Finally, the proposed algorithm has a time complexity of $O(n^2 * T)$. It can be seen that the time complexity is a quadratic term polynomial, and the algorithm execution efficiency is relatively simple and efficient. The specific algorithm pseudo-code is as follows:

---

**Algorithm BGFS**

---

**Input**: $X \in R^{n \times m}$, $Y \in R^{n \times 1}$
**Output**: the selected feature subset S

**Init**: Node $= \emptyset$, edge $= \emptyset$, S $= \emptyset$
1. **For** i to m **do**
2.      $v_i = I(X_i, Y)$
3.      Node = Node $\cup \{v_i\}$
4. **Endfor**
5. **For** i=1 **to** m **do**
6.      **For** j=i+1 **to** m **do**
7.          $e_{ij} = I(X_i, X_j)$
8.          **If** $e_{ij} > \theta$ **then**
9.              Edge = Edge $\cup \{e_{ij}\}$
10.    **Endfor**
11. **Endfor**
12. Construct_graph(Node, Edge)
13. **For** $i$ **to** m **do**
14.      Update($v_i$) using formula (7)
15. **End for**
16. **For** $i$ **to** m **do**
17.      flag $= 0$
18.      neibigours $= \text{Ner}(v_i)$
19.      **For each** $v_j \in$ nebigiours **do**
20.          **If** $\max(v_i, v_j) ==$ true **then**  //$v_i$ grater than $v_j$ return true
21.              flag = flag $+ 1$
22.      **Endfor**
23.      **If** flag $> \beta * $ len(neibigours) **then**
24.          S = S $\cup \{i\}$
25. **Endfor**
26. **Return S**

---

# 5  Experimental Study and Result Analysis

To verify the performance of the GBFS algorithm, this paper mainly evaluates the classification accuracy of the selected feature subsets on the SVM classifier. In the experiment, the performance of GBFS algorithm and SFS and ReliefF two classic filtering feature selection algorithms are compared.

## 5.1  Data Set Description

Eight data sets from UCI were selected in the experiment. Table 1 lists the relevant parameters of the data set. Where samples represent the number of samples in the data set, attributes represent the number of features, Discrete attributes represent the number of discrete attributes, Continuous attributes represent the number of continuous attributes, and classes represent the number of categories.

**Table 1.**  UCI data set.

| Dataset | Samples | Attributes | Discrete attributes | Continuous attributes | Classes |
|---|---|---|---|---|---|
| Dermatology | 366 | 34 | 33 | 1 | 6 |
| Ionosphere | 351 | 34 | 2 | 32 | 2 |
| Sonar | 208 | 60 | 0 | 60 | 2 |
| Wdbc | 569 | 30 | 0 | 30 | 2 |
| Wine | 178 | 13 | 0 | 13 | 3 |
| Parkinsons | 197 | 22 | 0 | 22 | 2 |
| Lung | 32 | 56 | 56 | 0 | 3 |
| Hill valley | 606 | 100 | 0 | 100 | 2 |

## 5.2  10-Fold Cross Validation

The experiment used a ten-fold cross-validation to estimate the performance of the model. The data set was divided into ten, and nine of them were taken as a training set and one was used as a test set. The corresponding accuracy is obtained for each experiment, and the average of the 10 results is used as an estimate of the performance of the algorithm. In the experiment, multiple ten-fold cross-validation was used, and the average value was calculated as an estimate of the accuracy of the algorithm. The reason for choosing to divide the data set into 10 is because a large number of researchers use a large number of data sets and use different algorithms to carry out continuous experiments, which shows that it is the best choice for obtaining the best error estimate, and there is also a theoretical proof.

## 5.3  Result Analysis

Because some UCI data sets are relatively neat, so before the experiment, the data set is disrupted, and then 10-fold cross-validation. The SFS algorithm and the ReliefF algorithm are sorting feature selection algorithms. Therefore, it is necessary to specify the number of features. In the experiment, the number of features selected according to the GBFS algorithm is selected, and the same number of features are specified in SFS and ReliefF. Compare the classification performance of the features selected by the three algorithms in the SVM classifier. Table 2 lists the experimental results of eight data sets, where SVM indicates that the original data set is directly trained using the SVM classifier, GBFS+SVM represents the feature selected by the GBFS feature, then put into the SVM trainer, and so on. And it can be seen that the number of features selected by the GBFS algorithm is about half of the original feature number, because the limit value of the algorithm in the experiment is the median, and the key point selection strategy is greater than half. If you want to continue to reduce features or increase features, you can adjust the parameters $\theta$ and $\beta$ in the pseudo-code. The data in Table 2 shows in discrete data sets, the proposed algorithm selects features better than other algorithms. For data-regulated data sets, the algorithm also performs well. For example, the second data set ionosphere, whose received signals were processed using an autocorrelation function before making data sets. But in some data sets with continuous value attributes, the proposed algorithms don't perform well, even worse than them. Because mutual information measures the continuous value data set is not very good, resulting in the construction of the topology map does not represent the relationship between features well. In the case of small sample size, the effect is not very good, because the essence of mutual information is still through probability statistics. The less the data, the less accurate the statistical probability. Therefore, the proposed method is applicable to the case where the attribute value is discrete and the sample size is large. From the final average accuracy, the GBFS algorithm is a bit higher than ReliefF and smaller than SFS.

**Table 2.** Classification results of data sets.

| Dataset | SVM | GBFS+SVM | RliefF+SVM | SFS+SVM | Features selected |
|---|---|---|---|---|---|
| Dermatology | 0.911 ± 0.024 | 0.959 ± 0.012 | 0.859 ± 0.025 | 0.884 ± 0.029 | 18 |
| Ionosphere | 0.942 ± 0.024 | 0.980 ± 0.018 | 0.977 ± 0.019 | 0.897 ± 0.026 | 18 |
| Sonar | 0.559 ± 0.11 | 0.581 ± 0.079 | 0.523 ± 0.064 | 0.826 ± 0.040 | 31 |
| Wdbc | 0.642 ± 0.038 | 0.690 ± 0.028 | 0.708 ± 0.046 | 0.730 ± 0.040 | 16 |
| Wine | 0.410 ± 0.072 | 0.792 ± 0.057 | 0.410 ± 0.052 | 0.968 ± 0.018 | 7 |
| Parkinsons | 0.759 ± 0.041 | 0.767 ± 0.040 | 0.761 ± 0.039 | 0.763 ± 0.046 | 12 |
| Lung | 0.433 ± 0.12 | 0.443 ± 0.12 | 0.452 ± 0.17 | 0.476 ± 0.17 | 29 |
| Hill valley | 0.478 ± 0.024 | 0.480 ± 0.020 | 0.476 ± 0.020 | 0.484 ± 0.027 | 51 |
| Average | 0.642 | 0.712 | 0.646 | 0.754 | – |

To verify the anti-interference of the model, the original data set was modified and were added with noise. The purpose of adding the noise dimension is to interfere with the correlation between the features and the labels, because the generated noise dimension may be more relevant to labels than the original features. As the noise increases, some of the algorithm's flaws are amplified. Suppose we have a data set (n samples, m features), add noise data, and represent noise data in gray. The first mode of adding noise, as shown in Fig. 2. According to the number of the original features, the same number of noise features are added, and the noise features generated by the random numbers are weakly correlated with each other. The experimental results are shown in Table 3. The horizontal anti-interference of the three algorithms is similar. The second noise-adding mode, as shown in Fig. 3, first copies the original data set so that the sample size is twice (2n), and then adds the noise features of the same feature number (m). The experimental results are shown in Table 4. Longitudinal expansion is equivalent to data enhancement, so accuracy is generally improved. In this noise mode, the proposed algorithm stability is similar to other algorithms.

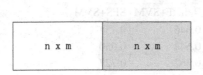

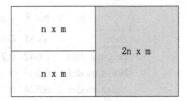

**Fig. 2.** Noise mode 1                         **Fig. 3.** Noise mode 2

Finally, the results of the original data set are averaged with the results of noise mode 1 and noise mode 2. As shown in Table 5, the anti-interference of GBFS is better than the other two algorithms. In other words, the accuracy of the various situations is combined and the proposed algorithm is more stable.

**Table 3.** The result of noise mode 1

| Dataset | SVM | GBFS+SVM | RliefF+SVM | SFS+SVM | Features selected |
|---|---|---|---|---|---|
| Dermatology | 0.911 ± 0.017 | 0.926 ± 0.022 | 0.909 ± 0.033 | 0.563 ± 0.064 | 35 |
| Ionosphere | 0.905 ± 0.036 | 0.939 ± 0.025 | 0.945 ± 0.025 | 0.895 ± 0.030 | 35 |
| Sonar | 0.523 ± 0.063 | 0.531 ± 0.055 | 0.537 ± 0.042 | 0.508 ± 0.083 | 61 |
| Wdbc | 0.627 ± 0.028 | 0.655 ± 0.030 | 0.655 ± 0.033 | 0.640 ± 0.029 | 31 |
| Wine | 0.531 ± 0.077 | 0.460 ± 0.068 | 0.411 ± 0.068 | 0.921 ± 0.044 | 14 |
| Parkinsons | 0.805 ± 0.048 | 0.785 ± 0.044 | 0.795 ± 0.036 | 0.835 ± 0.041 | 23 |
| Lung | 0.357 ± 0.14 | 0.381 ± 0.11 | 0.395 ± 0.12 | 0.343 ± 0.14 | 57 |
| Hill valley | 0.494 ± 0.045 | 0.498 ± 0.039 | 0.489 ± 0.036 | 0.491 ± 0.030 | 101 |
| Average | 0.644 | 0.647 | 0.642 | 0.650 | – |

**Table 4.** The result of noise mode 2

| Dataset | SVM | GBFS+SVM | RliefF+SVM | SFS+SVM | Features selected |
|---|---|---|---|---|---|
| Dermatology | 0.960 ± 0.011 | 0.967 ± 0.012 | 0.969 ± 0.010 | 0.918 ± 0.028 | 35 |
| Ionosphere | 0.976 ± 0.014 | 0.993 ± 0.0085 | 0.994 ± 0.012 | 0.953 ± 0.019 | 35 |
| Sonar | 0.587 ± 0.081 | 0.688 ± 0.033 | 0.690 ± 0.050 | 0.528 ± 0.029 | 61 |
| Wdbc | 0.884 ± 0.021 | 0.890 ± 0.019 | 0.891 ± 0.018 | 0.644 ± 0.027 | 31 |
| Wine | 0.873 ± 0.048 | 0.856 ± 0.049 | 0.826 ± 0.042 | 0.970 ± 0.022 | 14 |
| Parkinsons | 0.923 ± 0.015 | 0.931 ± 0.034 | 0.939 ± 0.028 | 0.901 ± 0.036 | 23 |
| Lung | 0.591 ± 0.12 | 0.729 ± 0.118 | 0.787 ± 0.085 | 0.388 ± 0.12 | 57 |
| Hill valley | 0.802 ± 0.019 | 0.790 ± 0.029 | 0.800 ± 0.024 | 0.511 ± 0.026 | 101 |
| Average | 0.824 | 0.856 | 0.862 | 0.727 | – |

**Table 5.** Anti-interference test results

| Data | SVM | GBFS+SVM | RliefF+SVM | SFS+SVM |
|---|---|---|---|---|
| Original data | 0.642 | 0.712 | 0.646 | 0.754 |
| Noise model 1 | 0.644 | 0.647 | 0.642 | 0.650 |
| Noise model 2 | 0.824 | 0.856 | 0.862 | 0.727 |
| Average | 0.703 | 0.738 | 0.717 | 0.710 |

# 6  Conclusion

This paper proposes a new feature selection algorithm based on the feature cluster theory. Different from the traditional selection scheme based on feature cluster theory, the proposed algorithm does not need to find feature clusters. The selection process of feature clusters is embedded into these two steps through the MPNN message delivery mechanism and node information update formula in the third section. Compared with other sorting feature selection methods, such as SFS and ReliefF algorithm, the proposed algorithm does not need to specify the number of selected features. Besides, the advantage of the proposed algorithm is that it can be used to increase the computational speed with distributed system calculations. In the experiment, we use matrix to represent the structure of the graph, so we can use the distributed parallel computing of the matrix to speed up the operation. The disadvantage of the proposed algorithm is that as the data dimension increases, the amount of calculation of the graph becomes larger and larger, and the number of edges can only be reduced by increasing the threshold, but this will bring about loss of information. The proposed algorithm does not trade well between performance and amount of computation. Future work will improve the proposed algorithm to resolve the contradiction between high dimensional data and computational complexity. Secondly, there is no uniform standard for data preprocessing, and there is no good evaluation standard for continuous value discretization. However, the impact of continuous value discretization on the experiment is still very large. In the next research work, the correlation criteria between continuous value attribute and discrete value attribute will be also further explored and studied. Finally, to verify the anti-interference of the proposed model, noise data is added horizontally and vertically in the eight original UCI data sets. The anti-interference experiment results show that the proposed model is better robust than the other two.

# References

1. Cai, Z., Goebel, R., Salavatipour, M., Lin, G.: Selecting dissimilar genes for multi-class classification, an application in cancer subtyping. BMC Bioinform. **8**, 206 (2007). (IF: 3.428)
2. Cai, Z., Zhang, T., Wan, X.: A computational framework for influenza antigenic cartography. PLoS Comput. Biol. **6**(10), e1000949 (2010)
3. Cai, Z., Xu, L., Shi, Y., Salavatipour, M., Goebel, R., Lin, G.: Using gene clustering to identify discriminatory genes with higher classification accuracy. In: IEEE the 6th Symposium on Bioinformatics and Bioengineering (BIBE 2006) (2006)
4. Yang, K., Cai, Z., Li, J., Lin, G.: A stable gene selection in microarray data analysis. BMC Bioinform. **7**, 228 (2006)
5. Gilmer, J., Schoenholz, S.S., Riley, P.F., et al.: Neural message passing for quantum chemistry (2017)
6. Lowe, D.G.: Distinctive image features from scale-invariant keypoints. Int. J. Comput. Vision **60**(2), 91–110 (2004)
7. Caruana, R., De Sa, V.R.: Benefitting from the variables that variable selection discards. J. Mach. Learn. Res. **3**, 1245–1264 (2003)
8. Kira, K., Rendell, L.A.: A practical approach to feature selection. In: Proceedings of the Ninth International Workshop on Machine Learning, pp. 249–256 (1992)
9. Kononenko, I.: Estimating attributes: analysis and extensions of RELIEF. In: Bergadano, F., De Raedt, L. (eds.) ECML 1994. LNCS, vol. 784, pp. 171–182. Springer, Heidelberg (1994). https://doi.org/10.1007/3-540-57868-4_57
10. Pal, S.K., Wang, P.P.: Genetic Algorithms for Pattern Recognition. CRC Press Inc., Boca Raton (1996)
11. Kennedy, J.: Particle swarm optimization. In: Sammut, C., Webb, G.I. (eds.) Encyclopedia of Machine Learning, pp. 760–766. Springer, Boston (2010). https://doi.org/10.1007/978-0-387-30164-8
12. Chuang, L.Y., Chang, H.W., Tu, C.J., et al.: Improved binary PSO for feature selection using gene expression data. Comput. Biol. Chem. **32**(1), 29–38 (2008)
13. Mirjalili, S., Mirjalili, S.M., Lewis, A.: Grey wolf optimizer. Adv. Eng. Softw. **69**(3), 46–61 (2014)
14. Emary, E., Zawbaa, H.M., Hassanien, A.E.: Binary grey wolf optimization approaches for feature selection. Neurocomputing **172**, 371–381 (2016)
15. Mafarja, M.M., Mirjalili, S.: Hybrid binary ant lion optimizer with rough set and approximate entropy reducts for feature selection. Soft. Comput. **23**(15), 6249–6265 (2019)
16. Al-Tashi, Q., Kadir, S.J.A., Rais, H.M., et al.: Binary optimization using hybrid grey wolf optimization for feature selection. IEEE Access **7**, 39496–39508 (2019)
17. Mafarja, M., Aljarah, I., Faris, H., et al.: Binary grasshopper optimisation algorithm approaches for feature selection problems. Expert Syst. Appl. **117**, 267–286 (2019)
18. Li, W., Chao, X.Q.: Improved particle swarm optimization method for feature selection. J. Front. Comput. Sci. Technol. **13**(6), 990–1004 (2019)
19. Efron, B., Hastie, T., Johnstone, I., et al.: Least angle regression. Ann. Stat. **32**(2), 407–451 (2004)

20. Wen, C., Zhang, A., Quan, S., et al.: BeSS: an R package for best subset selection in linear, logistic and CoxPH models (2017)
21. Wang, L., Jiang, S.: Novel feature selection method based on feature clustering. Appl. Res. Comput. **32**(5), 1305–1308 (2015)
22. Song, Q., Ni, J., Wang, G.: A fast clustering-based feature subset selection algorithm for high-dimensional data. IEEE Trans. Knowl. Data Eng. **25**(1), 1–14 (2013)

# Algorithms and Hardness Results for the Maximum Balanced Connected Subgraph Problem

Yasuaki Kobayashi[✉], Kensuke Kojima, Norihide Matsubara, Taiga Sone, and Akihiro Yamamoto

Kyoto University, Yoshida-honmachi, Sakyo-ku, Kyoto 606-8501, Japan
kobayashi@iip.ist.i.kyoto-u.ac.jp

**Abstract.** The Balanced Connected Subgraph problem (BCS) was recently introduced by Bhore et al. (CALDAM 2019). In this problem, we are given a graph $G$ whose vertices are colored by red or blue. The goal is to find a maximum connected subgraph of $G$ having the same number of blue vertices and red vertices. They showed that this problem is NP-hard even on planar graphs, bipartite graphs, and chordal graphs. They also gave some positive results: BCS can be solved in $O(n^3)$ time for trees and $O(n + m)$ time for split graphs and properly colored bipartite graphs, where $n$ is the number of vertices and $m$ is the number of edges. In this paper, we show that BCS can be solved in $O(n^2)$ time for trees and $O(n^3)$ time for interval graphs. The former result can be extended to bounded treewidth graphs. We also consider a weighted version of BCS (WBCS). We prove that this variant is weakly NP-hard even on star graphs and strongly NP-hard even on split graphs and properly colored bipartite graphs, whereas the unweighted counterpart is tractable on those graph classes. Finally, we consider an exact exponential-time algorithm for general graphs. We show that BCS can be solved in $2^{n/2}n^{O(1)}$ time. This algorithm is based on a variant of Dreyfus-Wagner algorithm for the Steiner tree problem.

**Keywords:** Balanced connected subgraph · Exact exponential time algorithm · Interval graph · Tree · Treewidth

## 1 Introduction

*Fairness* is one of the most important concepts in recent machine learning studies and numerous researches concerning "fair solutions" have been done so far such as fair bandit problem [16], fair clustering [9], fair ranking [8], and fair regression [7]. This brings us to a new question: Is it easy to find "fair solutions" in classical combinatorial optimization problems? Chierichetti et al. [10]

This work is partially supported by JSPS KAKENHI Grant Number JP17H01788 and JST CREST JPMJCR1401.

Y. Li et al. (Eds.): COCOA 2019, LNCS 11949, pp. 303–315, 2019.
https://doi.org/10.1007/978-3-030-36412-0_24

recently addressed a fair version of matroid constrained optimization problems and discussed polynomial time solvability, approximability, and hardness results for those problems.

In this paper, we study the problem of finding a "fair subgraph". Here, our goal is to find a maximum cardinality connected "fair subgraph" of a given bicolored graph. To be precise, we are given a graph $G = (B \cup R, E)$ in which the vertices in $B$ are colored by blue and those in $R$ are colored by red. We say that a subgraph is *balanced* if it contains an equal number of blue and red vertices. The objective of the problem is to find a balanced connected subgraph with the maximum number of vertices. This problem is called the balanced connected subgraph problem (BCS), recently introduced by Bhore et al. [3]. Although finding a maximum size connected subgraph is trivially solvable in linear time, they proved that BCS is NP-hard even on bipartite graphs, on chordal graphs, and on planar graphs. They also gave some positive results on some graph classes: BCS is solvable in polynomial time on trees, on split graphs, and on properly colored bipartite graphs. In particular, they gave an $O(n^3)$ time algorithm for trees, where $n$ is the number of vertices of the input tree.

BCS can be seen as a special case of the graph motif problem in the following sense. We are given a vertex (multi)colored graph $G = (V, E, c)$ with coloring function $c : V \rightarrow \{1, 2, \ldots, q\}$ and a multiset $M$ of colors $\{1, 2, \ldots, q\}$. The objective of the graph motif problem is to find a connected subgraph $H$ of $G$ that agrees with $M$: the multiset $c(H) = \{c(v) : v \in V(H)\}$ coincides with $M$. If $M$ is given as a set of $k/2$ red colors and $k/2$ blue colors, a feasible solution of the graph motif problem is a balanced connected subgraph of $k$ vertices. Björklund et al. [1] proved that there is an $O^*(2^{|M|})$ time randomized algorithm for the graph motif problem, where the $O^*$ notation suppresses the polynomial factor in $n$. This allows us to find a balanced connected subgraph of $k$ vertices in time $O^*(2^k)$ and hence BCS can be solved in $\max\{O^*(2^{0.773n}), O^*(2^{H(1-0.773)n})\} \subseteq O(1.709^n)$ time by using this $O^*(2^k)$ time algorithm for $k \leq 0.773n$ or by guessing the complement of an optimal solution for otherwise, where $H(x) = -x \log_2 x - (1 - x) \log_2 (1 - x)$ is the binary entropy function.

In this paper, we improve the previous running time $O(n^3)$ to $O(n^2)$ for trees and also give a polynomial time algorithm for interval graphs, which is in sharp contrast with the hardness result for chordal graphs. The algorithm for trees can be extended to bounded treewidth graphs. These results are given in Sects. 3 and 4. For general graphs, we show in Sect. 6 that BCS can be solved in $O^*(2^{n/2}) = O(1.415^n)$ time. The idea of this exponential-time algorithm is to exploit the Dreyfus-Wagner algorithm [14] for the Steiner tree problem. Let $R$ be the set of red vertices of $G$. Then, for each $S \subseteq R$ and for each $v$ in $G$, we first compute a tree $T$ that contains all the vertices $S \cup \{v\}$ but excludes all the vertices $R \setminus (S \cup \{v\})$. This can be done in $O^*(2^{|R|})$ time by the Dreyfus-Wagner algorithm and its improvement due to Björklund et al. [2]. Once we have such a tree for each $S$ and $v$, we can in linear time compute a balanced connected subgraph $H$ with $V(H) \cap R = S$. We also consider a weighted counterpart of BCS, namely WBCS: the input is vertex-weighted bicolored graph and the goal is to find a maximum weight connected subgraph $H$ in which the total weights of red vertices and of blue vertices are equal. If every vertex has a unit weight, the problem exactly corresponds to the normal BCS, and hence the hardness results for BCS on bipartite, chordal, and planar graphs also hold for WBCS. In contrast to the unweighted case, WBCS is particularly hard. In Sect. 5, we show that WBCS is (weakly) NP-hard even on properly colored star graphs and strongly NP-hard even on split and properly colored bipartite graphs. The hardness result for stars is best possible in the sense that WBCS on trees can be solved in pseudo-polynomial time.

## 2 Preliminaries

Throughout the paper, all the graphs are simple and undirected. Let $G$ be a graph. We denote by $V(G)$ and $E(G)$ the set of vertices and edges in $G$, respectively. We use $n$ to denote the number of vertices of the input graph. We say that a vertex set $U \subseteq V(G)$ is *connected* if its induced subgraph $G[U]$ is connected. A graph is *bicolored* if every vertex is colored by blue or red. Note that this coloring is not necessarily proper, that is, there may be adjacent vertices having the same color. We denote by $B$ (resp. $R$) the set of blue (resp. red) vertices of the input graph. The problems we consider are as follows.

**The Balanced Connected Subgraph Problem (BCS)**
**Input:** A bicolored graph $G = (B \cup R, E)$.
**Output:** A maximum size connected induced subgraph $H$ of $G$ such that $|V(H) \cap B| = |V(H) \cap R|$.

**The Weighted Balanced Connected Subgraph Problem (WBCS)**
**Input:** A bicolored vertex weighted graph $G = (B \cup R, E, w)$, $w : B \cup R \to \mathbb{N}$.
**Output:** A maximum weight connected induced subgraph $H$ of $G$ such that

$$\sum_{v \in V(H) \cap B} w(v) = \sum_{v \in V(H) \cap R} w(v).$$

Here, the size and the weight of a subgraph are measured by the number of vertices and the sum of the weight of vertices in the subgraph, respectively.

## 3 Trees and Bounded Treewidth Graphs

This section is devoted to showing that BCS can be solved in $O(n^2)$ time for trees, which improves upon the previous running time $O(n^3)$ of [3]. We also give an algorithm for BCS on bounded treewidth graphs whose running time is $O(n^2)$ as well.

The essential idea behind our algorithm is the same as one in [3]. Let $T$ be a bicolored rooted tree. For each $v \in V(T)$, we denote by $T_v$ the subtree of $T$ rooted at $v$. For the sake of simplicity, we convert the input tree $T$ into a rooted

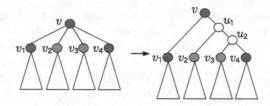

**Fig. 1.** Binarizing a degree-$p$ vertex with $p > 2$.

binary tree by adding uncolored vertices as follows. For each $v \in V(T)$ having $p > 2$ children $v_1, v_2, \ldots, v_p$, we introduce a path of $p-2$ vertices $u_1, u_2, \ldots, u_{p-2}$ that are all uncolored and make $T$ a rooted binary tree $T'$ as in Fig. 1.

We also assume that each internal node has exactly two children by appropriately adding uncolored children. This conversion can be done in $O(n)$ time. It is not hard to see that $T$ has a balanced connected subgraph of size $2k$ if and only if $T'$ has a connected subgraph with $k$ blue vertices and $k$ red vertices. Moreover, $T'$ has $O(n)$ vertices. Thus, in the following, we seek a connected subgraph with $k$ blue vertices and $k$ red vertices, where $k$ is as large as possible. For each $v \in V(T'_v)$ and each integer $-|V(T'_v)| \le d \le |V(T'_v)|$, we say that $S \subseteq V(T')$ is *feasible for* $(v, d)$ if it satisfies

- if $S$ is not empty, $S$ must contain $v$ and
- $|S \cap B| - |S \cap R| = d$.

We denote by $\mathrm{bcs}(v, d)$ the maximum size of $S$ that is feasible for $(v, d)$, where the size is measured by the number of colored vertices only. Let us note that for any $v \in V(T')$, the empty set is feasible for $(v, d)$ when $d = 0$. Given this, our goal is to compute $\mathrm{bcs}(v, d)$ for all $v$ and $d$.

Let $f : V(T') \to \{-1, 0, 1\}$ be a function, where $f(v) = 1$ if $v$ is a blue vertex, $f(v) = -1$ if $v$ is a red vertex, and $f(v) = 0$ if $v$ is an uncolored vertex. Suppose $v$ is a leaf of $T'$. Then, $\mathrm{bcs}(v, f(v)) = |f(v)|$, $\mathrm{bcs}(v, 0) = 0$, and $\mathrm{bcs}(v, d) = -\infty$ for $d \notin \{0, f(v)\}$. Suppose otherwise that $v$ is an internal node. Let $v_l$ and $v_r$ be the children of $v$. Observe that every feasible solution for $(v, d)$ can be split by $v$ into two feasible solutions for $(v_l, d_l)$ and $(v_r, d_r)$ for some $d_l$ and $d_r$. Conversely, for every pair of feasible solutions for $(v_l, d_l)$ and $(v_r, d_r)$, we can construct a feasible solution for $(v, d_l + d_r + f(v))$. Thus, we have the following straightforward lemma.

**Lemma 1.** *Let $d$ be an integer with $-|V(T'_v)| \le d \le |V(T'_v)|$. Then,*

$$\mathrm{bcs}(v, d) = \max_{\substack{d_l, d_r \\ d_l + d_r + f(v) = d}} (\mathrm{bcs}(v_l, d_l) + \mathrm{bcs}(v_r, d_r)) + |f(v)|,$$

*where the maximum is taken among all pairs $d_l, d_r$ with $d_l + d_r + f(v) = d$.*

The running time of evaluating this recurrence may be estimated to be $O(n^3)$ in total to compute $\mathrm{bcs}(v, d)$ for all $v \in V(T')$ and $-|V(T'_v)| \le d \le |V(T'_v)|$

since there are $O(n^2)$ subproblems and solving each subproblem may take $O(n)$. However, for a node $v$ with two children $v_l$ and $v_r$ the evaluation can be done in $O(|V(T'_{v_l})||V(T'_{v_r})|)$ time in total for all $d$ by simply joining all the pairs bcs$(v_l, d_l)$ and bcs$(v_r, d_r)$. Therefore, the total running time is

$$\sum_{v \in V(T')} O(|V(T'_{v_l})||V(T'_{v_r})|) = O(n^2).$$

This upper bound follows from the fact that the left-hand side can be seen as counting the number of edges in the complete graph on $V(T'_v)$.

**Theorem 1.** *BCS on trees can be solved in $O(n^2)$ time.*

This algorithm can be extended for bounded treewidth graphs. *Treewidth* is a well-known graph invariant, measuring "tree-likeness" of graphs. A *tree decomposition* of a graph $G = (V, E)$ is a rooted tree $T$ that satisfies the following properties: (1) for each $v \in V(T)$, some vertex set $X_v$, called a *bag*, is assigned and $\bigcup_{v \in V(T)} X_v = V$; (2) for each $e \in E$, there is $v \in V(T)$ such that $e \subseteq X_v$; (3) for each $w \in V$, the set of nodes $v \in V(T)$ containing $w$ (i.e. $w \in X_v$) induces a subtree of $T$. The *width* of $T$ is the maximum size of its bag minus one. The *treewidth* of $G$, denoted by tw$(G)$, is the minimum integer $k$ such that $G$ has a tree decomposition of width $k$.

The algorithm is quite similar to dynamic programming algorithms based on tree decompositions for connectivity problems [13], such as the Steiner tree problem and the Hamiltonian cycle problem. It is worth noting that the property of being balanced may not be able to be expressed by a formula in the Monadic Second Order Logic (MSO) with bounded length. Thus, we cannot directly apply the famous Courcelle's theorem [11,12] to our problem.

Here, we only sketch an overview of the algorithm and the proof is almost the same as those for connectivity problems. Let $T$ be a tree decomposition of $G$ whose width is $O(\text{tw}(G))$. Such a decomposition can be obtained in $2^{O(\text{tw}(G))} n$ time by the algorithm of Bodlaender et al. [5]. We can assume that $T$ is rooted. For each bag $X$ of $T$, we denote by $T_X$ the subtree rooted at $X$ and by $V_X$ the set of vertices appeared in some bag of $T_X$. For each bag $X$ of $T$, integer $d$ with $-|V_X| \leq d \leq |V_X|$, $S \subseteq X$, and a partition $\mathcal{S}$ of $S$, we compute the maximum size bcs$(X, d, S, \mathcal{S})$ of a set of vertices $U \subseteq V_X$ such that $|U \cap B| - |U \cap R| = d$, $X \cap U = S$, and $u, v \in S$ is connected in $U$ if and only if $u$ and $v$ belong to the same block in the partition $\mathcal{S}$. We can compute bcs$(X, d, S, \mathcal{S})$ guided by the tree decomposition in $2^{O(\text{tw}(G) \log \text{tw}(G))} n^2$ time. To improve the running time, we can apply the rank-based approach of Bodlaender et al. [5] to this dynamic programming in the same way as the Steiner tree problem. The running time is still quadratic in $n$ but the exponential part can be improved to $2^{O(\text{tw}(G))}$.

**Theorem 2.** *BCS can be solved in $2^{O(\text{tw}(G))} n^2$ time.*

We can extend the algorithm for trees to the weighted case, namely WBCS. For a tree $T$, instead of computing bcs$(v, d)$, we compute wbcs$(v, d)$; For $v \in$

$V(T)$ and for $d$ with $-\sum_{u \in V(T_v)} w(u) \le d \le \sum_{u \in V(T_v)} w(u)$, wbcs$(v, d)$ is the maximum total weight of a subtree $T_v'$ of $T_v$ that contains $v$ and satisfies $\sum_{u \in V(T_v') \cap B} w(u) - \sum_{w \in V(T_v') \cap R} w(u) = d$. The algorithm itself is almost the same with the previous one, but the running time analysis is slightly different. Let $W$ be the total weight of the vertices of $T$. The straightforward evaluation is that for each $v \in V(T)$, the values wbcs$(v, *)$ are computed by a dynamic programming algorithm, which runs in $O(p_v W^2)$ time, where $p_v$ is the number of children of $v$. Therefore, the overall running time is upper bounded by $O(nW^2)$. To improve the quadratic dependency of $W$, we can exploit the heavy-light recursive dynamic programming technique [17]. They proved that, given a tree whose vertex contains an item and each item has a weight and a value, the problem, called *tree constrained knapsack problems*, of maximizing the total value of items that induces a subtree subject to the condition that the total weight is upper bounded by a given budget can be solved in $O(n^{\log 3} W) = O(n^{1.585} W)$ time, where $W$ is the total weight of items. WBCS can be seen as this tree constrained knapsack problem and then almost the same algorithm works as well. Therefore, WBCS can be solved in $O(n^{1.585} W)$ time.

**Theorem 3.** *WBCS on trees can be solved in* $\min\{O(nW^2), O(n^{1.585} W)\}$ *time, where $W$ is the total weight of the vertices.*

## 4   Interval Graphs

In this section, we show that BCS can be solved in $O(n^3)$ time on interval graphs. Very recently, another polynomial time algorithm for interval graphs has been developed by [4].

A graph $G = (V, E)$ is *interval* if it has an interval representation: an *interval representation* of $G$ is a set of intervals that corresponds to its vertex set $V$, such that two vertices $u, v \in V$ are adjacent to each other in $G$ if and only if the corresponding intervals have a non-empty intersection. We denote by $I_v$ the interval corresponding to vertex $v$ and by $l_v$ and $r_v$ the left and right end points of $I_v$, respectively. Hence, in what follows, we do not distinguish between vertices and intervals and interchangeably use them. Given an interval graph, we can compute an interval representation in linear time [6]. Moreover, we can assume that, in the interval representation, every end point of intervals has a unique integer coordinate between 1 and $2n$.

First, we sort the input intervals in ascending order of their left end points, that is, for any $I_{v_i} = [l_i, r_i]$ and $I_{v_j} = [l_j, r_j]$ with $i < j$, it holds that $l_i < l_j$. The following lemma is crucial for our dynamic programming.

**Lemma 2.** *Let $S$ be a non-empty subset of $V$ such that $G[S]$ is connected and let $v$ be the vertex in $S$ whose index is maximized. Then $G[S \setminus \{v\}]$ is connected.*

*Proof.* Suppose for contradiction that $G[S \setminus \{v\}]$ has at least two connected components, say $C_a$ and $C_b$. An important observation is that an interval graph is connected if and only if the union of their intervals forms a single interval.

Thus, $C_a$ and $C_b$ respectively induce intervals $\mathcal{I}_a$ and $\mathcal{I}_b$ that have no intersection with each other. Without loss of generality, we may assume that $\mathcal{I}_a$ is entirely to the left of $\mathcal{I}_b$, i.e., the right end of $\mathcal{I}_a$ is strictly to the left of the left end of $\mathcal{I}_b$. Since $G[S]$ is connected, $I_v$ must have an intersection with both $\mathcal{I}_a$ and $\mathcal{I}_b$. This contradicts the fact that $l_u < l_v$ for every $u \in S \setminus \{v\}$. $\qquad\square$

For $0 \leq i \leq n$, $1 \leq k \leq 2n$, and $-n \leq d \leq n$, we say that a non-empty set $S \subseteq \{v_1, v_2, \ldots, v_i\}$ is *feasible for* $(i, k, d)$ if $S$ induces a connected subgraph of $G$, $\max_{v \in S} r_v = k$, and $|S \cap B| - |S \cap R| = d$. We denote by $\mathrm{bcs}(i, k, d)$ the maximum cardinality set that is feasible for $(i, k, d)$. We also define as $\mathrm{bcs}(i, k, d) = -\infty$ if there is no feasible subset for $(i, k, d)$. In particular, $\mathrm{bcs}(0, k, d) = -\infty$ for all $1 \leq k \leq 2n$ and $-n \leq d \leq n$. Let $f : V \to \{1, -1\}$ such that $f(v) = 1$ if and only if $v \in B$. The algorithm is based on the following recurrences.

**Lemma 3.** *For $i > 0$, $\mathrm{bcs}(i, k, d) =$*

$$
\begin{cases}
\max\{\mathrm{bcs}(i - 1, k, d - f(v_i)) + 1, \mathrm{bcs}(i - 1, k, d)\} & (k > r_i) \\
\displaystyle\max_{l_i < k' < r_i} \mathrm{bcs}(i - 1, k', d - f(v_i)) + 1 & (k = r_i \text{ and } d \neq f(v_i)) \\
\max\{1, \displaystyle\max_{l_i < k' < r_i} \mathrm{bcs}(i - 1, k', d - f(v_i)) + 1\} & (k = r_i \text{ and } d = f(v_i)) \\
\mathrm{bcs}(i - 1, k, d) & (\text{otherwise}).
\end{cases}
$$

*Proof.* We first show that the left-hand side is at most the right-hand side in all cases. Let $S \subseteq \{v_1, v_2, \ldots, v_i\}$ be feasible for $(i, k, d)$ with $|S| = \mathrm{bcs}(i, k, d)$. Suppose first that $v_i \notin S$. This implies that $k \neq r_i$. In this case, $S$ is also feasible for $(i-1, k, d)$ and hence we have $\mathrm{bcs}(i, k, d) \leq \mathrm{bcs}(i-1, k, d)$. Suppose otherwise that $v_i \in S$. By the definition of feasibility, it holds that $k \geq r_i$. If $S = \{v_i\}$, it holds that $\mathrm{bcs}(i, k, d) = 1$ if and only if $k = r_i$ and $d = f(v_i)$. Thus, the left-hand side is at most the right-hand side in the third recurrence. Let $S' = S \setminus \{v_i\}$ be non-empty. Suppose moreover that $k > r_i$, that is, $\max_{v \in S'} r_v = k$. Since $v_i$ has the maximum index among $S$, by Lemma 2, $G[S']$ is connected. Moreover, $|S' \cap R| - |S' \cap B| = |S \cap R| - |S \cap B| - f(v_i)$ holds. Therefore, $S'$ is feasible for $(i - 1, k, d - f(v_i))$, and then $\mathrm{bcs}(i, k, d) \leq \mathrm{bcs}(i - 1, k, d - f(v_i)) + 1$ follows. Finally, if $k = r_i$, it holds that $l_i < \max_{v \in S'} r_v < r_i$. This follows from the fact that $S$ is connected and there are no intervals that share end points. Similar to the case where $k > r_i$, $S'$ is feasible for $(i - 1, \max_{v \in S'} r_v, d - f(v_i))$. Hence, $\mathrm{bcs}(i, k, d) \leq \max_{l_i < k' < r_i} \mathrm{bcs}(i - 1, k', d - f(v_i)) + 1$.

For the converse direction, we assume that $k \geq r_i$ since otherwise $\mathrm{bcs}(i - 1, k, d) = \mathrm{bcs}(i, k, d)$. Suppose first that $k > r_i$. If there is a feasible set for $(i - 1, k, d)$, this is also feasible for $(i, k, d)$ and $\mathrm{bcs}(i, k, d) \geq \mathrm{bcs}(i - 1, k, d)$ follows. Let $S'$ be feasible for $(i - 1, k, d - f(v_i))$ with $|S'| = \mathrm{bcs}(i - 1, k, d - f(v_i))$. Since the intervals are sorted in their left end and $k > r_i$, $S'$ contains an interval that entirely covers the interval $I_{v_i}$. This means that $S := S' \cup \{v_i\}$ is connected and then feasible for $(i, k, d - f(v_i))$. Therefore, we have $\mathrm{bcs}(i, k, d) \geq |S'| + 1$.

Suppose otherwise that $k = r_i$. Let $S'$ be feasible for $(i - 1, k', d - f(v_i))$ with $l_i < k' < r_i$ and let $S := S' \cup \{v_i\}$. As $S'$ contains an interval whose right end is strictly in between $l_i$ and $r_i$, $S$ is connected, and hence feasible for $(i, k, d)$. Therefore, $\mathrm{bcs}(i, k, d) \geq |S'| + 1$. Finally, suppose that $k = r_i$, $d = f(v_i)$. Even if there is no feasible set for $(i - 1, k', d - f(v_i))$ with $l_i < k' < r_i$, the singleton $\{v_i\}$ can be feasible and hence $\mathrm{bcs}(i, k, d) \geq 1$.

Overall, the right-hand side is at most the left-hand side in all cases.    □

**Theorem 4.** *Given an $n$-vertex interval graph, BCS can be solved in $O(n^3)$ time.*

*Proof.* For each $i > 0$, we can evaluate the recurrence in Lemma 3 in time $O(n^2)$ using dynamic programming and hence the theorem follows.    □

As a special case of the results for interval graphs and trees, BCS on paths can be solved in linear time. Let $v_1, v_2, \ldots, v_n$ be a path in which $v_i$ and $v_{i+1}$ are adjacent to each other for $1 \leq i < n$. First, we compute $\mathrm{left}(d)$ that is the minimum integer $i$ such that $|\{v_1, v_2, \ldots, v_i\} \cap B| - |\{v_1, v_2, \ldots, v_i\} \cap R| = d$ for all $d$ with $-n \leq d \leq n$ and $\mathrm{pref}(i) = |\{v_1, v_2, \ldots, v_i\} \cap B| - |\{v_1, v_2, \ldots, v_i\} \cap R|$ for all $1 \leq i \leq n$. We can compute these values in $O(n)$ time and store them into a table. Note that $\mathrm{left}(0) = 0$ and some $\mathrm{left}(d)$ is defined to be $\infty$ when there is no $i$ satisfying the above condition. Using these value, for each $1 \leq i \leq n$, the maximum size of a balanced subpath whose rightmost index is $i$ can be computed by $i - \mathrm{left}(\mathrm{pref}(i))$. Therefore, BCS on paths can be solved in linear time.

**Theorem 5.** *Given an $n$-vertex path, BCS can be solved in $O(n)$ time.*

## 5    Hardness for WBCS

In this section, we discuss the hardness of the weighted counterpart of BCS, namely WBCS. Bhore et al. [3] proved that BCS is respectively solvable in polynomial time on trees, split graphs, and properly colored bipartite graphs. However, we prove in this section that WBCS is hard even on those graph classes.

**Theorem 6.** *WBCS is NP-hard even on properly colored star graphs.*

*Proof.* We can easily encode the subset sum problem into WBCS on star graphs. The subset sum problem asks for, given a set of integers $S = \{s_1, s_2, \ldots, s_n\}$ and an integer $B$, a subset $S' \subseteq S$ whose sum is exactly $B$, which is known to be NP-complete [15]. We take a blue vertex of weight $B$, add a red vertex of weight $s_i$ for each $s_i \in S$, and make adjacent each red vertex to the blue vertex. It is easy to see that the obtained graph has a feasible solution if and only if the instance of the subset sum problem has a feasible solution.    □

Let us note that WBCS can be solved in pseudo-polynomial time on trees using the algorithm described in Sect. 3. However, WBCS is still hard on properly colored bipartite graphs and split graphs even if the total weight is polynomially upper bounded.

**Theorem 7.** *WBCS is strongly NP-hard even on properly colored bipartite graphs.*

*Proof.* The reduction is performed from the Exact 3-Cover problem, where given a finite set $E$ and a collection of three-element subsets $\mathcal{F} = \{S_1, S_2, \ldots, S_n\}$ of $E$, the goal is to find a subcollection $\mathcal{F}' \subseteq \mathcal{F}$ such that $\mathcal{F}'$ is mutually disjoint and entirely covers $E$. This problem is known to be NP-complete [15].

For an instance $(E, \mathcal{F})$ of the Exact 3-Cover problem, we construct a bipartite graph $G = (V_E \cup V_{\mathcal{F}} \cup \{w\}, E_{\mathcal{F}} \cup E_w)$ as: $V_E = \{v_e : e \in E\}$, $V_{\mathcal{F}} = \{V_S : S \in \mathcal{F}\}$, $E_{\mathcal{F}} = \{\{v_e, v_S\} : e \in E, S \in \mathcal{F}, v_e \in S\}$, and $E_w = \{\{w, v_S\} : S \in \mathcal{F}\}$. We color the vertices of $V_{\mathcal{F}}$ with red and the other vertices with blue. Indeed, the graph obtained is bipartite and properly colored. We may assume that $n = 3k$ for some integer $k$. We assign weight one to each $v_e \in V_E$, weight $n^2$ to each $v_S \in V_{\mathcal{F}}$, and weight $k(n^2 - 3)$ to $w$. In the following, we show that $\mathcal{F}$ has a solution if and only if $G$ has a solution of total weight at least $2kn^2$.

Let $\mathcal{F}' \subseteq \mathcal{F}$ be a solution of Exact 3-Cover. We choose $w$, all the vertices of $V_E$, and $v_S$ for each $S \in \mathcal{F}$. Clearly, the chosen vertices have total weight $2kn^2$. As $\mathcal{F}'$ covers $E$, every vertex in $V_E$ is adjacent to some $v_S$, which is chosen as our solution. Moreover, every vertex in $V_{\mathcal{F}}$ is adjacent to $w$. This implies that the chosen vertices are connected. Therefore, $G$ has a feasible solution of total weight at least $2kn^2$.

Conversely, let $U \subseteq V_E \cup V_{\mathcal{F}} \cup \{w\}$ be connected in $G$ with total weight at least $2kn^2$. Since the total weight of the blue vertices in $G$ is $kn^2$, we can assume that the total weight of $U$ is exactly $2kn^2$. This means that $U$ contains exactly $k$ vertices of $V_{\mathcal{F}}$. Let $\mathcal{F}' \subseteq \mathcal{F}$ be the subsets corresponding to $U \cap V_{\mathcal{F}}$. We claim that $\mathcal{F}'$ is a solution of Exact 3-Cover. To see this, suppose that $\mathcal{F}'$ does not cover $e \in E$. Since $U$ is connected, $v_e$ has a neighbor $v_S$ in $U \cap V_{\mathcal{F}}$, contradicting that $e$ is not covered by $\mathcal{F}'$. □

**Theorem 8.** *WBCS is strongly NP-hard even on split graphs.*

*Proof.* Recall that a graph is *split* if the vertex set can be partitioned into a clique and an independent set. The proof is almost the same with Theorem 7. In the proof of Theorem 7, we construct a bipartite graph $G$ that has a solution of total weight $2kn^2$ if and only if the instance of the Exact 3-Cover problem has a solution. We construct a split graph $G'$ from $G$ by adding an edge for each pair of vertices of $V_{\mathcal{F}}$. Analogously, we can show that $G'$ has a solution of total weight $2kn^2$ if and only if the instance of the Exact 3-Cover problem has a solution. □

# 6   General Graphs

Since BCS is NP-hard [3], efficient algorithms for general graphs are unlikely to exist. From the viewpoint of exact exponential-time algorithms, the problem can be solved in time $O^*(1.709^n)$ using the algorithm due to Björklund et al. [1], discussed in Sect. 1. In this section, we improve this running time to $O^*(2^{n/2})$ by modifying the well-known Dreyfus-Wagner algorithm for the minimum Steiner tree problem [14].

Before describing our algorithm, we briefly sketch the Dreyfus-Wagner algorithm and its improvement by Björklund et al. [2]. The minimum Steiner tree problem asks for, given a graph $G = (V, E)$ and a terminal set $T \subseteq V$, a connected subgraph of $G$ that contains all the vertices of $T$ having the least number of edges. The Dreyfus-Wagner algorithm solves the minimum Steiner tree problem in time $O^*(3^{|T|})$ by dynamic programming. For $S \subseteq T$ and $v \in V$, we denote by $\text{opt}(S, v)$ the minimum number of edges in a connected subgraph of $G$ that contains $S \cup \{v\}$. Assume that $|S \cup \{v\}| \geq 3$ as otherwise, $\text{opt}(S, v)$ can be computed in polynomial time. Let $F$ be a connected subgraph that contains $S \cup \{v\}$ with $|E(F)| = \text{opt}(S, v)$. Note that $F$ must be a tree as otherwise we can delete at least one edge from $F$ without being disconnected. A key observation for applying the below algorithm is that every leaf of $F$ belongs to $S \cup \{v\}$. This enables us to decompose $F$ into (at most) three parts. Suppose first that $v$ is a leaf of $F$. Then, there is $w \in F$ such that the edge set of $F$ can be partitioned into three edge disjoint subtrees $F_1$, $F_2$, and $F_3$: $F_1$ is a shortest path between $v$ and $w$, $F_2$ and $F_3$ induce a non-trivial partition of $S \cup \{w\}$, that is, $V(F_2) \cap (S \cup \{w\})$ and $V(F_3) \cap (S \cup \{w\})$ are both non-empty. Suppose otherwise that $v$ is an interval vertex of $F$. We can also the edges of $F$ partition into two edge disjoint subtrees $F_1$ and $F_2$ such that $F_1$ contains $S' \cup \{w\}$ and $F_2$ contains $(S \cup \{w\}) \setminus S'$ for some non-empty proper subset $S'$ of $S$. This leads to the following recurrence.

$$\text{opt}(S, v) = \min_{w \in V} \left\{ d(v, w) + \min_{\substack{S' \subset S \\ S' \neq \emptyset}} (\text{opt}(S', w) + \text{opt}(S \setminus S', w)) \right\}.$$

Note that if $v$ is an internal vertex, the minimum is attained when $v = w$ in the above recurrence. A naive evaluation of this recurrence takes $O^*(3^{|T|})$ time in total. Björklund et al. [2] proposed a fast evaluation technique for the above recurrence known as the *fast subset convolution*, described in Theorem 9, which allows us to compute $\text{opt}(S, v)$ for all $S \subseteq T$ and $v \in V$ in total time $O^*(2^{|T|})$.

**Theorem 9** ([2]). *Let $U$ be a finite set. Let $M$ be a positive integer and let $f, g : 2^U \to \{0, 1, \ldots, M, \infty\}$. Then, the subset convolution over the min-sum semiring*

$$(f * g)(X) = \min_{Y \subseteq X} (f(Y) + g(X \setminus Y))$$

*can be computed in $2^{|U|}(|U| + M)^{O(1)}$ time in total for all $X \subseteq U$.*

For our problem, namely BCS, we first solve a variant of the minimum Steiner tree problem defined as follows. Let $G = (V, E)$ be the instance of BCS. Without loss of generality, we assume that $|R| \leq |B|$. For $S \subseteq R$ and $v \in V \setminus S$, we compute $\text{opt}'(S, v)$ the minimum number of edges of a tree connecting all the vertices of $S \cup \{v\}$ and *excluding any vertex of $R \setminus (S \cup \{v\})$*. The recurrence for $\text{opt}'$ is quite similar to one for the ordinary minimum Steiner tree problem, but an essential difference from the above recurrence is that the shortest path between $v$ and $w$ does not contain any red vertex other than $S \cup \{v\}$.

**Lemma 4.** *For $S \subseteq R$ and for $v \in V \setminus (R \setminus S)$,*

$$\text{opt}'(S, v) = \min_{w \in V \setminus (R \setminus (S \cup \{v\}))} \left\{ d'(v, w) + \min_{\substack{S' \subset S \\ S' \neq \emptyset}} (\text{opt}'(S', w) + \text{opt}'(S \setminus S', w)) \right\},$$

*where $d'(v, w)$ is the number of edges in a shortest path between $v$ and $w$ excluding red vertices except for its end vertices. If there is no such a path, $d'(v, w)$ is defined to be $\infty$.*

*Proof.* The idea of the proof is analogous to the Dreyfus-Wagner algorithm for the ordinary Steiner tree problem. Suppose that $F$ is an optimal solution that contains every vertex in $S \cup \{v\}$ and does not contain every vertex in $R \setminus (S \cup \{v\})$. Similarly, we can assume that every leaf of $F$ belongs to $S \cup \{v\}$ as otherwise such a leaf can be deleted without losing feasibility. Then, the edge set of $F$ can be partitioned into three edge disjoint subtrees $F_1$, $F_2$, and $F_3$ such that $F_1$ is a shortest path between $v$ and $w$, $F_2$ and $F_3$ induces a non-trivial bipartition of $S$. The only difference from the ordinary Steiner tree problem is that $F_1$, $F_2$ and $F_3$ should avoid any irrelevant red vertex. This can be done since $F_1$ does avoid such a vertex. ∎

Similar to the normal Steiner tree problem, we can improve the naive running time $O^*(3^{|R|})$ to $O^*(2^{|R|})$ by the subset convolution algorithm in Theorem 9.

Now, we are ready to describe the final part of our algorithm for BCS. We have already know $\text{opt}'(S, v)$ for all $S \subseteq V$ and $v \in V \setminus (R \setminus S)$. Suppose $\text{opt}'(S, v) < \infty$. Let $F$ be a tree with $|E(F)| = \text{opt}'(S, v)$ such that $V(F) \cap R = (S \cup \{v\}) \cap R$. Such a tree can be obtained in polynomial time using the standard traceback technique. Since $F$ is a tree, we know that $|V(F)| = \text{opt}'(S, v) + 1$. Let $k$ be the number of red vertices in $F$. If $F$ contains more than $k$ blue vertices, we can immediately conclude that there is no balanced connected subgraph $H$ with $V(H) \cap R = V(F) \cap R$. Suppose otherwise. The following lemma ensures that an optimal solution of BCS can be computed from some Steiner tree.

**Lemma 5.** *Let $R_F = V(F) \cap R$. If there is a balanced connected subgraph $H$ with $V(H) \cap R = R_F$, then there is a balanced connected subgraph $H'$ with $V(S)' \cap R = R_F$ such that $F$ is a subtree of $H'$. Moreover, such a subgraph $H'$ can be constructed in linear time from $F$.*

*Proof.* We prove this lemma by giving a linear time algorithm that constructs a balanced connected subgraph $H'$ when given $F$ as in the statement of the lemma. Suppose that there is a balanced connected subgraph $H$ with $V(H) \cap R = R_F$. Since $F$ is a minimum Steiner tree with $V(F) \cap R = R_F$, the number of blue vertices in $F$ is not larger than that of red vertices. We greedily add a blue vertex that has a neighbor in $F$ to $F$ as long as it is not balanced. We claim that it is

Y. Kobayashi et al.

possible to construct a balanced connected subgraph using this procedure. Let $H'$ be a maximal graph that is obtained by the above procedure. Suppose for contradiction that $|V(H')| < |V(H)|$. Let $v \in (V(H) \setminus V(H')) \cap B$ be a blue vertex and let $r$ be a red vertex in $F$. We choose $v$ and $r$ in such a way that the distance between $v$ and $r$ in $H$ is as small as possible. Consider a shortest path between $v$ and $r$ in $H$. By the choice of $v$ and $r$, there are no red vertices other than $r$ and no blue vertices of $V(H) \setminus V(H')$ on the path. Since $v \notin V(H')$ and $r \in V(H)$, there is a vertex $v'$ on the path such that $v'$ has a neighbor in $V(H')$ but not contained in $H'$, which contradicts the maximality of $V(H')$.

This greedy algorithm runs in linear time and hence the lemma follows. □

Overall, we have the following running time.

**Theorem 10.** *BCS can be solved in $O^*(2^{n/2})$ time.*

# References

1. Björklund, A., Kaski, P., Kowalik, Ł.: Constrained multilinear detection and generalized graph motifs. Algorithmica **74**(2), 947–967 (2016)
2. Björklund, A., Husfeldt, T., Kaski, P., Koivisto, M.: Fourier meets möbius: fast subset convolution. In: Proceedings of STOC 2007, pp. 67–74 (2007)
3. Bhore, S., Chakraborty, S., Jana, S., Mitchell, J.S.B., Pandit, S., Roy, S.: The balanced connected subgraph problem. In: Pal, S.P., Vijayakumar, A. (eds.) CALDAM 2019. LNCS, vol. 11394, pp. 201–215. Springer, Cham (2019). https://doi.org/10.1007/978-3-030-11509-8_17
4. Bhore, S., Jana, S., Mitchell, J. S., Pandit, S., Roy, S.: Balanced connected subgraph problem in geometric intersection graphs. ArXiv:1909.03872 (2019)
5. Bodlaender, H.L., Cygan, M., Kratsch, S., Nederlof, J.: Deterministic single exponential time algorithms for connectivity problems parameterized by treewidth. Inf. Comput. **243**, 86–111 (2015)
6. Booth, K.S., Lueker, G.S.: Testing for the consecutive ones property, interval graphs, and graph planarity using PQ-tree algorithms. J. Comput. Syst. Sci. **13**(3), 335–379 (1976)
7. Calders, T., Karim, A., Kamiran, F., Ali, W., Zhang, X.: Controlling attribute effect in linear regression. In: Proceedings of ICDM 2013, pp. 71–80. IEEE (2013)
8. Celis, L. E., Straszak, D., Vishnoi, N. K.: Ranking with fairness constraints. In: Proceedings of ICALP 2018, LIPIcs, pp. 28:1–28:15 (2018)
9. Chierichetti, F., Kumar, R., Lattanzi, S., Vassilvitskii, S.: Fair clustering through fairlets. In: Proceedings of NIPS 2017, pp. 5029–5037 (2017)
10. Chierichetti, F., Kumar, R., Lattanzi, S., Vassilvitskii, S.: Matroids, matchings, and fairness. In: Proceedings of AISTATS 2019, pp. 2212–2220 (2019)
11. Courcelle, B.: The monadic second-order logic of graphs. I. Recognizable sets of finite graphs. Inf. Comput. **85**(1), 12–75 (1990)
12. Courcelle, B., Mosbah, M.: Monadic second-order evaluations on tree-decomposable graphs. Theor. Comput. Sci. **109**(1), 49–82 (1993)
13. Cygan, M., et al.: Parameterized Algorithms. Springer, Cham (2015). https://doi.org/10.1007/978-3-319-21275-3
14. Dreyfus, S.E., Wagner, R.A.: The Steiner problems. Network **1**(3), 195–207 (1971)

15. Garey, M.R., Johnson, D.S.: Computers and Intractability: A Guide to the Theory of NP-Completeness. W. H. Freeman & Co., New York (1979)
16. Joseph, M., Kearns, M., Morgenstern, J.H., Roth, A.: Fairness in learning: classic and contextual bandits. In: Proceedings of NIPS 2016, pp. 325–333 (2016)
17. Kumabe, S., Maehara, T., Sin'ya, R.: Linear pseudo-polynomial factor algorithm for automaton constrained tree knapsack problem. In: Das, G.K., Mandal, P.S., Mukhopadhyaya, K., Nakano, S. (eds.) WALCOM 2019. LNCS, vol. 11355, pp. 248–260. Springer, Cham (2019). https://doi.org/10.1007/978-3-030-10564-8_20

# A Fast Exact Algorithm for Airplane Refueling Problem

Jianshu Li[1,2]($\boxtimes$), Xiaoyin Hu[1,2], Junjie Luo[1,2], and Jinchuan Cui[1]

[1] Academy of Mathematics and Systems Science, Chinese Academy of Sciences,
Beijing 100190, China
{ljs,hxy,luojunjie,cjc}@amss.ac.cn
[2] School of Mathematical Sciences, University of Chinese Academy of Sciences,
Beijing 100049, China

**Abstract.** We consider the airplane refueling problem, where we have a fleet of airplanes that can refuel each other. Each airplane is characterized by specific fuel tank volume and fuel consumption rate, and the goal is to find a drop out order of airplanes that last airplane in the air can reach as far as possible. This problem is equivalent to the scheduling problem $1||\sum w_j(-\frac{1}{C_j})$. Based on the dominance properties among jobs, we reveal some structural properties of the problem and propose a recursive algorithm to solve the problem exactly. The running time of our algorithm is directly related to the number of schedules that do not violate the dominance properties. An experimental study shows our algorithm outperforms state of the art exact algorithms and is efficient on larger instances.

**Keywords:** Scheduling · Dominance properties · Branch and bound

## 1 Introduction

The *airplane refueling problem*, originally introduced by Gamow and Stern [8], is a special case of single machine scheduling problem. Consider a fleet of several airplanes with certain fuel tank volume and fuel consumption rate. Suppose all airplanes travel at the same speed and they can refuel each other in the air. An airplane will drop out of the fleet if it has given out all its fuel to other airplanes. The goal is to find a drop out order so that the last airplane can reach as far as possible.

**Problem Definition.** In the original description of airplane refueling problem, all airplanes are defaulted to be identical. Woeginger [1] generalized this problem that each airplane $j$ can have arbitrary tank volume $w_j$ and consumption rate $p_j$. Denote the set of all airplanes by $J$, a solution is a drop out order $\sigma$ :

This work is supported by Key Laboratory of Management, Decision and Information Systems, CAS.

$\{1, 2, \ldots, |J|\} \mapsto J$, where $\sigma(i) = j$ if airplane $j$ is the $i^{\text{th}}$ airplane to leave the fleet. As a result, the objective value of the drop out order $\sigma$ is:

$$\sum_{j=1}^{n} \left( w_{\sigma(j)} / \sum_{k=j}^{n} p_{\sigma(k)} \right).$$

As pointed out by Vásquez [18], we can rephrase the problem as a single machine scheduling problem, which is equivalent to finding a permutation $\pi$ (the reverse of $\sigma$) that minimizes:

$$\sum_{j=1}^{n} \left( - w_{\pi(j)} / \sum_{k=j}^{n} p_{\pi(k)} \right) = \sum_{j=1}^{n} -w_j / C_j,$$

where $C_j$ is the completion time of job $j$, and $p_j, w_j$ correspond to its processing time and weight, respectively. This scheduling problem is specified as $1 || \sum w_j (-\frac{1}{C_j})$ using the classification scheme introduced by Graham et al. [10].

In the rest of this paper, we study the equivalent scheduling problem instead.

**Dominance Properties.** Since the computational complexity status of airplane scheduling problem is still open [1], existing algorithms find the optimal solution with branch and bound method. While making branching decisions in a branch and bound search, it would be much more useful if we know the dominance relation among jobs. For example, if we know job $i$ always precedes job $j$ in an optimal solution, we can speed up the searching process by pruning all the branches with job $i$ processed after job $j$. Let the start time of job $j$ be $t_j$, we refer to [7] for the definition of *local dominance* and *global dominance* :

- *local dominance:* Suppose job $j$ starts at time $t$ and is followed directly by job $i$ in a schedule. If exchanging the positions of jobs $i, j$ strictly improves the cost, we say that *job $i$ locally dominants job $j$ at time $t$* and denote this property by $i \prec_{l(t)} j$. If $i \prec_{l(t)} j$ holds for all $t \in [a, b]$, we denote it by $i \prec_{l[a,b]} j$.
- *global dominance:* Suppose schedule $S$ satisfies $a \le t_j \le t_i - p_j \le b$. If exchanging the positions of jobs $i, j$ strictly improves the cost, we say that *job $i$ globally dominants job $i$ in time interval $[a, b]$* and denote this property by $i \prec_{g[a,b]} j$. If it holds that $i \prec_{g[0,\infty)} j$, we denote this property by $i \prec_g j$.

We call the schedule that do not violate the dominance properties as *potential schedule*. The effect of dominance properties is to narrow the search space to the set of all potential schedules, whose cardinality is much smaller than $n!$, the number of all job permutations.

**Related Work.** Airplane refueling problem is a special case of a more general scheduling problem $1 || \sum w_j C_j^\beta$. For most problems of this form, including airplane refueling problem, no polynomial algorithm has been found. Existing methods resort to approximation algorithms or branch and bound schemes.

For approximations, several constant factor approximations and polynomial time approximation scheme (PTAS) have been devised for different cost functions [3,5,13,15]. Recently, Gamzu and Segev [9] gave the first polynomial-time approximation scheme for airplane refueling problem.

The focus of exact methods is to find stronger dominance properties. Following a long series of improvements [2,6,12,16,17], Dürr and Vásquez [7] conjectured that for all cost functions of the form $f_j(t) = w_j t^\beta, \beta > 0$ and all jobs $i, j$, $i \prec_l j$ implies $i \prec_g j$. Latter, Bansal et al. [4] confirmed this conjecture, and they also gave a counter example of the generalized conjecture that $i \prec_{l[a,b]} j$ implies $i \prec_{g[a,b]} j$. For airplane refueling problem, Vásquez [18] proved that $i \prec_{l[a,b]} j$ implies $i \prec_{g[a,b]} j$. The establish dominance properties are commonly incorporated into a branch and bound scheme, such as Algorithm $A^*$ [11], to speed up the searching process.

**Our Contribution.** Existing branch and bound algorithms search for potential schedules in a trail and error manner. Specifically, when making branching decisions, it is unknown whether current branch contains any potential schedule unless we exhaust the entire branch. So if we can prune all the branches that do not contain potential schedule, it will considerably improve the efficiency of the searching process.

In this paper we give an exact algorithm with the merit above for airplane refueling problem. Specifically, every branch explored by our algorithm is guaranteed to contain at least one potential schedule, and the time to find each potential schedule is bounded by $\mathcal{O}(n^2)$. Numerical experiments exhibit empirical advantage of our algorithm over the $A^*$ algorithms proposed by former studies, and the advantage is more significant on instances with more jobs.

The main difference between previous methods and our algorithm is that instead of branching on possible succeeding jobs, we branch on the possible start times of a certain job. To this end, we introduce a prefixing technique to determine the possible start times of a certain job in a potential schedule. In addition, the relative order of other jobs regarding that certain job is also decided. Thus, each branch divides the original problem into subproblems with fewer jobs and we can solve the problem recursively.

**Organization.** The rest of the paper is organized as follows. In Sect. 2, we introduce a new auxiliary function and give a concise representation of dominance property. Section 3 establishes some useful lemmas. We present our algorithm in Sect. 4 and experimental results in Sect. 5. Finally, Sect. 6 concludes the paper. Due to the lack of space, many proofs of results with ($\star$) are deferred to the full version of this paper [14].

## 2 Preliminaries

A *dominance rule* stipulates the local and global dominance properties among jobs. We can refer to a dominance rule by the set $R := \cup_{i,j \in J} \{r_{ij}\}$, where $r_{ij}$

represents the dominance properties between jobs $i, j$ specified by the rule. Given a dominance rule $R$, we formally define the potential schedule as follows.

**Definition 1 (Potential Schedule).** *We call $S$ a potential schedule with respect to dominance rule $R$ if for all $r_{ij} \in R$, start times of jobs $i, j$ in $S$ do not violate relation $r_{ij}$.*

**Auxiliary Function.** For each job $j$, we introduce an auxiliary function $\varphi_j(t)$, where $t \geq 0$ represents the possible start time: $\varphi_j(t) = \frac{w_j}{p_j(p_j+t)}$. We remark that $\varphi_j(t)$ is well defined since the processing time $p_j$ is positive. With the help of this auxiliary function, we can obtain a concise characterization of local precedence between two jobs.

**Lemma 1 ($\star$).** *For any two jobs $i, j$ and time $t \geq 0$, $i \prec_{l(t)} j$ if and only if $\varphi_i(t) > \varphi_j(t)$.*

**Dominance Rule.** Vásquez [18] proved the following dominance properties of airplane refueling problem.

**Theorem 1 (Vásquez).** *For all jobs $i, j$ and time points $a, b$, the dominance property $i \prec_{l[a,b]} j$ implies $i \prec_{g[a,b]} j$.*

Based on Theorem 1 and Lemma 1 we obtain the following dominance rule.

**Corollary 1.** *For any two jobs $i, j$ with $w_i > w_j$:*

1. *If $\varphi_i(t) \geq \varphi_j(t)$ for $t \in [0, \infty)$, $i \prec_g j$;*
2. *If $\varphi_j(t) \geq \varphi_i(t)$ for $t \in [0, \infty)$, $j \prec_g i$;*
3. *else $\exists\, t_{ij}^* = \frac{w_j p_i^2 - w_i p_j^2}{w_i p_j - w_j p_i} > 0$ with:*
   - *$\varphi_i(t) > \varphi_j(t)$ for $t \in [0, t_{ij}^*)$, $i \prec_{g[0,t_{ij}^*)} j$;*
   - *$\varphi_j(t) \geq \varphi_i(t)$ for $t \in [t_{ij}^*, \infty)$, $j \prec_{g[t_{ij}^*,\infty)} i$.*

Actually, this rule is equivalent to the rule given in [18], whereas we use $\varphi$ to indicate the dominance relation between any jobs $i, j$. Figure 1 gives an illustrative example of the scenario where $t_{ij}^* = \frac{w_j p_i^2 - w_i p_j^2}{w_i p_j - w_j p_i} > 0$.

In this paper, we care about potential schedules with respect to Corollary 1. We refer to these schedules as *potential schedules* for short.

## 3   Technical Lemmas

In this section, we establish several important properties concerning potential schedules.

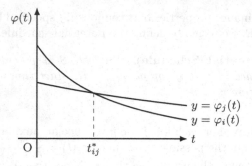

**Fig. 1.** Illustration of dominance property between job $i, j$. For $t \in [0, t^*_{ij})$, $\varphi_i(t) > \varphi_j(t)$, $i \prec_{g[0, t^*_{ij})} j$; for $t \in [t^*_{ij}, \infty)$, $\varphi_i(t) \leq \varphi_j(t)$, $j \prec_{g[t^*_{ij}, \infty)} i$.

### 3.1  Relative Order Between Two Jobs

To begin with, we show that the dominance rule makes it impossible for a job to start within some time intervals in a potential schedule. Let $T := \sum_{j \in J} p_j$ be the total processing time, we have:

**Lemma 2 ($\star$).** *For two jobs $i, j \in J$ with $\varphi_i(0) > \varphi_j(0)$ and $t^*_{ij} \in (0, T)$, if in a complete schedule $S$ job $i$ starts in time interval $[t^*_{ij}, t^*_{ij} + p_j)$, then $S$ is not a potential schedule.*

According to Lemma 2, job $i$ starts either in $[0, t^*_{ij})$ or in $[t^*_{ij} + p_j, T - p_i)$ in a potential schedule. Next lemma shows that if we fix the start time of job $i$ to one of these intervals, the relative order of job $i, j$ in a potential schedule is already decided.

**Lemma 3 ($\star$).** *For two jobs $i, j \in J$ with $\varphi_i(0) > \varphi_j(0)$ and $t^*_{ij} \in (0, T)$, suppose $S$ is a schedule with $t_j \notin [t^*_{ij}, t^*_{ij} + p_j)$.*

1. *If $t_i < t^*_{ij}$, job $i$ should be processed before job $j$, otherwise $S$ is not a potential schedule;*
2. *If $t_i \geq t^*_{ij} + p_j$, job $i$ should be processed after job $j$, otherwise $S$ is not a potential schedule.*

Conditioning on the start time of job $i$, Lemmas 2 and 3 provide a complete characterization of the relative order between jobs $i, j$ in a potential schedule. See Fig. 2 for an overview of these relations.

### 3.2  Possible Start Times in a Potential Schedule

In this subsection we consider a general *sub-problem* scenario. Let $J' \subseteq J$ be a set of jobs to be processed consecutively at start time $t_o$, where $t_o \in [0, T - \sum_{j \in J'} p_j]$. For convenience we denote the completion time $t_o + \sum_{j \in J'} p_j$ by $t_e$. Notice that when $J' = J$, we will have $t_o = 0$ by definition, the setting above describes the

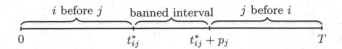

**Fig. 2.** The relative order of job $i$ and job $j$ in a potential schedule conditioned on the start time of job $i$ with $t_{ij}^* \in (0, T)$ and $\varphi_i(0) > \varphi_j(0)$. When $t_i \in [0, t_{ij}^*)$, job $i$ should precede job $j$; when $t_i \in [t_{ij}^*, t_{ij}^* + p_j)$, the resulting schedule will violate dominance property; when $t_i \in [t_{ij}^* + p_j, T)$, job $j$ should precede job $i$.

original problem. We further denote the partial schedule of $J'$ by $S'$. If $S'$ does not violate any dominance rule, we call it as a *partial potential schedule*. Now consider job $\alpha \in J'$ such that: $\alpha = \arg\max_i \varphi_i(t_o)$. When there are more than one job with the maximum $\varphi(t_o)$, the tie breaking rule is to choose the job with the maximum $\varphi(t_e)$.

Our goal is to study the possible start times of job $\alpha$ in a partial potential schedule $S'$. We start with analyzing the precedence relation of job $\alpha$ with other jobs in a partial potential schedule. To that end, we divide time interval $[t_o, t_e]$ into consecutive subintervals with the time set:

$$C_{J'}^\alpha := \{c_{\alpha j} | c_{\alpha j} = t_{\alpha j}^* + p_j, c_{\alpha j} \in (t_o, t_e), j \in J' \setminus \{\alpha\}\} \cup \{t_o, t_e\}.$$

We re-index the set $C_{J'}^\alpha$ according to their value rank, that is, if a time point ranks the $q^{\text{th}}$ in the set, it will be denoted by $c_q$. Besides, we use mapping $M_{J'}^\alpha : J' \mapsto \mathbb{Z}^+$ to maintain job information of the index. The mapping is defined as:

$$M_{J'}^\alpha(j) = \begin{cases} 1, & \text{if } j = \alpha; \\ \text{the rank of } c_{\alpha j} \text{ in } C_{J'}^\alpha, & \text{if } t_{\alpha j}^* \in (t_o, t_e); \\ |J'| + 1, & \text{if } t_{\alpha j}^* \notin (t_o, t_e). \end{cases}$$

We consider the start time of job $\alpha$ in each subinterval $[c_q, c_{q+1})$. Next lemma shows that once we fix $t_\alpha$ to a subinterval, the positions of all remaining jobs in a potential schedule relative to $\alpha$ are determined.

**Lemma 4 ($\star$).** *Suppose $S'$ is partial potential schedule that job $\alpha$ starts within time interval $[c_q, c_{q+1})$, then all $j$ with $M_{J'}^\alpha(j) \le q$ should proceed $\alpha$, and the rest jobs with $M_{J'}^\alpha(j) > q$ should come after $\alpha$.*

At last, for the possible start times of job $\alpha$ in a partial potential schedule, we have:

**Lemma 5 ($\star$).** *For every time interval $[c_q, c_{q+1})$, there is at most one possible start time for job $\alpha$ in $[c_q, c_{q+1})$. Thus, there are at most $|J'|$ possible start times of job $\alpha$ in a partial potential schedule $S'$.*

## 4   Algorithm

In this section we devise an exact algorithm for airplane refueling problem and analyze its running time.

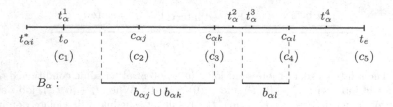

Fig. 3. Possible start times of job $\alpha$.

## 4.1   An Exact Algorithm for Air Plane Refueling Problem

We start with some notations. For each job pair $i, j$ of Lemma 2, we name the time interval $b_{ij} = [t^*_{ij}, t^*_{ij} + p_j)$ as the *banned interval* of job $i$ imposed by job $j$, and denote $B_i := \cup_{j \in J \setminus i} b_{ij}$ as the union of all banned intervals of job $i$.

Given an instance of airplane refueling problem, we find the job $\alpha$ and try to start $\alpha$ in each interval induced by $C^\alpha_{J'}$. For an interval $[c_q, c_{q+1})$ and corresponding $J_l, J_r$, the possible start time of $\alpha$ will be $t_\alpha = \sum_{j \in J_l} p_j$. If $t_\alpha \in [c_q, c_{q+1})$ and $t_\alpha \notin B_\alpha$, then we have found a potential start time of job $\alpha$.

Figure 3 provides an illustrative example of the above procedure, where $J' = \{i, j, k, l, \alpha\}$. As shown in the figure, the set $C^\alpha_{J'}$ divided time interval $[t_o, t_e)$ into 4 subintervals. Job $i$ should always behind job $\alpha$ since $t^*_{\alpha i} < t_o$. For other jobs in $J'$, condition on the start time of job $\alpha$ we have:

1. $t_\alpha \in [c_1, c_2)$: Job $\alpha$ serves as the first job and $t^1_\alpha = t_o$.
2. $t_\alpha \in [c_2, c_3)$: Job $j$ precedes job $\alpha$ in a potential schedule. As $t^2_\alpha = t_o + p_j > c_3$, job $\alpha$ can not start in this interval.
3. $t_\alpha \in [c_3, c_4)$: Job $j, k$ precedes job $\alpha$ in a potential schedule. As $t^3_\alpha = t_o + p_j + p_k$ is in banned intervals of job $\alpha$, job $\alpha$ can not start in this interval.
4. $t_\alpha \in [c_4, c_5)$: Job $j, k, l$ precedes job $\alpha$, $t^4_\alpha = t_o + p_j + p_k + p_l$.

Once we find a possible start time of job $\alpha$, we can solve $J_l$ and $J_r$ recursively with the procedure above until the subproblem has only one job. See Algorithm 1 for the pseudo code of our algorithm, where the initial inputs are the complete job set $J$ and original start time $t = 0$.

Next, we prove the correctness of Algorithm 1.

**Theorem 2.** *Algorithm 1 returns the optimal solution of airplane refueling problem.*

*Proof.* We only need to show that Algorithm 1 does not eliminate any potential schedules. While locating possible start times of job $\alpha$, our algorithm drops out schedules with $t_\alpha$ in banned intervals or $t_\alpha$ exceeds current interval. According to Lemmas 2 and 5, all the excluded schedules violate Corollary 1. Each recursive call also drops out schedules that have jobs in $J_l$ processed after job $\alpha$ or jobs in $J_r$ processed before job $\alpha$. By Lemma 3, these schedules are not potential schedules either. $\qquad \square$

---

**Algorithm 1.** Fast Schedule

---

**Input:** Job set $J$ and start time $t$.
**Output:** The optimal schedule and the optimal cost.
1: **function** FASTSCHEDULE($J, t$)
2:     **if** $J$ is empty **then**
3:         **return** $0, []$
4:     **else if** $J$ contains only one job $j$ **then**
5:         **return** $\frac{w_j}{p_j + t}, [j]$
6:     **end if**
7:     Find job $\alpha$
8:     $J_l, J_r, opt \leftarrow [], J, 0$
9:     **for** $c_q \in C_J^\alpha$ **do**
10:         Find job $i$ with $M_J^\alpha(i) = q$       ▷ $i = \alpha$ for $q = 1$, i.e. job $\alpha$ is the first job
11:         $J_l, J_r \leftarrow J_l \cup \{i\}, J_r \setminus \{i\}$       ▷ update jobs processed before and after $\alpha$
12:         **if** $c_q \le t + \sum_{i \in J_l} p_i < c_{q+1}$ & $(t + \sum_{i \in J_l} p_i) \notin B_\alpha$ **then**       ▷ new branch
13:             $opt_l, seq_l \leftarrow$ FASTSCHEDULE($J_l \setminus \{\alpha\}, t$)
14:             $opt_r, seq_r \leftarrow$ FASTSCHEDULE($J_r, t + \sum_{i \in J_l} p_i + p_\alpha$)
15:             **if** $opt_l + opt_r + \frac{w_\alpha}{p_\alpha + \sum_{i \in J_l} p_i} > opt$ **then**
16:                 $opt, seq \leftarrow (opt_l + opt_r + \frac{w_\alpha}{p_\alpha + \sum_{i \in J_l} p_i}), [seq_l, \alpha, seq_r]$
17:             **end if**
18:         **end if**
19:     **end for**
20:     **return** opt, seq
21: **end function**

---

## 4.2 Running Time of Algorithm 1

In this section we analyze the running time of Algorithm 1 with respect to the number of potential schedules. Our main result can be stated as follows.

**Theorem 3.** *Algorithm 1 finds the optimal solution of airplane refueling problem in $\mathcal{O}(n^2(\log n + K))$ time, where $K$ is the number of potential schedules with respect to Corollary 1.*

Before proving this theorem, we need to establish some properties of Algorithm 1. We start by showing that for any job set $J'$ that starts at time $t$, there is at least one partial potential schedule.

**Lemma 6 ($\star$).** *Suppose $J' \subseteq J$ is a set of job that starts at time $t$, then there exists a procedure that can find one partial potential schedule for $J'$.*

While Lemma 6 concerns about a single job set, the following lemma describes the relationship between two job sets $J_l$ and $J_r$.

**Lemma 7 ($\star$).** *Given start time $t_\alpha \in [c_q, c_{q+1})$ of job $\alpha$ with $t_\alpha \in [t_o, t_e - p_\alpha]$ and $t_\alpha \notin B_\alpha$, for any two jobs $j \in J_l$ and $k \in J_r$ with $M_J^\alpha(j) \le q < M_J^\alpha(k)$, if job $j$ is scheduled before job $\alpha$ and job $k$ is scheduled after job $\alpha$, then no matter where job $j$ and job $k$ start, dominance properties among $\alpha, j$ and $k$ will not be violated.*

Combining Lemmas 6 and 7, we can identify a strong connection between Algorithm 1 and potential schedules.

**Lemma 8** ($\star$). *Whenever Algorithm 1 adds a new branch for an instance $(J, t)$, there is at least one potential schedule on that branch.*

Now we are ready to prove Theorem 3.

*Proof (Proof of Theorem 3).* We only need to generate all $t_{ij}^*$ and sorted them once, which takes $\mathcal{O}(n^2 \log n)$ time. While finding the possible start times of $\alpha$, with the already calculated $t_{ij}^*$, we can construct the set $C_{\mathcal{J}}^\alpha$ and the mapping $M_{\mathcal{J}}^\alpha$ in $\mathcal{O}(n)$ time. The iteration over each subinterval $[c_q, c_{q+1})$ also needs at most $\mathcal{O}(n)$ time. Therefore, it takes at most $\mathcal{O}(n)$ time to find a potential start time of a job.

By Lemma 8 there exists at least one potential schedule with job $\alpha$ starting at that time. For any potential schedule there are $n$ start times to be decided, as a consequence, Algorithm 1 finds each potential schedule in at most $\mathcal{O}(n^2)$ time. To find all the potential schedules it will take $\mathcal{O}(n^2(\log n + K))$ time. $\square$

## 5   Experimental Study

We code our algorithm with Python 3 and perform the experiment on a Linux machine with one Intel Core i7-9700K @ 3.6 GHz × 8 processor and 16 GB RAM. Notice that our implementation only invokes one core at a time.

For experimental data set we adopt the method introduced by Höhn and Jacobs [12] to generate random instances. For an instance with $n$ jobs, the processing time $p_i$ of job $i$ is an integer generated from uniform distribution ranging from 1 to 100, whereas the priority weight $w_i = 2^{N(0,\sigma^2)} \cdot p_i$ with $N$ being normal distribution. Therefore, a random instance is characterized by two parameters $(n, \sigma)$. According to previous results [12,18], instances generated with smaller $\sigma$ are more likely to be harder to solve, which means we can roughly tune the hardness of instances by adjusting the value $\sigma$.

At first we compare our algorithm with the $A^*$ algorithm given by Vásquez [18]. This part of experiment is conducted on data set $S_1$, which is generated with $\sigma = 0.1$ and job size ranges from $\{10, 20, \dots, 140\}$, where for each configuration there are 50 instances. Secondly, we evaluate the empirical performance of Algorithm 1 on data set $S_2$. In detail, for each job size $n$ in $\{100, 500, 1000, 2000, 3000\}$ and for each $\sigma$ value $\{0.1, 0.101, 0.102, \dots, 1\}$, there are 5 instances, that is, $S_2$ has 4505 instances in total. At last, we generate data set $S_3$ to examine the relations of instance hardness with the number of potential schedules and the value of $\sigma$. This data set has 5 instances of 500 jobs for each $\sigma$ from $\{0.100, 0.101, 0.102, \dots, 1\}$.

**Comparison with $A^*$.** We consider the ratio between the average running time of $A^*$ and Algorithm 1 on data set $S_1$. As shown in Fig. 4, our algorithm outperforms $A^*$ on all sizes and the speed up is more significant on instances with larger size. For instances with 140 jobs, our algorithm is over 100 times faster.

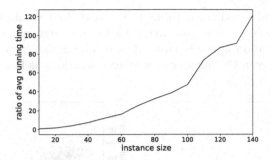

**Fig. 4.** Speed up factor on dependence of instance size, on data set $S_1$.

**Empirical Performance.** Table 1 presents the running time of Algorithm 1 on data set $S_2$. We set a 10000 seconds timeout while performing the experiment. When our algorithm does not solve all the instances within this time bound, we present the percentage of the solved instances. For all instances with less than 2000 jobs as well as most instances with 2000 and 3000 jobs, our algorithm returns the optimal solution within the time bound. While for hard instances of 2000 and 3000 jobs generated with $\sigma$ less than 0.2, our algorithm can solve 92% and 79% of the instances within the time bound respectively.

**Table 1.** Running time on data set $S_2$.

| $|J|$ | 100 | | 500 | | 1000 | | 2000 | | 3000 | |
|---|---|---|---|---|---|---|---|---|---|---|
| | avg. | std. | avg. | std. | avg. | std. | avg. | std. | avg. | std. |
| $[0.1, 0.2)$ | 0.37 | 0.16 | 62.26 | 128 | 645.4 | 2223 | 92.40% | – | 79.60% | – |
| $[0.2, 0.3)$ | 0.30 | 0.05 | 13.38 | 8.5 | 65.08 | 43.23 | 328.6 | 210.2 | 955.4 | 871.5 |
| $[0.3, 0.4)$ | 0.27 | 0.03 | 7.11 | 9.27 | 41.91 | 12.46 | 222.7 | 128.0 | 538.6 | 339.4 |
| $[0.4, 0.5)$ | 0.26 | 0.02 | 8.16 | 0.91 | 35.74 | 5.00 | 172.8 | 38.70 | 421.5 | 87.15 |
| $[0.5, 0.6)$ | 0.25 | 0.01 | 7.64 | 0.59 | 33.27 | 3.24 | 159.9 | 28.54 | 392.0 | 53.67 |
| $[0.6, 0.7)$ | 0.25 | 0.01 | 7.37 | 0.38 | 31.84 | 2.74 | 151.0 | 14.37 | 367.3 | 41.37 |
| $[0.7, 0.8)$ | 0.25 | 0.01 | 7.18 | 0.35 | 30.52 | 1.52 | 142.1 | 12.97 | 349.4 | 22.09 |
| $[0.8, 0.9)$ | 0.24 | 0.01 | 7.07 | 0.32 | 30.10 | 1.62 | 144.8 | 9.07 | 341.4 | 22.79 |
| $[0.9, 1.0]$ | 0.24 | 0.01 | 6.94 | 0.23 | 29.28 | 1.00 | 141.5 | 7.41 | 327.3 | 16.93 |

**Instance Hardness.** Figure 5 depicts relations of different hardness indicators. The chart on the left shows the number of potential schedules and the running time of Algorithm 1 on data set $S_3$. Since both variables cover a large range, we present the figure as a log-log plot. As indicated by Theorem 3, there is a strong correlation between the number of potential schedules and the running time of Algorithm 1. Therefore, the number of potential schedules can serve as a rough measure of the instance hardness. For the relation between the number of potential schedules and the $\sigma$ value, the second chart exhibits that instances generated with small $\sigma$ are more likely to have a large number of potential schedules. The relation above is more obvious on smaller $\sigma$, while on larger $\sigma$ the difference between the number of potential schedules is less significant.

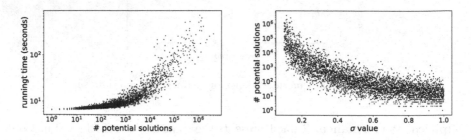

**Fig. 5.** Different indicators of instance hardness, on data set $S_3$.

## 6   Conclusions

We devise an efficient exact algorithm for airplane refueling problem. Based on the dominance properties of the problem, we propose a method that can prefix some jobs' start times and determine the relative orders among jobs in a potential schedule. This technique enables us to solve airplane refueling problem in a recursive manner. Our algorithm outperforms the state of the art exact algorithm on random generated data sets. For large instances with hard configurations that can not be tackled by previous algorithms, our algorithm can solve most of them in a reasonable time.

The empirical efficiency of our algorithm can be attributed to two factors. First, our algorithm explores only the branches that contain potential schedules. Second, on the root node of each branch, the problem is further divided into smaller subproblems, which can also speedup the searching process. Theoretically, we prove that the running time of our algorithm is upper bounded by the number of potential schedules times a polynomial overhead in the worst case.

## References

1. Albers, S., Baruah, S.K., Möhring, R.H., Pruhs, K.: Open problems - scheduling. In: Scheduling. No. 10071 in Dagstuhl Seminar Proceedings, Schloss Dagstuhl - Leibniz-Zentrum fuer Informatik, Germany, Dagstuhl, Germany (2010)
2. Bagga, P.C., Karlra, K.R.: A node elimination procedure for townsend's algorithm for solving the single machine quadratic penalty function scheduling problem. Manage. Sci. **26**(6), 633–636 (1980). https://doi.org/10.1287/mnsc.26.6.633
3. Bansal, N., Pruhs, K.: The geometry of scheduling. SIAM J. Comput. **43**(5), 1684–1698 (2014). https://doi.org/10.1137/130911317

4. Bansal, N., Dürr, C., Thang, N.K., Vásquez, O.C.: The local-global conjecture for scheduling with non-linear cost. J. Sched. **20**(3), 239–254 (2017). https://doi.org/10.1007/s10951-015-0466-5
5. Cheung, M., Mestre, J., Shmoys, D., Verschae, J.: A primal-dual approximation algorithm for min-sum single-machine scheduling problems. SIAM J. Discrete Math. **31**(2), 825–838 (2017). https://doi.org/10.1137/16M1086819
6. Croce, F.D., Tadei, R., Baracco, P., Tullio, R.D.: On minimizing the weighted sum of quadratic completion times on a single machine. In: 1993 Proceedings IEEE International Conference on Robotics and Automation, vol. 3, pp. 816–820 (1993). https://doi.org/10.1109/ROBOT.1993.292245
7. Dürr, C., Vásquez, O.C.: Order constraints for single machine scheduling with non-linear cost. In: 16th Workshop on Algorithm Engineering and Experiments (ALENEX 2014), pp. 98–111. SIAM (2014). https://doi.org/10.1137/1.9781611973198.10
8. Gamow, G., Stern, M.: Puzzle-Math. Viking, New York (1958)
9. Gamzu, I., Segev, D.: A polynomial-time approximation scheme for the airplane refueling problem. J. Sched. **22**(1), 119–135 (2019). https://doi.org/10.1007/s10951-018-0569-x
10. Graham, R.L., Lawler, E.L., Lenstra, J.K., Kan, A.H.G.R.: Optimization and approximation in deterministic sequencing and scheduling: a survey. In: Hammer, P.L., Johnson, E.L., Korte, B.H. (eds.) Annals of Discrete Mathematics, Discrete Optimization II, vol. 5, pp. 287–326. Elsevier (1979). https://doi.org/10.1016/S0167-5060(08)70356-X
11. Hart, P.E., Nilsson, N.J., Raphael, B.: A formal basis for the heuristic determination of minimum cost paths. IEEE Trans. Syst. Sci. Cybern. **4**(2), 100–107 (1968). https://doi.org/10.1109/TSSC.1968.300136
12. Höhn, W., Jacobs, T.: An experimental and analytical study of order constraints for single machine scheduling with quadratic cost. In: Bader, D.A., Mutzel, D. (eds.) 2012 Proceedings of the Fourteenth Workshop on Algorithm Engineering and Experiments (ALENEX), pp. 103–117. Society for Industrial and Applied Mathematics (2012). https://doi.org/10.1137/1.9781611972924.11
13. Höhn, W., Mestre, J., Wiese, A.: How unsplittable-flow-covering helps scheduling with job-dependent cost functions. Algorithmica **80**(4), 1191–1213 (2018). https://doi.org/10.1007/s00453-017-0300-x
14. Li, J., Hu, X., Luo, J., Cui, J.: A fast exact algorithm for airplane refueling problem (2019). arXiv:1910.03241
15. Megow, N., Verschae, J.: Dual techniques for scheduling on a machine with varying speed. In: Fomin, F.V., Freivalds, R., Kwiatkowska, M., Peleg, D. (eds.) ICALP 2013. LNCS, vol. 7965, pp. 745–756. Springer, Heidelberg (2013). https://doi.org/10.1007/978-3-642-39206-1_63
16. Mondal, S.A., Sen, A.K.: An improved precedence rule for single machine sequencing problems with quadratic penalty. Eur. J. Oper. Res. **125**(2), 425–428 (2000). https://doi.org/10.1016/S0377-2217(99)00207-6
17. Sen, T., Dileepan, P., Ruparel, B.: Minimizing a generalized quadratic penalty function of job completion times: an improved branch-and-bound approach. Eng. Costs Prod. Econ. **18**(3), 197–202 (1990). https://doi.org/10.1016/0167-188X(90)90121-W
18. Vásquez, O.C.: For the airplane refueling problem local precedence implies global precedence. Optim. Lett. **9**(4), 663–675 (2015). https://doi.org/10.1007/s11590-014-0758-2

# Approximation Algorithm and Incentive Ratio of the Selling with Preference

Pan Li[1], Qiang Hua[1], Zhijun Hu[2], Hing-Fung Ting[3], and Yong Zhang[4(✉)]

[1] College of Mathematics and Information Science, Hebei University,
Baoding, People's Republic of China
pan.li.hbu@outlook.com, huaq@hbu.cn
[2] School of Mathematics and Statistics, Guizhou University,
Guiyang, People's Republic of China
zjunhu75@163.com
[3] Department of Computer Science, The University of Hong Kong,
Hong Kong, People's Republic of China
hfting@cs.hku.hk
[4] Shenzhen Institutes of Advanced Technology, Chinese Academy of Sciences,
Beijing, People's Republic of China
zhangyong@siat.ac.cn

**Abstract.** We consider the market mechanism to sell two types of products, $A$ and $B$, to a set of buyers $I = \{1, 2, ..., n\}$. The amounts of products are $m_A$ and $m_B$ respectively. Each buyer $i$ has his information including the budget, the preference and the utility function. On collecting the information from all buyers, the market maker determines the price of each product and allocates some amount of product to each buyer. The objective of the market maker is design a mechanism to achieve the semi market equilibrium. In this paper, we show that maximizing the total utility of the buyers in satisfying the semi market equilibrium is NP-hard and give a 1.5-approximation algorithm for this optimization problem. Moreover, in the market, a buyer may get more utility by misreporting his information. We consider the situation that a buyer may misreport his preference and prove that the incentive ratio, the percentage of the improvement by misreporting the information, is upper bounded by 1.618.

## 1 Introduction

Market mechanism plays a very important role in Economy to match the product provider and consumer. In recent years, many new electronic commercial companies appear, e.g., Taobao, Uber, DiDi, AirBnB. Compared with the traditional models, such new market models are more complicated and involve more people. Thus, fairness and efficiency should be considered as the key features of the market mechanism.

Generally speaking, market contains two participants, the supplier and the buyer. The supplier has some product to be sold so as to maximize his revenue,

© Springer Nature Switzerland AG 2019
Y. Li et al. (Eds.): COCOA 2019, LNCS 11949, pp. 328–339, 2019.
https://doi.org/10.1007/978-3-030-36412-0_26

while the buyer wants to buy some products within his budget to maximize his utility. The supplier sets the price for each product and allocates the product to buyers. When the price goes higher, the demand of buyers will decrease accordingly. If the supply equals to the demands, such situation is called *the market equilibrium*. Arrow and Debreu formally defined the market equilibrium [2], which satisfies the following conditions.

- *market clearance*: the product is sold out.
- *budget bound*: the total payment of a buyer is no more than his budget.
- *individual optimal*: each buyer get the optimal utility w.r.t. the designated price.

Given all information of the supplier and the buyers, market equilibrium exist. However, if a buyer misreport his information, he may benefit and get some extra utility. To prevent such situation, the concept of truthful (or strategy-proof) was introduced. We say a mechanism is truthful is no buyer can get more utility by misreporting his information. In both economics and computer science, the truthful mechanisms have been well studied [3,8]. A famous truthful mechanism is the VCG auction [9–11].

On the other hand, if misreporting cannot be prevented, how much a buyer can benefit is also an interesting direction. To evaluate the extra benefit achieved via misreporting, the *incentive ratio* was introduced [5]. Formally, let $U_i(s_i, s_{-i})$ and $U_i(s_i', s_{-i})$ be the utility of buyer $i$ when he reports or misreports his true information respectively, where $s_i$ and $s_i'$ are buyer $i$'s true information and misreporting information respectively, $s_{-i}$ is other buyers' information. The incentive ratio is

$$\max_{i,s} \frac{U_i(s_i', s_{-i})}{U_i(s_i, s_{-i})}$$

In recent years, some well-adopted utility functions have been analyzed [4,8]. Adsul et al. showed that truthful market equilibrium can be achieved if and only if all buyers have the same utility function [1]. Chen et al. proved in [5–7] that for the linear utility function, the incentive ratio is upper bounded by 2, for Cobb-Douglas utility function, the incentive ratio is $e^{1/e} \approx 1.445$, while for the weak gross substitute utility, the incentive ratio is 2.

In this paper, we consider the generalization of the market equilibrium where each buyer may have an optional preference and can be satisfied by any product in his preferences. Formally, the supplier has two types of products, $A$ and $B$. Each buyer has a preference which denote the product the buyer may buy. The preference may be a single product, $\{A\}$ or $\{B\}$, or optional $\{A, B\}$. In the latter case, both products may fit the buyer, but he can only buy one of them. The supplier decides the price of each product and assigns some amount of product to the buyers according to their preferences and budgets. In a fair market, each buyer has no interest switching to another product, i.e., he cannot get more utility via switching. Thus, the prices of the product must be determined carefully and these two products may not be both sold out (an example is shown in the latter part of this section).

The problem we studied in this paper is described as follows.

**Selling with Preference:** A supplier has two types of products, $A$ and $B$, to be sold to a set of buyers $I = \{1, 2, ..., n\}$. The amounts of them are $m_A$ and $m_B$ respectively. The preference of each buyer is either product $A$ or product $B$, or both, i.e., each buyer's preference is one of the three forms, $\{A\}$, $\{B\}$, $\{A, B\}$. Each buyer can only choose one type of product to buy. The budget of buyer $i$ is $b_i$. When buyer $i$ buys some products, he get some utility. In this paper, we consider the linear utility function. Precisely speaking, the utility of buyer $i$ is

$$u_i = \begin{cases} c_{iA} \cdot x_{iA} \text{ if buyer } i \text{ buys } x_{iA} \text{ amount of product } A \\ c_{iB} \cdot x_{iB} \text{ if buyer } i \text{ buys } x_{iB} \text{ amount of product } B \end{cases}$$

where $c_{iA}$ and $c_{iB}$ are user $i$'s utility coefficients for buying product $A$ and product $B$ respectively.

The supplier decides the price of each product and assigns some amount of products to each buyer. We say the outcome is `semi market equilibrium` if the following conditions hold.

- *market semi clearance.* At least one type of product is fully assigned, i.e., either $\sum_{i \in I} x_{iA} = m_A$, or $\sum_{i \in I} x_{iB} = m_B$.
- *budget bound.* The total cost of each buyer is upper bounded by his budget, i.e., $x_{iX} \cdot p_X \leq b_i$ for $i \in I$ and $X \in \{A, B\}$.
- *individual optimality.* Each buyer's utility is maximized w.r.t. the designated price, i.e., $c_{iX} \cdot b_i/p_X \geq c_{iX'} \cdot b_i/p_{X'}$ if the supplier assigns product $X$ to buyer $i$, where $\{X, X'\} = \{A, B\}$.

Consider a simple example. The amount of two products are 1 and 1.5 respectively. The budget of two buyers are both 1, the utility functions are $u_{1A} = u_{1B} = u_{2A} = u_{2B} = x$, where $x$ is the assigned amount to the buyer. The supplier sets $p_A = p_B = 1$ and allocates 1 product $A$ to buyer 1 and 1 product $B$ to buyer 2. This configuration is semi market equilibrium, product $A$ is sold out while product $B$ has some remaining amount. The supplier cannot decrease the price to have another semi market equilibrium.

The remaining part of the paper is organized as follows. In Sect. 2, we show that in satisfying the condition of semi market equilibrium, maximizing the total utility among buyers is NP-hard, then give a 1.5-approximation algorithm for handling this problem. In Sect. 3, we prove that the incentive ratio of the selling with preference problem is 1.68. We conclude this paper in Sect. 4.

## 2 Approximation Algorithm

We first show it is NP-hard to find the maximum total utility for the selling with preference problem.

**Theorem 1.** *Maximizing the total utility among buyers is NP-hard in satisfying the semi market equilibrium.*

*Proof.* We prove the NP-hardness by the reduction from the partition problem. In the partition problem, given a set of values $V = \{v_1, v_2, ...v_n\}$, we are asked whether $V$ can be partitioned into two disjoint sets $V_1$ and $V_2$ such that $\sum_{i \in V_1} v_i = \sum_{i \in V_2} v_i$.

Now we construct an instance of the selling problem from the partition problem. The amount of two products are both $m_A = m_B = \sum_{i \in V} v_i/2$. There are $n$ buyers and each buyer's preference is $\{A, B\}$. For buyer $i$, the budget is $v_i$ and the utility function is $u_i = x_{iA}$ if assigned with product $A$, or $u_i = x_{iB}$ if assigned with product $B$. Thus, the maximum possible total utility is $\sum_{i \in V} v_i$.

According to the condition of individual optimality, if $p_A < p_B$ or $p_A > p_B$, all buyers will prefer the product with the lower price. In such situation, all amount of either product $A$ or product $B$ remains unassigned, which is apparently not optimal. Thus, in the constructed instance, the supplier must set price $p_A = p_B$.

In one direction, if the answer of the partition problem is *yes*, i.e., $V = V_1 + V_2$ and $\sum_{i \in V_1} v_i = \sum_{i \in V_2} v_i = \sum_{i \in V} v_i/2$. Then for the constructed selling problem, set $p_A = p_B = 1$ and assign $v_i$ product $A$ (or $B$) to buyer $i$ if $i \in V_1$ (or $i \in V_2$). In this case, the total utility is $\sum_{i \in V} v_i$ and get the maximum value.

In another direction, assume that the constructed selling problem can be solved with the total utility $\sum_{i \in V} v_i$. Since the amount of each product is $\sum_{i \in V} v_i/2$, product $A$ and product $B$ are both fully assigned. Note that $p_A = p_B$ and the coefficient of each buyer on each product is 1, the designated prices are $p_A = p_B = 1$. Each buyer $i$ is assigned with the amount of $v_i$. Therefore, the solution of the selling problem leads to the *yes* answer of the partition problem.

Therefore, we conclude that maximizing the total utility among buyers is NP-hard in satisfying the semi market equilibrium. ∎

We analyze the property of the assignment satisfying the semi market equilibrium. Let $r_i = \frac{c_{iA}}{c_{iB}}$, if the preference of buyer $i$ is $\{A\}$ (or $\{B\}$), then $r_i = +\infty$ (or 0).

**Lemma 1.** *In an assignment satisfying the semi market equilibrium, $r_i \geq r_j$ for any buyer $i$ assigned at $A$ and any buyer $j$ assigned at $B$.*

*Proof.* Suppose buyer $i$ is assigned at $A$ with price $p_A$ and buyer $j$ is assigned at $B$ with price $p_B$. Due to individual optimality, we have $\frac{b_i}{p_A} \cdot c_{iA} \geq \frac{b_i}{p_B} \cdot c_{iB}$ and $\frac{b_j}{p_B} \cdot c_{jB} \geq \frac{b_j}{p_A} \cdot c_{jA}$, thus, we have $\frac{p_A}{p_B} \leq \frac{c_{iA}}{c_{iB}}$ and $\frac{p_A}{p_B} \geq \frac{c_{jA}}{c_{jB}}$ Therefore,

$$r_i = \frac{c_{iA}}{c_{iB}} \geq \frac{p_A}{p_B} \geq \frac{c_{jA}}{c_{jB}} = r_j.$$

∎

Since each buyer will select the product with the higher utility, we have the following fact.

**Fact 1.** *For two assignments $C$ and $C'$ with prices $(p_A, p_B)$ and $(p'_A, p'_B)$ respectively, if $p_A \geq p'_A$ and $p_B \geq p'_B$, then the total utility of $C$ is no more than the total utility of $C'$.*

We analyze the property of the assignment satisfying the budget bound and the individual optimality, so as to find the relationship between the assignment with the maximum total utility in terms of the amount of products. We say an assignment is *fair* if it satisfies the *budget bound* and the *individual optimality*.

**Lemma 2.** *In a fair assignment with price $p_A$ and $p_B$, if the assigned amount of product A and product B are at least $m_A$ and $m_B$ respectively, the total utility is at least the maximum assignment satisfying the semi market equilibrium.*

With the help of Lemma 2, we give the algorithm *Product-Assignment* to maximize the total utility. Let $S_A$ and $S_B$ be the set of buyers assigned to $A$ and $B$ respectively.

---

**Algorithm 1.** Product-Assignment

1: Initially, $S_A = S_B = \emptyset$.
2: Set $p_A = \frac{\sum_i b_i}{m_A}$ for all buyers interested in product $A$.
3: Set $p_B = \frac{\sum_i b_i}{m_B}$ for all buyers interested in product $B$.
4: **for** $i = 1$ to $n$ **do**
5:      **if** $\frac{c_{iA}}{c_{iB}} \geq \frac{p_A}{p_B}$ **then**
6:          Assign $\frac{b_i}{p_A}$ product $A$ to buyer $i$, $S_A = S_A \cup \{i\}$.
7:      **else**
8:          Assign $\frac{b_i}{p_B}$ product $B$ to buyer $i$, $S_B = S_B \cup \{i\}$.
9:      **end if**
10: **end for**
11: **while** both product $A$ and $B$ are not fully assigned **do**
12:      Update-Assignment
13: **end while**

---

**Procedure: Update-Assignment**

1: Decrease $p_A$ and $p_B$ with the same ratio until $\frac{\sum_{i \in A} b_i}{p_A} = m_A$ or $\frac{\sum_{i \in B} b_i}{p_B} = m_B$.
2: **if** $\frac{\sum_{i \in X} b_i}{p_X} = m_X$ and $\frac{\sum_{i \in Y} b_i}{p_Y} \leq m_Y$ **then**     $\triangleright \{X, Y\} = \{A, B\}$
3:      Decrease $p_Y$ until some buyer $i \in X$ has the same utility in $Y$ w.r.t. $p_Y$.
4:      **if** more than one buyer satisfy the above condition **then**
5:          Select the one, say buyer $j$, with the lowest budget.
6:          **if** $\frac{\sum_{i \in Y} b_i + b_j}{p_Y} < m_Y$, i.e., switch buyer $j$ to $Y$ not reach the amount $m_Y$
    **then**
7:              $S_X = S_X - \{j\}$, $S_Y = S_Y + \{j\}$.
8:          **end if**
9:      **end if**
10: **end if**

---

**Theorem 2.** *The approximation ratio of Product-Assignment is 1.5.*

*Proof.* Consider the final configuration $\mathcal{C}$ of the assignment from the algorithm, either $A$ or $B$ is fully assigned. W.l.o.g., assume that product $A$ is fully assigned to buyers in $S_A$ and some amount of product $B$ remains unassigned. Let $\delta \cdot m_B$ be the free product in $B$. There must exist some buyer $i \in S_A$ such that assigning $i$ to $B$ with price $p_B$ has the same utility. Otherwise, the price of $B$ can be further decreased. However, switching buyer $i$ from $A$ to $B$ will either reach or beyond the amount $m_B$, i.e., $\frac{b_i}{p_B} \geq \delta \cdot m_B$.

If buyer $i$ can be partially assigned to both $A$ and $B$, and the occupation ratio in both products are exactly equal, we may decrease the prices $p_A$ and $p_B$ with the same ratio until both product fully utilized. According to Lemma 2, the total utility is at least the optimal solution. We use this idea to upper bound the approximation ratio.

Assume buyer $i$ can be split to buyer $i_A$ and $i_B$ with budget $b_{iA}$ and $b_{iB}$ respectively. Thus, $b_i = b_{iA} + b_{iB}$. Further assume that from the configuration $\mathcal{C}$, switching buyer $i_B$ to product $B$ lead to a configuration $\mathcal{C}'$ in which the occupation ratio of each product are equal. That means

$$\frac{\delta \cdot m_B - \frac{b_{iB}}{p_B}}{m_B} = \frac{\frac{b_{iB}}{p_A}}{m_A}.$$

Thus,

$$b_{iB} = \frac{\delta}{\frac{1}{p_A \cdot m_A} + \frac{1}{p_B \cdot m_B}}$$

The approximation is upper bounded by

$$\frac{m_A}{m_A - \frac{b_{iB}}{p_A}} = \frac{m_A}{m_A - \frac{\delta}{\frac{1}{m_A} + \frac{p_A}{p_B \cdot m_B}}} = \frac{m_A \cdot p_A + m_B \cdot p_B}{m_A \cdot p_A + (1 - \delta) \cdot m_B \cdot p_B} = \frac{1 + k}{1 + k - \delta}$$

where $k = \frac{m_A \cdot p_A}{m_B \cdot p_B}$.

From the above analysis, the higher value of $\delta$, the higher of the approximation ratio. Since $b_i \geq \delta \cdot m_B \cdot p_B$, large $\delta$ means the higher budget of buyer $i$. We use this fact to give another bound of the approximation ratio. If assign buyer $i$ to product $A$, the price of product $A$ is at least $\frac{b_i}{m_A}$ and all amount of product $A$ is assigned to buyer $i$. If assign buyer $i$ to product $A$ with price $p_A$, the assigned amount is $\frac{b_i}{p_A}$. In this case, the approximation ratio is at most $\frac{m_A \cdot p_A}{b_i}$. Similarly, if assign buyer $i$ to product $B$ with price $p_B$, the approximation ratio is at most $\frac{m_B \cdot b_B}{b_i}$. Thus, if only consider buyer $i$ with the price $p_A$ or $p_B$, the approximation ratio is upper bounded by

$$\max\{\frac{m_A \cdot p_A}{b_i}, \frac{m_B \cdot p_B}{b_i}\} \leq \frac{\max\{m_A \cdot p_A, m_B \cdot p_B\}}{\delta \cdot m_B \cdot p_B} = \frac{\max\{k, 1\}}{\delta}.$$

Combining the above analysis, the approximation ratio is at most

$$\min\{\frac{1 + k}{1 + k - \delta}, \frac{\max\{k, 1\}}{\delta}\}$$

- If $k \geq 1$, the approximation ratio is at most $\min\{\frac{1+k}{1+k-\delta}, \frac{k}{\delta}\}$. These two values are equal when $\delta = \frac{k^2+k}{2k+1}$. Thus, the ratio is $\frac{2k+1}{k+1} \leq 1.5$.
- If $k < 1$, the approximation ratio is at most $\min\{\frac{1+k}{1+k-\delta}, \frac{1}{\delta}\}$. These two values are equal when $\delta = \frac{k+1}{k+2}$. Thus, the ratio is $\frac{k+2}{k+1} \leq 1.5$.

■

## 3  Incentive Ratio

In this section, we consider the case that if a buyer misreports his preference, how much extra utility he can achieve. For example, a buyer $i$'s preference is $\{A, B\}$, he may claim his preference is $\{B\}$ to the supplier. Note that in this case, buyer $i$ cannot be assigned to $A$ even if he can get more utility on $A$.

**Example:** Consider an instance that the amount of products are $m_A = (\frac{1+\sqrt{5}}{2}) \cdot x$ (where $x \leq \frac{1+\sqrt{5}}{2}$) and $m_B = \frac{1+\sqrt{5}}{2}$, respectively. There are two buyers with budget $b_1 = 1$ and $b_2 = \frac{1+\sqrt{5}}{2}$, respectively. The preferences of these two buyers are both $\{A, B\}$ and the preference parameters are $c_{1A} = c_{1B} = c_{2A} = c_{2B} = 1$.

In assignment 1, $p_A = p_B = 1$, buyer 1 is assigned with 1 amount of product $A$ while buyer 2 is assign with $\frac{1+\sqrt{5}}{2}$ amount of product $B$. The utility of buyer 1 and buyer 2 are 1 and $\frac{1+\sqrt{5}}{2}$ respectively. Since $x$ is upper bounded by $\frac{1+\sqrt{5}}{2}$, the amount of product $A$ is at most $\frac{3+\sqrt{5}}{2}$. In this case, it can be verified that assignment 1 is semi market equilibrium.

If buyer 2 claims his preference is $\{A\}$ instead of $\{A, B\}$, the assignment will be changed. In assignment 2, $p_A = \frac{1}{x}$, $p_B = \frac{2}{\sqrt{5}+1}$. Buyer 1 is assigned with $\frac{1+\sqrt{5}}{2}$ amount of product $B$ while buyer 2 is assigned with $(\frac{1+\sqrt{5}}{2}) \cdot x$ amount of product $A$. The utility of buyer 1 and buyer 2 are changed to $\frac{1+\sqrt{5}}{2}$ and $(\frac{1+\sqrt{5}}{2}) \cdot x$ respectively. Again, assignment 2 is semi market equilibrium w.r.t. the claimed preference. However, it is not semi market equilibrium w.r.t. the previous preference if $x < \frac{1+\sqrt{5}}{2}$. In this way, the utility of buyer 2 is increased from $\frac{1+\sqrt{5}}{2}$ to $(\frac{1+\sqrt{5}}{2}) \cdot x$, the incentive ratio of buyer 2 is at most $\frac{1+\sqrt{5}}{2} \approx 1.618$.

In this problem, the amount of two products may not be equal. To make the analysis simple and easy to read, the amount can be normalized to be 1 as follows.

**Normalization:** For each buyer $i$, let $\tilde{C}_{iA} = C_{iA} \cdot m_A$, $\tilde{C}_{iB} = C_{iB} \cdot m_B$, $\tilde{p}_A = p_A \cdot m_A$, $\tilde{p}_B = p_B \cdot m_B$. The normalization can be justified as follows.

- The total assigned amount in product $X$ is bounded by 1, i.e.,

$$\frac{\sum_{i \in S_X} b_i}{\tilde{p}_X} = \frac{\sum_{i \in S_X} b_i}{p_X \cdot m_X} \leq 1 \quad \text{if} \quad \frac{\sum_{i \in S_X} b_i}{p_X} \leq m_X$$

– The utility of each buyer is the same as the original one, i.e.,

$$\frac{b_i}{\tilde{p}_X} \cdot \tilde{C}_{iX} = \frac{b_i}{p_X \cdot m_X} \cdot (C_{iX} \cdot m_X) = \frac{b_i}{p_X} \cdot C_{iX}$$

where $X$ is $A$ or $B$.

We consider a general case to show the bound of the incentive ratio. Let $E$ and $E'$ be the semi market equilibrium assignments such that buyer $i$ lies or not. The designated prices are $\tilde{p}_A$ and $\tilde{p}_B$ in $E$, while the designated prices are $\tilde{p}'_A$ and $\tilde{p}'_B$ in $E'$.

Let $\{X, Y\} = \{A, B\}$. W.l.o.g., assume that $Y$ is fully utilized in $E$. Let $S_{XY}$ denote the buyers who are in $X$ w.r.t. $E$ and in $Y$ w.r.t. $E'$. Let $S_{YX}$ denote the buyers who are in $Y$ w.r.t. $E$ and in $X$ w.r.t. $E'$. Let $M_{XY}$ and $M_{YX}$ be the assigned amounts of $S_{XY}$ and $S_{YX}$ in $E$ respectively.

Consider a virtual assignment $\tilde{E}$ such that buyers in $S_{XY}$ are switched to $Y$ with the price $\tilde{p}_Y$ and buyers in $S_{YX}$ are switched to $X$ with the price $\tilde{p}_X$.

**Lemma 3.** *In the virtual assignment $\tilde{E}$, if either $X$ or $Y$ is fully assigned or over assigned, the lying buyer cannot get more utility.*

*Proof.* In the virtual assignment, the utility of any buyer cannot increase. If the lying buyer is assigned at the fully (or over) assigned product, the corresponding price cannot be decreased. If the lying buyer is assigned at the non-fully utilized product, the price on this side cannot be decreased, too. Otherwise, some buyer on the other product may be switched to this product to get more utility.  ∎

Thus, we only need to consider both $X$ and $Y$ are not fully assigned in $\tilde{E}$. Let $M'_{XY}$ and $M'_{YX}$ be the assigned amount of $S_{XY}$ and $S_{YX}$ w.r.t. $\tilde{E}$. Since $Y$ is fully assigned in $E$, the amount $M'_{XY}$ in $Y$ w.r.t. $\tilde{E}$ is upper bounded by $M_{YX}$. Thus, we have $M_{YX} \geq M'_{XY}$ and $M_{XY} \leq M'_{YX}$.

---

**Procedure: Achieving the assignment $E'$**
In the virtual assignment $\tilde{E}$, decrease the prices of $X$ and $Y$ with the same ratio until one of them is fully utilized.

---

In assignment $E$, assume that the free amount of $X$ is $\Delta$. Now we consider the assignment $E'$ and analyze the decreased ratio of the price.

**Case 1:** If $Y$ is fully utilized in $E'$, i.e., $\frac{1}{1-\Delta-M_{XY}+M'_{YX}} \geq \frac{1}{1-M_{YX}+M'_{XY}}$.
From the assumption, we have

$$\Delta \geq (M_{YX} - M'_{XY}) + (M'_{YX} - M_{XY}). \tag{1}$$

Note that the prices of $X$ and $Y$ are decreased with the same ratio, which is

$$\frac{\tilde{p}_Y}{\tilde{p}'_Y} = \frac{\tilde{p}_X}{\tilde{p}'_X} = \frac{1}{1 - M_{YX} + M'_{XY}}.$$

Consider two cases $\tilde{p}_X \leq \tilde{p}_Y$ and $\tilde{p}_X > \tilde{p}_Y$.

- $\tilde{p}_X \leq \tilde{p}_Y$.

In this case, $M_{YX} \leq M'_{YX}$ and $(M'_{YX} - M_{XY}) \geq (M_{YX} - M'_{XY})$. According to Eq. 1,

$$\Delta \geq (M_{YX} - M'_{XY}) + (M'_{YX} - M_{XY}) \geq 2 \cdot (M_{YX} - M'_{XY}).$$

If $\Delta > M'_{YX}$, we can slightly decrease $\tilde{p}_X$ to achieve an semi market equilibrium, which is a contradiction. So, $M'_{YX} \geq \Delta$, which leads to

$$\Delta - (M'_{YX} - M_{XY}) \leq M'_{YX} - (M'_{YX} - M_{XY}) = M_{XY}.$$

Therefore

$$
\begin{aligned}
1 &\geq \Delta + M_{XY} \\
&\geq 2\Delta - (M'_{YX} - M_{XY}) \\
&\geq \Delta + (M_{YX} - M'_{XY}) \\
&\geq 3(M_{YX} - M'_{XY})
\end{aligned}
$$

The decreased ratio of the price is $\frac{1}{1 - M_{YX} + M'_{XY}} \leq \frac{3}{2}$, which means the utility is increased at most $\frac{3}{2}$ times to the original one.

- $\tilde{p}_X > \tilde{p}_Y$.

In this case, $M_{YX} \geq M'_{YX}$, $M'_{XY} \geq M_{XY}$ and $\frac{M'_{YX} - M_{XY}}{M_{YX} - M'_{XY}} = \frac{\tilde{p}_Y}{\tilde{p}_X}$. Let $k = \frac{\tilde{p}_Y}{\tilde{p}_X}$, from the assumption, $k < 1$. According to Eq. 1,

$$\Delta \geq (1 + k) \cdot (M_{YX} - M'_{XY}).$$

Note that $M'_{YX} \geq \Delta$, we have $M'_{XY} \geq \frac{1}{k} \cdot (M_{YX} - M'_{XY})$. Since $\Delta + M_{XY} \leq 1$ and $M_{XY} = k \cdot M'_{XY}$, we have

$$
\begin{aligned}
1 &\geq \Delta + M_{XY} \\
&\geq (1 + k) \cdot (M_{YX} - M'_{XY}) + (M_{YX} - M'_{XY}) \\
&= (2 + k) \cdot (M_{YX} - M'_{XY})
\end{aligned}
$$

Therefore,

$$(M_{YX} - M'_{XY}) \leq \frac{1}{2 + k} \tag{2}$$

Since $M'_{XY} \geq \frac{1}{k} \cdot (M_{YX} - M'_{XY})$, we have

$$
\begin{aligned}
1 &\geq M_{YX} \\
&= (M_{YX} - M'_{XY}) + M'_{XY} \\
&\geq (1 + \frac{1}{k}) \cdot (M_{YX} - M'_{XY})
\end{aligned}
$$

Therefore,

$$(M_{YX} - M'_{XY}) \leq \frac{1}{1 + 1/k} \tag{3}$$

Note that the decreased ratio of the price is $\frac{1}{1-M_{YX}+M'_{XY}}$, which is higher when the value of $M_{YX} - M'_{XY}$ gets higher. Combining (2) and (3), the highest value of $(M_{YX} - M'_{XY})$ is $\frac{3-\sqrt{5}}{2}$ when $\frac{1}{2+k} = \frac{1}{1+1/k}$. In this case, the decreased ratio of the price is

$$\frac{1}{1 - M_{YX} + M'_{XY}} = \frac{1 + \sqrt{5}}{2} \approx 1.618.$$

**Case 2:** If $X$ is fully utilized, $\frac{1}{1-\Delta-M_{XY}+M'_{YX}} \leq \frac{1}{1-M_{YX}+M'_{XY}}$

From the assumption, we have

$$\Delta \leq (M_{YX} - M'_{XY}) + (M'_{YX} - M_{XY}). \tag{4}$$

Again, we consider two cases, $\tilde{p}_X \leq \tilde{p}_Y$ and $\tilde{p}_X > \tilde{p}_Y$.

– $\tilde{p}_X \leq \tilde{p}_Y$.
Since $\frac{M'_{YX}-M_{XY}}{M_{YX}-M'_{XY}} = \frac{\tilde{p}_Y}{\tilde{p}_X} \geq 1$, $M'_{YX} - M_{XY} \geq M_{YX} - M'_{XY}$ and thus

$$\Delta - (M'_{YX} - M_{XY}) \leq M_{YX} - M'_{XY} \leq M'_{YX} - M_{XY}.$$

Since $\Delta \leq M'_{YX}$, we have

$$\Delta - (M'_{YX} - M_{XY}) \leq M'_{YX} - (M'_{YX} - M_{XY}) = M_{XY}.$$

Combining with the fact that $\Delta + M_{XY} \leq 1$,

$$\Delta - (M'_{YX} - M_{XY}) \leq 1/3.$$

Thus, the decreased ratio of the price is at most

$$\frac{1}{1 - \Delta - M_{XY} + M'_{YX}} \leq \frac{3}{2}.$$

– $\tilde{p}_X > \tilde{p}_Y$.
In this case, $M_{YX} < M'_{YX}$ and $M'_{XY} > M_{XY}$. Let $k = \frac{\tilde{p}_Y}{\tilde{p}_X} < 1$. Since $\frac{M'_{YX}-M_{XY}}{M_{YX}-M'_{XY}} = \frac{\tilde{p}_Y}{\tilde{p}_X}$, we have $M'_{YX} - M_{XY} = k \cdot (M_{YX} - M'_{XY})$. According to Eq. 4,

$$\Delta \leq (1 + k) \cdot (M_{YX} - M'_{XY}).$$

and

$$k \cdot (\Delta - (M'_{YX} - M_{XY})) \leq k \cdot (M_{YX} - M_{XY'}) = M'_{YX} - M_{XY}$$

Since $\Delta \leq M'_{YX}$, we have

$$M_{XY} \geq \Delta - (M'_{YX} - M_{XY}).$$

Thus,

$$1 \geq (\Delta - (M'_{YX} - M_{XY}) + (M'_{YX} - M_{XY}) + M_{XY}$$
$$\geq (2 + k) \cdot (\Delta - (M'_{YX} - M_{XY}))$$

Therefore,

$$\Delta - (M'_{YX} - M_{XY}) \leq \frac{1}{2 + k}. \tag{5}$$

Since $M_{XY} \geq \Delta - (M'_{YX} - M_{XY})$, we have $M'_{XY} \geq \frac{1}{k} \cdot (\Delta - (M'_{YX} - M_{XY}))$. Thus,

$$1 \geq (M_{YX} - M'_{XY}) + M'_{XY}$$
$$\geq (1 + 1/k) \cdot (\Delta - (M'_{YX} - M_{XY}))$$

Therefore,

$$\Delta - (M'_{YX} - M_{XY}) \leq \frac{1}{1 + 1/k} \tag{6}$$

Similar to the analysis in Case 1, combining the inequalities (5) and (6), the highest value of $\Delta - (M'_{YX} - M_{XY})$ is $\frac{3-\sqrt{5}}{2}$ when $\frac{1}{2+k} = \frac{1}{1+1/k}$. In this case, the decreased ratio of the price is

$$\frac{1}{1 - \Delta - M'_{YX} + M_{XY}} = \frac{1 + \sqrt{5}}{2} \approx 1.618.$$

According to the above analysis, we have the following conclusion.

**Theorem 3.** *The decreased ratio of the price is at most 1.618, moreover, this bound is tight for the problem.*

Due to the bound on the decreased price, the incentive ratio can be also bounded.

**Theorem 4.** *The incentive ratio of the selling with preference problem is 1.618.*

*Proof.* There are four possible cases of the lying buyer $i$: in $S_{XY}$, in $S_{YX}$, always assigned at $X$ and always assigned at $Y$.

1. If the lying buyer $i$ is in $S_{XY}$.
   In this case, the incentive ratio is at most the decreased ratio of the price, which is 1.618.
2. If the lying buyer $i$ is in $S_{YX}$.
   In this case, the price of $X$ cannot be further decreased. Otherwise, it will violate the individual optimality. Thus, from the assignment $E$ to $E'$, the incentive ratio is at most 1.618.
3. If the lying buyer always in $X$.
   The incentive ratio is the same as case 1, which is at most 1.618.
4. If the lying buyer always in $Y$.
   The incentive ratio is the same as case 2, which is at most 1.618.

From the example shown in the beginning of Sect. 3, we can see that the incentive ratio 1.618 is tight for the selling with preference problem. ∎

# 4    Concluding Remark

With the great development of information technology, market economy has changed radically. Many revolutionary market models appear and improve our daily life. We should pay more attention in analyzing and evaluating the existed and the new coming models. In this work, we study the market equilibrium by considering the buyer's preference. It is a simple try on two types of products. An interesting future work might be considering many types of product and more complicated buyer behaviors, e.g., buyer's interested bundle contains multiple types of products and the optional preference may be the subset of his interested bundle.

**Acknowledgement.** This research is supported by Major Scientific and Technological Special Project of Guizhou Province (20183001), China's NSFC grants (No. 61433012, 71361003, 71461003), Shenzhen research grant (No. KQJSCX20180330170311901, JCYJ 20180305180840138 and GGFW2017073114031767), Hong Kong GRF-17208019, Shenzhen Discipline Construction Project for Urban Computing and Data Intelligence, Natural Science Foundations of Guizhou Province (No. [2018] 3002) and the Key Science and Technology Foundation of the Education Department of Hebei Province, China (ZD2019021)

# References

1. Adsul, B., Babu, C.S., Garg, J., Mehta, R., Sohoni, M.: Nash equilibria in fisher market. In: Proceedings of SAGT 2010, pp. 30–41 (2010)
2. Arrow, K.J., Debreu, G.: Existence of an equilibrium for a competitive economy. Econometrica **22**(3), 265–290 (1954)
3. Barberá, S., Jackson, M.O.: Strategy-proof exchange. Econometrica **63**(1), 51–87 (1995)
4. Brânzei, S., Chen, Y., Deng, X., Filos-Ratsikas, A., Frederiksen, S.K.S., Zhang, J.: The fisher market game: equilibrium and welfare. In: Proceedings of AAAI 2014, pp. 587–593 (2014)
5. Chen, N., Deng, X., Zhang, J.: How profitable are strategic behaviors in a market? In: Demetrescu, C., Halldórsson, M.M. (eds.) ESA 2011. LNCS, vol. 6942, pp. 106–118. Springer, Heidelberg (2011). https://doi.org/10.1007/978-3-642-23719-5_10
6. NinChen, N., Deng, X., Zhang, H., Zhang, J.: Incentive ratios of fisher markets. In: Proceedings of ICALP 2012, pp. 464–475 (2012)
7. Chen, N., Deng, X., Tang, B., Zhang, H.: Incentives for strategic behavior in fisher market games. In: Proceedings of AAAI 2016, pp. 453–459 (2016)
8. Cheng, Y., Deng, X., Scheder, D.: Recent studies of agent incentives in internet resource allocation and pricing. 4OR **16**, 231–260 (2018)
9. Clarke, E.H.: Multipart pricing of public goods. Public Choice **11**(1), 17–33 (1971)
10. Groves, T.: Incentives in teams. Econometrica **41**(4), 617–631 (1973)
11. Vickrey, W.: Counterspeculation, auctions, and competitive sealed tenders. J. Financ. **16**(1), 8–37 (1961)

# Car-Sharing Problem: Online Scheduling with Flexible Advance Bookings

Haodong Liu[1], Kelin Luo[1($\boxtimes$)], Yinfeng Xu[1,2], and Huili Zhang[1]

[1] School of Management, Xi'an Jiaotong University, Xi'an 710049, China
HaodongLiu0515@outlook.com, luokelin@stu.xjtu.edu.cn,
{yfxu,zhang.huilims}@xjtu.edu.cn
[2] The State Key Lab for Manufacturing Systems Engineering, Xi'an 710049, China

**Abstract.** We study an online scheduling problem that is motivated by applications such as car-sharing for trips among a number of locations. Users submit ride requests, and the scheduler aims to accept as many requests as $k$ servers (cars) could. Each ride request specifies the pick-up time, the pick-up location, and the drop-off location. A request can be submitted at any time during a certain time interval that precedes the pick-up time. The scheduler has to decide whether or not to accept a request immediately at the time when the request is submitted (booking time). For the general case in which requests may have arbitrary lengths, $L$, the ratio of the longest to the shortest request, is an important parameter. We give an algorithm with competitive ratio $O(L)$, where the ratio $L$ is known in advance. It is shown that no $O(L^{\alpha})$-competitive algorithm exists for the fixed booking variant, where $0 < \alpha < 1$. It is also proved that no $O(L)$-competitive algorithm exists for the variable booking variant.

**Keywords:** Car-sharing system · Online scheduling · Competitive analysis · General network

## 1 Introduction

Recently, car-sharing becomes one of the popular transportation services; customers can book their requests online over time while the car-sharing company offers cars to serve these requests. It could contribute in response to the growing demand for mobility as well as urban space requirements. In addition to this, shared cars also help to push more private vehicles out of the mobility system, and it will alleviate the current traffic congestion and environment issues [1]. Based on this, the car-sharing problem (CSP) has become one of the most important areas of study within the field of operations research. In this paper, we consider an online car-sharing problem that the goal is to maximize the number of satisfied requests (or customers).

© Springer Nature Switzerland AG 2019
Y. Li et al. (Eds.): COCOA 2019, LNCS 11949, pp. 340–351, 2019.
https://doi.org/10.1007/978-3-030-36412-0_27

## 1.1  Related Work

The online car-sharing problem has been studied in several papers with different networks (two locations [9–11], star network [12]), different numbers of servers (one server [9], two servers [10], and $k$ servers [11,12]) and different booking variants (fixed bookings and variable bookings), and the lower bound and upper bound are matched in most cases. More specifically, in the car-sharing problem with two locations and $k$ servers, where $k$ can be arbitrarily large [11], the authors considered both fixed booking times and variable booking times. Moreover, the authors in [12] studied an online scheduling problem that is motivated by applications such as car-sharing for trips between an airport and a group of hotels. Both the unit travel time variant and the arbitrary travel time variant are considered. In the unit travel time variant, the travel time between the airport and any hotel is a fixed value $t$. They gave a 2-competitive algorithm in the case that the length of booking interval (pick-up time minus booking time) is at least $t$ and the number of servers is even. In the arbitrary travel time variant, the travel time between the airport and a hotel is bounded by $t$ and $Lt$, where $L \geq 1$ denotes the ratio of the longest to the shortest distance(travel time). They presented an algorithm with competitive ratio $O(\log L)$ if the number of servers is at least $\log L$. For both variants, they showed matching lower bounds on the competitive ratio of any deterministic online algorithm. We will extend the problem to the version with a more general network in this paper.

The offline car-sharing problem in a general network is investigated by Böhmová et al. [4]. They showed that if all customer requests are known in advance, the problem of maximizing the number of accepted requests is solvable in polynomial time. Furthermore, they considered the problem variant with two locations where each customer requests two rides (in opposite directions) aiming to maximize the number of satisfied customers. This variant is NP-hard and APX-hard.

A well-studied problem that is closest to our setting is the online dial-a-ride problem (OLDARP). The only difference is that the requests should be served 'as soon as possible' in OLDARP instead of at a specific time in the online car-sharing problem. It offers that all of arrived requests must be served. The objective contains the minimum makespan or the minimum maximum flow time to serve all the requests [2,3,6]. The interested readers can refer the work of [7] for more details.

Another related problem is the online interval scheduling. The online car-sharing problem can be phrased in terms of interval scheduling if all the pick-up and drop-off locations are the same. Let a car be a machine and each interval is associated to a request, one for each car in the instance. The starting time of an interval corresponds to the pick-up time of a particular request, and the ending time of an interval corresponds to the drop-off time of that request. A feasible solution consists of a schedule of intervals that the selected intervals in a machine are disjoint. Lipton and Tomkins [8] considered this problem on one machine. They showed that no (randomized) algorithm can achieve competitive ratio $O(\log \Delta)$ (where $\Delta$ denotes the ratio between the longest and the shortest interval, and $\Delta$ is unknown to the algorithm), and gave an $O((\log \Delta)^{1+\varepsilon})$-competitive randomized algorithm.

## 1.2   Problem Description and Preliminaries

We consider a setting with $m$ locations and $k$ servers (denoted by $s_1, s_2, \ldots, s_k$). The $k$ servers are initially located at one location. Let $R$ denote a sequence of requests that are released over time. Customer requests for car bookings arrive over time, and the decision about each request (to accept or reject this request) must be made immediately, without any knowledge of future requests. The requests with the same release time are released one by one in an arbitrary order. The $i$-th request is denoted by $r_i = (\tilde{t}_{r_i}, t_{r_i}, p_{r_i}, \dot{p}_{r_i})$ and specifies the booking time or release time $\tilde{t}_{r_i}$, the start time or pick-up time $t_{r_i}$, the pick-up location $p_{r_i}$, and the drop-off location $\dot{p}_{r_i}$. Let $t(p, q)$ denote the travel time between location $p$ and location $q$, which is decided by both two locations. We assume that the service time satisfy $t \leq t(p, q) \leq Lt$ for all $1 \leq i \leq m$ and $1 \leq j \leq m$, where $t$ is the travel time of the shortest request. Note that the drop-off time $\dot{t}_{r_i}$ equals to $t_{r_i} + t(p_{r_i}, \dot{p}_{r_i})$.

An empty movement between two locations $p$ and $q$ also takes time $t(p, q)$. Each server can only serve one request at a time. A car can serve two consecutive requests only if the drop-off location of the first request is the same as the pick-up location of the second one, or if there is enough time to travel between the two locations. If two requests are such that they can not both be served by the same server, we say that the requests are *in conflict*. We denote the set of requests accepted by an algorithm by $R'$. The goal of the car-sharing problem is to accept a set of requests $R'$ that maximizes the number of satisfied requests $|R'|$. We refer to this problem as the online car-sharing problem.

With respect to constraints on the booking time, one can consider the fixed booking time variant and the variable booking time variant of the online car-sharing problem. Customers can book a request at a specific time (fixed booking time variant) or at any time of an interval (variable booking time variant) before the start time of the request. The problem for the fixed booking time variant in which $t_{r_i} - \tilde{t}_{r_i} = a$ holds for all requests $r_i$, where $a \geq Lt$ is a constant, is called the CSP-F problem. For the variable booking time variant, the booking time $\tilde{t}_{r_i}$ of any request $r_i$ must satisfy $t_{r_i} - b_u \leq \tilde{t}_{r_i} \leq t_{r_i} - b_l$, where $b_l$ and $b_u$ are constants, with $Lt \leq b_l < b_u$, that specify the minimum and maximum travel time, respectively, of the time interval between the booking time and the start time. The problem for the variable booking time variant is called the CSP-V problem. We do not require that the algorithm assigns an accepted request to a specific server immediately, provided that it ensures that some servers could serve that request. In some settings, however, it is not necessary for an algorithm to use this flexibility. Furthermore, we forbid unprompted moves. It means that the server is only allowed to make an empty move to other location which is the pick-up location of an accepted request.

The competitive ratio is used to measure the performance of an online algorithm [5]. For any request sequence $R$, let $P^A$ denote the objective value produced by an online (randomized or deterministic) algorithm $A$, and $P^*$ that obtained by an optimal scheduler $OPT$ that has full information about the request sequence in advance. The competitive ratio of algorithm $A$ is defined as $\rho^A = \sup_R \frac{P^*}{P^A}$. We say that $A$ is $\rho$-competitive if $P^* \leq \rho P^A$ (for the maximum objective) holds for all request sequences $R$. Let $ON$ be the set of all online algorithms for a problem. A value $\beta$ is a lower bound on the best possible competitive ratio if $\rho^A \geq \beta$ (for the maximum objective) for all $A$ in $ON$.

## 1.3   Contribution

In this paper, we consider the online car-sharing problem with $k$ servers and $m$ locations in a general network with both fixed booking times and variable booking times. An overview of our results is shown in Table 1. In Sect. 2, we prove the lower bounds. In Sect. 3, we propose a greedy algorithm that achieves the best possible competitive ratio.

**Table 1.** Lower and upper bounds for the online car-sharing problem

| Problem | Lower bound | Upper bound |
|---------|-------------|-------------|
| CSP-F | $O(L)$ | $O(L)$ |
| CSP-V | $O(L^\alpha)$  $(0 < \alpha < 1)$ | $O(L)$ |

## 2   Lower Bounds

In this section, we present lower bounds for the car-sharing problem. We use $ALG$ to denote any online algorithm and $OPT$ to denote an optimal scheduler. The set of requests accepted by $ALG$ is denoted by $R'$, and the set of requests accepted by $OPT$ by $R^*$.

Recall that two requests are in conflict if they can not be served by one server. Our basic strategy to obtain a lower bound is in place to ensure that most of the servers serve a request which is conflict with all other requests. For this purpose, the adversary releases requests phase by phase, and each phase includes many groups where each group contains $k$ identical requests. To implement this process, we place three rules for releasing requests: (1) Any two requests in one phase are in conflict; (2) In any two consecutive phases, any request in the last group of preceding phase and a request in the subsequent phase are not in conflict; (3) In any two phases, any request in the preceding phase, except for that in the last group, is in conflict with the request in the subsequent phase. Once an algorithm accepts less than a number of requests $(\max\{\lfloor \frac{k}{m^\alpha} \rfloor, 1\}$ for CSP-F, and 1 for CSP-V) in current request group, the adversary releases requests in the next phase; otherwise, it will release requests of next group in the current phase. Owing to the above rules, we will observe that most of the servers (the servers who does not serve a request in the last group of any phase) serve at most one request.

In the CSP-F and CSP-V problem, consider a line network with $m + 1$ locations and $t(0, i) = it$ for $1 \leq i \leq m$ (Fig. 1 shows an example). Note that $L = m$. Recall that a request $r$ is denoted by $r = (\tilde{t}_r, t_r, p_r, \dot{p}_r)$ and is specified by the booking time $\tilde{t}_r$, the start time $t_r$, the pick-up location $p_r$, and the drop-off location $\dot{p}_r$.

**Fig. 1.** Illustration of a line

The adversary presents requests in $\gamma$ phases ($\gamma = m$ in variable booking variant), where phase $i$, for $1 \leq i \leq \gamma$, consists of $l_i$ groups of requests, Let $R_{i,j}$ denote the set of requests in group $j$ of phase $i$. $R_{i,j}$ consists of $k$ identical requests. Let $k_{i,j}$ be the number of requests accepted by $ALG$ in $R_{i,j}$. Since any two requests in $R_{i,j}$ are in conflict, $k_{i,j}$ also is the number of servers that $ALG$ used in group $j$ of phase $i$.

## 2.1   Lower Bound for CSP-F

**Theorem 1.** *No deterministic online algorithm for the car-sharing problem with fixed bookings can achieve a competitive ratio smaller than $O(L^\alpha)$ for any $0 < \alpha < 1$.*

*Proof.* The adversary releases requests based on the releasing rule shown in Algorithm 1. For each phase, if $ALG$ accepts no more than $\max\{\lfloor \frac{k}{m^\alpha} \rfloor, 1\}$ requests in a group, without loss of generality, suppose phase $i$ group $j$ , the adversary presents the requests in the first group of the next phase, phase $i + 1$ group 1; otherwise, the adversary presents the requests in next group, phase $i$ group $j + 1$.

---

**Algorithm 1.** Releasing rule for car-sharing problem with fixed bookings

---

*Initialization*: The adversary presents the requests in phase 1 group 1, $R_{1,1}$.
$i = 1, j = 1$.
While $\dot{p}_{r_{i,j}} \leq m$
    if $k_{i,j} \leq \max\{\lfloor \frac{k}{m^\alpha} \rfloor, 1\}$, then $l_i = j$, $i = i + 1$, $j = 1$ and the adversary releases $R_{i,j}$;
    else  $j = j + 1$, and the adversary releases $R_{i,j}$;
*Output*: $\gamma = i$ and $l_h$ for all $1 \leq h \leq \gamma$; $k_{h,j}$ for all $1 \leq h \leq \gamma$ and $1 \leq j \leq l_h$.

---

Let $a$ be the advance time of a request in the the fixed booking variant, i.e., the booking time of each request is the start time of each request minus $a$. If we know the start time of $r_{i,j}$, we may ignore the booking time when we specify $r_{i,j}$. Specifically, the requests in phase $i$ group $j$ use the following notation.

- $R_{1,1}$ consists of $k$ copies of the request $r_{1,1}$ with $r_{1,1} = (\nu \cdot t - a, \nu \cdot t, 0, 1)$ for some $\nu$ such that $\nu \in \mathbb{N}$ and $\nu \cdot t - a \geq Lt$;
- $R_{i,1}$ ($i > 1$ and $l_{i-1} > 1$) consists of $k$ copies of the request $r_{i,1}$ with start time $t_{r_{i,1}} = t_{r_{i-1,l_{i-1}}} + t + \varepsilon_{i,1}$, pick-up location $p_{r_{i,1}} = \sum_{h=1}^{i-1} l_h$, and drop-off location $\dot{p}_{r_{i,1}} = \sum_{h=1}^{i-1} l_h + 1$, where $\varepsilon_{i,1} = \frac{\varepsilon_{i-1,l_{i-1}} - \varepsilon_{i-1,l_{i-1}-1}}{m^{i-1}}$;
- $R_{i,1}$ ($i > 1$ and $l_{i-1} = 1$) consists of $k$ copies of the request $r_{i,1}$ with start time $t_{r_{i,1}} = t_{r_{i-1,l_{i-1}}} + t + \varepsilon_{i,1}$, pick-up location $p_{r_{i,1}} = \sum_{h=1}^{i-1} l_h$, and drop-off location $\dot{p}_{r_{i,1}} = \sum_{h=1}^{i-1} l_h + 1$, where $\varepsilon_{i,1} = \varepsilon_{i-1,1}$;
- $R_{i,j}$ ($j > 1$) consists of $k$ copies of the request $r_{i,j}$ with start time $t_{r_{i,j}} = t_{r_{i,1}} + (j-1)t - \varepsilon_{i,j}$, pick-up location $p_{r_{i,j}} = p_{r_{i,1}} + j - 1$, and drop-off location $\dot{p}_{r_{i,j}} = p_{r_{i,1}} + j$, where $0 < \varepsilon_{i,j-1} < \varepsilon_{i,j} < \varepsilon_{i,1}$. Set $\varepsilon_{1,1} = t$.

Firstly, we state that the requests in phase $i$ group $j$ are reasonable. We claim that the requests start in order of phases, and the requests in each phase start in order of groups. Observe that each request $r_{i,j}$ satisfies $\dot{p}_{r_{i,j}} - p_{r_{i,j}} = 1$ and the booking time of each request is $\hat{t}_{r_{i,j}} = t_{r_{i,j}} - a$.

Based on the definition of $\varepsilon_{i,j}$ in $R_{i,1}$ ($i > 1$ and $l_{i-1} = 1$), we have $\varepsilon_{i,1} = \varepsilon_{i-1,1}$. Based on the definition of $\varepsilon_{i,j}$ in $R_{i,1}$ ($i > 1$ and $l_{i-1} > 1$) and $R_{i,j}$ ($j > 1$), we know that $\varepsilon_{i,1} = \frac{\varepsilon_{i-1,l_{i-1}} - \varepsilon_{i-1,l_{i-1}-1}}{m^{i-1}}$ and $\varepsilon_{i-1,l_{i-1}} - \varepsilon_{i-1,l_{i-1}-1} < \varepsilon_{i-1,1}$, we have $0 < \varepsilon_{i,1} \leq \varepsilon_{i-1,1}$. Since $\varepsilon_{1,1} = t$, by induction, for any $i, h \in \{1, 2, ..., \gamma\}$ with $i < h$, we have $0 < \varepsilon_{h,1} \leq \varepsilon_{i,1} \leq t$.

In phase $i$ group $j$ and phase $i$ group $f$ ($1 \leq f < j$), since $\varepsilon_{i,j} < t$ and $t_{r_{i,j}} = t_{r_{i,1}} + (j-1)t - \varepsilon_{i,j} > t_{r_{i,1}} + (f-1)t > t_{r_{i,f}}$, we have $t_{r_{i,j}} > t_{r_{i,f}}$ for any $i$. It means the requests in one phase start in order of groups.

For any two consecutive phase $i-1$ and $i$ ($1 < i \leq \gamma$), since $t_{r_{i,1}} = t_{r_{i-1,l_{i-1}}} + t + \varepsilon_{i,1}$ with $\varepsilon_{i,1} > 0$, we have $t_{r_{i,1}} > t_{r_{i-1,l_{i-1}}}$ for any $1 < i \leq \gamma$. That means the requests start in order of phases.

Next, we state that the requests satisfy the releasing rule.

(1) In phase $i$, for any two groups $j, f \in \{1, 2, ..., l_i\}$ with $j > f$, we have $t_{r_{i,j}} = \hat{t}_{r_{i,f}} + (j-f-1)t + \varepsilon_{i,f} - \varepsilon_{i,j} < \hat{t}_{r_{i,f}} + (j-f-1)t$, $p_{r_{i,j}} = \dot{p}_{r_{i,f}} + j - f - 1$. Therefore, $t_{r_{i,j}} - \hat{t}_{r_{i,f}} < (p_{r_{i,j}} - \dot{p}_{r_{i,f}}) \cdot t$ for phase $i$ group $j, f \in \{1, 2, ..., l_i\}$ with $i > 1$ and $j > f$. That means request $r_{i,f}$ and request $r_{i,j}$ are in conflict (all requests in phase 1 in Fig. 2 show an example).

(2) In phase $i$ group $j$ ($1 < j \leq l_i$), we have $t_{r_{i,j}} = t_{r_{i,1}} + (j-1)t - \varepsilon_{i,j} = t_{r_{i-1,l_{i-1}}} + t + \varepsilon_{i,1} + (j-1)t - \varepsilon_{i,j} \geq t_{r_{i-1,l_{i-1}}} + j \cdot t$, $p_{r_{i,j}} = \dot{p}_{r_{i-1,l_{i-1}}} + j - 1$. Therefore, $t_{r_{i,j}} \geq \hat{t}_{r_{i-1,l_{i-1}}} + j \cdot t - t$, we have $t_{r_{i,j}} - \hat{t}_{r_{i-1,l_{i-1}}} \geq (p_{r_{i,j}} - \dot{p}_{r_{i-1,l_{i-1}}}) \cdot t$ for any $i > 1$ and $j \leq l_i$. That means any request in the last group of phase $i-1$ and a request in phase $i$ are not in conflict (all red dotted requests in Fig. 2 show an example).

(3) In phase $i$ group $j$ and phase $h$ group $f$ ($h > i$, $j < l_i$), according to the notion of request in mentioned cases, we have $p_{r_{h,f}} = \dot{p}_{r_{i,j}} + (l_i - j) + \sum_{\nu=i+1}^{h-1} l_\nu + f - 1$, $t_{r_{h,f}} = t_{r_{h-1,l_{h-1}}} + t + \varepsilon_{h,1}$ if $f = 1$, and $t_{r_{h,f}} = t_{r_{h,1}} + (f-1)t - \varepsilon_{h,f}$ if $f > 1$.

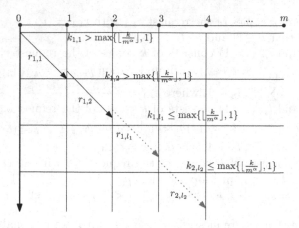

**Fig. 2.** Illustration of the request sequences $R$: $ALG$ accepts no more than $\max\{\lfloor\frac{k}{m^\alpha}\rfloor, 1\}$ requests in last groups $R_{1,l_1}$ and $R_{2,l_2}$; a request in $R_{1,j}$ ($j < l_1$) and a request in $R_{1,f}$ ($f \neq j$) are in conflict; a request in $R_{1,l_1}$ and a request in $R_{i,j}$ ($i > 1$) are not in conflict; a request in $R_{1,j}$ ($j < l_1$) and a request in $R_{i,j}$ ($i > 1$) are in conflict.

If $f = 1$, we have $t_{r_{h,f}} = t_{r_{h-1,l_{h-1}}} + t + \varepsilon_{h,1} = t_{r_{i,1}} + \sum_{\nu=i}^{h-1} l_\nu \cdot t + \sum_{\nu=i+1}^{h} \varepsilon_{\nu,1} - \sum_{\nu=i}^{h-1} \varepsilon_{\nu,l_\nu} = t_{r_{i,j}} - (j-1)t + \varepsilon_{i,j} + \sum_{\nu=i}^{h-1} l_\nu \cdot t + \sum_{\nu=i+1}^{h} \varepsilon_{\nu,1} - \sum_{\nu=i}^{h-1} \varepsilon_{\nu,l_\nu}$. Since $\varepsilon_{i+1,1} < \varepsilon_{i,l_i} - \varepsilon_{i,l_i-1} \leq \varepsilon_{i,l_i} - \varepsilon_{i,j}$, and $\varepsilon_{\nu,1} < \varepsilon_{\nu-1,l_{\nu-1}}$, we have $t_{r_{i,j}} - (j-1)t + \varepsilon_{i,j} + \sum_{\nu=i}^{h-1} l_\nu \cdot t + \sum_{\nu=i+1}^{h} \varepsilon_{\nu,1} - \sum_{\nu=i}^{h-1} \varepsilon_{\nu,l_\nu} < t_{r_{i,j}} - (j-1)t + \sum_{\nu=i}^{h-1} l_\nu \cdot t$, and then $t_{r_{h,f}} - \dot{t}_{r_{i,j}} < (p_{r_{h,f}} - \dot{p}_{r_{i,j}}) \cdot t$.

If $f > 1$, we have $t_{r_{h,f}} = t_{r_{h,1}} + (f-1)t - \varepsilon_{h,f}$. Similarly, since $(f-1)t - \varepsilon_{h,f} < (p_{r_{h,f}} - p_{r_{h,1}}) \cdot t$, we have $t_{r_{h,f}} - \dot{t}_{r_{i,j}} < ((p_{r_{h,f}} - p_{r_{h,1}}) + (p_{r_{h,1}} - \dot{p}_{r_{i-1,l_{i-1}}})) \cdot t < (p_{r_{h,f}} - \dot{p}_{r_{i,j}}) \cdot t$. That means any other requests (except the requests in the last group of phase $i$) in phase $i$ and a request in subsequent phase $h$ ($h > i$) are in conflict (the black requests in the first phase and red dotted requests in the second phase in Fig. 2 show an example).

Finally, we divide the requests accepted by $ALG$ into two parts: (1) $R_1'$, the requests accepted by $ALG$ in $R_{i,j}$, where $1 \leq i \leq \gamma$ and $j < l_i$; (2) $R_2'$, the requests accepted by $ALG$ in $R_{i,j}$, where $1 \leq i \leq \gamma$ and $j = l_i$.

Observe that in each group $R_{i,j}$ ($1 \leq i \leq \gamma$ and $j < l_i$), $ALG$ uses more than $\lfloor\frac{k}{m^\alpha}\rfloor$ servers. We know that the number of groups such that $R_{i,j}$ ($1 \leq i \leq \gamma$ and $j < l_i$) is no more than $m^\alpha$. Otherwise, $ALG$ uses more than $(\lfloor\frac{k}{m^\alpha}\rfloor+1) \cdot m^\alpha > k$ servers, which derives a contradiction. From this it follows that $\gamma \geq m - m^\alpha$.

Since in each group $R_{i,l_i}$ ($1 \leq i \leq \gamma$), $ALG$ uses no more than $\lfloor\frac{k}{m^\alpha}\rfloor$ servers, we have $|R_2'| \leq \lfloor\frac{k}{m^\alpha}\rfloor \cdot \gamma$. Suppose there are $k_1 \leq k$ servers who accept a request in $R_{i,j}$ ($1 \leq i \leq \gamma$ and $j < l_i$). Therefore, $|R'| = |R_1'| + |R_2'| \leq k + \lfloor\frac{k}{m^\alpha}\rfloor \cdot \gamma$.

$OPT$ accepts all requests in $R_{i,l_i}$ where $1 \leq i \leq \gamma$ (based on releasing rule 2, any two requests in the last group of two different phases are not in conflict, i.e., a request in $R_{i,l_i}$ and a request in $R_{h,l_h}$ are not in conflict), we have $|R^*| = k \cdot \gamma$. Since $|R'| \leq k + \lfloor \frac{k}{m^\alpha} \rfloor \cdot \gamma$, we get $|R^*|/|R'| = \frac{k \cdot \gamma}{k + \lfloor \frac{k}{m^\alpha} \rfloor \cdot \gamma} = O(m^\alpha)$.

## 2.2   Lower Bound for CSP-V

**Theorem 2.** *No deterministic online algorithm for the car-sharing problem with variable bookings can achieve a competitive ratio smaller than $O(L)$.*

*Proof.* The adversary releases requests based on the releasing rule shown in Algorithm 2. For each phase, if $ALG$ accepts no requests in a group, without loss of generality, suppose phase $i$ group $j$, the adversary presents the requests in the first group of the next phase, phase $i+1$ group 1; otherwise, the adversary presents the requests in next group, phase $i$ group $j + 1$.

---

**Algorithm 2.** Releasing Rule for car-sharing problem with variable bookings

---

*Initialization*: The adversary presents the requests in phase 1 group 1, $R_{1,1}$.
$i = 1$, $j = 1$.
While $i \leq m$ do
    While $j \leq k + 1$ do
        if $k_{i,j} > 0$, then $j = j + 1$ and the adversary releases the requests in $R_{i,j}$;
        else $l_i = j$, $i = i + 1$, $j = 1$ and the adversary releases the requests in $R_{i,j}$;
*Output*: $\gamma = i$ and $l_h$ for all $1 \leq h \leq \gamma$; $k_{h,j}$ for all $1 \leq h \leq \gamma$ and $1 \leq j \leq l_h$.

---

Specifically, the requests in phase $i$ group $j$ uses the following notation:

- $R_{1,1}$ consists of $k$ copies of the request $r_{1,1}$ with $(\nu \cdot t - b_u, \nu \cdot t, 0, 1)$ for some $\nu$ such that $\nu \in \mathbb{N}$ and $\nu \cdot t - b_u \geq L \cdot t$;
- $R_{i,1}$ ($i > 1$ and $j = 1$) consists of $k$ copies of the request $r_{i,1}$ with booking time $t_{r_{i,1}} - b_u$, start time $t_{r_{i-1,l_{i-1}}} + t + \frac{k}{k+1} \cdot \frac{\min\{b_u - b_l, t\}}{(k+1)^{i-1}}$, pick-up location $i - 1$, and drop-off location $i$;
- $R_{i,j}$ ($i \geq 1$ and $j > 1$) consists of $k$ copies of the request $r_{i,j}$ with booking time $\tilde{t}_{r_{i,1}}$, start time $t_{r_{i,1}} - (j - 1) \cdot \frac{\min\{b_u - b_l, t\}}{(k+1)^i}$, pick-up location $i - 1$, and drop-off location $i$.

Observe that $l_i \leq k + 1$ for all $1 \leq i \leq m$.

Firstly, we state that the requests in phase $i$ group $j$ are reasonable. Observe that the time between the start time and the booking time of each request is between $b_l$ and $b_u$. We claim that the requests are presented in order of phases, and the requests in one phase are presented in order of groups.

In phase $i$ group $j$ (recall that $j \leq l_i \leq k+1$), because of the booking interval $t_{r_{i,j}} - \tilde{t}_{r_{i,j}} = t_{r_{i,1}} - \tilde{t}_{r_{i,1}} - (j - 1) \cdot \frac{\min\{b_u - b_l, t\}}{(k+1)^i}$ and $t_{r_{i,1}} - \tilde{t}_{r_{i,1}} = b_u$, we know that $b_l \leq t_{r_{i,j}} - \tilde{t}_{r_{i,j}} \leq b_u$, i.e., all requests satisfying the variable booking times, the booking interval is not greater than $b_u$ and is not less than $b_l$.

Observe that the booking time of all requests in phase $i$ ($1 \leq i \leq m$) are equal. We know that in one phase, the requests are booked in order of non-decreasing groups. For any $i$, the booking time of requests in phase $i$ is $\tilde{t}_{r_{i,1}} = t_{r_{i,1}} - b_u = t_{r_{i-1,l_{i-1}}} + t + \frac{k}{k+1} \cdot \frac{\min\{b_u - b_l, t\}}{(k+1)^{i-1}} - b_u \geq \tilde{t}_{r_{i-1,1}} - b_u$. From this we know that all requests are booked in order of non-decreasing phases.

Next, we state that the requests satisfy the releasing rule.

(1) In phase $i$, since all requests start at the same location and the start time of any two requests are not greater than $2t$ (because $l_i \leq k+1$, $\max_j t_{r_{i,j}} - \min_j t_{r_{i,j}} = t_{r_{i,1}} - t_{r_{i,l_i}} \leq (k+1)\frac{\min\{b_u - b_l, t\}}{(k+1)^i} \leq t$), we know that any two requests in one phase are in conflict (all requests in phase 1 in Fig. 3 show an example).

(2) In phase $i$ group $j$, since $j \leq l_i \leq k+1$, $\frac{k}{k+1} \cdot \frac{\min\{b_u - b_l, t\}}{(k+1)^{i-1}} - (j-1) \cdot \frac{\min\{b_u - b_l, t\}}{(k+1)^i} \geq 0$, we have $t_{r_{i,j}} = t_{r_{i,1}} - (j-1) \cdot \frac{\min\{b_u - b_l, t\}}{(k+1)^i} = t_{r_{i-1,l_{i-1}}} + t + \frac{k}{k+1} \cdot \frac{\min\{b_u - b_l, t\}}{(k+1)^{i-1}} - (j-1) \cdot \frac{\min\{b_u - b_l, t\}}{(k+1)^i} \geq t_{r_{i-1,l_{i-1}}}$, $p_{r_{i,j}} = \dot{p}_{r_{i-1,l_{i-1}}}$. Therefore, $t_{r_{i,j}} - \dot{t}_{r_{i-1,l_{i-1}}} \geq (p_{r_{i,j}} - \dot{p}_{r_{i-1,l_{i-1}}})t = 0$ for any $i > 1$ and $1 \leq j \leq l_i$. That means any request in the last group of phase $i - 1$ and a request in phase $i$ are not in conflict (all red dotted requests in Fig. 3 show an example).

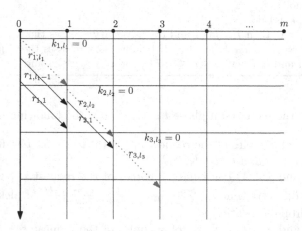

**Fig. 3.** Illustration of the request sequences $R$: $ALG$ accepts no requests in last groups $R_{1,l_1}$ and $R_{2,l_2}$; a request in $R_{1,j}$ ($j < l_1$) and a request in $R_{1,f}$ ($f \neq j$) are in conflict; a request in $R_{1,l_1}$ and a request in $R_{i,j}$ ($i > 1$) are not in conflict; a request in $R_{1,j}$ ($j < l_1$) and a request in $R_{i,j}$ ($i > 1$) are in conflict.

(3) In phase $i$ group $j$ ($i > 1$, $l_{i-1} > 1$ and $1 < j < l_{i-1}$), because of $t_{r_{i-1,j}} = t_{r_{i-1,1}} - (j-1) \cdot \frac{\min\{b_u - b_l, t\}}{(k+1)^{i-1}}$, we have $t_{r_{i-1,j}} \leq t_{r_{i-1,1}} - \frac{\min\{b_u - b_l, t\}}{(k+1)^{i-1}}$, and then $\dot{t}_{r_{i-1,j}} \leq t_{r_{i,1}} - \frac{k}{k+1} \cdot \frac{\min\{b_u - b_l, t\}}{(k+1)^{i-1}} - \frac{\min\{b_u - b_l, t\}}{(k+1)^{i-1}}$ (based on the definition of $t_{r_{i,1}}$). For any $1 < f \leq l_i$, $t_{r_{i,f}} = t_{r_{i,1}} - (f-1) \cdot \frac{\min\{b_u - b_l, t\}}{(k+1)^i} \geq t_{r_{i,1}} - \frac{k}{k+1} \cdot \frac{\min\{b_u - b_l, t\}}{(k+1)^{i-1}} - \frac{\min\{b_u - b_l, t\}}{(k+1)^{i-1}}$ ($f \leq l_i \leq k+1$), we have $\dot{t}_{r_{i-1,j}} \leq t_{r_{i,f}}$ and thus $t_{r_{i-1,j}} \leq \dot{t}_{r_{i,f}}$.

Since $t_{r_{i-1,j}} = t_{r_{i-1,1}} - (j-1) \cdot \frac{\min\{b_u - b_l, t\}}{(k+1)^{i-1}}$, we have $t_{r_{i-1,j}} \geq t_{r_{i-1,l_{i-1}}} + \frac{\min\{b_u - b_l, t\}}{(k+1)^{i-1}} = t_{r_{i-1,1}} - (l_{i-1} - 1) \cdot \frac{\min\{b_u - b_l, t\}}{(k+1)^{i-1}} + \frac{\min\{b_u - b_l, t\}}{(k+1)^{i-1}}$. From this it follows that $\dot{t}_{r_{i-1,j}} \geq t_{r_{i,1}} - \frac{k}{k+1} \cdot \frac{\min\{b_u - b_l, t\}}{2(k+1)^{i-1}} - (l_{i-1} - 1) \cdot \frac{\min\{b_u - b_l, t\}}{(k+1)^{i-1}} + \frac{\min\{b_u - b_l, t\}}{(k+1)^{i-1}} > t_{r_{i,1}} - 2t$ ($f \leq l_i \leq k+1$). For all $1 < f \leq l_i$, $t_{r_{i,f}} \leq t_{r_{i,1}}$, then we have $t_{r_{i,f}} - \dot{t}_{r_{i-1,j}} < 2t$.

Note that $p_{r_{i,f}} = i - 1 = \dot{p}_{r_{i-1,j}}$ for all $1 \leq f \leq l_i$ and $1 \leq j \leq l_{i-1}$. Since $t_{r_{i-1,j}} - t_{r_{i,f}} < 0 = (\dot{p}_{r_{i-1,j}} - p_{r_{i,f}})t$ and $t_{r_{i,f}} - \dot{t}_{r_{i-1,j}} < 2t = (\dot{p}_{r_{i,f}} - p_{r_{i-1,j}})t$, we know that request $r_{i-1,j}$ and request $r_{i,f}$ are in conflict (the black requests in the first phase and all requests in the second phase in Fig. 3 show an example).

Similar to the proof in fixed booking variant, we can prove that for any phase $i < \gamma = m$, if a server accept a request $r_{i,j} \in R_{i,j}$ ($j \neq l_i$), this server can not accept any more request in a subsequent phase $h$ ($h > i$). Since no $ALG$ server accepts request in the last group of each phase, each server of $ALG$ can accept at most one request. From this it follows that $ALG$ can accept at most $k$ requests in $R$, and $|R'| \leq k$.

Observe that for any two consecutive phases $i$ and $i+1$, a request in $R_{i,l_i}$ and a request in $R_{i+1,l_{i+1}}$ are not in conflict (based on releasing rule 2). $OPT$ accepts all requests in the last group of each phase, i.e., phase $i$ group $l_i$ for all $1 \leq i \leq m$ (by mathematical induction). That means each server accepts $m$ requests (all red requests in Fig. 3 show an example). Thus the requests accepted by $OPT$ is $|R^*| = mk$. Since $|R'| \leq k$, we get $|R^*|/|R'| \geq m$. The claimed lower bound of $O(L)$ follows.

# 3 Upper Bound for the Online Car-Sharing Problem with Fixed Booking or Variable Booking

We propose a Greedy Algorithm (GA) for the CSP-F and CSP-V problem, shown in Algorithm 3. It can be stated in a simple way: When a request $r_i$ arrives, if it is acceptable to any server, accept $r_i$ with that server; otherwise, reject it. For an arbitrary request sequence $R = \{r_1, r_2, r_3, \ldots, r_n\}$, note that we have $\tilde{t}_{r_i} \leq \tilde{t}_{r_{i+1}}$ for $1 \leq i < n$. Denote the requests accepted by $OPT$ by $R^* = \{r_1^*, r_2^*, \ldots, r_{|R^*|}^*\}$ and the requests accepted by GA by $R' = \{r_1', r_2', \ldots, r_{|R'|}'\}$ indexed in order of non-decreasing booking times.

**Theorem 3.** *The Greedy algorithm is $O(L)$-competitive for CSP-V (or CSP-F) problem.*

*Proof.* For each server $s_e^* \in S^*$ in $OPT$ solution, let $R_e^*$ be the set of requests accepted by $s_e^*$ and $\bar{R}_{ej}^*$ be the set of requests accepted by $s_e^*$ that are not accepted by server $s_j'$ in $ALG$. We claim that $|\bar{R}_{ej}^*| \leq \alpha \cdot L \cdot |R_j'|$ for some constant $\alpha$ and any server $s_e^* \in S^*$, $s_j' \in S'$. If this claim holds, we get that $|R_e^*| \leq |R_j'| + |\bar{R}_{ej}^*| \leq |R_j'| + \alpha \cdot L \cdot |R_j'| = O(L) \cdot |R_j'|$, proving the lemma.

It remains to prove the claim. Consider any request $r_h$ in $\bar{R}_{ej}^*$. As $s_j'$ did not accept $r_h$, $s_j'$ must have accepted another request $r_c$ with start time in

---

**Algorithm 3.** Greedy Algorithm (GA)

---

*Input*: $k$ servers, requests arrive over time.

*Step*: When request $r_i$ arrives, accept $r_i$ if $r_i$ is acceptable;

*Note 1.* $R'_j$ is the list of requests accepted by server $s_j$ before $r_i$ is released.

*Note 2.* $r_i$ is acceptable to a server $s_j$ if and only if $r_i$ is not in conflict with the requests in $R'_j$, i.e., $\forall r'_q \in R'_j, t_{r_i} - t_{r'_q} \geq t(p_{r_i}, \dot{p}_{r'_q})$ if $t_{r_i} \geq t_{r'_q}$, and $t_{r'_q} - t_{r_i} \geq t(p_{r'_q}, \dot{p}_{r_i})$ if $t_{r_i} < t_{r'_q}$.

---

$(t_{r_h} - 2Lt, t_{r_h}]$ or $[t_{r_h}, t_{r_h} + 2Lt)$ (resp. $(t_{r_h} - 2Lt, t_{r_h}]$ in the problem with fixed bookings); otherwise, the $2Lt$ time units is sufficient for $s'_j$ to serve the previous or later request and make an empty movement to the pick-up location $p_{r_h}$. We charge $r_h$ to $r_c$. In this way, every request in $\bar{R}^*_{ej}$ is charged to a request in $R'_j$.

We bound the number of requests that can be charged to a single request $r_c$ in $R'_j$. By the above charging scheme, every request that is accepted by $s^*_e$ and charges $r_c$ has a start time in $(t_{r_c} - 2Lt, t_{r_c} + 2Lt)$. As all requests have travel time at least $t$, the start times of any two consecutive requests accepted by $s^*_e$ differ by at least $t$. A half-open interval of length $4Lt$ can contain at most $\frac{4Lt}{t} = 4L$ request start times, and hence $r_c$ is charged by at most $4L$ requests from $\bar{R}^*_{ej}$. This establishes the claim, with $\alpha = 4$. For any servers $s^*_e \in S^*$ and $s'_j \in S'$, $|R^*_{ej}| \leq O(L) \cdot |R'_j|$. Since $|R^*| = \sum_{i=1}^{k*} |R^*_i|$ and $|R'| = \sum_{i=1}^{k} |R'_i|$, we have $|R^*|/|R'| \leq O(L)$. The theorem is proved.

# 4  Conclusion

We have studied the online car-sharing problem with both fixed booking times and variable booking times that is motivated by applications such as car-sharing and taxi dispatching. The problem is an extension of online car-sharing problems researched by Luo et al. For both variants we have given the matching lower and upper bounds on the competitive ratio. The upper bounds are all achieved by the same greedy algorithm (GA). As mentioned in the introduction, we only consider the deterministic algorithms. The main contribution is that we proved the no deterministic online algorithm for CSP-F (resp. CSP-V) can achieve a competitive ratio smaller than $O(L^\alpha)$, $0 < \alpha < 1$ (resp. $O(L)$) when the goal is to maximize the number of satisfied requests.

In the future research, it would be interesting to study this problem from the stochastic viewpoint. It would also be interesting to determine how the constraints on the booking time affect the competitive ratio for CSP-F and CSP-V.

**Acknowledgments.** This work was partially supported by the China Postdoctoral Science Foundation (Grant No. 2016M592811), and the National Nature Science Foundation of China (Grant No. 71701048).

# References

1. The future of driving: Seeing the back of the car. The Economist (2012)
2. Ascheuer, N., Krumke, S.O., Rambau, J.: Online dial-a-ride problems: minimizing the completion time. In: Reichel, H., Tison, S. (eds.) STACS 2000. LNCS, vol. 1770, pp. 639–650. Springer, Heidelberg (2000). https://doi.org/10.1007/3-540-46541-3_53
3. Bjelde, A., et al.: Tight bounds for online TSP on the line. In: Proceedings of the 28th Annual ACM-SIAM Symposium on Discrete Algorithms (SODA 2017), pp. 994–1005. SIAM (2017). https://doi.org/10.1137/1.9781611974782.63
4. Böhmová, K., Disser, Y., Mihalák, M., Šrámek, R.: Scheduling transfers of resources over time: towards car-sharing with flexible drop-offs. In: Kranakis, E., Navarro, G., Chávez, E. (eds.) LATIN 2016. LNCS, vol. 9644, pp. 220–234. Springer, Heidelberg (2016). https://doi.org/10.1007/978-3-662-49529-2_17
5. Borodin, A., El-Yaniv, R.: Online Computation and Competitive Analysis. Cambridge University Press, Cambridge (1998)
6. Krumke, S.O., de Paepe, W.E., Poensgen, D., Lipmann, M., Marchetti-Spaccamela, A., Stougie, L.: On minimizing the maximum flow time in the online dial-a-ride problem. In: Erlebach, T., Persinao, G. (eds.) WAOA 2005. LNCS, vol. 3879, pp. 258–269. Springer, Heidelberg (2006). https://doi.org/10.1007/11671411_20
7. Christman, A., Forcier, W., Poudel, A.: From theory to practice: maximizing revenues for on-line dial-a-ride. J. Comb. Optim. **35**(2), 512–529 (2018)
8. Lipton, R.J., Tomkins, A.: Online interval scheduling. In: Proceedings of the Fifth Annual ACM-SIAM Symposium on Discrete Algorithms, 23–25 January 1994, Arlington, Virginia, USA, pp. 302–311 (1994). http://dl.acm.org/citation.cfm?id=314464.314506
9. Luo, K., Erlebach, T., Xu, Y.: Car-sharing between two locations: online scheduling with flexible advance bookings. In: Wang, L., Zhu, D. (eds.) COCOON 2018. LNCS, vol. 10976, pp. 242–254. Springer, Cham (2018). https://doi.org/10.1007/978-3-319-94776-1_21
10. Luo, K., Erlebach, T., Xu, Y.: Car-sharing between two locations: online scheduling with two servers. In: 43rd International Symposium on Mathematical Foundations of Computer Science, MFCS 2018, 27–31 August 2018, Liverpool, UK. LIPIcs, vol. 117, pp. 50:1–50:14. Schloss Dagstuhl - Leibniz-Zentrum fuer Informatik (2018). https://doi.org/10.4230/LIPIcs.MFCS.2018.50
11. Luo, K., Erlebach, T., Xu, Y.: Online scheduling of car-sharing requests between two locations with many cars and flexible advance bookings. In: ISAAC 2018: The 29th International Symposium on Algorithms and Computation. LIPIcs, vol. 123. Schloss Dagstuhl - Leibniz-Zentrum fuer Informatik (2018, to appear)
12. Luo, K., Erlebach, T., Xu, Y.: Car-sharing on a star network: on-line scheduling with k servers. In: 36th International Symposium on Theoretical Aspects of Computer Science, STACS 2019, 13–16 March 2019, Berlin, Germany, pp. 51:1–51:14 (2019). https://doi.org/10.4230/LIPIcs.STACS.2019.51

# Improved Approximation Algorithm for Minimum Weight $k$-Subgraph Cover Problem

Pengcheng Liu[1], Xiaohui Huang[2], and Zhao Zhang[1]($\boxtimes$)

[1] College of Mathematics and Computer Science, Zhejiang Normal University,
Jinhua 321004, Zhejiang, China
hxhzz@sina.com
[2] Library and Information Center, Zhejiang Normal University,
Jinhua 321004, Zhejiang, China

**Abstract.** For a graph $G$, a vertex subset $C$ is a Connected $k$-Subgraph Cover (VCC$_k$) if every connected subgraph on $k$ vertices of $G$ contains at least one vertex from $C$. Using local ratio method, a $(k-1)$-approximation algorithm was given in [14] for the minimum weight VCC$_k$ problem under the assumption that the girth $g(G)$ (the length of a shortest cycle of $G$) is at least $k$. In this paper, we prove that a $(k-1)$-approximation can be achieved when $k \geq 5$ and $g(G) > 2k/3$. Although our algorithm also employs the local ratio method, the analysis has a big difference from that in [14], this is why the girth constraint can be relaxed from $k$ to $2k/3$.

**Keywords:** Connected $k$-subgraph cover · Approximation algorithm · Girth

## 1 Introduction

*Wireless sensor networks* (WSNs) have wide applications in the real world, including battlefield monitoring, traffic control, disaster detection, and home automation, etc. [4,8,12]. These applications bring several important research issues into studies, such as coverage, connectivity, and security [8,9,11]. In a WSN, a sensor has limited capabilities and is vulnerable to be captured, which motivates many optimization models with an attempt to achieve higher security. For example, Zhang *et al.* proposed the *minimum weight connected k-subgraph cover problem* (MinWVCC$_k$). For an undirected graph $G = (V, E)$ and an integer $k$, a vertex subset $C$ is a *connected k-subgraph cover* (VCC$_k$) if every connected subgraph on $k$ vertices of $G$ contains at least one vertex from $C$. Given a weight function $w : V \to \mathbb{R}^+$ on the vertex set, where $\mathbb{R}^+$ is the set of non-negative real numbers, the MinWVCC$_k$ problem is to find a VCC$_k$ with the minimum weight

This research work is supported in part by NSFC (11771013, 11531011, 61751303), and ZJNSFC (LD19A010001, LY19A010018).

© Springer Nature Switzerland AG 2019
Y. Li et al. (Eds.): COCOA 2019, LNCS 11949, pp. 352–361, 2019.
https://doi.org/10.1007/978-3-030-36412-0_28

$w(C) = \sum_{u \in C} w_u$. The study on such a model is motivated by the consideration of security and supervisory control. For example, in a WSN, if an attacker knows at least $k$ related information fragments (where relation is measured by connectivity), then he can decode the whole information. Therefore, every connected $k$-vertex set must have at least one protected vertex to ensure security.

## 1.1  Related Works

Also motivated by a security consideration, a $k$-generalized Canvas scheme was proposed by Novotný [9] with the attempt to guarantee data integrality by requiring that every $k$-path (a path on $k$ vertices) contains at least one protected vertex. Since protected vertices cost more, it is desirable that the number of protected vertices is as small as possible. In graph theoretical language, the problem can be abstracted as a *minimum $k$-path vertex cover problem (MinVCP$_k$)* [1,13,14], the goal of which is to select a minimum weight vertex set $C$ containing at least one vertex from every $k$-path. When $k = 2$, both MinVCP$_2$ and MinVCC$_2$ problems are the *minimum vertex cover problem* (MinVC) which is a classic topic in combinatorial optimization and approximation algorithm. When $k = 3$, the MinVCP$_3$ problem coincides with the MinVCC$_3$ problem. For $k \geq 4$, previous studies show that there are some essential theoretical differences between them.

   In some other literatures [5,6], the MinVCC$_k$ problem is also called the *minimum $k$-vertex separator problem*. These studies are motivated by applications in graph partition. Finding a $k$-separator is often served as a subroutine to build algorithms based on divide-and-conquer methods [10].

## 1.2  Contribution

From the above subsection, it can be seen that studies on MinVCC$_k$ are much less than studies on MinVCP$_k$. In this paper, we study MinVCC$_k$. The main contribution of this paper is a $(k-1)$-approximation algorithm for MinWVCC$_k$ when $k \geq 5$ and $g(G) > 2k/3$, where $g(G)$ is the girth of graph $G$. In a previous work, Zhang et al. [14] obtained approximation ratio $k-1$ under the assumption that the graph has girth at least $k$. Our result improves the girth constraint from $k$ to $2k/3$ for $k \geq 5$. The algorithm uses the same local ratio method, but more refined analysis is used to improve the girth requirement. Recall that the algorithm presented by Lee in [6] has approximation ratio $O(\log k)$ which runs in time $n^{O(1)} + 2^{O(k)}n$. It should be noticed that Lee's algorithm is for the cardinality version while our algorithm applies for the weighted version. Furthermore, in a worst case, the constant in the big $O$ of Lee's paper is at least 8. So, for $k \leq 44$, our ratio $k-1$ is better than Lee's ratio.

## 2  Approximation Algorithm for MinWVCC$_k$

We first introduce some notations. For a graph $G = (V, E)$ with vertex weight $w : V \rightarrow \mathbb{R}^+$, denote by $V_{>0}(w)$ the set of vertices with positive weights. The degree of vertex $v$ in $G$ is denoted as $d_G(v)$. A weight function $w$ is a *degree-weight function* if there exists a constant $c$ such that $w(v) = c \cdot d_G(v)$ for every vertex $v \in V$. For two vertex sets $V_1$ and $V_2$, let $E(V_1, V_2)$ be the set of edges with one end in $V_1$ and the other end in $V_2$. For a vertex set $U \subseteq V(G)$, the symbol $G[U]$ represents the subgraph of $G$ induced by $U$, that is, the subgraph of $G$ with vertex set $U$ and edge set $\{uu' \in E(G) \colon u, u' \in U\}$. For a subgraph $\widetilde{G}$ of $G$ and a vertex set $U$ (not necessarily a subset of $V(\widetilde{G})$), the symbol $\widetilde{G}|_U$ represents the subgraph of $\widetilde{G}$ restricted on $U$, that is, $\widetilde{G}[V(\widetilde{G}) \cap U]$. The *order* of a graph $G$ is the number of vertices contained in $G$.

The algorithm uses local ratio method. The general theory for the local ratio method is that if the weight function $w$ can be decomposed into $w = w_1 + \cdots + w_t$ and one can find a feasible solution which is an $\alpha$-approximation w.r.t. every $w_i$, then the feasible solution is also an $\alpha$-approximation w.r.t. $w$. Our algorithm uses degree-weight functions to serve as those $w_i$'s. The details are described in Algorithm 1. In the algorithm, $w_i$ is the degree-weight function which is peeled off in the $i$-th iteration, $w^{(i)} = w - w_1 - \cdots - w_i$ is the residual weight, and $R_i$ is the set of vertices whose residual weight is reduced to zeros in the $i$-th iteration. After taking the vertices in $R_i$ temperately into the solution, the next iteration considers the residual subgraph $G_{i+1}$ which is induced by those vertices with positive residual weights. The process continues until there is no connected $k$-subgraph in the residual graph, and thus $\emptyset$ is an optimal solution for the residual instance. Then the algorithm adds back $R_i$'s in a backward manner, leaving off some redundant vertices to ensure the minimality while feasibility is kept. The sets obtained in the $i$-th iteration of the for loop is denoted by $F_i$.

---

**Algorithm 1.** Local Ratio Algorithm for MinWVCC$_k$ (LR-MinWVCC$_k$)

---

**Input:** A connected graph $G = (V, E)$ and a weight function $w$ on $V$.
**Output:** A vertex set $F$ which is a VCC$_k$ of $G$.

1: $i \leftarrow 0$, $G_0 \leftarrow G$, $w^{(i)} \leftarrow w$, $R_i \leftarrow \emptyset$.
2: **while** $G[V_{>0}(w^{(i)})]$ contains a connected component with at least $k$ vertices **do**
3:     $i \leftarrow i + 1$.
4:     $c_i \leftarrow \min\limits_{v \in V(G_{i-1})} \left\{ \dfrac{w^{(i-1)}(v)}{d_{G_{i-1}}(v)} \right\}$.
5:     $w_i(v) \leftarrow \begin{cases} c_i \cdot d_{G_{i-1}}(v), & v \in V(G_{i-1}) \\ 0, & v \in V \setminus V(G_{i-1}) \end{cases}$
6:     $w^{(i)}(v) = w^{(i-1)}(v) - w_i(v)$ for $v \in V$.
7:     $G_i \leftarrow G_{i-1}[V_{>0}(w^{(i)})]$.
8:     $R_i \leftarrow (V_{>0}(w^{(i-1)}) - V_{>0}(w^{(i)}))$
9: **end while**
10: $t \leftarrow i$, $F_t \leftarrow \emptyset$.
11: **for** $i = t-1, t-2, \ldots, 0$ **do**
12:     $F_i \leftarrow F_{i+1} \cup R'_{i+1}$, where $R'_{i+1}$ is a *minimal subset* of $R_{i+1}$ such that $F_{i+1} \cup R'_{i+1}$ is a VCC$_k$ of $G_i$.
13: **end for**
14: **return** $F \leftarrow F_0$.

---

By induction on $i = 0, \ldots, t$, it can be proved that $F_i$ is a minimal $VCC_k$ of $G_i$. As a consequence, we have the following result, which proves the correctness of the algorithm.

**Lemma 1.** *The output $F$ of Algorithm 1 is a minimal $VCC_k$ of graph $G$.*

The following lemma reveals a relation between the girth and the number of edges of a graph.

**Lemma 2.** *Suppose $G$ is a connected graph on $n$ vertices which has girth $g$. If $g > 2(n+1)/3$, then $G$ has at most $n$ edges.*

*Proof.* Since $g > 2(n+1)/3$ implies that $n \leq 2g - 2$. So, if $G$ has at least two cycles, then there exists two cycles $C_1, C_2$ which share a common path (see Fig. 1). Denote the common path by $P$, and let the paths $Q_1 = C_1 - P$ and $Q_2 = C_2 - P$. Since the union of $Q_1, Q_2$ and the two ends of $P$ is also a cycle in $G$, and because every cycle in $G$ has at least $g$ vertices, we have

$$|V(Q_1)| + |V(P)| \geq g,$$
$$|V(Q_2)| + |V(P)| \geq g,$$
$$|V(Q_1)| + |V(Q_2)| + 2 \geq g.$$

Adding the above three inequalities together, and by noticing that $|V(Q_1)| + |V(Q_2)| + |V(P)| \leq n$, we have

$$3g \leq 2(|V(Q_1)| + |V(Q_2)| + |V(P)|) + 2 \leq 2(n+1).$$

So, if $g > 2(n+1)/3$, then $G$ has at most one cycle. Then the lemma follows from the fact that a connected graph on $n$ vertices with at most one cycle has at most $n$ edges.

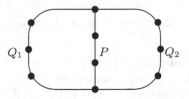

**Fig. 1.** Illustration for the proof of Lemma 2.

To analyze the approximation ratio of Algorithm 1, we first consider its effect on a degree-weight function.

**Lemma 3.** *Suppose $k \geq 5$, $G$ is a graph whose $g(G) > 2k/3$, $w$ is a degree-weight function on $G$, $F$ is a minimal $VCC_k$ of $G$, and $F^*$ is a minimum weight $VCC_k$ of $G$ w.r.t. $w$. Then $w(F) \leq (k-1)w(F^*)$.*

*Proof.* For a vertex subset $U \subseteq V$, denote $d_G(U) = \sum_{v \in U} d_G(v)$. Since $w$ is a degree-weight function, we have

$$\frac{w(F)}{w(F^*)} = \frac{c \sum_{v \in F} d_G(v)}{c \sum_{v \in F^*} d_G(v)} = \frac{d_G(F)}{d_G(F^*)}. \tag{1}$$

Rewrite $d_G(F)$ and $d_G(F^*)$ in the following way:

$$\begin{aligned} d_G(F) =& d_G(F \cap F^*) + d_G(F \setminus F^*) \\ =& 2|E(F \setminus F^*, F \cap F^*)| + (d_G(F \cap F^*) - |E(F \setminus F^*, F \cap F^*)|) \\ &+ (d_G(F \setminus F^*) - |E(F \setminus F^*, F \cap F^*)|), \end{aligned} \tag{2}$$

$$\begin{aligned} d_G(F^*) =& d_G(F \cap F^*) + d_G(F^* \setminus F) \\ =& |E(F \setminus F^*, F \cap F^*)| + (d_G(F \cap F^*) - |E(F \setminus F^*, F \cap F^*)|) \\ &+ d_G(F^* \setminus F). \end{aligned} \tag{3}$$

Comparing expressions (2) and (3), we have

$$\frac{d_G(F)}{d_G(F^*)} \leq \max\left\{2, \frac{d_G(F \setminus F^*) - |E(F \setminus F^*, F \cap F^*)|}{d_G(F^* \setminus F)}\right\}.$$

So, to prove the lemma, it suffices to prove

$$\begin{aligned} &\frac{\sum_{v \in F \setminus F^*}(d_G(v) - |E(v, F \cap F^*)|)}{d_G(F^* \setminus F)} \\ =& \frac{d_G(F \setminus F^*) - |E(F \setminus F^*, F \cap F^*)|}{d_G(F^* \setminus F)} \leq k - 1. \end{aligned} \tag{4}$$

**Proof Idea:** We prove (4) by a charging method. The idea is to establish a mapping $\varphi$ from $F \setminus F^*$ to subsets of $E(F^* \setminus F, V \setminus F^*)$, assign $(k-1)$ coins to every edge of $E(F^* \setminus F, V \setminus F^*)$, and for every vertex $v \in F \setminus F^*$, charge $d_G(v) - |E(v, F \cap F^*)|$ coins from $\varphi(v)$. Since $d_G(F^* \setminus F) \geq |E(F^* \setminus F, V \setminus F^*)|$, so if the number of coins are sufficient for such a charging, then inequality (4) is proved.

**Construction of the Mapping $\varphi$:** For any vertex $v \in F \setminus F^*$, by the minimality of $F$, graph $G[(V \setminus F) \cup \{v\}]$ has a component of size at least $k$. Denote this component as $G_v$. As a consequence, $E(v, V \setminus F) \neq \emptyset$. If $E(v, F^* \setminus F) \neq \emptyset$, we call such $v$ as a *type I vertex* and let $\varphi(v) = E(v, F^* \setminus F)$. If $E(v, F^* \setminus F) = \emptyset$, we call such $v$ as a *type II vertex*.

Next, we consider how to map a type II vertex $v$. Notice that $V(G_v) \cap F^* \neq \emptyset$. In fact, if $V(G_v) \cap F^* = \emptyset$, then $G_v$ is contained in a component of $G[V \setminus F^*]$, and thus $G_v$ will have at most $k - 1$ vertices, a contradiction. As a consequence, the subgraph $G_v$ has the following structure (see Fig. 2):

(i) $G_v$ has an edge $uu'$ such that $u' \in F^*$, $u \in V \setminus F^*$;

(ii) $v$ is adjacent with a vertex $\tilde{v} \in V \setminus (F \cup F^*)$, and there is a path in $G[V \setminus (F \cup F^*)]$ connecting $\tilde{v}$ to $u$.

In this case, we define $\varphi(v) = \{uu'\}$. Furthermore, if there are more than one choices, we choose $\varphi(v)$ to be a new edge as long as possible. Consider Fig. 2 for an example, both $uu'$ and $xx'$ can serve as $\varphi(v')$. But since $\varphi(v) = uu'$, we let $\varphi(v') = xx'$.

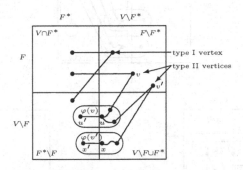

**Fig. 2.** An illustration for the construction of the mapping $\varphi$.

**Proof That the Number of Coins are Sufficient for Charging:** For each vertex $v \in F \setminus F^*$, since the component of $G[V \setminus F^*]$ containing $v$ has at most $k - 1$ vertices, we have $d_{G[V \setminus F^*]}(v) \leq k - 2$. Hence for each type I vertex $v$,

$$d_G(v) - |E(v, F \cap F^*)| \leq |E(v, F^* \setminus F)| + k - 2 \leq (k-1)|\varphi(v)|.$$

In this case, the number of coins on the edges of $\varphi(v)$ is sufficient for the charging of $v$.

Next, we show that

$$\text{the coins on the edges of } \bigcup_{v:\text{type II vertex}} \varphi(v) \text{ are sufficient} \tag{5}$$
$$\text{to pay for the charging of type II vertices.}$$

Since a type II vertex $v$ has $d_G(v) - |E(v, F \cap F^*)| \leq k - 2$ and $\varphi(v)$ has $k - 1$ coins, so if $\varphi(v_1) \neq \varphi(v_2)$ holds for any type II vertices $v_1 \neq v_2$, then (5) is true. The trouble arises when more than one type II vertices are mapped to the same edge.

For an edge $uu'$ in the image of $\varphi$, let $\varphi^{-1}(uu')$ be the set of vertices mapped onto $uu'$ by $\varphi$. By the construction of $\varphi$, every vertex $v \in \varphi^{-1}(uu')$ is connected to $u$ by a path in $G[V \setminus F^*]$, so all vertices in $\varphi^{-1}(uu')$ belong to a same component of $G[V \setminus F^*]$. Denote this component as $\widetilde{G}_{uu'}$. Suppose $\widetilde{G}_{uu'}$ has $l$ vertices. Since $l \leq k - 1$, and the girth of $G$ satisfies $g(G) > 2k/3 \geq 2(l+1)/3$, by Lemma 2,

$$\widetilde{G}_{uu'} \text{ has at most } l \leq k - 1 \text{ edges.} \tag{6}$$

Denote by $V_{uu'} = V(\widetilde{G}_{uu'}) \cap (F \setminus F^*)$. Notice that $\varphi^{-1}(uu')$ might be a proper subset of $V_{uu'}$ (see Fig. 3). We distinguish two cases in the following.

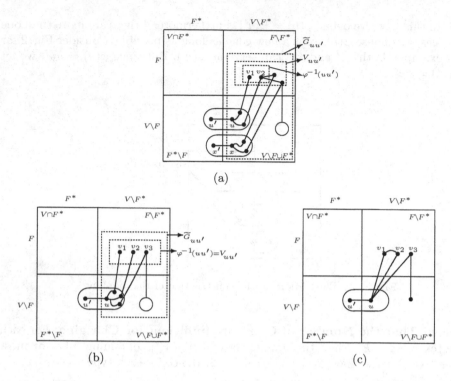

**Fig. 3.** An illustration for the proof that the number of coins are sufficient for type II vertices. (a) illustrates Case 1, (b) illustrates Case 2, and an extremal structure which occurs in Case 2 is depicted in (c).

**Case 1.** $|\varphi(V_{uu'})| \geq 2$ (see Fig. 3(a)).

In this case, by property (6), we have

$$\sum_{v \in V_{uu'}} (d_G(v) - |E(v, F \cap F^*)|) \leq 2|E(\widetilde{G}_{uu'})| \leq (k-1)|\varphi(V_{uu'})|,$$

and thus the coins on the edges of $\varphi(V_{uu'})$ are sufficient to pay for the charging of the vertices in $V_{uu'}$.

**Case 2.** $|\varphi(V_{uu'})| = 1$ (see Fig. 3(b)).

In this case, $\varphi^{-1}(uu') = V_{uu'}$. If $v_1, v_2 \in V_{uu'}$ are two adjacent vertices in $G$, then the concatenation of the path from $v_1$ to $u$, the path from $u$ to $v_2$, and edge $v_i v_j$ will contain a cycle of $\widetilde{G}_{uu'}$. Since a connected graph on $l$ vertices and at most $l$ edges has at most one cycle, we see that there is at most one edge between vertices of $V_{uu'}$, that is, $|E(G[V_{uu'}])| \leq 1$. Then

$$\sum_{v \in V_{uu'}} (d_G(v) - |E(v, F \cap F^*)|)$$

$$= 2|E(G[V_{uu'}])| + |E(V_{uu'}, V \setminus (F \cup F^*))|$$

$$= |E(G[V_{uu'}])| + |E(\widetilde{G}_{uu'})| - |E(\widetilde{G}_{uu'}|_{V \setminus (F \cap F^*)})| \tag{7}$$

$$\leq 1 + |E(\widetilde{G}_{uu'})| = 1 + l \leq k.$$

If $\sum_{v \in V_{uu'}} (d_G(v) - |E(v, F \cap F^*)|) \leq k - 1$, then the $k-1$ coins on edge $uu'$ are sufficient to pay for the charging of vertices in $V_{uu'}$. In the following, consider the case when $\sum_{v \in V_{uu'}} (d_G(v) - |E(v, F \cap F^*)|) = k$. By inequality (7), this is possible only when $|E(G[V_{uu'}])| = 1$, $E(\widetilde{G}_{uu'}|_{V \setminus (F \cap F^*)}) = \emptyset$, and $|V(\widetilde{G}_{uu'})| = k - 1$. Then we arrive at a structure depicted in Fig. 3(c). In particular, the structure (ii) in the construction of mapping $\varphi$ combined with the property that $E(\widetilde{G}_{uu'}|_{V \setminus (F \cap F^*)}) = \emptyset$ implies that every vertex $v \in V_{uu'}$ is adjacent with vertex $u$. Then the existence of an edge in $G[V_{uu'}]$ implies that the girth $g = 3$. By the girth condition $g(G) > 2k/3$, we have $k \leq 4$, which contradicts the assumption that $k \geq 5$.

Notice that the edge sets we use to pay for the charging of both type I vertices and type II vertices are disjoint. Hence the lemma is proved.

**Theorem 1.** *Suppose $k \geq 5$. Then the approximation ratio of Algorithm 1 on graphs with girth $g(G) > 2k/3$ is at most $k - 1$.*

*Proof.* Suppose $F^*$ is an optimal solutions on $G$ w.r.t. $w$, and for $i = 1, \ldots, t$, $F_i^*$ is an optimal solution on $G_i$ w.r.t. $w_i$. Notice that for any $i \in \{1, \ldots, t\}$ and any vertex $v$, we have $w(v) = \sum_{j=1}^{i} w_i(v) + w^{(i)}(v)$ and $w^{(i)}(v) \geq 0$. As a consequence, $w(v) \geq \sum_{j=1}^{t} w_j(v)$ for any vertex $v \in V$, and equality holds for any vertex $v \in F$. It follows that

$$\frac{w(F)}{w(F^*)} \leq \frac{\sum_{i=1}^{t} w_i(F)}{\sum_{i=1}^{t} w_i(F^*)} \leq \max_{i \in \{1, \ldots, t\}} \left\{ \frac{w_i(F)}{w_i(F^*)} \right\}. \tag{8}$$

Observe that

$$w^{(i)}(v) = 0 \text{ (and thus } w_i(v) = 0) \text{ holds for any vertex } v \in V \setminus V(G_i). \tag{9}$$

In particular, $w_i(v) = 0$ holds for any vertex $v \in F \setminus F_i$. Hence

$$w_i(F) = w_i(F_i). \tag{10}$$

Also by (9), it can be seen that $F_i^* \cup (V \setminus V(G_i))$ is an optimal solution on $G$ w.r.t. $w_i$ and $w_i(F_i^* \cup (V \setminus V(G_i))) = w_i(F_i^*)$. Since $F^*$ is a feasible solution on $G$, we have

$$w_i(F^*) \geq w_i(F_i^* \cup (V \setminus V(G_i))) = w_i(F_i^*). \tag{11}$$

Combining inequalities (8), (10), and (11),

$$\frac{w(F)}{w(F^*)} \leq \max_{i \in \{1, \ldots, t\}} \left\{ \frac{w_i(F_i)}{w_i(F_i^*)} \right\}. \tag{12}$$

By Lemma 1, $F_i$ is a minimal $VCC_k$ of $G_i$. Then it follows from Lemma 3 that $w_i(F_i) \leq (k-1)w_i(F_i^*)$ holds for any $i = 1, \ldots, t$. The theorem is proved.

## 3    Conclusion

This paper presents a local ratio algorithm for the $MinWVCC_k$ problem which achieves approximation ratio $k - 1$ under the assumption that $k \geq 5$ and girth $g(G) > 2k/3$. The result improves the one in [14] by relaxing the girth assumption from $g(G) \geq k$ to $g(G) > 2k/3$. Unfortunately, this result does not apply to the case when $k = 4$. An interesting question is whether some other method can yield 3-approximation for the $MinWVCC_4$ problem without any girth assumption. From a communication aspect of view, there are also a lot of works studying the *minimum connected $VCC_k$ problem* ($MinCVCC_k$) in which the computed $VCC_k$ should induce a connected subgraph [2,3,7]. In particular, Li *et al.* presented a $k$-approximation for $MinCVCC_k$ without any girth assumption [7]. It is for the unweighted case. Studying weighted case seems to be a challenging and interesting problem.

## References

1. Brešar, B., Kardoš, F., Katrenič, J., Semanišin, G.: Minimum $k$-path vertex cover. Discrete Appl. Math. **159**, 1189–1195 (2011)
2. Chen, L., Huang, X., Zhang, Z.: A simpler PTAS for connected $k$-path vertex cover in homogeneous wireless sensor network. J. Combin. Optim. **36**(1), 35–43 (2018)
3. Fujito, T.: On approximability of connected path vertex cover. In: Solis-Oba, R., Fleischer, R. (eds.) WAOA 2017. LNCS, vol. 10787, pp. 17–25. Springer, Cham (2018). https://doi.org/10.1007/978-3-319-89441-6_2
4. Khan, I., Belqasmi, F., Glitho, R., Crespi, N., Morrow, M., Polakos, P.: Wireless sensor network virtualization: a survey. IEEE Commun. Surv. Tutor. **18**(1), 553–576 (2016)
5. Guruswami, V., Lee, E.: Inapproximability of $H$-transversal/packing. SIAM J. Discrete Math. **31**(3), 1552–1571 (2017)
6. Lee, E.: Partitioning a graph into small pieces with applications to path transversal. In: Proceedings of the Twenty-Eighth Annual ACM-SIAM Symposium on Discrete Algorithms (SODA17), pp. 1546–1558. SIAM, Barcelona (2017)
7. Li, X., Zhang, Z., Huang, X.: Approximation algorithms for minimum (weight) connected $k$-path vertex cover. Discrete Appl. Math. **205**, 101–108 (2016)
8. Li, Y., Thai, M., Wu, W.: Wireless Sensor Networks and Applications. Springer, Boston (2008). https://doi.org/10.1007/978-0-387-49592-7
9. Novotný, M.: Design and analysis of a generalized canvas protocol. In: Samarati, P., Tunstall, M., Posegga, J., Markantonakis, K., Sauveron, D. (eds.) WISTP 2010. LNCS, vol. 6033, pp. 106–121. Springer, Heidelberg (2010). https://doi.org/10.1007/978-3-642-12368-9_8
10. Shmoys, D.: Cut problems and their application to divide-and-conquer. In: Hochbaum, D. (ed.) Approximation Algorithms for NP-hard Problems pp 192–235. PWS Publishing, Boston (1996)

11. Wan, P., Alzoubi, K. M., Frieder, O.: Distributed construction of connected dominating set in wireless ad hoc networks. In: International conference on computer communications(INFOCOM), pp. 1597–1604. IEEE, New York (2002)
12. Yick, J., Mukherjee, B., Ghosal, D.: Wireless sensor network survey. Comput. Netw. **52**(12), 2292–2330 (2008)
13. Zhang, Z., Li, X., Shi, Y., Nie, H., Zhu, Y.: PTAS for minimum $k$-path vertex cover in ball graph. Inform. Process. Lett. **119**, 9–13 (2017)
14. Zhang, Y., Shi, Y., Zhang, Z.: Approximation algorithm for the minimum weight connected $k$-subgraph cover problem. Theoret. Comput. Sci. **535**, 54–58 (2014)

# A True $O(n \log n)$ Algorithm for the All-k-Nearest-Neighbors Problem

Hengzhao Ma and Jianzhong Li$^{(\boxtimes)}$

Harbin Institute of Technology, Harbin 150001, Heilongjiang, China
hzma@stu.hit.edu.cn, lijzh@hit.edu.cn

**Abstract.** In this paper we examined an algorithm for the All-k-Nearest-Neighbor problem proposed in 1980s, which was claimed to have an $O(n \log n)$ upper bound on the running time. We find the algorithm actually exceeds the so claimed upper bound, and prove that it has an $\Omega(n^2)$ lower bound on the time complexity. Besides, we propose a new algorithm that truly achieves the $O(n \log n)$ bound. Detailed and rigorous theoretical proofs are provided to show the proposed algorithm runs exactly in $O(n \log n)$ time.

**Keywords:** Computation geometry · Nearest neighbor problems · All-k-Nearest-Neighbors

## 1 Introduction

The All-k-Nearest-Neighbors problem, or All-kNN for short, is an important problem that draws intensive research efforts. Early works about All-kNN date back to 1980s [4], and there are still some new results about this problem published in recent years [2,15–17,20,21]. The reason why the All-kNN problem has been continuously studied is that many applications invoke All-kNN as an important sub-procedure, such as data base [14,18], classification [22], agglomerative clustering [8], computational biology [33], image retrieval [28], and recommendation systems [11]. For many of these applications, solving All-kNN is reported as the main bottleneck [8].

The All-kNN problem can be briefly defined as follows. Let $(X, D)$ be a metric space, where $X$ is a point set and $D(\cdot, \cdot)$ is a distance function. The input of All-kNN is a point set $P \subseteq X$, and the output is the $k$-nearest-neighbor for all points $p \in P$, where the $k$-nearest-neighbor of a point $p$ is a set of $k$ points in $P$ that are closest to $p$ under the distance function $D(\cdot, \cdot)$. The formal definition will be given in Sect. 2.

There is an obvious brute-force solution for All-kNN, which is to compute the pairwise distances for all points in $P$, and select the $k$ points with smallest

This work was supported by the National Natural Science Foundation of China under grants 61732003, 61832003 and U1811461.

Y. Li et al. (Eds.): COCOA 2019, LNCS 11949, pp. 362–374, 2019.
https://doi.org/10.1007/978-3-030-36412-0_29

distance to $p$ for each point $p$. This solution takes $O(n^2)$ time, which is unacceptable when the size $n$ of the input is large. There have been a lot of algorithms proposed to efficiently solve the All-kNN problem, which can be categorized into the following three classes.

The first class of algorithms uses different techniques to accelerate the empirical running time of the algorithm, while the theoretical $O(n^2)$ upper bound is unchanged. There are basically three different kinds of techniques. The first kind is based on tree-like spacial indexes, such as $k$-$d$ tree [9] and Voronoi diagram [7] based index. The second kind is based on space filling curves, including Hilbert curve [10] and Z-order curves [19]. The space filling curve is a useful method to build an one-dimensional index on multidimensional data. There is an important property about the index based on the space filling curves, that is, the elements near to each other tend to be indexed into near entries. This property helps to reduce the number of distance computation to solve the All-kNN problem, as reported in [5,21,29]. The third kind is based on the idea of *neighborhood propagation*, which uses the intuition that the neighbors of neighbors are also likely to be neighbors. The NN-descent algorithm proposed in [6] is the seminal work on neighborhood propagation, and it is still one of the best algorithms for All-kNN problem. Other works use the neighborhood propagation technique to refine the primary result returned by some preprocessing step. For example, the authors of [26] use a multiple random divide and conquer approach as preprocessing, and in [32] the Locality Sensitive Hashing method is used.

The second class of algorithms turns to solve the problem on parallel systems. The theoretical work [3] states that there is an optimal solution for All-kNN which needs $O(\log n)$ time and $O(n)$ processors on CREW PRAM model. Other works try to solve All-kNN on different parallel platforms, such as MapReduce platform [23,27] and GPU environment [12].

Different from most algorithms in the above two classes which do not reduce the theoretical $O(n^2)$ upper bound on running time, the third class of algorithms is proved to have lower upper bounds. And they are all serial algorithms, different with the work in [3]. For example, Bentley gives an multidimensional divide-and-conquer framework that can solve the All-kNN problem in $O(n(\log n)^{d-1})$ time [1], where $d$ is the number of dimension. Besides, the algorithm given in [4] takes $O(n(\log \delta))$ time, where $\delta$ is the ratio of the maximum and minimum distance between any two points in the input. Finally, the algorithm proposed by Vaidya [25] is claimed to have upper bound of $O(kd^d n \log n)$ on the running time.

After all, it can be summarized that most of the works about the All-kNN problem focus on improving the empirical running time of the algorithms, but few of them succeed in reducing the $O(n^2)$ worst case upper bound. To the best of our knowledge, the algorithms proposed by [1,4,25] are the only ones that have lower upper bound than $O(n^2)$. We will not consider the parallel situation so that the work in [3] is excluded. Unfortunately, it has been as long as 30 years since the theoretical results are published. There is an urgent demand to renew and improve these historical results, and that is exactly the main work in this paper.

Among the three theoretical works, the one proposed by Vaidya [25] has the best upper bound on $n$, which is $O(kd^d n \log n)$. We have carefully examined the algorithm and the proofs in [25], and unfortunately a major mistake is found, which is that some part of the algorithm proposed in [25] actually exceeds the $O(n \log n)$ bound. Since the work has been cited over 300 times combined with the conference version [24], it is necessary to point out the mistake and fix it. Here the contributions of this paper are listed as follows.

1. We point out that the algorithm proposed in [25] needs $\Omega(n^2) + O(kd^d n \log n)$ time, contradicting with the claimed $O(n \log n)$ upper bound.
2. A new algorithm for All-kNN is proposed whose running time is proved to be up-bounded by $O(k(k + (\sqrt{d})^d) \cdot n \log n)$. The algorithm only applies to the Euclidean space.
3. While the algorithm proposed in [25] is designed for All-1NN and needs non-trivial modifications to generalize to All-kNN, the algorithm in this paper can directly solve the All-kNN problem for arbitrary integral value of $k$.

The rest of the paper is organized as follows. In Sect. 2 the definition of the problem and some prerequisite knowledge are introduced. Then the algorithm proposed in [25] is analyzed in Sect. 3. The modified algorithm is proposed in Sect. 4, and the correctness and the upper bound on running time is formally proved in Sect. 5. Finally we conclude the paper in Sect. 6.

## 2    Problem Definitions and Preliminaries

The problem studied in this paper is the All-k-Nearest-Neighbors problem under Euclidean Space. In the following discussions we will assume that the input is a set $P$ of points where each $p \in P$ is a d-dimensional vector $(p^{(1)}, p^{(2)}, \cdots, p^{(d)})$. The distance between two points $p$ and $p'$ is measured by the Euclidean distance, which is $D(p, p') = \sqrt{\sum_{i=1}^{d} (p^{(i)} - p'^{(i)})^2}$. The formal definition of the All-k-Nearest-Neighbors problem is given below.

**Definition 1 (k-NN).**  *Given the input point set $P \subseteq R^d$ and a query point $q \in R^d$, define $kNN(q, P)$ to be the set of $k$ points in $P$ that are nearest to $q$. Formally,*

*1. $kNN(q, P) \subseteq P$, and $|kNN(q, P)| = k$;*
*2. $D(p, q) \le D(p', q)$ for $\forall p \in kNN(q, P)$ and $\forall p' \in P \setminus kNN(q, P)$.*

**Definition 2 (All-k-Nearest-Neighbors, All-kNN).**  *Given the input point set $P \subseteq R^d$, find $kNN(p, P \setminus \{p\})$ for all $p \in P$.*

Next we introduce some denotations that will be used in the algorithm proposed in Sects. 3 and 4. Define a rectangle $\mathfrak{r}$ in $R^d$ to be the product of $d$ intervals, i.e., $I_1 \times I_2 \times \cdots \times I_d$, where each interval $I_i$ can be open, closed or semi-closed for $1 \leq i \leq d$. For a rectangle $\mathfrak{r}$, let $Lmax(\mathfrak{r}), Lmin(\mathfrak{r})$ be the longest and shortest length of the intervals defining $\mathfrak{r}$, respectively. When $Lmax(\mathfrak{r}) = Lmin(\mathfrak{r})$, $\mathfrak{r}$ is called a d-cube, and denote the side length of the d-cube $\mathfrak{r}$ as $Len(\mathfrak{r}) = Lmax(\mathfrak{r}) = Lmin(\mathfrak{r})$. For a point set $P$, $\mathfrak{r}$ is called the bounding rectangle of $P$ if $\mathfrak{r}$ is the smallest rectangle containing all the points in $P$, then let $|\mathfrak{r}|$ to be equivalent to $|P|$ which is the number of points in $P$. Besides, a d-cube $\mathfrak{r}$ is a Minimal Cubical Rectangle (MCR) for a given point set $P$ iff $\mathfrak{r}$ contains all the points in $P$ and has the minimal side length. Note that for a specific point set $P$, its bounding rectangle is unique but its MCR may not.

On the other hand, define a d-ball to be the set $B(c,r) = \{x \in R^d \mid D(x,c) \leq r\}$, where $c$ is the center and $r$ is the radius of the d-ball. For a point set $P$, define the Minimum Enclosing Ball (MEB) of $P$ to be the minimum radius ball containing all the points in $P$, denoted as $MEB(P)$. It is known that the there exists one unique $MEB(P)$ for a given $P$, which can be computed by solving a quadratic-programming problem [30]. From now on let $\mathcal{C}_P$ and $\mathcal{R}_P$ denote the unique center and radius of $MEB(P)$, respectively. Besides the exact MEB, the approximate MEB is equally useful and easier to compute. A d-ball $B(c_P, r_P)$ is an $\epsilon$-MEB of a point set $P$ iff $P \subseteq B(c_P, r_P)$ and $r_P \leq \epsilon \cdot \mathcal{R}_P$. The following algorithm can compute a $\frac{3}{2}$-MEB in linear time, which is proposed in [31].

---

**Algorithm 1.** Compute $\frac{3}{2}$-MEB

---

**Input:** $B(c_P, r_P)$ which is a $\frac{3}{2}$-MEB of $P$, and another point set $Q$
**Output:** a $\frac{3}{2}$-MEB of $P \cup Q$

1   **if** $P = \emptyset$ **then**
2      $c_0 \leftarrow$ a random point in $Q$;
3      $r_0 \leftarrow 0$;
4   **else**
5      $c_0 \leftarrow c_P, r_0 \leftarrow r_p$;
6   **end**
7   **while** $\exists q \in Q$ **do**
8      **if** $d(q, c_0) > r_0$ **then**
9         $\delta \leftarrow \frac{1}{2}(D(q, c_0) - r_0)$;
10        $r_1 \leftarrow r_0 + \delta$;
11        $c_1 \leftarrow c_0 + \frac{\delta}{D(q, c_0)}(q - c_0)$;
12        $c_0 \leftarrow c_1, r_0 \leftarrow r_1$;
13      **end**
14      $P \leftarrow P \cup \{q\}, Q \leftarrow Q \setminus \{q\}$;
15 **end**

---

The main characteristic of Algorithm 1 is that it can be viewed as a dynamic algorithm, where the $\frac{3}{2}$-MEB of $P$ is precomputed, and the set $Q$ is an update to $P$. The algorithm runs in $O(|Q|)$ time to compute the $\frac{3}{2}$-MEB of $P \cup Q$ based on the precomputed $\frac{3}{2}$-MEB of $P$. This characteristic will play an important part in the algorithm proposed in Sect. 4.

## 3    The Analysis of the Algorithm in [25]

In this section we point out the main reason why the algorithm given in [25] exceeds the claimed $O(n \log n)$ upper bound.

The algorithm in [25] organizes the points in the input set $P$ into a collection $S$ of d-cubes. And for each d-cube $\mathfrak{r}$, a set $Nbr(\mathfrak{r})$ of d-cubes is maintained based on the $Est(\mathfrak{r})$ value, which is defined below.

For any two d-cubes $\mathfrak{r}$ and $\mathfrak{r}'$ in $S$, define $D_{max}(\mathfrak{r}), D_{min}(\mathfrak{r}, \mathfrak{r}'), D_{max}(\mathfrak{r}, \mathfrak{r}')$ as follows:

$$D_{max}(\mathfrak{r}) = \max_{p,p' \in \mathfrak{r}} \{D(p,p')\}$$

$$D_{min}(\mathfrak{r}, \mathfrak{r}') = \min_{p \in \mathfrak{r}, p' \in \mathfrak{r}'} \{D(p,p')\}$$

$$D_{max}(\mathfrak{r}, \mathfrak{r}') = \max_{p \in \mathfrak{r}, p' \in \mathfrak{r}'} \{D(p,p')\}$$

Then define $Est(\mathfrak{r})$ by the following equation:

$$Est(\mathfrak{r}) = \begin{cases} D_{max}(\mathfrak{r}), & if\ |\mathfrak{r}| \geq 2 \\ \min_{\mathfrak{r}' \in S}\{D_{max}(\mathfrak{r}, \mathfrak{r}')\}, & otherwise \end{cases} \tag{1}$$

Now the definitions of the $Nbr$ and $Frd$ sets can be given using the denotations above.

**Definition 3** ([25]). $Nbr(\mathfrak{r}) = \{\mathfrak{r}' \in S \mid D_{min}(\mathfrak{r}, \mathfrak{r}') \leq Est(\mathfrak{r})\}$.

**Definition 4** ([25]). $Frd(\mathfrak{r}) = \{\mathfrak{r}' \in S \mid \mathfrak{r} \in Nbr(\mathfrak{r}')\}$.

The computation of the $Est$ value is exactly where the algorithm exceeds the $O(n \log n)$ bound, as is shown in the following lemma.

**Lemma 1.** *The time to create and maintain the $Est(\mathfrak{r})$ value for all d-cubes is $\Omega(n^2)$.*

*Proof.* Please pay attention to the first term of Eq. 1. When $|\mathfrak{r}| \geq 2$, $Est(\mathfrak{r})$ is defined by $Est(\mathfrak{r}) = D_{max}(\mathfrak{r})$, where $D_{max}(\mathfrak{r}) = \max_{p_1,p_2 \in \mathfrak{r}} D(p_1, p_2)$. We assure that this is the original definition in [25]. According to this definition, it will exceed the $O(n \log n)$ bound merely to compute $Est(\mathfrak{r}_0)$, where $\mathfrak{r}_0$ is the MCR of the input point set $P$. Actually, it needs $\Theta(n^2)$ time to compute $Est(\mathfrak{r}_0)$, because it needs to compute the pairwise distance between each pair of points in $P$, and there are obviously $\frac{n(n-1)}{2}$ pairs. Then the time to compute $Est(\mathfrak{r})$ for all d-cubes is obviously $\Omega(n^2)$. □

## 4    The Proposed Algorithm

According to the analysis in Sect. 3, the main defect of the algorithm in [25] is that the $Est$ value defined in Eq. 1 can not be efficiently computed. In this section we propose our algorithm that solves this problem elegantly.

The algorithm is divided into tree parts, which will be introduced one by one in the rest of this section.

## 4.1  Constructing the Rectangle Split Tree

The first part of our algorithm is to build the Rectangle Split Tree (RST). There are three important points of the algorithm that need to be clarified. First, our algorithm represents the subsets of the input point set $P$ by bounding rectangles. Second, the algorithm will recursively split the rectangles and organize them into the RST tree structure. Third, the method to split a rectangle $\mathfrak{r}$ is to cut the longest edge of $\mathfrak{r}$ into two equal halves. Let $\mathfrak{r}_{large}$ denote the one in the two sub-rectangles of $\mathfrak{r}$ that contains more points, and $\mathfrak{r}_{small}$ denote the other. Then the Rectangle Split Tree is defined in the following Definition 5.

**Definition 5 (Rectangle Split Tree, RST).** *Given a point set $P$, an RST based on $P$ is a tree structure satisfying:*

1. *the root of $T$ is the bounding rectangle of $P$,*
2. *each node in $T$ represents a rectangle, which is the bounding rectangle of a subset of $P$, and*
3. *there is an edge from $\mathfrak{r}$ to $\mathfrak{r}_{large}$ and $\mathfrak{r}_{small}$.*

*Besides, $T$ is called fully built if all the leaf nodes contain only one point.*

Next we give the algorithm to build the RST. The complexity of Algorithm 2 will be proved to be $O(dn\log n)$ in Sect. 5.

---

**Algorithm 2.** Constructing the RST

---

**Input:** a point set $P$, and the bounding rectangel $\mathfrak{r}_P$ of $P$
**Output:** an RST $T$

1  Create a tree $T$ rooted at $\mathfrak{r}_P$;
2  $S_0 \leftarrow \{\mathfrak{r}\}, S_1 \leftarrow \emptyset$;
3  **while** $|S_1| < |P|$ **do**
4  $\quad$ $\mathfrak{r} \leftarrow$ an arbitrary rectangle in $S_0$;
5  $\quad$ Split $\mathfrak{r}$ into $\mathfrak{r}_{large}$ and $\mathfrak{r}_{small}$;
6  $\quad$ **foreach** $\mathfrak{r}_i \in \{\mathfrak{r}_{large}, \mathfrak{r}_{small}\}$ **do**
7  $\quad\quad$ Create a node for $\mathfrak{r}_i$ and hang it under the node for $\mathfrak{r}$ in $T$;
8  $\quad\quad$ **if** $|\mathfrak{r}_i| > 1$ **then**
9  $\quad\quad\quad$ $S_0 \leftarrow S_0 \cup \{\mathfrak{r}_i\}$;
10 $\quad\quad$ **else**
11 $\quad\quad\quad$ $S_1 \leftarrow S_1 \cup \{\mathfrak{r}_i\}$;
12 $\quad\quad$ **end**
13 $\quad$ **end**
14 $\quad$ $S_0 \leftarrow S_0 \setminus \{\mathfrak{r}\}$;
15 **end**

---

## 4.2  Computing the Approximate MEB

The second part of the proposed algorithm is to compute the approximate MEB for each node in the RST, which is given as Algorithm 3. This algorithm receives the constructed RST $T$ as input, and traverse $T$ with post-root order, where Algorithm 1 will be invoked at each node. It will be shown in Sect. 5 that the algorithm takes $O(dn \log n)$ time.

---

**Algorithm 3.** Compute the approximate MEB

---

**Input:** a RST $T$

**Output:** Compute an $\frac{3}{2}$-MEB for $\forall \mathfrak{r} \in T$

1  Invoke ComputeMEB $(root(T))$, where $root(T)$ is the root of $T$;

2  **Procedure** ComputeMEB($\mathfrak{r}$):

3  $\quad$ **if** $|\mathfrak{r}| = 1$ **then**

4  $\quad\quad$ $c_{\mathfrak{r}} = p$, where $p$ is the only point in $\mathfrak{r}$;

5  $\quad\quad$ $r_{\mathfrak{r}} = 0$;

6  $\quad\quad$ return;

7  $\quad$ **end**

8  $\quad$ ComputeMEB($\mathfrak{r}_{large}$);

9  $\quad$ ComputeMEB($\mathfrak{r}_{small}$);

10 $\quad$ Invoke Algorithm 1, where the parameters are set to :

$\quad\quad$ $c_{\mathfrak{r}_{large}}, r_{\mathfrak{r}_{large}}, \mathfrak{r}_{small}$;

11 **end**

---

### 4.3   Computing All-KNN

Based on the algorithm for constructing the RST and computing MEB, the algorithm for All-kNN is given as Algorithm 4. It is worthy to point out that the algorithm naturally applies to all integer $k \geq 1$, which is an advantage against the algorithm in [25].

The algorithm first invokes Algorithm 2 on $P$ to construct an RST $T$ (Line 1). Then Algorithm 3 is invoked at Line 2 to compute the approximate MEB for each node in $T$. The rest of the algorithm aims to traverse $T$ and construct $kNbr(\mathfrak{r})$ and $kFrd(\mathfrak{r})$ sets for each $\mathfrak{r} \in T$. The formal definition of the two sets along with two auxiliary definitions are given below. In these definitions, assume that a set $H$ of rectangles is given, and the approximate MEB $B(c_{\mathfrak{r}}, r_{\mathfrak{r}})$ of each $\mathfrak{r} \in H$ is precomputed.

**Definition 6.** *Given two rectangles $\mathfrak{r}$ and $\mathfrak{r}' \in H$, define the relative quality of $\mathfrak{r}'$ against $\mathfrak{r}$ as follows: $Qlty(\mathfrak{r}', \mathfrak{r}) = \sqrt{(D(c_{\mathfrak{r}'}, c_{\mathfrak{r}}) + \frac{1}{2}r_{\mathfrak{r}'})^2 + r_{\mathfrak{r}'}^2}$.*

**Definition 7.** *Let $\min\limits_{k}\{x \in S \mid f(x)\}$ be the k-th smallest value in the set $\{x \in S \mid f(x)\}$. Define $kThres(\mathfrak{r})$ for each $\mathfrak{r} \in H$ as follow:*

$$kThres(\mathfrak{r}) = \begin{cases} 2 \cdot r_{\mathfrak{r}}, & if |\mathfrak{r}| \geq k+1 \\ \min\limits_{k}\{\mathfrak{r}' \in H \setminus \{\mathfrak{r}\} \mid Qlty(\mathfrak{r}', \mathfrak{r})\}, & if |\mathfrak{r}| < k+1 \end{cases} \quad (2)$$

**Definition 8.** $kNbr(\mathfrak{r}) = \{\mathfrak{r}' \in H \mid D(c_{\mathfrak{r}'}, c_{\mathfrak{r}}) \leq r_{\mathfrak{r}} + r_{\mathfrak{r}'} + kThres(\mathfrak{r})\}$.

**Definition 9.** $kFrd(\mathfrak{r}) = \{\mathfrak{r}' \in H \mid \mathfrak{r} \in kNbr(\mathfrak{r}')\}$.

The algorithm will visit all the rectangles in the RST $T$ by a descending order on the radius of the approximate MEB. To do so, a heap $H$ is maintained to store the rectangles in $T$ ordered by the radius of the approximate MEB of them. The main part of Algorithm 4 is a *While* loop. Each time the *While* loop is executed, the top element $\mathfrak{r}$ of $H$, which is the one with the largest approximate MEB radius, will be popped out of $H$. Then the algorithm will determine the $\mathcal{S}_{son}$ set, push all the rectangles in $\mathcal{S}_{son}$ into $H$ and process the rectangles in $\mathcal{S}_{son}$ by invoking the $MntnNbrFrd$ process (Algorithm 5). The *While* loop will terminate when $|H| = |P|$ and that is when all the rectangles in $T$ are processed.

The set $\mathcal{S}_{son}$ is determined by the following criterion. First, $\mathcal{S}_{son}$ is set to be $\{\mathfrak{r}_{large}, \mathfrak{r}_{small}\}$. Then, if any $\mathfrak{r}_i \in \mathcal{S}_{son}$ satisfies $|\mathfrak{r}_i| < k+1$, $\mathfrak{r}_i$ will be replaced by the set of the leaf nodes in the subtree of $T$ rooted at $\mathfrak{r}_i$. Such work is done by Line 9 to 16 in Algorithm 4.

---

**Algorithm 4.** All-kNN

**Input:** a point set $P$
**Output:** All-kNN on $P$

1  Invoke Algorithm 2 on $P$, and construct an RST $T$;
2  Invoke `ComputeMEB`$(root(T))$ where $root(T)$ is the root of $T$;
3  $kNbr(root(T)) = \emptyset, kFrd(root(T)) = \emptyset$;
4  Initialize a heap $H = \{root(T)\}$, which is order by the radius of the approximate MEB;
5  **while** $|H| < |P|$ **do**
6  $\quad$ $\mathfrak{r} \leftarrow H.pop()$;
7  $\quad$ $\mathcal{S}_{son} \leftarrow \{\mathfrak{r}_{large}, \mathfrak{r}_{small}\}$;
8  $\quad$ **foreach** $\mathfrak{r}_i \in \mathcal{S}_{son}$ **do**
9  $\quad\quad$ **if** $|\mathfrak{r}_i| < k+1$ **then**
10 $\quad\quad\quad$ $\mathcal{S}_{leaf} \leftarrow$ the set of the leaf nodes in the subtree rooted at $\mathfrak{r}_i$;
11 $\quad\quad\quad$ $\mathcal{S}_{son} \leftarrow \mathcal{S}_{son} \setminus \{\mathfrak{r}_i\} \cup \mathcal{S}_{leaf}$;
12 $\quad\quad\quad$ Delete the non-leaf nodes in the sub-tree;
13 $\quad\quad\quad$ Hang the leaf nodes directly under $\mathfrak{r}$;
14 $\quad\quad$ **end**
15 $\quad\quad$ Push $\mathfrak{r}_i$ into $H$;
16 $\quad$ **end**
17 $\quad$ `MntnNbrFrd`$(\mathfrak{r}, \mathcal{S}_{son})$;
18 **end**

---

After the $\mathcal{S}_{son}$ is determined, Algorithm 4 invokes the $MntnNbrFrd$ process (Algorithm 5) to construct the $kNbr$ and $kFrd$ sets for each $\mathfrak{r}_i \in \mathcal{S}_{son}$. Algorithm 5 relies on four sub-procedures, which are $AddInNbr$, $DelFromNbr$, $DelFromFrd$ and $TruncateNbr$, respectively. The name of these sub-procedures intuitively shows their functionality, and the pseudo codes are given in Algorithm 6.

Algorithm 6 shows the functionalities of the sub-procedures mentioned in Algorithm 5. In the pseudo codes of the $AddInNbr$ sub-procedure, it can be

---

**Algorithm 5.** The $MntnNbrFrd$ process

---

```
1  Procedure MntnNbrFrd(t, S_son):
2      foreach t' ∈ kNbr(t) do
3          DelFromFrd(t, t');                        // Delete t from kFrd(t')
4          foreach t_i ∈ S_son do
5          |   AddInNbr (t', t_i);                    // Add t' into kNbr(t_i)
6          end
7      end
8      foreach t' ∈ kFrd(t) do
9          DelFromNbr(t, t');                        // Delete t from kNbr(t')
10         foreach t_i ∈ S_son do
11         |   AddInNbr (t_i, t');                    // Add t_i into kNbr(t')
12         end
13     end
14     foreach t_i ∈ S_son do
15         for t_j ∈ S_son \ {t_i} do
16         |   AddInNbr(t_i, t_j);                    // Add t_i into kNbr(t_j)
17         end
18     end
19     foreach t' ∈ S_son ∪ kFrd(t) do
20         if |t'| = 1 then
21         |   TruncateNbr(t');
22         end
23     end
24 end
```

---

seen that an extra set $Cand(t)$ is used in the special case of $|t| = 1$, which stores the candidate rectangles that might contain the k-NN of the only point in $t$. The $Cand$ set is formally defined as follows.

**Definition 10.** *For $t$ where $|t| = 1$, the $Cand(t)$ set satisfies:*

1. $Cand(t) \subseteq kNbr(t)$, $|Cand(t)| = k$, and
2. $Qlty(t', t) \leq Qlty(t'', t)$ for $\forall t' \in Cand(t)$ and $\forall t'' \in kNbr(t) \setminus Cand(t)$.

Recall that $kThres(t)$ is defined to be $\min_k \{t' \in H \setminus \{t\} \mid Qlty(t', t)\}$ when $|t| < k + 1$. And then with $Cand(t)$ defined, $kThres(t)$ can be equivalently defined as $\max_{t' \in Cand(t)} \{Qlty(t', t)\}$, as is shown at Line 15 in Algorithm 6.

Last but not least, these sub-procedures above require the $kNbr$, $kFrd$ and $Cand$ sets to be implemented by specific data structures. Specifically, $kNbr$ sets should be implemented by B+ trees, $kFrd$ sets should be implemented by red-black trees, and *Cand* sets should be implemented by heaps. The detailed discussion is omitted due to space limitation.

---

**Algorithm 6.** The sub-procedures

```
1  Procedure DelFromNbr(r′,r):
2  |   Delete r′ from kNbr(r);
3  end
4  Procedure DelFromFrd(r′,r):
5  |   Delete r′ from kFrd(r);
6  end
7  Procedure AddInNbr(r′,r):
8  |   if |r| = 1 then
9  |   |   if |Cand(r)| < k then
10 |   |   |   Add r′ into Cand(r);
11 |   |   else
12 |   |   |   if Qlty(r′,r) < kThres(r) then
13 |   |   |   |   Add r′ into Cand(r);
14 |   |   |   |   Pop the top element out of Cand(r);
15 |   |   |   |   kThres(r) ←  max  {Qlty(r″,r)};
                      r″∈Cand(r)
16 |   |   |   end
17 |   |   end
18 |   end
19 |   if D(c_{r′},c_r) < r_{r′} + r_r + kThres(r) then
20 |   |   Add r′ into kNbr(r);
21 |   |   Add r into kFrd(r′);
22 |   end
23 end
24 Procedure TruncateNbr(r):
25 |   Find the first element r′ ∈ kNbr(r) such that
   |       D(c_{r′},c_r) − r_{r′} > r_r + kThres(r);
26 |   Truncate kNbr(r) by deleting all elements from r′ to the last one;
27 end
```

---

## 5  Analysis

Here only the main theorems are proposed, and a lot of lemmas needed to prove these theorems are omitted due to space limitation. The detailed proofs can be found in the online full version of this paper [13].

### 5.1  Correctness

**Theorem 1.** *When Algorithm 4 terminates, $kNbr(r) = Cand(r) = kNN(p, P \setminus \{p\})$ holds for any leaf node $r$, where $p = c_r$ is the only point in $r$.*

## 5.2  Complexities

**Theorem 2.** *The time complexity of Algorithm 2 is $O(dn \log n)$.*

**Theorem 3.** *The time complexity of Algorithm 3 is $O(dn \log n)$.*

**Theorem 4.** *The time complexity of Algorithm 4 is $O(k(k + (\sqrt{d})^d) \cdot n \log n)$.*

## 6  Conclusion

In this paper the All-k-Nearest-Neighbors problem is considered. An algorithm proposed in [25] is the inspiring work of this paper. In [25] the author claimed that the algorithm has an upper bound of $O(kd^d n \log n)$ on the time complexity. However, we find that this bound is unachievable according to the descriptions in [25]. We give formal analysis that the algorithm needs at least $\Omega(n^2)$ time. On the other hand, we propose another algorithm for the All-k-Nearest-Neighbor problem whose time complexity is truly up-bounded by $O(k(k + (\sqrt{d})^d)n \log n)$. After all, we have renewed an result that has an history of over 30 years and has been cited more than 300 times. Considering the importance of the All-k-Nearest-Neighbor problem, this work should be considered valuable.

## References

1. Bentley, J.L.: Multidimensional divide-and-conquer. Commun. ACM **23**(4), 214–229 (1980)
2. Cai, Z., Miao, D., Li, Y.: Deletion propagation for multiple key preserving conjunctive queries: approximations and complexity. In: 35th IEEE International Conference on Data Engineering, ICDE 2019, Macao, China, 8–11 April 2019, pp. 506–517. IEEE (2019). http://ieeexplore.ieee.org/xpl/mostRecentIssue.jsp?punumber=8725877
3. Callahan, P.B.: Optimal parallel all-nearest-neighbors using the well-separated pair decomposition. In: Proceedings of the 1993 IEEE 34th Annual Foundations of Computer Science, SFCS 1993, pp. 332–340. IEEE Computer Society, Washington (1993)
4. Clarkson, K.L.: Fast algorithms for the all nearest neighbors problem. In: 24th Annual Symposium on Foundations of Computer Science (SFCS 1983), vol. 16, pp. 226–232. IEEE, November 1983
5. Connor, M., Kumar, P.: Fast construction of k-nearest neighbor graphs for point clouds. IEEE Trans. Vis. Comput. Graph. **16**(4), 599–608 (2010)
6. Dong, W., Moses, C., Li, K.: Efficient k-nearest neighbor graph construction for generic similarity measures. In: Proceedings of the 20th international conference on World wide web - WWW 2011, p. 577. ACM Press, New York (2011)
7. Edelsbrunner, H.: Algorithms in Combinatorial Geometry, 1st edn. Springer, Heidelberg (2012)
8. Franti, P., Virmajoki, O., Hautamaki, V.: Fast agglomerative clustering using a k-nearest neighbor graph. IEEE Trans. Pattern Anal. Mach. Intell. **28**(11), 1875–1881 (2006)

9. Friedman, J.H.: An algorithm for finding best matches in logarithmic expected time. **3**(3), 209–226 (1977)
10. Hilbert, D.: Über die stetige Abbildung einer Linie auf ein Flächenstück. In: Hilbert, D. (ed.) Dritter Band: Analysis · Grundlagen der Mathematik · Physik Verschiedenes, pp. 1–2. Springer, Heidelberg (1935). https://doi.org/10.1007/978-3-662-38452-7_1
11. Karypis, G.: Evaluation of item-based top- N recommendation algorithms. In: Proceedings of the Tenth International Conference on Information and Knowledge Management - CIKM 2001, p. 247. ACM Press, New York (2001)
12. Komarov, I., Dashti, A., D'Souza, R.: Fast $k$-NNG construction with GPU-based quick multi-select. 1–20 (2013)
13. Ma, H., Li, J.: A true $o(n \log n)$ algorithm for the all-k-nearest-neighbors problem (2019). https://arxiv.org/abs/1908.00159
14. Miao, D., Cai, Z., Li, J.: On the complexity of bounded view propagation for conjunctive queries. IEEE Trans. Knowl. Data Eng. **30**(1), 115–127 (2018)
15. Miao, D., Cai, Z., Li, Y.: SEF view deletion under bounded condition. Theor. Comput. Sci. **749**, 17–25 (2018)
16. Miao, D., Cai, Z., Yu, J., Li, Y.: Triangle edge deletion on planar glasses-free rgb-digraphs. Theor. Comput. Sci. **788**, 2–11 (2019)
17. Miao, D., Liu, X., Li, J.: On the complexity of sampling query feedback restricted database repair of functional dependency violations. Theor. Comput. Sci. **609**, 594–605 (2016)
18. Miao, D., Liu, X., Li, Y., Li, J.: Vertex cover in conflict graphs. Theor. Comput. Sci. **774**, 103–112 (2019)
19. Morton, G.M.: A computer oriented geodetic data base and a new technique in file sequencing. Technical report, IBM Ltd, Ottawa, Canada (1966)
20. Park, Y., Lee, S.G.: A novel algorithm for scalable k-nearest neighbour graph construction. J. Inf. Sci. **42**(2), 274–288 (2016)
21. Sieranoja, S.: High dimensional k NN-graph construction using space filling curves. Ph.D. thesis, University of Eastern Finland (2015)
22. Szummer, M., Jaakkola, T.: Partially labeled classification with markov random walks. In: Proceedings of the 14th International Conference on Neural Information Processing Systems: Natural and Synthetic, NIPS 2001, pp. 945–952. MIT Press, Cambridge (2001)
23. Trad, M.R., Joly, A., Boujemaa, N.: Distributed KNN-graph approximation via hashing. In: Proceedings of the 2nd ACM International Conference on Multimedia Retrieval - ICMR 2012, p. 1. No. section 3. ACM Press, New York (2012)
24. Vaidya, P.M.: An optimal algorithm for the all-nearest-neighbors problem. In: Intergovernmental Panel on Climate Change (ed.) 27th Annual Symposium on Foundations of Computer Science (SFCS 1986), pp. 117–122. IEEE, Cambridge, October 1986
25. Vaidya, P.M.: An O(n logn) algorithm for the all-nearest-neighbors problem. Discrete Comput. Geom. **4**(2), 101–115 (1989)
26. Wang, J., Wang, J., Zeng, G., Tu, Z., Gan, R., Li, S.: Scalable k-NN graph construction for visual descriptors. In: 2012 IEEE Conference on Computer Vision and Pattern Recognition, pp. 1106–1113. IEEE (2012)
27. Warashina, T., Aoyama, K., Sawada, H., Hattori, T.: Efficient K-nearest neighbor graph construction using MapReduce for large-scale data sets. IEICE Trans. Inform. Syst. **97**–**D**(12), 3142–3154 (2014)

28. Yang, X., Latecki, L.J.: Affinity learning on a tensor product graph with applications to shape and image retrieval. In: Proceedings of the 2011 IEEE Conference on Computer Vision and Pattern Recognition, pp. 2369–2376. CVPR 2011. IEEE Computer Society, Washington (2011)
29. Yao, B., Li, F., Kumar, P.: K nearest neighbor queries and kNN-Joins in large relational databases (almost) for free. In: 2010 IEEE 26th International Conference on Data Engineering (ICDE 2010), pp. 4–15. IEEE (2010)
30. Yildirim, E.A.: Two algorithms for the minimum enclosing ball problem. SIAM J. Optim. **19**(3), 1368–1391 (2008)
31. Zarrabi-Zadeh, H., Chan, T.: A simple streaming algorithm for minimum enclosing balls. In: Proceedings of 18th Annual Canadian Conference Computing, pp. 14–17 (2006)
32. Zhang, Y.-M., Huang, K., Geng, G., Liu, C.-L.: Fast $k$NN Graph Construction with Locality Sensitive Hashing. In: Blockeel, H., Kersting, K., Nijssen, S., Železný, F. (eds.) ECML PKDD 2013. LNCS (LNAI), vol. 8189, pp. 660–674. Springer, Heidelberg (2013). https://doi.org/10.1007/978-3-642-40991-2_42
33. Zhou, J., Sander, J., Cai, Z., Wang, L., Lin, G.: Finding the nearest neighbors in biological databases using less distance computations. IEEE/ACM Trans. Comput. Biol. Bioinform. **7**(4), 669–680 (2010)

# Approximation Algorithms for Some Minimum Postmen Cover Problems

Yuying Mao, Wei Yu[✉], Zhaohui Liu, and Jiafeng Xiong

Department of Mathematics, East China University of Science and Technology,
Shanghai 200237, China
{y30170158,y20170003}@mail.ecust.edu.cn,
{yuwei,zhliu}@ecust.edu.cn

**Abstract.** In this work we study the Minimum Rural Postmen Cover Problem (MRPCP) and the Minimum Chinese Postmen Cover Problem (MCPCP). The MRPCP aims to cover a given subset $R$ of edges of an undirected weighted graph $G = (V, E)$ by a minimum size set of closed walks of bounded length $\lambda$. The MCPCP is a special case of the MRPCP with $R = E$. We give the first approximation algorithms for these two problems with approximation ratios 5 and 4, respectively.

**Keywords:** Approximation algorithms · Traveling Salesman Problem · Rural Postman Problem · Chinese Postman Problem · Postmen Cover

## 1 Introduction

In the Minimum Rural Postmen Cover Problem (MRPCP), we are given an undirected graph $G = (V, E)$, which may be multi-graph, with vertex set $V$ and edge set $E$, a subset $R \subseteq E$ of *required edges* and a nonnegative integer $\lambda$. Each edge $e$ is associated with a nonnegative integral weight $w(e)$. The objective is to find a set of closed walks of length at most $\lambda$ which collectively traverse all the edges in $R$ such that the number of closed walks used is a minimum. When $R = E$ in the MRPCP, i.e., all the edges are required edges, we have the Minimum Chinese Postmen Cover Problem (MCPCP).

It can be seen that deciding if the optimal value of the MRPCP equals one is exactly the decision version of the classical Rural Postman Problem (RPP), which asks if there is a closed walk of length no more than $\lambda$ that traverses all the required edges of a graph. It is well known that the decision version of the RPP is NP-Complete [5], which implies that the MRPCP cannot be approximated within the ratio 2, unless P = NP. By a similar reduction in [19], one can show that deciding if the optimal value of the MCPCP is no more than 2 is NP-Complete even if $G$ is a tree graph. Therefore, the MCPCP cannot have an algorithm with approximation ratio less than $3/2$, unless P = NP.

The RPP and its variants have a wide range of applications that include street sweeping, garbage collection, mail delivery, school bus routing, etc. (see Eiselt

© Springer Nature Switzerland AG 2019
Y. Li et al. (Eds.): COCOA 2019, LNCS 11949, pp. 375–386, 2019.
https://doi.org/10.1007/978-3-030-36412-0_30

et al. [8]). Recently, a special case of the RPP, called the Subpath Planning Problem (SPP), has found new applications in both the artificial intelligence community and the robotics research community [20,21]. Since the MRPCP/MCPCP addresses a more practical scenario, where a crew of postmen is available, than the RPP/CPP in which only a single postman is considered, they may occur more frequently in applications.

Although many heuristics and exact algorithms have been proposed for some variants of the RPP/CPP with multiple postmen [5], we focus on approximability results in this work. Arkin et al. [1] gave the first approximation algorithm for the Minimum Rural Postmen Walk Cover Problem (MRPWCP), which is a variant of the MRPCP with closed walks replaced by (open) walks. Their algorithm has an approximation ratio of 4. The authors also considered a similar variant of the MCPCP, called the Minimum Chinese Postmen Walk Cover Problem (MCPWCP), and derived a 3-approximation algorithm. Moreover, Arkin et al. [1] developed a 6-approximation algorithm for the Minimum Cycle Cover Problem (MCCP), which is a special case of the MRPCP with $R$ consisting of all $|V|$ zero-length self-loops. After the improvements in [16,24], Yu et al. [25] gave the current best approximation ratio 32/7 for the MCCP. On the other hand, the MCCP clearly cannot be approximated within the ratio 2 since deciding if the optimal value of the MCCP equals one is equivalent to the decision version of the well-known Traveling Salesman Problem, which is NP-Complete.

A closely related problem is the Min-Max Rural Postmen Cover Problem (MMRPCP) where the input is the same as the MRPCP except that $\lambda$ is replaced by a positive integer $k$ and the goal is to find at most $k$ closed walks to minimize the length of the longest closed walks. Analogously, by replacing closed walks with (open) walks in the MMRPCP we have the Min-Max Rural Postmen Walk Cover Problem (MMRPWCP). Arkin et al. [1] devised a 7-approximation algorithm for the MMRPWCP. This was improved to a 5-approximation algorithm by Yu et al. [26], who also proposed the first constant-factor approximation algorithms for the MMRPCP as well as the corresponding Chinese Postman variant. Frederickson et al. [12] obtained some approximation algorithms for the MMRPCP/MMCPCP with a single depot vetex, which has to be contained in all the closed walks.

Notice that in the MRPCP/MRPWCP, the walks may have repeated vertices and edges. If the closed (open) walks are replaced by edge-disjoint cycles (vertex-disjoint paths) and the graph is unweighted, we have the $\lambda$-cycle partition problem ($\lambda$-path partition problem) studied in [3,6,13,18,23].

In this paper we give the first approximation algorithms for the MRPCP and the MCPCP. For the MRPCP, a 7-approximation algorithm can be obtained using a similar approach in [1] tailored for the MRPWCP. To improve the approximation ratio, we provide a generalized version of the tree-splitting procedure in [9,16], which we call graph decomposition (see Sect. 2 for details). Based on this graph decomposition, we devise two 5-approximation algorithms. The first one has a simple analysis on the approximation ratio. The second one has a faster running time while the analysis is more complicated. For the MCPCP, we develop a 4-approximation algorithm. Note that a direct application of the approach for the MCPWCP in [1] can only derive a 5-approximation algorithm.

The remainder of the paper is organized as follows. We formally describe the problems and derive the graph decomposition algorithm in Sect. 2. In Sect. 3 we develop two 5-approximation algorithms for the MRPCP, which is followed by the result on the MCPCP in Sect. 4.

## 2  Preliminaries

Given an undirected weighted graph $G = (V, E)$ with vertex set $V$ and edge set $E$, $w(e)$ denotes the weight or length of edge $e$. If $e = (u, v)$, we also use $w(u, v)$ to denote the weight of $e$. For $R \subseteq E$ and $B > 0$, $G_R[B]$ denotes the subgraph of $G$ obtained by removing all the edges in $E \setminus R$ with weight greater than $B$. A walk in $G$ is a sequence $W = v_0 e_1 v_1 \ldots v_{l-1} e_l v_l$ such that $v_i \in V$ for $i = 0, 1, \ldots, l$ and $e_i = (v_{i-1}, v_i) \in E$ for $i = 1, \ldots, l$. If $v_0 = v_l$, $W$ is referred to as a closed walk. If all $v_i$'s are distinct except that $v_0$ may be identical to $v_l$, then $W$ is called a path. A cycle is a closed path. We simply write $W = e_1 \ldots e_l$ when $v_0, v_1, \ldots, v_l$ is clear in the context.

For a subgraph $G'$ of $G$, the graph obtained by adding some copies of the edges in $G'$ is called a multi-subgraph of $G$. For a (multi-)subgraph $H$ (e.g. path, cycle, walk) of $G$, let $V(H), E(H)$ be the vertex set and edge set of $H$, respectively. The weight of $H$ is defined as $w(H) = \sum_{e \in E(H)} w(e)$. If $H$ is a multi-subgraph, $E(H)$ is a multi-set of edges and the edges appearing multiple times contribute multiply to $\sum_{e \in E(H)} w(e)$. By doubling $H$ we mean that we add another copy of each edge in $E(H)$ to $H$ to derive a multi-subgraph of weight $2w(H)$. The distance between $H$ and another (multi-)subgraph $H'$ is given by

$$d(H, H') = \min\{w(e) \mid e = (u, v) \in E \text{ such that } u \in V(H), v \in V(H')\}.$$

The weight of a path (cycle, walk) is also called its length. A cycle $C$ is also called a tour on $V(C)$. A path (cycle, walk) containing only one vertex and no edges is a trivial path (cycle, walk) and its length is defined as zero.

For an edge subset $E' \subseteq E$, a set $\{W_1, \ldots, W_k\}$ of closed walks (some of them may be trivial walks) is called a *rural postmen cover* of $E'$ if $E' \subseteq \bigcup_{i=1}^{k} E(W_i)$. And the cost of this rural postmen cover is defined as $\max_{1 \le i \le k} w(W_i)$, i.e., the maximum length of the closed walks. By replacing closed walks with (open) walks we can define *rural postmen walk cover* and its cost similarly. A rural postmen (walk) cover of $E$ is simply called a postmen (walk) cover. Now we can formally state the following problems.

**Definition 1.** *In the Minimum Rural Postmen Cover Problem (MRPCP), we are given an undirected graph $G = (V, E)$ with a nonnegative integral weight function $w$ on $E$, an edge subset $R \subseteq E$ and a nonnegative integer $\lambda$, the goal is to find a rural postmen cover of $R$ of cost at most $\lambda$ such that the number of closed walks used is a minimum.*

**Definition 2.** *In the Minimum Chinese Postmen Cover Problem (MCPCP), we are given an undirected graph $G = (V, E)$ with a nonnegative integral weight function $w$ on $E$ and a nonnegative integer $\lambda$, the goal is to find a postmen cover of cost at most $\lambda$ such that the number of closed walks used is a minimum.*

If we replace the rural postmen cover of $R$ in Definition 1 with a rural postmen walk cover of $R$ we have the Minimum Rural Postmen Walk Cover Problem (MRPWCP).

Given an instance $I$ of MRPCP/MRPWCP defined on $G = (V, E)$ with required edge set $R$, we can derive a new instance $I'$ with exactly the same input except that the graph $G$ is replaced by a graph $G' = (V', E' \cup R)$ such that: (i) $V' = V(R)$, i.e., each vertex in $V'$ is an end vertex of some edge in $R$; (ii) For any $u, v \in V'$, there is an edge $(u, v) \in E'$ whose length equals the length of the shortest path between $u$ and $v$ in $G$; (iii) For any $e = (u, v) \in R$, there is a parallel edge between $u$ and $v$ with the same length as in $G$. We call $G'$ the *reduced graph* of $G$. Similar to the RPP (see [8]), one can verify that $I'$ is equivalent to $I$. Therefore, we focus on the instances of MRPCP/MRPWCP in which $G$ is the reduced graph of some graph, say $\bar{G}$, in the following discussion.

Given an instance $I$ of MRPCP/MRPWCP/MCPCP, we use $n$ to denote the number of vertices of $G$ and $OPT(I)$ indicates its optimal value as well as the corresponding optimal solution. Each (closed) walk in $OPT(I)$ is called an optimum (closed) walk. Note that the optimum (closed) walks may be neither edge-disjoint nor vertex-disjoint. If the instance $I$ is clear in the context, we write $OPT$ for $OPT(I)$.

The following result on decomposing a tree of large weight into a series of trees of small weight is very useful to design and analyze the algorithms for the Minimum Cycle Cover Problem.

**Lemma 1** [9,16]. *Given $B > 0$ and a tree $T$ with $\max_{e \in E(T)} w(e) \le B$, we can decompose $T$ into $k \le \max\{\lfloor \frac{w(T)}{B} \rfloor, 1\}$ edge-disjoint trees $T_1, T_2, \ldots, T_k$ with $w(T_i) \le 2B$ for each $i = 1, 2, \ldots, k$ in $O(|V(T)|)$ time.*

However, this result cannot be used directly to deal with the MRPCP, which needs a more general version where the upper bound on edge length is removed and the graph to be decomposed is connected but not necessarily a tree.

**Lemma 2.** *Given $B > 0$ and a connected graph $H$, we can decompose $H$ into $k \le \max\{\lceil \frac{w(H)}{B} \rceil, 1\}$ edge-disjoint connected subgraphs $H_1, H_2, \ldots, H_k$ such that either $w(H_i) \le 2B$ or $|E(H_i)| = 1$ for each $i = 1, 2, \ldots, k$ in $O(|E(H)|)$ time.*

*Proof.* We prove the lemma by **Algorithm Graph Decomposition** described in Fig. 1. We explain the steps of this algorithm by giving an example in Fig. 2. The input consists of $B > 0$ and a connected graph $H$. We see the tree $T$ generated in Step 2 as a rooted tree with an arbitrary root vertex $r$. A subtree of $T$ is called light (medium, or heavy) if its weight is smaller than $B$ (in the range $[B, 2B)$, or no less than $2B$). Let $T_v$ be the subtree induced by $v$ and all its descendants. For $e = (u, v) \in E(T)$ with $v$ a child of $u$, $T_e$ denotes the subtree consisting of $e$ and $T_v$.

**Algorithm** *Graph Decomposition*

Step 1. Find an arbitrary spanning tree $T_H$ of $H$.

Step 2. For each edge $e = (u, v) \in E(H) \setminus E(T)$, we add an edge $e' = (u, v_e)$ with $w(e') = w(e)$ to $T$, where $v_e$ is a new vertex corresponding to $e$. Let $E' = \{e' = (u, v_e) \mid e \in E(H) \setminus E(T)\}$. Then $T = T_H \cup E'$ is a tree with $|E(H)|$ edges and $w(T) = w(H)$. We see $T$ as a rooted tree with an arbitrary root vertex $r$. Set $\mathcal{T} = \emptyset$.

Step 3. Split all subtrees $T_v$ or $T_e$ of weight in interval $[B, 2B)$ from the current tree and put them into $\mathcal{T}$. If no edges is left, turn to Step 5; otherwise, we are left with a tree $T_{left}$ whose subtrees $T_v$ and $T_e$ are either light or heavy.

Step 4. For $T_{left}$, do the following:

(i) If $w(T_{left} < B)$, then put $T_{left}$ into $\mathcal{T}$ and turn to Step 5;

(ii) If $w(T_{left}) \geq 2B$, find a vertex $u$ with children $v_1, v_2, \ldots, v_l$ such that $T_u$ is heavy and $T_{v_i}$ is light and $T_{e_i}$ is either light or heavy for $i = 1, 2, \ldots, l$, where $e_i = (u, v_i)$. We assume w.l.o.g that $T_{e_1}, \ldots, T_{e_h}$ are heavy.

**Case 1.** $h \geq 1$. For $i = 1, \ldots, h$, if $w(T_{left}) - w(T_{v_i}) \leq 2B$, split $T_{v_i}$ and the whole leftover tree (both subtrees have a weight of no more than $2B$ by definition), put them into $\mathcal{T}$ and turn to Step 5; otherwise, split $T_{v_i}$ and a subtree consisting of a single edge $e_i$, put them into $\mathcal{T}$.

**Case 2.** $h = 0$, i.e. all the subtrees $T_{e_i}$ $(i = 1, 2, \ldots, l)$ are light. Then we split a sequence of subtrees $T^1 = \bigcup_{i=i_0+1}^{i_1}, \ldots, T^t = \bigcup_{i=i_{t-1}+1}^{i_t}$ (set $i_0=0$ for convenience) such that $i_j$ $(j = 1, \ldots, t)$ is the minimum index with $\sum_{i=i_{j-1}+1}^{i_j} w(T_{e_i}) \geq B$. Since $w(T_{e_j}) < B$, we have $w(T^j) < 2B$, i.e., $T^j$ is medium. Put $T^1, \ldots, T^t$ into $\mathcal{T}$ (Note that $t \geq 1$ since $T_u$ is heavy).

(iii) Removing the edges in the subtrees split will result in a set of subtrees. Update the current tree as the only tree containing $r$ and turn to Step 3.

Step 5. Suppose $\mathcal{T} = \{T_1, T_2, \ldots, T_k\}$. For subtree $T_i$ $(i = 1, 2, \ldots, k)$, replace each edge $e' = (u, v_e) \in E(T_i) \cap E'$ by $e \in E(H) \setminus E(T)$ and delete $v_e$ to obtain a connected subgraph $H_i$. Return $H_1, H_2, \ldots, H_k$.

**Fig. 1.** Graph decomposition

The main idea of decomposition of $T$ is similar to [9,16], i.e., to "split" medium subtrees or subtrees consisting of a single edge until we are left with a tree of total length less than $2B$. For a vertex $v$ connected to its children with edges $e_1, e_2, \ldots, e_l$, splitting subtree $T' = \bigcup_{i=p}^{q} T_{e_i} (1 \leq p \leq q \leq l)$ means removing all the edges in $\bigcup_{i=p}^{q} E(T_{e_i})$ and all the vertices in $\bigcup_{i=p}^{q} V(T_{e_i}) \setminus \{v\}$ from $T$ and putting $T'$ into our decomposition $\mathcal{T}$.

By construction the subtrees in $\mathcal{T}$ are connected, edge-disjoint and contain all the edges of $T$, which implies that the subgraphs $H_1, H_2, \ldots, H_k$, are connected, edge-disjoint and contain all the edges of $E(H)$. Since we always split medium subtrees or subtrees consisting of a single edge, either $w(H_i) \leq 2B$ or $|E(H_i)| = 1$ for each $i = 1, 2, \ldots, k$. Moreover, one can see that when adding subtrees to $\mathcal{T}$ we split either a medium subtree (Step 3, Step 4(ii) Case 2) or two subtrees of total weight at least $2B$ (Step 4(ii) Case 1) except for the last subtree. Therefore, $k \leq \max\{\lceil \frac{w(T)}{B} \rceil, 1\} = \max\{\lceil \frac{w(H)}{B} \rceil, 1\}$.

Using a standard bottom-up data structure of a tree, the algorithm can be implemented in $O(|E(H)|)$ time. This completes the proof.     □

# 3   Minimum Rural Postmen Cover

In this section we give two 5-approximation algorithms for the MRPCP. The first is a relatively simple 5-approximation algorithm. The second is a faster 5-approximation algorithm, which adopts the approach in [25] designed for the MCCP.

## 3.1   A Simple 5-Approximation Algorithm

Next we analyze the following algorithm with inputs $G, R, \lambda$.

**Algorithm MRPCP1**

For each $l = n_R, n_R-1, \ldots, 1$, where $n_R$ is the number of connected components of the subgraph induced by the edges in $R$, do the following:

Step 1. Find a minimum weight subgraph, say $G^l$, of $G$ that includes all the edges of $R$ and has exactly $l$ connected components, denoted by $G_1^l, G_2^l, \ldots, G_l^l$. Set $S_l := \emptyset$.

Step 2. For $i = 1, 2, \ldots, l$, run Algorithm *Graph Decomposition* on $G_i^l$ with $B = \frac{\lambda}{4}$ to obtain a series of connected subgraphs. For each subgraph $H$ of these connected subgraphs, we distinguish the following three cases:

**Case 1.** If $w(H) \le 2B = \frac{\lambda}{2}$, we double $H$ to derive a closed walk of length at most $4B = \lambda$ and put it into $S_l$.

**Case 2.** If $H$ consists of a single edge $e \in R$, we derive a closed walk by connecting the two end vertices of $e$ with a shortest path in $G$.

**Case 3.** If $H$ consists of a single edge $e \in E \setminus R$, we simply discard $H$.

The algorithm returns one of the $S_l$ with minimum number of closed walks.

We show the performance of this algorithm in the following lemma.

**Theorem 1.** *Algorithm MRPCP1 is an $O(n^2 \log n + n_R|E|)$ time 5-approximation algorithm for the MRPCP.*

**Remark 1.** In Arkin et al.'s algorithm for the MRPWCP, they used a similar framework to Algorithm *MRPCP1*. However, in Step 2 they chose a different approach to generate a rural postmen walk cover of $R$. The authors first double the graph $G_i^l$ to produce a closed walk covering the edges in $R \cap E(G_i^l)$ and then cut it into walks of length at most $\lambda$ using a greedy procedure similar to the cycle-splitting technique in [12,22]. Since this procedure may throw away some edges, they have to copy the edges in $R \cap E(G_i^l)$ to avoid being removed and hence generate a feasible rural postmen walk cover of $R$. This approach can be modified to deal with the MRPCP and achieve an approximation ratio of 7 by a careful examination.

One can see that Algorithm *MRPCP1* adopts a more direct approach, i.e., Algorithm *Graph Decomposition*, which avoid doubling $G_i^l$ and copying the edges in $R \cap E(G_i^l)$. This is the reason why Algorithm *MRPCP1* has a better approximation ratio.

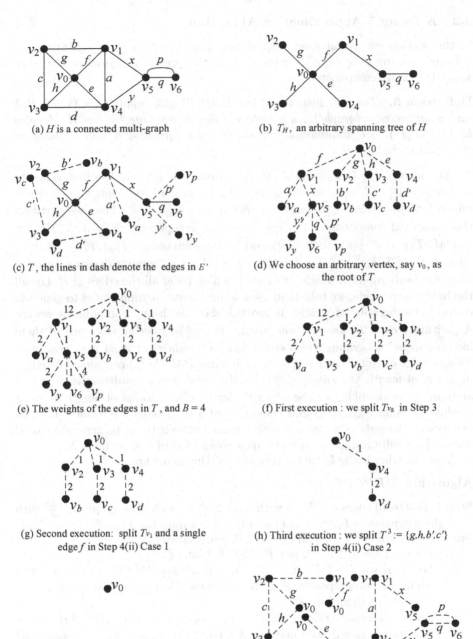

(a) $H$ is a connected multi-graph

(b) $T_H$, an arbitrary spanning tree of $H$

(c) $T$, the lines in dash denote the edges in $E'$

(d) We choose an arbitrary vertex, say $v_0$, as the root of $T$

(e) The weights of the edges in $T$, and $B = 4$

(f) First execution : we split $T_{v_5}$ in Step 3

(g) Second execution: split $T_{v_1}$ and a single edge $f$ in Step 4(ii) Case 1

(h) Third execution : we split $T^3 := \{g, h, b', c'\}$ in Step 4(ii) Case 2

(i) Fourth execution : we split the $T_{left}$ with weight less than $B$

(j) All the subgraphs (five) generated by Algorithm Graph Decomposition on $H$

**Fig. 2.** An example for Algorithm *Graph Decomposition*

## 3.2   A Faster 5-Approximation Algorithm

In this section we give another 5-approximation algorithm. This algorithm has a faster running time $O(|E| + n \log n)$ than the previous algorithm while the analysis is more complicated.

**Definition 3.** *Given an instance of the MRPCP with inputs $G = (V, E), R, \lambda$ and a connected subgraph $H$, a pseudo minimum spanning tree on $H$, denoted by $PMST(H)$, is a minimum weight connected subgraph of $H$ that contains all the edges in $E(H) \cap R$.*

The idea of the 5-approximation algorithm is as follows. We first generate a graph $G_R[\frac{\lambda}{A}]$ ($A$ is an integer greater than one and its value will be determined later) by deleting all edges $e \in E \setminus R$ with $w(e) > \frac{\lambda}{A}$. Then we partition the connected components of $G_R[\frac{\lambda}{A}]$ into light components $F_1, F_2, \ldots, F_l$ with $w(PMST(F_i)) \leq \frac{\lambda}{2}(i = 1, 2, \ldots, l)$ and heavy components $F_{l+1}, F_{l+2}, \ldots, F_{l+h}$ with $w(PMST(F_i)) > \frac{\lambda}{2}(i = l + 1, l + 2, \ldots, l + h)$. Next we derive a series of connected subgraphs of weight at most $\frac{\lambda}{2}$, which cover all the edges in $R$. For all the light components, we take their pseudo minimum spanning trees to generate connected subgraphs of weight at most $\frac{\lambda}{2}$. For the heavy components we use Algorithm *Graph Decomposition* (taking $B = \frac{\lambda}{4}$) to decompose each of them into connected subgraphs which either have a weight of at most $\frac{\lambda}{2}$ or consist of a single edge $e$ (Note that $e$ has to be in $R$ since $PMST(F_i)$ contains only edges in $E \setminus R$ of length at most $\frac{\lambda}{A} \leq \frac{\lambda}{2}$). Lastly, for connected subgraphs of weight at most $\frac{\lambda}{2}$ we doubling all the edges to derive closed walks of length at most $\lambda$ while for connected subgraphs consisting of a single edge we connect the end vertices of the only edge by a shortest path between them to generate closed walks. This will produce a rural postmen cover of $R$ of cost at most $\lambda$.

The following is the formal description of the algorithm.

## Algorithm *MRPCP2*

Step 1. Delete all edges $e \in E \setminus R$ with $w(e) > \frac{\lambda}{A}$ to derive the graph $G_R[\frac{\lambda}{A}]$ with light components $F_1, F_2, \ldots, F_l$ and heavy components $F_{l+1}, F_{l+2}, \ldots, F_{l+h}$.

Step 2. Set $T := \emptyset$, $S := \emptyset$ and do the following substeps:

(i) For each $i = 1, 2, \ldots, l$, put $PMST(F_i)$ into $T$.

(ii) For each $s = l + 1, l + 2, \ldots, l + h$, decompose $PMST(F_s)$ into a set of connected subgraphs by Algorithm *Graph Decomposition* (taking $B = \frac{\lambda}{4}$) and put them into $T$.

(iii) For each connected subgraph $H$ in $T$, if $w(H) \leq \frac{\lambda}{2}$ we double all the edges to obtain a closed walk and put it into $S$; otherwise, i.e., $H$ is composed of a single edge $e \in R$, we connect the end vertices of $e$ by a shortest path between them to derive a closed walk and put it into $S$.

Step 4. Return $S$.

By the steps of the algorithm one can verify that $S$ is a rural postmen cover of $R$ of cost no more than $\lambda$. We proceed to show that the number of closed walks in $S$ is at most $5OPT$.

Assume that $OPT = k$ and $C_1^*, C_2^*, \ldots, C_k^*$ are the optimum closed walks. We have similar definitions about "incident", "light closed walk", "bad closed walk" and "heavy closed walk", "long edge" and "internal long edge" in [25].

**Definition 4.** *Given a subgraph $H$ of $G$, if $V(F_i) \cap V(H) \neq \emptyset$ for some $i$ with $1 \leq i \leq l + h$ we call $F_i$ is incident to $H$ or $H$ is incident to $F_i$.*

**Definition 5.** *Given an optimum closed walk $C$, it is called a light closed walk (heavy closed walk) if $C$ is incident to only light components (heavy components). When $C$ is incident to at least one light component and at least one heavy component, it is called a bad closed walk.*

Let $k_l, k_h, k_b$ be the number of light, heavy and bad closed walks, respectively. It is obvious that $k = k_l + k_h + k_b$.

**Definition 6.** *An edge $e$ of an optimum closed walk with weight $w(e) > \frac{\lambda}{A}$ is called a long edge. A long edge $e$ incident to only one connected component of $G_R[\frac{\lambda}{A}]$ is called an internal long edge. If $e$ is incident to two distinct connected components of $G_R[\frac{\lambda}{A}]$, it is referred to as an external long edge.*

The number of external long edges of an optimum closed walk $C$ is denoted by $Ex(C)$. Since $w(C) \leq \lambda$, we have $Ex(C) \leq A - 1$, $i = 1, 2, \ldots, k$. Since $C$ is a closed walk it cannot include a single external long edge. So we have

*Property 1.* For any optimum closed walk $C$, $Ex(C) \in \{0, 2, 3, \ldots, A - 1\}$.

A similar argument to that in [25] leads to

*Property 2.* For any optimum closed walk $C$, it is incident to at most $A - 1$ components in $\{F_1, F_2, \ldots, F_{l+h}\}$.

By these two properties, we can classify all the bad closed walk into $B_1, B_2, \ldots, B_{A-1}$, where $B_j$ denotes the set of bad closed walks incident to exactly $j$ light components. If we define $k_j = |B_j|$, it is clear that $k_b = \sum_{j=1}^{A-1} k_j$.

Now we can bound the number $l$ of connected subgraphs derived in Step 2(i) of Algorithm $MRPCP2$.

**Lemma 3.** $l \leq (A - 1)k_l + \sum_{j=1}^{A-1} jk_j$.

By this lemma, if $h = 0$ Algorithm $MRPCP2$ returns a $(A - 1)$-approximate solution because we have $k = k_l$ in this case, which implies that Step 2(ii) does not generate any connected subgraph. Therefore, we assume $h \geq 1$ in the following discussion.

To bound the number of connected subgraphs generated in Step 2(ii) of Algorithm $MRPCP2$, we give an upper bound on the total weight of the pseudo minimum spanning trees of the heavy components.

**Lemma 4.** $\sum_{s=l+1}^{l+h} w(PMST(F_s)) \leq k_b \lambda + (1 + \frac{1}{A})k_h \lambda - \frac{\lambda}{A} \sum_{j=1}^{A-1} jk_j - h\frac{\lambda}{A}$.

**Lemma 5.** *The Step 2(ii) of Algorithm MRPCP2 generates at most*

$$\begin{cases} 4\left(k_b + \frac{A+1}{A}k_h - \frac{1}{A}\sum_{j=1}^{A-1} jk_j\right), & A = 2,3, \\ \frac{6A}{A+2}k_b + \frac{6(A+1)}{A+2}k_h - \frac{6}{A+2}\sum_{j=1}^{A-1} jk_j, & A \geq 4, \end{cases}$$

*connected subgraphs.*

Using Lemmas 3 and 5 with $A \geq 4$, the number of connected subgraphs generated by Algorithm $MRPCP2$ is at most

$$l + \frac{6A}{A+2}k_b + \frac{6(A+1)}{A+2}k_h - \frac{6}{A+2}\sum_{j=1}^{A-1} jk_j$$

$$\leq (A-1)k_l + \frac{6A}{A+2}k_b + \frac{6(A+1)}{A+2}k_h + \frac{A-4}{A+2}\sum_{j=1}^{A-1} jk_j$$

$$\leq (A-1)k_l + \frac{A^2+8}{A+2}k_b + \frac{6(A+1)}{A+2}k_h$$

where the last inequality follows from $\sum_{j=1}^{A-1} jk_j \leq (A-2)k_b$. So we have

**Lemma 6.** *If we take $A = 4$, it holds that $|S| = |T| \leq 4k_l + 4k_b + 5k_h \leq 5k$.*

As for the time complexity of Algorithm $MRPCP2$, in Step 1 we can obtain the graph $G_R[\frac{\lambda}{A}]$ in $O(|E|)$ time. To compute the $PMST$ on $F_i$ $(i = 1, 2, \ldots, l+h)$ we use the Prim's Algorithm implemented with Fibonacci heap [17], which takes $O(|E(F_i)| + |V(F_i)|\log|V(F_i)|)$ time. Therefore, the overall time complexity of Step 1 is $O(|E| + n\log n)$. In Step 2, Algorithm *Graph Decomposition* runs in $O(|E|)$ time. Above all, Algorithm $MRPCP2$ has a time complexity of $O(|E| + n\log n)$. Combining this with Lemma 6, we have

**Theorem 2.** *There is an $O(|E| + n\log n)$ time 5-approximation algorithm for the MRPCP.*

## 4    Minimum Chinese Postmen Cover

In this section we deal with the MCPCP and devise a 4-approximation algorithm.

Given an instance of the MCPCP with inputs $G$ and $\lambda$, if $G$ is composed of connected components $G_1, G_2, \ldots, G_p$, then the optimum closed walks in $OPT$ must be composed of $p$ groups and the $i$th group is a postmen cover of $E(G_i)$ of cost at most $\lambda$ that only use edges in $G_i$. If we define $p$ subinstances of the MCPCP with the $i$th subinstance consisting of a connected graph $G_i$ and $\lambda$, then any $p$ $\rho$-approximate solutions for the corresponding subinstances can be merged to obtain a $\rho$-approximate solution for the original instance. This is because $OPT = \sum_{i=1}^{p} OPT_i$, where $OPT_i$ is the optimal value of the $i$th subinstance. In a word, we may assume w.l.o.g that $G$ is connected in the rest of this section.

We give the following algorithm for the MCPCP. The input consists of a connected graph $G = (V, E)$ and $\lambda \geq 0$.

**Algorithm** *MCPCP*

Step 1. We solve the Chinese Postman Problem defined on $G$ to obtain a optimum closed walk $C_0$ containing all the edges in $E$. Set $C := C_0$, $T := \emptyset$, $S := \emptyset$ and $X := \emptyset$.

Step 2. If $E(C) = \emptyset$, turn to Step 4.

Step 3. If $w(C) \leq \frac{\lambda}{2}$, let $W_{left} := C$ be the sub-walk of $C_0$ containing all the edges in $C$. Put $W_{left}$ into $T$ and put a closed walk obtained by doubling $W_{left}$ into $S$. After that turn to Step 4. Otherwise, i.e. $w(C) \leq \frac{\lambda}{2}$, we assume that $C = e_1 e_2 \cdots e_l$. We can find the minimum index $j$ with $\sum_{i=1}^{j} w(e_i) > \frac{\lambda}{2}$.

    **Case 1.** $j = 1$. Let $W_j = e_1$ be the sub-walk of $C_0$ consisting of a single edge $e_1$. $C_j$ is defined as the closed walk consisting of $e_1$ and a shortest path $P_1$ between the two end vertices of $e_1$. Update $C$ by removing the edge $e_1$.

    **Case 2.** $j > 1$. Let $W_j = e_1 \cdots e_{j-1} e_j$ be the sub-walk of $C_0$ consisting of the first $j$ edges of $C$. Define $C_j$ as the closed walk obtained by doubling the edges in the sub-walk $e_1 \cdots e_{j-1}$. Update $C$ by removing the edges $e_1, \ldots, e_{j-1}$. Set $X := X \cup \{e_j\}$.

Put $W_j$ and $C_j$ into $T$ and $S$, respectively. Then turn to Step 2.

Step 4. Return the set $S$ of closed walks.

We analyze the approximation ratio in the following theorem.

**Theorem 3.** *There is an $O(n^3)$ time 4-approximation algorithm for the MCPCP.*

**Acknowledgements.** This research is supported by the National Natural Science Foundation of China under grants number 11671135, the Natural Science Foundation of Shanghai under grant number 19ZR1411800 and the Fundamental Research Fund for the Central Universities under grant number 22220184028.

# References

1. Arkin, E.M., Hassin, R., Levin, A.: Approximations for minimum and min-max vehicle routing problems. J. Algorithms **59**, 1–18 (2006)
2. van Bevern, R., Hartung, S., Nichterlein, A., Sorge, M.: Constant-factor approximations for capacitated arc routing without triangle inequality. Oper. Res. Lett. **42**, 290–292 (2014)
3. Chen, Y., Goebel, R., Lin, G., Su, B., Xu, Y., Zhang, A.: An improved approximation algorithm for the minimum 3-path partition problem. J. Comb. Optim. **38**, 150–164 (2019)
4. Christofides, N.: Worst-case analysis of a new heuristic for the traveling salesman problem. Technical report, Graduate School of Industrial Administration, Carnegie-Mellon University, Pittsburgh (1976)

5. Corberan, A., Laporte, G. (eds.): Arc Routing: Problems, Methods, and Applications. SIAM, Philadelphia (2014)
6. Dyer, M., Frieze, A.: On the complexity of partitioning graphs into connected subgraphs. Discret. Appl. Math. **10**, 139–153 (1985)
7. Edmonds, J., Johnson, E.L.: Matching, Euler tours and the Chinese postman. Math. Program. **5**(1), 88–124 (1973)
8. Eiselt, H.A., Gendreau, M., Laporte, G.: Arc routing problems, part II: the rural postman problem. Oper. Res. **43**, 399–414 (1995)
9. Even, G., Garg, N., Koemann, J., Ravi, R., Sinha, A.: Min-max tree covers of graphs. Oper. Res. Lett. **32**, 309–315 (2004)
10. Farbstein, B., Levin, A.: Min-max cover of a graph with a small number of parts. Discret. Optim. **16**, 51–61 (2015)
11. Frederickson, G.N.: Approximation algorithms for some postman problems. J. ACM **26**(3), 538–554 (1979)
12. Frederickson, G.N., Hecht, M.S., Kim, C.E.: Approximation algorithms for some routing problems. SIAM J. Comput. **7**(2), 178–193 (1978)
13. Gutin, G., Muciaccia, G., Yeo, A.: Parameterized complexity of $k$-Chinese postman problem. Theor. Comput. Sci. **513**, 124–128 (2013)
14. Holyer, I.: The NP-completeness of some edge-partition problem. SIAM J. Comput. **10**, 713–717 (1981)
15. Karpinski, M., Lampis, M., Schmied, R.: New inapproximability bounds for TSP. J. Comput. Syst. Sci. **81**, 1665–1677 (2015)
16. Khani, M.R., Salavatipour, M.R.: Improved approximation algorithms for min-max tree cover and bounded tree cover problems. Algorithmica **69**, 443–460 (2014)
17. Korte, B., Vygen, J.: Combinatorial Optimization: Theory and Algorithms. Springer, Heidelberg (2007)
18. Monnot, J., Toulouse, S.: The path partition problem and related problems in bipartite graphs. Oper. Res. Lett. **35**, 677–684 (2007)
19. Nagarajan, V., Ravi, R.: Approximation algorithms for distance constrained vehicle routing problems. Networks **59**(2), 209–214 (2012)
20. Safilian, M., Hashemi, S.M., Eghbali, S., Safilian, A.: An approximation algorithm for the subpath planning. In: The Proceedings of the 25th International Joint Conference on Artificial Intelligence, pp. 669–675 (2016)
21. Sumita, H., Yonebayashi, Y., Kakimura, N., Kawarabayashi, K.: An improved approximation algorithm for the subpath planning problem and its generalization. In: The Proceedings of the 26th International Joint Conference on Artificial Intelligence, pp. 4412–4418 (2017)
22. Xu, Z., Xu, L., Li, C.-L.: Approximation results for min-max path cover problems in vehicle routing. Naval Res. Log. **57**, 728–748 (2010)
23. Yan, J., Chang, G., Hedetniemi, S., Hedetniemi, S.: $k$-path partitions in trees. Discret. Appl. Math. **78**, 227–233 (1997)
24. Yu, W., Liu, Z.: Improved approximation algorithms for some min-max cycle cover problems. Theor. Comput. Sci. **654**, 45–58 (2016)
25. Yu, W., Liu, Z., Bao, X.: New approximation algorithms for the minimum cycle cover problem. Theor. Comput. Sci. **793**, 44–58 (2019)
26. Yu, W., Liu, Z., Bao, X.: Approximation algorithms for some min-max postmen cover problems (2019, Manuscript submitted for publication)

# New Results on a Family of Geometric Hitting Set Problems in the Plane

Joseph S. B. Mitchell[1] and Supantha Pandit[2(✉)]

[1] Stony Brook University, Stony Brook, NY, USA
joseph.mitchell@stonybrook.edu
[2] Dhirubhai Ambani Institute of Information and Communication Technology,
Gandhinagar, Gujarat, India
pantha.pandit@gmail.com

**Abstract.** We study some geometric optimal hitting set (stabbing) problems involving certain classes of objects in the plane. The objects include axis-parallel line segments, red/blue sets of pseudo-segments, axis-parallel 2-link "L" chains, pairs of line segments, etc. We examine cases in which the objects are constrained so that at least one endpoint of each object is on an inclined line (a line with slope −1). We prove that stabbing a set of vertical segments using a minimum number of horizontal segments is NP-hard when the input segments are each touching the inclined line, possibly from both sides of the line. Previously, the problem was known to be NP-hard for the general version, stabbing vertical segments with a minimum number of horizontal segments in the plane [9], and for a constrained version of this problem, in which all of the vertical segments intersect a horizontal line [3]. We provide some constant factor approximation algorithms as well. In particular, we present a PTAS for this problem using the local search technique. In contrast, if both vertical and horizontal segments are touching the inclined line from exactly one side, then the problem can be solved in polynomial time. We prove that stabbing a class of 2-link chains ("I-chains") by horizontal segments is NP-hard, when both the chains and the segments have an endpoint on an inclined line and lie on one side (say, the right side) of the line. Finally, we prove that stabbing pairs of segments (each pair contains either two vertical segments or one vertical and one horizontal segments) by horizontal segments is NP-hard, when the segments are touching the inclined line from only one side.

**Keywords:** Stabbing segments · Geometric hitting set · Set cover · NP-hard · Local search · PTAS

J. S. B. Mitchell—Partially supported by the National Science Foundation (CCF-1526406), the US-Israel Binational Science Foundation (project 2016116), and DARPA (Lagrange).

Y. Li et al. (Eds.): COCOA 2019, LNCS 11949, pp. 387–399, 2019.
https://doi.org/10.1007/978-3-030-36412-0_31

# 1    Introduction

Geometric stabbing and covering problems with line segments, rays, and lines in the plane are well-studied [9]. Recently some variations are also studied [3]. We are given two sets of objects $R$ and $B$. We say that an object $r \in R$ *stabs* (or *hits*) an object $b \in B$ if $r \cap b \neq \emptyset$. In the stabbing problem [9], the goal is to find a minimum cardinality set $R' \subseteq R$ such that each object in $B$ is hit. In the orthogonal segment stabbing problem, the sets $R$ and $B$ are the subsets of axis-parallel segments in the plane. This problem was studied by Katz et al. [9]. They proved that the problem is NP-hard when the set $R$ contains horizontal segments and $B$ contains vertical segments in the plane. In [3], Bandyapadhyay et al. studied a constrained version of the orthogonal segment stabbing problem where the every horizontal segment intersects a vertical line. They provide polynomial time algorithms and several NP-hardness results based on the sets $R$ and $B$ contains different subsets of axis-parallel segments.

We study some special cases of the segment stabbing problem, with various geometric objects. In particular, we consider geometric stabbing problems in the plane with line segments, 2-link chains, and pairs of line segments. Further, we assume that each input segment has at least one endpoint on a given inclined line (e.g., a line with slope $-1$, wlog); refer to Fig. 1. We define this problem in general as follows.

> **Geometric Hitting Set:** We are given two sets, a red set $R$ and a blue set $B$ of anchored geometric objects in the plane (e.g., line segments, polygonal chains, pseudo-segments, pairs of segments, etc), with each object being "anchored" (having an endpoint) on an inclined line. The objective is to find a minimum-cardinality collection $R' \subseteq R$ of red objects that hit (intersect) each blue object in $B$.

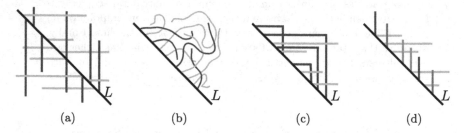

(a)                (b)                (c)                (d)

**Fig. 1.** Different problem instances. (a) Stabbing vertical segments by horizontal segments, with each segment touching an inclined line. (b) Stabbing blue pseudo-segments with red pseudo-segments. (c) Stabbing 2-link chains by horizontal segments, with each object touching an inclined line from the right side of the line. (d) Stabbing pairs of vertical segments by horizontal segments, with each segment touching an inclined line from the right side of the line. (Color figure online)

## 1.1    Previous Work

Katz et al. [9] have studied a variety of orthogonal segment stabbing problems, on line segments, rays, and lines in the plane. They show that the problem is NP-complete for objects involving line segments. For the versions involving axis-parallel rays or lines they provided polynomial-time algorithms. Recently, a constrained version of the segment covering problem was studied by Bandya-padhyay and Mehrabi [3]. Given two sets of segments, $S$ and $D$, they define the $(S, D)$-stabbing problem as the problem of stabbing all of the line segments in $D$ with the minimum number of line segments in $S$. They considered orthogonal line segments and assumed that all of the horizontal line segments intersect a vertical line. Based on whether $S$ and $D$ contain either horizontal or vertical, or both, segments, they provided polynomial time algorithms and NP-hardness results. For some of the variations they provided PTAS's and constant-factor approximation algorithms.

Optimization problems involving objects intersecting an inclined line were first studied by Chepoi and Felsner [7]. They considered the Independent Set and Piercing Set problems with rectangles such that the rectangles are inter-secting an axis-monotone curve. For both problems they provided constant fac-tor approximation algorithms. Subsequently, Correa et al. [8] studied both of these problems, in which the rectangles are intersecting an inclined line. They showed that the Independent Set problem is NP-hard and provided constant factor approximation algorithms for both problems. They left open the compu-tational complexity of the Piercing Set problem. In [10], the authors answered this open question by showing an NP-hardness result for the Piercing Set prob-lem. Apart from rectangles they also considered other objects such as squares, unit-height rectangles. They studied the Set Cover, and Hitting Set problems as well. In [12], the Dominating Set problem with rectangles and unit squares is considered where the objects are touching an inclined line. He showed that the problem is NP-hard for rectangles and can be solved polynomial time for unit squares. Recently, Bandyapadhyay et al. [2], consider this problem and provided a $(2 + \epsilon)$ approximation algorithm for rectangles.

## 1.2    Our Contributions

- **Stabbing vertical segments**
  We show that stabbing vertical segments by horizontal segments can be solved in polynomial time where the segments are touching an inclined line $L$ from exactly one side (say right side) of $L$. If both vertical and horizontal segments are touching $L$ from both sides, then we prove that this problem is NP-hard. We show that this problem has a 2- and a 4-factor approximation algorithm. Further, we provide a PTAS using local search.

- **Stabbing pseudo-segments**
  We generalize from horizontal/vertical straight line segments that are touch-ing $L$ from one side to the case of two sets of pseudo-segments touching $L$ from one side: a blue set $B$ of disjoint simple curves and a red set $R$ of disjoint

simple curves, with the goal to compute a minimum cardinality subset of the red curves to hit all of the blue curves. (The pseudo-segment property of $B, R$ requires that any red curve crosses a blue curve at most once.)

- **Stabbing 2-link chains**
  We prove that stabbing 2-link chains (⌐-shapes) by horizontal segments is NP-hard where both the chains and segments are touching an inclined line $L$ from the same side.
- **Stabbing pairs of segments**
  We prove (in the full paper[1]) that stabbing pairs of vertical segments by horizontal segments is NP-hard where the segments are touching an inclined line $L$ from one side. It is also NP-hard for the case in which each pair includes exactly one vertical and one horizontal segment, even if all segments are of unit length.

## 1.3    Preliminaries

In the Rectilinear Planar Monotone 3SAT (RPM3SAT) problem, we are given a 3SAT formula $\phi$ with $n$ variables $x_1, x_2, \ldots, x_n$ and $m$ clauses $C_1, C_2, \ldots, C_m$, each containing exactly 3 literals, all of which are either positive (positive clause) or negative (negative clause). Each variable corresponds to a horizontal segment; all variable segments lie on a horizontal line, e.g., $y = 0$. Each clause also corresponds to a horizontal segment; positive clause segments lie above $y = 0$, negative clause segments lie below. Clause segments are connected to the corresponding three variables by vertical line segments. The entire construction is planar; see Fig. 2(a). We are to decide if there is a truth assignment for the variables that satisfies $\phi$.

We also consider a modified version of the RPM3SAT problem, the Mod-RPM3SAT problem. We place the variable segments on an inclined line $L$. Let $C$ be a positive clause in the RPM3SAT problem (Fig. 2(a)) that contains variables $x_i$, $x_j$, and $x_k$. We directly connect the clause segment of $C$ with the variable segment of $x_i$. Further, we extend the vertical segments corresponding to $x_j$ and $x_k$ such that they connect to the variable segments of $x_j$ and $x_k$ respectively. A symmetric construction is done for the negative clauses. See Fig. 2(b) for the modified instance of the RPM3SAT problem instance in Fig. 2(a). de Berg and Khosravi [4] proved that the RPM3SAT problems is NP-complete. This implies that the Mod-RPM3SAT problem is also NP-complete.

We now define some terminology that is used in this paper. Let us consider a variable $x_i$. Order the positive clauses left to right that connect to $x_i$. Now let $C_\ell$ be a positive clause that contains $x_i$, $x_j$, and $x_k$, and assume that the variables are also ordered from left to right. Then, by the above ordering of the positive clauses, we sat that, $C_\ell$ is a $\ell_1$-th, $\ell_2$-th, and $\ell_3$-th clause for the variables $x_i$, $x_j$, and $x_k$, respectively. This is true for both the RPM3SAT and Mod-RPM3SAT problems. A similar ordering can be done for the negative clauses.

---

[1] Several proofs are deferred to the full paper, because of page limitations here.

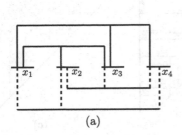 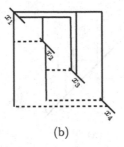

(a) (b)

**Fig. 2.** (a) An instance of RPM3SAT. (b) After modification of Fig. 2(a).

Another variation of the 3SAT problem is the linear-SAT (LSAT) problem that is introduced by Arkin et al. [1]. They proved that this problem is NP-complete. In this problem the literals are placed on a horizontal line and they have a left to right ordering. The clauses also have a left to right ordering and they connect to the literals from above. Any two consecutive clauses share at most one literal. See Fig. 3 for an instance $\phi$ of the LSAT problem.

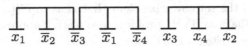

**Fig. 3.** An LSAT problem instance $\phi = C_1 \wedge C_2 \wedge C_3$, where $C_1 = (x_1 \vee \bar{x}_2 \vee \bar{x}_3)$, $C_2 = (\bar{x}_3 \vee \bar{x}_1 \vee x_4)$, and $C_3 = (x_3 \vee \bar{x}_4 \vee x_2)$

### 1.4 Segments Touching an Inclined Line from One Side

We show that the HVH1 problem can be solved in polynomial time when the segments are touching the inclined line from one side. Let $h_1, h_2, \ldots, h_n$ be a set of $n$ horizontal and $v_1, v_2, \ldots, v_m$ be a set of $m$ vertical segments such that they touch a given line $L$ from right side of $L$. We reduce this problem to the hitting set problem with rectangles where the rectangles are anchored on a vertical line. We take a infinite vertical line $\ell$ at the extreme right of all given segments such that no segment intersects $\ell$. Now for each vertical segment $v$, we take a rectangle $r_v$ that is anchored on $\ell$ and whose left boundary is $v$. For each horizontal segment $h$, take a point $p_h$ as the right endpoint of $h$. Now clearly, finding a minimum number of points stabbing all the anchored rectangles is equivalent to finding a minimum number of horizontal segments hit all the vertical segments. Since the hitting set problem on anchored rectangle can be solved in $O(\min(m, n)^2(m+n) + n \log n + m \log m)$ time [5] (Katz et al. [9] also considers the problem of stabbing segments with rays and gave an algorithm with running time $O(n^2(m + n)))$, the HVH1 problem with segments touching an inclined line from one side also solved in the same time.

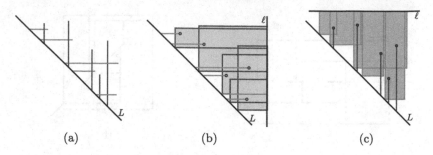

**Fig. 4.** (a) Given input. (b) A hitting set instance. (c) A set cover instance.

**Approximation Algorithm:** The HVH1 problem is equivalent to the set cover problem with rectangles such that the rectangles are anchored on a horizontal line. We take an infinite horizontal line $\ell$ at the top of all the given segments such that no segment intersects $\ell$. Now for each horizontal segment $h$, we take a rectangle $r_h$ that is anchored on $\ell$ and whose bottom boundary is $h$. For each vertical segment $v$, take a point $p_v$ as the top endpoint of $v$. See Fig. 4(c) for this construction. Katz et al. [9] provided a factor 2 approximation algorithm for this anchored version of the set cover problem that runs in $O((n + m) \log(n + m))$ time. This implies that the HVH1 problem also has a 2 approximation algorithm that runs in $O((n + m) \log(n + m))$ time.

## 2    Pseudo-Segments Touching from One Side

Consider now the case in which the input objects are sets of pseudo-segments that touch an inclined line from one side. Note that, in this pseudo-segment setting, the inclination of the line is not relevant: Thus, we assume that the "inclined line" is simply the $x$-axis, with each input curve lying above the axis (in the halfplane $y \geq 0$), with one of its endpoints being on the $x$-axis. Let $R$ denote the input set of $n$ pairwise-disjoint red pseudo-segments, $r_1, \ldots, r_n$, indexed according to the order of their endpoints on the $x$-axis. Let $B$ denote the input set of $m$ pairwise-disjoint blue segments, $b_1, \ldots, b_m$, indexed according to the order of their endpoints on the $x$-axis. The pseudo-segment property tells us that if $r_i$ intersects $b_j$, then the two curves intersect in a single point, either a shared endpoint or a proper crossing point; it suffices to assume, without loss of generality, that if two curves intersect they do so in a proper crossing point. For computational purposes, we assume quite general forms of curves; we have only to assume that one has a means of testing intersection and computing a crossing point of two curves (one red, one blue), if it exists, in $O(1)$ time.

Our goal is to compute a smallest cardinality subset, $R' \subseteq R$, such that $R'$ forms a hitting set for $B$ (i.e., every $b_j$ is intersected by some red curve in $R'$). We give a dynamic programming algorithm to solve the problem exactly in polynomial time ($O(n^3 m)$).

A pair of red pseudo-segments, $r_{i_1}$ and $r_{i_2}$, defines *subproblem* $(i_1, i_2)$, in which we seek to find the minimum number, $f(i_1, i_2)$, of red pseudo-segments from $\{r_{i_1+1}, \ldots, r_{i_2-1}\}$ needed to hit all blue segments, not already intersected by $r_{i_1}$ or $r_{i_2}$, whose endpoint on the $x$-axis lies between the endpoint of $r_{i_1}$ and $r_{i_2}$. We augment the set $R$ with a red curve $r_0$ (disjoint from curves $R$ and $B$) that touches the $x$-axis to the left of all red curves $\{r_1, \ldots, r_n\}$ and all blue curves $B$, and a red curve $r_{n+1}$ (disjoint from curves $R$ and $B$) that touches the $x$-axis to the right of all red curves $\{r_1, \ldots, r_n\}$ and all blue curves $B$. Our overall goal, then, is to compute $f(0, n+1)$.

Let $R_{i_1, i_2} = \{r_{i_1+1}, r_{i_1+2}, \ldots, r_{i_2-1}\}$ denote the set of red curves available to use in solving subproblem $(i_1, i_2)$. Then, our algorithm is based on a dynamic program in which we evaluate the functions:

$$f(i_1, i_2) = \begin{cases} 0 & \text{if all blue curves anchored between} \\ & r_{i_1} \text{ and } r_{i_2} \text{ intersect either } r_{i_1} \text{ or } r_{i_2} \\ \min_{i_1 < k < i_2} \{f(i_1, k) + 1 + f(k, i_2)\} & \text{otherwise.} \end{cases}$$

In order to justify the correctness of the dynamic program, we argue that, for any nonempty optimal set of red curves hitting all blue curves in a subproblem $(i_1, i_2)$, there must be a red curve $r_k$, with $i_1 < k < i_2$, that *partitions* the subproblem, in the following sense: any blue curve anchored between $r_{i_1}$ and $r_k$ is hit by a red curve $r_j$ with $i_1 \le j \le k$, and any blue curve anchored between $r_k$ and $r_{i_2}$ is hit by a red curve $r_{j'}$ with $k \le j' \le i_2$.

In order to make the partitioning argument, we define a partial order on red curves: We say that $r_i \succ r_{i'}$ if and only if there exists a blue curve $b_j \in B$ crossing $r_i$ at a point $w$ such that the curve $r_{i'}$ lies inside the (bounded) simply connected region bounded by the closed Jordan curve defined by the portion of $r_i$ from the $x$-axis to $w$, the portion of $b_j$ from $w$ to the $x$-axis, and the portion of the $x$-axis between the anchor points of $r_i$ and $b_j$. Then, in a nonempty optimal set of red curves for subproblem $(i_1, i_2)$, the claimed partitioning red curve $r_k$ is any red curve with $i_1 < k < i_2$ in the optimal solution for which $r_k$ is maximal in the partial order $\succ$, so that any blue curve $b_j$ anchored between $r_{i_1}$ and $r_{i_2}$ must be hit by a red curve of the optimal solution on the same side of $r_k$ as is $b_j$ (if $b_j$ were only hit by red curves on the opposite side of $r_k$, then one such red curve, $r'$, together with $b_j$, shows that $r' \succ r_k$, contradicting the fact that $r_k$ is undominated (maximal).

## 3  Segments Touching an Inclined Line from Both Sides

In this section we prove that hitting vertical segments by horizontal segments such that the segments touch a given inclined line $L$ from both sides of the line (HVH problem) is NP-hard. We give a reduction from the RPM3SAT problem.

**Reduction:** The gadget for the variable $x_i$ is shown in Fig. 5(a). We take $4m$ vertical segments $\{v_1, v_2, \ldots, v_{4m}\}$ and $4m$ horizontal segments $\{h_1, h_2, \ldots, h_{4m}\}$

in a circular fashion such that one endpoint of each segment touches $L$. The horizontal segment $h_i$ hits the vertical segments $v_{i-1}$ and $v_i$, for $2 \leq i \leq 4m$ and $h_1$ hits $v_1$ and $v_{4m}$. Clearly, there are two optimal sets of horizontal segments $H_1^i = \{h_1, h_3, \ldots, h_{4m-1}\}$ and $H_2^i = \{h_2, h_4, \ldots, h_{4m}\}$ containing $2m$ segments that hit all the vertical segments.

For each clause $C_\ell$, the gadget consists of a single vertical segment $v^\ell$. Now assume that $C_\ell$ contains variable $x_i$, $x_j$ and $x_k$. Let $C_\ell$ be a $\ell_1$-, $\ell_2$-, and $\ell_3$-th clause for $x_i$, $x_j$ and $x_k$ respectively (see preliminaries). There are three cases.

**Case 1:** $x_k$ is a rightmost variable in $C_\ell$. We place the segment $v^\ell$ such that it intersects $h_{2\ell_3}$ in the gadget of $x_k$.

**Case 2:** $x_j$ is a middle variable in $C_\ell$. We extend the segment $h_{2\ell_2}$ horizontally rightward such that it intersects $v^\ell$.

**Case 3:** $x_i$ is a leftmost variable in $C_\ell$. We extend the segment $h_{2\ell_1}$ horizontally rightward such that it intersects $v^\ell$.

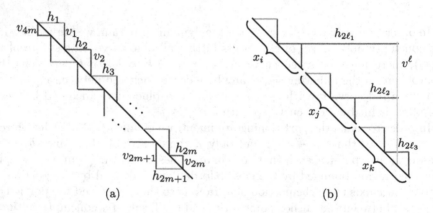

(a)                                    (b)

**Fig. 5.** (a) Structure of a variable gadget. (b) A clause gadget and its interaction with the variable gadgets.

Refer to the construction in Fig. 5(b). Clearly this construction takes polynomial time with respect to $n$ and $m$. We conclude (proof in the full paper):

**Theorem 1.** *The HVH problem is NP-hard.*

**Approximation Algorithm:** We show that there is a 2-approximation algorithm for the HVH problem. Let $H = \{h_1, h_2, \ldots, h_n\}$ be a set of horizontal segments and $V = \{v_1, v_2 \ldots, v_m\}$ be a set of horizontal segments. We first partition the set $V$ into $V'$ and $V''$, where $V' \subseteq V$ and $V'' \subseteq V$ are the sets of vertical segments to the right and left of $L$, respectively. We solve two HVH1 problems; one is to stab all the segments in $V'$ using $H$ and another one is to stab all the segments in $V''$ using $H$. Finally, we return the union of these two solutions.

Since the HVH1 problem can be solved in $O(n^2(m+n))$ time exactly, we get a 2-approximation for the HVH problem in the same time bound.

Note that, since there is a 2-approximation algorithm for the HVH1 problem that takes $O((n+m)\log(n+m))$ time, the HVH problem has a 4-approximation algorithm that runs in $O((n+m)\log(n+m))$ time.

**PTAS for the HVH Problem:** We present a PTAS for the HVH problem using the local search technique independently proposed by Mustafa and Ray [11] and Chan and Har-peled [6]. Let $H$ and $V$ be given sets of horizontal and vertical segments. We first preprocess the given input as follows: If there are two horizontal segments in $H$ having the same $y$-coordinates, then we keep the longer one among them and remove the shorter one from $H$; we do similarly for vertical segments in $V$, but we keep the shorter segment and remove the longer one. Let $H$ and $V$ be the updated sets, after this preprocessing.

Now we describe the local search algorithm to get a PTAS. The algorithm has the following steps.

**Initial Step:** We begin with any subset $H' \subset H$ that is a feasible solution (stabs all of the elements in $V$). Actually, it suffices to use $H' = H$.

**Local Improvement Step:** If any set $H_k \subseteq H'$ of $k$ segments can be replaced by a set $H_{k-1} \subseteq H$ of $k-1$ segments such that $(H' \setminus H_k) \cup H_{k-1}$ still stabs all the elements in $V$, then we set $H' = (H' \setminus H_k) \cup H_{k-1}$.

**Halting Step:** If no such local improvement is possible, then the algorithm stops and returns the solution $H'$.

The running time of the algorithm is $n^{O(k)}$. Let $O$ be the set of horizontal segments in an optimal solution, and let $P$ be the set of horizontal segments returned by the above local search algorithm (i.e., $P = H'$). Assume that $O \cap P = \emptyset$. We now show that the given input satisfies the following locality condition:

**Locality Condition:** There exists a planar bipartite graph $G = (O, P, E)$ such that there is an edge $(o, p) \in E$ if and only if there exists a vertical segment $v \in V$ for which $o \cap v \neq \emptyset$ and $p \cap v \neq \emptyset$.

Now, by an analysis similar to that in [11], we obtain the following lemma.

**Lemma 1.** *For a given set system $(H, V)$ satisfying the locality condition, there is a $(1 + \epsilon)$-approximation algorithm that takes $n^{O(1/\epsilon^2)}$ time.*

Now the only thing that remains is to show the existence of a graph $G$ that is planar and bipartite.

**Construction of a Planar Bipartite Graph $G = (O, P, E)$:** For each horizontal segment $s \in O \cup P$, take a vertex at the intersection of $s$ with the inclined line $L$. Now we connect these vertices using edges. Note that any vertical segment $v$ is stabbed by at least one segment $o$ from $O$ and at least one segment $p$ from $P$. Assume that $v$ is to the right of the inclined line $L$. There are two ways to stab $v$: (i) $p$ and $o$ are on the opposite sides of $L$ (Fig. 6(a)), and (ii) $p$ and $o$ are on the same side of $L$ (Fig. 6(b)).

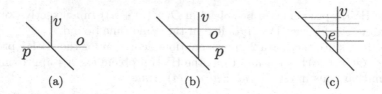

**Fig. 6.** (a) $v$ is stabbed by both $p$ and $o$, on opposite sides of $L$. (b) $v$ is stabbed by both $p$ and $o$ on the same side of $L$. (c) Generation of an edge $e$.

Observe that any horizontal segment to the left of $L$ can stab a vertical segment to the right of $L$ only at its bottom-most point on $L$. We add edges to the graph $G$ as follows. First we consider the right side of $L$. We sort the vertical segments by decreasing order of their $x$-coordinates. Take the segments one-by-one in this order. Consider the first vertical segment, say $v$, in this order. Take the two segments, one, say $o$, from $O$, and the other, say $p$, from $P$ having the smallest $y$-coordinates. Here we mention that if no such $o$ and $p$ is found (in this case either $o$ or $p$ is to the left of $L$), we handle such cases at a later stage. We take a curve (treated as an edge) say $e$ that connects the two intersection points of the two segments $o$ and $p$ with $L$ (see Fig. 6(c)). Take this curve to be to the right of $L$.

The next vertical segment has three options: (i) it intersects both $o$ and $p$, in which case we take the same curve $e$ in $G$; (ii) it intersects only one of $o$ and $p$ (say, $o$), in which case we take the segment $p'$ in $P$ with smallest $y$-coordinate and take a curve joining the two intersection points of the two segments $p'$ and $o$ with $L$; and, (iii) it intersects none of $o$ and $p$, in which case we repeat the same process as we did for $v$. Note that, the curves generated at the end of this process does not intersect with each other, they may touch at their end points on $L$. Further, there is no edge between two blue or two red vertices.

A similar process is performed for the vertical segments left of $L$.

After adding the edges using the above process we consider the vertical segments that are not stabbed yet. Note that these segments are of type (i) (see Fig. 6(a)) i.e., stabbed by $o$ and $p$ that are on both sides of $L$. Before adding a curve between the two vertices corresponding to $o$ and $p$ on $L$ we do the following. We perturb all the vertices slightly upwards (resp. downwards) on $L$ that are the end points of some curves on the right (resp. left) side of $L$. We do this perturbation in such a way that the ordering of all the vertices (the touching point of all horizontal lines with $L$) on $L$ should preserved. This perturbation ensures that no two curves; one from the right of $L$ and another from the left of $L$ do not coincides on two different vertices (corresponding to the two different horizontal lines; one to the right and another to the left of $L$) that have the same $x$- and $y$-coordinates. As a result all the curves to the right of $L$ are disjoint from all the curves to the left of $L$.

Now come back to add a curve between the two vertices corresponding to $o$ and $p$ on $L$. Observe that these two vertices are either coinciding (none of these vertices are the end points of a curve and hence not perturbed) or not coinciding (at least one of the two vertices is an end point of some curve and hence perturbed). In the first case, we perturb the two vertices corresponding to $o$ and $p$ on $L$ and add one curve either side of $L$ between them such that the curve does not intersect any other existing curve. In the second case, we add one curve, on either side of $L$, between the two vertices corresponding to $o$ and $p$ on $L$ such that it does not intersect any other curve.

We now argue that the graph $G$ generated by the above process is planar and bipartite. Note that there is no edges between two red or two blue vertices on $L$. Also no two curves intersect (they can touch at their end points on $L$). Hence, the resulting graph is planar and bipartite. Hence we have the following theorem.

**Theorem 2.** *The HVH problem admits a PTAS.*

## 4    Chains Touching an Inclined Line from One Side

We prove that hitting 2-link chains by horizontal segments (H2LSH) is NP-hard where the chains and segments touch a given inclined line $L$ from the right side of $L$. We give a reduction from the LSAT problem. Here we assume that the shape of a 2-link chain is an ⌐-chain. Each ⌐-chain has two segments, one horizontal and one vertical, and two endpoints, left and bottom. Let $\phi$ be an instance of the LSAT problem. We generate an instance $I_\phi$ of the H2LSH problem as follows.

**Reduction:** Let $L$ be the given inclined line. Let $0, 1, 2, \ldots, 2n-1$ be the number associated with the literals in the order they present in $\phi$.

Now for each variable $x_i$, we take a ⌐-chain $s_i$. Let the two literals $x_i$ and $\overline{x}_i$ are associated with the numbers $q$ and $q'$, respectively. If $q < q'$ (resp., $q' < q$), then the left (resp., bottom) endpoint of $s_i$ is at $(q, 2n-q)$, and the bottom (resp., left) endpoint is at $(q', 2n - q')$. If the literal corresponding to $q$ is positive, we place a horizontal segment $h_i$ whose left endpoint is at $(q, 2n-q)$ and a horizontal segment $\overline{h}_i$ whose left endpoint is at $(q', 2n - q')$. Otherwise, we place the left endpoint of $h_i$ at $(q', 2n - q')$ and the left endpoint of $\overline{h}_i$ at $(q, 2n - q)$. The length of each $h_i$ and $\overline{h}_i$ is 0.2. Refer to Fig. 7 for this construction.

Now for each clause $C_\ell$, we take 5 ⌐-chain s $\{t_1, t_2, t_3, t_4, t_5\}$ and additional 4 horizontal segments $\{g_1, g_2, g_3, g_4\}$. The lengths of each of $g_i$'s is 0.2 unit. Let $C_\ell$ contains three literals whose corresponding numbers are $q_1$, $q_2$, and $q_3$ in increasing order. Note that these numbers are consecutive. The left endpoint of $t_1$ is at $(q_1, 2n - q_1)$ and the bottom endpoint is at $(q_2 + .4, 2n - q_2 - .4)$. The left endpoint of $t_3$ is at $(q_2 + .2, 2n - q_2 - .2)$ and the bottom endpoint is at $(q_3 + .3, 2n - q_3 - .3)$. The left endpoint of $t_5$ is at $(q_3, 2n - q_3)$ and the bottom endpoint is at $(q_3 + .5, 2n - q_3 - .5)$. The left endpoint of $t_2$ is at $(q_3 + .25, 2n - q_3 - .25)$ and the bottom endpoint is at $(q_3 + .7, 2n - q_3 - .7)$. Finally, the left endpoint of $t_4$ is at $(q_3, 2n - q_3)$ and the bottom endpoint is at $(q_3 + .9, 2n - q_3 - .9)$. The left endpoints of $g_1, g_2, g_3$, and $g_4$ is at $(q_2 + .2, 2n - q_2 - .2)$, $(q_3 + .3, 2n - q_3 - .3)$,

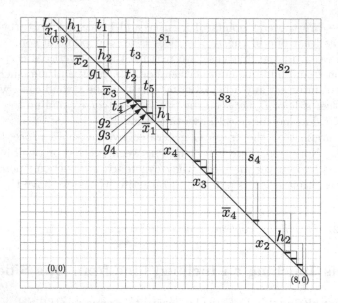

**Fig. 7.** Construction for the formula $\phi = C_1 \wedge C_2 \wedge C_3$, where $C_1 = (x_1 \vee \overline{x}_2 \vee \overline{x}_3)$, $C_2 = (\overline{x}_3 \vee \overline{x}_1 \vee x_4)$, and $C_3 = (x_3 \vee \overline{x}_4 \vee x_2)$.

$(q_3+.5, 2n-q_3-.5)$, and $(q_3+.7, 2n-q_3-.7)$ respectively. The overall construction for the instance of $\phi$ in Fig. 3 is shown in Fig. 7.

The construction can be done in polynomial time with respect to $n$ and $m$. This leads to the following theorem (whose proof appears in the full paper):

**Theorem 3.** *The H2LSH problem is NP-hard.*

# References

1. Arkin, E.M., Banik, A., Carmi, P., Citovsky, G., Katz, M.J., Mitchell, J.S.B., Simakov, M.: Selecting and covering colored points. Discret. Appl. Math. **250**, 75–86 (2018)
2. Bandyapadhyay, S., Maheshwari, A., Mehrabi, S., Suri, S.: Approximating dominating set on intersection graphs of rectangles and L-frames. In: MFCS, pp. 37:1–37:15 (2018)
3. Bandyapadhyay, S., Mehrabi, S.: Constrained orthogonal segment stabbing. CoRR abs/1904.13369 (2019). Proceedings of the CCCG 2019, Edmonton, Canada, pp. 195–202
4. de Berg, M., Khosravi, A.: Optimal binary space partitions for segments in the plane. Int. J. Comput. Geom. Appl. **22**(03), 187–205 (2012)
5. Chan, T.M., Grant, E.: Exact algorithms and APX-hardness results for geometric packing and covering problems. Comput. Geom. **47**(2), 112–124 (2014)
6. Chan, T.M., Har-Peled, S.: Approximation algorithms for maximum independent set of pseudo-disks. Discret. Comput. Geom. **48**(2), 373–392 (2012)

7. Chepoi, V., Felsner, S.: Approximating hitting sets of axis-parallel rectangles intersecting a monotone curve. Comput. Geom. **46**(9), 1036–1041 (2013)
8. Correa, J.R., Feuilloley, L., Pérez-Lantero, P., Soto, J.A.: Independent and hitting sets of rectangles intersecting a diagonal line: algorithms and complexity. Discret. Comput. Geom. **53**(2), 344–365 (2015)
9. Katz, M.J., Mitchell, J.S.B., Nir, Y.: Orthogonal segment stabbing. Comput. Geom. Theory Appl. **30**(2), 197–205 (2005)
10. Mudgal, A., Pandit, S.: Covering, hitting, piercing and packing rectangles intersecting an inclined line. In: Lu, Z., Kim, D., Wu, W., Li, W., Du, D.-Z. (eds.) COCOA 2015. LNCS, vol. 9486, pp. 126–137. Springer, Cham (2015). https://doi.org/10.1007/978-3-319-26626-8_10
11. Mustafa, N.H., Ray, S.: Improved results on geometric hitting set problems. Discret. Comput. Geom. **44**(4), 883–895 (2010)
12. Pandit, S.: Dominating set of rectangles intersecting a straight line. In: Canadian Conference on Computational Geometry, CCCG, pp. 144–149 (2017)

# Two-Machine Flow Shop Scheduling Problem Under Linear Constraints

Kameng Nip[1(✉)] and Zhenbo Wang[2]

[1] School of Mathematics (Zhuhai), Sun Yat-sen University, Zhuhai, China
niejm3@mail.sysu.edu.cn
[2] Department of Mathematical Sciences, Tsinghua University, Beijing, China

**Abstract.** We introduce a two-machine flow shop scheduling problem under linear constraints (2-FLC problem in short), in which the processing times of two stages of jobs are also decision variables and satisfy a system of linear constraints. The goal is to determine the processing times of each job, and to schedule the jobs to the two-machine flow shop such that the makespan, i.e., the completion time of all the jobs is minimized. This problem can find applications in various areas, such as industrial production and advertising planning. We study the computational complexity and algorithms for the 2-FLC problem. Particularly, we show that although the two-machine flow shop scheduling problem can be solved in polynomial time, the 2-FLC problem is generally NP-hard in the strong sense. Then we consider the design and analysis of algorithms on various settings of the 2-FLC problem. In particular, we propose a polynomial time algorithm for the 2-FLC problem when there is a fixed number of constraints. For the general case, we first propose a simple 2-approximation algorithm, and then design a polynomial time approximation scheme (PTAS).

**Keywords:** Flow shop scheduling · Computational complexity · Linear programming · Approximation algorithm

## 1 Introduction

This paper studies a generalization of the two-machine flow shop scheduling problem, in which the processing times of jobs are not given in advance, but can be determined by a system of linear constraints. Given a set of jobs and two machines, each job must be executed on the first machine, and then executed on the second machine, without interruption. The classical problem is to find the schedule which has minimum makespan, i.e., the completion time of the last job. It is usually denoted by $F2||C_{\max}$ [1]. In our problem, the processing times of jobs are not fixed and exogenously given, but are decision variables that satisfy a set of linear constraints. We call the problem a two-machine flow shop scheduling problem under linear constraints, and 2-FLC for short. The goal of this problem

This work has been supported by NSFC No. 11801589, No.11771245 and No. 11371216.

Y. Li et al. (Eds.): COCOA 2019, LNCS 11949, pp. 400–411, 2019.
https://doi.org/10.1007/978-3-030-36412-0_32

is to determine the processing times of the jobs, as well as finding the schedule that has minimum makespan among all the feasible choices.

The current work can be viewed as an extension of the existing works on the scheduling problem under linear constraints [2,3]. In these works, several single-stage scheduling problems under linear constraints, including parallel machine scheduling [2], related machine scheduling [3] have been introduced and studied. Under this framework, some parameters of the scheduling problem, such as the processing times or machine speeds must satisfy a set of linear constraints. The scheduling problem under linear constraints can be simpler or harder than the original scheduling problem. These problems may have different computational complexities, and require different techniques in designing and analyzing algorithms. For instance, the parallel machine scheduling where the processing times satisfying linear constraints [2] can be solved in polynomial time if both the number of constraints and machines are fixed constants, whereas the parallel machine scheduling problem itself is NP-hard even when there are only two machines. These results suggest us to explore various scheduling models, and to study their complexities and design of algorithms under linear constraints. Apart from scheduling problems, these are also some other researches on various combinatorial optimization problems under linear constraints in the literature, such as bin packing problem [4] and knapsack problem [5].

Next we describe some application scenarios that motivate the study of our 2-FLC problem. We remark that the scenarios are actually the extensions of those introduced in [2,3]. In fact, we can naturally replace the machine requirement from parallel machines to flow-shop machines in their examples.

1. *Industrial Production Problem.* Consider a problem arises in the steel industry, in which the decision maker requires certain amounts of different raw metals (say, iron, copper, aluminum, and etc.). The raw metals are obtained by extracting from several different types of steel. The extraction of each type of steel can be seen as a job. This is a typical application of flow shop scheduling problem on the industrial production (see, e.g., [6]). For example, each job should undergo wire-drawing first, and then annealing, which can be regarded as the two stages of flow-shop machines. The goal is to find a schedule of the jobs. e.g., to finish the entire production as early as possible. In practice, the processing time of each job depends on the processing quantities of steel that needs to be processed. Moreover, those quantities are usually determined by the demands of the raw metals and can be formulated as a blending problem [7,8]. Table 1 is a concrete example.

Let $x_i$ be the processing time of steel $i$ to be processed in the first machine (e.g., wire-drawing), and $y_i$ be the processing time of steel $i$ to be processed in the second machine (e.g., annealing). In the example shown in Table 1, the demand of iron is 56. If we assign $x_1 + y_1$ units of processing time on steel 1 in total, then we can produce $24(x_1 + y_1)$ unit of iron. Similarly, if we assign $x_2 + y_2$ units of processing time on steel 2 in total, then we can produce $8(x_2 + y_2)$ unit of iron. Then the requirement on the demand of iron can be represented as a linear inequality

$24(x_1 + y_1) + 8(x_2 + y_2) + 3(x_3 + y_4) + \cdots + 2(x_n + y_n) \geq 56$. Moreover, the processing time of a job in the second stage always depend on the processing time and (amounts of steel) first stage. For example, if the processing time (amounts) of steel 1 the first stage is $x_1$ units, then we require exactly $x_1/2$ units of processing time on the second machine, which can be represented as a linear equality $x_1 = 2y_1$. Due to the environment of the machines, there are usually some limits on the processing times of the steels on both machines. For example, the maximum processing time of steel 1 on the first machine is 10, and the processing time of steel 1 on the second machine should be within $[2, 7]$. Therefore, we have constraints $x_1 \leq 10$ and $5 \leq y_1 \leq 7$. Similarly, we can write linear constraints for the demand of other required metals and other constraints. In this problem, the decision maker needs to determine the non-negative job processing times $x_1, \ldots, x_n, y_1, \ldots, y_n$ satisfying the above linear constraints, and then assign these jobs to the two-machine flow shop machines such that the last completion time is minimized. This problem can be viewed as a minimum makespan two-machine flow shop scheduling problem, where the processing times of jobs satisfy several linear constraints.

2. *Advertising Media Production and Planning Problem.* The flow shop scheduling problem can find application in the area of media and Internet (see, e.g., [9,10]). For example, deciding the sequences of the media production and broadcast can be regarded as two stages of the flow shop machine scheduling problem. Consider a scenario that a media production company has several potential advertisements that need to be produced and then be broadcast. Let $x_i$ be the entire production time (say, e.g., casting, filming and post production) and $y_i$ be the duration of the advertisement $i$. Each advertisement $i$ must be broadcast after its production is finished, and the goal is to finish the entire broadcast of all the advertisements as early as possible. The whole process can be viewed as a two-machine flow shop scheduling problem. Furthermore, the decision maker is required to decide the production time and duration for the advertisements, that is, the processing times of the jobs. It can always be represented as several linear constraints. An example of such a problem is given in Table 2.

In Table 2, there is a budget of the production of all advertisements. Each unit production time of advertisement 1 costs 100, each unit production time of advertisement 2 costs 200, and so forth. The total budget constraint is no more than 10000, and hence can be represented as $100x_1 + 200x_2 + \cdots + 150x_n \leq 10000$. Moreover, the company will gain revenue $2000y_1$ for $y_1$ units of broadcasting time of advertisement 1. The company requires a total revenue at least 15000, which can be represented as $1000x_1 + 2000x_2 + \cdots + 1500x_n \geq 150000$. The production time always depending on the duration (broadcast time) of the advertisement. For example, each unit of broadcast time of advertisement 1 requires at least 10 units of production time, which can be represented as $x_1 \geq 10y_1$. Similar to the first example, we can formulate the other constraints as several linear constraints. The above-described problem can be naturally

**Table 1.** Example for the industrial production problem

| Composition | Steel | | | | Demand |
|---|---|---|---|---|---|
| | 1 | 2 | 3 | $\cdots$ $n$ | |
| Iron | $24(x_1 + y_1)$ | $8(x_2 + y_2)$ | $3(x_3 + y_3)$ | $\cdots$ $2(x_n + y_n)$ | $\geq 56$ |
| Copper | $3(x_1 + y_1)$ | $3(x_2 + y_2)$ | $3(x_3 + y_3)$ | $\cdots$ $1(x_n + y_n)$ | $\geq 30$ |
| $\vdots$ | $\vdots$ | $\vdots$ | $\vdots$ | $\vdots$ $\vdots$ | $\vdots$ |
| Aluminum | $4(x_1 + y_1)$ | $33(x_2 + y_2)$ | $13(x_3 + y_3)$ | $\cdots$ $100(x_n + y_n)$ | $\geq 100$ |
| Relations | | | | | |
| Steel 1 | $x_1 - 2y_1$ | 0 | 0 | $\cdots$ 0 | $= 0$ |
| Steel 2 | 0 | $2x_2 - y_2$ | 0 | $\cdots$ 0 | $= 0$ |
| $\vdots$ | $\vdots$ | $\vdots$ | $\vdots$ | $\vdots$ $\vdots$ | $\vdots$ |
| Steel $n$ | 0 | 0 | 0 | $\cdots$ $x_n - 10y_n$ | $= 0$ |
| Limits | | | | | |
| Max. of 1(Stage 1) | $x_1$ | 0 | 0 | $\cdots$ 0 | $\leq 10$ |
| Min. of 1(Stage 2) | $y_1$ | 0 | 0 | $\cdots$ 0 | $\geq 2$ |
| Max. of 1(Stage 2) | $y_1$ | 0 | 0 | $\cdots$ 0 | $\leq 7$ |
| $\vdots$ | $\vdots$ | $\vdots$ | $\vdots$ | $\vdots$ $\vdots$ | $\vdots$ |
| Max. of $n$ (Stage 2) | 0 | 0 | 0 | $\cdots$ $y_n$ | $\leq 15$ |

formulated as a two-machine flow shop scheduling problem scheduling problem in which the processing times (running times of the advertisements) are determined by a system of linear constraints.

In this paper, we study the computational complexity and algorithms for the 2-FLC problem. First, we show that the problem is generally NP-hard in the strong sense. It is well-known that the original two-machine flow shop scheduling problem can be solved in $O(n \log n)$ time by Johnson's rule [11], where $n$ is the number of jobs. This result suggests that the 2-FLC problem with arbitrary number of constraints has a very different complexity from the original two-machine flow shop scheduling problem. Then we consider the design of algorithms on various settings of the 2-FLC problem. In particular, we design a polynomial time algorithm for the 2-FLC problem when there is a fixed number of constraints. For the general case, we first propose a simple 2-approximation algorithm, and then design a polynomial time approximation scheme (PTAS).

The remainder of this paper is organized as follows: In Sect. 2, we give a formal definition of the 2-FLC problem studied in this paper, and briefly review some related literature. In Sect. 3, we study the computational complexity of the 2-FLC problem. In Sect. 4, we discuss the case with a fixed number of constraints. In Sect. 5, we study the approximation algorithm for the general case. We provide some concluding remarks in Sect. 6.

**Table 2.** Example for the advertising media production and planning problem

| | Each unit time broadcast provides | | | | | |
|---|---|---|---|---|---|---|
| | ad 1 | ad 2 | ad 3 | $\cdots$ | ad $n$ | |
| Production budget | $100x_1$ | $200x_2$ | $100x_3$ | $\cdots$ | $150x_n$ | $\leq 10000$ |
| Broadcast revenue | $2000y_1$ | $3500x_2$ | $2000y_3$ | $\cdots$ | $1500y_n$ | $\geq 150000$ |
| Attractions to women | $20y_1$ | $100y_2$ | $100y_3$ | $\cdots$ | $10y_n$ | $\geq 500$ |
| Attractions to men | $15y_1$ | $10y_2$ | $0y_3$ | $\cdots$ | $80y_n$ | $\geq 500$ |
| Attractions to teens | $30y_1$ | $0y_2$ | $30y_3$ | $\cdots$ | $100y_n$ | $\geq 200$ |
| $\vdots$ | $\vdots$ | $\vdots$ | $\vdots$ | $\vdots$ | $\vdots$ | $\vdots$ |
| Relation of ad 1 | $x_1 - 10y_1$ | 0 | 0 | $\cdots$ | 0 | $\geq 0$ |
| Relation of ad 2 | 0 | $x_2 - 15y_2$ | 0 | $\cdots$ | 0 | $\geq 0$ |
| $\vdots$ | $\vdots$ | $\vdots$ | $\vdots$ | $\vdots$ | $\vdots$ | $\vdots$ |
| Max time for ad 1 | $y_1$ | 0 | 0 | $\cdots$ | 0 | $\leq 20$ |
| Min time for ad 1 | $y_1$ | 0 | 0 | $\cdots$ | 0 | $\geq 10$ |
| Max time for ad 2 | 0 | $y_2$ | 0 | $\cdots$ | 0 | $\leq 35$ |
| $\vdots$ | $\vdots$ | $\vdots$ | $\vdots$ | $\vdots$ | $\vdots$ | $\vdots$ |

## 2   Problem Description

In this paper, we consider the following the two-machine flow shop scheduling problem under linear constraints (2-FLC) problem stated below:

**Definition 1.** *There are $n$ jobs and two flow-shop machines. Each job has to be processed on the first machine and then on the second machine without interruption. The processing times of job $i$ on the first machine and the second machine are $x_i$ and $y_i$ respectively, which are determined by $k$ linear constraints. The goal of the two-machine flow shop scheduling under linear constraints (2-FLC) problem is to determine the processing times of the jobs such that they satisfy the linear constraints and to schedule the jobs to the two flow-shop machines such that the makespan is minimized.*

Let $A, C \in \mathbb{R}^{k \times n}$, $b \in \mathbb{R}^{k \times 1}$, the processing times $x = \{x_1, ..., x_n\}$, $y = \{y_1, ..., y_n\}$ should be feasible solutions to $Ax + Cy \geq b$, $x, y \geq 0$. It is well-known that the two-machine flow shop scheduling problem $F2||C_{\max}$ has an optimal schedule which is a permutation schedule (see, e.g., [9]), i.e., the orders of jobs in both machines are identical. Therefore, we can denote a schedule of our problem as $\sigma = \{\sigma(1), ..., \sigma(n)\}$, where $\sigma(i)$ indicates the job that is assigned on position $i$ in the schedule. The makespan $C_{\max}$ of a schedule is thus the completion time of job $\sigma(n)$ on the second machine.

Here we give a brief literature review on the flow shop scheduling and related problems. Flow shop scheduling is one of the three basic models (open shop, flow shop, job shop) of multi-stage scheduling problems. The flow shop scheduling which minimizes the makespan is usually denoted by $Fm||C_{max}$, where $m$ is the number of machines. Garey et al. proved that $Fm||C_{max}$ is strongly NP-hard for $m \geq 3$ [12], and Hall proposed a PTAS algorithm for $Fm||_{max}$ [13]. Particularly, the two-machine flow shop scheduling problem $F2||C_{max}$ can be solved by Johnson's algorithm in $O(n \log n)$ time [11]. If all the jobs are processed in the same order, then we call this schedule a permutation schedule. It has been shown in [14] that $F2||C_{max}$ or $F3||C_{max}$ has an optimal permutation schedule.

# 3   Computational Complexity of 2-FLC Problem

We first study the computation complexity of 2-FLC problem. We show that this problem is NP-hard in the strong sense, even though the original $F2||C_{max}$ problem can be solved in polynomial time by Johnson's algorithm.

**Theorem 1.** *The 2-FLC problem with two machines is NP-hard in the strong sense.*

*Proof.* We reduce the Exact Cover by 3-Sets problem (X3C) to the 2-FLC problem. Given a ground set $S$ with $|S| = 3n$ ($n \geq 1$) and a collection $C$ of 3-element subsets of $S$. The X3C problem is to decide if there exists a subset $C'$ of $C$ with size $n$, in which every element of $S$ occurs in $C'$ exactly once. Let $m = |C|$ be the total number of subsets and assume that $m > n$ without loss of any generality. We construct an instance of 2-FLC with $m + 1$ jobs. Each subset $S_i \in C$ is associated with a job $i$ that has processing times $x_i$ and $y_i$ in the two stages, respectively. The remaining job $m + 1$ has processing times $x_{m+1} = n$ in the first stage and $y_{m+1} = m - n$ in the second stage. The processing times are determined by the following linear constraints:

$$\sum_{i:j\in S_i} y_i = 1 \qquad\qquad \forall\, j \in S \qquad\qquad (1a)$$

$$x_j = 1 - y_j \qquad\qquad \forall\, i = 1,\ldots,m \qquad (1b)$$

$$x_{m+1} = n,\ y_{m+1} = m - n \qquad\qquad\qquad (1c)$$

$$\boldsymbol{x}, \boldsymbol{y} \geq 0.$$

We show that the instance for X3C is YES if and only if there exists a schedule of 2-FLC in which the processing times are feasible to (1) and has makepan at most $m$.

First, suppose that theres exist an exact cover $C'$ of $S$. Let $x_i = 0$, $y_i = 1$ if $S_i \in C'$ and $x_i = 1$, $y_i = 0$ otherwise, and $x_{m+1} = n, y_{m+1} = m - n$ . Since $C'$ is an exact cover, it can be seen that the linear constraints (1) are satisfied. By Johnson's rule, the jobs correspond to $S_i \in C'$ are scheduled first, then the job $m + 1$, and last the jobs correspond to $S_i \notin C'$. Note that the number of subsets in every exact cover is $n$, therefore the makespan of the schedule is exactly $n + (m - n) = m$.

Now we prove the opposite direction. Note that the job $m+1$ has total processing time $n+(m-n)=m$, hence the makespan of this schedule is exactly $m$. The constraints in (1b) guarantees that all the $x_i$s and $y_i$s are binary, as otherwise the makespan of the schedule must be more than $m$. To see this, if there are some job $k$ which has processing time $x_k \geq y_k > 0$ (the case with $0 < x_k < y_k$ is analogous), then job $k$ must be assigned after job $m+1$ by Johnson's rule, which leads to a schedule with makespan at least $n+(m-n)+y_k > m$. This contradicts to the condition that the schedule has makespan at most $m$. Denote $J'$ as the jobs having processing time $x_i = 0$, $y_i = 1$ and $\bar{J}$ as the remaining jobs, that is, the jobs have processing time $x_i = 1$, $x_i = 0$. Again, by Johnson's rule and the makespan is $m$, it can be seen that there is exactly $n$ jobs in $J'$. By (1a), the subsets in $\mathcal{C}$ corresponds to $J'$ constitute an exact cover of $S$.    □

## 4    Fixed Number of Constraints

In this section, we study the case when the number of constraints is fixed.

**Lemma 1.** *The 2-FLC problem has an optimal solution, in which each machine has at most $k$ jobs with nonzero processing time.*

*Proof.* We first show that given any optimal solution to the 2-FLC problem, we can construct an optimal solution that at most $k$ jobs have nonzero processing time in the second machine. Now we fix the permutation of jobs in an optimal solution. Note that the jobs have same orders in both machines. To simplicity of notation, we denote the order of the jobs as $1, \ldots, n$. We can construct the following linear program:

$$\min \quad t \tag{2a}$$

$$\text{s.t.} \sum_{i=1}^{l} x_i + \sum_{i=l}^{n} y_i \leq t \qquad \forall\, l = 1, \ldots, n \tag{2b}$$

$$\sum_{i=1}^{n} a_{ji} x_i + c_{ji} y_i \geq b_j \qquad \forall\, j = 1, \ldots, k \tag{2c}$$

$$t, x_i, y_i \geq 0 \qquad \forall\, i = 1, \ldots, n.$$

It can be seen that any optimal solution to (2) is also optimal to the 2-FLC problem. To see this, let $(\boldsymbol{x}^*, \boldsymbol{y}^*)$ be the processing times of the optimal solution to the 2-FLC problem. Then by the $(\boldsymbol{x}, \boldsymbol{y})$ is the optimal solution to (2), we have $t = \max_{l=1,\ldots,n}\left\{\sum_{i=1}^{l} x_i + \sum_{i=l}^{n} y_i\right\} \geq \left\{\sum_{i=1}^{l} x_i^* + \sum_{i=l}^{n} y_i^*\right\} = C_{\max}^*$, where $C_{\max}^*$ is the optimal makespan to the 2-FLC problem.

Introducing slack variables to (2b), we obtain a system of $n$ constraints:

$$\sum_{i=1}^{l} x_i + \sum_{i=l}^{n} y_i + z_i = t \quad \forall\, l = 1, \ldots, n. \tag{3}$$

Applying Gaussian elimination to (3), we obtain

$$x_1 = t - \sum_{i=1}^{n} y_i - z_1 \tag{4}$$

$$x_i = y_{i-1} + z_{i-1} - z_i \quad \forall \, i = 2, \ldots, n.$$

Therefore, (2a) is equivalent to the following linear program:

$$\min \quad t \tag{5a}$$

$$\text{s.t.} \quad x_1 + z_1 = t - \sum_{i=1}^{n} y_i \tag{5b}$$

$$x_i + z_i = y_{i-1} + z_{i-1} \quad \forall \, i = 2, \ldots, n \tag{5c}$$

$$\sum_{i=1}^{n} a_{ji} x_i + c_{ji} y_i \geq b_j \quad \forall \, j = 1, \ldots, k \tag{5d}$$

$$t, x_i, y_i, z_i \geq 0 \qquad \forall \, i = 1, \ldots, n,$$

Now we find the optimal basic feasible solution to (5), which is also an optimal solution to the 2-FLC problem. Note that there are at most $n + k$ variables among $x_i$, $y_i$, $z_i$, and $t$ can have nonzero processing time. If $t = 0$, then all $x_i$s and $y_i$s are zero and the lemma is proved. Therefore, we assume that $t > 0$ and thus at most $n + k - 1$ of $x_i$, $y_i$, and $z_i$ have nonzero processing time. Let $S = \{i \in \{2, ..., n\} \mid x_i + z_i > 0\}$ be the subset of indices that with $x_i + z_i$ is positive. If $|S| = n - 1$, then at most $k$ of the $y_i$s have nonzero processing times, since the total number of remaining variables which can have nonzero processing times it at most $n + k - 1 - (n - 1) = k$. Otherwise, $|S| < n - 1$ and there exists some $i$ such that $x_i + z_i = 0$. Note that by constraint (5c), the corresponding $y_{i-1} = z_{i-1} = 0$. Now we fix the processing times of all those $x_i$, $z_i$ and $y_{i-1}$ to be zero, and consider the linear program without these variables:

$$\min \quad t \tag{6a}$$

$$\text{s.t.} \quad x_1 + z_1 = t - \sum_{i \in S} y_{i-1} \tag{6b}$$

$$x_i + z_i = y_{i-1} + z_{i-1} \qquad \forall \, i \in S \tag{6c}$$

$$\sum_{i \in \{1\} \cup S} a_{ji} x_i + \sum_{i \in S} c_{j,i-1} y_{i-1} \geq b_j \qquad \forall \, j = 1, \ldots, k \tag{6d}$$

$$t, x_1, z_1, x_i, y_{i-1}, z_i \geq 0 \qquad \forall \, i \in S.$$

Note that the optimal solution to (5) is feasible to (6), hence the optimal solution to (6) is still optimal to 2-FLC. We repeat the above procedure, that is, find an optimal basic feasible solution to (6), which has at most $|S| + k$ variables with nonzero processing time. If at some point $x_i + z_i$ is strictly greater than zero for all $i \in S$, then we obtain an optimal basic feasible solution in which at most $|S| + k - |S| = k$ of the $y_i$s have nonzero processing times; otherwise we can reduce the size of $|S|$ and find the optimal basic feasible solution to the reduced linear program similar to (6), and repeat the above procedure. It can be observed that the size of $S$ is strictly decreasing. Therefore if the procedure continues, we must have $|S| \leq k$ at some point. In that case, we obtain an optimal basic feasible solution which has at most $k$ of the $x_i + z_i$s have nonzero processing times (note that the $x_i + z_i$ for all $i \notin S$ have been fixed to be zero before), and so for the $x_i$s and $y_i$s. Therefore, there exists an optimal solution that the second machine must have at most $k$ nonzero processing time jobs. Fixed the processing times on the second machine and applied the same method on the first machine, we obtain a solution that each machine has at most $k$ nonzero processing time jobs.

□

Based on Lemma 1, we can propose a polynomial time algorithm for the 2-FLC problem when the number of constraints is a fixed number of constraints. We summarize the algorithm as Algorithm 1 and Theorem 2.

---

**Algorithm 1.** Enumeration algorithm for 2-FLC problem with fixed $k$

---

**Input:** $n$ jobs, $k$ linear constraints $Ax + Cy \geq b$
**Output:** the schedule to the two flow-shop machines that has the minimum makespan, and the corresponding processing times of jobs
1: **for** each subset $J'$ of $J$ with $k$ jobs **do**
2:     **for** each permuation of the jobs in $J'$ **do**
3:         Let $\sigma(1), \ldots, \sigma(k)$ be the permutation, i.e., the schedule of these $k$ jobs. Solve the following LP while setting $y_i = 0$ for $i \notin J'$.

$$\min \quad t \tag{7a}$$

$$\text{s.t.} \sum_{i=1}^{l} x_{\sigma(i)} + \sum_{i=l}^{k} y_{\sigma(i)} \leq t \quad \forall\, l = 1, \ldots, k \tag{7b}$$

$$\sum_{i=1}^{n} x_i \leq t \tag{7c}$$

$$\sum_{i=1}^{n} a_{ji}x_i + c_{ji}y_i \geq b_j \quad \forall\, j = 1, \ldots, k \tag{7d}$$

$$y_i = 0 \qquad\qquad \forall\, i \notin J' \tag{7e}$$

$$t, x_i, y_i \geq 0 \qquad\qquad \forall\, i = 1, \ldots, n.$$

4:     **if** (7) is infeasible **then**
5:         Let the processing times of jobs be the optimal solution to (7), and record the schedule and the makespan.
6: **return** the schedule with the smallest makespan among all these iterations and its corresponding processing times.

---

**Theorem 2.** *The 2-FLC problem when the number of constraints is fixed has a polynomial time algorithm with complexity $O(n^{k+3}L)$.*

The proofs of the subsequent lemmas/theorems will be provided in Appendix.

## 5    General Case: Approximation Algorithms

First we propose a simple 2-approximation algorithm for the problem. We solve the linear program with minimum total progressing time subject to $Ax = b$ to obtain the processing times $x$, $y$. Then we assign the jobs with processing times $x$, $y$ by Johnson's rule.

**Theorem 3.** *There is a 2-approximation algorithm for the 2-FLC problem.*

Now we propose a PTAS for the problem. First, we run the 2-approximation algorithm and obtain a value $C_{\max}$. By Theorem 1, $C_{\max}/2$ and $C_{\max}$ are lower bound and upper bound of the optimal makespan of the 2-FLC problem, respectively. We denote $LB = C_{\max}/2$ and $UB = C_{\max}$.

Given $\epsilon > 0$, we denote the job sets (with respect to $x$ and $y$) as follow:

$$J_1 = \{i \in J \mid x_i \leq y_i\}, \quad L_1 = \{i \in J \mid x_i \leq y_i,\ y_i > \epsilon C_{\max}^*(x, y)\},$$
$$J_2 = \{i \in J \mid x_i \geq y_i\}, \quad L_2 = \{i \in J \mid x_i \geq y_i,\ x_i > \epsilon C_{\max}^*(x, y)\}, \qquad (8)$$

where $C_{\max}^*(x, y)$ is the makespan of optimal schedule of $F2||C_{\max}$ with processing time $x$ and $y$. Note that we have the sizes of $L_1$ and $L_2$ are $|L_1| \leq 1/\epsilon$ and $|L_2| \leq 1/\epsilon$ as otherwise we have $\sum_{i=1}^n x_i > C_{\max}^*(x, y) \geq C_{\max}^*$ or $\sum_{i=1}^n y_i > C_{\max}^*(x, y) \geq C_{\max}^*$, which leads to a contradiction. The algorithm first guesses the value of the optimal makespan, then guesses the set $L_1$ and $L_2$ in an optimal solution (w.r.t. $x^*$ and $y^*$), as well as enumerating all the possible permutations of these jobs. During each guess, we obtain the values of the processing times of jobs by checking the feasibility of a specific linear program. The jobs are then scheduled by the Johnson's rule. The algorithm finally returns the best schedule among all these iterations. We summarize the details as Algorithm 2.

To establish the approximation performance result of Algorithm 2, we first state a property of a specified schedule in the classical two-machine flow shop scheduling problem $F2||C_{\max}$, which has makespan at most $1 + \epsilon$ of the optimal solution to $F2||C_{\max}$ and allows us to design a PTAS for the 2-FLC problem. Fixed any processing times $x$ and $y$ for the jobs, $J_1$, $J_2$, $L_1$, $L_2$, $C_{\max}^*(x, y)$ are defined as (8), $S_1 = J_1 \setminus L_1 = \{i \in J \mid x_i \leq y_i,\ y_i \leq \epsilon C_{\max}^*(x, y)\}$ and $S_2 = J_2 \setminus L_2 = \{i \in J \mid x_i \geq y_i,\ x_i \leq \epsilon C_{\max}^*(x, y)\}$. The lemma is described as below.

**Lemma 2.** *Let $C_{\max}'(x, y)$ be the makespan of schedule obtained by assigning the jobs in the order $S_1$, $L_1$, $L_2$, $S_2$ (identical for both machines), where the jobs in $L_1$ and $L_2$ are scheduled according to Johnson's rule, $S_1$ and $S_2$ are scheduled arbitrarily. Then, $C_{\max}'(x, y) \leq (1 + \epsilon)C_{\max}^*(x, y)$.*

Based on this lemma, we have the following result.

**Theorem 4.** *Algorithm 2 is a PTAS for the 2-FLC problem.*

---

**Algorithm 2.** PTAS for 2-FLC problem

---

**Input:** $n$ jobs, $k$ linear constraints $Ax + Cy \geq b$

**Output:** the schedule to the two flow-shop machines that has the minimum makespan, and the corresponding processing times of jobs

1: Given $\epsilon \in (0,1)$, let $h = \lceil 1/\epsilon \rceil$ and $UB$, $LB$ defined as before
2: **for** each subset $L_1^h$ and each subset $L_2^h$ of $J$, with $L_1^h \cup L_2^h = \emptyset$ and $|L_1^h| = |L_2^h| = h$ **do**
3:     **for** each permutation $\sigma_1 = (\sigma_1(1), \ldots, \sigma_1(h))$ and each permutation $\sigma_2 = (\sigma_2(1), \ldots, \sigma_2(h))$ of $(1, \ldots, h)$ **do**
4:         Denote $(\sigma_1(1), \ldots, \sigma_1(h))$ and $(\sigma_2(1), \ldots, \sigma_2(h))$ as the jobs and the schedule in $L_1^h$ and $L_2^h$, respectively.
5:         Set $S^h = J \setminus (L_1^h \cup L_2^h)$, $T = LB$.
6:         **while** $T < (1+\epsilon)UB$ **do**
7:             Check the feasibility of the following linear program (9):

$$Ax + Cy \geq b \tag{9a}$$

$$\sum_{i=1}^{n} x_i \leq (1+\epsilon)T, \quad \sum_{i=1}^{n} y_i \leq (1+\epsilon)T \tag{9b}$$

$$\sum_{i \in S^h} t_i + \sum_{i=1}^{l} x_{\sigma_1(i)} + \sum_{i=l}^{h} y_{\sigma_1(i)} + \sum_{i=1}^{h} y_{\sigma_2(i)} \leq (1+\epsilon)^2 T \quad \forall\, l \in 1, \ldots, h \tag{9c}$$

$$\sum_{i \in S^h} t_i + \sum_{i=1}^{h} x_{\sigma_1(i)} + \sum_{i=1}^{l} x_{\sigma_2(i)} + \sum_{i=l}^{h} y_{\sigma_2(i)} \leq (1+\epsilon)^2 T \quad \forall\, l = 1, \ldots, h \tag{9d}$$

$$t_i \geq x_i, \quad t_i \geq y_i \qquad \forall\, i = 1, \ldots, n \tag{9e}$$

$$y_j \leq y_i, \quad x_i \leq y_i \qquad \forall\, j \in S^h, i \in L_1^h \tag{9f}$$

$$x_j \leq x_i, \quad y_i \leq x_i \qquad \forall\, j \in S^h, i \in L_2^h \tag{9g}$$

$$x_i, y_i, t_i \geq 0, \quad \forall\, i = 1, \ldots, n.$$

8:         **if** (9) is infeasible **then**
9:             Set $T = (1+\epsilon)T$, continue to Step 6.
10:        **else**
11:            Record the processing times $x$ and $y$ and $T$ of a feasible solution to 9, continue to Step 3.
12: **return** the solution with the smallest $T$ among all these iterations and its corresponding processing times, and schedule the jobs by Johnson's rule.

---

## 6    Conclusions

We study the two-machine flow shop scheduling problem under linear constraints in this paper. The problem is NP-hard in the strong sense, and we propose several optimal or approximation algorithms for various settings of it. An immediate research question is to study the FLC problem where the number of machines is more than 2. Furthermore, it is also interesting to consider the shop scheduling problem under linear constraints where the machine environment is open shop or job shop, and the shop scheduling problem with additional constraints, such as the no-wait case.

# References

1. Graham, R., Lawler, E., Lenstra, J., Kan, A.: Optimization and approximation in deterministic sequencing and scheduling: a survey. Ann. Discret. Math. $5(1)$, 287–326 (1979)
2. Nip, K., Wang, Z., Wang, Z.: Scheduling under linear constraints. Eur. J. Oper. Res. $253(2)$, 290–297 (2016)
3. Zhang, S., Nip, K., Wang, Z.: Related machine scheduling with machine speeds satisfying linear constraints. In: Kim, D., Uma, R.N., Zelikovsky, A. (eds.) COCOA 2018. LNCS, vol. 11346, pp. 314–328. Springer, Cham (2018). https://doi.org/10.1007/978-3-030-04651-4_21
4. Wang, Z., Nip, K.: Bin packing under linear constraints. J. Comb. Optim. $34(5)$, 1198–1209 (2017)
5. Nip, K., Wang, Z., Wang, Z.: Knapsack with variable weights satisfying linear constraints. J. Global Optim. $69(3)$, 713–725 (2017)
6. Nagar, A., Heragu, S.S., Haddock, J.: A branch-and-bound approach for a two-machine flowshop scheduling problem. J. Oper. Res. Soc. $46(6)$, 721–734 (1995)
7. Danø, S.: Linear Programming in Industry, Theory and Applications: An Introduction. Springer, Vienna (1960). https://doi.org/10.1007/978-3-7091-3644-7
8. Eiselt, H.A., Sandblom, C.L.: Linear Programming and its Applications. Springer, Heidelberg (2007). https://doi.org/10.1007/978-3-540-73671-4
9. Emmons, H., Vairaktarakis, G.: Flow Shop Scheduling: Theoretical Results, Algorithms, and Applications. Springer, New York (2013). https://doi.org/10.1007/978-1-4614-5152-5
10. Allahverdi, A., Al-Anzi, F.S.: Using two-machine flowshop with maximum lateness objective to model multimedia data objects scheduling problem for WWW applications. Comput. Oper. Res. $29(8)$, 971–994 (2002)
11. Johnson, S.M.: Optimal two- and three-stage production schedules with setup times included. Naval Res. Logistics Q. $1$, 61–68 (1954)
12. Garey, M.R., Johnson, D.S., Sethi, R.: The complexity of flowshop and jobshop scheduling. Math. Oper. Res. $1$, 117–129 (1976)
13. Hall, L.A.: Approximability of flow shop scheduling. Math. Program. $82$, 175–190 (1998)
14. Conway, R.W., Maxwell, W., Miller, L.: Theory of Scheduling. Addison-Wesley, Reading (1967)

# Some Graph Optimization Problems with Weights Satisfying Linear Constraints

Kameng Nip[1], Zhenbo Wang[2($\boxtimes$)], and Tianning Shi[2]

[1] School of Mathematics (Zhuhai), Sun Yat-sen University, Zhuhai, China
[2] Department of Mathematical Sciences, Tsinghua University, Beijing, China
wangzhenbo@tsinghua.edu.cn

**Abstract.** In this paper, we study several graph optimization problems in which the weights of vertices or edges are variables determined by several linear constraints, including the maximum matching problem under linear constraints (max-MLC), the minimum perfect matching problem under linear constraints (min-PMLC), the shortest path problem under linear constraints (SPLC) and the vertex cover problem under linear constraints (VCLC). The objective of these problems is to decide the weights that are feasible to the linear constraints, and to find the optimal solution of the graph optimization problems among all the feasible choices of weights. Even though all the original graph optimization problems can be solved in polynomial time or be approximated within a fixed constant factor, we find that these problems under linear constraints are intractable in general. In particular, we show that the max-MLC problem is NP-hard, while the min-PMLC, SPLC, VCLC problems are all NP-hard and do not even have any polynomial-time algorithms unless $P = NP$. These findings suggest us to explore the special cases of these problems which are tractable. Particularly, we show that when the number of constraints is a fixed constant, all these problems under linear constraints are polynomially solvable. Moreover, if there are fixed number of distinct weights, then the max-MLC, min-PMLC and SPLC are polynomially solvable, and the VCLC has 2-approximation algorithm. In addition, we propose several approximation algorithms for max-MLC.

**Keywords:** Graph optimization · Polynomial-time algorithm · Approximation algorithm · Linear programming · Computational complexity

## 1 Introduction

Optimization problem over graphs is a large class of important problems in the field of combinatorial optimization. In Karp's famous paper [1] on the reducibility of NP-compete problems, 10 of 21 are decision problems of graph optimization problems, and some of the others can also be naturally formulated or solved

This work has been supported by NSFC No. 11801589, No. 11771245 and No. 11371216.

Y. Li et al. (Eds.): COCOA 2019, LNCS 11949, pp. 412–424, 2019.
https://doi.org/10.1007/978-3-030-36412-0_33

as problems over graph. A traditional graph optimization problem is basically stated as follows. Given an undirected or directed graph $G$ with vertices (or nodes) $V$ and edges (or arcs) $E$, the vertices/edges (or both) are associated with a (usually nonnegative) weight vector $w \in \mathbb{R}^{|V|}/w \in \mathbb{R}^{|E|}$, which is often deterministic and exogenously given. The goal is to select a subset of vertices/edges satisfying some specific structure, e.g., a path, a spanning tree, or a covering, such that the total weight of the selected vertices/edges are maximized or minimized. It includes numerous widely-studied graph optimization problems, such as minimum spanning tree problem, shortest path problem, matching problem, vertex cover problem, maximum independent set problem and traveling salesman problem (see, e.g., [2,3]). These problems are motivated from various real-world applications, for example, in transportation, telecommunications, computer science and social networks. The design of efficient algorithms or approximation algorithms and the studies of computational complexity for these problems have long been a core issue of combinatorial optimization.

In a series of recent works [4–7], several combinatorial optimization problems are investigated in which some of the parameters are also variables and must satisfy several linear constraints, but not deterministic and given in advance. For instance, in the knapsack problem with weights satisfying linear constraints studied in [5], the weights of items are also variables and must satisfy a system of linear constraints, while the capacity of knapsack and the values (of value-to-weight ratios) of items are given and known. The goal is to determine the weight of each item and to find a subset of items, such that the total weight is no more than the capacity and the total value is maximized. Such problems are motivated from industrial production, advertising, transportation and resource allocation. These works suggest that the combinatorial optimization problems under linear constraints are generally hard to solve or be approximated, as they can be naturally formulated as mixed integer programming problems [8,9]. Despite of this, under many cases, these problems can be polynomially solved or be efficiently approximated within a fixed constant factor. For instance, if the number of constraints is a fixed constant, the parallel machine scheduling problem with fixed number of machines [4] and the knapsack problem under linear constraints [5] can be solved in polynomial time. If the number of constraints is arbitrary, the problems may have various computational complexities, which could be a similar or harder problem than the original problems. For example, the parallel machine scheduling problem under linear constraints [4] and bin packing problem under constraints [6] also admit approximation algorithms similar to List Scheduling algorithm and Next Fit algorithm respectively, and have the same approximation factors. On the contrary, the knapsack problem under linear constraints might be even harder in the sense that it is unlikely to have any constant-factor approximation algorithms [5], while the knapsack problem has a well-known fully polynomial time approximation scheme (FPTAS) (see, e.g., [3]). Following this line of research, it is interesting to study the graph optimization problems under linear constraints, in which the weights satisfy a system of linear constraints.

Below we briefly illustrate a potential application scenario of these graph optimization problems under linear constraints. Suppose that a decision maker needs to assign $n_e$ employees to $n_t$ different tasks, with each employee is assigned to at most one task and each task is conducted by at most one employee. She/he makes the decision based on the evaluation score for each pair of the employees and tasks, which may be assessed by the capability and availability of the employees, and etc. The goal is to find an assignment that has maximum total evaluation scores. This is a typical bipartite matching (assignment) problem, in which each vertex indicates the employee or task, and each edge indicates the score of an employee to a task. In the classic problem, these scores are deterministic values that are given exogenously to the entire decision. However, it is always the case that the evaluation scores are assessed by the same decision maker or department (e.g., in the human resources department of a company) through various evaluation processes, such as interview or examination. Performance evaluation is one of the powerful applications of linear programming in the field of organizational management and educations, which has been studied extensively in the literature (see, e.g., [10–12]). Here we adapt the applications in these works and state a simple model for this scenario. Let $x_{ij}$ be the evaluation score of employee $i$ to task $j$. For example, the total score of each employee $i$ is at most 100, and is at least some nonnegative value $s_i$ that is based on her/his general performance in the general interview process. Therefore, thus we have linear constraints $s_i \leq \sum_{j=1}^{n_t} x_{ij} \leq 100$, for $i = 1, ..., n_e$. Based on the employees past experience and other evaluation, each score has a rough estimation, say $l_{ij} \leq x_{ij} \leq u_{ij}$ for each $i = 1, ..., n_e$ and $j = 1, ..., n_t$, where $l_{ij}$ and $u_{ij}$ are values obtained by some processes, such as through resume review process. Then after more specialized evaluation processes of different departments, the decision maker have more detailed informations about the comparisons between different employees for each task. Suppose that in the evaluation for the task 1, the entire evaluations of all the employees is recorded as $(a_{11}, a_{21}, ..., a_{n_e 1})$ where $a_{ij}$ is some nonnegative value. Then the scores should satisfy the following linear constraints: $x_{11} = \frac{a_{11}}{a_{21}} x_{21}$, $x_{11} = \frac{a_{11}}{a_{31}} x_{31}, ..., x_{11} = \frac{a_{11}}{a_{n_e 1}} x_{n_e 1}$. Similarly, the scores of comparisons for task $j$ can also be written as $x_{1j} = \frac{a_{1j}}{a_{2j}} x_{2j}$, $x_{1j} = \frac{a_{1j}}{a_{3j}} x_{3j}, ..., x_{1j} = \frac{a_{1j}}{a_{n_e j}} x_{n_e j}$. Overall, the decision maker needs to decide the evaluation scores for each employee to each task, which should satisfy all the above linear constraints, and also find the matching between the employees to the tasks, such that the total evaluation scores of the selected matching is maximized. This leads to the maximum matching problem with weights satisfying linear constraints as above. Analogously, we can find some potential applications on other graph optimization problems under linear constraints.

In this research, we concentrate on studying three types of classic graph optimization problems under linear constraints, including the matching problem under linear constraints (MLC), the shortest path problem under linear constraints (SPLC) and the vertex cover problem under linear constraints (VCLC). Especially for the MLC problem, we study both the maximum matching problem under linear constraints (max-MLC) and minimum perfect matching problem

under linear constraints (min-PMLC). All of these graph optimization problems are fundamental and well-studied problems in combinatorial optimization,
and have widespread real-world applications. It is well-known that the matching
problem and shortest path problem with nonnegative weights have polynomial-
time algorithm (see, e.g, [13]), while the vertex cover problem is NP-hard [14] but
has several 2-approximation algorithms. However, to our surprise, all the above
problems under linear constraints are NP-hard in the general case. Furthermore, all the minimization problems, namely, min-PMLC, SPLC and VCLC are
unlikely to have any polynomial-time approximation algorithms, unless $P = NP$.
These findings suggest us to explore the special cases of these problems that have
polynomial-time algorithms or approximation algorithms. Particularly, we show
that when the number of constraints is a fixed constant, all these problems are
polynomially solvable. Moreover, if the vertices/edges of the problems are preliminarily grouped as a fixed number of mutually disjoint sets in which each set
of vertices/edges must have identical weights, then the max-MLC, min-PMLC
and SPLC are polynomially solvable, and the VCLC has 2-approximation algorithms. In addition, we propose several approximation algorithms for the max-
MLC problem in the general case and some special cases.

The remainder of this paper is organized as follows: In Sect. 2, we formally
describe the problems that are studied in this paper, as well as some related
literature reviews. In Sect. 3, we study the computational complexities of these
problems. In Sect. 4, we discuss the cases where the number of constraints is fixed.
In Sect. 5, we study the cases where the number of distinct weights/groups is
fixed. In Sect. 6, we propose some approximation algorithms for the max-MLC
problem. Finally, we provide several concluding remarks in Sect. 7.

## 2   Problem Description

### 2.1   Matching Problem Under Linear Constraints

**Definition 1 (max-MLC).** *There is an undirected graph[1] $G = (V, E)$, where
$V$ is the set of vertices, and $E$ is the set of edges. Let $e = (i, j)$ be the edge
between vertex $i$ and vertex $j$. The weight of edge $e$ is denoted as $x_e$, which is
determined by $k$ linear constraints. The goal of the maximum matching problem
under linear constraints (max-MLC) problem is to determine the weights of edges
that satisfying the linear constraints, and to find matching of $G$, i.e., a set of
edges $E' \subset E$ without common vertices such that the total weight is maximized.*

Throughout this paper, we denote by $|V| = n$ and $|E| = m$. Let $A \in \mathbb{R}^{k \times n}$,
$b \in \mathbb{R}^{k \times 1}$, the weights should satisfy $Ax \le b$ (or $Ax \ge b$ for other minimization
problems). We note that it is without loss of generality to stipulate that the
edges have non-negative weights, since the weights with negative weights will
not be included in any optimal maximum matching. By the standard integer

---

[1] All the results on this paper holds for $G$ is a directed graph.

programming formulation of the matching problem, the max-MLC problem can be formulated as follows:

$$(\text{max-MLC-NMIP}) \qquad \max \sum_{e \in E} x_e y_e \tag{1a}$$

$$\text{s.t.} \sum_{e \in \delta(i)} y_e \leq 1 \qquad\qquad \forall i \in V \tag{1b}$$

$$Ax \leq b \tag{1c}$$

$$x_e \geq 0, \ y_e \in \{0, 1\} \qquad\qquad \forall e \in E, \tag{1d}$$

where $y_e = 1$ indicates that the edge $e$ is selected into the matching, and $\delta(i)$ denotes the set of edges that one of the endpoints is $i$ for each $i \in V$. To avoid trivial cases, we assume that the linear inequalities (1c) always has a feasible solution, and the optimal solution of the problem is bounded throughout this work. Note that this is a bilinear mixed integer programming problem on $x$ and $y$, which is very hard to solve and approximate in general [8,9].

We also study the minimum version of the matching problem which weights satisfy linear constraints. If there are both positive and negative weights, then the minimum matching can be simply transformed to a maximum matching problem, and the problem under linear constraints is basically identical to (max-MLC). Therefore, we focus on the case of minimum perfect matching under linear constraints (min-PMLC). The definition of PMLC is similar to that in Definition 1, in which the max in the objective function (1a) is replaced by min and the matching is required to be a perfect matching. A matching is said to be a perfect matching if each vertex is an endpoint of an edge in the matching, or equivalently, the $\leq$ in (1b) is replaced by $=$. We can similarly formulate this problem as (min-MLC-NMIP).

**Definition 2 (min-PMLC).** *There is an undirected graph $G = (V, E)$, where $V$ is the set of vertices, and $E$ is the set of edges. Let $e = (i, j)$ be the edge between vertex $i$ and vertex $j$. The weight of edge $e$ is denoted as $x_e$, which is determined by $k$ linear constraints. The goal of the minimum perfect matching problem under linear constraints (min-PMLC) problem is to determine the weights of edges that satisfy the linear constraints, and to find a perfect matching of $G$ such that the total weight is minimized.*

It is well-known that the maximum/minimum matching problem can be solved in polynomial time, since the constraint matrix of the matching problem (1b) is a totally unimodular matrix (see, e.g., [3,15]). The matching problem that the graph $G$ is a bipartite graph is usually referred to assignment problem. Kuhn proposed the famous Hungarian algorithm [16] for the assignment problem, which is regarded as one of the beginnings of combinatorial optimization algorithms. There are numerous improvements and studies on the algorithms and variants of the assignment problem, and we refer the interested readers to [17] for more details. For the general maximum matching problem, Edmond's blossom algorithm [18,19] can solve it with worst-case time complexity $O(|V|^2|E|)$, which has better time complexity than solving the linear program of matching problem directly. Moreover, Edmond's blossom algorithm can be generalized to solve the minimum perfect matching problem (see, e.g., [3, Section 11.2]).

## 2.2   Shortest Path Problem Under Linear Constraints

Note that we focus on the shortest path problem with nonnegative weights, since the shortest path with negative cycles is known to be NP-hard and cannot admit any approximation algorithms with a fixed constant factor [14]. For ease of exposition, we still say en edge rather than an arc of a graph even if the problem is defined in directed graphs.

**Definition 3 (SPLC).** *There is a directed graph $G = (V, E)$, where $V$ is the set of vertices with two specific vertices $s, t \in V$, and $E$ is the set of edges. Let $e = (i, j)$ be the edge between vertex $i$ and vertex $j$. The weight of edge $e$ is denoted as $x_e$, which is nonnegative and determined by $k$ linear constraints. The goal of the shortest path problem under linear constraints (SPLC) problem is to determine the weights of edges that satisfying the linear constraints, and to find a directed path from $s$ to $t$ such that the total weight is minimized.*

By the integer programming formulation of the shortest path problem with [15], the SPLC problem can also be formulated as a bilinear mixed integer program:

$$(\text{SPLC-NMIP}) \qquad \min \sum_{e \in E} x_e y_e \tag{2a}$$

$$\text{s.t.} \quad \sum_{e:e \in H_i} y_e - \sum_{e:e \in T_i} y_e = \begin{cases} 1, & i = s, \\ -1, & i = t, \\ 0, & i \neq s, t. \end{cases} \tag{2b}$$

$$Ax \geq b \tag{2c}$$

$$x_e \geq 0, \ y_e \in \{0, 1\} \qquad\qquad \forall e \in E, \tag{2d}$$

where $y_e = 1$ indicates that edge $e$ is selected into the path, $H_i = \{e \in E \mid e = (i, j) \in E, j \in V\}$ is the set of edges starting from vertex $i$ and $T_i = \{e \in E \mid e = (j, i) \in E, j \in V\}$ is the set of edges ending at vertex $i$.

Similar to the matching problem, the shortest path problem with nonnegative weights can be solved in polynomial time, since the constraint matrix of the matching problem (2b) is a totally unimodular matrix (see, e.g., [3,15]). Shortest path problem is one of the basic network optimization problems (see, e.g., [15]). If all the weights of the shortest path problem are nonnegative, then it can be solved in time $O(|V|^2)$ by the famous Dijkstra's algorithm [20]. The time complexity can be further improved to $O(|E| + |V| \log |V|)$ by implementing Fibonacci heap [21]. For more results on the shortest path, we refer the interested readers to [15].

## 2.3   Vertex Cover Problem Under Linear Constraints

**Definition 4 (VCLC).** *There is an undirected graph $G = (V, E)$, where $V$ is the set of vertices, and $E$ is the set of edges. The weight of vertex $i$ is denoted as*

$x_i$, which is nonnegative and determined by $k$ linear constraints. The goal of the vertex cover problem under linear constraints (VCLC) problem is to determine the weights of vertices that satisfy the linear constraints and to find a vertex cover of $G$, i.e., a subset $V'$ of vertices $V$ such that each edge has at least one endpoint in $V'$ such that the total weight is minimized.

Analogously, the VCLC problem can also be formulated as a bilinear mixed integer program:

$$\text{(VCLC-NMIP)} \qquad \min \quad \sum_{i \in V} x_i y_i \tag{3a}$$

$$\text{s.t.} \quad y_i + y_j \geq 1 \qquad\qquad \forall (i,j) \in E \qquad \text{(3b)}$$

$$A\boldsymbol{x} \geq \boldsymbol{b} \tag{3c}$$

$$x_i \geq 0, y_i \in \{0,1\} \qquad\qquad \forall j \in V, \qquad \text{(3d)}$$

where $y_i = 1$ indicates that the vertex $i$ is selected in the vertex cover. Note that we can replace the vertex cover problem in Definition 4 and the constraints (3) by more general covering problems and their corresponding constraints, such as the hitting set problem, set cover problem [1,3] and obtain other covering problems under linear constraints. The results of VCLC problem can be directly applied to these problems, therefore we only discuss VCLC problem in this research for ease of exposition.

The decision problem of vertex cover is one of the six fundamental NP-complete problems introduced by Garey and Johnson [14], as well as one of Karp's 21 NP-complete problems [1]. There are several standard techniques that can be applied to design 2-approximation algorithms for the vertex cover, such as the linear programming relaxation, local ratio and primal-dual method (see, e.g., [22]). On the other hand, Dinur and Safra proved that unless $P = NP$, vertex cover does not have any approximation algorithm with performance factor better than 1.36 [23]. Under a stronger assumption Unique Game Conjecture, Khot and Regev showed that it is unlikely to obtain an approximation algorithm with performance factor better than 2 [24]. For more details on the vertex cover and related problems, we refer the interested readers to [22].

## 3   Computational Complexities

Although the maximum matching problem is polynomally solvable, we prove that both the maximum and minimum versions of the MLC problem are NP-hard. The hardness result is established by a reduction from the famous NP-hard partition problem [14]. The proofs of this and all subsequent lemmas/theorems are omitted, and will be provided in the full version.

**Theorem 1.** *The max-MLC problem is NP-hard, even if the graph is a bipartite graph.*

We can also show that the minimum version of the perfect-matching problem under linear constraints (min-PMLC) is NP-hard. Furthermore, we can show that it is unlikely to design polynomial-time approximation algorithm for this problem.

**Theorem 2.** *The min-PMLC problem is NP-hard and does not admit any approximation algorithm unless $P = NP$, even if the graph is a bipartite graph.*

We can also prove that the SPLC problem and VCLC problem are NP-hard and do not admit any approximation algorithm unless $P = NP$.

**Theorem 3.** *The SPLC problem is NP-hard and does not admit any approximation algorithm unless $P = NP$.*

**Theorem 4.** *The VCLC problem is NP-hard and does not admit any approximation algorithm unless $P = NP$.*

# 4 Fixed Number of Constraints

As we show in Sect. 3, it is extremely hard to find polynomial-time approximation algorithms for the minimum version of our problems. Therefore, we consider several special cases of the problems that can be optimally solved in polynomial time. In this section, we first consider the problems where the total number of constraints $k$ is a fixed constant. Even though all the previous problems are NP-hard or inapproximable, they can be solved in polynomial time if the number of constraints $k$ is fixed.

**Lemma 1.** *The max-MLC problem has an optimal solution, in which there are at most $k$ edges that has nonzero weights.*

By Lemma 1, we can find the optimal solution to max-MLC problem by enumeration. The idea is to first guess all the edges that have nonzero weights and are selected in an optimal matching. Since $k$ is fixed, the enumeration can be done in polynomial time $O(n^k)$. We then solve the corresponding (max-MLC-LP) to obtain the weights of the edges. Due to the limit of space, we will only describe the high-level ideas of the algorithms, and the pseudocodes of the algorithms stated in the theorems in this paper are provided in the of full version.

**Theorem 5.** *The max-MLC problem can be solved in polynomial time if the total number of constraints $k$ is fixed.*

A direct corollary of Lemma 1 is that the min-PMLC problem also has the same property, since we can replace the max in (max-MLC-LP) by min.

**Corollary 1.** *The min-PMLC problem has an optimal solution, in which there are at most $k$ edges that has nonzero weights.*

The enumeration algorithm to solve min-PMLC problem is analogous, even though the general case of this problem cannot have any approximation algorithm. However, we require some extra effort to find the feasible solution, i.e., a perfect matching. The main difference with the max-MLC problem is that since this is a minimization problem, if we select the edges that have nonzero weights but are not selected in the optimal solution, we might obtain a suboptimal solution (which is still optimal to the max-MLC problem).

**Theorem 6.** *The min-PMLC problem can be solved in polynomial time if the total number of constraints $k$ is fixed.*

For the SPLC problem with a fixed number of constraints, we have a similar result with Lemma 1. Based on Lemma 2, we can also use an enumeration algorithm to find the optimal solution when the number of constraints is fixed.

**Lemma 2.** *The SPLC problem has an optimal solution, in which there are at most $k$ edges that has nonzero weights (note necessarily in the path).*

**Theorem 7.** *The SPLC problem can be solved in polynomial time if the total number of constraints $k$ is fixed.*

The VCLC problem has the same result as Lemmas 1 and 2.

**Lemma 3.** *The VCLC problem has an optimal solution, in which there are at most $k$ edges that has nonzero weights.*

Note that the case for VCLC problem is easier than the min-PMLC problem and SPLC problem, since we can simply select all the vertices that have zero weights and obtain a vertex cover.

**Theorem 8.** *The VCLC problem can be solved in polynomial time if the total number of constraints $k$ is fixed.*

## 5    Fixed Number of Distinct Weights/Groups

In this section, we focus on the problems when the total number of different groups $h$ is a fixed constraint. Let $X$ be the set of all vertices/edges of the problem, we assume that the vertices/edges are preliminarily grouped as $X = X_1 \cup X_2 \cup \cdots \cup X_h$, in which the vertices/edges in each group $X_g$ have the same value of weights and $X_i \cap X_j = \emptyset$ for any $i$, $j$. If $h = n$, then each $X_i$ is a singleton and it is the original problem. In this section, we focus on the case when $h$ is a fixed constant. We note that this case is equivalent to adding the constraints $x_i = x_j, \forall i, j \in X_g, \forall g = 1, ..., h$ to the original problem, which has $O(n)$ constraints. Therefore, the results for the fixed number of constraints stated in Sect. 4 cannot be directly applied to the cases studied in this section.

The framework for all the problems are analogous, therefore we describe the algorithm in terms of the max-MLC problem for clearness. Let $\boldsymbol{y}^* \in \{0,1\}^m$ be the selection of edges in the optimal solution, where $y_e^* = 1$ indicates that edge

$e$ is selected into the matching. Then by the discussion in the proof of Lemma 1, the optimal weights can be obtained by solving the linear program:

$$(\text{max-MLC-LP}) \qquad \max \quad \sum_{e \in E} y_e^* x_e$$

$$\text{s.t.} \quad Ax \leq b$$

$$x_e \geq 0 \quad \forall e \in E,$$

Now note that in this case, the vertices are preliminarily grouped by their weights. Let $z_g$ be the weight of vertices in group $X_g$, for $g = 1, ..., h$. Then we can rewrite (max-MLC-LP) in terms of $z_g$:

$$(\text{max-MLC-LP2}) \qquad \max \quad \sum_{g=1}^{h} \left( \sum_{e \in X_g} y_e^* \right) z_g$$

$$\text{s.t.} \quad \sum_{g=1}^{h} \sum_{e \in X_g} a_{ie} z_g \leq b_i \quad \forall i = 1, ..., k$$

$$z_g \geq 0 \qquad\qquad\qquad \forall g = 1, ..., h,$$

It can be seen that the optimal weights of the max-MLC problem can also be obtained by solving the linear program (max-MLC-LP2). By the fundamental theorem of linear programming (see e.g, [25]), since the problem has a bounded optimal solution by assumption, there must have an optimal solution to (max-MLC-LP2) which is an extreme point of the polytope defined by $P_z = \{z \in \{0,1\}^h : \sum_{g=1}^{h} \sum_{e \in X_g} a_{ie} z_g \leq b_i, \forall i = 1, ..., k; z_g \geq 0, g = 1, ..., h\}$. Note that the polytope $P_z$ has $h + k$ constraints and $h$ variables, we can enumerate all the extreme points, which are at most $\binom{k+h}{h} = O((k+h)^h)$ (see, e.g, [26]). If $h$ is a fixed constraint, then the number of all extreme points is a polynomial on the size of input instance of the problem. After the weights are determined, we can find a maximum matching by the Edmond's blossom algorithm or other matching algorithm. Note that the above discussion stands for any problem studied in this paper, since the linear program obtained after the selection of vertices/edges are similar to (max-MLC-LP2) (see Sect. 4).

**Theorem 9.** *There are polynomial time algorithms for the max-MLC, min-PMLC, SPLC problems, when the total number of groups $h$ is fixed.*

**Theorem 10.** *There is a 2-approximation algorithm for the VCLC problem, where the total number of groups $h$ is a fixed constant.*

## 6    Approximation Algorithms for the Max-MLC Problem

In this section, we study the approximation algorithms for the max-MLC problem. In particular, we propose an $1/\min\{m, k\}$-approximation algorithm for the general case, and a PTAS for the special case that the total number of groups is a fixed constant (that is, the case studied in Sect. 5).

For each $e' \in E$, we solve the linear program:

$$\max \quad x_{e'} \tag{6}$$
$$\text{s.t.} \quad Ax \leq b$$
$$x \geq 0$$

and select $e'$ as the matching. Since a single edge is always a feasible matching, we can return the largest solution and corresponding matching among all these $|E| = m$ linear programs.

**Theorem 11.** *The max-MLC problem has an $\frac{1}{\min\{m,k\}}$-approximation algorithm.*

Based on the previous approximation algorithm, we can propose a PTAS for the max-MLC problem where the total number of groups is fixed. If the solution returned by the $1/\min\{m,k\}$-approximation is unbounded, then this algorithm already returns an optimal solution of the problem. Therefore we can assume w.l.o.g. that the returned solution is bounded and denote it as $UB$. Since (6) searches the maximum possible value of weight for each edge, $UB$ is clearly an upper bound of the weight of each edge in an optimal solution. Moreover, $UB$ is a lower bound of the optimal value as every single edge is a matching. The idea of the algorithm is that we can approximately guess the values of vertices/edges in each set $V_g$ for $g = 1, ..., h$. After the values are determined, we find a feasible solution to the problem which total cost is at most $(1 - \epsilon)$ of the optimal solution, for any $\epsilon \in (0, 1)$. We note that the running time of the algorithm stated in Theorem 12 does not depend on the number of constraints $k$, which is different from the algorithm in Theorem 9.

**Theorem 12.** *Given any $\epsilon \in (0, 1)$, there is a PTAS for the max-MLC problem when the number of groups $h$ is fixed.*

# 7   Conclusions

We study several graph optimizations problem under linear constraints. These problems are NP-hard in general, and most of them are unlikely to have any approximation algorithms. We propose polynomial time algorithms and approximation algorithms for several special cases of them. A possible future direction is to see if there is any other nontrivial special cases that are tractable. It is also interesting to study other graph optimizations problem under linear constraints.

# References

1. Karp, R.M.: Reducibility among combinatorial problems. In: Complexity of Computer Computations, pp. 85–103 (1972)
2. Papadimitriou, C.H., Steiglitz, K.: Combinatorial Optimization: Algorithms and Complexity. Courier Dover Publications, New York (1998)
3. Korte, B., Vygen, J.: Combinatorial Optimization: Theory and Algorithms, 4th edn. Springer, Heidelberg (2012)
4. Nip, K., Wang, Z., Wang, Z.: Scheduling under linear constraints. Eur. J. Oper. Res. **253**(2), 290–297 (2016)
5. Nip, K., Wang, Z., Wang, Z.: Knapsack with variable weights satisfying linear constraints. J. Global Optim. **69**(3), 713–725 (2017)
6. Wang, Z., Nip, K.: Bin packing under linear constraints. J. Comb. Optim. **34**(5), 1198–1209 (2017)
7. Zhang, S., Nip, K., Wang, Z.: Related machine scheduling with machine speeds satisfying linear constraints. In: Kim, D., Uma, R.N., Zelikovsky, A. (eds.) COCOA 2018. LNCS, vol. 11346, pp. 314–328. Springer, Cham (2018). https://doi.org/10.1007/978-3-030-04651-4_21
8. Burer, S., Letchford, A.N.: Non-convex mixed-integer nonlinear programming: a survey. Surv. Oper. Res. Manag. Sci. **17**(2), 97–106 (2012)
9. Köppe, M.: On the complexity of nonlinear mixed-integer optimization. In: Lee, J., Leyffer, S. (eds.) Mixed-Integer Nonlinear Programming. IMA, vol. 154, pp. 533–558. Springer, Berlin (2011). https://doi.org/10.1007/978-1-4614-1927-3_19
10. Fagoyinbo, I., Ajibode, I.: Application of linear programming techniques in the effective use of resources for staff training. J. Emerg. Trends Eng. Appl. Sci. **1**(2), 127–132 (2010)
11. Gupta, K.M.: Application of linear programming techniques for staff training. Int. J. Eng. Innov. Technol. **3**(12), 132–135 (2014)
12. Uko, L.U., Lutz, R.J., Weisel, J.A.: An application of linear programming in performance evaluation. Acad. Educ. Lead. J. **21**, 1–7 (2017)
13. Kleinberg, J., Tardos, É.: Algorithm Design. Pearson Education, Incorporated, New York (2006)
14. Garey, M.R., Johnson, D.S.: Computers and Intractability: A Guide to the Theory of NP-Completeness. Freeman, New York (1979)
15. Ahuja, R.K., Magnanti, T.L., Orlin, J.B.: Network Flows: Theory, Algorithms, and Applications. Prentice Hall, New Jersey (1993)
16. Kuhn, H.W.: The hungarian method for the assignment problem. Nav. Res. Logist. Q. **2**(1 & 2), 83–97 (1955)
17. Burkard, R., Dell'Amico, M., Martello, S.: Assignment Problems. Society for Industrial and Applied Mathematics, Philadelphia (2009)
18. Edmonds, J.: A glimpse of heaven. In: Lenstra, J., Kan, A.R., Schrijver, A. (eds.) History of Mathematical Programming - A Collection of Personal Reminiscences, pp. 32–54. Centrum Wiskunde & Informatica, Amsterdam (1961)
19. Edmonds, J.: Paths, trees, and flowers. Can. J. Math. **17**, 449–467 (1965)
20. Dijkstra, E.W.: A note on two problems in connexion with graphs. Numer. Math. **1**, 269–271 (1959)
21. Fredman, M.L., Tarjan, R.E.: Fibonacci heaps and their uses in improved network optimization algorithms. J. ACM **34**(3), 596–615 (1987)
22. Williamson, D.P., Shmoys, D.B.: The Design of Approximation Algorithms. Cambridge University Press, New York (2011)

23. Dinur, I., Safra, S.: On the hardness of approximating minimum vertex cover. Ann. Math. **162**(1), 439–485 (2005)
24. Khot, S., Regev, O.: Vertex cover might be hard to approximate to within 2-$\epsilon$. J. Comput. Syst. Sci. **74**(3), 335–349 (2008)
25. Luenberger, D.G., Ye, Y.: Linear and Nonlinear Programming. Springer, New York (2015)
26. Schrijver, A.: Theory of Linear and Integer Programming. Wiley, Incorporated, New York (1986)

# On the Hardness of Some Geometric Optimization Problems with Rectangles

Supantha Pandit[✉]

Dhirubhai Ambani Institute of Information and Communication Technology,
Gandhinagar, Gujarat, India
pantha.pandit@gmail.com

**Abstract.** We study the Set Cover, Hitting Set, Piercing Set, Independent Set, Dominating Set problems, and discrete versions (Discrete Independent Set and Discrete Dominating Set) for geometric instances in the plane. We focus on certain restricted classes of geometric objects, including axis-parallel lines, strips, and rectangles. For rectangles, we consider the cases in which the rectangles are (i) anchored on a horizontal line, (ii) anchored on two lines (either two parallel lines or one vertical and one horizontal line), and (iii) stabbed by a horizontal line. Some versions of these problems have been studied previously; we focus here on the open cases, for which no complexity results were known.

**Keywords:** Discrete Dominating Set · Discrete Independent Set ·
NP-hard · Anchored rectangles · Stabbed by a horizontal line

## 1 Introduction

We study several special cases of various optimization problems in the plane, including the Set Cover (SC), Hitting Set (HS), Piercing Set (PS), Independent Set (IS), and Dominating Set (DS) problems. In addition, we consider *discrete* versions of the IS and DS problems, the Discrete Independent Set (DIS) and Discrete Dominating Set (DDS) problems. The inputs of these two problems are a set of objects $O$ and a set of points $P$. In the DIS problem the objective is to select maximum cardinality subset $O' \subseteq O$ of objects such that any two objects in $O'$ do not share a point in $P$. On the other hand, in the DDS problem the objective is to select a minimum collection $O' \subseteq O$ of objects such that the intersection of any object in $O \setminus O'$ and an object in $O'$ contains a point in $P$.

In this paper we study various optimization problems on various types of geometric objects as follows (see Fig. 1 for some types of objects):

⇢ **Line:** Axis parallel lines.
⇢ **Strip:** Axis-parallel strips.
⇢ **R-AHL:** Rectangles anchored on a horizontal line (Fig. 1(a)).
⇢ **R-ATL:** Rectangles anchored on two lines (Fig. 1(b)).

© Springer Nature Switzerland AG 2019
Y. Li et al. (Eds.): COCOA 2019, LNCS 11949, pp. 425–436, 2019.
https://doi.org/10.1007/978-3-030-36412-0_34

⇒ **R-ATOL:** Rectangles anchored on two orthogonal lines (Fig. 1(c)).
⇒ **R-SHL:** Rectangles stabbing a horizontal line.

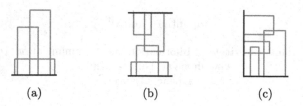

(a)                    (b)                    (c)

**Fig. 1.** (a) Rectangles anchored on a horizontal line. (b) Rectangles anchored on two lines. (c) Rectangles anchored on two orthogonal lines.

## 1.1    Previous Work

The optimization problems considered in this paper are NP-hard for simple geometric objects like unit squares [9], unit disks [9], rectangles [9], etc. The SC problem admits a PTAS for weighted unit squares [8]. The SC and HS problems are APX-hard even for axis-parallel strips [5]. For rectangles anchored on a horizontal line both the SC and HS problems can be solved in polynomial time [5,11]. When the rectangles are anchored on two lines both these problems are NP-hard [18]. Chepoi and Felsner [6] considered the IS and PS problems with rectangles where the rectangles are intersecting an axis-monotone curve. Correa et al. [7] studied the same problem, where the objects intersect a diagonal line. In [17], their results were extended. They also considered the SC and HS problems with other geometric objects as well.

In [19], the author studied the DS problem using axis-parallel rectangles and unit squares where the objects are intersecting a straight line which makes an angle with the $x$-axis. Recently, Bandyapadhyay et al. [3] gave a PTAS for the DS problem with axis-parallel rectangles touching a line with slope -1 from exactly one side of the line. Recently, Madireddy et al. [16] studied the DDS and DIS problem with arbitrary radii disks and arbitrary length squares in the plane and provide PTASes. They also showed that both DDS and DIS problems are NP-hard for unit disks intersecting a horizontal line and for axis-parallel unit squares intersecting a straight line with slope $-1$. In [20], the author studied the SC, HS, PS, and IS problems with right angled triangles such that the triangles intersect a straight line (either a horizontal line or a line with slope of $-1$).

## 1.2    Our Contributions

We list our contributions in Table 1.

**Table 1.** Our contributions are shown in colored text (P-> polynomial time, H-> NP-hard). The results in non-colored text for which no references are given are either trivial to show or can be derived from the other problems easily.

| Problems | IS | SC | HS | PS | DS | DDS | DIS |
|---|---|---|---|---|---|---|---|
| LINE | P | P | P [10] | P | P | *H* Th 1 | *P* Th 4 |
| STRIP | P | H [5] | H [5] | P | P | *H* Cor 1 | H [5] |
| R-AHL | P | P [11,5] | P [11,5] | P | P | ?? | P [5] |
| R-ATL | P [12,14,1,2] | H [18] | H [18] | ?? | ?? | *H* Th 2 | *H* Th 5 |
| R-ATOL | P [12,14,1,2] | H | H | ?? | ?? | *H* | H |
| R-SHL | P | *H* | *H* | P | P | *H* Th 3 | *H* Th 6 |

### 1.3 Prerequisites

In a planar 3-SAT (P-3-SAT) problem we are given a 3-CNF formula $\phi$ with $n$ variables $x_1, x_2, \ldots, x_n$ and $m$ clauses $C_1, C_2, \ldots, C_m$ such that each clause contains exactly 3 literals. For each variable or clause take a vertex in the plane. A literal is present in a clause if and only if there is an edge between the two vertices corresponding to the variable and the clause. Moreover, the formula should be such that the resulting graph must be planar. The goal is now to decide whether there is a truth assignment to the variables such that $\phi$ is satisfiable. Lichtenstein [15] proved that this problem is NP-complete. Later on Knuth and Raghunathan [13] considered a restricted version of the P-3-SAT problem, the rectilinear planar 3-SAT (R-P-3-SAT) problem can be defined as follows. For each variable or clause we take a horizontal line *segment*. The variable segments are placed on a horizontal line and clause segments are connected to these variable segments either from above or below by vertical line segments called *connections* such that none of these line segments and connections intersect. The goal is to decide whether there is a truth assignment to the variables such that $\phi$ is satisfiable. Figure 2 shows an instance of the R-P-3-SAT problem. Knuth and Raghunathan [13] showed that every P-3-SAT problem instance has an equivalent R-P-3-SAT problem instance and hence the later problem is also NP-complete. Note that we can order the variable segments in increasing $x$ direction. Let $C_t = (x_i \vee x_j \vee x_k)$ be a clause that connects the variables from above, where $x_i, x_j, x_k$ are ordered in the above ordering. Then we say that, $x_i$ is a *left*, $x_j$ is a *middle*, $x_k$ is a *right* variable.

## 2 Discrete Dominating Set

### 2.1 Axis-Parallel Lines

We prove that the DDS-LINE problem is NP-hard. The reduction is from the minimum dominating set problem on bipartite graphs that is NP-complete [4].

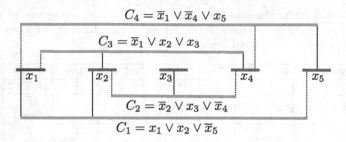

$$C_4 = \overline{x}_1 \vee \overline{x}_4 \vee x_5$$

$$C_3 = \overline{x}_1 \vee x_2 \vee x_3$$

$$C_2 = \overline{x}_2 \vee x_3 \vee \overline{x}_4$$

$$C_1 = x_1 \vee x_2 \vee \overline{x}_5$$

**Fig. 2.** An instance of R-P-3-SAT problem. Solid (resp. dotted) clause vertical segments represent that the variable is positively (resp. negatively) present in the corresponding clauses. For clause $C_4$, $x_1$ is a left, $x_4$ is a middle, and $x_5$ is a right variable.

Let $G(A, B, E)$ be a bipartite graph where the vertices of $A$ are on a vertical line and the vertices or $B$ are on another vertical line to the right of $A$. Let $1, 2, \ldots, \tau$ be the number of the vertices of $A$ from top to bottom. Similarly, let $1, 2, \ldots, \kappa$ be the number of the vertices of $B$ from top to bottom. Now for each vertex $i \in A$, take a horizontal line $h_i$ as $y = i$ and for each vertex $j \in B$ take a vertical line $v_j$ as $x = j$. Let $e_{ij} = (i, j)$ such that $i \in A$ and $j \in B$ be an edge in $E$. Take a point $p_{ij}$ with coordinate $(i, j)$ corresponding to $e_{ij}$ at the intersection of $h_i$ and $v_j$. See Fig. 3 for this construction. This construction takes polynomial time. It is observed that finding a minimum set of vertices in $G$ that dominates all the vertices in $G$ is equivalent to finding a minimum set of axis-parallel lines that dominates all the lines. Hence we conclude:

**Theorem 1.** DDS-LINE *is* NP-*hard.*

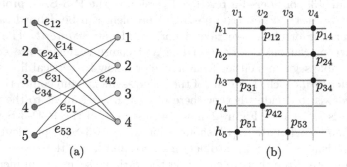

**Fig. 3.** (a) An instance of a bipartite graph. (b) An instance of the DDS-LINE problem constructed from the instance in (a).

**Corollary 1.** *The* NP-*hardness of the* DDS-LINE *problem directly implies the* NP-*hardness of the* DDS-STRIP *problem by replacing each horizontal (resp vertical) line by a thin horizontal (resp vertical) strip. The* NP-*hardness of the* DDS-STRIP *problem shows the* NP-*hardness of the* DDS-R-ATOL *problem. We just take a vertical and a horizontal line both at* $-\infty$. *Clearly these lines restrict the horizontal and vertical strips at* $-\infty$.

## 2.2   Rectangles Anchored on Two Lines

We prove that the DDS-R-ATL problem is NP-hard by a reduction from the R-P-3-SAT problem (see prerequisites).

**Reduction:** To represent a variable gadget of the DDS-R-ATL problem, we assume the graph $G$ in Fig. 4. The following result on $G$ can be proved easily.

**Lemma 1.** *There are exactly two optimal dominating sets, $D_0 = \{v_4, v_8, \ldots, v_{8\alpha}\}$ and $D_1 = \{v_2, v_6, \ldots, v_{8\alpha-2}\}$, of vertices each with cost exactly $2\alpha$ for graph $G$.*

We choose $\alpha$ to be the maximum number of clause vertical connections connecting from clause segments to a single variable segment either from above or from below. We encode the graph in Fig. 4 as a variable gadget of the DDS-R-ATL problem. For each vertex, we take a rectangle and for each edge, we take a point that is contained in exactly the rectangle corresponding to the two vertices that form the edge. The gadget for the variable $x_i$ is shown in Fig. 5. We take $8\alpha$ rectangles $R_i$ and $10\alpha$ points $P_i$ in two sides of a horizontal line $L$. The $4\alpha$ rectangles $\{s_1^i, s_2^i, \ldots, s_{4\alpha}^i\}$ and $5\alpha$ points $\{p_1^i, p_2^i, \ldots, p_{5\alpha}^i\}$ are one side of $L$ and the $4\alpha$ rectangles $\{s_{4\alpha+1}^i, s_{4\alpha+2}^i, \ldots, s_{8\alpha}^i\}$ and $5\alpha$ points $\{p_{5\alpha+1}^i, p_{5\alpha+2}^i, \ldots, p_{10\alpha}^i\}$ are another side $L$. Therefore, by Lemma 1 we conclude that for each variable gadget there are exactly two optimal dominating set of rectangles $S_i^0 = \{s_4^i, s_8^i, \ldots, s_{8\alpha}^i\}$ and $S_i^1 = \{s_2^i, s_6^i, \ldots, s_{8\alpha-2}^i\}$; each with size exactly $2\alpha$. This represents the truth value of the variable $x_i$.

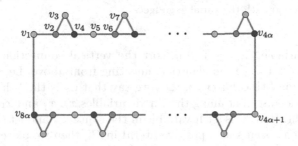

**Fig. 4.** Structure of the graph $G$.

The construction of the clause gadgets in the above and below are independent, and hence we describe the clause gadgets only for the above. For a clause $C_t$ that contains three variables $x_i$, $x_j$ and $x_k$ in this order from left to right, we take a rectangle $r^t$ and three points $p^{t_i}, p^{t_j}, p^{t_k}$. The bottom boundary of $r^t$, say $b^t$, are on the horizontal segment of $C_t$. In Fig. 6, we give a schematic diagram of the clause rectangles and positions of the points corresponding to the clauses. We now describe how $r^t, p^{t_i}, p^{t_j}, p^{t_k}$ interact with the variable gadgets.

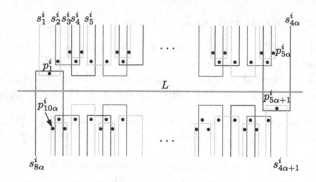

**Fig. 5.** Structure of a variable gadget.

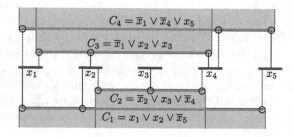

**Fig. 6.** Schematic diagram of the clause rectangles and position of the points (circles) and their interaction with the variable gadget.

For each variable $x_i$, $1 \leq i \leq n$, sort the vertical connections from left to right that connect to $x_i$ from clauses connecting from above. Let the clause $C_t$ connects to $x_i$ via $l$-th connection, then we say that $C_t$ is the $l$-th clause for $x_i$.

Let $C_t$ be a clause containing the three variables $x_i$, $x_j$ and $x_k$ in this order from left to right. Here $x_i$ is a left variable in the clause $C_t$ and let $C_t$ be the $l_1$-th clause for $x_i$. If $x_i$ occurs as a positive literal in $C_t$, then we place the point $p_{t_i}$ on $b^t$ and inside the rectangle $s^i_{4l_1+4}$ only. Otherwise, we place the point $p_{t_i}$ on $b^t$ and inside the rectangle $s^i_{4l_1+2}$ only. The interaction is similar for $x_j$ (middle variable) and $x_k$ (right variable) by replacing $l_1$ with $l_2$ and $l_3$ respectively. See Fig. 7 for the above construction. Clearly, the above construction can be done in polynomial time. We now prove the correctness of the above construction.

**Lemma 2.** *The formula $\phi$ is satisfiable iff there exists a solution to $\mathcal{D}$, an instance of the* DDS-R-ATL *problem constructed from $\phi$, with cost at most $2\alpha n$.*

*Proof.* Assume that $\phi$ is satisfiable and let $A : \{x_1, x_2, \ldots, x_n\} \rightarrow \{true, false\}$ be a satisfying assignment. For the $i$-th variable gadget, take the solution $S^0_i$, if $A(x_i) = true$. Otherwise take $S^1_i$. Clearly, we choose a total of $2\alpha n$ rectangles and these rectangles dominate all the variable and clause rectangles.

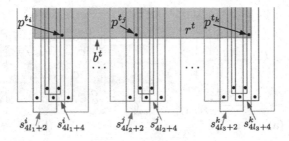

**Fig. 7.** Structure of a clause gadget and its interaction with variable gadgets.

On the other hand, suppose that there is a solution to $\mathscr{D}$ with cost at most $2\alpha n$. To dominate all the half-strips in a variable gadget requires at least $2\alpha$ rectangles (see Claim 1). Note that all the variable gadgets are disjoint. Therefore, from each variable gadget we must choose exactly $2\alpha$ rectangles (either set $S_i^0$ or set $S_i^1$). Set a variable to *true* if $S_i^0$ is chosen in its variable gadget, otherwise set it to *false*. Note that there are three points in a clause rectangle. Since the clause rectangle is dominated, at least one of these three points is covered by the solution. Such a point is either in solution $S_i^0$ or in solution $S_i^1$ of the corresponding variable gadget based on whether the variable is positively or negatively present in the clause. Hence, the assignment is a satisfying assignment.  □

**Theorem 2.** *The* DDS-R-ATL *problem is* NP-*hard.*

## 2.3 Rectangles Stabbed by a Horizontal Line

We prove that the DDS-R-SHL problem is NP-hard. The reduction is similar to the reduction given in Sect. 2.2. Here also we encode the graph $G$ in Fig. 4 as a variable gadget (see Fig. 8(a)). Note that in Fig. 8(a), all the rectangles are anchored on $L$ except $s_{4\alpha}^i$ and $s_{8\alpha}^i$ that is stabbed in the middle by $L$. Clearly, using Lemma 1, we say that for each variable gadget there are exactly two optimal dominating sets of rectangles each with size $\alpha$: $S_i^0 = \{s_4^i, s_8^i, \ldots, s_{8\alpha}^i\}$ and $S_i^1 = \{s_2^i, s_6^i, \ldots, s_{8\alpha-2}^i\}$. These two sets represent the truth value of $x_i$.

The clause gadgets are exactly the same as the clause gadgets in Sect. 2.2. However, here the interaction between the clause gadgets and the variable gadgets is different. We first reverse the connection of the clauses in the R-P-3-SAT problem instance i.e., the clauses those connect the variables from above (resp below) are now connect the variables from below (resp above). Now observe that the description of the variable clause connection for the clauses that connects to the variable from above in Sect. 2.2 are true here for the variable clause connection for the clauses that connects to the variable from below. In Fig. 8, we give a schematic diagram of the clause rectangle and position of the points corresponding to the clauses. A similar proof of Theorem 2 concludes:

**Theorem 3.** *The* DDS-R-SHL *problem is* NP-*hard.*

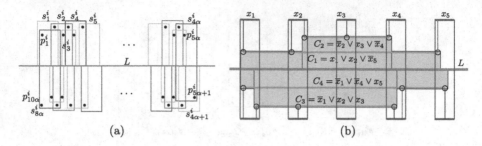

**Fig. 8.** (a) A variable gadget. (b) Schematic diagram of the variable and clause gadgets and their interaction. Blue rectangles are schematically represent the variable gadgets. (Color figure online)

## 3   Discrete Independent Set

### 3.1   Axis-Parallel Lines

We show that the DIS-LINE problem can be solved in polynomial time. We consider the reduction in the Sect. 2.1 in the reverse reduction, which ensures that finding a solution to the DIS-LINE problem is equivalent to finding a solution to the maximum independent set problem in a bipartite graph. Since the later problem can be solved in polynomial time, the DIS-LINE problem can also be solved in polynomial time.

**Theorem 4.** *The* DIS-LINE *problem can be solved in polynomial time.*

### 3.2   Rectangles Anchored on Two Lines

We prove that the DIS-R-ATL problem is NP-hard. We give a reduction from the R-P-3-SAT problem (see prerequisites).

**Reduction:** Note that, we choose $\alpha$ to be the maximum number of clause vertical connections connecting from clause segments to a single variable segment either from above or from below. The gadget for the variable $x_i$ is shown in Fig. 9(a). We take $16\alpha$ rectangles and $16\alpha$ points in two sides of a horizontal line $L$. The $8\alpha$ rectangles $\{s_1^i, s_2^i, \ldots, s_{8\alpha}^i\}$ and $8\alpha$ points $\{p_1^i, p_2^i, \ldots, p_{8\alpha}^i\}$ are one side of $L$ and the $8\alpha$ rectangles $\{s_{8\alpha+1}^i, s_{8\alpha+2}^i, \ldots, s_{16\alpha}^i\}$ and $8\alpha$ points $\{p_{8\alpha+1}^i, p_{8\alpha+2}^i, \ldots, p_{16\alpha}^i\}$ are another side of $L$. Each pair of consecutive rectangles have a point in common. Since these rectangles forms a cycle graph, where rectangles corresponding to vertices and two rectangles share a point if and only if there is an edge between the corresponding vertices of these two rectangles.

**Observation 1.** *For each variable gadget there are exactly two optimal independent sets of rectangles $H_i^0 = \{s_2^i, s_4^i, \ldots, s_{16\alpha}^i\}$ and $H_i^1 = \{s_1^i, s_3^i, \ldots, s_{16\alpha-1}^i\}$ each with size exactly $8\alpha$.*

The construction of the clause gadgets in above and below are independent, and hence we describe the clause gadgets only for above. Let $C_t$ be a clause that contains variables $x_i$, $x_j$ and $x_k$ in this order from left to right. For $C_t$, we take 5 rectangles $\{s_1^t, s_2^t, s_3^t, s_4^t, s_5^t\}$ and 6 points; 1 point $p^{t_i}$ corresponding to $x_i$, 4 points $p_1^{t_j}, p_2^{t_j}, p_3^{t_j}, p_4^{t_j}$ corresponding to $x_j$, and 1 point $p^{t_k}$ corresponding to $x_k$. The rectangle $s_1^t$ covers the points $\{p^{t_i}, p_1^{t_j}, p_4^{t_j}\}$, $s_2^t$ covers $\{p^{t_i}, p_1^{t_j}, p_2^{t_j}\}$, $s_3^t$ covers $\{p_2^{t_j}, p_3^{t_j}\}$, $s_4^t$ covers $\{p_1^{t_j}, p_2^{t_j}, p_4^{t_j}, p^{t_k}\}$, and $s_5^t$ covers $\{p_1^{t_j}, p_4^{t_j}, p^{t_k}\}$. See Fig. 10 for this construction. We now describe the placement of the points and rectangles with respect to the variable gadget. We take a rectangle $r^t$. The bottom boundary of $r^t$, say $b^t$, are on the horizontal segment of $C_t$ and it can extends to the infinity (actually we can take horizontal line in a far enough distance such that the top boundaries of all such rectangles touch it) in the upward direction. We take a horizontal thin rectangular **region** along the top edge of $r^t$. We place points corresponding to the clause $C_t$ inside this region. The rectangles corresponding to $C_t$ are exclusively in $r^t$ and their top boundaries are inside the region. In Fig. 9(b), we give a schematic diagram of the clause rectangles, regions, and positions of the points corresponding to the clauses. We now describe how the rectangles and points corresponding to the clauses interact with the rectangles and points corresponding to variables.

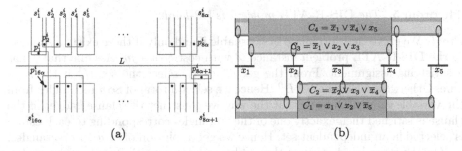

**Fig. 9.** (a) Structure of a variable gadget. (b) Position of the rectangles and points (empty ellipses) corresponding to the clauses.

For each variable $x_i$, $1 \leq i \leq n$, sort the vertical connections from left to right that connect to $x_i$ from clauses connecting from above. Let clause $C_t$ connect to $x_i$ via $l$-th connection, then we say that $C_t$ is the $l$-th clause for the variable $x_i$. Assume that the clause $C_t$ contains three variables $x_i$, $x_j$, and $x_k$ in this order from left to right. We now have the following cases.

➤ Here $x_i$ is a left variable in the clause $C_t$ and let $C_t$ be the $l_1$-th clause for $x_i$. If $x_i$ occurs as a positive literal in $C_t$, then we place the point $p^{t_i}$ inside the rectangle $s_{8l_1-5}^i$. Otherwise, we place $p^{t_i}$ inside the rectangle $s_{8l_1-4}^i$.

➤ Here $x_j$ is a middle variable in the clause $C_t$ and let $C_t$ be the $l_2$-th clause for $x_j$. If $x_j$ occurs as a positive literal in $C_t$, then we place the point $p_1^{t_j}$, $p_2^{t_j}$,

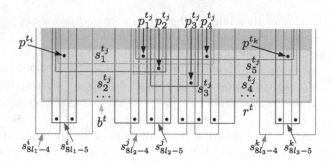

**Fig. 10.** Structure of a clause gadget and its interaction with variable gadgets.

$p_3^{t_j}$, and $p_4^{t_j}$ inside the rectangle $s_{8l_2-6}^j$, $s_{8l_2-5}^j$, $s_{8l_2-3}^j$, and $s_{8l_2-2}^j$ respectively. Otherwise, we shift all the points one rectangle to the right.

➤ Here $x_k$ is a right variable in the clause $C_t$ and let $C_t$ be the $l_3$-th clause for $x_k$. If $x_k$ occurs as a positive literal in $C_t$, then we place the point $p^{t_k}$ inside the rectangle $s_{8l_3-5}^k$. Otherwise, we place $p^{t_k}$ inside the rectangle $s_{8l_3-4}^k$.

See Fig. 10 for the above construction. Clearly, the construction described above can be done in polynomial time.

**Theorem 5.** *The* DIS-R-ATL *problem is* NP-*hard.*

*Proof.* We prove that formula $\phi$ is satisfiable if and only if there exists a solution to the DIS-R-ATL problem instance $\mathscr{D}$ with cost $8\alpha n + m$. Assume that $\phi$ has a satisfying assignment. From the gadget of $x_i$, select the set $H_i^1$ if the $x_i$ is true. Otherwise select the set $H_i^0$. Hence we select a total of $8\alpha n$ rectangles from the variable gadget. Observe that the way we construct the clause gadget, if the clause is satisfied then exactly one of the rectangles corresponding to each clause is selected in an independent set. Hence we get a solution of $8\alpha n + m$ rectangles.

On the other hand, assume that $\mathscr{D}$ has a solution with $8\alpha n + m$ rectangles. From the gadget of $x_i$ we select $8\alpha$ rectangles either $H_i^0$ or $H_i^1$. We set $x_i$ to be true if $H_i^1$ is selected otherwise set $x_i$ to be false if $H_i^0$ is selected. We now argue that this is a satisfying assignment of $\phi$ i.e., every clause is satisfied. Consider a clause $C_t = (x_i \vee x_j \vee x_k)$ (a similar argument can be applied for other clauses as well). If $C_t$ is not satisfied, then we select the sets $H_i^0$, $H_j^0$, and $H_k^0$ from the corresponding variable gadget. These rectangles prevent in selecting any rectangle from the set of rectangles corresponding to $C_t$. This contradicts the fact that the size of the solution is $8\alpha n + m$. However if one of $H_i^1$, $H_j^1$, and $H_k^1$ is selected then from the set of rectangles of $C_t$, exactly one rectangle is selected in a solution. Therefore, the above assignment is a satisfying assignment.    □

### 3.3    Rectangles Stabbed by a Horizontal Line

We prove that the DIS-R-SHL problem is NP-hard. The reduction is from the R-P-3-SAT problem and is a composition of the two reductions in Sects. 3.2 and

2.3. The way the gadget in Figure 8 is constructed from the gadget in Fig. 5, the similar way we construct the variable gadget here from the gadget in Fig. 9(a). See Fig. 11(a) for the structure of a variable gadget. Clearly, Observation 1 is true for any variable gadget.

The clause gadgets are exactly the same as the clause gadgets in Sect. 3.2. However here the interaction between the clause gadgets and the variable gadgets is different. We first reverse the connection of the clauses in the R-P-3-SAT problem instance i.e., the clauses those connect the variables from above (resp below) are now connect the variables from below (resp above). Now observe that the description of the variable clause connection for the clauses that connects to the variable from above in Sect. 3.2 are true here for the variable clause connection for the clauses that connects to the variable from below. In Fig. 11(b), we give a schematic diagram of the clause rectangle and position of the points corresponding to the clauses. Hence a proof similar to the proof of Theorem 5 concludes:

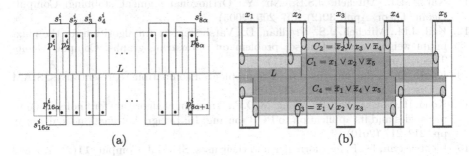

**Fig. 11.** (a) Structure of a variable gadget. (b) Schematic diagram of the variable and clause gadgets and their interaction.

**Theorem 6.** *The* DIS-R-SHL *problem is* NP-*hard.*

**Acknowledgements.** We would like to thank Joseph S. B. Mitchell for fruitful discussions in the early stages of this paper.

# References

1. Ahmadinejad, A., Zarrabi-Zadeh, H.: Finding maximum disjoint set of boundary rectangles with application to PCB routing. IEEE Trans. Comput.-Aided Des. Integr. Circuits Syst. **36**(3), 412–420 (2017)
2. Ahmadinejad, A., Assadi, S., Emamjomeh-Zadeh, E., Yazdanbod, S., Zarrabi-Zadeh, H.: On the rectangle escape problem. Theor. Comput. Sci. **689**, 126–136 (2017)
3. Bandyapadhyay, S., Maheshwari, A., Mehrabi, S., Suri, S.: Approximating dominating set on intersection graphs of rectangles and L-frames. Comput. Geom. **82**, 32–44 (2019)

4. Bertossi, A.A.: Dominating sets for split and bipartite graphs. Inf. Process. Lett. **19**(1), 37–40 (1984)
5. Chan, T.M., Grant, E.: Exact algorithms and APX-hardness results for geometric packing and covering problems. Comput. Geom. **47**(2, Part A), 112–124 (2014)
6. Chepoi, V., Felsner, S.: Approximating hitting sets of axis-parallel rectangles intersecting a monotone curve. Comput. Geom. **46**(9), 1036–1041 (2013)
7. Correa, J., Feuilloley, L., Pérez-Lantero, P., Soto, J.A.: Independent and hitting sets of rectangles intersecting a diagonal line: algorithms and complexity. Discrete Comput. Geom. **53**(2), 344–365 (2015)
8. Erlebach, T., van Leeuwen, E.J.: PTAS for weighted set cover on unit squares. In: Serna, M., Shaltiel, R., Jansen, K., Rolim, J. (eds.) APPROX/RANDOM -2010. LNCS, vol. 6302, pp. 166–177. Springer, Heidelberg (2010). https://doi.org/10.1007/978-3-642-15369-3_13
9. Fowler, R.J., Paterson, M.S., Tanimoto, S.L.: Optimal packing and covering in the plane are NP-complete. Inf. Process. Lett. **12**(3), 133–137 (1981)
10. Hassin, R., Megiddo, N.: Approximation algorithms for hitting objects with straight lines. Discrete Appl. Math. **30**(1), 29–42 (1991)
11. Katz, M.J., Mitchell, J.S.B., Nir, Y.: Orthogonal segment stabbing. Comput. Geom. Theor. Appl. **30**(2), 197–205 (2005)
12. Keil, J.M., Mitchell, J.S., Pradhan, D., Vatshelle, M.: An algorithm for the maximum weight independent set problem on outerstring graphs. Comput. Geom. Theor. Appl. **60**(C), 19–25 (2017)
13. Knuth, D.E., Raghunathan, A.: The problem of compatible representatives. SIAM J. Discrete Math. **5**(3), 422–427 (1992)
14. Kong, H., Ma, Q., Yan, T., Wong, M.D.F.: An optimal algorithm for finding disjoint rectangles and its application to PCB routing. In: Design Automation Conference, pp. 212–217 (2010)
15. Lichtenstein, D.: Planar formulae and their uses. SIAM J. Comput. **11**(2), 329–343 (1982)
16. Madireddy, R.R., Mudgal, A., Pandit, S.: Hardness results and approximation schemes for discrete packing and domination problems. In: Kim, D., Uma, R.N., Zelikovsky, A. (eds.) COCOA 2018. LNCS, vol. 11346, pp. 421–435. Springer, Cham (2018). https://doi.org/10.1007/978-3-030-04651-4_28
17. Mudgal, A., Pandit, S.: Covering, hitting, piercing and packing rectangles intersecting an inclined line. In: Lu, Z., Kim, D., Wu, W., Li, W., Du, D.-Z. (eds.) COCOA 2015. LNCS, vol. 9486, pp. 126–137. Springer, Cham (2015). https://doi.org/10.1007/978-3-319-26626-8_10
18. Mudgal, A., Pandit, S.: Geometric hitting set, set cover and generalized class cover problems with half-strips in opposite directions. Discrete Appl. Math. **211**, 143–162 (2016)
19. Pandit, S.: Dominating set of rectangles intersecting a straight line. In: CCCG, pp. 144–149 (2017)
20. Pandit, S.: Covering and packing of triangles intersecting a straight line. In: Pal, S.P., Vijayakumar, A. (eds.) CALDAM 2019. LNCS, vol. 11394, pp. 216–230. Springer, Cham (2019). https://doi.org/10.1007/978-3-030-11509-8_18

# On Vertex-Edge and Independent Vertex-Edge Domination

Subhabrata Paul[1]([⊠]) and Keshav Ranjan[2]

[1] Department of Mathematics, IIT Patna, Bihta, India
subhabrata@iitp.ac.in
[2] Department of Computer Science and Engineering, IIT Madras, Chennai, India
keshav@cse.iitm.ac.in

**Abstract.** Given a graph $G = (V, E)$, a vertex $u \in V$ *ve-dominates* all edges incident to any vertex of $N_G[u]$. A set $S \subseteq V$ is a *ve-dominating set* if for all edges $e \in E$, there exists a vertex $u \in S$ such that $u$ ve-dominates $e$. Lewis [Ph.D. thesis, 2007] proposed a linear time algorithm for ve-domination problem for trees. In this paper, first we have constructed an example where the proposed algorithm fails. Then we have proposed a linear time algorithm for ve-domination problem in block graphs, which is a superclass of trees. We have also proved that finding minimum ve-dominating set is NP-complete for undirected path graphs. Finally, we have characterized the trees with equal ve-domination and independent ve-domination number.

**Keywords:** Vertex-edge domination · Independent vertex-edge domination · NP-completeness

## 1 Introduction

Domination and its variants are one of the classical problems in graph theory. Let $G = (V, E)$ be a graph and $N_G(v)$ (or $N_G[v]$) be the *open* (respectively, *closed*) neighborhood of $v$ in $G$. A set $D \subseteq V$ is called a *dominating set* of a graph $G = (V, E)$ if $|N_G[v] \cap D| \geq 1$ for all $v \in V$. Our goal is to find a dominating set of minimum cardinality which is known as *domination number* of $G$ and denoted by $\gamma(G)$. For details the readers are refered to [5,6].

In this paper, we have studied one variant of domination problem, namely *vertex-edge domination* problem, also known as *ve-domination* problem. Given a graph $G = (V, E)$, a vertex $u \in V$ *ve-dominates* all edges incident to any vertex of $N_G[u]$. A set $S \subseteq V$ is a *vertex-edge dominating set* (or simply a *ve-dominating set*) if for all edges $e \in E$, there exists a vertex $u \in S$ such that $u$ ve-dominates $e$. The minimum cardinality among all the ve-dominating sets of $G$ is called the *vertex-edge domination number* (or simply *ve-domination number*),

---

K. Ranjan—This work was done when the second author was persuing his M.Tech. at IIT Patna.

and is denoted by $\gamma_{ve}(G)$. A set $S$ is called an *independent ve-dominating set* if $S$ is both an independent set and a ve-dominating set. The *independent ve-domination number* of a graph $G$ is the minimum cardinality of an independent ve-dominating set and is denoted by $i_{ve}(G)$.

The vertex-edge domination problem was introduced by Peters [12] in his PhD thesis in 1986. However, it did not receive much attention until Lewis [9] in 2007 introduced some new parameters related to it and established many new results in his PhD thesis. In his PhD thesis, Lewis has given some lower bound on $\gamma_{ve}(G)$ for different graph class like connected graphs, $k$-regular graphs, cubic graphs etc. On the algorithmic side, Lewis has also proved that the ve-domination problem is NP-Complete for bipartite, chordal, planar and circle graphs and independent ve-domination problem is NP-Complete even when restricted to bipartite and chordal graph. Also approximation algorithm and approximation hardness results are proved in [9]. In [10], the authors have characterized the trees with equal domination and vertex-edge domination number. In [8], both upper and lower bounds on ve-domination number of a tree have been proved. Some upper bounds on $\gamma_{ve}(G)$ and $i_{ve}(G)$ and some relationship between ve-domination number and other domination parameters have been proved in [2]. In [13], Żyliński has shown that for any connected graph $G$ with $n \geq 6, \gamma_{ve}(G) \leq n/3$. Other variations of ve-dominations have also been studied [1,7].

In [9], Lewis proposed a linear time algorithm for ve-domination problem for trees. Basically, he proposed a linear time algorithm for finding minimum distance-3 dominating set of a weighted tree. A set $D \subseteq V$ is called a *distance-3 dominating set* of a graph $G = (V, E)$ if every vertex in $V$ is at most distance 3 from some vertex in $D$. In case of a weighted graph, the goal is to find *distance-3 dominating set* with minimum weight. Lewis claimed that "Given any tree $T = (V, E)$,

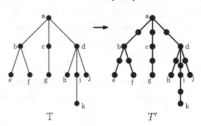

**Fig. 1.** Counter example

define a new tree $T' = (V', E')$ by subdividing each edge of $E$. Then place a weight of one on each of $V$ and a weight of $\infty$ on each of $V' \setminus V$. Now solve the weighted distance-3 dominating set problem for $T'$, the result will give the $\gamma_{ve}(T)$." But, we have found a counter example of this claim. In Fig. 1, it is easy to see that $\gamma_{ve}(T) = 2$. Now, in the new weighted tree $T' = (V', E')$, it is not possible to find any distance-3 dominating set whose weight is 2.

This motivates us to study ve-domination problem in trees and other graph classes. The rest of the paper is organized as follows. In Sect. 2, we have proposed a linear time algorithm for finding minimum ve-dominating set in block graphs, which is a superclass of trees. Section 3 deals with NP-completeness of this problem in undirected path graphs. In Sect. 4, we have characterized the trees having equal ve-domination number and independent ve-domination number. Finally Sect. 5 concludes the paper.

## 2   Block Graph

In this section, we propose a linear time algorithm for block graphs to solve ve-domination problem. A vertex $v \in V$ is called a *cut vertex* of $G = (V, E)$ if removal of $v$ increases the number of components in $G$. A maximal connected induced subgraph without a cut vertex of $G$ is called a *block* of $G$. A graph $G$ is called a *block graph* if each block of $G$ is a complete subgraph. The intersection of two distinct blocks can contain at most one vertex. Two blocks are called *adjacent blocks* if they contain a common cut vertex of $G$. A block graph with one or more cut vertices contains at least two blocks, each of which contains exactly one cut vertex. Such blocks are called *end blocks*. The *distance between two blocks* $B_i$ and $B_j$ is defined as $dist_G(B_i, B_j) = max\{dist(v_i, v_j)|v_i \in B_i, v_j \in B_j\} - 1$. The *distance between a vertex $v$ and a block $B$* of a block graph $G$ is denoted as $dist_G(v, B) = max\{dist_G(v, u)|u \in B\} - 1$.

Our proposed algorithm is a labeling based algorithm. Let $G = (V, E)$ be a block graph where each vertex $v \in V$ is associated with a label $l(v)$ and each edge $e = (xy) \in E$ is associated with a label $m(xy)$, where $l(v), m(xy) \in \{0, 1\}$. We call such graph a *labelled block graph*.

**Definition 1.** *Given a labelled block graph $G = (V, E)$ with labels $l$ and $m$, we first define an optional ve-dominating set as a subset $D \subseteq V$ such that*

*(i) if $l(v) = 1$, then $v \in D$,*
*(ii) $D$ ve-dominates every edge $e = (xy)$ with $m(xy) = 1$.*

*The optional ve-domination number, denoted by $\gamma_{opve}(G)$, is the minimum cardinality among all the optional ve-dominating sets of $G$.*

Note if $l(v) = 0$ for all $v \in V$ and $m(xy) = 1$ for all $e = (xy) \in E$, then the minimum optional ve-dominating set is nothing but a minimum ve-dominating set of $G$. Given a labelled block graph $G$ with labels $l(v) = 0$ for each $v \in V$ and $m(xy) = 1$ for each $xy \in E$, our proposed algorithm basically outputs a minimum optional ve-dominating set of $G$.

Next we present the outline of the algorithm. Let $B_0$ be an end block of a block graph $G = (V, E)$. Since, block graph has a tree like structures, we can view $G$ as a graph rooted at the end block $B_0$. The *height* of $G$ is defined as $h(G) = max\{dist_G(B_0, B)|B \text{ is end block of G}\}$. At each step, the algorithm processes one of the end blocks at maximum height. Moreover, out of all the blocks at the same maximum height, the blocks with more number of edges with $m(xy) = 1$, are processed first. Based on some properties of the edges having label 1 in an end block $B$, we decide whether to take some vertices from $B$ in the optional ve-dominating set or not and then delete that block $B$ (except the cut vertex) from $G$. We also modify the labels of some of the vertices and edges of the new graph. In the next iteration, we process another end block and the process continues till we are left with the root block $B_0$. For the root block, we directly calculate the optional ve dominating set. The outline of the algorithm is given in Algorithm 1. In Algorithm 1, for an end block $B, t_B$ denotes the number of edges with $m(xy) = 1$ in $B$ and $P$ denotes the set of non-cut vertices of $B$,

---

**Algorithm 1.** MIN_OPT_VEDom(G)

---

**Input:** A labelled block graph $G$ with $l(v) = 0, \forall v \in V$ and $m(xy) = 1, \forall xy \in E$

**Output:** Minimum optional ve-dominating set $S$ of $G$

1: $S = \phi$
2: **for** i=$h(G)$ **to** 1 **do**
3:    $B$ is an end block at level $i$ and $c$ is the cut vertex of $B$
4:    **while** $(t_B \geq 2)$ **do**
5:       $l(c) = 1$
6:       $G = G \setminus P$
7:       $m(xy) = 0, \forall x \in N_G(c)$
8:    **while** $((t_B = 1$ such that $m(uv) = 1)$ AND $(u \neq c)$ AND $(v \neq c))$ **do**
9:       $l(c) = 1$
10:      $G = G \setminus P$
11:      $m(xy) = 0, \forall x \in N_G(c)$
12:    **while** $(t_B = 1)$ **do**
13:      $l(F(c)) = 1$
14:      $G = G \setminus P$
15:      $m(xy) = 0, \forall x \in N_G(F(c))$
16:    **while** $(t_B = 0)$ **do**
17:      $S = S \cup \{x | x \in P$ and $l(x) = 1\}$
18:      $G = G \setminus P$
19: **if** $h(G) = 0$ **then**
20:    **if** $t_B > 0$ **then**
21:      select any $v \in V(G)$
22:      $S = S \cup \{v\}$ and **Return S**
23:    **else**
24:      $S = S \cup \{v | v \in G$ AND $l(v) = 1\}$ and **Return S**

---

i.e., $P = V(B) \setminus \{c\}$, where $V(B)$ is the set of vertices of $B$ and $c$ is the cut vertex of $B$. Also let $F(c)$ denote the unique cut vertex of $G$ in $N_G[c]$ which has the minimum distance from $B_0$. Note that $F(c) = c$ if and only if $c$ is the cut vertex of $B_0$.

Next we prove the correctness of Algorithm 1. Note that, the modification of the labels of the reduced graph is done in such a way that if $l(v) = 1$ for some $v \in V$, then $m(xy) = 0$ for all edges incident to any vertex $x \in N_G(v)$. Hence we have the following claim:

**Claim 2.** *After each iteration, in any block $B$, the set of edges with label 1 forms a clique.*

*Proof.* We will prove this by showing that it is not possible to have an edge $uv$ with $m(uv) = 0$ and $m(ux) = m(vy) = 1$ in any block. Suppose a block $B$ is having an edge $uv$ with $m(uv) = 0$. Since $m(uv) = 0$, there must be a vertex $p \in N[u]$ or $p \in N[v]$ with $l(p) = 1$.

**Case 1.** *(When $p \in N[u]$):* In this case all the edges incident to $u$ must be labeled 0 since they are being ve-dominated by the vertex $p$. But $m(ux) = 1$ for the edge $ux$, which is a contradiction.

**Case 2.** *(When $p \in N[v]$):* In this case all the edges incident to $v$ must be labeled 0 since they are being ve-dominated by the vertex $p$. But $m(vy) = 1$ for the edge $vy$, which is a contradiction.

Hence, if $m(uv) = 0$ then either all edges incident to $u$ is labelled 0 or all edges incident to $v$ is labelled 0. And hence whenever an edge is labelled 0 all other edges incident to at least one of its end point is also labelled 0. It reduces the size of clique (formed by label-1 edges) by at least 1.                                    □

**Lemma 3.** *Let $G$ be a block graph with an end block $B_0$ as root and $B$ be another end block such that $dist_G(B_0, B) = h(G)$. Also assume that $P = V(B) \setminus \{c\}$, where $c$ is the cut vertex of $B$ and $t_B$ denotes the number of edges with label 1 in $B$. Then followings are true.*

*(a) If $t_B \geq 2$, and $G'$ is new block graph results from $G$ by relabelling $c$ as $l(c) = 1$, deleting all $v \in P$ and relabelling all edges $xy$ as $m(xy) = 0 \; \forall x \in N(c)$. Then $\gamma_{opve}(G) = \gamma_{opve}(G')$.*

*(b) If $t_B = 1$ but the edge with label-1 is not incident to $c$ and $G'$ is new block graph results from $G$ by relabelling $c$ as $l(c) = 1$, deleting all $v \in P$ and relabelling all edges $xy$ as $m(xy) = 0 \; \forall x \in N(c)$. Then $\gamma_{opve}(G) = \gamma_{opve}(G')$.*

*(c) Let the conditions in (a) and (b) are not satisfied but $t_B = 1$ and $G'$ is new block graph results from $G$ by relabelling $F(c)$ as $l(F(c)) = 1$, deleting all $v \in P$ and relabelling all edges $xy$ as $m(xy) = 0 \; \forall x \in N_{G'}(F(c))$. Then $\gamma_{opve}(G) = \gamma_{opve}(G')$.*

*(d) If $t_B = 0$ and $B$ has $k$ many vertices with $l(v) = 1$ and $G'$ is new block graph results from $G$ by deleting all $v \in P$. Then $\gamma_{opve}(G) = \gamma_{opve}(G') + k$.*

*Proof.* (a) Let $S$ be $\gamma_{opve}$-set of $G$. If $\exists v \in P \cap S$ then $(S \setminus \{v\}) \cup \{c\}$ is also optional ve-dominating set of $G$, where $c$ is considered as a vertex with label 1. So assume $S \cap P = \phi$. Now pick any edge $e$ with $m(e) = 1$ in $G'$. There must be some $v \in S$ such that $v$ ve-dominates $e$. Also $v \notin P$. Therefore $S$ is optional ve-dominating set of $G'$. Hence, $\gamma_{opve}(G') \leq \gamma_{opve}(G)$.

Conversely, let $S'$ be $\gamma_{opve}$-set of $G'$. Since $l(c) = 1, c \in S'$. Pick any edge $e$ with $m(e) = 1$ from $G$. If $e \notin B$ then obviously some $v \in S'$ ve-dominates $e$. If $e \in B$ then $c$ ve-dominates $e$. So, $S'$ is also optional ve-dominating set of $G$. Hence $\gamma_{opve}(G) \leq \gamma_{opve}(G')$.

(b) The proof is same as the proof in (a).

(c) Let $S$ be $\gamma_{opve}$-set of $G$. If $\exists v \in P \cap S$ then $(S \setminus \{v\}) \cup \{F(c)\}$ is also optional ve-dominating set of $G$. Since all edges private to $v$ are also ve-dominated by $F(c)$. So assume $S \cap P = \phi$. Now pick any edge $e$ with $m(e) = 1$ from $G'$. Since $S \cap P = \phi$ there must be some $u \in S$ such that $u$ ve-dominates $e$. Also $u \notin P$. Therefore, $S$ is optional ve-dominating set of $G'$. Hence, $\gamma_{opve}(G') \leq \gamma_{opve}(G)$.

Conversely, let $S'$ be $\gamma_{opve}$-set of $G'$. Since $l(F(c)) = 1, F(c) \in S'$. So, all edges incident to $c$ is ve-dominated by $F(c)$. In block $B$ only one edge is labelled 1 and is incident to $c$. So, it is ve-dominated by $F(c)$ and $S'$ is optional ve-dominating set of $G$. Hence, $\gamma_{opve}(G) \leq \gamma_{opve}(G')$.

(d) Let $S$ be $\gamma_{opve}$-set of $G$. $Q = \{p | p \in P$ and $l(p) = 1\}$. So $Q \subseteq S$. There are two cases $Q = \phi$ and $Q \neq \phi$.

**Case 1.** *(Q = φ i.e. (k = 0)) Pick any edge e with m'(e) = 1 from G'. Since S is $\gamma_{opve}$-set of G. There must exist some $v \in S$ such that v ve-dominates e and this $v \notin P$. Therefore, $\gamma_{opve}(G') \leq \gamma_{opve}(G)$ and hence $\gamma_{opve}(G') \leq \gamma_{opve}(G)$.*

*Conversely, let S' be $\gamma_{opve}$-set of G'. All the edges of G, except the edges of the newly added block B is ve-dominated by S' and none of the edges of block B needs to be ve-dominated. Hence $\gamma_{opve}(G) \leq \gamma_{opve}(G')$.*

**Case 2.** *(Q ≠ φ i.e. (k > 0)) Pick any edge e with m'(e) = 1 from G'. Since $Q \neq \phi$, e is not incident to c. There must be some vertex $v \in S \setminus Q$ to ve-dominate e. Hence, $S \setminus Q$ is optional ve-dominating set of G'. Hence $\gamma_{opve}(G') \leq \gamma_{opve}(G) - k$.*

*Conversely, let S' be $\gamma_{opve}$-set of G'. All the edges of G, except the edges of the newly added block B is ve-dominated by S' and none of the edges of block B needs to be ve-dominated. Since Q contains vertices with label 1, $S' \cup Q$ is optional ve-dominating set of G. Hence, $\gamma_{opve}(G) \leq \gamma_{opve}(G') + k$.* ☐

**Lemma 4.** *Let G be a complete graph, i.e., G = B. If $t_B \geq 1$, then $\gamma_{opve}(G) = 1$. Otherwise, $\gamma_{opve}(G) = k$, where k is the number of vertices of B with l(v) = 1.*

*Proof.* When $t_B \geq 1$ then B does not have any vetex with label-1. So, we need at least one vertex from block B to ve-dominate all the edges with m(e) = 1 and only one vertex is sufficient to ve-dominate all the edges. Hence, $\gamma_{opve}(G) = 1$.

When $t_B = 0$, none of the edges needs to be ve-dominated. So, all the vertices with l(v) = 1 forms an optional ve-dominating set and there are k many such vertices. Hence, $\gamma_{opve}(G) = k$. ☐

Lemmas 3 and 4 shows that the output of Algorithm 1 is minimum optional ve-dominating set. At each iteration, we are taking $\mathcal{O}(deg(c))$ time. Hence the running time of Algorithm 1 is $\mathcal{O}((n + m)$. Thus, we have the following theorem.

**Theorem 5.** *The ve-domination problem can be solved in $\mathcal{O}(n + m)$ time for block graphs.*

*Remark 1.* We can also find the minimum independent ve-dominating set of a given block graph in a similar approach. The algorithm is similar to Algorithm 1, but with little modification, to ensure the output is an independent set. Thus, for block graphs, the independent ve-domination problem can also be solved in linear time.

## 3   Undirected Path Graphs

In this section, we prove that the ve-domination problem for undirected path graphs is NP-complete by showing a polynomial time reduction from 3 - dimensional matching problem which is a well-known NP-complete problem [3]. A graph G is called an *undirected path graph* if G is the intersection graphs of a family of paths of a tree. In [4], Gavril proved that a graph $G = (V, E)$ is an undirected path graph if and only if there exists a tree T whose vertices are the maximal cliques of G and the set of all maximal cliques containing a particular vertex $v$ of $V$ forms a path in $T$. This tree is called the clique tree of the undirected path graph $G$. The 3-dimensional matching problem is as follows: given a set $M \subseteq U \times V \times W$, where $U, V$ and $W$ are disjoint set with $|U| = |V| = |W| = q$, does $M$ contains a matching $M'$, i.e., a subset $M' \subseteq M$ such that $|M'| = q$ and no two elements of $M'$ agree in any coordinate?

**Theorem 6.** *The ve-domination problem is NP-complete for undirected path graphs.*

*Proof.* It is easy to see that ve-domination problem is in NP. Now, we describe polynomial reduction form 3-dimensional matching problem to ve-domination problem in undirected path graph. Let $U = \{u_r | 1 \leq r \leq q\}, V = \{v_s | 1 \leq s \leq q\}, W = \{w_t | 1 \leq t \leq q\}$, and $M = \{m_i = (u_r, v_s, w_t) | 1 \leq i \leq p, u_r \in U, v_s \in V, w_t \in W\}$ be an instance of 3-dimensional matching problem. Now we construct a tree $T$ having $8p + 6q + 1$ vertices that becomes the clique tree of an undirected path graph $G$. The vertices of the tree $T$ are maximal cliques of $G$. The vertex set and the edge set are as follows:

For $1 \leq i \leq p$, each $m_i = (u_r, v_s, w_t) \in M$ corresponds to 8 cliques which are vertices of $T$, namely $\{A_i, B_i, C_i, D_i\}, \{A_i, B_i, D_i, F_i\}, \{C_i, D_i, G_i\}, \{A_i, B_i, E_i\}, \{C_i, G_i, K_i\}, \{A_i, E_i, H_i\}, \{B_i, E_i, I_i\}$, and $\{B_i, I_i, J_i\}$. These vertices depend only on the triple $m_i$ itself but not on the elements within the triple. These eight vertices induces a subtree corresponding to $m_i$ as illustrated in Fig. 2. Further, for each $u_r \in U, 1 \leq r \leq q$, we take two cliques $\{R_r\} \cup \{A_i | u_r \in m_i\}$ and $\{R_r, X_r\}$ which are vertices of $T$ forming a subtree as shown in Fig. 2. Similarly, for each $v_s \in V, 1 \leq s \leq q$ and $w_t \in W, 1 \leq t \leq q$, we add the cliques $\{S_s\} \cup \{B_i | v_s \in m_i\}, \{S_s, Y_k\}$ and $\{T_t\} \cup \{C_i | w_t \in m_i\}, \{T_t, Z_t\}$, respectively to the tree $T$ as shown in Fig. 2. Finally, $\{A_i, B_i, C_i | 1 \leq i \leq p\}$ is the last vertex of tree $T$. The construction of $T$ is illustrated in Fig. 2.

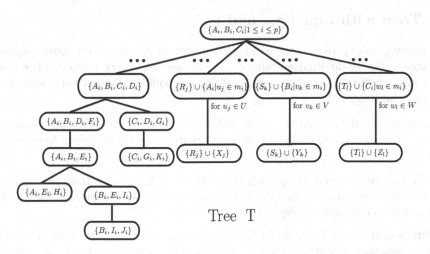

Tree $T$

**Fig. 2.** The clique tree of the undirected path graphs

Hence, $T$ is the clique tree of the undirected path graph $\mathcal{G}$ whose vertex set is $\{A_i, B_i, C_i, D_i, E_i, F_i, G_i, H_i, I_i, J_i, K_i | 1 \leq i \leq p\} \cup \{R_j, S_j, T_j, X_j, Y_j, Z_j : 1 \leq j \leq q\}$.

**Claim 7.** *The graph $\mathcal{G}$ has a ve-dominating set of size $2p + q$ if and only if 3-dimensional matching has a solution.*

*Proof.* Let $\mathcal{D}$ be a ve-dominating set of $\mathcal{G}$ of size $2p+q$. For any $i \in \{1, 2, \ldots, p\}$, the only way to ve-dominate the edge-set of the subgraph induced by the vertex set $\{A_i, B_i, C_i, D_i, E_i, F_i, G_i, H_i, I_i, J_i, K_i\}$ corresponding to $m_i$ with two vertices is to choose $D_i$ and $E_i$. Hence, to ve-dominate the edge-set of that induced subgraph by any larger vertex set, at least three vertices has to be taken. Note that, the set $\{A_i, B_i, C_i\}$ ve-dominates the edge-set of that induced subgraph. So, without loss of generality, we assume that $\mathcal{D}$ consists of $A_i, B_i, C_i$ for $t$ many $m_i$'s and $D_i, E_i$ for $(p-t)$ many other $m_i$'s. Also to ve-dominate the edges of the form $R_r X_r, S_s Y_s$ and $T_t Z_t$, $\mathcal{D}$ contains at least $max\{3(q-t), 0\}$ many vertices (namely, $R_j$ or $X_j$, $S_k$ or $Y_k$, $T_l$ or $Z_l$). Hence, we have,

$$2p + q = |\mathcal{D}| \geq 3t + 2(p-t) + 3(q-t) = 2p + 3q - 2t.$$

So, $t \geq q$. i.e. $\mathcal{D}$ must contain at least $q$ many $A_i, B_i, C_i$. Picking the corresponding $m_i$'s form a matching $M'$ of size $q$.

Conversely, let $M'$ be the solution of the 3-dimensional matching problem of size $q$. Then we can form the ve-dominating set $\mathcal{D}$ as $\mathcal{D} = \{A_i, B_i, C_i : m_i \in M'\} \cup \{D_i, E_i : m_i \notin M'\}$. Clearly, $\mathcal{D}$ is a ve-dominating set of $\mathcal{G}$ of size $2p + q$. $\square$

Hence, the ve-domination problem is NP-complete for undirected path graph. $\square$

## 4    Trees with Equal $\gamma_{ve}$ and $i_{ve}$

For every graph $G$, the independent ve-domination number is obviously at least as large as the ve-domination number. In this section, we characterize the trees for which these two parameters $\gamma_{ve}$ and $i_{ve}$ are equal. We start with some pertinent definitions.

**Definition 8.** *An atom $A$ is a tree with at least 3 vertices with a vertex, say $c$, designated as center of the atom such that distance of every vertex from $c$ is at most 2.*

We denote an atom along with its center by $(A, c)$. Note that the center $c$ ve-dominates all edges of the atom $A$. Next, we define an operation for joining two atoms to construct a bigger tree.

**Definition 9.** *Let $(A', c')$ and $(A, c)$ be two atoms along with their centers $c'$ and $c$, respectively. For some $i, j \in \{0, 1, 2\}$, we define $(i-j)$-join, denoted by $\otimes$, as the addition of an edge $(x_{c'}, x_c)$ between the vertices $x_{c'} \in V(A')$ and $x_c \in V(A)$ such that $dist_{A'}(c', x_{c'}) = i$ and $dist_A(c, x_c) = j$.*

With slight abuse of notation, $T' \bigotimes (A, c)$ denotes the tree obtained by $(i-j)$-join between two atoms $(A', c')$ and $(A, c)$, where $(A', c')$ is an atom in $T'$. Given a subset $S \subseteq V$, a vertex $v \in S$ has a *private edge* $e \in E$ with respect to the set $S$ if $v$ ve-dominates the edge $e$ and no other vertex in $S$ ve-dominates the edge $e$. An edge $e = xy \in E$ is called a *distance-1 private edge* of $v \in S$ with respect to the set $S$ if $e$ is a private edge of $v$ with respect to $S$ and $\min\{dist(v, x), dist(v, y)\} = 1$. Next, we give the recursive definition of a family of trees, say $\mathcal{T}$, using the notion of atom and $(i - j)$-join.

**Definition 10.** *The recursive definition of the family $\mathcal{T}$ of trees is as follows:*

1. *every atom $(A, c) \in \mathcal{T}$ and*
2. *Let $T' \in \mathcal{T}$ and $(A', c')$ be an atom in $T'$ and $S'$ be the set of all atom centers in $T'$. Then $T = T' \bigotimes (A, c) \in \mathcal{T}$ if one of the following cases hold*
   (i) *$\bigotimes$ is a $(0-1)$-join such that*
      (a) *$c'$ has a neighbour $y$ such that all edges incident to $y$, except $yc'$, are pendent edges. Also, $c$ has no distance-1 edge and at least two edges of $(A, c)$ is incident to $c$.*
      (b) *$c'$ has distance-1 private edges with respect to $S'$ and at least one distance-1 edge of $c$ is not incident to $x_c$.*
      (c) *$c'$ has no distance-1 private edge with respect to $S'$ and at least one distance-1 edge of $c$ is not incident to $x_c$.*
   (ii) *$\bigotimes$ is a $(1-0)$-join such that*
      (a) *$c$ has distance-1 edges and $c'$ has a neighbour $y$ such that all edges incident to $y$, except $c'y$, are pendent and private edges of $c'$ with respect to $S'$.*
      (b) *$c$ has distance-1 edges and $c'$ has a neighbour $y$ such that $y$ is a leaf vertex and $y \neq x_{c'}$.*
   (iii) *$\bigotimes$ is a $(1-1)$-join when at least one distance-1 private edge of $c'$ with respect to $S'$ is not incident to $x_{c'}$ and at least one distance-1 edge of $c$ is not incident to $x_c$.*
   (iv) *$\bigotimes$ is a $(2-1)$-join when all distance-1 private edges of $c'$ with respect to $S'$ are not incident to $y$, where $y \in N_{T'}(c') \cap N_{T'}(x_{c'})$, and at least one distance-1 edge of $c$ is not incident to $x_c$.*

Now, we show that if $T \in \mathcal{T}$, then *ve-domination number* and *independent ve-domination number* are same.

**Lemma 11.** *If $T \in \mathcal{T}$, then the set of all atom centers of $T$ forms a minimum ve-dominating set.*

*Proof.* We prove this by induction on the number of atoms in $T$. Clearly, when $T$ is an atom, the hypothesis is true. Let $T \in \mathcal{T}$ be a tree containing $k$ atoms and $T$ is obtained from $T' \in \mathcal{T}$ by joining the atom $(A, c)$ with an atom $(A', c')$ of $T'$ satisfying the joining rules. Let $S$ and $S'$ be the atom centers of $T$ and $T'$, respectively. Clearly, $S = S' \cup \{c\}$. By induction hypothesis, $S'$ is a $\gamma_{ve}$-set of $T'$. We show that $S$ is a $\gamma_{ve}$-set of $T$ for all the seven types of joining defined in Definition 10.

$(\mathbf{0-1})(\mathbf{a})$: In this type of joining, $(A,c)$ is star and $c'$ has a neighbour $y$ such that all edges incident to $y$, except $yc'$, are pendent edges. Since $S'$ is a ve-dominating set of $T'$, $S$ is obviously a ve-dominating set of $T$. If possible, let us assume that $S$ is not a $\gamma_{ve}$-set of $T$. Let $D$ be a $\gamma_{ve}$-set of $T$ such that $|D| < |S|$. Note that $D$ contains exactly one vertex from $(A,c)$, say $p$, to ve-dominate all edges of $(A,c)$. Also, to ve-dominate the pendent edges which are incident to $y$, $D$ must contain one vertex, say $q$, from $\{x,y,c'\}$, where $x$ is any leaf adjacent to $y$. Since the set of edges that are ve-dominated by $\{p,q\}$ can also be ve-dominated by $\{c,c'\}$, $D' = (D \setminus \{p,q\}) \cup \{c,c'\}$ is also a $\gamma_{ve}$-set of $T$. It is easy to see that $D' \setminus \{c\}$ is a $\gamma_{ve}$-set of $T'$. This contradicts the fact that $S'$ is a ve-dominating set of $T'$ of minimum cardinality. Hence, $S$ is a $\gamma_{ve}$-set of $T$.

$(\mathbf{0-1})(\mathbf{b})$: If possible, let us assume that $S$ is not a $\gamma_{ve}$-set of $T$. Let $D$ be a $\gamma_{ve}$-set of $T$ such that $|D| < |S|$. Clearly, $D$ contains exactly one vertex, say $p$, from $(A,c)$. Since the set of edges that are ve-dominated by $p$ can also be ve-dominated by $c$, $D' = (D \setminus \{p\}) \cup \{c\}$ is also a $\gamma_{ve}$-set of $T$. It is easy to see that $D' \setminus \{c\}$ is a $\gamma_{ve}$-set of $T'$. This is a contradiction. Hence, $S$ is a $\gamma_{ve}$-set of $T$.

$(\mathbf{0-1})(\mathbf{c})$: If possible, let us assume that $S$ is not a $\gamma_{ve}$-set of $T$. Let $D$ be a $\gamma_{ve}$-set of $T$ such that $|D| < |S|$. Clearly, $D$ contains exactly one vertex, say $p$, from $(A,c)$. Since the set of edges that are ve-dominated by $p$ can also be ve-dominated by $c$, $D' = (D \setminus \{p\}) \cup \{c\}$ is also a $\gamma_{ve}$-set of $T$. It is easy to see that $D' \setminus \{c\}$ is a $\gamma_{ve}$-set of $T'$. This is a contradiction. Hence, $S$ is a $\gamma_{ve}$-set of $T$.

$(\mathbf{1-0})(\mathbf{a})$: In this type of joining, $c'$ has a neighbour $y$ such that all edges incident to $y$, except $(yc')$, are pendent edges. If possible, let us assume that $S$ is not a $\gamma_{ve}$-set of $T$. Let $D$ be a $\gamma_{ve}$-set of $T$ such that $|D| < |S|$. Note that $D$ contains exactly one vertex from $(A,c)$, say $p$, to ve-dominate all edges of $(A,c)$. Suppose $x$ is a leaf node adjacent to $y$. So, to ve-dominate the pendent edge $xy$, $D$ must contain one vertex, say $q$, from $\{x,y,c'\}$. Since the set of edges that are ve-dominated by $\{p,q\}$ can also be ve-dominated by $\{c,c'\}$, $D' = (D \setminus \{p,q\}) \cup \{c,c'\}$ is also a $\gamma_{ve}$-set of $T$. It is easy to see that $D' \setminus \{c\}$ is a $\gamma_{ve}$-set of $T'$. This is a contradiction. Hence, $S$ is a $\gamma_{ve}$-set of $T$.

$(\mathbf{1-0})(\mathbf{b})$: In this case, $c'$ has a neighbour $y \neq x_{c'}$ such that $y$ is leaf vertex. If possible, let us assume that $S$ is not a $\gamma_{ve}$-set of $T$. Let $D$ be a $\gamma_{ve}$-set of $T$ such that $|D| < |S|$. Note that $D$ contains exactly one vertex from $(A,c)$, say $p$, to ve-dominate all edges of $(A,c)$. Since $p$ cannot ve-dominate the edge $c'y$, $D$ must contain a vertex, say $q$, to ve-dominate this edge. Note that $q \in N[c']$ because $y$ is leaf vertex and $q$ also ve-dominate the edge $c'x_{c'}$. Since the set of edges that are ve-dominated by $p$ can also be ve-dominated by $c$, $D' = (D \setminus \{p\}) \cup \{c\}$ is also a $\gamma_{ve}$-set of $T$. It is easy to see that $D' \setminus \{c\}$ is a $\gamma_{ve}$-set of $T'$. This is a contradiction. Hence, $S$ is a $\gamma_{ve}$-set of $T$.

$(\mathbf{1-1})\textbf{-join}$: Since $S'$ is a ve-dominating set of $T'$, $S$ is obviously a ve-dominating set of $T$. If possible, let us assume that $S$ is not a $\gamma_{ve}$-set of $T$. Let $D$ be a $\gamma_{ve}$-set of $T$ such that $|D| < |S|$. Clearly, $D$ contains exactly one vertex, say $p$, from $(A,c)$. Since the set of edges that are ve-dominated by $p$ can also be ve-dominated by $c$, $D' = (D \setminus \{p\}) \cup \{c\}$ is also a $\gamma_{ve}$-set of $T$. It is easy to see that $D' \setminus \{c\}$ is a $\gamma_{ve}$-set of $T'$. This is a contradiction. Hence, $S$ is a $\gamma_{ve}$-set of $T$.

$(2-1)$-**join:** Since $S'$ is a ve-dominating set of $T'$, $S$ is obviously a ve-dominating set of $T$. If possible, let us assume that $S$ is not a $\gamma_{ve}$-set of $T$. Let $D$ be a $\gamma_{ve}$-set of $T$ such that $|D| < |S|$. Clearly, $D$ contains exactly one vertex, say $p$, from $(A, c)$. Since the set of edges that are ve-dominated by $p$ can also be ve-dominated by $c$, $D' = (D \setminus \{p\}) \cup \{c\}$ is also a $\gamma_{ve}$-set of $T$. It is easy to see that $D' \setminus \{c\}$ is a $\gamma_{ve}$-set of $T'$. This is a contradiction. Hence, $S$ is a $\gamma_{ve}$-set of $T$. $\qquad \square$

**Theorem 12.** *For all $T \in \mathscr{T}$, $\gamma_{ve}(T) = i_{ve}(T)$.*

*Proof.* By the definition of $\mathscr{T}$, the distance between any two atom centers in $T$ is at least 2. Hence, the set of these atom centers, say $S$, forms an independent set. In Lemma 11, $S$ is also a $\gamma_{ve}$-set. Hence, $\gamma_{ve}(T) = i_{ve}(T)$ for all $T \in \mathscr{T}$. $\quad \square$

Next, we show that the converse of Theorem 12 is also true. For that, first we prove following lemmas that allow us to construct an *independent ve-dominating set* of a tree with some desirable properties.

**Lemma 13.** *For any tree $T$ $(n \geq 3)$, there exists an $i_{ve}$-set which does not contain any leaf.*

*Proof.* Let $S$ be an $i_{ve}$-set of $T$. If $S$ does not contain any leaf, then we are done. Otherwise assume that $S$ contains a leaf, say $x$. Let the neighbour of $x$ is $y$ and $N(y) = \{z_1, z_2, \ldots, z_q, x\}$. Since $S$ is independent set, $y \notin S$. Also, none of the $z_i$ are in $S$, because if any of $z_i \in S$, then $S \setminus \{x\}$ is also an independent ve-dominating set. Hence by replacing $x$ by $y$, we get another $i_{ve}$-set. Repeating this process, we can form an $i_{ve}$-set of $T$ which does not contain any leaf. $\quad \square$

**Lemma 14.** *Let $S$ be an $i_{ve}$-set of a rooted tree $T$ having depth $h$, which does not have any leaf. If the vertex $u \in S$ is at $(h-1)^{th}$-level and $v$ is the closest vertex to $u$ such that $v \in S$ and $dist(u, v) \geq 3$, then $S' = (S \setminus \{u\}) \cup \{w\}$ is also an $i_{ve}$-set without having any leaf, where $w$ is the parent of $u$ in $T$.*

*Proof.* It is easy to see that all the edges that are ve-dominated by $u$ can also be ve-dominated by $w$. So, $S'$ is $\gamma_{ve}$-set. Also, since the minimum distance between $u$ and $v$ is at least 3, $S'$ is independent set. Hence, $S'$ is an $i_{ve}$-set without having any leaf. $\qquad \square$

**Theorem 15.** *If $\gamma_{ve}(T) = i_{ve}(T)$ for a tree $T$ with $n \geq 3$, then $T \in \mathscr{T}$.*

For the proof of the theorem, please refer to the proof of Theorem 15 in [11]. Combining Theorems 12 and 15, we have the main result of this section.

**Theorem 16.** *For a tree $T$ with at least 3 vertices, $\gamma_{ve}(T) = i_{ve}(T)$ if and only if $T \in \mathscr{T}$.*

# 5    Conclusions

We proposed a linear time algorithm for ve-domination problem in block graphs and also pointed out that independent ve-domination problem can also be solved using similar technique. Further, we proved that finding minimum ve-dominating set is NP-complete for undirected path graphs. Finally, we characterized the trees for which $\gamma_{ve} = i_{ve}$. It would be interesting to study this problem in other subclasses like interval graphs, directed path graphs etc. Also characterization of graphs having equal ve-domination parameters is another interesting problem.

# References

1. Boutrig, R., Chellali, M.: Total vertex-edge domination. Int. J. Comput. Math. **95**(9), 1820–1828 (2018)
2. Boutrig, R., Chellali, M., Haynes, T.W., Hedetniemi, S.T.: Vertex-edge domination in graphs. Aequ. Math. **90**(2), 355–366 (2016)
3. Garey, M.R., Johnson, D.S.: Computers and Intractability: A Guide to the Theory of NP-Completeness. W. H. Freeman & Co., New York (1990)
4. Gavril, F.: A recognition algorithm for the intersection graphs of directed paths in directed trees. Discrete Math. **13**(3), 237–249 (1975)
5. Haynes, T., Hedetniemi, S., Slater, P.: Domination in Graphs: Advanced Topics. Marcel Dekker, New York (1998)
6. Haynes, T., Hedetniemi, S., Slater, P.: Fundamentals of Domination in Graphs. Marcel Dekker, New York (1998)
7. Krishnakumari, B., Chellali, M., Venkatakrishnan, Y.B.: Double vertex-edge domination. Discrete Math. Algorithms Appl. **09**(04), 1750045 (2017)
8. Krishnakumari, B., Venkatakrishnan, Y.B., Krzywkowski, M.: Bounds on the vertexedge domination number of a tree. C.R. Math. **352**(5), 363–366 (2014)
9. Lewis, J.: Vertex-edge and edge-vertex parameters in graphs. Ph.D. thesis, Clemson, SC, USA (2007)
10. Lewis, J.R., Hedetniemi, S.T., Haynes, T.W., Fricke, G.H.: Vertex-edge domination. Util. Math. **81**, 193–213 (2010)
11. Paul, S., Ranjan, K.: On vertex-edge and independent vertex-edge domination. CoRR, abs/1910.03635 (2019)
12. Peters Jr., K.W.: Theoretical and Algorithmic Results on Domination and Connectivity (Nordhaus-gaddum, Gallai Type Results, Max-min Relationships, Linear Time, Series-parallel). Ph.D. thesis, Clemson, SC, USA (1986)
13. Żyliński, P.: Vertex-edge domination in graphs. Aequ. Math. **93**(4), 735–742 (2019)

# The Balanced Connected Subgraph Problem: Complexity Results in Bounded-Degree and Bounded-Diameter Graphs

Benoit Darties, Rodolphe Giroudeau, König Jean-Claude,
and Valentin Pollet[(✉)]

LIRMM - CNRS UMR 5506, Montpellier, France
{benoit.darties,rgirou,konig,pollet}@lirmm.fr

**Abstract.** We present new complexity results for the Balanced Connected Subgraph (BCS) problem. Given a graph whose vertices are colored either blue or red, find the largest connected subgraph containing as many red vertices as blue vertices. We establish the NP-completeness of the decision variant of this problem in bounded-diameter and bounded-degree graphs: bipartite graphs of diameter four, graphs of diameter three and bipartite cubic graphs. BCS being polynomially solvable in graphs of diameter two and maximum degree two, our results close some of the existing gaps in the complexity landscape.

**Keywords:** Complexity · Bounded diameter · Bounded degree

## 1 Introduction

Many well-studied combinatorial optimization problems consist in finding induced subgraphs with a given property. For instance, finding a maximum clique or a maximum independent set are one of the 21 NP-complete problems classified by Karp [10]. Garey and Johnson [7] describe a general version of these problems (GT21-22): maximum induced (connected) subgraph with property $\Pi$. If $\Pi$ is hereditary and non-trivial then the problem is NP-complete and some approximability results hold.

In this article, we investigate the Balanced Connected Subgraph problem (BCS) as introduced by Bhore et al. [3]. Given a 2-colored graph (using colors red and blue), find the largest connected subgraph containing as many vertices of each color. Notice that the property of being color-balanced is far from being hereditary, hence the need for an *ad-hoc* study.

BALANCED CONNECTED SUBGRAPH (BCS)
**Input:** $G = (V, E)$ a 2-colored graph, $k \in \mathbb{N}$
**Question:** Does $G$ have a connected subgraph of size at least $2k$ containing as many vertices of each color ?

## 1.1  Related Work

Bhore et al. [3] showed that BCS remains NP-complete in bipartite graphs, chordal graphs and planar graphs. They also gave polynomial algorithms solving BCS in trees, splits, graphs of diameter 2 and properly colored bipartite graphs.

As they point out, BCS is strongly related to the Maximum Weight Connected Subgraph (MWCS) problem mentioned by Johnson [9] in one of his columns. Note that BCS is neither a special case nor a generalization of MWCS. In MWCS, the goal is to find a connected subgraph of maximum weight. If BCS were to be formulated as a MWCS with weights say +1 for red vertices and −1 for blue vertices, we would search for the largest subgraph of weight exactly 0. Exact approaches, like the one proposed by Álvarez-Miranda et al. [1], for solving MWCS hence cannot be used to solve BCS directly.

BCS is also related to the Steiner Tree problem. In fact, assume that you are given a graph $G$ along with a 2-coloration (using colors red and blue) of its vertices (less red vertices than blue vertices). Asking whether there exists a BCS in $G$ containing all the red vertices can be seen as a special case of Steiner Tree: the red vertices are the terminals, and we search for a Steiner Tree of size roughly twice the number of terminals. In that case, one can determine the existence of a BCS containing all the red vertices by using efficient exact algorithms for Steiner.

The Graph Motif (GM) problem is tied to BCS as well. GM consists, given a colored graph $G$ and a multiset of colors $M$, in finding a connected subgraph such that the multiset of colors assigned to its vertices is exactly $M$. Finding a balanced connected subgraph of size at least $2k$ can be reduced to a polynomial number of motif searches in a 2-colored graph: all one has to do is to search for the motif {red,...,red, blue,..., blue} with $k$ occurrences of red and blue, then $k + 1$ occurrences of each, and so on, upon either finding a balanced connected subgraph or having proved that none exists.

GM was first introduced by Lacroix et al. [11] in the context of metabolic networks. They showed that GM is NP-complete even if the input graph is restricted to be a tree. Fellows et al. [5,6] further proved that GM remains NP-complete in trees of maximum degree 3, and even if the input graph is a 2-colored bipartite graph of maximum degree 4. As a positive result, they gave an FPT algorithm for GM parameterized by the size of the motif in the general case. Since BCS can be solved by solving a polynomial number of instances of GM, using their FPT algorithm to do so would result in an FPT algorithm for BCS parameterized by the size of the solution.

Dondi et al. [4] described an optimization variant of GM: find the largest connected subgraph which multiset of colors is included in the given motif. Dondi et al. [4] proved this variant to be APX-hard in trees of maximum degree 3.

Bhore et al. [3] give a rather comprehensive list of applications for MWCS, and claim that BCS may be better-suited in some cases. While we do not motivate BCS with further practical applications, we believe it may prove to be useful in network design applications (where the colors represent roles assigned to the nodes), social data-mining (colors represent classes of individuals), or even electoral applications (see [2] for a study of *gerrymandering* as a graph partitioning problem with a red-blue colored graph as input).

On the other hand, BCS appears of interest in a purely theoretical point of view. The problem is quite hard complexity-wise, and can be generalized in a lot of different ways. For instance, one can loosen the "balanced" constraint and ask for a connected subgraph minimizing the ratio between the number of red vertices and blue vertices. Other generalizations would be increasing the number of colors in the input coloration, enforcing the subgraph being looked for to have additional properties (being a path, tree, 2-connected ...), or coloring the edges instead of the vertices.

## 1.2 Our Contribution

We study the computational complexity of BCS in three restricted cases, namely bipartite graphs of diameter 4, graphs of diameter 3 and bipartite cubic graphs. In each case, we establish the NP-completeness of BCS by polynomial-time reduction from well-known problems.

**Organization of the Paper.** Section 2 recaps the notations and problems that will be referred to in the rest of the paper. Section 3 is dedicated to NP-completeness proofs in bounded-diameter graphs: bipartite graphs of diameter 4 and graphs of diameter 3. Section 4 focuses on bipartite cubic graphs. Perspectives are given in Sect. 5.

## 2 Pre-requisites

In this paper, the graphs we consider are undirected and connected. For a graph $G$, we denote $\Delta(G)$ (or simply $\Delta$ if there is no ambiguity) the maximum degree of $G$ and $D(G)$ (or simply $D$) its diameter. For any $X \subset V(G)$ (resp. $T \subset E(G)$), we denote $G[X]$ (resp. $G[T]$) the subgraph of $G$ induced by $X$ (resp. $T$).

We assume that a 2-coloration of $G$ uses colors red and blue, and we denote $V_r$ (resp. $V_b$) the set of vertices receiving color red (resp. blue). Without loss of generality, we assume that $|V_r| \leq |V_b|$.

In the rest of the paper, several well-known problems are referred to. The following is their statement along with a reference for their NP-completeness.

**Dominating Set (DS):** Given a graph $G$ and an integer $k$, find $D \subset V(G)$ s.t. $|D| \leq k$ and for all $x \in V(G)$, either $x \in D$ or $x \in N(y)$ for some $y \in D$ [10].

**Exact Cover by 3-sets (X3C):** Given a universe $X$ and a collection $C$ of triples of $X$, find $C' \subset C$ s.t. each element of $X$ appears in exactly one triple of $C'$ [7].

**Exact Cover by 3-sets with occurrence 3 (X3C3):** Same as above, but the each element in $X$ belongs to exactly three triples of $C$ [8].

# 3   Bounded-Diameter Graphs

## 3.1   Bipartite Graphs of Diameter 4

In the following, we prove that BCS remains NP-complete in graphs of diameter 4. The reduction is based on Dominating Set in graphs of diameter 2, which is NP-complete [12]. The following construction transforms any graph of diameter 2 into a 2-colored bipartite graph of diameter 4.

**Construction 1.** *Let $G = (V, E)$ be a graph on $n$ vertices and $k \in \mathbb{N}$. We build $G' = (V', E')$ an instance of BCS as follows:*

- *add $2n$ blue vertices $V_1 = \{v_1^1, \ldots, v_n^1\}$ and $V_2 = \{v_1^2, \ldots, v_n^2\}$;*
- *add $n + k$ red vertices $Q = \{q_1, \ldots, q_n\}$ and $P = \{p_1, \ldots, p_k\}$;*
- *for all $i \in [n]$, add the edge $v_i^2 q_i$;*
- *for all $i \in [n]$, for all $w \in N[v]$ add the edge $v_i^1 w^2$;*
- *for all $i \in [n]$, for all $j \in [k]$ add the edge $v_i^1 p_j$.*

Construction 1 is clearly done in polynomial time. If the base graph $G$ has diameter 2, then the graph $G'$ obtained after transformation has diameter 4. Indeed, each couple $(x, y) \in V_1 \times V_1$ or $V_2 \times V_2$ can be connected by a path of length 2 between $x$ and $y$. It follows that each couple $(x, y) \in V_1 \times V_2$ can be connected by a path of length at most 3. Finally, since all the red vertices have a neighbor in $V_1$ or $V_2$, and $G'[P \cup V_1]$ is a complete bipartite graph, any pair of red vertices can be connected by a path of length at most 4.

**Theorem 1.** BALANCED CONNECTED SUBGRAPH *remains* NP-*complete on bipartite graphs of diameter* 4.

*Proof.* Let $G$ be a graph of diameter 2 and $k \in \mathbb{N}$. Let $G'$ be the graph obtained from $G$ using Construction 1. Let us prove that $G$ contains a dominating set of size $k$ if and only if $G'$ has a BCS of size $2(n + k)$.

⇒ if $G$ contains a dominating set $D$ of size $k$, then let $D_1 \subset V'$ be the vertices of $V_1$ in $G'$ corresponding to $D$. $B = \{v \in V' : v$ is red $\} \cup V_2 \cup D_1$ contains $n + k$ red vertices and $n + k$ blue vertices. Since $D$ is a dominating set in $G$, every vertex in $V_2$ is connected to a vertex of $D_1$. $G'[B]$ is thus connected and balanced.

⇐ if $G'$ has a BALANCED CONNECTED SUBGRAPH $S$ of size $2(n+k)$, then it has to include all the red vertices. For the pendant red vertices to be connected in $S$, $S$ must include $V_2$. Since $V_2$ is of size $n$, $S$ contains exactly $k$ other blue vertices, and those vertices belong to $V_1$. Moreover, $V_2$ being an independent set and $S$ being connected, every vertex of $V_2$ must be connected to at least one vertex in $V_1 \cap S$. The vertices of $G$ corresponding to $V_1 \cap S$ in $G'$ thus form a dominating set of size $k$.

Since DOMINATING SET is NP-complete in graphs of diameter 2 [12], and BCS being in NP, the discussion above proves the theorem.     □

## 3.2   Graphs of Diameter Three

To prove that BALANCED CONNECTED SUBGRAPH remains NP-hard in graphs of diameter three, we design a reduction from COLORFUL CONNECTED SUBGRAPH which is stated as follows.

COLORFUL CONNECTED SUBGRAPH (CCS)
**Input:** $G = (V, E)$ a $p$-colored graph, $k \in \mathbb{N}$ s.t. $k \leq p$.
**Question:** Does $G$ contain a connected subgraph of size at least $k$ which each color appears at most once in ?

To the best of our knowledge, the complexity of CCS – as stated above – is not clearly established. To show that CCS is NP-complete, we use a result given by Fellows et al. [5] on the Graph Motif problem. Recall that GM consists, given a colored graph $G = (V, E)$ and a multiset of colors $\mathcal{M}$, in finding a connected subgraph of $G$ which multiset of colors is exactly $\mathcal{M}$.

Fellows et al. [5] show in their Theorem 1 that GM remains NP-hard even if the motif $\mathcal{M}$ is colorful, that is each color appears at most once in $\mathcal{M}$. In their reduction, the instance of GM they obtain is such that the motif $\mathcal{M}$ is exactly the set of all colors. Therefore, GM remains NP-complete even if $\mathcal{M}$ is the set containing each color once. Now, one can observe that COLORFUL CONNECTED SUBGRAPH is NP-hard because in case $p$ is equal to $k$, then CCS is equivalent to GM in the aforementioned case. The following lemma holds, since COLORFUL CONNECTED SUBGRAPH is clearly in NP.

**Lemma 1.** COLORFUL CONNECTED SUBGRAPH *is* NP-*complete*.

We now reduce COLORFUL CONNECTED SUBGRAPH to BALANCED CONNECTED SUBGRAPH in graphs of diameter 3. The idea of the construction is to create a clique containing one red vertex and "a lot" of blue ones for each color. The cliques are then interconnected, making sure that any pair of cliques is connected by at least one edge.

**Construction 2.** *Let* $G = (V, E)$ *be a graph and* $c : V \to \{1, \ldots, p\}$ *a $p$-coloring of its vertices. We build* $G' = (V', E')$ *a 2-colored graph in the following way (refer to Fig. 1).*

- $V' = C_1 \cup C_2 \cdots \cup C_p$ *with* $C_i = \{x \in V : c(x) = i\} \cup \{r_i\}$.
  - *For all* $i \in \{1, \ldots, p\}$, $G'[C_i]$ *is connected as a clique, and each one of its vertices, except* $r_i$, *is blue* .
- *For all* $uv \in E$, *add the corresponding edge to* $G'$.
- *For all* $C_i, C_j$ *such that there is no edge between* $C_i$ *and* $C_j$, *add a blue vertex* $x_i^j$ *to* $C_i$ *(connected to every vertex in* $C_i$*) and a blue vertex* $x_j^i$ *to* $C_j$ *(connected to every vertex in* $C_j$*), as well as the edge* $x_i^j x_j^i$.

Construction 2 can be applied in polynomial time. The resulting graph has diameter three since it is composed of pairwise connected cliques.

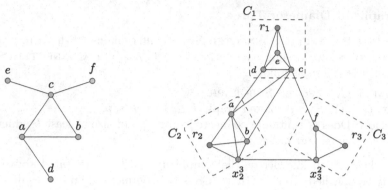

(a) 3-colored graph.        (b) Graph obtained after applying construction 2.

**Fig. 1.** Applying Construction 2 to a 3-colored graph. The clique $C_1$ corresponds the green vertices $c, d$ and $e$, the clique $C_2$ to the pink vertices $a$ and $b$, and the clique $C_3$ to the orange vertex $f$. Vertices $x_2^3$ and $x_3^2$ originate from the absence of a yellow-pink edge and ensure $r_2$ and $r_3$ are connected by a path of length 3. (Color figure online)

**Theorem 2.** BALANCED CONNECTED SUBGRAPH *is NP-complete in graphs of diameter three.*

*Proof.* Let $G = (V, E)$ be a graph and $c : V \rightarrow \{1, \ldots, p\}$ a $p$-coloring of its vertices. Let $G' = (V', E')$ be the 2-colored graph obtained by applying Construction 2 to $G$. Let $k \in \mathbb{N}$, $k \geq 3$. We claim that $G$ has a CCS of size at least $k$ if and only if $G'$ has a BCS of size at least $2k$.

$\Rightarrow$ If $G$ has a CCS, $S$, of size at least $k$, then the corresponding vertices in $G'$ are blue and induce a connected subgraph intersecting each clique at most once. For each clique that $S$ intersects, just add said clique's red vertex to $S$. Doing so, we build a balanced connected subgraph in $G'$ of size at least $2k$.

$\Leftarrow$ If $G'$ has a BCS, $S$, of size at least $2k$ then its contains at least $k$ red vertices belonging to at least $k$ different cliques. Since the neighborhoods of red vertices are pairwise disjoint, each red vertex must have exactly one blue neighbor (in its clique) belonging to $S$. Assume that a vertex $x_i^j$ belongs to $S$, then it is the sole neighbor of $r_i$. In order to connect $r_i$ to other red vertices, $x_j^i$ has to belong to $S$. Now since $x_j^i$ is assumed to belong to $S$, $r_j$ cannot have any other blue neighbor in $S$. Under those assumptions, $S$ cannot be of size greater than 4 which absurd because we assumed $k \geq 3$. Therefore, vertices of type $x_i^j$ cannot belong to $S$, and every blue vertex in $S$ corresponds to a vertex in $G$. Since the red vertices have degree 1 in $S$, removing them from $S$ does not break the connectivity and thus the set of blue vertices in $S$ corresponds to a connected subgraph in $G$. Since $S$ contains at most one blue vertex per clique, the set of blue vertices of $S$ is a CCS of size at least $k$ in $G$.

The previous discussion concludes the polynomial-time reduction from COL- ORFUL CONNECTED SUBGRAPH to BALANCED CONNECTED SUBGRAPH. Since the instances of BALANCED CONNECTED SUBGRAPH obtained through Construction 2 have diameter three, the theorem holds.    □

## 4  Bounded-Degree Graphs

We prove BCS to be NP-complete in cubic bipartite graphs by reduction from Exact Cover by 3-Sets with occurrence 3 (X3C3). The construction consists in encoding each set by a blue subgraph (a set-gadget) and the elements by red vertices. The graph is then completed by making sure there are less red vertices than blue vertices. In the end, the starting X3C3 instance is positive if and only if there is a BCS containing all the red vertices in the constructed graph.

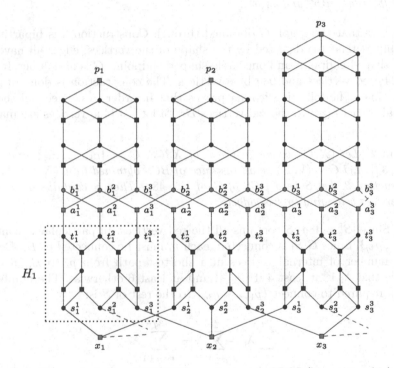

**Fig. 2.** Encoding an instance of X3C3 into an instance of BCS (bipartite cubic). From bottom to top – $3q$ red element-vertices; $3q$ blue set-gadgets on 14 vertices; $3q$ red connectivity-gadgets on 6 vertices connected as an accordion; $3q$ chains of 2 to 3 blue 2-regular bipartite graphs on 6 vertices; and $3q$ red terminal vertices. (Color figure online)

**Construction 3.** *Let* $(X, C)$ *be an instance of X3C3,* $C = \{c_1, \ldots, c_{3q}\}$ *and* $X = \{x_1, \ldots, x_{3q}\}$, *with* $q$ *even. Let* $G = (V, E)$ *be an instance of BCS obtained from* $(X, C)$ *as follows (see Fig. 2).*

- *For each $x_i \in X$ add a vertex $x_i$ (an element-vertex);*
- *for each $c_i \in C$, $c_i = \{x_j, x_k, x_l\}$:*
  - *add a blue gadget $H_i$ on 14 vertices. Denote $t_i^1$, $t_i^2$ and $t_i^3$ (resp. $s_i^1$, $s_i^2$ and $s_i^3$) the vertices of degree 2 at the top (resp. bottom) of $H_i$;*
  - *add the edges $s_i^1 x_j$, $s_i^2 x_k$ and $s_i^3 x_l$;*
  - *add 6 red vertices partitionned into $A_i = \{a_i^1, a_i^2, a_i^3\}$ and $B_i = \{b_i^1, b_i^2, b_i^3\}$. Add the edges $a_i^1 t_i^1$, $a_i^2 t_i^2$, $a_i^3 t_i^3$ $a_i^1 b_i^2$, $a_i^3 b_i^2$, $a_i^2 b_i^1$ and $a_i^2 b_i^3$;*
  - *add 2 (if $i \le \frac{q}{2}$) or 3 (elsewise) 2-regular bipartite graphs on 6 blue vertices connected by matchings. Connect the first of these graphs to $B_i$ by a matching. Enforce 3-regularity by adding a red terminal vertex $p_i$ connected to the remaining 3 vertices of degree 2;*
- *for each $i \in \{1, \dots, 3q - 1\}$, add the edges $a_i^3 b_{i+1}^1$ and $b_i^3 a_{i+1}^1$;*
- *add the edges $a_1^1 b_1^1$ and $a_{3q}^3 b_{3q}^3$.*

Observe that the graph $G$ obtained through Construction 3 is bipartite: in Fig. 2 bipartiteness is depicted by the shape of the vertices, edges all have one square-shaped endpoint and one circle-shaped endpoint. $G$ is also cubic. It contains $24q$ red vertices and $93q$ blue vertices. The construction is done in polynomial time. The following lemma proves that in order to connect all the red terminal vertices ($p_i$) to the red vertices $B_i$, "a lot" of blue vertices are mandatory.

**Lemma 2.** *Let $(X, C)$ be an instance of X3C3, $C = \{c_1, \dots, c_{3q}\}$ and $X = \{1, \dots, 3q\}$. Let $G = (V, E)$ be an instance of BCS obtained from $(X, C)$ through Construction 3. Let $S$ be a BCS of $G$ of size $48q$. Then $S$ contains at most $7q$ blue vertices belonging to set-gadgets.*

*Proof.* Since $|S| = 48q$, $S$ contains all the red vertices. In particular $S$ contains $\{p_1, \dots, p_{3q}\}$. Since $G[S]$ is connected, each $p_i$ must be connected to $B_i$. Denote $\lambda_i$ the number of internal vertices in a shortest path from $p_i$ to $B_i$ in $G[S]$. Observe that $\lambda_i$ is at least 4 if $i \le \frac{q}{2}$, and at least 6 otherwise. The number of vertices required to connect $\{p_1, \dots, p_{3q}\}$ to the rest of $S$ is:

$$\sum_{i=1}^{3q} \lambda_i = \sum_{i=1}^{\frac{q}{2}} \lambda_i + \sum_{i=\frac{q}{2}+1}^{3q} \lambda_i \tag{1}$$

$$\ge 4 \times \frac{q}{2} + 6 \times \frac{5q}{2} \tag{2}$$

$$\ge 17q \tag{3}$$

Since $S$ contains $24q$ blue vertices and at least $17q$ form paths connecting $\{p_1, \dots, p_{3q}\}$ to $\{B_1, \dots, B_{3q}\}$, the number of vertices belonging to set-gadgets cannot exceed $7q$ thus the lemma holds. $\qquad\square$

We now prove the existence and unicity of a solution to an integer linear program that will describe how "expensive" (in the number of blue vertices) it is to connect all the element-vertices to the other red vertices.

**Lemma 3.** *Let $q \in \mathbb{N}$ and $(L)$ be the following integer linear program:*

$$min \quad 7u_1 + 5u_2 + 3u_3$$
$$s.t.$$
$$3u_1 + 2u_2 + u_3 = 3q \quad (1)$$
$$u_1, u_2, u_3 \in \mathbb{N}$$

$u_2 = u_3 = 0$ *and* $u_1 = q$ *is the unique optimal solution to* $(L)$.

*Proof.* Let $(u_1, u_2, u_3)$ be an optimal solution to $(L)$.

- If $u_2 > 0$, then $u_3 = 0$ because if $u_3 > 0$ then $(u_1 + 1, u_2 - 1, u_3 - 1$ has better cost which is absurd. Assuming $u_3 = 0$, the constraint (1) becomes $3u_1 + 2u_2 = 3q$ which implies $u_2 \geq 3$ since $u_2 > 0$. Now $(u_1 + 2, u_2 - 3, u_3)$ has lower cost, this is absurd therefore $\underline{u_2 = 0}$.
- Assume $u_3 > 0$. With $u_2 = 0$, the constraint (1) becomes $3u_1 + u_3 = 3q$ which implies $u_3 \geq 3$. Now $(u_1 + 1, u_2, u_3 - 3)$ has lower cost, this is absurd therefore $\underline{u_2 = 0}$.

We now have $u_2 = u_3 = 0$ and necessarily $u_1 = q$ to satisfy (1) and the lemma holds.                                                                    □

**Theorem 3.** *BCS is NP-complete in bipartite (sub)cubic graphs.*

*Proof.* Let $(X, C)$ be an instance of X3C3 and $G = (V, E)$ the graph obtained from $(X, C)$ through Construction 3. Let us prove that $(X, C)$ is positive if and only if $G$ contains a BCS of size $48q$.

$\Leftarrow$ If $G$ contains a BCS, $S$, of size $48q$ we claim that $S$ contains exactly $q$ paths of length 7, spread across $q$ distinct set gadgets. To prove this, we reason on how the set-gadgets are used to connect elements. Denote $R \subset V$ the set of red vertices in $G$, and $R^\uparrow \subset R$ the red vertices that are not element-vertices.

Since $|S| = 48q$, $S$ contains all the red vertices. In particular, $S$ contains all the element-vertices. For an element $x$ to be connected to $R^\uparrow$, $S$ must contain a path from $x$ to $R^\uparrow$ and that path goes through vertices of set-gadgets. Either the vertices of this path belong to a unique set-gadget, or they belong to several. In case the path spreads across several set-gadgets, it must go through other element-vertices, one of them being connected to $R^\uparrow$ by a path belonging to a unique set-gadget.

Now, see Fig. 3. We classify the gadgets in 3 types, depending on how they connect elements in $S$. We abuse notation by saying that $S$ contains a gadget when we actually mean that $S$ contains some vertices of said gadget.

- Type-1 gadgets connect 3 elements to $R^\uparrow$. At least 7 vertices of those gadgets must belong to $S$.
- Type-2 gadgets connect 2 elements. Here, we have three cases:
  - the two elements are connected to $R^\uparrow$ by a structure reaching one of the top vertices of the gadget (see Fig. 3b and d);

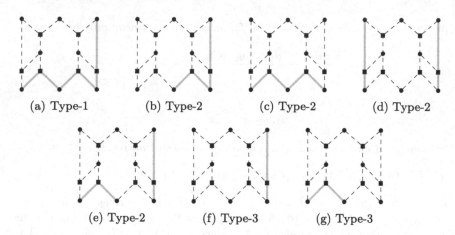

(a) Type-1          (b) Type-2          (c) Type-2          (d) Type-2

(e) Type-2          (f) Type-3          (g) Type-3

**Fig. 3.** Different types of gadget usage in a BCS. Type-1 gadgets connect 3 element to the top using at least 7 vertices. Type-2 gadgets connect 2 elements to the top using at least 5 vertices. Type-3 gadgets connect 1 element to the top using at least 3 vertices.

- one of them directly to the top and the other to another element (Fig. 3e)
- the two elements are connected to another element (Fig. 3c)

In all cases, $S$ contains at least 5 vertices per type-2 gadget.
- Type-3 gadgets connect 1 element. Either directly to $R^\uparrow$ or to another element somehow connected. In both cases, $S$ contains at least 3 vertices per type-2 gadget.

Denote $u_1$ the number of type-1 gadgets in $S$, $u_2$ the number of type-2 gadgets and $u_3$ the number of type-3 gadgets. By construction, $3u_1 + 2u_2 + u_3 = 3q$. In addition, the number of blue vertices in $S$ belonging to set-gadgets is at least $7u_1 + 5u_2 + 3u_3$. $(u_1, u_2, u_3)$ is thus a solution to the linear program $(L)$ described in Lemma 3 and $7u_1 + 5u_2 + 3u_3$ is the objective function of $(L)$. Since $(L)$ has a unique optimal solution costing $7q$, $(u_1, u_2, u_3)$ must be optimal because otherwise $7u_1 + 5u_2 + 3u_3 > 7q$ which is absurd by Lemma 2. We thus have $u_1 = q$ and $u_2 = u_3 = 0$ i.e. $S$ contains exactly $q$ type-1 gadgets, and for each of those gadgets, exactly 7 blue vertices belong to $S$. Since each type-1 gadget connects 3 red vertices, the set of type-1 gadgets in $S$ corresponds to an exact cover in $(X, C)$, therefore if $G$ contains a BCS of size $48q$ then $(X, C)$ is positive.

$\Rightarrow$ If $(X, C)$ is positive, then there exists $q$ sets in $C$ covering $X$ exactly. Take all the red vertices in $G$, and for each set in the exact cover pick a path of length 7 in the corresponding set gadget (see Fig. 3a). Connect the red accordion to the sinks $p_i$ using exactly $17q$ blue vertices. The structure thus obtained is a BCS of size $48q$ in $G$.

BCS belonging to NP, and X3C3 being NP-complete, the discussion above proves the NP-completeness in cubic bipartite graphs. Observe that some edges incident to the terminal vertices $p_i$ can be removed, and these deletions do not impact the reduction, therefore the problem remains NP-complete on subcubic graphs, the theorem thus holds.                                                                              □

# 5 Conclusion

In this paper, we improved the complexity results for BCS. We gave a proof of NP-completeness in bipartite cubic graphs, graphs of diameter 3 and bipartite graphs of diameter 4. Our results nicely complement the ones of Bhore et al. [3]. Indeed, they proved BCS to be polynomially solvable in graphs of diameter 2 and graphs of maximum degree 2.

Despite remaining computationally difficult in restrictive settings, BCS parameterized by the size of the solution belongs to FPT. Indeed, as we stated in the introduction, BCS can be Turing-reduced to the graph motif problem. Since this problem parameterized by the size of the motif is FPT, it implies BCS parameterized by the size of the solution is FPT as well. On the negative side, our results in graphs of bounded degree and bounded diameter imply that BCS parameterized by the diameter or the maximum degree of the input graph is in fact not FPT.

A short-term perspective of ours would be determining the complexity in the case of bipartite graphs of diameter 3 to achieve a complete dichotomy between polynomiality and NP-completeness in bipartite graphs.

Longer-term perspectives include studying other restrictive settings (block graphs, interval graphs) and several variants of the problem: weighted BCS; $k$-colored graph as input; finding balanced 2-connected subgraphs; and so on.

# References

1. Álvarez-Miranda, E., Ljubić, I., Mutzel, P.: The maximum weight connected subgraph problem. In: Jünger, M., Reinelt, G. (eds.) Facets of Combinatorial Optimization, pp. 245–270. Springer, Heidelberg (2013). https://doi.org/10.1007/978-3-642-38189-8_11
2. Apollonio, N., Becker, R., Lari, I., Ricca, F., Simeone, B.: Bicolored graph partitioning, or: gerrymandering at its worst. Discrete Appl. Math. **157**(17), 3601–3614 (2009)
3. Bhore, S., Chakraborty, S., Jana, S., Mitchell, J.S.B., Pandit, S., Roy, S.: The balanced connected subgraph problem. In: Pal, S.P., Vijayakumar, A. (eds.) CALDAM 2019. LNCS, vol. 11394, pp. 201–215. Springer, Cham (2019). https://doi.org/10.1007/978-3-030-11509-8_17
4. Dondi, R., Fertin, G., Vialette, S.: Maximum motif problem in vertex-colored graphs. In: Kucherov, G., Ukkonen, E. (eds.) CPM 2009. LNCS, vol. 5577, pp. 221–235. Springer, Heidelberg (2009). https://doi.org/10.1007/978-3-642-02441-2_20
5. Fellows, M.R., Fertin, G., Hermelin, D., Vialette, S.: Sharp tractability borderlines for finding connected motifs in vertex-colored graphs. In: Arge, L., Cachin, C., Jurdziński, T., Tarlecki, A. (eds.) ICALP 2007. LNCS, vol. 4596, pp. 340–351. Springer, Heidelberg (2007). https://doi.org/10.1007/978-3-540-73420-8_31

6. Fellows, M.R., Fertin, G., Hermelin, D., Vialette, S.: Upper and lower bounds for finding connected motifs in vertex-colored graphs. J. Comput. Syst. Sci. **77**(4), 799–811 (2011)
7. Garey, M.R., Johnson, D.S.: Computers and Intractability; A Guide to the Theory of NP-Completeness. W. H. Freeman & Co., New York (1990)
8. Gonzalez, T.F.: Clustering to minimize the maximum intercluster distance. Theoret. Comput. Sci. **38**, 293–306 (1985)
9. Johnson, D.S.: The np-completeness column: an ongoing guide. J. Algorithms **6**(1), 145–159 (1985)
10. Karp, R.M.: Reducibility among combinatorial problems. In: Proceedings of a Symposium on the Complexity of Computer Computations, IBM Thomas J. Watson Research Center, pp. 85–103 (1972)
11. Lacroix, V., Fernandes, C.G., Sagot, M.F.: Motif search in graphs: application to metabolic networks. IEEE/ACM Trans. Comput. Biol. Bioinform. **3**(4), 360–368 (2006)
12. Lokshtanov, D., Misra, N., Philip, G., Ramanujan, M.S., Saurabh, S.: Hardness of $r$-DOMINATING SET on graphs of diameter (r+1). In: Gutin, G., Szeider, S. (eds.) IPEC 2013. LNCS, vol. 8246, pp. 255–267. Springer, Cham (2013). https://doi.org/10.1007/978-3-319-03898-8_22

# Card-Based Secure Ranking Computations

Ken Takashima[1]($\boxtimes$) (iD), Yuta Abe[1], Tatsuya Sasaki[1], Daiki Miyahara[1,4] (iD),
Kazumasa Shinagawa[2,4] (iD), Takaaki Mizuki[3] (iD), and Hideaki Sone[3]

[1] Graduate School of Information Sciences, Tohoku University,
6–3–09 Aramaki-Aza-Aoba, Aoba-ku, Sendai 980–8578, Japan
`ken.takashima.q4@dc.tohoku.ac.jp`
[2] Graduate School of Information Sciences and Engineering,
Tokyo Institute of Technology, 2–12–1 Ookayama, Megro, Tokyo 152–8552, Japan
[3] Cyberscience Center, Tohoku University,
6–3 Aramaki-Aza-Aoba, Aoba-ku, Sendai 980–8578, Japan
`mizuki+lncs@tohoku.ac.jp`
[4] National Institute of Advanced Industrial Science and Technology,
2–3–26, Aomi, Koto-ku, Tokyo 135-0064, Japan

**Abstract.** Consider a group of people who want to know the "rich list" among them, namely the ranking in terms of their total assets, without revealing any information about the actual value of their assets. This can be achieved by a "secure ranking computation," which was first considered by Jiang and Gong (CT-RSA 2006); they constructed a secure ranking computation protocol based on a public-key cryptosystem. In this paper, instead of using a public-key cryptosystem, we use a deck of physical cards to provide secure ranking computation protocols. Therefore, our card-based protocols do not rely on computers, and they are simple and easy for humans to implement. Specifically, we design four protocols considering tradeoffs between the number of cards and the number of shuffles required to execute protocols. We also present a guide to choose an appropriate protocol according to the number of people participating in the protocol and the size of the input range.

**Keywords:** Card-based protocols · Secure ranking computation · Secure multiparty computations · Millionaire problem · Deck of cards

## 1 Introduction

Assume that there are $n$ players $P_1, \ldots, P_n$ such that each player $P_i$, $1 \le i \le n$, holds $x_i$ dollars with $1 \le x_i \le m$. They want to know the "rich list" $(r_1, \ldots, r_n)$ among them, namely the ranking in terms of their total assets, without revealing any information about the actual values of $x_1, \ldots, x_n$. More formally, they want to securely compute the *ranking function* rk : $\{1, 2, \ldots, m\}^n \to \{1, 2, \ldots, n\}^n$ defined as

$$\text{rk} : (x_1, \ldots, x_n) \mapsto (r_1, \ldots, r_n)$$

Y. Li et al. (Eds.): COCOA 2019, LNCS 11949, pp. 461–472, 2019.
https://doi.org/10.1007/978-3-030-36412-0_37

such that $r_i = 1 + \left| \{x_j \mid x_j > x_i, 1 \leq j \leq n\} \right|$ for every $i, 1 \leq i \leq n$,

where $|X|$ denotes the cardinality of a set $X$. For example,

$$\mathsf{rk}(3, 1, 3, 5) = (2, 4, 2, 1),$$

which means that $P_4$ is the richest, $P_2$ is the poorest, and both $P_1$ and $P_3$ are tied; all the players want to know only the ranking list $(2, 4, 2, 1)$ without revealing the amounts of their individual money $(3, 1, 3, 5)$.

This is a *secure ranking computation* problem, which was first considered by Jiang and Gong [5]. They constructed a secure ranking computation protocol based on a public-key cryptosystem. That is, their protocol allows $P_1 \cdots, P_n$ whose secret properties are $(x_1, \ldots, x_n)$ to know only the ranking $\mathsf{rk}(x_1, \ldots, x_n) = (r_1, \ldots, r_n)$ without revealing any information about the inputs $(x_1, \ldots, x_n)$. Note that this falls into the category of secure multiparty computations (MPCs) [12].

## 1.1  Card-Based Protocols

It is well known that MPCs can be conducted by using a deck of physical cards (such as ♣ ♣ ♡ $\cdots$ ), and such a research area is called *card-based cryptography* (e.g. [1,7]). The goal of this paper is to construct efficient card-based protocols for solving the secure ranking computation problem. Compared to the use of a public-key cryptosystem, there are certain advantages in using a deck of cards; their computation, correctness, and secrecy are easier to understand.

As elementary card-based computations, efficient NOT, AND, OR, XOR, and COPY protocols have been proposed [2,8]. Therefore, for any $n$-variable function $f$, we can construct a card-based protocol that securely computes $f$ by combining these elementary protocols. Thus, if one created a logical circuit representing the ranking function $\mathsf{rk}$, then one could construct a card-based protocol securely computing $\mathsf{rk}$. However, such a circuit would be huge and complicated, and hence, the resulting protocol would no longer be practical.

While generic constructions as above tend to be impractical, a protocol specialized for a specific problem is usually more efficient. For example, consider the millionaire problem [12]: Two players, Alice and Bob, want to know who is richer. Indeed, efficient card-based millionaire protocols were proposed [3,6,9]. The goal of this paper is to design specialized card-based protocols that efficiently solve the ranking computation problem.

One may notice that the secure ranking computation can be realized by repeatedly executing a millionaire protocol. However, our proposed protocols perform the secure ranking computation more efficiently than by repeating the existing millionaire protocol, as seen later.

We use a deck of four types of cards: black cards ♣, red cards ♡, a joker Jo, and number cards 1, 2, ..., m, which satisfy the following properties:

1. Cards with the same symbol on their faces cannot be distinguished from each other.

2. All cards have identical backs $\boxed{?}$, and the pattern of backs is vertically asymmetric so that we can distinguish between a face-down card that is right-side up $\boxed{?}$ and one that is upside down $\boxed{¿}$.

We call black cards, red cards, and the joker as *color cards*.

## 1.2  Contribution

In this study, we propose four secure ranking computation protocols using a deck of cards. The protocols are named as follows: two-phase, one-phase, shuffle-efficient, and card-efficient protocols. Table 1 summarizes their performance. The two-phase protocol and the one-phase protocol are not finite-runtime, but they are expected finite-runtime protocols that use an expected finite number of shuffles, i.e., they are Las Vegas algorithms. By contrast, both the shuffle-efficient protocol and the card-efficient protocol are finite-runtime; they terminate with a finite number of shuffles. More details along with the organization of this paper are as follows.

**Table 1.** The performance of our proposed protocols.

| Protocol | Finite-runtime? | # of color cards | # of num. cards | (Expected) # of shuffles |
|---|---|---|---|---|
| Two-phase | No | $n(2m+1)$ | 0 | $k + 2 + (m - k + 1)\sum_{i=1}^{k}\frac{1}{i}(= S)$ |
| One-phase | No | $n(m+1)$ | 0 | $\leq S + 1 + (m+1)\sum_{i=1}^{m-k}\frac{1}{i}$ |
| Shuffle-efficient | Yes | $(n+k)(m+1)$ | $m+1$ | $k+3$ |
| Card-efficient | Yes | $n(m+1)$ | $m+1$ | $3k+1$ |

In Sect. 2, we review fundamental shuffling operations in card-based cryptography.

In Sect. 3.1, we propose the two-phase protocol. The first phase of this protocol computes the "equality" of input $(x_1, \ldots, x_n)$, i.e., it reveals the quotient set $\{1, \ldots, n\}/\{(i,j) \mid x_i = x_j\}$; each equivalent class corresponds to a group of players who have the same amount of money. We often use $k$ to denote

$$k = m - |\{1, \ldots, n\}/\{(i,j) \mid x_i = x_j\}|$$

throughout this paper (including Table 1); in other words, $k$ is the number of amounts of money that nobody has. Making the use of $k$, the second phase of this protocol outputs $\mathsf{rk}(x_1, \ldots, x_n)$. Note that knowing the ranking $\mathsf{rk}(x_1, \ldots, x_n)$ implies learning the "equality," and hence, revealing such a quotient set will not leak any information beyond the ranking.

While the two-phase protocol mentioned above has two phases, we propose the one-phase protocol in Sect. 3.2. That is, we get rid of the "equality check" phase so that the number of required cards is $n(m+1)$, which is the smallest among all the proposed protocols.

Remember that both the two-phase and one-phase protocols are not finite-runtime. In Sect. 4, we focus our attention on constructing finite-runtime protocols that always terminate with a fixed number of shuffles. We first propose the finite-runtime shuffle-efficient protocol, which uses only $k + 3$ shuffles although it requires a relatively large number of cards. We can modify this protocol to reduce the number of required cards by increasing the number of shuffles. Carrying this technique further, we next propose the finite-runtime card-efficient protocol.

In Sect. 5, we compare the two finite-runtime protocols with the existing millionaire protocol and discuss which protocol is appropriate in practical situations.

In Sect. 6, we summarize this paper and mention future work.

## 2 Preliminaries

In this section, we review three types of shuffles that are often used in card-based protocols. As seen later, we can conceal input values by applying a shuffle operation to a sequence of face-down cards.

### 2.1 Random Cut

Consider a sequence of $\ell$ face-down cards $\boxed{?}\,\boxed{?}\cdots\boxed{?}$ denoted by $(c_1, c_2, \ldots, c_\ell)$. A *random cut* randomly and cyclically shifts a sequence of cards. Applying a random cut to the sequence $(c_1, c_2, \ldots, c_\ell)$ results in $(c_{1+r}, c_{2+r}, \ldots, c_{\ell+r})$, where $r$ is uniformly randomly generated from $\mathbb{Z}/\ell\mathbb{Z}$ and is unknown to everyone. Note that a random cut can be implemented securely by human hands [11].

### 2.2 Pile-Scramble Shuffle

Consider a pile of face-down cards $\boxed{?}$. Assume that we have $\ell$ such piles $\boxed{?}\,\boxed{?}\cdots\boxed{?}$, all of whose sizes are the same; we denote it by $(\boldsymbol{p}_1, \boldsymbol{p}_2, \ldots, \boldsymbol{p}_\ell)$. A *pile-scramble shuffle* [4] is a shuffle operation for such a sequence of piles. Applying a pile-scramble shuffle to the sequence $(\boldsymbol{p}_1, \boldsymbol{p}_2, \ldots, \boldsymbol{p}_\ell)$ results in $(\boldsymbol{p}_{\pi^{-1}(1)}, \boldsymbol{p}_{\pi^{-1}(2)}, \ldots, \boldsymbol{p}_{\pi^{-1}(\ell)})$, where $\pi \in S_\ell$ is a uniformly distributed random permutation and $S_\ell$ is the symmetric group of degree $\ell$. We can easily implement a pile-scramble shuffle by making piles with rubber bands or envelopes.

Alternatively, we can reduce the implementation of a pile-scramble shuffle to random cuts. We make a sequence of piles by using upside down cards as follows:

$$\underbrace{\boxed{\raisebox{0pt}{\scriptsize ¿}}\boxed{?}\cdots\boxed{?}}_{\boldsymbol{p}_1}\underbrace{\boxed{\raisebox{0pt}{\scriptsize ¿}}\boxed{?}\cdots\boxed{?}}_{\boldsymbol{p}_2}\cdots\underbrace{\boxed{\raisebox{0pt}{\scriptsize ¿}}\boxed{?}\cdots\boxed{?}}_{\boldsymbol{p}_l}.$$

Then, after applying a random cut to it, we take out the first pile, which is randomly chosen from $\ell$ piles. Next, do the same operation for the remaining $\ell - 1$ piles. Repeating this $\ell - 1$ times in total, we have a randomly permuted sequence of $\ell$ piles.

## 2.3  Pile-Shifting Shuffle

A *pile-shifting shuffle*[10] cyclically and randomly shifts a sequence of piles. Applying a pile-shifting shuffle to $(p_1, p_2, \ldots, p_\ell)$ results in $(p_{1+r}, p_{2+r}, \ldots, p_{\ell+r})$, where $r$ is uniformly randomly generated from $\mathbb{Z}/\ell\mathbb{Z}$ and is unknown to everyone.

Similar to the implementation of a pile-scramble shuffle based on upside down cards, as discussed in Sect. 2.2, we can implement a pile-shifting shuffle by applying a random cut once.

## 3  Simple Protocols

In this section, we propose two simple protocols. The first is the fundamental one, which has two phases. The second is obtained by removing one phase from the two-phase protocol. The numbers of cards and shuffles required to execute the two protocols were shown in Table 1; note that there is a tradeoff.

### 3.1  Proposal of Two-Phase Protocol

This simple ranking protocol consists of two phases, namely, the Check-equality phase and the Create-ranking phase, after the setup.

*Setup.* Each player $P_i$ holds one red card $\boxed{\heartsuit}$ and $m-1$ black cards $\boxed{\clubsuit}$s, and places a sequence of $m$ cards face down, denoted by $\alpha_i$, that encodes their private value $x_i \in \{1, \ldots, m\}$, as follows:

$$\alpha_i \quad \underbrace{\boxed{?}}_{\clubsuit} \underbrace{\boxed{?}}_{\clubsuit} \cdots \underbrace{\boxed{?}}_{\clubsuit} \underbrace{\boxed{?}}_{\heartsuit} \underbrace{\boxed{?}}_{\clubsuit} \cdots \underbrace{\boxed{?}}_{\clubsuit} .$$
$$\text{1-st} \qquad \qquad x_i\text{-th} \qquad \qquad m\text{-th}$$

That is, $P_i$ puts $\boxed{\heartsuit}$ at the $x_i$-th position and $\boxed{\clubsuit}$s at the remaining positions with their faces down. Likewise, each player $P_i$ makes the following sequence of cards $\beta_i$:

$$\beta_i \quad \underbrace{\boxed{?}}_{\heartsuit} \underbrace{\boxed{?}}_{\heartsuit} \cdots \underbrace{\boxed{?}}_{\heartsuit} \underbrace{\boxed{?}}_{\clubsuit} \underbrace{\boxed{?}}_{\heartsuit} \cdots \underbrace{\boxed{?}}_{\heartsuit} .$$
$$\text{1-st} \qquad \qquad x_i\text{-th} \qquad \qquad m\text{-th}$$

*Check-Equality Phase.* In this phase, we reveal the "equality" of the inputs $(x_1, \ldots, x_n)$. Formally speaking, we want to know the quotient set

$$\{1, \ldots, n\}/\{(i,j) \mid x_i = x_j\}.$$

Note that, as mentioned before, knowing $\mathsf{rk}(x_1, \ldots, x_n)$ implies learning the quotient set, and hence, revealing the equality does not leak any information beyond the ranking.

1. Take the $i$-th $(i = 1, \ldots, m)$ cards from all players' input sequences $\alpha_1, \ldots, \alpha_n$ (while keeping the order unchanged) to make a pile $\boldsymbol{p}_i$ of size $n$:

$$
\begin{array}{c}
\alpha_1 \quad \boxed{?} \; \boxed{?} \; \cdots \; \boxed{?} \\
\alpha_2 \quad \boxed{?} \; \boxed{?} \; \cdots \; \boxed{?} \\
\vdots \quad \vdots \;\; \vdots \quad \quad \vdots \\
\alpha_n \quad \boxed{?} \; \boxed{?} \; \cdots \; \boxed{?} \\
\downarrow \quad \downarrow \quad\quad \downarrow \\
\boxed{?} \; \boxed{?} \; \cdots \; \boxed{?} \;. \\
\boldsymbol{p}_1 \;\; \boldsymbol{p}_2 \quad\quad \boldsymbol{p}_m
\end{array}
\tag{1}
$$

Note that if the $j$-th and $\ell$-th cards of the pile $\boldsymbol{p}_i$ (for some $j, \ell$) are red:

$$
\boldsymbol{p}_i \quad
\overset{1}{\boxed{?}} \overset{2}{\boxed{?}} \cdots \overset{j}{\boxed{?}} \cdots \overset{\ell}{\boxed{?}} \cdots \overset{m}{\boxed{?}} ,
$$

then both $P_j$ and $P_\ell$ hold $i$ dollars, i.e., $x_j = x_\ell = i$, and hence, they are tied.

2. Apply a pile-scramble shuffle to $(\boldsymbol{p}_1, \boldsymbol{p}_2, \ldots, \boldsymbol{p}_m)$.

3. Reveal all the piles $(\boldsymbol{p}_{\pi^{-1}(1)}, \boldsymbol{p}_{\pi^{-1}(2)}, \ldots, \boldsymbol{p}_{\pi^{-1}(m)})$. Then, two types of piles should appear: a pile consisting of only ♣s (which we call an *empty pile*) or a pile containing ♡s (which we call a *non-empty pile*).

   If a non-empty pile contains ♡s at $j$-th and $\ell$-th positions for some $j, \ell$, we have $x_j = x_\ell$, but we cannot know the actual value of $x_j = x_\ell$. Therefore, each non-empty pile tells us an equivalent class, and hence, all the non-empty piles reveal the equality, namely $\{1, \ldots, n\}/\{(i, j) \mid x_i = x_j\}$.

*Create-Ranking Phase.* Let $k$ be the number of empty piles, i.e., $k = m - |\{1, \ldots, n\}/\{(i, j) \mid x_i = x_j\}|$.

1. Every player $P_i$ adds a face-down joker[1] to the last of $\beta_i$:

$$
\beta_i \quad
\underbrace{\boxed{?}}_{\text{1-st}} \; \boxed{?} \; \cdots \; \boxed{?} \; \underbrace{\boxed{?}}_{x_i\text{-th}} \; \boxed{?} \; \cdots \; \boxed{?} \; \underbrace{\boxed{?}}_{m+1\text{-st}} \;.
\tag{2}
$$

2. Create a sequence $(\boldsymbol{p}_1, \boldsymbol{p}_2, \ldots, \boldsymbol{p}_{m+1})$ in the same way as in (1) in Sect. 3.1.

3. Repeat the following until $k$ empty piles[2] are removed.
   (a) Apply a pile-shifting shuffle to the sequence.
   (b) Choose (any) one pile and reveal all the cards of the pile.
       - If it is an empty pile, remove it from the sequence.
       - If it is not an empty pile, turn over all the face-up cards and return the pile to the sequence.

4. Apply a pile-shifting shuffle to the sequence.

5. Reveal all the remaining piles. Then, we obtain the ranking by cyclically shifting the revealed piles so that the pile consisting of only jokers is at the end of the sequence.

---

[1] $\boxed{\text{Jo}}$ can be substituted by $\boxed{♣}$; we use $\boxed{\text{Jo}}$ for the purpose of easy understanding.

[2] An empty pile consists of only ♡ (and a joker) here.

## 3.2  Proposal of One-Phase Protocol

In this subsection, we construct a protocol without relying on checking equality.

Recall that in the Create-ranking phase of the two-phase protocol, we remove a pile if it is an empty pile. If it is not an empty pile, it goes back to the sequence. Notice that each non-empty pile is unique and contains information on players having the same amount of money, namely the corresponding equivalent class. Therefore, even if we do not know the "equality" in advance, we can find all empty piles by searching them until all piles are revealed. Players can memorize revealed piles so that they can determine when they finish.

Thus, we can skip the Check-equality phase, and our one-phase protocol proceeds as follows.

*Setup.* $P_i$ creates a sequence of cards $\beta_i$ in the same way as in (2) in Sect. 3.1.

*Create-ranking phase.*

1. Create $(\boldsymbol{p}_1, \boldsymbol{p}_2, \ldots, \boldsymbol{p}_{m+1})$ in the same way as in (1) in Sect. 3.1.
2. Repeat the following until all $m + 1$ piles are revealed.
   (a) Apply a pile-shifting shuffle to the sequence.
   (b) Reveal (any) one pile.
       – If it is an empty pile, remove it from the sequence.
       – If it is not an empty pile, memorize it, turn over all the face-up cards, and return the pile to the sequence.
3. Apply a pile-shifting shuffle to the sequence.
4. Reveal all the remaining piles. We obtain the ranking by shifting the sequence of piles based on the pile consisting of only jokers.

# 4  Finite-Runtime Protocols

The two proposed protocols in the previous section are not finite-runtime (because we have to repeat the shuffle until all empty piles are found). In this section, we solicit finite-runtime protocols that perform the secure ranking computation with a fixed number of shuffles. Specifically, we design two protocols considering the tradeoff between the number of cards and the number of shuffles. First, we propose the finite-runtime shuffle-efficient protocol specialized for reducing the number of shuffles. Next, we introduce a technique for reducing the number of cards, and propose the finite-runtime card-efficient protocol.

## 4.1  Proposal of Shuffle-Efficient Protocol

In this protocol, we first find all empty piles at once, add markers to them, and search the empty piles with the help of the markers.

1. $P_i$ creates a sequence of cards $\beta_i$ in the same way as in (2) in Sect. 3.1.
2. Create $(\boldsymbol{p}_1, \boldsymbol{p}_2, \ldots, \boldsymbol{p}_{m+1})$ in the same way as in (1) in Sect. 3.1.

3. For every $j$, $1 \leq j \leq m+1$, add a number card $j$ to the top of the pile $p_j$ so that we have a new pile $p'_j$ of size $n+1$.

4. Apply a pile-scramble shuffle to the sequence of piles:

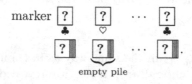

5. Reveal all the cards except for the number cards. Now, we find all the empty piles (whose positions are randomly permuted). Let $k$ be the number of empty piles.

6. For each found empty pile, add a marker which consists of $m$ ♣s and one ♡ being placed on the piles so that the red card ♡ indicates the empty pile, as follows:

marker | ? | ? | ⋯ | ? |
        | ♣ | ♡ |   | ♣ |

| ? | ? | ⋯ | ? |

empty pile

One marker indicates one empty pile, and hence, we stack $k$ markers on the sequence of piles. We denote the resulting sequence by $(p''_1, p''_2, \ldots, p''_{m+1})$.

7. Apply a pile-scramble shuffle to the sequence of piles:

| ? | ? | ⋯ | ? |

$p''_{\pi_2^{-1}(1)}$ $p''_{\pi_2^{-1}(2)}$ $p''_{\pi_2^{-1}(m+1)}$

8. Reveal only the number cards, each of which is the $(k+1)$-st card from the top of the corresponding pile, and rearrange the order of the piles so that the number cards are in ascending order; remove all the number cards:

| ? | ? | ⋯ | ? |

$p'''_1$ $p'''_2$ $p'''_{m+1}$

9. We remove all empty piles without anyone getting to know the "distance" between any pair of non-empty piles: Repeat the following until $k$ empty piles are removed.
   (a) Apply a pile-shifting shuffle to the sequence.
   (b) Reveal one marker and remove the corresponding empty pile.

10. Apply a pile-shifting shuffle to the sequence.

11. Reveal all the remaining piles. We obtain the ranking by shifting the sequence of piles based on the pile consisting of only jokers.

## 4.2   Proposal of Card-Efficient Protocol

In the finite-runtime shuffle-efficient protocol presented in Sect. 4.1, we use $k$ markers, i.e., we use $k(m+1)$ cards as markers. We consider a method of reducing the number of cards used for these markers. If we arrange only one marker in Step 6 of the finite-runtime shuffle-efficient protocol, we can compute the ranking by repeating the process from Step 3 to Step 9 $k$ times. Then, the number of cards used for markers can be reduced from $k(m+1)$ to $m+1$. However, the number of required shuffles increases from $k+3$ to $3k+1$.

By extending this idea further, no marker is needed. We construct a finite-runtime card-efficient protocol that does not rely on markers. The procedure is shown below.

1. $P_i$ creates a sequence of cards $\beta_i$ in the same way as in (2) in Sect. 3.1.
2. Create $(p_1, p_2, \ldots, p_{m+1})$ in the same way as in (1) in Sect. 3.1.
3. Repeat the following until $k$ empty piles are removed (while repeating, the number of piles is decreasing from $m+1$ to $m+1-k$):
   (a) Apply a random cut to a sequence consisting of number cards:

   $$\boxed{?}\ \boxed{?}\ \cdots\ \boxed{?}\ \rightarrow\ \boxed{?}\ \boxed{?}\ \cdots\ \boxed{?}\ .$$
   $$\ _{1}\quad _{2}\qquad\ _{m+1}\qquad _{1+r}\ _{2+r}\qquad\ _{m+1+r}$$

   (b) For each pile $p_j$, add the number card $j+r$ to the pile so that we have a new pile $p'_j$.
   (c) Apply a pile-scramble shuffle to the sequence of piles:

   $$\underbrace{\boxed{?}}_{p'_{\pi_1^{-1}(1)}}\ \underbrace{\boxed{?}}_{p'_{\pi_1^{-1}(2)}}\ \cdots\ \underbrace{\boxed{?}}_{p'_{\pi_1^{-1}(m+1)}}\ .$$

   (d) Reveal all the cards except for the number cards.
   (e) Remove one empty pile along with the number card. Note that because the removed number card does not indicate the position of the pile (because of the random permutation $\pi_1$), information on the input is not leaked.
   (f) Turn all the cards face-down and apply a pile-scramble shuffle to the sequence of piles:

   $$\underbrace{\boxed{?}}_{p'_{\pi_1^{-1}\left(\pi_2^{-1}(1)\right)}}\ \underbrace{\boxed{?}}_{p'_{\pi_1^{-1}\left(\pi_2^{-1}(2)\right)}}\ \cdots\ \underbrace{\boxed{?}}_{p'_{\pi_1^{-1}\left(\pi_2^{-1}(m+1)\right)}}\ .$$

   (g) Reveal only the number cards, each of which is the top of the corresponding pile, and rearrange the order of the piles so that the number cards are in ascending order. Remove all the number cards.
4. Apply a pile-shifting shuffle to the sequence of piles.
5. Reveal all the remaining piles. We obtain the ranking by shifting the sequence of piles based on the pile consisting of only jokers.

## 5  Discussion

In this section, we discuss comparison of protocols and real use cases.

**Comparing Protocols.** We compare the two finite-runtime protocols (presented in Sect. 4) with the repetition of the existing millionaire protocol [6] in terms of the number of cards and shuffles, and the input range $m$. We fix the number of players at $n = 5$, the number of color cards at 52, and the number of number cards at 10.

Figure 1 shows the number of shuffles required to execute the three protocols. One can confirm that our protocols use fewer shuffles compared to the case of repeating the millionaire protocol.

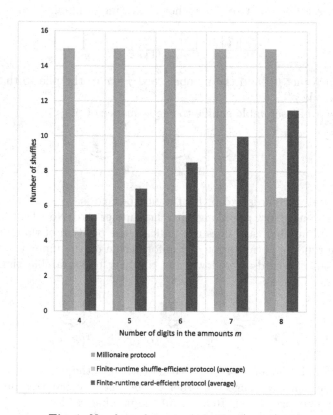

**Fig. 1.** Number of required shuffles $(n = 5)$

**Use Case.** Next, we discuss how our protocols can be used. To reduce the number of cards, it is reasonable to compare digits in the amounts instead of direct amounts of money. For example, if the amount of money that any player holds is between 10 and $10^{y+1} - 1$ for some $y \in \mathbb{N}$, we can set the input range to $\{1, \ldots, y\}$.

Assume that the number of players is $n = 5$ and the number of digits in the amounts is $m = 6$. For example, if a player $P_i$ holds more than $10^3$ dollars and less than $10^4$ dollars, she sets to $x_i = 3$, if she holds $10^6$ dollars, she sets to $x_i = 6$, and so on. In this case, 42 cards are required to execute our card-efficient protocol. The breakdown is as follows: 10 ♣ s, 25 ♡ s, and 7 number cards. Assume that, for instance, the number of empty piles is $k = 3$, and the number of required shuffles is 10. Both the numbers of cards and shuffles are within a feasible range in practice.

If there are two or more players at the same rank, it is possible to compare the amounts of money between the players in finer units. If there are three or more players to compare, using our finite-runtime shuffle-efficient protocol or finite-runtime card-efficient protocol is efficient; if there are only two players, using the existing millionaire protocol would be a good choice.

## 6   Conclusion

In this paper, we first proposed card-based secure ranking computation protocols. While our two simple protocols are Las Vegas algorithms, our shuffle-efficient protocol and card-efficient protocol always terminate with a finite number of shuffles. Future work will be to devise a protocol that works with a standard deck of playing cards. Finding lower bounds on the number of shuffles for a secure ranking computation protocol is also interesting.

**Acknowledgement.** This work was supported by JSPS KAKENHI Grant Numbers JP17K00001, JP17J01169, and JP19J21153. We would like to thank the anonymous reviewers for their fruitful comments.

## References

1. Boer, B.: More efficient match-making and satisfiability the five card trick. In: Quisquater, J.-J., Vandewalle, J. (eds.) EUROCRYPT 1989. LNCS, vol. 434, pp. 208–217. Springer, Heidelberg (1990). https://doi.org/10.1007/3-540-46885-4_23
2. Crépeau, C., Kilian, J.: Discreet solitary games. In: Stinson, D.R. (ed.) CRYPTO 1993. LNCS, vol. 773, pp. 319–330. Springer, Heidelberg (1994). https://doi.org/10.1007/3-540-48329-2_27
3. Hibiki Ono, Y.M.: Efficient card-based cryptographic protocols for the millionaires' problem using private input operations. In: 2018 13th Asia Joint Conference on Information Security (AsiaJCIS), pp. 23–28. IEEE (2018)
4. Ishikawa, R., Chida, E., Mizuki, T.: Efficient card-based protocols for generating a hidden random permutation without fixed points. In: Calude, C.S., Dinneen, M.J. (eds.) UCNC 2015. LNCS, vol. 9252, pp. 215–226. Springer, Cham (2015). https://doi.org/10.1007/978-3-319-21819-9_16
5. Jiang, S., Gong, G.: A round and communication efficient secure ranking protocol. In: Pointcheval, D. (ed.) CT-RSA 2006. LNCS, vol. 3860, pp. 350–364. Springer, Heidelberg (2006). https://doi.org/10.1007/11605805_22

6. Miyahara, D., Hayashi, Y., Mizuki, T., Sone, H.: Practical and easy-to-understand card-based implementation of Yao's millionaire protocol. In: Kim, D., Uma, R.N., Zelikovsky, A. (eds.) COCOA 2018. LNCS, vol. 11346, pp. 246–261. Springer, Cham (2018). https://doi.org/10.1007/978-3-030-04651-4_17

7. Mizuki, T., Shizuya, H.: Computational model of card-based cryptographic protocols and its applications. IEICE Trans. Fundam. Electron. Commun. Comput. Sci. **E100.A**(1), 3–11 (2017). https://doi.org/10.1587/transfun.E100.A.3

8. Mizuki, T., Sone, H.: Six-card secure AND and four-card secure XOR. In: Deng, X., Hopcroft, J.E., Xue, J. (eds.) FAW 2009. LNCS, vol. 5598, pp. 358–369. Springer, Heidelberg (2009). https://doi.org/10.1007/978-3-642-02270-8_36

9. Nakai, T., Tokushige, Y., Misawa, Y., Iwamoto, M., Ohta, K.: Efficient card-based cryptographic protocols for millionaires' problem utilizing private permutations. In: Foresti, S., Persiano, G. (eds.) CANS 2016. LNCS, vol. 10052, pp. 500–517. Springer, Cham (2016). https://doi.org/10.1007/978-3-319-48965-0_30

10. Nishimura, A., Hayashi, Y., Mizuki, T., Sone, H.: Pile-shifting scramble for card-based protocols. IEICE Trans. Fundam. Electron. Commun. Comput. Sci. **E101.A**(9), 1494–1502 (2018). https://doi.org/10.1587/transfun.E101.A.1494

11. Ueda, I., Nishimura, A., Hayashi, Y., Mizuki, T., Sone, H.: How to implement a random bisection cut. In: Martín-Vide, C., Mizuki, T., Vega-Rodríguez, M.A. (eds.) TPNC 2016. LNCS, vol. 10071, pp. 58–69. Springer, Cham (2016). https://doi.org/10.1007/978-3-319-49001-4_5

12. Yao, A.C.: Protocols for secure computations. In: Proceedings of the 23rd Annual Symposium on Foundations of Computer Science, FOCS 1982, pp. 160–164. IEEE Computer Society, Washington, DC (1982). https://doi.org/10.1109/SFCS.1982.88

# Improved Stretch Factor of Delaunay Triangulations of Points in Convex Position

Xuehou Tan[1,2], Charatsanyakul Sakthip[2], and Bo Jiang[1(✉)]

[1] Dalian Maritime University, Linghai Road 1, Dalian, China
bojiang@dlmu.edu.cn
[2] Tokai University, 4-1-1 Kitakaname, Hiratsuka 259-1292, Japan

**Abstract.** Let $S$ be a set of $n$ points in the plane, and let $DT(S)$ be the planar graph of the Delaunay triangulation of $S$. For a pair of points $a, b \in S$, denote by $|ab|$ the Euclidean distance between $a$ and $b$. Denote by $DT(a,b)$ the shortest path in $DT(S)$ between $a$ and $b$, and let $|DT(a,b)|$ be the total length of $DT(a,b)$. Dobkin *et al.* were the first to show that $DT(S)$ can be used to approximate the complete graph of $S$ in the sense that the stretch factor $\frac{|DT(a,b)|}{|ab|}$ is upper bounded by $((1+\sqrt{5})/2)\pi \approx 5.08$. Recently, Xia improved this factor to 1.998. Amani *et al.* have also shown that if the points of $S$ are in convex position, then a planar graph with these vertices can be constructed such that its stretch factor is 1.88. A set of points is said to be in *convex position*, if all points form the vertices of a convex polygon.

In this paper, we prove that if the points of $S$ are in convex position, then the stretch factor of $DT(S)$ is less than 1.83, improving upon the previously known factors, for either the Delaunay triangulation or planar graph of a point set. Our result is obtained by investigating some geometric properties of $DT(S)$ and showing that there exists a convex chain between $a$ and $b$ in $DT(S)$ such that it is either contained in a semicircle of diameter $ab$, or enclosed by segment $ab$ and a simple (convex) chain that consists of a circular arc and two or three line segments.

**Keywords:** Computational geometry · Delaunay triangulation · Stretch factor · Convex polygons

## 1 Introduction

Let $S$ be a set of $n$ points in the plane, and let $G(S)$ be such a graph that each vertex corresponds to a point in $S$ and the weight of an edge is the Euclidean distance between its two endpoints. For a pair of points $u, v$ in the plane, denote

The work by Tan was partially supported by JSPS KAKENHI Grant Number 15K00023, and the work by Jiang was partially supported by National Natural Science Foundation of China under grant 61702242.

by $uv$ the line segment connecting $u$ and $v$, and $|uv|$ the Euclidean distance between $u$ and $v$. For a pair of points $a, b \in S$, denote by $G(a, b)$ the shortest path in $G(S)$ between $a$ and $b$, and let $|G(a, b)|$ be the total length of path $G(a, b)$. The graph $G(S)$ is said to approximate the complete graph of $S$ if $\frac{|G(a,b)|}{|ab|}$, called the *stretch factor* of $G(S)$, is upper bounded by a constant, independent of $S$ and $n$. It is then desirable to identify classes of graphs that approximate complete graphs well and have only $O(n)$ edges (in comparison with $O(n^2)$ edges of complete graphs), as these graphs have potential applications in geometric network design problems [2,5,6].

Denote by $DT(S)$ the planar graph of the Delaunay triangulation of $S$. Dobkin *et al.* [4] were the first to give a stretch factor $((1 + \sqrt{5})/2)\pi \approx 5.08)$ of Delaunay triangulations to complete graphs, which was later improved to $2\pi/(3\cos(\pi/6)) \approx 2.42$ by Keil and Gutwin [7]. Recently, this factor has been improved to 1.998 by Xia [9]. On the other hand, Xia and Zhang [10] gave a lower bound 1.5932 on the stretch factor of $DT(S)$. Determining the worst possible stretch factor of the Delaunay triangulation has been a long standing open problem in computational geometry.

Cui *et al.* [3] have also studied the stretch factor of $DT(S)$ for the points in convex position. A set of points is said to be in *convex position*, if all points form the vertices of a convex polygon. The currently best known stretch factor in this special situation is 1.88, due to a work of Amani *et al.* on the stretch factor of planar graphs [1]. Notice that the planar graph studied by Amani *et al.* is *not* the Delaunay triangulation of the given point set. Although studying the convex case may not lead to improved upper bounds for the general case, it shows a large possibility in obtaining a better upper bound on the stretch factor of $DT(S)$ and may give some intelligent hints for the general case.

In this paper, we prove that $\frac{|DT(a,b)|}{|ab|} < 1.83$ holds for a set $S$ of points in convex position. This improves upon the previously known factor 1.998 for the delaunay triangulation of a point set (although it is in general position), as well as the stretch factor 1.88 for a planar graph with a set of vertices in convex position. Our result is obtained by investigating some geometric properties of $DT(S)$ and showing that there exists a convex chain between $a$ and $b$ in $DT(S)$ such that it is either contained in a semicircle of diameter $ab$, or enclosed by segment $ab$ and a simple (convex) chain that consists of a circular arc and two or three line segments. The total length of the simple chain is less than $1.83|ab|$.

## 2    Preliminary

Without loss of generality, assume that no four points of $S$ are on the boundary of a circle in the plane. The *Voronoi diagram* for $S$, denoted by $Vor(S)$, is a partition of the plane into regions, each containing exactly one point in $S$, such that for each point $p \in S$, every point within its corresponding region, denoted by $Vor(p)$, is closer to $p$ than to any other point of $S$. The boundaries of these Voronoi regions form a planar graph. The *Delaunay triangulation* of $S$, denoted by $DT(S)$, is the straight-line dual of the Voronoi diagram for $S$; that is, we connect a pair of points in $S$ if and only if they share a Voronoi boundary. Since $DT(S)$ is a planar graph, it has $O(n)$ edges.

For a pair of points $a, b \in S$, denote by $DT(a, b)$ the shortest path in $DT(S)$ between $a$ and $b$, and $|DT(a, b)|$ the total length of path $DT(a, b)$. The *stretch factor* of $DT(S)$ is then the maximum ratio $\frac{|DT(a,b)|}{|ab|}$ among all point pairs $(a, b)$.

We now briefly review an important idea of Dobkin *et al.*'s work [4]. Let $a = a_0, a_1, \ldots, a_m = b$ be the sequence of the points of $S$, whose Voronoi regions intersect segment $ab$ (Fig. 1). If a Voronoi edge happens to be on segment $ab$, either of the points defining that Voronoi edge can be chosen as the one on the direct path from $a$ to $b$. The path obtained in this way is called the *direct path from $a$ to $b$* [4].

The direct path from $a$ to $b$ is said to be *one-sided* if all points of the path are to the same side of the line through $a$ and $b$. See Fig. 1. If the direct path from $a$ to $b$ is one-sided, then it has length at most $\pi|ab|/2$.

**Lemma 1** *(Dobkin et al. [4]). If the direct path from $a$ to $b$ is one-sided, then it has length at most $\pi|ab|/2$.*

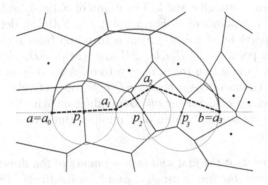

**Fig. 1.** A one-sided, direct path from $a$ to $b$.

Let $p_i$ be the intersection point of $ab$ with the Voronoi edge between $Vor(a_{i-1})$ and $Vor(a_i)$, for $1 \leq i \leq m$. It follows from the definition of the Voronoi diagram that $p_i$ is the center of a circle that passes through $a_{i-1}$ and $a_i$ but contains no points of $S$ in its *interior*, see Fig. 1. All points of the direct path from $a$ to $b$ are thus contained in the circle of diameter $ab$, no matter whether the path is one-sided or not.

A more general result than Lemma 1 is the following.

**Lemma 2** *(Dobkin et al. [4]). Let $C_1, C_2, \ldots, C_k$ be the circles all centered on a same line such that $C = \bigcup_{1 \le i \le k} C_i$ is connected. The boundary of $C$ has length at most $\pi|ab|$ and is contained in the circle of diameter $ab$, where $a$ and $b$ are two extreme endpoints of $C$ on the line.*[1]

## 3   The Main Result

Assume that the set $S$ of given points is in convex position. For a point $p$ in the plane, denote the coordinates of $p$ by $p(x)$ and $p(y)$, respectively. Without loss of generality, assume that both $a$ and $b$ lie on the $x$-axis (i.e., $a(y) = b(y) = 0$), with $a(x) < b(x)$. The *bisector* of two points $u$ and $v$, denoted by $B_{u,v}$, is the perpendicular line through the middle point of segment $uv$.

We say segment $ab$ *properly* intersects a Delaunay triangle if it intersects the interior of the triangle (i.e., segment $ab$ does not intersect only at a vertex of the triangle). Note that if a Delaunay triangle does not properly intersect $ab$, then at least one of its vertices (and two edges incident to that vertex) can be deleted from $DT(S)$, without affecting the value of $\frac{|DT(a,b)|}{|ab|}$. Hence, we assume below that $ab$ properly intersects all triangles of $DT(S)$.

Denote by $SA[a, b]$ (resp. $SB[a, b]$) the portion of the *convex chain* of $S$ above (resp. below) the line through $a$ and $b$. The union of $SA[a, b]$ and $SB[a, b]$ is then the convex hull of the points of $S$. For a point $p \in SA[a, b]$, denote by $SA[a, p]$ and $SA[p, b]$ two portions of $SA[a, b]$ from $a$ to $p$ and from $p$ to $b$, respectively. Analogously, for a point $q \in SB[a, b]$, $SB[a, q]$ (resp. $SB[q, b]$) represents the portion of $SB[a, b]$ from $a$ to $q$ (resp. from $q$ to $b$). Also, denote by $SA(a, b)$ and $SB(a, b)$ the open chains of $SA[a, b]$ and $SB[a, b]$, respectively.

Let $C$ be the circle of diameter $ab$. Depending on whether the direct path from $a$ to $b$ intersects segment $ab$ an odd number times or not, we can obtain the following results.

**Lemma 3.** *Suppose that the first and last segments of the direct path from $a$ to $b$ are below and above the line through $a$ and $b$, respectively. Then, there exists an angle $\alpha$ such that $|DT(a, b)|/|ab| \le \sin(\alpha) + \cos(\alpha)(\cos(\alpha) + \alpha)$, $0 < \alpha < \pi/2$, or $|DT(a, b)|/|ab| \le \sin(\alpha) + \cos(\alpha)(2\sin(\alpha) + \pi/2 - 2\alpha)$, $0 < \alpha < \pi/6$.*

**Proof.** Assume that neither $SA[a, b]$ nor $SB[a, b]$ is not completely contained in $C$; otherwise, $|DT(a, b)|/|ab| \le \pi/2$ and we are done. Denote by $ac$ and $bd$ the first and last segments of the direct path from $a$ to $b$ respectively, as viewed from $a$. Then, both $ac$ and $bd$ are contained in $C$, see Fig. 2. Extend segments $ac$ and $bd$ until they touch the boundary of $C$, say, at points $c'$ and $d'$ respectively. Since $\angle bc'a = \angle ad'b = \pi/2$, either $\angle c'ad'$ or $\angle d'bc'$ is at least $\pi/2$. In the following, we assume that $\angle d'bc' \ge \pi/2$, or equivalently, $\angle dbc' \ge \pi/2$.

---

[1] It simply follows from the proof of Lemma 2 of [4] that the boundary of $C$ is contained in the circle of diameter $ab$.

Let $i$ be the intersection point of $C$ with $B_{b,d}$, which is vertically below segment $ab$. Since $B_{b,d}$ is perpendicular to $bd$, and since $\angle bc'a = \pi/2$ and $\angle dbc' \geq \pi/2$, $B_{b,d}$ intersects segment $ac'$. Thus, segment $bi$ intersects $ac'$, and point $i$ is outside of the convex hull of $S$, see Fig. 2.

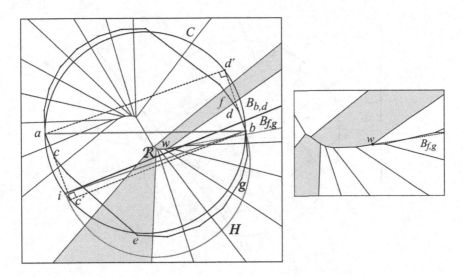

**Fig. 2.** Illustration of the proof of Lemma 3.

Denote by $H$ the semicircle of diameter $bi$, which is vertically below $bi$. In the following, we first show that $SB[a,b]$ is contained in the union of $H$ and a semicircle of $C$, see Fig. 2. To bound the length of $DT(a,b)$, we further introduce one or two tangents to the convex chain of $S$ contained in $H$. As a final result, $SB[a,b]$ is completely contained in the region bounded by segment $ab$ and a simple (convex) chain that consists of a circular arc of diameter $bi$ and two or three line segments.

Let $e$ be the first point of $SB[a,b]$ outside of $C$, as viewed from $a$, and $f$ the last point of $SA[a,b]$ such that $Vor(e)$ and $Vor(f)$ are adjacent. Denote by $\mathcal{R}$ the chain formed by the common edges of $Vor(p)$ and $Vor(q)$, $p \in SB[e,b)$ and $q \in SA[f,b)$. See Figs. 2 and 3 for some examples, where $\mathcal{R}$ is emphasized in dotted (and solid) line.

We first claim that $\mathcal{R}$ is vertically above $B_{b,d}$. It is clear that $Vor(d)$ is vertically above $B_{b,d}$. Assume that $f$ is not identical to $d$ (otherwise, our claim is trivially true). If $f$ is adjacent to $d$, then there exists a point $g \in SB[e,b)$, such that $Vor(f)$, $Vor(d)$ and $Vor(g)$ share a Voronoi vertex, say $w$ (see Fig. 2). So, $w$ is on $B_{f,g}$ and the *rightmost* (finite) vertex of $Vor(f)$. From the convexity of Voronoi regions, $B_{f,g}$ intersects $Vor(d)$, if $Vor(d)$ is *not* a triangular region (see the right of Fig. 2); otherwise, $w$ is on $B_{b,d}$ and our claim is trivially true. Hence, $B_{f,g}$ intersects $B_{b,d}$ at a point, whose $x$-coordinate is larger than $w(x)$. Since

$Vor(f)$ is vertically above $B_{f,g}$, it is vertically above $B_{b,d}$, too. In the case that $f$ is not adjacent to $d$, a similar argument on each pair of consecutive points of $SA[f, b]$ can show that all the regions $Vor(q)$, $q \in SA[f, b]$, are vertically above $B_{b,d}$. Therefore, our claim is proved.

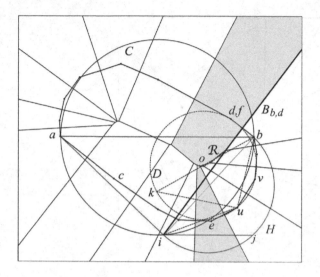

**Fig. 3.** Illustration of the proof of Lemma 3 (continued).

Let $u$, $v$ be two adjacent points on $SB[e, b]$ such that $u$ is immediately before $v$ on $SB[e, b]$. Then, $u \neq b$. Since it has been assumed that every triangle of $DT(S)$ properly intersects $ab$, the Delaunay triangle with an edge $uv$ has its third vertex on $SA[f, b]$. Denote by $D$ the circumcircle of the Delaunay triangle with edge $uv$, centered at a Voronoi vertex $o$ of $\mathcal{R}$ (Fig. 3). It follows from our claim that point $o$ is vertically above $B_{b,d}$, and $bi$ as well.

We show below that $\angle bui > \pi/2$. By definition of $DT(S)$, point $b$ is on or outside of $D$. Let $k$ be the intersection point of $D$ with the line through $b$ and $o$, such that $k$ is not contained in segment $ob$, see Fig. 3. Since no point of $S$ is contained in the interior of $D$, point $k$ is contained in the convex hull of $S$. Moreover, since $i$ is outside of the convex hull of $S$ and $o$ is vertically above $bi$ and below $ab$, point $k$ is contained in the triangle $\triangle_{a,b,i}$. Hence, segment $bi$ intersects $uk$. Therefore, $\angle bui > \angle buk \geq \pi/2$. From the convexity of $S$ and the definition of points $i$ and $e$, $SB[a, b]$ is completely contained in the region bounded by $ab$, $ai$ and $H$, see Figs. 2 and 3.

Let us now give a method to bound the length of $DT(a, b)$. Let $\alpha = \angle abi$, $0 < \alpha < \pi/2$. Denote by $j$ the intersection point of $H$ with the horizontal line through $i$, see Figs. 3 and 4. So, $ij$ is parallel to $ab$. Since $\angle bji = \pi/2$, we have $\angle abj = \pi/2$. Thus, the line through $b$ and $j$ is tangent to $C$. If the whole chain $SB[a, b]$ is vertically above the line through $i$ and $j$, then $SB[a, b]$ is contained in the convex region bounded by $ba$, $ai$, $ij$ and the arc $\widehat{jb}$ of diameter $bi$ and angle $\angle bij$ ($= \alpha$). See Fig. 3. Clearly, $|ai| = \sin(\alpha)|ab|$, $|bi| = \cos(\alpha)|ab|$, $|ij| = cos^2(\alpha)|ab|$ and $|\widehat{jb}| = \alpha \cos(\alpha)|ab|$. A simple argument (as in [8]) then shows that the length of $SB[a,b]$, denoted by $|SB[a,b]|$, is less than $(\sin(\alpha) + \cos(\alpha)(\cos(\alpha) + \alpha))|ab|$. Thus, $|DT(a,b)| \leq |SB[a,b]| \leq (\sin(\alpha) + \cos(\alpha)(\cos(\alpha) + \alpha))|ab|$.

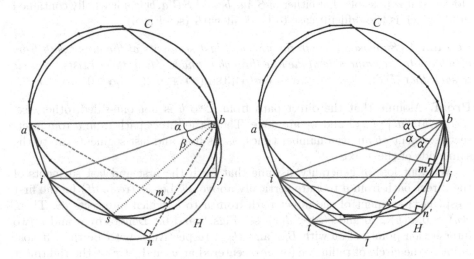

**Fig. 4.** Two shortcuts $it$ and $tl$ are introduced in $H$.

Consider the situation in which a portion of $SB[a,b]$ is below $ij$. To bound the length of $SB[a,b]$, we draw a tangent from point $i$ to the portion of $SB[a,b]$ contained in $H$. (Recall that $i$ is outside of the convex hull of $S$.) The tangent intersects $H$ at a point, say, $n$ ($\neq i$). Since a portion of $SB[a,b]$ is below $ij$, segment $bn$ intersects $C$ at a point, say, $m$ ($\neq b$). See the left of Fig. 4.

Let $\beta = \angle ibm$. Since $n$ and $m$ are on $H$ and $C$ respectively, $\angle bni = \angle bma = \pi/2$. Two segments $am$ and $in$ are thus parallel. Since $in$ is tangent to $SB[a,b]$, it intersects $C$ at a point, say, $s$ ($\neq i$). Thus, two circular arcs $\widehat{ai}$ and $\widehat{ms}$ of $C$ are of the same length. Hence, we have $\angle sbm = \alpha$, and $\beta \geq \alpha$. Since $\angle jib = \alpha$ and $\angle ibj > \beta \geq \alpha$, we obtain $\alpha < \pi/6$.

Let $t$ be the point on $H$ such that $\angle ibt = \alpha$, see the right of Fig. 4. Since $\beta \geq \alpha$, segment $jt$ does not intersect $SB[a,b]$. Moreover, since $\alpha < \pi/6$, we can draw another tangent from $t$ to the portion of $SB[a,b]$ contained in $H$. Denote by $n'$ ($\neq t$) the intersection point of $H$ with that tangent. As discussed above, we have $\angle tbn' \geq \alpha$. Again, let $l$ be the point on $H$ such that $\angle tbl = \alpha$. Also, segment $tl$ does not intersect $SB[a,b]$.

In summary, $SB[a,b]$ is contained in the region bounded by $ba$, $ai$, $it$, $tl$ and the circular arc $\widehat{lb}$ of diameter $bi$ and angle $\pi/2 - 2\alpha$ (see the right of

Fig. 4). Since $|it| = |tl| = \cos(\alpha)\sin(\alpha)|ab|$, we have $|DT(a,b)| \le |SB[ab]| \le (\sin(\alpha)+\cos(\alpha)(2\sin(\alpha)+\pi/2-2\alpha))|ab|$, $0 < \alpha < \pi/6$. The proof is complete. $\square$

Another part of our work handles the situation in which the direct path from $a$ to $b$ intersects $ab$ an even number times. In most case, either $SA[a,b]$ or $SB[a,b]$ is completely contained in $C$, and thus, we can obtain $|DT(a,b)|/|ab| \le \pi/2$. However, it is possible that either of $SA[a,b]$ and $SB[a,b]$ is not wholly contained in $C$, which is the difficult case to be dealt with (see Fig. 6).

**Lemma 4.** *Suppose that both the first and last segments of the direct path from $a$ to $b$ are to the same side of the line through $a$ and $b$. Then, there exists an angle $\alpha$ such that $|DT(a,b)|/|ab| \le \alpha + \cos(\alpha)(3\sin(\alpha) + \pi/2 - 3\alpha)$, $0 < \alpha < \pi/6$.*

**Proof.** Assume that the direct path from $a$ to $b$ is not one-sided; otherwise, $|DT(a,b)|/|ab| \le \pi/2$ and we are done. Then, the direct path from $a$ to $b$ intersects segment $ab$ an even number times, as its first and last segments are to the same side of the $x$-axis.

Without loss of generality, assume that both the first and last segments of the direct path from $a$ to $b$ are vertically above $ab$. Let $ce$ (resp. $df$) be the first (resp. last) segment of the direct path from $a$ to $b$, which intersects $ab$. Then, $c(y) > 0$, $d(y) > 0$ and $c(x) < d(x)$, see Figs. 5 and 6.[2] Denote by $u$ and $v$ two intersection points of $ab$ with $B_{c,e}$ and $B_{d,f}$, respectively. Let $l$ be the leftmost point of the circle of radius $uc$ (or $ue$), centered at $u$, and let $r$ be the rightmost point of the circle of radius $vd$ (or $vf$), centered at $v$. Since both $ce$ and $df$ belong to the direct path from $a$ to $b$, segment $lr$ is completely contained in $ab$, see Fig. 5.

Let $c = p_1, p_2, \ldots, p_{k+1} = d$ be the sequence of points on $SA[c,d]$, $k \ge 2$. Here, $SA[c,d]$ denotes the portion of $SA[a,b]$ between $c$ and $d$. From our assumption that the direct path from $a$ to $b$ is not one-sided, some Voronoi edges defined by point pairs $(p_i, p_{i+1})$ $(1 \le i \le k)$ do no intersect $ab$. Denote by $q_1, q_2, \ldots, q_k$ the points on the $x$-axis such that $|q_i p_i| = |q_i p_{i+1}|$, for all $1 \le i \le k$. Since point $u$ is the rightmost among the intersection points of $Vor(c)$ with $ab$ and since the edge of $Vor(c)$ on $B_{p_1,p_2}$ does not intersect $ab$, we have $u(x) < q_1(x)$, see Figs. 5 and 6. Analogously, $q_k(x) < v(x)$ holds.

We first claim that if $q_1(x) \le q_k(x)$, then $|SA[a,b]|/|ab| \le \pi/2$ and $SA[a,b]$ is wholly contained in $C$. Since $|q_1 p_1| < |q_1 u| + |u p_1|$, the leftmost point of the circle of radius $q_1 p_1$, centered at $q_1$, is to the right of point $l$ on $ab$. Also, since $|q_k p_{k+1}| < |q_k v| + |v p_{k+1}|$, the rightmost point of the circle of radius $q_k p_{k+1}$, centered at $q_k$, is to the left of point $r$ on $ab$. If $u(x) < q_1(x) \le q_2(x) \le \cdots \le q_k(x) < v(x)$ holds, then by an analogous argument, the leftmost point of the circle of radius $q_j p_{j+1}$ $(2 \le j \le k)$, centered at $q_j$, on $ab$ is to the right of, or identical to the leftmost point of the circle of radius $q_{j-1} p_j$, centered at $q_{j-1}$. Hence, the leftmost points of all circles of radius $q_i p_{i+1}$, centered at $q_i$ for all

---

[2] There are some instances in which the direct path from $a$ to $b$ intersects $ab$ more than four times.

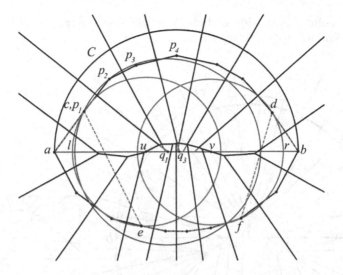

**Fig. 5.** Illustration of the proof of Lemma 4.

$1 \leq i \leq k$, are to the right of $l$ on $ab$. Analogously, the rightmost points of all circles of radius $q_i p_i$, centered at $q_i$ for all $1 \leq i \leq k$, are to the left of $r$ on $ab$. Since all the points of $SA[a, c] \cup SA[d, b]$ belong to the direct path from $a$ to $b$, it then follows from Lemma 2 that $|SA[a, b]|/|ab| \leq \pi/2$ and $SA[a, b]$ is wholly contained in $C$.

Consider the situation in which $u(x) < q_1(x) \leq q_2(x) \leq \cdots \leq q_k(x) < v(x)$ does *not* hold. For ease of presentation, let $q_0(x) = u(x)$ and $q_{k+1}(x) = v(x)$. Assume that $[i, j]$ is a *maximal* interval such that $1 \leq i < j \leq k$ and $q_i(x) > q_l(x)$, for all $l = i + 1, i + 2, \ldots, j$. So, $q_{i-1}(x) \leq q_i(x)$ and $q_j(x) \leq q_{j+1}(x)$. Clearly, it suffices to consider the situation in which $[i, j]$ is the first (or leftmost) maximal interval on $[1, k]$. Since $q_i(x) > q_{i+1}(x)$, both $q_i$ and $p_{i+2}$ are to the same side of $B_{p_{i+1}, p_{i+2}}$, and thus $|q_i p_{i+2}| < |q_i p_{i+1}|$. Analogously, since both $q_i$ and $p_{i+l}$, $l \geq 3$ and $i + l \leq j + 1$, are to the same side of $B_{p_{i+l-1}, p_{i+l}}$, we have $|q_i p_{i+l}| < |q_i p_{i+l-1}| < \cdots < |q_i p_{i+1}|$. This implies that $p_{i+2}, \ldots p_{j+1}$ are all contained in the circle of radius $q_i p_{i+1}$, centered at $q_i$. From the convexity of $S$, all points $q_{i+1}, \ldots, q_j$ can be ignored in our discussion. For the instance of Fig. 5, the maximal interval we considered is $[1, 2]$, and thus, point $q_2$ is ignored. This process can repeatedly be performed, until an $x$-monotone sequence of the points $q_n$, which are *not* ignored, is obtained. The rest discussion is the same as that in the case $u(x) < q_1(x) \leq q_2(x) \leq \cdots \leq q_k(x) < v(x)$. Again, $|SA[a, b]|/|ab| \leq \pi/2$, and $SA[ab]$ is wholly contained in $C$. Our claim is thus proved.

Following from the above claim, it suffices to consider only the situation in which $q_1(x) > q_k(x)$. Let $t$ be the intersection point of $B_{p_1, p_2}$ and $B_{p_k, p_{k+1}}$. (Recall that $p_1 = c$ and $p_{k+1} = d$.) From the convexity of $Vor(p_1)$ and

$Vor(p_{k+1})$, point $t$ is outside of both $Vor(p_1)$ and $Vor(p_{k+1})$. Since $q_1(x) >$ $q_k(x)$, we also have $t(y) > 0$, see Fig. 6.

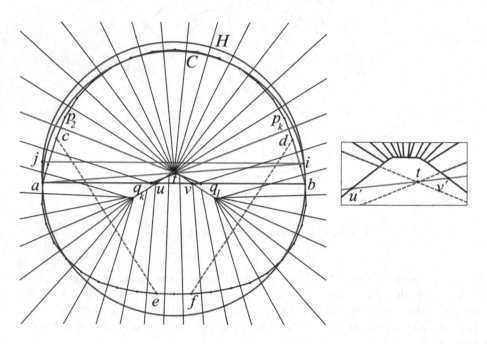

**Fig. 6.** A situation in which $q_1(x) > q_k(x)$.

Denote by $i$ ($\neq a$) the intersection point of the line through $a$ and $t$ with $C$. Both $B_{p_1,p_2}$ and $B_{p_k,p_{k+1}}$ then intersect $ai$ at $t$. Denote by $u'$ (resp. $v'$) the rightmost (resp. leftmost) intersection point of $Vor(p_1)$ (resp. $Vor(p_{k+1})$) with segment $ai$, see the right of Fig. 6. Let $q_1', q_2', \ldots, q_k'$ be the points on the line through $a$ and $i$ such that $|q_i'p_i| = |q_i'p_{i+1}|$, for all $1 \leq i \leq k$. Since $q_1'$ is identical to $q_k'$, from the proof of the above claim, $SA[c,d]$ is wholly contained in the circle of diameter $l'r'$, where $l'$ and $r'$ are the leftmost and rightmost points of the circles of radii $u'c$ and $v'd$, centered at $u'$ and $v'$, on segment $ai$ respectively. Since $l'r'$ is contained in $ai$ and $SA[a,c]$ is contained in $C$, $SA[a,d]$ is wholly contained in the circle of diameter $ai$, too.

Denote by $H$ the semicircle of diameter $ai$, which is vertically above $ai$. Notice that the chain $SA[d,b]$ is contained in $C$ and intersects segment $ai$. Thus, all points of $SA[a,b]$ are contained in the convex region bounded by $ab$, $H$ and the arc $\widehat{ib}$ of $C$ (Fig. 6). Denote by $j$ the intersection point of $H$ with the horizontal line through $i$. As discussed in the proof of Lemma 3, the line through $a$ and $j$ is tangent to $C$. Hence, segment $aj$ does not intersect $SA[a,c]$ (nor $SA[a,b]$), excluding point $a$.

Let $\alpha = \angle bai$. As shown in the proof of Lemma 3, $\alpha < \pi/6$ holds and two line segments of length $\cos(\alpha)\sin(\alpha)|ab|$ can be used to cut off their corresponding arcs of $H$. Since $aj$ does not intersect $SA(a, b]$, we have $|DT(a, b)|/|ab| \leq \alpha + \cos(\alpha)(3\sin(\alpha) + \pi/2 - 3\alpha)$, $0 < \alpha < \pi/6$. $\square$

We can now give the main result of this paper.

**Theorem 1.** *Suppose that the set $S$ of given points is in convex position, and $a$ and $b$ are two points of $S$. In the Delaunay triangulation of $S$, there is a path from $a$ to $b$ such that its length is less than $1.83|ab|$.*

**Proof.** Suppose that the direct path from $a$ to $b$ is not one-sided; otherwise, $|DT(a, b)| \leq \pi|ab|/2$ and we are done. Let $f_1(\alpha) = \sin(\alpha) + \cos(\alpha)(\cos(\alpha) + \alpha)$, $\alpha \in (0, \pi/2)$, $f_2 = \sin(\alpha) + \cos(\alpha)(2\sin(\alpha) + \pi/2 - 2\alpha)$, $\alpha \in (0, \pi/6)$, $f_3 = \alpha + \cos(\alpha)(3\sin(\alpha) + \pi/2 - 3\alpha)$, $\alpha \in (0, \pi/6)$. It then follows from Lemmas 3 and 4 that $|DT(a, b)|/|ab| \leq \max\{\pi/2, f_1(\alpha), f_2(\alpha), f_3(\alpha)\}$. Since the function $f_i(\alpha)$ is convex, $i = 1, 2$ or $3$, we can obtain $f_i(\alpha) < 1.83$ by letting $f_i'(\alpha) = 0$. $\square$

## 4   Concluding Remarks

We have shown that the stretch factor of the Delaunay triangulation of a set of points in convex position is less than 1.83. We believe that the same stretch factor also holds for the set of points in general position. Again, the direct path from $a$ to $b$ may be used to distinguish two different situations. For instance, if the direct path from $a$ to $b$ intersects $ab$ an even number times, then as in the proof of Lemma 4, we can find a path between $a$ and $b$ such that it is wholly above or below $ab$ and contained in the region bounded by $ab$, an arc, say, $\overset{\curvearrowright}{ai}$ of $C$, and a semicircle of diameter $ai$. A new difficulty is that the considered path between $a$ and $b$ is non-convex, which needs be dealt with. Also, it is a challenging open problem to reduce the stretch factor of $DT(S)$ further, so as to close the gap to its lower bound (roughly about 1.60).

## References

1. Amani, M., Biniaz, A., Bose, P., De Carufel, J.-L., Maheshwair, A., Smid, M.: A plane 1.88-spanner for points in convex position. J. Comput. Geom. **7**, 520–539 (2016)
2. Chew, L.P.: There are planar graphs almost as good as the complete graph. J. Comput. Syst. Sci. **39**, 205–219 (1989)
3. Cui, S., Kanji, I.A., Xia, G.: On the stretch factor of Delaunay triangulations of points in convex position. Comput. Geom. **44**, 104–109 (2011)
4. Dobkin, D.P., Friedman, S.T., Supowit, K.J.: Delaunay graphs are almost as good as complete graphs. Discret. Comput. Geom. **5**, 399–407 (1990)
5. Eppstein, D.: Spanning trees and spanners. In: Sack, J.-R., Urrutia, J., (eds.), Handbook of Computational Geometry, pp. 425–462. North-Hollard Press (2000)
6. Narasimhan, G., Smid, M.: Geometric Spanner Networks. Cambridge University Press, Cambridge (2007)
7. Keil, J.M., Gutwin, C.A.: The Delaunay triangulation closely approximates the complete graph. Discret. Comput. Geom. **7**, 13–28 (1992)

8. Hershberger, J., Suri, S.: Practical methods for approximating shortest paths on a convex polytope in $R^3$. Comput. Geom. **10**, 31–46 (1998)
9. Xia, G.: The stretch factor of the Delaunay triangulation is less than 1.998. SIAM J. Comput. **42**, 1620–1659 (2013)
10. Xia, G., Zhang, L.: Towards the tight bound of the stretch factor of Delaunay triangulations. In: Proceedings of CCCG (2011)

# Solving $(k - 1)$-Stable Instances
# of k-terminal cut with Isolating Cuts

Mark Velednitsky$^{(\boxtimes)}$ (iD)

University of California, Berkeley, USA
marvel@berkeley.edu

**Abstract.** The k-terminal cut problem, also known as the Multiway
Cut problem, is defined on an edge-weighted graph with $k$ distinct vertices called "terminals." The goal is to remove a minimum weight collection of edges from the graph such that there is no path between any
pair of terminals. The problem is NP-hard.

Isolating cuts are minimum cuts which separate one terminal from the
rest. The union of all the isolating cuts, except the largest, is a $(2 - 2/k)$-
approximation to the optimal k-terminal cut. This is the only currently-
known approximation algorithm for k-terminal cut which does not
require solving a linear program.

An instance of k-terminal cut is $\gamma$-stable if edges in the cut can be
multiplied by up to $\gamma$ without changing the unique optimal solution. In
this paper, we show that, in any $(k - 1)$-stable instance of k-terminal
cut, the source sets of the isolating cuts are the source sets of the
unique optimal solution of that k-terminal cut instance. We conclude
that the $(2 - 2/k)$-approximation algorithm returns the optimal solution
on $(k - 1)$-stable instances. Ours is the first result showing that this
$(2 - 2/k)$-approximation is an exact optimization algorithm on a special
class of graphs.

We also show that our $(k - 1)$-stability result is tight. We construct $(k - 1 - \epsilon)$-stable instances of the k-terminal cut problem which
only have trivial isolating cuts: that is, the source set of the isolating
cut for each terminal is just the terminal itself. Thus, the $(2 - 2/k)$-
approximation does not return an optimal solution.

## 1 Introduction

The k-terminal cut problem, also known as the Multiway Cut problem, is
defined on an edge-weighted graph with $k$ distinct vertices called "terminals."
The goal is to remove a minimum weight collection of edges from the graph
such that there is no path between any pair of terminals. The k-terminal cut
problem is known to be APX-hard [6].

In [3], Bilu and Linial introduced the concept of stability for graph cut problems. An instance is said to be $\gamma$-stable if the optimal cut remains uniquely optimal when every edge in the cut is multiplied by a factor up to $\gamma$. The concept

---

The author is a Fellow of the National Physical Science Consortium.

© Springer Nature Switzerland AG 2019
Y. Li et al. (Eds.): COCOA 2019, LNCS 11949, pp. 485–495, 2019.
https://doi.org/10.1007/978-3-030-36412-0_39

of robustness in linear programming is closely related [2,9]. In [8], the authors showed that, for 4-stable instances of k-terminal cut, the solution to a certain linear programming relaxation of the problem will necessarily be integer. The result was later improved to $(2 - 2/k)$-stable instances using the same linear programming technique [1].

In an instance of k-terminal cut, isolating cuts are minimum cuts which separate one terminal from the rest of the terminals. They can give useful information about the optimal solution: the source set of a terminal's isolating cut is a subset of that terminal's source set in an optimal solution [6]. Furthermore, the union of all the isolating cuts, except for the cut with largest weight, is a $(2 - 2/k)$-approximation for the k-terminal cut problem [6]. This algorithm is the only currently-known approximation algorithm for k-terminal cut which does not require solving a linear program [4,5,7,10]. Thanks to their relative simplicity, isolating cuts are easily put into practice [11]. It is natural to wonder how the $(2 - 2/k)$-approximation performs on non worst-case instances.

In this paper, we establish a connection between isolating cuts and stability. We show that in $(k - 1)$-stable instances of k-terminal cut, the source sets of the isolating cuts equal the source sets of the unique optimal solution of that k-terminal cut instance. It follows that the simple $(2 - 2/k)$-approximation of [6] returns the optimal solution on $(k-1)$-stable instances. Ours is the first result showing that this $(2 - 2/k)$-approximation is an exact optimization algorithm on a special class of graphs.

Our result is tight. For $\epsilon > 0$, we construct $(k - 1 - \epsilon)$-stable instances of the k-terminal cut problem which only have trivial isolating cuts: that is, the source set of the isolating cut for each terminal is just the terminal itself. In these $(k - 1 - \epsilon)$-stable instances, the $(2 - 2/k)$-approximation does not return an optimal solution.

In Sect. 2, we introduce definitions and notation. In Sect. 3, we prove the main structural result, that in $(k - 1)$-stable instance of k-terminal cut the source sets of the isolating cuts equal the source sets of the optimal k-terminal cut. In Sect. 4, we construct a $(k - 1 - \epsilon)$-stable graph in which the source set of the isolating cut for each terminal is just the terminal itself.

## 2    Preliminaries

The notation $\{G = (V, E), w, T\}$ refers to an instance of the k-terminal cut problem, where $G = (V, E)$ is an undirected graph with vertices $V$ and edges $E$. $T = \{t_1, \ldots, t_k\} \subseteq V$ is a set of $k$ terminals. The weight function $w$ is a function from $E$ to $\mathbb{R}^+$.

For a subset of edges $E' \subseteq E$, the notation $w(E')$ is the total weight of edges in $E'$:

$$w(E') = \sum_{e \in E'} w(e).$$

For two disjoint subsets of vertices $V_1 \subseteq V$, $V_2 \subseteq V$, the notation $w(V_1, V_2)$ is the total weight of edges between $V_1$ and $V_2$:

$$w(V_1, V_2) = \sum_{\substack{(v_1, v_2) \in E \\ v_1 \in V_1 \\ v_2 \in V_2}} w((v_1, v_2)).$$

We can further generalize this notation to allow for several disjoint subsets of vertices $V_1, \ldots, V_m \subseteq V$. In this case, we calculate the total weight of edges that go between two distinct subsets:

$$w(V_1, \ldots, V_m) = \sum_i \sum_{j > i} w(V_i, V_j).$$

For an instance $\{G = (V, E), w, T\}$ of the k-terminal cut problem, we can refer to optimal solution in two equivalent ways. The first is in terms of the edges that are cut and the second is in terms of the source sets.

Referring to the optimal cut in terms of edges, we use the notation $E_{\text{OPT}}$: the subset of $E$ of minimum total weight whose removal ensures that there is no path between any pair of terminals.

Source sets are a partition of $V$ into $S_1, S_2, \ldots, S_k$ such that $t_i \in S_i$. We say that $S_i$ is the source set corresponding to $t_i$. We denote the optimal source sets $S_1^*, S_2^*, \ldots, S_k^*$.

The set of edges in the optimal cut is precisely the set of edges which go between distinct elements of the optimal partition $(S_1^*, \ldots, S_k^*)$. Combining the notation introduced in this section,

$$w(E_{\text{OPT}}) = w(S_1^*, \ldots, S_k^*).$$

## 2.1   Stability

**Definition 1 ($\gamma$-Perturbation).** *Let $G = (V, E)$ be a weighted graph with edge weights $w$. Let $G' = (V, E)$ be a weighted graph with the same set of vertices $V$ and edges $E$ and a new set of edge weights $w'$ such that, for every $e \in E$ and some $\gamma > 1$,*

$$w(e) \leq w'(e) \leq \gamma w(e).$$

*Then $G'$ is a $\gamma$-perturbation of $G$.*

Stable instances are instances where the optimal solution remains uniquely optimal for any $\gamma$-perturbation of the weighted graph.

**Definition 2 ($\gamma$-Stability).** *An instance $\{G = (V, E), w, T\}$ of k-terminal cut is $\gamma$-stable ($\gamma > 1$) if there is an optimal solution $E_{\text{OPT}}$ which is uniquely optimal for k-terminal cut for every $\gamma$-perturbation of $G$.*

Note that the optimal solution need not be $\gamma$ times as good as *any* other solution, since two solutions may share many edges. Given an alternative feasible solution, $E_{\text{ALT}}$, to the optimal cut, $E_{\text{OPT}}$, in a $\gamma$-stable instance, we can make a statement about the relative weights of the edges where the cuts differ. The following equivalence was first noted in [8]:

**Lemma 1 ($\gamma$-Stability).** *Let $\{G = (V, E), w, T\}$ be an instance of k-terminal cut with optimal cut $E_{\text{OPT}}$. Let $\gamma > 1$. $G$ is $\gamma$-stable iff for every alternative feasible k-terminal cut $E_{\text{ALT}} \neq E_{\text{OPT}}$, we have*

$$w(E_{\text{ALT}} \setminus E_{\text{OPT}}) > \gamma w(E_{\text{OPT}} \setminus E_{\text{ALT}}).$$

*Proof.* Note that $E_{\text{ALT}}$ cannot be a strict subset of $E_{\text{OPT}}$ (since $E_{\text{OPT}}$ is optimal) and the claim is trivial if $E_{\text{OPT}}$ is a strict subset of $E_{\text{ALT}}$. Thus, we can assume that both $E_{\text{ALT}} \setminus E_{\text{OPT}}$ and $E_{\text{OPT}} \setminus E_{\text{ALT}}$ are non-empty.

For the "if" direction, consider an arbitrary $\gamma$-perturbation of $G$ in which the edge $e$ is multiplied by $\gamma_e$. We first derive the following two inequalities,

$$\sum_{e \in E_{\text{OPT}}} \gamma_e w(e) = \sum_{e \in E_{\text{OPT}} \cap E_{\text{ALT}}} \gamma_e w(e) + \sum_{e \in E_{\text{OPT}} \setminus E_{\text{ALT}}} \gamma_e w(e)$$
$$\leq \sum_{e \in E_{\text{OPT}} \cap E_{\text{ALT}}} \gamma_e w(e) + \gamma w(E_{\text{OPT}} \setminus E_{\text{ALT}}),$$

and

$$\sum_{e \in E_{\text{ALT}}} \gamma_e w(e) = \sum_{e \in E_{\text{OPT}} \cap E_{\text{ALT}}} \gamma_e w(e) + \sum_{e \in E_{\text{ALT}} \setminus E_{\text{OPT}}} \gamma_e w(e)$$
$$\geq \sum_{e \in E_{\text{OPT}} \cap E_{\text{ALT}}} \gamma_e w(e) + w(E_{\text{ALT}} \setminus E_{\text{OPT}}).$$

Since we have the inequality

$$w(E_{\text{ALT}} \setminus E_{\text{OPT}}) > \gamma w(E_{\text{OPT}} \setminus E_{\text{ALT}}),$$

we conclude that

$$\sum_{E_{\text{OPT}}} \gamma_e w(e) < \sum_{E_{\text{ALT}}} \gamma_e w(e).$$

Hence, $E_{\text{OPT}}$ remains uniquely optimal in any $\gamma$-perturbation.

For the "only if" direction, if $G$ is $\gamma$-stable, then we can multiply each edge in $E_{\text{OPT}}$ by $\gamma$ and $E_{\text{OPT}}$ will still be uniquely optimal:

$$w(E_{\text{ALT}} \setminus E_{\text{OPT}}) + \gamma w(E_{\text{ALT}} \cap E_{\text{OPT}}) > \gamma w(E_{\text{OPT}} \setminus E_{\text{ALT}}) + \gamma w(E_{\text{ALT}} \cap E_{\text{OPT}}).$$

Thus,

$$w(E_{\text{ALT}} \setminus E_{\text{OPT}}) > \gamma w(E_{\text{OPT}} \setminus E_{\text{ALT}}).$$

$\square$

We make a few observations about $\gamma$-stability:

**Fact 1.** *Any `k-terminal cut` instance that is stable with $\gamma > 1$ must have a unique optimal solution.*

*Proof.* By Definition 1, any graph is a $\gamma$-perturbation of itself. Thus, by Definition 2, the optimal solution must be unique. □

**Fact 2.** *Any `k-terminal cut` instance that is $\gamma_2$-stable is also $\gamma_1$-stable for any $1 < \gamma_1 < \gamma_2$.*

*Proof.* The set of $\gamma_1$-perturbations is a subset of the set of $\gamma_2$-perturbations, since

$$w(e) \leq w'(e) \leq \gamma_1 w(e) \implies w(e) \leq w'(e) \leq \gamma_2 w(e).$$

□

Thus, for example, every instance which is 4-stable is necessarily 2-stable, but not the other way around.

## 2.2 Isolating Cuts

**Definition 3 ($t_i$-Isolating Cut).** *The $t_i$-isolating cut is a minimum $(s,t)$-cut which separates source terminal $s = t_i$ from all the other terminals (shrunk into a single sink terminal $t = T \setminus \{t_i\}$).*

We will use the notation $Q_i$ to denote the source set of this isolating cut (the set of vertices which remain connected to $t_i$). We use $E_i$ to denote the set of edges which are cut. Let $E_{\text{ISO}}$ be the union of the $E_i$ except for the $E_i$ with largest weight. The following two lemmas are due to [6]:

**Lemma 2.** $E_{\text{ISO}}$ *is a $(2 - 2/k)$-approximation for the optimal $k$-terminal cut.*

**Lemma 3.** *Let $\{G, w, T\}$ be an instance of `k-terminal cut`. Let $i \in \{1, \ldots, k\}$. Then there exists an optimal solution $(S_1^*, \ldots, S_k^*)$ in which*

$$Q_i \subseteq S_i^*.$$

The condition that "there exists" an optimal solution can make the implication of Lemma 3 somewhat complicated when there are multiple optimal solutions, since the equation $Q_i \subseteq S_i^*$ need not be simultaneously true for all $i$. Conveniently, when an instance is $\gamma$-stable ($\gamma > 1$), it has a unique optimal solution (Fact 1). Thus, the condition $Q_i \subseteq S_i^*$ will be simultaneously true for all $i$.

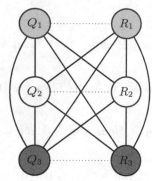 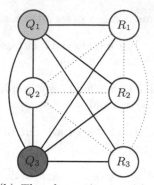

(a) The optimal partition, in which each $R_i$ is in a source set with its respective $Q_i$.

(b) The alternative partition used in Theorem 1 when $i = 2$, where all of the $R_i$ are in a source set with $Q_2$.

**Fig. 1.** The sets $Q_1, Q_2, Q_3$ and $R_1, R_2, R_3$ defined in Theorem 1 when $k = 3$. Solid lines represent edges which are cut. Dashed lines represent edges which are not cut.

## 3    Proof of Main Result

**Theorem 1.** *Let $\{G, w, T\}$ be a $(k-1)$-stable instance of $k$-terminal cut. Then, for all $i$, $Q_i = S_i^*$.*

*Proof.* We will primarily be working with the $k$ vertex sets $Q_1, \ldots, Q_k$ and the $k$ vertex sets $S_1^* \setminus Q_1, \ldots S_k^* \setminus Q_k$. For convenience, we will use the notation $R_i = S_i^* \setminus Q_i$. As a consequence of Lemma 3, $S_i^* = Q_i \cup R_i$. We will assume, for the sake of contradiction, that at least one $R_i$ is non-empty.

Since $Q_i$ is the source set for the isolating cut for terminal $t_i$:

$$w(Q_i, V \setminus Q_i) \leq w(S_i^*, V \setminus S_i^*)$$
$$w(Q_i, V \setminus Q_i) \leq w(R_i, V \setminus S_i^*) + w(Q_i, V \setminus S_i^*)$$
$$-w(Q_i, V \setminus S_i^*) + w(Q_i, V \setminus Q_i) \leq w(R_i, V \setminus S_i^*)$$
$$w(Q_i, R_i) \leq w(R_i, V \setminus S_i^*)$$
$$w(Q_i, R_i) \leq \sum_{\{j \mid j \neq i\}} w(R_i, R_j) + \sum_{\{j \mid j \neq i\}} w(R_i, Q_j)$$

Summing these inequalities over all the $i$:

$$\sum_i w(Q_i, R_i) \leq \sum_i \sum_{\{j \mid j \neq i\}} w(R_i, R_j) + \sum_i \sum_{\{j \mid j \neq i\}} w(R_i, Q_j)$$
$$\sum_i w(Q_i, R_i) \leq 2w(R_1, \ldots, R_k) + \sum_i \sum_{\{j \mid j \neq i\}} w(R_i, Q_j) \tag{1}$$

Next, we will consider alternatives to the optimal cut $(S_1^*, \ldots, S_k^*)$ and apply Lemma 1. The optimal cut can be written as

$$(S_1^*, \ldots, S_k^*) = (Q_1 \cup R_1, \ldots, Q_k \cup R_k).$$

We will consider alternative cuts $E_{\text{ALT}}^{(i)}$ where all the $R_j$ are in the same set of the partition, associated with $Q_i$. That is, we will consider

$$\left( S_1, \ldots, S_{i-1}, S_i, S_{i+1}, \ldots, S_k \right)$$
$$= \left( Q_1, \ldots, Q_{i-1}, Q_i \cup (R_1 \cup \ldots \cup R_k), Q_{i+1}, \ldots, Q_k \right).$$

See Fig. 1 for an illustration. We assumed that at least one of the $R_i$ is non-empty, so at least $k - 1$ of these alternative cuts are distinct from the optimal one[1]. In order to apply Lemma 1, we need to calculate $w(E_{\text{OPT}} \setminus E_{\text{ALT}}^{(i)})$ and $w(E_{\text{ALT}}^{(i)} \setminus E_{\text{OPT}})$.

To calculate $w(E_{\text{ALT}}^{(i)} \setminus E_{\text{OPT}})$, consider the edges in $E_{\text{ALT}}^{(i)}$ with one endpoint in $Q_j$ ($j \neq i$). The only edges which are *not* counted in $E_{\text{OPT}}$ are those which go to $R_j$. Thus,

$$w(E_{\text{ALT}}^{(i)} \setminus E_{\text{OPT}}) = \sum_{\{j | j \neq i\}} w(R_j, Q_j).$$

To calculate $w(E_{\text{OPT}} \setminus E_{\text{ALT}}^{(i)})$, we must consider the set of edges which are in $E_{\text{OPT}}$ but not in $E_{\text{ALT}}^{(i)}$. For an edge not to be in $E_{\text{ALT}}^{(i)}$, it must be internal to one of the $Q_j$ ($j \neq i$) or internal to $Q_i \cup (R_1 \cup \ldots \cup R_k)$. None of the internal edges of the $Q_j$ are in $E_{\text{OPT}}$, so we need only consider the internal edges of $Q_i \cup (R_1 \cup \ldots \cup R_k)$:

$$w(E_{\text{OPT}} \setminus E_{\text{ALT}}^{(i)}) = w(R_1, \ldots, R_k) + \sum_{\{j | j \neq i\}} w(R_j, Q_i).$$

We apply Lemma 1, with $\gamma = k - 1$:

$$(k - 1) \cdot w(E_{\text{OPT}} \setminus E_{\text{ALT}}^{(i)}) < w(E_{\text{ALT}}^{(i)} \setminus E_{\text{OPT}}) \quad (2)$$
$$(k - 1) \cdot w(R_1, \ldots, R_k) + (k - 1) \cdot \sum_{\{j | j \neq i\}} w(R_j, Q_i) < \sum_{\{j | j \neq i\}} w(R_j, Q_j).$$

Averaging over the $k$ inequalities (one for each $i$) (See footnote 1):

$$(k - 1) \cdot w(R_1, \ldots, R_k) + \frac{k - 1}{k} \sum_i \sum_{\{j | j \neq i\}} w(R_j, Q_i) < \frac{k - 1}{k} \sum_i w(R_i, Q_i).$$
$$(3)$$

---

[1] If only one $R_i$ is non-empty, then $E_{\text{OPT}} = E_{\text{ALT}}^{(i)}$ for this $i$. The corresponding inequality in Eq. 2 is not strict (both sides are 0), but the other $k - 1$ inequalities are strict and so the average (Eq. 3) is still a strict inequality.

We combine this with the inequality derived in Eq. 1:

$$(k-1) \cdot w(R_1, \ldots, R_k) + \frac{k-1}{k} \sum_i \sum_{\{j|j \neq i\}} w(R_j, Q_i)$$

$$< 2\frac{k-1}{k} w(R_1, \ldots, R_k) + \frac{k-1}{k} \sum_i \sum_{\{j|j \neq i\}} w(R_i, Q_j).$$

Notice that

$$\sum_i \sum_{\{j|j \neq i\}} w(R_j, Q_i) = \sum_i \sum_{\{j|j \neq i\}} w(R_i, Q_j).$$

Therefore,

$$(k-1) \cdot w(R_1, \ldots, R_k) < 2\frac{k-1}{k} w(R_1, \ldots, R_k).$$

This is a contradiction, so it must be the case that $R_i = \emptyset$ for all $i$. Thus, $Q_i = S_i^*$ for all $i$. □

**Corollary 1.** Let $\{G, w, T\}$ be a $(k-1)$-stable instance of *k-terminal cut*. Then $E_{ISO}$ is the unique optimal solution to *k-terminal cut* on this instance.

## 4     Tightness of Main Result

**Theorem 2.** There exists a $(k-1-\epsilon)$-stable instance of *k-terminal cut* for which $Q_i = \{t_i\} \neq S_i^*$ for all $i \in \{1, \ldots, k\}$.

*Proof.* Consider a graph with $2k$ vertices. There are $k$ terminals $(t_1, \ldots, t_k)$ and $k$ other vertices $(s_1, \ldots, s_k)$. The $\binom{k}{2}$ edges between $s_i$ and $s_j$ $(i \neq j)$ have weight $a \in \mathbb{R}^+$. The $k$ edges from $t_i$ to $s_i$ have weight $b \in \mathbb{R}^+$. The $k(k-1)$ edges from $t_i$ to $s_j$ $(i \neq j)$ have weight $c \in \mathbb{R}^+$. Call this graph $G_k$. See Fig. 2a for a drawing of $G_k$ when $k = 3$. We will show that this graph has the desired properties for appropriate choices of $a, b, c \in \mathbb{R}^+$.

Consider an arbitrary $\gamma$. We would like to chose values of $a$, $b$, and $c$ such that the unique, $\gamma$-stable optimal cut is the one in which each $s_i$ remains connected to the corresponding $t_i$. Equivalently, zero edges with weight $b$ are cut. All the other edges *are* cut. Thus, the optimal cut should have weight

$$\binom{k}{2} a + k(k-1)c.$$

To verify that the optimal cut is $\gamma$-stable, we need to consider every other possible cut. We are helped by the symmetry of the construction. Consider an alternative cut, $E_{ALT}^{(p)}$, in which **exactly $p$ edges with weight $b$ are in the cut** (see Fig. 2b). Equivalently, exactly $k-p$ of the $s_i$ remain connected to the corresponding $t_i$. The optimal cut is the unique cut with $p = 0$, so we need only

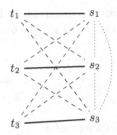 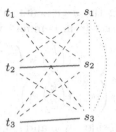

(a) Dotted lines have weight $a$. Solid have weight $b$. Dashed have weight $c$.

(b) When $p = 1$. The red edges are in $E_{\text{ALT}}^{(1)}$ and the blue edge are not.

**Fig. 2.** The construction used in Theorem 2 when $k = 3$. (Color figure online)

consider alternative cuts where $p \in \{1, \ldots, k\}$. We would like to construct an inequality for the alternative cut of the form in Lemma 1.

Consider $w(E_{\text{ALT}}^{(p)} \setminus E_{\text{OPT}})$. By construction, the number of edges of weight $b$ which are in $E_{\text{ALT}}^{(p)}$ but not $E_{\text{OPT}}$ is exactly $p$. Thus,

$$w(E_{\text{ALT}}^{(p)} \setminus E_{\text{OPT}}) = pb.$$

Calculating $w(E_{\text{OPT}} \setminus E_{\text{ALT}}^{(p)})$ is more difficult. In order to create the tightest possible inequality in Lemma 1, we want to include as few edges of weight $a$ and $c$ as possible in $E_{\text{ALT}}^{(p)}$ in order to maximize $w(E_{\text{OPT}} \setminus E_{\text{ALT}}^{(p)})$. We will consider the edges of weight $c$ and $a$ in the next two paragraphs.

Consider the edges of weight $c$. Notice that every $s_i$ is adjacent to all the terminals. Thus, if $s_i$ is one of the $k - p$ which remains connected to $t_i$, then all of the $k - 1$ edges between $s_i$ and $t_j$ $(j \neq i)$ must be in $E_{\text{ALT}}^{(p)}$. On the other hand, if the edge between $s_i$ and $t_i$ is in $E_{\text{ALT}}^{(p)}$, then at most one of the edges of weight $c$ adjacent to $s_i$ can be excluded from $E_{\text{ALT}}^{(p)}$. Thus, the number of edges of weight $c$ in $w(E_{\text{OPT}} \setminus E_{\text{ALT}}^{(p)})$ is at most $p$.

Consider the edges of weight $a$. Recall that $k - p$ of the edges with weight $b$ are *not* in $E_{\text{ALT}}^{(p)}$, which means that there are $k - p$ vertices $s_i$ connected to the corresponding $t_i$. Between these $k - p$ vertices, all of the edges of weight $a$ must be in $E_{\text{ALT}}^{(p)}$. Thus, $\binom{k-p}{2}$ edges of weight $a$ must be in $E_{\text{ALT}}^{(p)}$. Of the $p$ vertices $s_j$ which are not connected to the corresponding $t_j$, each one can remain connected to at most one of the aforementioned $k - p$ vertices. When $p < k$, this gives an additional $p(k - p - 1)$ edges which must be in $E_{\text{ALT}}^{(p)}$.

Combining the arguments in the two preceding paragraphs, the strongest inequality we get from Lemma 1 when exactly $p$ edges of weight $b$ are cut is

$$pb > \gamma \left( pc + (\binom{k}{2}) - \binom{k-p}{2}) - (k-p-1)p)a \right) \qquad \text{if } p < k$$

$$pb > \gamma \left( pc + (\binom{k}{2})a \right) \qquad \text{if } p = k.$$

Dividing both sides by $p$ and simplifying the coefficient of $a$:

$$b > \gamma \left( c + \frac{p+1}{2}a \right) \qquad \text{if } p < k$$

$$b > \gamma \left( c + \frac{k-1}{2}a \right) \qquad \text{if } p = k.$$

We need only consider the case $p = k-1$ to get the strongest possible inequality:

$$b > \gamma(c + \frac{k}{2}a). \tag{4}$$

Computing the condition for the isolating cuts to have trivial source sets is easier. Knowing that $t_i$ and $s_i$ are connected in the optimal cut, we know that the optimal isolating cut can have source set either $\{t_i\}$ or $\{t_i, s_i\}$. The source set is $\{t_i\}$ if

$$b + (k-1)c < (k-1)a + 2(k-1)c.$$
$$b < (k-1)(a+c). \tag{5}$$

In summary, if inequality 4 is satisfied then $E_{\text{OPT}}$ is the $\gamma$-stable optimal **k-terminal cut** and if inequality 5 is satisfied then the isolating cuts have trivial source sets. When $\gamma = k - 1 - \epsilon$, the following values simultaneously satisfy inequalities 4 and 5:

$$a = 2\epsilon$$
$$b = k(k-1)(k-1-\epsilon)$$
$$c = k(k-1-\epsilon) - \epsilon.$$

□

## 5   Conclusions

In this paper, we proved that, in $(k-1)$-stable instances of **k-terminal cut**, the source sets of the isolating cuts are the source sets of the unique optimal solution to that **k-terminal cut** instance. As an immediate corollary, we concluded that the well-known $(2 - 2/k)$-approximation algorithm for **k-terminal cut** is optimal for $(k-1)$-stable instances.

We also showed that the factor of $k-1$ is tight. We constructed $(k-1-\epsilon)$-stable instances of **k-terminal cut** in which the source set of the isolating cut for a terminal is just the terminal itself. In those instances, the $(2 - 2/k)$-approximation algorithm does not return an optimal solution.

# References

1. Angelidakis, H., Makarychev, K., Makarychev, Y.: Algorithms for stable and perturbation-resilient problems. In: Proceedings of the 49th Annual ACM SIGACT Symposium on Theory of Computing, pp. 438–451. ACM (2017)
2. Ben-Tal, A., Nemirovski, A.: Robust solutions of linear programming problems contaminated with uncertain data. Math. Program. **88**(3), 411–424 (2000)
3. Bilu, Y., Linial, N.: Are stable instances easy? Comb. Probab. Comput. **21**(5), 643–660 (2012)
4. Buchbinder, N., Naor, J.S., Schwartz, R.: Simplex partitioning via exponential clocks and the multiway cut problem. In: Proceedings of the Forty-Fifth Annual ACM Symposium on Theory of Computing, pp. 535–544. ACM (2013)
5. Călinescu, G., Karloff, H., Rabani, Y.: An improved approximation algorithm for multiway cut. In: Proceedings of the Thirtieth Annual ACM Symposium on Theory of Computing, pp. 48–52. ACM (1998)
6. Dahlhaus, E., Johnson, D.S., Papadimitriou, C.H., Seymour, P.D., Yannakakis, M.: The complexity of multiterminal cuts. SIAM J. Comput. **23**(4), 864–894 (1994)
7. Karger, D.R., Klein, P., Stein, C., Thorup, M., Young, N.E.: Rounding algorithms for a geometric embedding of minimum multiway cut. Math. Oper. Res. **29**(3), 436–461 (2004)
8. Makarychev, K., Makarychev, Y., Vijayaraghavan, A.: Bilu-Linial stable instances of max cut and minimum multiway cut. In: Proceedings of the Twenty-Fifth Annual ACM-SIAM Symposium on Discrete Algorithms, pp. 890–906. Society for Industrial and Applied Mathematics (2014)
9. Robinson, S.M.: A characterization of stability in linear programming. Oper. Res. **25**(3), 435–447 (1977)
10. Sharma, A., Vondrák, J.: Multiway cut, pairwise realizable distributions, and descending thresholds. In: Proceedings of the Forty-Sixth Annual ACM Symposium on Theory of Computing, pp. 724–733. ACM (2014)
11. Velednitsky, M., Hochbaum, D.S.: Isolation branching: a branch and bound algorithm for the k-Terminal cut problem. In: Kim, D., Uma, R.N., Zelikovsky, A. (eds.) COCOA 2018. LNCS, vol. 11346, pp. 624–639. Springer, Cham (2018). https://doi.org/10.1007/978-3-030-04651-4_42

# A Task Assignment Approach
# with Maximizing User Type Diversity
# in Mobile Crowdsensing

Ana Wang[1,2,3], Lichen Zhang[1,2,3]($\boxtimes$), Longjiang Guo[1,2,3]($\boxtimes$), Meirui Ren[1,2,3],
Peng Li[1,2,3], and Bin Yan[1,2,3]

[1] Key Laboratory of Modern Teaching Technology, Ministry of Education,
Xi'an 710062, China
[2] Engineering Laboratory of Teaching Information Technology of Shaanxi Province,
Xi'an 710119, China
[3] School of Computer Science, Shaanxi Normal University,
Xi'an 710119, China
{zhanglichen,longjiangguo}@snnu.edu.cn

**Abstract.** Mobile crowdsensing (MCS) employs numerous mobile users
to perform sensing tasks, in which task assignment is a challenging issue.
Existing researches on task assignment mainly consider spatial-temporal
diversity and capacity diversity, but not focus on the type diversity of
users, which may lead to low quality of tasks. This paper formalizes a
novel task assignment problem in MCS, where a task needs the cooperation of various types of users, and the quality of a task is highly related
to the various types of the recruited users. Therefore, the goal of the
problem is to maximize the user type diversity subject to limited task
budget. This paper uses three heuristic algorithms to try to resolve this
problem, so as to maximize user type diversity. Through further simulations, the proposed algorithm UR-GAT (Unit Reward-based Greedy
algorithm by type) obviously improves the user type diversity.

**Keywords:** Mobile crowdsensing · Task assignment · User type
diversity

## 1 Introduction

Mobile crowdsensing (MCS) [1–4] is a new paradigm and widely applied in many
areas, which mainly utilizes the sensing, computing, storage and communication
functions of mobile smart devices. It regards mobile smart devices carried by
mobile users as wireless sensors with powerful functions and completes the sensing task together through their cooperation. In MCS, a sensing task is assigned
to amounts of normal users, who use their carried smart devices to sense data
dynamically without interference of users [5–7].

As we know, a sensing task usually needs the cooperation of numerous users
to sense data. In daily life, users may have different roles or types, such as worker,

Y. Li et al. (Eds.): COCOA 2019, LNCS 11949, pp. 496–506, 2019.
https://doi.org/10.1007/978-3-030-36412-0_40

teacher, student, etc. The sensed data obtained from a user with a specific type may be quite different from users with other types. Therefore, it is beneficial to improve the quality of a task if assigning the task to all kinds of types of users more uniformly. For example, we want to understand the satisfaction degree of a new kind of food. The results may be reliable if the collected data are submitted by various types of customers, such as women, men, children, etc. On the contrary, if only the children are recruited to submit their satisfaction, the results will be one-sided and thus has low quality. Therefore, it is shown that user type diversity has great impact on the quality of task in MCS.

However, the current researches mainly focuses on spatial-temporal diversity [8–10] and capability diversity [11–14], without considering the user type diversity. Furthermore, the existing task assignment mechanisms mainly focus on the optimization of some goals, such as minimizing cost [15–18], maximizing profit [19], maximizing social welfare [20,21], maximizing user coverage quality [22–26], and so on. In [12], Peng et al. consider multi-skill spatial crowdsourcing and maximize workers benefit under the budget constraint. In [13], Wang et al. study the location-aware and location diversity based on dynamic crowd-sensing system, where workers move over time and tasks arrive stochastically. In [14], Li et al. focus on the room allocation problem with capacity diversity and budget constraints. The above-related work does not take into account the type diversity, which may be lead to a low quality.

Motivated by the above observation, this paper formalizes a novel task assignment problem in MCS, where a task needs the cooperation of various types of users, and the quality of a task is highly related to the types of the recruited users. The main contributions are given as follows.

1. The paper proposes a novel type diversity-oriented task assignment problem under the constraints of budget, the latest ending time, the required types and the required number of users, in which the concept of entropy is introduced to measure the optimization goal of the problem.
2. In order to maximize the type diversity of the selected users and meet the type requirements of the task, this paper proposes three heuristic algorithms: Cost-based Greedy Algorithm (C-GA), Cost-based Greedy Algorithm by Type (C-GAT), and Unit Reward-based Greedy Algorithm by Type (UR-GAT).
3. This paper conducts extensive simulations to show that Unit Reward-based Greedy Algorithm by Type has better entropy and total profit than the other algorithms.

## 2  Model and Problem Formulation

In this section, we first describe the system model for maximizing type diversity task assignment in MCS. Then, we formulate the task assignment problem when all users and task-related information are known in advance. For easy reference, we list the important notations used in this paper in Table 1.

**Table 1.** Definition of notation.

| Notation | Description |
|---|---|
| $U, T$ | User and type sets |
| $u_i, \tau$ | $u_i$ is the $i$th user and task $\tau$ |
| $c_i, r_i$ | The cost and reward of the user $u_i$ |
| $s_i$ | The latest start time of the user $u_i$ |
| $t_i$ | The type set of the user $u_i$ |
| $d_\tau$ | The latest ending time of the task $\tau$ |
| $T_\tau$ | The required types set of the task $\tau$ |
| $n_\tau$ | The required number of users for the task $\tau$ |
| $b_\tau$ | The budget of task $\tau$ |
| $t_i^j$ | Whether the user $u_i$ belongs to the type $j$ |
| $t_\tau^j$ | Whether type $j$ is required by the task $\tau$ |
| $k$ | The number of types |
| $m$ | The number of users |
| $x_i$ | Whether user $u_i$ is assigned to task $\tau$ |
| $T_j$ | A set of users with type $j$ by cost in ascending order |
| $count_j$ | The number of selected users with type $j$ |
| $p_j$ | The probability is $count_j$ account for $n_\tau$ |

## 2.1 Model

In MCS, a sensing task is denoted by $\tau = < d_\tau, b_\tau, T_\tau, n_\tau >$, in which the four attributes refer to the latest ending time, the budget, the required types and the required number of users, respectively. Note that, the task $\tau$ will expire after time $d_\tau$. We further use $T = \{1, 2, ..., k\}$ to denote the set of types and $T_\tau \subseteq T$ to denote the required types of task $\tau$ with $|T_\tau| > 1$.

Assume that a crowd of mobile users $U = \{u_1, ..., u_i, ..., u_m\}$ are interested in participating sensing task. A mobile user is denoted by $u_i = < c_i, r_i, s_i, t_i >$, in which the four attributes refer to the cost, the reward, the latest start time and the type of user $u_i$, respectively. Specifically, user $u_i$ is not available before this time $s_i$, so that he/she cannot perform any task. We use $t_i$ to denote the type of user $u_i$ and each user belongs to a type. In our model, a sensing task usually needs cooperation of multiple users with the required types of the task.

## 2.2 Problem Formulation

In order to improve the quality of a task, the type distribution of recruited users should be as uniform as possible. Motivated by the above analysis, we introduce the concept of entropy to measure the type uniformity of the recruited users, with the optimization goal of maximizing the type diversity while satisfying the task budget. Therefore, we formalize the task assignment problem as follows.

$$\text{Max}: -\sum_{j\in T} p_j \cdot \log(p_j)$$

$$
\begin{aligned}
&s.t. \ 1) \ \sum_{i=1}^{m} r_i \cdot x_i \leq b_\tau \\
&2) \ s_i \cdot x_i \leq d_\tau, \forall i \in \{1,...,m\} \\
&3) \ \sum_{i=1}^{m} x_i \leq n_\tau \\
&4) \ \sum_{i=1}^{m} x_i \cdot t_i^j \geq t_\tau^j, \forall j \in \{1,...,k\} \\
&5) \ x_i \in \{0,1\}, \forall i \in \{1,...,m\}
\end{aligned}
\tag{1}
$$

In Eq. 1, $p_j = count_j/n_\tau$ indicates the ratio of the number of selected users in this type $j$ over the required number of users for the task, and $count_j = \sum_{i\in T_j} x_i$ is the number of selected users with type $j$, and $T_j$ is a set of users with type $j$ in ascending order according to cost (see C-GAT algorithm for details).

Here, Constraint 1 implies the total reward of all the assigned users should not exceed the budget of the task. Constraint 2 means that the latest start time for all assigned users does not exceed the latest ending time of the task. Constraint 3 indicates that the number of users performing the task should not exceed the required number of users.

Constraint 4 indicates that if task $\tau$ requires type $j(t_\tau^j = 1)$, at least one user of type $j$ is selected to perform the task. Where $t_i^j = 1$ shows that user $u_i$ belongs to type $j$, $t_\tau^j = 1$ means that type $j$ is required by task $\tau$. In Constraint 5, if user $u_i$ is assigned to task $\tau$, $x_i = 1$; otherwise, $x_i = 0$.

In this case, the above-mentioned problem can be converted into 0–1 integer programming problem [5,19], which leads to that the mentioned problem is also NP-hard.

## 3 The Proposed Task Assignment Approach

For solving the proposed problem efficiently, this paper proposes three heuristic algorithms, named C-GA (Cost-based Greedy Algorithm), C-GAT (Cost-based Greedy Algorithm by Type), and UR-GAT (Unit Reward-based Greedy Algorithm by Type), respectively.

### 3.1 Cost-Based Greedy Algorithm (C-GA)

As shown in Algorithm 1, C-GA is depicted as follows. We sort users by cost in ascending order (line 1). If user $u_i$ satisfies the required types of task $\tau$ while considering the latest ending time and the budget constraints, the task $\tau$ will be assigned to user $u_i$, $x_i = 1$ (line 2–4). Next, we update the budget and the required number of users, if the required number of task is not more than zero, then the loop will be broken (line 5–11). Finally, the algorithm returns the result of task assignment (line 12).

---

**Algorithm 1.** Cost-based Greedy Algorithm (C-GA).

---

**Require:** task $\tau$, user set $U$.
**Ensure:** the task assignment results: $\{x_i | u_i \in U\}$.
 1: Sort users by cost in ascending order;
 2: **for all** $u_i \in U$ **do**
 3:     **if** $s_i < d_\tau$ and $b_\tau \geq r_i$ and $n_\tau > 0$ and $t_i \in T_\tau$ **then**
 4:         set $x_i \leftarrow 1$;
 5:         set $b_\tau \leftarrow b_\tau - r_i$;
 6:         set $n_\tau \leftarrow n_\tau - 1$;
 7:         **if** $n_\tau \leq 0$ **then**
 8:             break;
 9:         **end if**
10:     **end if**
11: **end for**
12: return $x_i$.

---

## 3.2   Cost-Based Greedy Algorithm by Type (C-GAT)

In order to maximize the type diversity, we consider the relationship between the required types of task and the type of users. As shown in Algorithm 2, C-GAT is depicted as follows. We sort users with type $j(j \leq k)$ by cost in ascending order as $T_j$ (line 1–3). Next, we select user $u_i$ from $T_j$ (line 5). Then, if user $u_i$ satisfies the required types of task $\tau$ while considering the latest ending time and the budget constraints, the task $\tau$ will be assigned to user $u_i$, $x_i = 1$. The algorithm goes to the next loop when none of the constraints are satisfied (line 6–18). Finally, the algorithm returns the result of task assignment (line 19).

## 3.3   Unit Reward-Based Greedy Algorithm by Type (UR-GAT)

Our algorithm maximizes the total profit while meeting the budget of task. Profit equals the user's reward minus cost, unit reward is equal to the ratio of rewards to costs. The intuition is that the greater the unit reward of users, the greater the user's profit, and vice versa. Compared with Algorithm 2, this algorithm comprehensively considers the costs and rewards of selected users, selects users with higher unit rewards of each type, and calculates the total profit and type diversity of selected users.

On the basis of Algorithm 2, the implementation of the Algorithm 3 is modified. Line 2 is modified to sort users with type $j$ by unit reward in descending order as $T_j$.

---

**Algorithm 2.** Cost-based Greedy Algorithm by Type (C-GAT).

---

**Require:** task $\tau$, user set $U$.
**Ensure:** the task assignment results: $\{x_i | u_i \in U\}$.
 1: **for all** type $j$ **do**
 2:     Sort users with type $j$ by cost in ascending order as $T_j$;
 3: **end for**
 4: **while** $j$ **do**
 5:     select the lowest cost user $u_i$ from $T_j$;
 6:     **if** $s_i < d_\tau$ and $b_\tau \geq r_i$ and $n_\tau > 0$ and $t_i \in T_\tau$ **then**
 7:         set $x_i \leftarrow 1$;
 8:         set $b_\tau \leftarrow b_\tau - r_i$;
 9:         set $n_\tau \leftarrow n_\tau - 1$;
10:         **if** $n_\tau \leq 0$ **then**
11:             break;
12:         **end if**
13:     **end if**
14:     $j \leftarrow j + 1$;
15:     **if** $j == k + 1$ **then**
16:         $j \leftarrow 1$;
17:     **end if**
18: **end while**
19: return $x_i$.

---

---

**Algorithm 3.** Unit Reward-based Greedy Algorithm by Type (UR-GAT).

---

**Require:** task $\tau$, user set $U$.
**Ensure:** the task assignment results: $\{x_i | u_i \in U\}$.
 1: **for all** type $j$ **do**
 2:     Sort users with type $j$ by unit reward in descending order as $T_j$;
 3: **end for**
 4: **while** $j$ **do**
 5:     select user $u_i$ with the highest unit reward from $T_j$;
 6:     **if** $s_i < d_\tau$ and $b_\tau \geq r_i$ and $n_\tau > 0$ and $t_i \in T_\tau$ **then**
 7:         set $x_i \leftarrow 1$;
 8:         set $b_\tau \leftarrow b_\tau - r_i$;
 9:         set $n_\tau \leftarrow n_\tau - 1$;
10:         **if** $n_\tau \leq 0$ **then**
11:             break;
12:         **end if**
13:     **end if**
14:     $j \leftarrow j + 1$;
15:     **if** $j == k + 1$ **then**
16:         $j \leftarrow 1$;
17:     **end if**
18: **end while**
19: return $x_i$.

---

# 4   Evaluation

In this section, we evaluate the performance of the proposed algorithms with various number of users, budget of task and the required types of task in terms of total profit and type diversity. In order to reduce the influence from random progress, we compare three algorithms under the same parameters, and finally analyze the type diversity and total profit. The simulation parameters are shown in Table 2.

Table 2. Simulation parameters.

| Parameter | Value |
|-----------|-------|
| $U$ | $\{20, 30, 40, 50, 60\}$ |
| $c_i$ | $[5, 10]$ |
| $r_i$ | $[15, 20]$ |
| $s_i$ | $[15, 24]$ |
| $d_\tau$ | $[21, 24]$ |
| $T_\tau$ | $\{2, 3, 4, 5, 6\}$ |
| $n_\tau$ | $[4, 6]$ |
| $b_\tau$ | $\{100, 105, 110, 115, 120\}$ |

## 4.1   Type Diversity

Figure 1(a) shows that the type diversity of all algorithms with a varying budget of task, when the required type of task and the number of users are fixed. As seen, the C-GAT and UR-GAT achieve better type diversity with the increase of the budget of task, whereas the C-GA is insensitive to the budget of task. This observation can be easily interpreted as follows. For the C-GAT and UR-GAT, they select users in turn for each required type of task, so they have the same trend. However, the C-GA achieves the worst performance in terms of type diversity, and the result is relatively unstable with the increase of the budget of task, this is because the algorithm randomly selects the lowest cost user.

When the budget of the task and the number of users are fixed, the type diversity increases substantially with the increase of the required type of task, as shown in Fig. 1(b). As expected, we can see that the C-GAT and UR-GAT achieve the highest performance in terms of the type diversity, and observe that the result is relatively stable with the increase of required type of task, This is because the value of type diversity is closely related to the required type by task.

As the required type of task increases, more choices will be available. In Fig. 1(c), we change the number of users to test the type diversity of all algorithms. Clearly, C-GAT and UR-GAT are far superior to the C-GA.

It can be seen from the below three figures that the change trend of entropy in C-GAT and UR-GAT is the same, because they choose users evenly in each type. Conversely, the entropy of C-GA is always lower than the other two algorithms, because the algorithm selects users at random.

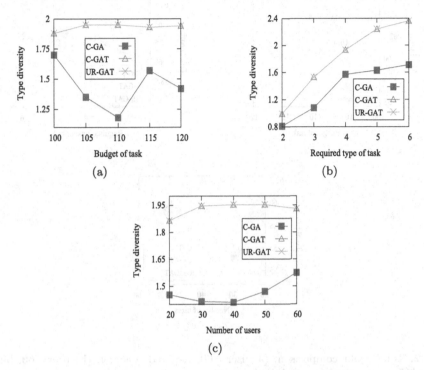

**Fig. 1.** Type diversity comparison. (a) user = 60, required type = 4. (b) user = 60, budget = 115. (c) required type = 4, budget = 115.

## 4.2    Total Profit

Total profit refers to the sum of the profit of all selected users. Figure 2(a) shows that compared with the C-GAT and C-GA, UR-GAT achieves the greatest total profit. It is because that UR-GAT sorts users by unit reward, and the other two algorithms sort users by cost. Moreover, unit reward is related to the rewards and costs of users.

With the increase of task requirement types, the overall trend of the three algorithms is the same, as shown in Fig. 2(b). It is because that total profit is unrelated to the types of users. Clearly, UR-GAT achieves a higher profit.

In Fig. 2(c), we change the number of users to test the total profit of all algorithms. Obviously, UR-GAT algorithm is far superior to the C-GA and C-GAT.

According to the below three figures, the total profit of UR-GAT is always higher than the other two algorithms, because this algorithm comprehensively considers the costs and rewards.

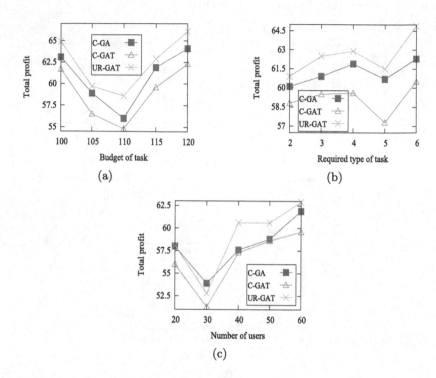

**Fig. 2.** Total profit comparison. (a) user = 60, required type = 4. (b) user = 60, budget = 115. (c) required type = 4, budget = 115.

## 5   Conclusion

In this paper, we propose a novel type diversity-oriented task assignment problem under the constraints, such as the budget, the latest ending time, the required types and the required number of users. Then, we formalize the task assignment problem. Moreover, we propose three heuristic algorithms to maximize the type diversity with budget constraint. If the users selected to perform the task are distributed with equal probability in each type required for the task, then the goal of task assignment optimization is to maximize the user type diversity. Through extensive simulation to evaluate the performances of our algorithms, the result shows that the proposed UR-GAT algorithm is superior to C-GA and C-GAT, in terms of total profit and type diversity.

**Acknowledgement.** This work is partly supported by the National Natural Science Foundation of China (No. 61872228), the Natural Science Basis Research Plan in Shaanxi Province of China (No. 2017JM6060), the Fundamental Research Funds for the Central Universities of China (Nos. GK201903090, GK201801004).

# References

1. Ganti, R.K., Ye, F., Lei, H.: Mobile crowdsensing: current state and future challenges. IEEE Commun. Mag. **49**(11), 32–49 (2011)
2. Wei, G., Zhang, B., Cheng, L.: Task assignment in mobile crowdsensing: present and future directions. IEEE Network **32**(4), 100–107 (2018)
3. Zhu, S., Cai, Z., Hu, H., Li, Y., Li, W.: zkCrowd: a hybrid blockchain-based crowdsourcing platform. IEEE Trans. Ind. Inform. (TII) (2019, online). https://doi.org/10.1109/TII.2019.2941735
4. Duan, Z., Li, W., Zheng, X., Cai, Z.: Mutual-preference driven truthful auction mechanism in mobile crowdsensing. In: The 39th IEEE International Conference on Distributed Computing Systems (ICDCS), Dallas, Texas, USA. IEEE (2019)
5. Li, J., Cai, Z., Wang, J., Han, M., Li, Y.: Truthful incentive mechanisms for geographical position conflicting mobile crowdsensing systems. IEEE Trans. Comput. Soc. Syst. **5**(2), 324–334 (2018)
6. Wang, Y., Cai, Z., Tong, X., Gao, Y., Yin, G.: Truthful incentive mechanism with location privacy-preserving for mobile crowdsourcing systems. Comput. Netw. **135**, 32–43 (2018)
7. Li, J., Cai, Z., Yan, M., Li, Y.: Using crowdsourced data in location-based social networks to explore influence maximization. In: The 35th Annual IEEE International Conference on Computer Communications (INFOCOM), San Francisco, CA, USA, pp. 1–9. IEEE (2016)
8. Cheng, P., Lian, X., Chen, Z., Chen, L., Han, J., Zhao, J.: Reliable diversity-based spatial crowdsourcing by moving workers. Proc. VLDB Endowment **8**(10), 1022–1033 (2015)
9. Cranshaw, J., Toch, E., Hong, J., Kittur, A., Sadeh, N.: Bridging the gap between physical location and online social networks. In: Proceeding of the 12th ACM International Conference on Ubiquitous Computing, Copenhagen, Denmark, pp. 119–128. ACM (2010)
10. Kazemi, L., Shahabi, C.: GeoCrowd: enabling query answering with spatial crowdsourcing. In: International Conference on Advances in Geographic Information Systems, Redondo Beach, CA, USA, pp. 189–198. ACM (2012)
11. Cohen, S., Yashinski, M.: Crowdsourcing with diverse groups of users. In: Proceedings of the 20th International Workshop on the Web and Databases, WebDB 2017, Chicago, IL, USA, pp. 7–12. ACM (2017)
12. Cheng, P., Lian, X., Chen, L., Han, J., Zhao, J.: Task assignment on multi-skill oriented spatial crowdsourcing. IEEE Trans. Knowl. Data Eng. **28**(8), 2201–2215 (2015)
13. Wang, X., Jia, R., Tian, X., Gan, X.: Dynamic task assignment in crowdsensing with location awareness and location diversity. In: IEEE Conference on Computer Communications, IEEE INFOCOM 2018, Honolulu, HI, USA, pp. 2420–2428. IEEE (2018)
14. Li, Y., Jiang, Y., Wu, W., Jiang, J., Fan, H.: Room allocation with capacity diversity and budget constraints. IEEE Access **7**, 42968–42986 (2019)

15. Yu, J., Xiao, M., Gao, G., Hu, C.: Minimum cost spatial-temporal task allocation in mobile crowdsensing. In: Yang, Q., Yu, W., Challal, Y. (eds.) WASA 2016. LNCS, vol. 9798, pp. 262–271. Springer, Cham (2016). https://doi.org/10.1007/978-3-319-42836-9_24

16. Wang, L., Yu, Z., Han, Q., Guo, B., Xiong, H.: Multi-objective optimization based allocation of heterogeneous spatial crowdsourcing tasks. IEEE Trans. Mob. Comput. **17**(7), 1637–1650 (2018)

17. Zhang, Y., Zhang, D., Li, Q., Wang. D.: Towards optimized online task allocation in cost-sensitive crowdsensing applications. In: IEEE 37th International Performance Computing and Communications Conference (IPCCC), Orlando, FL, USA, pp. 1–8. IEEE (2018)

18. Duan, Z., Li, W., Cai, Z.: Distributed auctions for task assignment and scheduling in mobile crowdsensing systems. In: IEEE 37th International Conference on Distributed Computing Systems (ICDCS), Atlanta, GA, USA, pp. 635–644. IEEE Computer Society (2017)

19. Li, L., Zhang, L., Wang, X., Yu, S., Wang, A.: An efficient task allocation scheme with capability diversity in crowdsensing. In: Shen, S., Qian, K., Yu, S., Wang, W. (eds.) CWSN 2018. CCIS, vol. 984, pp. 12–20. Springer, Singapore (2019). https://doi.org/10.1007/978-981-13-6834-9_2

20. Wang, Y., Cai, Z., Zhan, Z., Gong, Y., Tong, X.: An optimization and auction based incentive mechanism to maximize social welfare for mobile crowdsourcing. IEEE Trans. Comput. Soc. Syst. **6**(3), 414–429 (2019)

21. Wang, Y., Cai, Z., Ying, G., Gao, Y., Tong, X., Wu, G.: An incentive mechanism with privacy protection in mobile crowdsourcing systems. Comput. Netw. **102**, 157–171 (2016)

22. Abououf, M., Mizouni, R., Singh, S., Otrok, H., Ouali, A.: Multi-worker multi-task selection framework in mobile crowd sourcing. J. Netw. Comput. Appl **130**, 52–62 (2019)

23. Zhang, M., et al.: Quality-aware sensing coverage in budget-constrained mobile crowdsensing networks. IEEE Trans. Veh. Technol. **65**(9), 7698–7707 (2016)

24. Yin, X., Chen, Y., Li, B.: Task assignment with guaranteed quality for crowd-sourcing platforms. In: IEEE/ACM 25th International Symposium on Quality of Service, Vilanova i la Geltru, Spain, pp. 1–10. IEEE (2017)

25. Gong, W., Zhang, B., Li, C.: Location-based online task assignment and path planning for mobile crowdsensing. IEEE Trans. Veh. Technol. **68**(2), 1772–1783 (2019)

26. Tao, X., Song, W.: Location-dependent task allocation for mobile crowdsensing with clustering effect. IEEE Internet Things J. **6**(1), 1029–1045 (2019)

# Cake Cutting with Single-Peaked Valuations

Chenhao Wang[1,2,3] and Xiaoying Wu[1,2(✉)]

[1] University of Chinese Academy of Sciences, Beijing, China
[2] AMSS, Chinese Academy of Sciences, Beijing, China
{wangch,xywu}@amss.ac.cn
[3] City University of Hong Kong, Hong Kong, China

**Abstract.** In the cake cutting problem, one allocates a heterogeneous divisible resource (cake) to $n$ participating agents. The central criteria of an allocation to satisfy and optimize is envy-freeness and efficiency. In this paper, we consider cake cutting with *single-peaked* preferences: each agent is assumed to have a favorite point in the cake; the further a piece of cake is from her favorite point, the less her valuation on this piece is. Under this assumption, agents can be considered as a point embedded in a metric space, and thus this setting models many practical scenarios. We present a protocol in the standard query model which outputs an envy-free allocation in linear running time.

**Keywords:** Cake cutting · Envy-freeness · Fair division · Multiagent resource allocation

## 1 Introduction

Cake cutting is a fundamental topic in fair division, and has received wide attention in economics, mathematics and theoretical computer science. It concerns the setting in which a heterogeneous divisible cake needs to be allocated to multiple agents with different preferences. The cake is represented by an interval $[0, 1]$. Each agent has her own valuation function over the cake, which determines the value of any part of the cake for her. The utility that each agent obtains from the allocated portion of cake depends on her valuation function.

In the standard Robertson-Webb model [13], one uses a protocol to allocate the cake, which obtains the information of the private valuation functions by two kinds of queries. The first one is the *cut* query, in which an agent is asked to mark the cake at a distance from a given starting point so that the piece between these two points is worth a given value for her. The other one is the *evaluate* query, in which an agent is asked to evaluate a given piece according to her valuation function. The most classical protocol is *Cut & Choose Protocol* for two players: one agent is asked to cut the cake as equally as possible into two pieces, and the other agent choose one of them.

© Springer Nature Switzerland AG 2019
Y. Li et al. (Eds.): COCOA 2019, LNCS 11949, pp. 507–516, 2019.
https://doi.org/10.1007/978-3-030-36412-0_41

A central question is how to design a protocol to divide the cake fairly. There are two commonly used notions that capture fairness in resource allocation: proportionality and envy-freeness. An allocation is called *proportional*, if each agent is allocated at least an average share with respect to her own valuation function. An allocation is called *envy-free*, if no agent envies any other agent, *i.e.*, each agent values her own portion in the allocation no less than the portion of any other. It is easy to see that envy-freeness is a stronger concept of fairness than proportionality: every envy-free allocation must be proportional, but not vice versa. Aziz and Mackenzie [1] propose a discrete and bounded envy-free protocol for any number of agents, requiring exponential queries. It is still algorithmically intractable to design envy-free protocol for general valuation functions.

In this paper, we focus on a special class of valuation functions, *single-peaked* valuation functions, which means that each agent has a favorite point in the interval that represents the cake, and the valuation of those points that are further from her favorite point is preferred less. In addition, we assume the valuation of one point is linearly decreasing with respect to the distance to the favorite point. Thus, the single-peaked valuation functions considered fall into the class of piecewise linear valuation functions. A natural interpretation is that the agents are specified as points in the interval, and they prefer the pieces of cake that are close to their own locations, than the pieces that are further away. This setting can be applied to allocation of land resource, clean water and mineral deposits. This is also partially motivated by the facility location games and social choice under metric preference.

**Related Work.** The cake cutting problem has been studied by Steinhaus [15] since World War II, possibly as a result of the division of the former German Empire among the allied powers. After that, a large body of work has been devoted to its study. Usually, the cake is allocated by a protocol, which describes an interactive procedure recommended to be followed. The protocol itself does not hold any information about the agents' valuation functions. However, the protocol may ask some agent to provide her valuation of a specific piece of cake. Based on the agent's answer, the protocol may then recommend how to continue (e.g., for this agent to either accept the piece, or to trim it if she valued it above a certain threshold).

This interaction between the protocol and the agents is formalized by Robertson and Webb [13] by distinguishing two kinds of requests (or queries):

- cut requests: given a start point and a certain valuation, the protocol asks an agent to return where is the end point such that she could receive this given valuation according to her valuation function;
- evaluation requests: given a start point and an end point, the protocol asks an agent to return the valuation of this piece of cake according to her valuation function.

There is a variety of cake-cutting protocols described in the literature (see, e.g., the textbooks [4, 14]). The most important criterion for cake cutting protocols is fairness, especially envy-freeness and proportionality. A well-known envy-free protocol is the Cut & Choose protocol, which applies to the cake cutting problem with two agents, and the idea behind has been transferred to generalizations for three or more agents [7–11,16] - all of them are proportional, but none is envy-free. How to obtain a bounded envy-free protocol for more than three agents has been an open question for a long time, until Aziz and Mackenzie [2] propose such a protocol for four agents in 2016. After that, they [1] further propose a discrete and bounded envy-free protocol for any number of agents, requiring exponentially many queries.

Another valuation criterion is efficiency, especially the Pareto efficiency, which means that there is no other division that would make at least one agent better off without making any of the other agents worse off. Weller [17] proves the existence of divisions that are both Pareto efficient and envy-free.

Furthermore, some literatures focus on the design of algorithms for optimizing the social welfare, which is defined as the total value received by the agents. Cohler et al. [6] give a tractable algorithm which takes the valuation functions as input, and outputs an approximating optimal and approximately envy-free allocation. Bei et al. [3] design a polynomial-time approximation scheme to compute a proportional and connected allocation with optimal social welfare.

The work most related to ours is the cake cutting problem with piecewise linear valuation functions, that is, the interval representing the cake can be partitioned into a finite number of subintervals such that the valuation function is linear on each interval. The single-peaked function we consider in this work is piecewise linear, and thus the related results can be applied to our setting. Kurokawa et al. [12] design a protocol (called Sandwich) that finds an envy-free allocation with at most $O(n^6 k \ln k)$ queries in the Robertson-Webb model, assuming there are at most $k$ subintervals in the piecewise linear functions. Moreover, Chen et al. [5] consider the truthfulness from the point view of mechanism design, and provide a randomized mechanism that is truthful-in-expectation and proportional.

**Our Results.** We study the cake cutting problem with single-peaked valuations in the Robertson-Webb model. We propose an envy-free (and thus proportional) protocol that first obtains the information on the valuation functions of agents by making some queries, and derives an envy-free allocation using a natural idea. Our protocol is simple to run, and requires at most $O(n)$ queries, in particular two evaluation queries for each agent. Due to the simplicity of single-peaked functions, it costs much less time than the Sandwich protocol in [12] for piecewise linear valuations, which requires at least $\Omega(n^6)$ queries.

## 2 Model

The cake is modeled as the real interval $[0, 1]$, and the set of agents is $N = \{1, \ldots, n\}$. A piece of cake or a portion of cake is a finite set of disjoint subintervals of the cake $[0, 1]$. Each agent $i \in N$ has a integrable, non-negative *value density* function $v_i$. For each piece of cake $X$, an agent's value on $X$ is denoted by $V_i(X)$ and defined by the integral of its density function, *i.e.*, $V_i(X) := \int_X v_i(x)dx$. Thus, $V_i : 2^{[0,1]} \to \mathbb{R}^+$ is called the *valuation function* of agent $i$. For simplicity, we abuse the notion by writing $V_i(x, y)$ instead of $V_i([x, y])$ for each interval $[x, y]$. The definition of the valuation function implies the additivity (*i.e.*, $V_i(A \cup B) = V_i(A) + V_i(B)$ for disjoint $A$ and $B$) and non-atomicity (*i.e.*, $V_i(x, x) = 0$). As usual, we assume that the agents' valuations are normalized so that $V_i(0, 1) = 1$ for each $i \in N$. This assumption is without loss of generality because one can scale the valuation function by a constant factor when considering the properties.

We consider a restricted class of valuations. We say that an agent $i$ has a *single-peaked* valuation when her value density function $v_i$ satisfies that there is a number $p_i \in [0, 1]$ such that $v_i(p_i) = \max_{x \in [0,1]} v_i(x)$ and $v_i(x) = \max\{0, v_i(p_i) - k_i|x - p_i|\}$ for all $x \in [0, 1]$ and some coefficient $k_i > 0$. We call $p_i$ and $v_i(p_i)$ the peak location and peak density of agent $i$, respectively. Notice that the coefficient $k_i$ as well as the density function is uniquely determined by the peak location and peak density as the normalized assumption says $\int_0^1 v_i(x)dx = 1$. Fixing the peak location, varied peak densities can give different $k_i$ and thus different valuation functions. See Fig. 1 for an example.

An *allocation* (or, *division*) $A = (X_1, \ldots, X_n)$ is a partition of the cake among the agents, that is, each agent $i \in N$ receives the piece of cake $X_i$, the pieces are disjoint, and $\cup_{i \in N} X_i = [0, 1]$. Two notions for determining the fairness of an allocation is proportionality and envy-freeness. An allocation is *proportional* if $V_i(X_i) \geq \frac{1}{n}$ for all $i \in N$, and is *envy-free* if $V_i(X_i) \geq V_i(X_j)$ for all $i, j \in N$. It is easy to see that envy-freeness implies proportionality.

A cake-cutting protocol describes an interactive procedure recommended to be followed to divide cake among the agents. We use the well-known Cut & Choose protocol as an example to show what a cake-cutting protocol looks like. Given a cake $X = [0, 1]$, two players $\{1, 2\}$ and their valuation functions $V_1(X), V_2(X)$. The protocol outputs an allocation. First, player 1 executes one cut to divide the cake into two pieces $S_1$ and $S_2$ which she considers equal halves: $V_1(S_1) = V_1(S_2) = 1/2$. It holds that $X = S_1 \cup S_2$. Then player 2 chooses her best piece, and player 1 receives the remaining piece.

**Robertson-Webb Model.** Throughout this paper, we operate in the standard Robertson-Webb query model: it models the interaction between the protocol and the agents using two types of queries:

1. $Eval_i(x, y)$: agent $i$ is asked to output a value $\alpha$ such that $v_i(x, y) = \alpha$.
2. $Cut_i(x, \alpha)$: agent $i$ is asked to output a point $y$ such that $v_i(x, y) = \alpha$.

The number of queries made during the execution is a well-established measurement to estimate the runtime of a cake cutting protocol.

Furthermore, we show how the Cut & Choose protocol is simulated in the Robertson-Webb query model. First, agent 1 receives a query $Cut_1(0, \frac{1}{2})$ and reports a point $w$ such that the interval $[0, w]$ (as well as the complement $[w, 1]$) worth $\frac{1}{2}$ to agent 1. Next, agent 2 receives a query $Eval_2(0, w)$. If she reports a value more (or less) than $\frac{1}{2}$, we assign the piece $[0, w]$ (or $[w, 1]$) to her, and the piece $[w, 1]$ (or $[0, w]$) to agent 1. It guarantees that the valuation agent 1 receives is exactly $\frac{1}{2}$ and the valuation agent 2 receives is at least $\frac{1}{2}$.

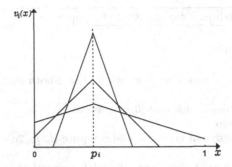

**Fig. 1.** The peak location of agent $i$ is $p_i$, and different peak densities lead to different density functions. The integral is always 1.

## 3    An Envy-Free Protocol

Let $U_i$ be the maximum subinterval such that agent $i$ has positive value on its every point, i.e., $v_i(x) > 0, \forall x \in U_i$ and $v_i(x) = 0, \forall x \in [0,1] \backslash U_i$. Define a triple $(l_i, p_i, r_i)$ for each agent $i \in N$, where $p_i$ is the peak location, $l_i$ is the left endpoint of $U_i$, and $r$ is the right endpoint of $U_i$. The triple may have overlap points, for example, the triple for the value density $v_i(x) = 2x$ is $(0,1,1)$. In the following procedure, we make two Cut queries for each agent to obtain her associated triple.

We define the computation of triples for different cases in this procedure as:

– If $a \leq 1/2$,

$$(l_i, p_i, r_i) = \begin{cases} (0, [-2c^2 - 7a^2/2 + 6ac]^{\frac{1}{2}}, 2c - a) & \frac{3a}{2} < c \leq \frac{1+a}{2}, \\ (3a - 2c, a, 2c - a) & c \leq \frac{3a}{2}, \\ (0, \left[\frac{a}{2} + \frac{2(1-c)(c-a)(1-2a)}{(4c-a-3)}\right]^{\frac{1}{2}}, 1) & c > \frac{1+a}{2}. \end{cases}$$

– If $a > 1/2$,

$$(l_i, p_i, r_i) = \begin{cases} (2b - a, a, 3a - 2b) & \frac{3a-1}{2} < b, \\ (2b - a, 1 + [(a - 2b + 1)/2 + 4(a - b)^2]^{\frac{1}{2}}, 1) & \frac{a}{2} \leq b < \frac{3a-1}{2}, \\ (0, 1 - [\frac{1}{2} + \frac{a}{2} - \frac{(4a-2)(b^2-a^2)}{4b-a}]^{\frac{1}{2}}, 1). & 0 < b < \frac{a}{2}. \end{cases}$$

The following lemma shows the correctness of the computation.

**Lemma 1.** *The procedure of Algorithm 1 as well as the above defined formulas correctly compute the triples of all agents.*

*Proof.* We only consider the case when $a \leq 1/2$, as the verification of the case when $a > 1/2$ is similar. We note that the peak $p_i$ must be located at $[0, a]$, since the integral of the value density function on interval $[0, 1]$ is 1.

---

**Algorithm 1.** Procedure for computing the triples

---

**Input**: Set of players $N = \{1, \ldots, n\}$
**Output**: triples of agents.

1: **for** each $i \in N$ **do**
2:    Let $a = Cut_i(0, \frac{1}{2})$, such that $[0, a]$ and $[a, 1]$ are both worth $\frac{1}{2}$ to agent $i$.
3:    **if** $a \leq \frac{1}{2}$ **then**
4:       Let $c = Cut_i(0, \frac{7}{8})$, such that $[0, c]$ is worth $\frac{7}{8}$.
5:       **if** $c \leq \frac{1+a}{2}$ **then**
6:          The right endpoint is $2c - a$, and compute $(l_i, p_i, 2c - a)$.
7:       **else**
8:          The left and right endpoints is 0 and 1, respectively. Compute $(0, p_i, 1)$.
9:       **end if**
10:   **else**
11:      $b = Cut_i(0, \frac{1}{8})$.
12:      **if** $b \geq \frac{a}{2}$ **then**
13:         The left endpoint is $2b - a$, and compute $(2b - a, p_i, r_i)$.
14:      **else**
15:         The left and right endpoint is 0 and 1, respectively. Compute $(0, p_i, 1)$.
16:      **end if**
17:   **end if**
18: **end for**
19: **return** triple $(l_i, p_i, r_i)$ for each $i \in N$.

---

**Case 1.** $c \leq \frac{1+a}{2}$. The condition $c \leq \frac{1+a}{2}$ indicates that the right break point $r_i$ is less than 1, *i.e.*, $v_i(r_i) = 0$. Then, $c$ is the mean between $a$ and $r_i$, which implies that $r_i = 2c - a$. As for the left break point $l_i$ and the peak $p_i$, we give the analysis as follows. If $a \geq \frac{2}{3}c$, it must have the peak point $p_i = a$ and the left break point $l_i = a - (r_i - a) = 3a - 2c$ (see Fig. 2(a)), which completes the computation.

If $a < \frac{2}{3}c$, it gives that the left break point $l_i = 0$ (see Fig. 2(b)).

To compute the peak $p_i$, denote the value density by

$$v_i(x) = \begin{cases} kx + b_1, & \text{if } 0 \leq x \leq p_i, \\ -kx + b_2, & \text{if } p_i < x \leq r_i. \end{cases}$$

Note that at the peak point $p_i$, it holds $v_i(p_i) = kp_i + b_1 = -kp_i + b_2$, which implies $b_1 = -2kp_i + b_2$. By the assumption of normality, we have

$$1 = \frac{(v_i(0) + v_i(p_i))p_i}{2} + \frac{v_i(p_i)(2c - a - p_i)}{2}$$
$$= -kp_i^2 + \frac{(-2kc + ka + b_2)p_i}{2} + \frac{(2c - a)b_2}{2}. \tag{1}$$

Recall that the right break point $r_i = 2c - a$, indicating $v_i(r_i) = -k(2c - a) + b_2 = 0$. Substituting it into (1), the peak location $p_i$ is equal to $[(-2 + (2c - a)b_2)/2k]^{1/2}$. Solving the two equations $v_i(r_i) = 0$ and $V_i(a, r_i) = 1/2$ with

**(a)**     **(b)**     **(c)**

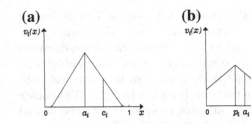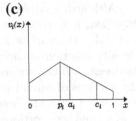

**Fig. 2.** (a) $a \leq \frac{1}{2}, c \leq \frac{3a}{2}$. (b) $a \leq \frac{1}{2}, \frac{3a}{2} < c \leq \frac{1+a}{2}$. (c) $a \leq \frac{1}{2}, c > \frac{1+a}{2}$.

respect to $k$ and $b_2$, we have

$$k = \frac{1}{4(c-a)^2}, \quad b_2 = (2c-a)k = \frac{2c-a}{4(c-a)^2}.$$

Now, we can compute the peak location as $p_i = [(-2 + (2c - a)b_2)/2k]^{1/2} = [-2c^2 - 7a^2/2 + 6ac]^{1/2}$.

Hence, when $a \leq 1/2$ and $c \leq (1+a)/2$, the triple is:

$$(l_i, p_i, r_i) = \begin{cases} (3a - 2c, a, 2c - a), & \text{if } c \leq \frac{3a}{2}, \\ (0, [-2c^2 - 7a^2/2 + 6ac]^{1/2}, 2c - a), & \text{if } \frac{3a}{2} < c < \frac{1+a}{2}, \end{cases}$$

**Case 2.** $c > \frac{1+a}{2}$. The right endpoint $r_i$ must be 1 in this case. Since the peak point $p_i$ is on the left side of $a$ and $a \leq 1/2$, we know that the left endpoint is $l_i = 0$. Figure 2(c) shows the value density function.

By the normality that $V_i(0,1) = 1$, we have

$$1 = \frac{(v_i(0) + v_i(p_i))p_i}{2} + \frac{(v_i(p_i) + v_i(1))(1 - p_i)}{2}$$
$$= \frac{(b_1 - kp_i + b_2)p_i}{2} + \frac{(-kp_i + 2b_2 - k)(1 - p_i)}{2}. \tag{2}$$

By simplifying Eq. (2), it is equivalent to $2kp_i^2 + k + 2b_2 - 2 = 0$, which implies $p_i = [(-2 - k + 2b_2)/2k]^{1/2}$. Solving the two equations $V(a, 1) = 1/2$ and $V(c, 1) = 1/8$ w.r.t. to $k$ and $b_2$, we have

$$k = \frac{1}{c-a}\left(\frac{1}{4(1-c)} - \frac{1}{1-a}\right), \quad b_2 = \frac{1}{2(1-a)} - \frac{1+a}{2}k.$$

Then, we can compute the peak location and the triple when $a \leq 1/2$ and $c > \frac{1+a}{2}$:

$$(l_i, p_i, r_i) = \left(0, \left[\frac{a}{2} + \frac{2(1-c)(c-a)(1-2a)}{(4c - a - 3)}\right]^{1/2}, 1\right).$$

$\square$

Although a triple does not uniquely determine a single-peaked value density function, it is sufficient for deriving an envy-free allocation. The basic idea is natural: arranging the points among all triples in a nondecreasing order, the value density function of each agent is linear (or even constant) over each subinterval between such two consecutive points; we partition each of such subintervals into $2n$ portions of equal length and label the portions from left to right. We allocate to agent 1 with the first and last portions, allocate to agent 2 with the second and second last portions, and so on. Algorithm 2 formally describes this protocol. To our knowledge, Chen *et al.* [5] firstly use this natural idea for the cake cutting problem with piecewise constant and piecewise linear valuations, from the point view of mechanism design, which requires each agent to explicitly report her value density function. We apply it to the Robertson-Webb query model, which asks the agents some queries.

---

**Algorithm 2.** Protocol for single-peaked valuations

---

**Input**: Set of agents $N = \{1, \ldots, n\}$
**Output**: An allocation.

1: Let $(l_i, p_i, r_i)$ be the triples obtained by Algorithm 1 for each $i \in N$
2: Let $M = \cup_{i \in N} \{l_i, p_i, r_i\} \cup \{0, 1\}$ be the set of points among all triples.
3: Denote by $M = \{z_1, \ldots, z_m\}$. Arrange the points in $M$ in a nondecreasing order $z_1 \leq \cdots \leq z_m$, interchanging the index if necessary.
4: **for** $j = 1, \ldots, m - 1$ **do**
5:     Uniformly divide the subinterval $[z_j, z_{j+1}]$ into $2n$ portions of equal length.
6:     Allocate the $i$-th and $(2n - i + 1)$-st portions to each agent $i \in N$.
7: **end for**
8: **return** The allocation derived above.

---

**Theorem 1.** *The protocol of Algorithm 2 outputs an envy-free allocation, using $2n$ queries in the Robertson-Webb model.*

*Proof.* In each sub-interval $[z_j, z_{j+1}]$, the value density functions of all agents must be linear or be constant 0. The valuation that agent $i$ receives in this subinterval is $\frac{V_i(z_j, z_{j+1})}{n}$, and the two portions that any other agent receives in this subinterval is also worth $\frac{V_i(z_j, z_{j+1})}{n}$ to agent $i$. Summing all subintervals, agent $i$ receives a valuation of $\frac{1}{n} \cdot \sum_{j=1}^{m-1} V_i(z_j, z_{j+1}) = \frac{1}{n} \cdot V_i(0, 1) = \frac{1}{n}$, and she does not envy anyone because the piece of cake allocated to any other agent is worth $\frac{1}{n}$ to her.

The procedure of Algorithm 2 asks two queries for each agent, and there are totally $2n$ queries. □

We use the following example to illustrate this protocol.

*Example 1.* Consider the two-agent cake cutting problem with single-peaked valuations shown in Fig. 1. The triples are $(0, \frac{1}{3}, \frac{2}{3})$ and $(\frac{1}{3}, \frac{2}{3}, 1)$, and the set of points among the triples is $M = \{0, \frac{1}{3}, \frac{2}{3}, 1\}$. We divide each of the three subintervals $[0, \frac{1}{3}], [\frac{1}{3}, \frac{2}{3}], [\frac{2}{3}, 1]$ into 4 portions of length $\frac{1}{12}$. Then, agent 1 receives the piece of cake $[0, \frac{1}{12}] \cup [\frac{1}{4}, \frac{5}{12}] \cup [\frac{7}{12}, \frac{3}{4}] \cup [\frac{11}{12}, 1]$, and agent 2 receives the complement.

Pareto-efficiency of an allocation says that there is no other allocation such that at least one agent is better off and no agent is worse off. Clearly, the allocation in this example is not Pareto-efficient: allocating $[0, \frac{1}{2}]$ to agent 1 and $[\frac{1}{2}, 1]$ to agent 2 improves both agents (Fig. 3).

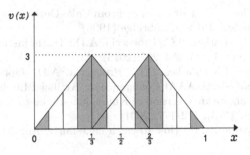

**Fig. 3.** A 2-agent example for the protocol of Algorithm 2. Agent 1 (who has peak location $\frac{1}{3}$) receives the gray pieces, while agent 2 (who has peak location $\frac{2}{3}$) receives the white pieces.

## 4 Conclusion

We consider the cake cutting problem with single-peaked valuations in the Robertson-Webb model, and present an envy-free protocol that makes $2n$ queries. But it is not Pareto efficient by Example 1. An immediate question is whether we can do better: can we design a protocol that satisfies envy-freeness and efficiency? Kurokawa *et al.* [12] proves that there is no Pareto-efficient protocol for piecewise constant (and thus piecewise linear) valuations, while Weller [17] proves the existence of allocations that are both Pareto-efficient and envy-free. It is open to explore the existence of Pareto-efficient and envy-free protocol for single-peaked valuations.

**Acknowledgements.** The authors thank Xiaodong Hu, Xujin Chen, and Minming Li for support and suggestions. They also thank all the reviewers of COCOA 2019 for comments.

# References

1. Aziz, H., Mackenzie, S.: A discrete and bounded envy-free cake cutting protocol for any number of agents. In: 2016 IEEE 57th Annual Symposium on Foundations of Computer Science (FOCS), pp. 416–427. IEEE (2016)
2. Aziz, H., Mackenzie, S.: A discrete and bounded envy-free cake cutting protocol for four agents. In: Proceedings of the Forty-Eighth Annual ACM Symposium on Theory of Computing, pp. 454–464. ACM (2016)
3. Bei, X., Chen, N., Hua, X., Tao, B., Yang, E.: Optimal proportional cake cutting with connected pieces. In: Twenty-Sixth AAAI Conference on Artificial Intelligence (2012)
4. Brams, S.J., Taylor, A.D.: Fair Division: From Cake-Cutting to Dispute Resolution. Cambridge University Press, Cambridge (1996)
5. Chen, Y., Lai, J.K., Parkes, D.C., Procaccia, A.D.: Truth, justice, and cake cutting. Games Econ. Behav. **77**(1), 284–297 (2013)
6. Cohler, Y.J., Lai, J.K., Parkes, D.C., Procaccia, A.D.: Optimal envy-free cake cutting. In: Twenty-Fifth AAAI Conference on Artificial Intelligence (2011)
7. Dawson, C.: An algorithmic version of Kuhns lone-divider method of fair division. Mo. J. Math. Sci. **13**(3), 172–177 (2001)
8. Dubins, L.E., Spanier, E.H.: How to cut a cake fairly. Am. Math. Mon. **68**(1P1), 1–17 (1961)
9. Even, S., Paz, A.: A note on cake cutting. Discret. Appl. Math. **7**(3), 285–296 (1984)
10. Fink, A.M.: A note on the fair division problem. Math. Mag. **37**, 341–342 (1964)
11. Kuhn, H.W.: On games of fair division. Essays in Mathematical Economics (1967)
12. Kurokawa, D., Lai, J.K., Procaccia, A.D.: How to cut a cake before the party ends. In: Twenty-Seventh AAAI Conference on Artificial Intelligence (2013)
13. Robertson, J., Webb, W.: Cake-Cutting Algorithms: Be Fair if You Can. AK Peters/CRC Press, Natick (1998)
14. Rothe, J.: Economics and Computation, vol. 4. Springer, Heidelberg (2015). https://doi.org/10.1007/978-3-662-47904-9
15. Steihaus, H.: The problem of fair division. Econometrica **16**, 101–104 (1948)
16. Steinhaus, H.: Mathematical Snapshots. Courier Corporation, Chelmsford (1999)
17. Weller, D.: Fair division of a measurable space. J. Math. Econ. **14**(1), 5–17 (1985)

# The One-Cop-Moves Game on Graphs of Small Treewidth

Lusheng Wang[1] and Boting Yang[2](✉)

[1] Department of Computer Science, City University of Hong Kong,
Kowloon Tong, Hong Kong
`cswangl@cityu.edu.hk`
[2] Department of Computer Science, University of Regina, Regina, SK, Canada
`boting.yang@uregina.ca`

**Abstract.** This paper considers the one-cop-moves game played on a graph. In this game, a set of cops and a robber occupy the vertices of the graph and move alternately along the graph's edges with perfect information about each other's positions. The goal of the cops is to capture the robber. At cops' turns, exactly one cop is allowed to move from his location to an adjacent vertex; at robber's turns, she is allowed to move from her location to an adjacent vertex or to stay still. We want to find the minimum number of cops to capture the robber. This number is known as the cop number. In this paper, we investigate the cop number of several classes of graphs, including graphs with treewidth at most 2, Halin graphs, and Cartesian product graphs. We also give a characterization of $k$-winnable graphs in the one-cop-moves game.

## 1 Introduction

Cops and Robbers was independently introduced by Nowakowski and Winkler [14] and Quilliot [16]. In [14,16], the game is played on a graph containing one cop and one robber, where the cop tries to capture the robber. They gave a characterization of graphs on which one cop can capture one robber. Aigner and Fromme [1] considered this game on a graph containing a set of cops and one robber. Both cops and the robber have perfect information, meaning that all of them know the whole graph and everyone's location at any moment. Initially, the cops choose a set of vertices to occupy; then the robber chooses a vertex to occupy. At even ticks of a clock (starting at 0), some subset of the cops move to adjacent vertices, and at odd ticks of the clock, the robber moves to an adjacent vertex or stays at her current vertex. Two consecutive turns, starting with a cops' turn, form a *round* of the game. Cops win if after some finite number of turns, one of the cops occupy the same vertex as the robber; this

L. Wang—Research supported by a grant for Hong Kong Special Administrative Region, China (CityU 11210119).
B. Yang—Research supported in part by an NSERC Discovery Research Grant, Application No.: RGPIN-2018-06800.

© Springer Nature Switzerland AG 2019
Y. Li et al. (Eds.): COCOA 2019, LNCS 11949, pp. 517–528, 2019.
https://doi.org/10.1007/978-3-030-36412-0_42

is called a *capture*. The robber wins if she can evade capture indefinitely. We want to find the minimum number of cops required to capture the robber in $G$. Such a number is called the *cop number* of $G$, denoted by $c(G)$. Hahn and MacGillivray [11] gave an algorithmic characterization of copwin finite digraphs. Clarke and MacGillivray [8] gave two characterizations of graphs $G$ with $c(G) \leq k$. A survey of results can be found in [2,7,10] and in the recent book of Bonato and Nowakowski [6].

In this paper, we study a variant of the Cops and Robbers game, known alternately as the *one-active-cop* game [15], *lazy cops and robbers* [4,5,17] or the *one-cop-moves* game [9,20]. Note that in a cops' turn of the standard Cops and Robbers game, any subset of cops can move to their neighbors. The only difference between the one-cop-moves game and the Cops and Robbers game is that in each cops' turn of the one-cop-moves game, only one of the cops is allowed to move from his present location to an adjacent vertex. The corresponding cop number of a graph $G$ in this game variant, denoted by $c_1(G)$, is called the *one-cop-moves cop number of* $G$, which is the minimum number of cops required to capture the robber in $G$.

The one-cop-moves cop number has been studied for hypercubes [4,15], random graphs [5] and Rook's graphs [17]. The capture time of the one-cop-moves game has been investigated for trees [20]. Little is known about the behaviour of the one-cop-moves cop number of planar graphs. Aigner and Fromme [1] proved that $c(G) \leq 3$ for every connected planar graph $G$, but Gao and Yang [9] showed that Aigner and Fromme's result does not generalize to the one-cop-moves game by constructing a connected planar graph whose structure is specifically designed for a robber to evade three cops indefinitely.

This paper is organized as follows. Section 2 gives definitions and notation. Section 3 shows that two cops can capture a robber on graphs of treewidth at most 2 in the one-cop-moves game. Section 4 deals with the one-cop-moves game on Halin graphs. Section 5 proves that the one-cop-moves cop number of the Cartesian product of two trees is equal to 2. Section 6 presents a relational characterization of $k$-winnable graphs in the one-cop-moves game. Finally, we conclude this paper with Sect. 7.

## 2     Preliminaries

For a graph $G = (V, E)$ with the vertex set $V$ and the edge set $E$, we use $\{u, v\}$ to denote an edge with endpoints $u$ and $v$. We also use $V(G)$ and $E(G)$ to denote the vertex set and edge set of $G$ respectively. The vertex set $\{u : \{u, v\} \in E\}$ is the *neighborhood* of $v$, denoted by $N_G(v)$, and the vertex set $N_G(v) \cup \{v\}$ is the *closed neighborhood* of $v$, denoted as $N_G[v]$. For a vertex set $U \subseteq V$, we define $N_G[U] = \cup_{v \in U} N_G[v]$. For a subset $V' \subseteq V$, we use $G[V']$ to denote the subgraph induced by $V'$, which consists of all vertices of $V'$ and all of the edges that connect vertices of $V'$ in $G$. For a vertex $v \in V$, we use $G - v$ to denote the subgraph induced by $V \setminus \{v\}$.

If $S$ is a collection of vertices of $G$ in which vertices are allowed to appear more than once, then $S$ is a *multiset* of vertices. For a graph $G$ and a positive integer $k$, we use $V_{G,k}$ to denote a multiset of vertices of $G$ such that each vertex of $G$ occurs exactly $k$ times (and so $V_{G,1} = V(G)$). The *cardinality* of $S$, denoted as $|S|$, is the total number of occurrences of vertices in $S$. So $|V_{G,k}| = k|V(G)|$. The notation of $S \subseteq V_{G,k}$ means that $S$ is a multisubset of $V_{G,k}$ such that each vertex of $S$ can occur at most $k$ times. If $v \in S$, we use $S \setminus \{v\}$ to denote a multisubset of $S$ such that the number of occurrences of $v$ is reduced by one (if this number is reduced to 0, then $v$ is removed from the multisubset). If $v \in V(G)$, we use $S \cup \{v\}$ to denote a multisuperset of $S$ such that the number of occurrences of $v$ is increased by one.

A *path* is a sequence $v_0 v_1 \ldots v_k$ of *distinct* vertices of $G$ with $\{v_{i-1}, v_i\} \in E$ for all $i$ ($1 \leq i \leq k$). The *length* of a path is the number of edges on the path. A cycle can be defined similarly. The *distance* between two vertices $u, v \in V$, denoted by $\mathrm{dist}_G(u, v)$, is the length of a shortest path between $u$ and $v$. The *distance* between a vertex $v$ and a subset $U$ of vertices in $G$, denoted by $\mathrm{dist}_G(v, U)$, is defined as

$$\mathrm{dist}_G(v, U) = \min\{\mathrm{dist}_G(v, u) : u \in U\}.$$

If there is no ambiguity, we use $N(v)$, $N[v]$, $N[U]$, $\mathrm{dist}(u, v)$ and $\mathrm{dist}(v, U)$ without subscripts. Definitions omitted here can be found in [19].

Given a graph $G = (V, E)$, a *tree decomposition* of $G$ is a pair $(T, W)$ with a tree $T = (I, F)$, $I = \{1, 2, \ldots, m\}$, and a family of non-empty subsets $W = \{W_i \subseteq V : i = 1, 2, \ldots, m\}$, such that

1. $\bigcup_{i=1}^{m} W_i = V$,
2. for each edge $\{u, v\} \in E$, there is an $i \in I$ with $\{u, v\} \subseteq W_i$, and
3. for all $i, j, k \in I$, if $j$ is on the path from $i$ to $k$ in $T$, then $W_i \cap W_k \subseteq W_j$.

The *width* of a tree decomposition $(T, W)$ is $\max\{|W_i| - 1 : 1 \leq i \leq m\}$. The *treewidth* of $G$, denoted by $\mathrm{tw}(G)$, is the minimum width over all tree decompositions of $G$.

**Definition 1 ($k$-winnable).** In the one-cop-moves game, a graph $G$ is called *$k$-winnable* if $c_1(G) \leq k$.

Note that "$k$-winnable" is an analog of "$k$-copwin" in the Cops and Robbers game. We use $\lambda_1, \ldots, \lambda_k$ to denote $k$ cops, and $\gamma$ to denote a robber. If there is no ambiguity, we also use $\lambda_1, \ldots, \lambda_k$ and $\gamma$ to denote the vertices they are occupying at the moment in question.

**Definition 2 (Protect).** We say that the cops *protect* a subgraph $G'$ of $G$ if for any sequence of robber moves leading to the robber moving to a vertex of $G'$, the robber is immediately captured by one of the cops when she enters $G'$.

## 3  Graphs with Treewidth at Most 2

The class of *two-terminal series-parallel graphs* are defined inductively as follows:

1. The graph with the single edge $\{s,t\}$ is a two-terminal series-parallel graph with terminals $\{s,t\}$.
2. Let $G_1$ and $G_2$ be two-terminal series-parallel graphs with terminals $\{s_1,t_1\}$ and $\{s_2,t_2\}$, respectively.
   (a) Identify $t_1 = s_2$ to create a new two-terminal series-parallel graph with terminals $\{s_1,t_2\}$. This is known as the *series composition* of $G_1$ and $G_2$.
   (b) Identify $s = s_1 = s_2$ and $t = t_1 = t_2$ to create a new two-terminal series-parallel graph with terminals $\{s,t\}$. This is known as the *parallel composition* of $G_1$ and $G_2$.

A graph $G$ is *series-parallel* if there is a pair of vertices $s$ and $t$ on $G$ such that $G$ is a two-terminal series-parallel graph with terminals $\{s,t\}$. Note that multiple edges are allowed in series-parallel graphs. A *block* of $G$ is a maximal connected subgraph of $G$ that has no cut vertex.

Theis [18] showed that series-parallel graphs are 2-copwin in the Cops and Robbers game. We extend this result to the one-cop-moves game by modifying the proof from [18].

**Theorem 1.** *If $G$ is a series-parallel graph, then $c_1(G) \leq 2$.*

*Proof.* Let $G$ be a series-parallel graph with terminals $\{s,t\}$. Initially, the two cops, $\lambda_1$ and $\lambda_2$, are placed on the terminals $s$ and $t$, respectively, and then the robber $\gamma$ is placed on a vertex. If $G$ is a graph with a single edge $\{s,t\}$, then the robber has been captured. If $G$ is a graph with at least two edges, the cops' strategy is to force the robber to move into a series-parallel subgraph $H$ with terminals $\{s_H,t_H\}$ such that $s_H$ and $t_H$ are occupied by the two cops. The cops can keep doing in this way until the subgraph $H$ becomes a graph with two vertices.

We will design a cops' strategy such that $\text{dist}(\lambda_1,s) \leq \text{dist}(\gamma,s)$ and $\text{dist}(\lambda_2,t) \leq \text{dist}(\gamma,t)$ after each of cops' turns. This guarantees that when the robber moves to $s$ or $t$, she will be captured immediately. If $G$ can be considered as a parallel composition of two series-parallel subgraphs, then the robber must be on one of them, say $H$. So we only need to consider the series-parallel subgraph $H$ whose terminals are occupied by cops. So the size of the problem is reduced. If $G$ can be considered as a series composition of two series-parallel subgraphs, let $s'$ be the closest cut vertex to $s$ and let $t'$ be the closest cut vertex to $t$ (note that $s'$ and $t'$ may be the same terminal shared by the two series-parallel subgraphs). Let $G_1$ be the series-parallel subgraph with terminals $\{s,s'\}$ and let $G_2$ be the series-parallel subgraph with terminals $\{t,t'\}$ (see Fig. 1). Without loss of generality, suppose that it is the cops' turn to move. We have the following cases.

CASE 1. $G_1$ or $G_2$ has only two vertices. Then $\lambda_1$ moves from $s$ to $s'$ or $\lambda_2$ moves from $t$ to $t'$. After that, we have a series-parallel subgraph whose terminals are occupied by cops. Thus the size of the problem is reduced.

CASE 2. $G_1$ and $G_2$ both have at least three vertices. If $\text{dist}(s,s') = 1$ and the robber is on $G_1$, then she cannot escape from $G_1$ because $\lambda_1$ protects $s$ and $s'$. So we can move $\lambda_2$ to $s'$ to reduce the size of the problem. If $\text{dist}(s,s') = 1$

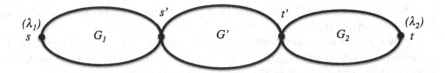

**Fig. 1.** $G_1$, $G_2$, $G'$ are series-parallel subgraphs, and $G_1$, $G_2$ are also blocks.

and the robber is not on $G_1$, then we can move $\lambda_1$ to $s'$ to reduce the size of the problem. If $\text{dist}(s, s') = 2$ and the robber is on $G_1$, then we can move $\lambda_1$ to the vertex that is adjacent to $s$ and $s'$. So the robber cannot escape from $G_1$. We move $\lambda_2$ to $s'$ and then move $\lambda_1$ back to $s$. Thus the size of the problem is reduced. If $\text{dist}(s, s') = 2$ and the robber is not on $G_1$, then we can move $\lambda_1$ to $s'$ in two rounds to reduce the size of the problem. We can argue similarly if $\text{dist}(t, t') \leq 2$. So assume that $\text{dist}(s, s') = k \geq 3$ and $\text{dist}(t, t') = \ell \geq 3$. We have the following subcases.

CASE 2.1. The robber is not on $G_1$ and $\text{dist}(\gamma, s') \geq k - 1$, or the robber is not on $G_2$ and $\text{dist}(\gamma, t') \geq \ell - 1$. Then $\lambda_1$ moves from $s$ to $s'$ along a shortest path or $\lambda_2$ moves from $t$ to $t'$ along a shortest path. So the size of the problem is reduced.

CASE 2.2. The robber is on $G_1$ and $\text{dist}(\gamma, t') \leq \ell - 2$. Let $P_1$ be a shortest path from $s$ to $s'$ and $P_2$ be a shortest path from $t$ to $t'$. Here is the cops' strategy: At the beginning of any one of cops' turns, if the robber is adjacent to a cop, then the robber is captured in this turn. At the end of each of cops' turns, both $\text{dist}(\lambda_1, s) \leq \text{dist}(\gamma, s)$ and $\text{dist}(\lambda_2, t) \leq \text{dist}(\gamma, t)$ hold. Namely, in the first phase, $\lambda_1$ moves from $s$ to $s'$ along $P_1$ until either $\text{dist}(\lambda_1, s) = \text{dist}(\gamma, s)$ or $\lambda_1$ arrives at $s'$. If $\lambda_1$ arrives at $s'$ and the robber is outside of $G_1$, then we have a smaller series-parallel subgraph containing the robber and both terminals are occupied by cops. If $\lambda_1$ arrives at $s'$ and the robber is on $G_1$, then $\lambda_1$ can protect both $s'$ and $s$ because he moves along a shortest path between them and can simultaneously ensure that he is closer to both $s$ and $s'$ than the robber. So $\gamma$ is trapped in $G_1$. When $\gamma$ either stays still or changes moving directions, $\lambda_2$ will move towards $t'$ along $P_2$. After $\lambda_2$ arrives at $t'$, $\lambda_1$ moves back to $s$. So after a finite number of turns, the size of the problem is reduced. If $\lambda_1$ has not arrived at $s'$ and we have a cops' turn such that $\text{dist}(\lambda_1, s) = \text{dist}(\gamma, s)$ at the end of this turn, then we start the second phase of the strategy. During the second phase, we have the following rules for the cops.

1. If $\gamma$ moves towards $s$ such that $\text{dist}(\lambda_1, s) = \text{dist}(\gamma, s) + 1$ at the end of her turn, then $\lambda_1$ also moves towards $s$ to maintain $\text{dist}(\lambda_1, s) = \text{dist}(\gamma, s)$.
2. If $\gamma$ moves towards $s$ such that $\text{dist}(\lambda_1, s) = \text{dist}(\gamma, s)$ at the end of her turn, then $\lambda_2$ moves towards $t'$ along $P_2$ while $\lambda_1$ stays still.
3. If $\gamma$ stays still with $\text{dist}(\lambda_1, s) = \text{dist}(\gamma, s)$, then $\lambda_2$ moves towards $t'$ along $P_2$ while $\lambda_1$ stays still.
4. If $\gamma$ moves towards $s'$ such that $\text{dist}(\lambda_1, s) = \text{dist}(\gamma, s) - 1$ at the end of her turn, then $\lambda_2$ moves towards $t'$ along $P_2$ while $\lambda_1$ stays still.

5. If $\gamma$ stays still with $\mathrm{dist}(\lambda_1, s) = \mathrm{dist}(\gamma, s) - 1$, then $\lambda_2$ moves towards $t'$ along $P_2$ while $\lambda_1$ stays still.
6. If $\gamma$ moves towards $s'$ such that $\mathrm{dist}(\lambda_1, s) = \mathrm{dist}(\gamma, s) - 2$ at the end of her turn and $\lambda_1$ is not adjacent to $s'$, then $\lambda_1$ also moves towards $s'$ such that $\mathrm{dist}(\lambda_1, s) = \mathrm{dist}(\gamma, s) - 1$ at the end of the cops' turn.
7. If $\gamma$ moves towards $s'$ such that $\mathrm{dist}(\lambda_1, s) = \mathrm{dist}(\gamma, s) - 2$ at the end of her turn and $\lambda_1$ is adjacent to $s'$, then $\lambda_2$ moves towards $t'$ along $P_2$ while $\lambda_1$ stays still. From then on, $\lambda_1$ will protect both $s'$ and $s$. So $\gamma$ is trapped in $G_1$.
8. If $\gamma$ moves away from $s'$ to an adjacent vertex outside $G_1$ such that $\mathrm{dist}(\lambda_1, s) = \mathrm{dist}(\gamma, s) - 2$ at the end of her turn, then $\lambda_1$ stays at the present vertex, which is adjacent to $s'$, and $\lambda_2$ moves back to $t$ along $P_2$ in the following rounds. After $\lambda_2$ arrives at $t$, $\lambda_1$ moves to $s'$, and thus, the size of the problem is reduced.

In the second phase using the above rules, if $\lambda_2$ arrives at $t'$, then $\lambda_1$ moves back to $s$ along $P_1$ so that the size of the problem is reduced.

CASE 2.3. The robber is on $G_2$ and $\mathrm{dist}(\gamma, s') \le k - 2$. By symmetry, we can have a cops' strategy similar to that in CASE 2.2.

CASE 2.4. The robber is not on $G_1$ or $G_2$, $\mathrm{dist}(\gamma, s') \le k - 2$ and $\mathrm{dist}(\gamma, t') \le \ell - 2$. We can easily modify the strategy in CASE 2.2 such that the robber is either captured or trapped in a smaller series-parallel subgraph whose terminals are occupied by cops.

In the above, we give explicit proofs for the cases where the robber makes reasonable moves. When the robber makes unreasonable moves (e.g., moving to a vertex occupied by $\lambda_1$ or $\lambda_2$), it is easy to see the robber would be caught immediately or trapped. So, after a finite number of turns, we always have a smaller series-parallel subgraph which contains the robber and whose both terminals are occupied by the cops. Therefore $c_1(G) \le 2$. □

A connected graph is called a *generalized series-parallel graph* if all its blocks are two-terminal series-parallel graphs such that any two blocks can share at most one vertex that is a terminal of both of them.

**Theorem 2.** *For a connected graph $G$, if $\mathrm{tw}(G) \le 2$, then $c_1(G) \le 2$.*

*Proof.* If $\mathrm{tw}(G) = 1$, then $G$ is a tree, and thus $c_1(G) = 1$. If $\mathrm{tw}(G) = 2$, then $G$ is a generalized series-parallel graph. So each block is a two-terminal series-parallel subgraph whose terminals are cut vertices of $G$ if they are also terminals of other blocks. Initially, we place the two cops $\lambda_1$ and $\lambda_2$ on a cut vertex $v_1$ of $G$. Suppose $G'$ is a connected component in $G - v_1$ that contains the robber. Let $G_1$ be a block of $G$ such that $v_1$ is a terminal of $G_1$ and the other terminal $v_2$ is in $G'$. Then $\lambda_1$ moves from $v_1$ to $v_2$ along a shortest path. If the robber is contained in $G_1$, then we can use the strategy described in the proof of Theorem 1 so that the two cops can capture the robber on $G_1$; otherwise, $\lambda_2$ also moves from $v_1$ to $v_2$. We can continue in this way until the robber is captured. □

**Corollary 1.** *For a connected graph $G$ with $\mathrm{tw}(G) = 2$, if $G$ contains an induced cycle of length at least 4, then $c_1(G) = 2$.*

*Proof.* Let $G$ be a connected graph of $n$ vertices such that $tw(G) = 2$ and $G$ contains an induced cycle of length at least 4. Assume that $c_1(G) = 1$. Then $G$ is copwin. We say that a vertex $x$ is a *corner* of $G$ if there is a vertex $y$ such that $N_G[x] \subseteq N_G[y]$. Since $G$ is copwin, from [14], the vertices of $G$ can be listed in a sequence $(u_1, u_2, \ldots, u_n)$ such that each $u_i$, $1 \leq i \leq n$, is a corner in the subgraph of $G$ induced by $\{u_i, u_{i+1}, \ldots, u_n\}$; this induced subgraph is denoted by $G_i$. Since $G$ contains an induced cycle of length at least 4, there is an index $i$, $1 \leq i \leq n$, such that $G_i$ contains an induced cycle $C$ of length at least 4 and $u_i$ is a vertex on $C$. Notice that $C$ is also an induced cycle in $G$. Since $u_i$ is a corner in $G_i$, there is a vertex $u_j$, $i + 1 \leq j \leq n$, such that $N_{G_i}[u_i] \subseteq N_{G_i}[u_j]$. Since $C$ is an induced cycle of length at least 4, we know that $u_j$ cannot be on this cycle. Note that $u_j$ is adjacent to $u_i$ and its two neighbors on $C$. Thus $K_4$ must be a minor in $G$, which contradicts $tw(G) = 2$.                                □

For a planar graph $G$, a *plane embedding* of $G$ is an embedding of $G$ on the plane so that distinct edges do not intersect except possibly at end vertices. A graph is *outerplanar* if it has a plane embedding such that all vertices of the graph are on the exterior face. Since all connected outerplanar graphs have treewidth at most 2, we have the following corollary.

**Corollary 2.** *For any connected outerplanar graph $G$, $c_1(G) \leq 2$.*

## 4    Halin Graphs

A *Halin graph* is a planar graph, which can be constructed from a plane embedding of a tree with at least 4 vertices and with no vertices of degree 2, by connecting all leaves with a cycle in the natural cyclic order defined by the embedding of the tree.

If $G$ is a wheel graph, which is a special Halin graph, then we can place a cop on the center of the wheel and this cop can capture the robber on any vertex of $H$. Thus we have the following lemma.

**Lemma 1.** *If $G$ is a wheel graph, then $c_1(G) = 1$.*

**Theorem 3.** *Let $H$ be a Halin graph. If $H$ is not a wheel graph, then $2 \leq c_1(H) \leq 3$.*

*Proof.* For the Halin graph $H$, we will also use $H$ to denote its plane embedding that consists of a tree $T$ and a cycle $C$, where $T$ has no vertex of degree two and has at least one vertex of degree three or more, and $C$ is a cycle connecting all the leaves of $T$ in the cyclic order.

Since $H$ is not a wheel graph, there must be two vertices $x$ and $y$ of $H$ such that $dist_C(x, y) = 1$ and $dist_T(x, y) \geq 3$. Assume that there is only one cop in the game. The robber can choose either $x$ or $y$ to occupy initially. If the robber just moves on the smallest cycle containing $x$ and $y$, then the cop cannot capture her. Hence $c_1(H) \geq 2$.

Pick an internal vertex $u$ of $T$ as the root. Let $N_T(u) = \{v_1, v_2, \ldots, v_\ell\}$ which are listed in the cyclic order, and let $T'$ be a subtree of $T$ which contains the root $u$, a proper subset of consecutive children of $u$ and all the descendants of these children. The subgraph $F$ of $H$ induced by $V(T')$ is called a *fan with apex* $u$. The leftmost leaf $w$ of $T'$ is called a *left corner* of $F$ and the rightmost leaf $w'$ is a *right corner* of $F$. The fan $F$ is denoted by $F(u, w, w')$. A vertex $x$ is a *center* of a tree $T$ if the radius of $T$ is equal to $\max\{\text{dist}_T(x, y) : y \in V(T)\}$.

We will give a strategy which guarantees that three cops $\lambda_1$, $\lambda_2$, $\lambda_3$ can capture the robber $\gamma$ on $H$. Let $a$ be a center of the tree $T$, and let $F(a, p, q)$ be a fan with apex $a$, left corner $p$ and right corner $q$. Let $P$ be the path on $T$ connecting two leaves $p$ and $q$ (which must contain $a$), and let $P'$ be the path on $C$ connecting $p$ and $q$. Initially, we place $\lambda_1$ on $p$, $\lambda_2$ on $q$ and $\lambda_3$ on $a$ (see Fig. 2). If $\gamma$ is not placed on a vertex of the fan $F(a, p, q)$, then she must be on a vertex of the neighboring fan $F(a, b, c)$, where $b \in N_H(q)$ and $c \in N_H(p)$ are leaves of $T$. Then we can move $\lambda_1$ to $c$ and $\lambda_2$ to $b$ in two rounds in such a way that $F(a, b, c)$ contains $\gamma$ and its apex and corners are occupied by cops. Without loss of generality, we suppose that $F(a, p, q)$ contains $\gamma$.

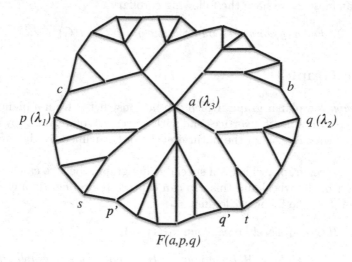

$F(a,p,q)$

**Fig. 2.** A fan $F(a, p, q)$ with apex $a$ occupied by $\lambda_3$, left corner $p$ occupied by $\lambda_1$, and right corner $q$ occupied by $\lambda_2$.

Note that $F(a, p, q)$ consists of a subtree $T'$ of $T$ and a path $P'$ on $C$ from $p$ to $q$. Since $a$ is on the path $P$ connecting $p$ and $q$, it must have at least two children in $T'$. Let $p'$ be the left corner of the maximal proper subfan $F(a, p', q)$ of $F(a, p, q)$, and let $q'$ be the right corner of the maximal proper subfan $F(a, p, q')$ of $F(a, p, q)$ (see Fig. 2). Without loss of generality, we assume that $a$ has at least three children in $T'$ (the case that $a$ has two children can be proved similarly). Let $s$ be a leaf of $T'$ adjacent to $p'$ on $C$ such that $F(a, p, s)$ is a fan not containing $p'$.

Let $t$ be a leaf of $T'$ adjacent to $q'$ on $C$ such that $F(a, t, q)$ is a fan not containing $q'$ (see Fig. 2).

Let $P_1$ be a shortest path in $F(a, p, s)$ from $p$ to $s$ and let $P_2$ be a shortest path in $F(a, t, q)$ from $q$ to $t$. If $\text{dist}_H(p, s) - 1 \leq \text{dist}_H(\gamma, s)$ and $\gamma$ is not on $F(a, p, s)$, then $\lambda_1$ moves from $p$ to $p'$ along $P_1$, and when $\lambda_1$ arrives at $p'$, the robber is trapped in $F(a, p', q)$, where $a, p', q$ are occupied by cops; similarly, if $\text{dist}_H(q, t) - 1 \leq \text{dist}_H(\gamma, t)$ and $\gamma$ is not on $F(a, t, q)$, then $\lambda_2$ moves from $q$ to $q'$ along $P_2$, and when $\lambda_2$ arrives at $q'$, the robber is trapped in $F(a, p, q')$, where $a, p, q'$ are occupied by cops; otherwise, we have the following situations for cops.

1. At the beginning of a cops' turn, if the robber is adjacent to one of the cops, then the robber is captured in this turn; if $\lambda_1$ is adjacent to $p'$, then $\lambda_1$ moves to $p'$; similarly, if $\lambda_2$ is adjacent to $q'$, then $\lambda_2$ moves to $q'$ in this turn.

2. If $\text{dist}_H(\lambda_1, p) < \text{dist}_H(\gamma, p) - 1$ at the beginning of a cops' turn, then $\lambda_1$ moves towards $p'$ along $P_1$. If at some point $\lambda_1$ arrives at $p'$, then in the following rounds, $\lambda_2$ moves back to $q$ along $P_2$ (if $\lambda_2$ is not on $q$) so that $\gamma$ is contained in $F(a, p', q)$ whose apex and corners are occupied by cops. Similarly, if $\text{dist}_H(\lambda_2, q) < \text{dist}_H(\gamma, q) - 1$ at the beginning of a cops' turn, then $\lambda_2$ moves towards $q'$ along $P_2$. If at some point $\lambda_2$ arrives at $q'$, then in the following rounds, $\lambda_1$ moves back to $p$ along $P_1$ so that $\gamma$ is contained in $F(a, p, q')$ whose apex and corners are occupied by cops. Break ties arbitrarily if both inequalities hold.

3. If $\text{dist}_H(\gamma, p) - 1 \leq \text{dist}_H(\lambda_1, p) \leq \text{dist}_H(\gamma, p)$ at the beginning of a cops' turn, then $\lambda_2$ moves towards $q'$ along $P_2$. If at some point $\lambda_2$ arrives at $q'$, then $\lambda_1$ moves back to $p$ along $P_1$ (if $\lambda_1$ is not on $p$) in the following rounds so that $\gamma$ is contained in $F(a, p, q')$ whose apex and corners are occupied by cops. Similarly, if $\text{dist}_H(\gamma, q) - 1 \leq \text{dist}_H(\lambda_2, q) \leq \text{dist}_H(\gamma, q)$ at the beginning of a cops' turn, then $\lambda_1$ moves towards $p'$ along $P_1$. If at some point $\lambda_1$ arrives at $p'$, then $\lambda_2$ moves back to $q$ along $P_2$ in the following rounds so that $\gamma$ is contained in $F(a, p', q)$ whose apex and corners are occupied by cops. Break ties arbitrarily if both inequalities hold.

4. If $\text{dist}_H(\lambda_1, p) = \text{dist}_H(\gamma, p) + 1$ (resp. $\text{dist}_H(\lambda_2, q) = \text{dist}_H(\gamma, q) + 1$) at the beginning of a cops' turn, then $\lambda_1$ (resp. $\lambda_2$) moves towards $p$ (resp. $q$) to maintain $\text{dist}_H(\lambda_1, p) \leq \text{dist}_H(\gamma, p)$ (resp. $\text{dist}_H(\lambda_2, q) \leq \text{dist}_H(\gamma, q)$) at the end of this turn.

In the searching process using the above strategy, at the end of each of cops' turn, both $\text{dist}(\lambda_1, p) \leq \text{dist}(\gamma, p)$ and $\text{dist}(\lambda_2, q) \leq \text{dist}(\gamma, q)$ hold. So the corners $p$ and $q$ are always protected. Thus, after a finite number of turns, we have a smaller subfan which contains the robber such that its left corner is occupied by $\lambda_1$, right corner is occupied by $\lambda_2$ and apex is occupied by $\lambda_3$ ($\lambda_3$ moves from $a$ to a new apex directly in some cases). We can repeat this process until the robber is captured. Therefore $c_1(H) \leq 3$. $\qquad\square$

## 5  Cartesian Product Graphs

Let $G_1$ and $G_2$ be graphs. The *Cartesian product* of $G_1$ and $G_2$, denoted $G_1 \,\square\, G_2$, has vertex set $V(G_1) \times V(G_2)$, where two vertices $(u_1, u_2)$ and $(v_1, v_2)$ are adjacent if and only if there exists $j$, $1 \le j \le 2$, such that $\{u_j, v_j\} \in E(G_j)$ and $u_i = v_i$ for $i \ne j$. Maamoun and Meyniel [13] studied the cop number of Cartesian products of trees in the Cops and Robbers game. The next result follows from Lemma 1 in [13].

**Theorem 4.** *Let $T_1$ and $T_2$ be two trees, each of which contains at least one edge. Then* $c_1(T_1 \,\square\, T_2) = 2$.

If we replace one of the trees by a clique, which is always 1-winnable, the above lemma still holds, i.e., $c_1(T \,\square\, K_n) = 2$, $n \ge 2$. However, this is not always true when we replace one of the trees by a 1-winnable graph. Here is an example. Let $H$ be an $n \times n$ grid with all diagonals $\{(i+1, j), (i, j+1)\}$, $1 \le i, j \le n - 1$, such that every face is a triangle except the exterior face. It is easy to see that $c_1(H) = 1$. If each side of $H$ has a large number of vertices, e.g., $n = 100$, then even if $T$ is a path $P$ with at least three vertices, we have $c_1(P \,\square\, H) \ge 3$.

On the other hand, if we replace one of the trees by a 2-winnable graph, Theorem 4 may still hold. Consider the Cartesian product of a tree $T$ and a cycle $C$. We have the following result.

**Lemma 2.** *Let $T$ be a tree with at least two vertices and $C$ be a cycle with at least three vertices. Then* $c_1(T \,\square\, C) = 2$.

## 6  Characterization of $k$-winnable Graphs in the One-Cop-Moves Game

The purpose of this section is to present a characterization of $k$-winnable graphs which extends the one given by Nowakowski and Winkler [14] and Quilliot [16]. The characterization involves a relation defined on the vertex set of a graph. The idea of this characterization is similar to that used by Clarke and MacGillivray [8], which has also be used in [3,11].

**Definition 3 (Relation $\mathbb{R}^i_{G,k}$).** Let $G$ be a finite connected graph and $k$ be an integer satisfying $1 \le k \le |V(G)|$. A relation $\mathbb{R}^0_{G,k}$ is defined as $\mathbb{R}^0_{G,k} = \{(u, S) : u \in S \subseteq V_{G,k} \text{ and } |S| = k\}$. Assume that the relation $\mathbb{R}^i_{G,k}$ has been defined for each integer $i \in [0, t]$. We now define $\mathbb{R}^{t+1}_{G,k}$ by saying $(u, S) \in \mathbb{R}^{t+1}_{G,k}$, where $u \in V(G)$, $S \subseteq V_{G,k}$ and $|S| = k$, if and only if for every $u' \in N[u]$, there are vertices $v \in S$ and $v' \in N[v]$ such that $(u', S') \in \mathbb{R}^i_{G,k}$ for some $i \in [0, t]$, where $S' = (S \setminus \{v\}) \cup \{v'\}$.

**Lemma 3.** *For any $i \ge 0$, $\mathbb{R}^i_{G,k} \subseteq \mathbb{R}^{i+1}_{G,k}$.*

**Lemma 4.** *For any $i \ge 0$, if $(u, S) \in \mathbb{R}^i_{G,k}$, then* $\mathrm{dist}_G(u, S) \le i$.

From Definition 3, we know that whenever we reach an index $t$ such that $\mathbb{R}_{G,k}^t = \mathbb{R}_{G,k}^{t-1}$, then $\mathbb{R}_{G,k}^i = \mathbb{R}_{G,k}^{t-1}$ for all $i \geq t$.

**Definition 4 (Relation $\mathbb{R}_{G,k}$).** From Lemma 3, there must be a smallest $t$ for which $\mathbb{R}_{G,k}^t = \mathbb{R}_{G,k}^{t-1}$. We then define the relation $\mathbb{R}_{G,k}$ to be $\mathbb{R}_{G,k}^{t-1}$. If $\mathbb{R}_{G,k} = V_{G,1} \times \{S : S \subseteq V_{G,k}$ and $|S| = k\}$, the relation $\mathbb{R}_{G,k}$ is called *complete*.

We have the following characterization of $k$-winnable graphs in the one-cop-moves game.

**Theorem 5.** *Let $G$ be a finite connected graph and $k$ be an integer satisfying $1 \leq k \leq |V(G)|$. $G$ is $k$-winnable in the one-cop-moves game if and only if the relation $\mathbb{R}_{G,k}$ is complete.*

**Corollary 3.** *For a connected graph $G$, if there is a multiset $S$ of vertices with $|S| = k$ and an integer $i$ such that $(u, S) \in \mathbb{R}_{G,k}^i$ for any vertex $u \in V(G)$, then $\mathbb{R}_{G,k}$ is complete.*

From the sufficiency proof of Theorem 5, the index in relation $\mathbb{R}_{G,k}^i$ gives us an upper bound on the optimal capture time of the one-cop-moves game [20].

**Corollary 4.** *For a connected graph $G$, if there is a multiset $S$ of vertices with $|S| = k$ and an integer $i$ such that $(u, S) \in \mathbb{R}_{G,k}^i$ for any vertex $u \in V(G)$, then in the one-cop-moves game on $G$ with $k$ cops and one robber, the robber can be captured in at most $i$ rounds.*

## 7  Conclusions

In the proof of $c_1(G) \leq 2$ for series-parallel graphs $G$, we recursively reduce the problem on a series-parallel graph to the problem on a series-parallel subgraph. The same technique is also used for Halin graphs. In the one-cop-moves game, since only one cop can move and all other cops act as guards in each round, we believe that this technique can conceivably be applied to other classes of graphs.

In [12], Joret et al. proved that $c(G) \leq \text{tw}(G)/2 + 1$. In Sect. 3, we proved $c_1(G) \leq \text{tw}(G)/2 + 1$ for any connected graph $G$ with $\text{tw}(G) \leq 2$, while in Sect. 4, we showed $c_1(H) \leq 3$ for any Halin graph $H$. Note that the treewidth of a Halin graph is at most 3. So we have the following conjecture for general graphs.

*Conjecture 1.* For any connected graph $G$, $c_1(G) \leq \lceil \text{tw}(G)/2 \rceil + 1$.

Similar to the one-cop-moves game, we can define the two-cops-move game. This game has the same setting as the Cops and Robbers game except that at cops' turns, at most two cops are allowed to move from their current locations to their adjacent vertices. Let $c_2(G)$ be the two-cops-move cop number of $G$. Aigner and Fromme [1] showed that $c(G) \leq 3$ for any connected planar graph $G$. Gao and Yang [9] showed that there are connected planar graphs $D$ such that $c_1(D) \geq 4$. We believe Aigner and Fromme's result can generalize to the two-cops-move game.

*Conjecture 2.* In the two-cops-move game, if $G$ is a connected planar graph, then $c_2(G) \leq 3$.

# References

1. Aigner, M., Fromme, M.: A game of cops and robbers. Discret. Appl. Math. **8**, 1–12 (1984)
2. Alspach, B.: Sweeping and searching in graphs: a brief survey. Matematiche **5**, 5–37 (2006)
3. Berarducci, A., Intrigila, B.: On the cop number of a graph. Adv. Appl. Math. **14**(4), 389–403 (1993)
4. Bal, D., Bonato, A., Kinnersley, W.B., Pralat, P.: Lazy cops and robbers on hypercubes. Comb. Probab. Comput. **24**(6), 829–837 (2015)
5. Bal, D., Bonato, A., Kinnersley, W.B., Pralat, P.: Lazy cops and robbers played on random graphs and graphs on surfaces. Int. J. Comb. **7**(4), 627–642 (2016)
6. Bonato, A., Nowakowski, R.J.: The Game of Cops and Robbers on Graphs. American Mathematical Society, Providence (2011)
7. Bonato, A., Yang, B.: Graph searching and related problems. In: Pardalos, P.M., Du, D.-Z., Graham, R.L. (eds.) Handbook of Combinatorial Optimization, pp. 1511–1558. Springer, New York (2013). https://doi.org/10.1007/978-1-4419-7997-1_76
8. Clarke, N.E., MacGillivray, G.: Characterizations of $k$-copwin graphs. Discret. Math. **312**, 1421–1425 (2012)
9. Gao, Z., Yang, B.: The one-cop-moves game on planar graphs. J. Comb. Optim. (2019). https://link.springer.com/article/10.1007/s10878-019-00417-x
10. Hahn, G.: Cops, robbers and graphs. Tatra Mt. Math. Publ. **36**, 163–176 (2007)
11. Hahn, G., MacGillivray, G.: A note on $k$-cop, $l$-robber games on graphs. Discret. Math. **306**, 2492–2497 (2006)
12. Joret, G., Kamiński, M., Theis, D.: The cops and robber game on graphs with forbidden (induced) subgraphs. Contrib. Discret. Math. **5**, 40–51 (2010)
13. Maamoun, M., Meyniel, H.: On a game of policemen and robber. Discret. Appl. Math. **17**, 307–309 (1987)
14. Nowakowski, R.J., Winkler, P.: Vertex-to-vertex pursuit in a graph. Discret. Math. **43**, 235–239 (1983)
15. Offner, D., Okajian, K.: Variations of cops and robber on the hypercube. Australas. J. Comb. **59**(2), 229–250 (2014)
16. Quilliot, A.: Jeux et pointes fixes sur les graphes, thèse de 3ème cycle, Université de Paris VI, pp. 131–145 (1978)
17. Sullivan, B.W., Townsend, N., Werzanski, M.: The $3 \times 3$ rooks graph is the unique smallest graph with lazy cop number 3. Preprint (2017). https://arxiv.org/abs/1606.08485
18. Theis, D.: The cops & robber game on series-parallel graphs. Preprint (2008). https://arxiv.org/abs/0712.2908
19. West, D.B.: Introduction to Graph Theory. Prentice Hall, Upper Saddle River (1996)
20. Yang, B., Hamilton, W.: The optimal capture time of the one-cop-moves game. Theor. Comput. Sci. **588**, 96–113 (2015)

# Bounded Degree Graphs Computed for Traveling Salesman Problem Based on Frequency Quadrilaterals

Yong Wang(✉) [ID]

North China Electric Power University, Beijing 102206, China
yongwang@ncepu.edu.cn

**Abstract.** Traveling salesman problem ($TSP$) is well-known in combinatorial optimization. Recent research demonstrates that there are competitive algorithms for $TSP$ on the bounded degree or genus graphs. We present an iterative algorithm to convert the complete graph of a $TSP$ instance into a bounded degree graph based on frequency quadrilaterals. First, the frequency of each edge is computed with $N$ frequency quadrilaterals. Second, $\frac{1}{3}$ of the edges having the smallest frequencies are cut for each vertex. The two steps are repeated until there are no quadrilaterals for most edges in the residual graph. At the $k^{th} \geq 1$ iteration, the maximum vertex degree is smaller than $\left\lceil \left(\frac{2}{3}\right)^k (n-1) \right\rceil$. Two theorems illustrate that the long edges are cut and the short edges will be preserved. Thus, the optimal solution or a good approximation will be found in the bounded degree graphs. The iterative algorithm is tested with the real-life $TSP$ instances.

**Keywords:** Traveling salesman problem · Bounded degree graph · Frequency quadrilateral · Algorithm

## 1 Introduction

We consider the symmetric traveling salesman problem ($TSP$). That is, we are given the complete graph $K_n$ on the vertices $\{1, \ldots, n\}$ such that there is a distance function $d$ such that for any vertices $x, y \in \{1, \ldots, n\}$ and $x \neq y$, $d(x, y) = d(y, x)$ is the distance between $x$ and $y$. An optimal Hamiltonian cycle ($OHC$) will be explored with respect to this distance function. That is, we want to find a permutation $\sigma = (\sigma_1 \ldots \sigma_n)$ of $1, \ldots, n$ such that $\sigma_1 = 1$ and the distance $d(\sigma) := d(\sigma_n, 1) + \sum_{i=1}^{n-1} d(\sigma_i, \sigma_{i+1})$ is as small as possible. The $TSP$ has been extensively studied to find the special classes of graphs where polynomial-time algorithms exist for either finding an $OHC$, or finding an approximate solution, i.e., a Hamiltonian cycle ($HC$) whose distance is close

The author acknowledge the funds supported by the Fundamental Research Funds for the Central Universities (No. 2018MS039 and No. 2018ZD09).

© Springer Nature Switzerland AG 2019
Y. Li et al. (Eds.): COCOA 2019, LNCS 11949, pp. 529–540, 2019.
https://doi.org/10.1007/978-3-030-36412-0_43

to that of the $OHC$. There are a number of special classes of graphs where one can find the $OHC$ in a reasonable computation time [1].

Karp [2] showed that $TSP$ is $NP$-complete in computation complexity. This means that there are no polynomial-time algorithms for $TSP$ unless $P = NP$. The computation time of the exact algorithms is generally $O(a^n)$ for some $a > 1$. For example, Held and Karp [3], and independently Bellman [4] gave the dynamic programming approach which required $O(n^2 2^n)$ time. Integer programming techniques, such as either branch-and-bound [5,6] or cutting-plane [7,8], are able to solve some examples of $TSP$ with thousands of nodes. In 2006, a VLSI instance (Euclidean $TSP$) with 85,900 nodes has been solved by the branch-and-cut method with a 128-node computer system [8].

Recently, researchers have developed polynomial-time algorithms for the $TSP$ on sparse graphs. In a sparse graph, the number of the $HC$s is greatly reduced. For example, Sharir and Welzl [9] proved that in a sparse graph of average degree $d = 3, 4$, the number of $HC$s is less than $e(\frac{d}{2})^n$ where $e$ is the base of the natural logarithm. Gebauer [10] gave the lower bound of the number of $HC$s roughly as $(\frac{4}{3})^n$ for a graph of average degree 3. In addition, Björklund [11] proved that the $TSP$ on a bounded degree graph could be solved in time $O((2 - \epsilon)^n)$, where $\epsilon$ depends on the maximum degree of a vertex in the graph. For the $TSP$ on a general bounded-genus graph, Borradaile, Demaine and Tazari [12] gave a polynomial-time approximation scheme. In the case of the asymmetric $TSP$ on a bounded-genus graph, Gharan and Saberi [13] designed a constant-factor approximation algorithm and the constant-factor is $22.51(1 + \frac{1}{n})$ for the $TSP$ on a planar graph. Thus, whether one is trying to find the exact solution or approximate solutions, one has a variety of more efficient algorithms available if one can reduce a given instance of $TSP$ to finding the $OHC$ in a sparse graph.

According to the $k - opt\ moves$, Hougardy and Schroeder [14] argued that some edges are impossibly contained in any $OHC$s. They designed a three-stage algorithm which is able to trim many edges out of the $OHC$. The experiments illustrate that the Concorde Solver is accelerated as the $TSP$ instance on a sparse graph is resolved. Wang [15] used the frequency graphs as another way to reduce the number of edges that one has to consider for finding an $OHC$. The basic idea of the frequency graphs is the following. Suppose that we are given $k$ vertices $\{v_1, v_2 \ldots, v_{k-1}, v_k\}$ in $K_n$ where $k \geq 4$, let $\boldsymbol{v}_\sigma = (v_{\sigma_1}, \ldots, v_{\sigma_k})$ be a permutation of $v_1, \ldots, v_k$ where $v_{\sigma_1} = v_1$ and $v_{\sigma_k} = v_k$ and let $d(\boldsymbol{v}_\sigma) = \sum_{i=1}^{k-1} d(v_{\sigma_i}, v_{\sigma_{i+1}})$. We assume that $d(\boldsymbol{v}_\sigma)$ all have different values, so there is an optimal path $\boldsymbol{v}_\sigma = (v_{\sigma_1}, \ldots, v_{\sigma_k})$ connecting $v_1$ and $v_k$ using the intermediate vertices $v_2, \ldots, v_{k-1}$ which makes $d(\boldsymbol{v}_\sigma)$ as small as possible. We will call such a path the *optimal k-vertex path* for $v_1$ and $v_k$. Given $k$ vertices $\{v_1, \ldots, v_k\}$, there are $\binom{k}{2}$ ways to pick the endpoints for a $k$-vertex path. Thus, there are $\binom{k}{2}$ optimal $k$-vertex paths that arise from the vertex set $\{v_1, \ldots, v_k\}$.

Let $\mathcal{OP}^k$ denote the set of optimal $k$-vertex paths in $K_n$. Then the frequency $f(u, v)$ of an edge $(u, v)$ is the number of the optimal $k$-vertex paths which contain $(u, v)$ as an edge. Our intuition is that if $(u, v)$ is an edge in the $OHC$, then its frequency is likely to be much higher than the average frequency. This has been born out by studying many real-life $TSP$ instances, see [15]. This suggests that we can safely eliminate the edges of small frequency below the average frequency and still keep the $OHC$. The hope is that by eliminating the edges of low frequency, we can be left with a sparse graph which has $O(n \log n)$ edges so that the techniques for either finding or approximating the $OHC$ for sparse graphs can be applied.

In the following, Wang and Remmel [16] compute the frequency of edges with the frequency quadrilaterals. As one chooses $N$ frequency quadrilaterals containing an edge to compute its frequency, a binomial distribution model illustrates the frequency of an $OHC$ edge is much higher than the average frequency. Thus, the edges of low frequency can be eliminated for computing a sparse graph for $TSP$. Moreover, Wang and Remmel [17] designed an iterative algorithm to compute a sparse graph for $TSP$ based on the binomial distribution model. At each iteration, all edges in a given graph $G = (V, E)$ are ordered according to their frequencies and the $\frac{1}{3}|E|$ edges of the lowest frequencies are eliminated. Since the low frequencies are not uniformly distributed to the edges containing each vertex, the degrees of vertices are different as the edges of low frequencies are cut. The maximum vertex degree is hard to estimate after edges elimination. In this research, we eliminate $\frac{1}{3}$ of the edges for each vertex according to their frequencies. At the iteration $k$, the maximum vertex degree of the preserved graph is less than $\left\lceil \left(\frac{2}{3}\right)^k (n-1) \right\rceil$. As $k$ is near to $\log_{1.5} \frac{n-1}{2}$, the algorithm will output a sparse bounded degree graph for $TSP$.

This paper is organized as follows. In Sect. 2, we shall briefly introduce the frequency quadrilaterals. In Sect. 3, an iterative algorithm is designed for eliminating edges for $TSP$. Two theorems are given in Sect. 4 to illustrate the difference between the eliminated edges and the preserved edges. In Sect. 5, we will do experiments to compute the sparse graphs for the real-world examples. In the last section, conclusions are drawn and the future research is proposed.

## 2    The Frequency Quadrilaterals

The frequency quadrilaterals have been computed in paper [16]. Given four vertices $\{A, B, C, D\}$ in $K_n$, they form a quadrilateral $ABCD$ as each pair of vertices are connected. The six edges are noted as $(A, B)$, $(A, C)$, $(A, D)$, $(B, C)$, $(B, D)$ and $(C, D)$, and the distances of the six edges are noted as $d(A, B)$, $d(A, C)$, $d(A, D)$, $d(B, C)$, $d(B, D)$ and $d(C, D)$, respectively. The corresponding frequency quadrilateral is computed with the six optimal four-vertex paths of given endpoints in $ABCD$. An example of the optimal four-vertex path for two vertices $A$ and $B$ is computed as follows. Fix endpoints $A$ and $B$, there are two four-vertex paths $P_1 = (A, C, D, B)$ and $P_2 = (A, D, C, B)$ in $ABCD$. Provided that the two paths have different distances, i.e., $d(P1) \neq d(P_2)$, the shorter path is taken as the optimal four-vertex path for $A$ and $B$. Since six pairs of endpoints originate from $\{A, B, C, D\}$, $ABCD$ contains the six optimal four-vertex paths. For an edge $e \in \{(A, B), (A, C), (A, D), (B, C), (B, D), (C, D)\}$, the frequency $f(e)$ is the number of the optimal four-vertex paths containing $e$.

The six optimal four-vertex paths are determined by the three sum distances $d(A,B) + d(C,D)$, $d(A,C) + d(B,D)$ and $d(A,D) + d(B,C)$ in $ABCD$. The bigger the sum distance is, the smaller the frequency of the related edges will be. For example, if $d(A,B) + d(C,D) < d(A,D) + d(B,C) < d(A,C) + d(B,D)$, the six optimal four-vertex paths are derived as $(A,B,C,D)$, $(B,C,D,A)$, $(C,D,A,B)$, $(D,A,B,C)$, $(B,A,C,D)$ and $(A,B,D,C)$. The frequencies of the six edges are computed as $f(A,B) = f(C,D) = 5$, $f(A,D) = d(B,C) = 3$ and $f(A,C) = f(B,D) = 1$, respectively. Since the three sum distances have six orderings, there are six sets of optimal four-vertex paths and six frequency quadrilaterals. The six frequency quadrilaterals are illustrated in Fig. 1(a)–(f). Below each frequency quadrilateral, the three sum distances are ordered according to the less than inequality sign "<". In each frequency quadrilateral, the number aside an edge is its frequency enumerated from a set of optimal four-vertex paths. The six optimal four-vertex paths are computed based on the orders of the three sum distances in the quadrilateral $ABCD$.

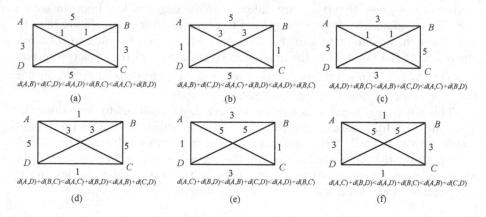

Fig. 1. The six frequency quadrilaterals for a quadrilateral $ABCD$

# 3   The Iterative Algorithm

Given a graph, $\frac{1}{3}$ of all edges with the smallest frequencies are cut in the paper [17]. The number of the preserved edges containing a vertex cannot be estimated after edges elimination. In this section, we provide the iterative algorithm for eliminating the edges for each vertex.

In a quadrilateral $ABCD$, every vertex associates three edges. For example, $A$ is contained in $(A,B)$, $(A,C)$ and $(A,D)$. Among the three edges, two of them are contained in the $OHC$ in $ABCD$. In each of the frequency quadrilaterals in Fig. 1(a)–(f), the edges with frequency 3 and 5 are the $OHC$ edges. The edges with frequency 1 can be eliminated as one searches the $OHC$ in $ABCD$. Given an edge $e$ containing a vertex $v$, such as $(A, B)$ containing $A$, it has each of the frequencies 1, 3 and 5 twice according to the six frequency quadrilaterals. Thus, the probability $p_1(e)$, $p_3(e)$, $p_5(e)$ that $e$ has 1, 3 and 5 in a frequency quadrilateral is $\frac{1}{3}$, respectively. The expected frequency of $e$ is 3. Moreover, $e$ (or $(A, B)$) is contained in four of the six $OHC$s in $ABCD$ and it has frequency 3 or 5 in the four frequency quadrilaterals. In the other two frequency quadrilaterals, $e$ (or $(A, B)$) has frequency 1 and it is not contained in the $OHC$. If we eliminate $e$ with the small frequency according to the frequency threshold 3, $e$ is eliminated two times. Let $p(e)$ denote the probability that $e$ is eliminated according to the frequency threshold 3. This probability is $p(e) = \frac{1}{3}$.

In $K_n$, a vertex $v$ is contained in $n - 1$ edges. Three of them form a quadrilateral $ABCD$. Given a $ABCD$, the corresponding frequency quadrilateral will be one of the six frequency quadrilaterals in Fig. 1(a)–(f). In average case, we assume an edge $e$ has the equal probability to be one of the edges containing $v$. In every frequency quadrilateral, the probability that $e$ has 1, 3 and 5 is $\frac{1}{3}$, respectively. According to the frequency threshold 3, the probability that $e$ is eliminated is $p(e) = \frac{1}{3}$. Considering the $n - 1$ edges containing $v$, $[\frac{1}{3}(n-1)]$ edges $e$ will be cut.

Given a common edge $e$, it has the probability $\frac{1}{3}$ to be cut according to 3 based on the six frequency quadrilaterals in Fig. 1. For an $OHC$ edge $e_o$ in $K_n$, this probability will decrease. Wang and Remmel [17] constructed $n - 3$ frequency quadrilaterals where $e_o$ has the frequency 3 or 5 rather than 1. Given a quadrilateral containing $e_o$, the probability that $e_o$ has 1, 3 and 5 is represented as Formula (1). According to the frequency threshold 3, the probability $p(e_o) = p_1(e_o)$ that $e_o$ is cut is smaller than $\frac{1}{3}$. As one chooses $N$ frequency quadrilaterals containing $e$ and $e_o$ respectively to compute their average frequency $\bar{f}(e)$ and $\bar{f}(e_o)$, $\bar{f}(e) = 3$ and $\bar{f}(e_o) = 3 + \frac{8}{3(n-2)} > 3$ hold. As $\frac{1}{3}$ of the edges $e$ with the smallest frequency below 3 are cut, the $OHC$ edges $e_o$ will be preserved. It mentions that the probability model (1) is very conservative for $e_o$ since only the $OHC$ edges are used to construct the quadrilaterals for $e_o$. $p_1(e_o)$ in Formula (1) can be taken as a upper bound for $e_o$.

$$p_5(e_o) = \frac{1}{3} + \frac{2}{3(n-2)}$$

$$p_3(e_o) = \frac{1}{3}$$

$$p_1(e_o) = \frac{1}{3} - \frac{2}{3(n-2)} \tag{1}$$

Based on the above analysis, the heuristic Algorithm 1 is given to compute a bounded degree graph for $TSP$, where $k$ and $d_{min}$ note the iteration and minimum vertex degree, respectively. First, the frequency of each edge is computed with $N$ frequency quadrilaterals. Second, $\frac{1}{3}$ of the edges containing each vertex is eliminated according to their frequencies. The maximum vertex degree of the preserved graphs decreases exponentially in proportion to $\frac{2}{3}$ according to $k$. At the $k^{th}$ iteration, the maximum vertex degree of the preserved graph is $\left\lceil \left(\frac{2}{3}\right)^k (n-1) \right\rceil$. Formula (1) guarantees that the $OHC$ edges will be preserved as $N$ is big enough. In the experiments, $N \geq 1$ for each preserved edge.

---

**Algorithm 1.** Algorithm for eliminating edges containing a vertex.

---

**Input**: $K_n = (V, E)$, the distance matrix and $N \geq 1$
**Output**: A residual graph of bounded degree
$k := 0$, $d_{min} := n - 1$;
**while** $d_{min} \geq 2$ *or 3* **do**
  Compute the average frequency of each edge with $N$ random
  frequency quadrilaterals;
  Order the edges containing an vertex according to their frequencies;
  Eliminate $\frac{1}{3}$ of the edges with the smallest frequencies for each vertex;
  $k := k + 1$ and $d_{min}$ are updated;
**end**

---

As the numbers of eliminated edges containing each vertex are equal, $N$ is equal to $\binom{\lfloor (\frac{2}{3})^k (n-2) \rfloor}{2}$ at iteration $k$. In applications, $N$ will be smaller than $\binom{\lfloor (\frac{2}{3})^k (n-2) \rfloor}{2}$ for most edges since the edges are not averagely eliminated for every vertex. In the best case, $N$ decreases exponentially in proportion to $(\frac{2}{3})^2$ for a preserved edge. At a certain iteration, there are few quadrilaterals for the preserved edges. The algorithm cannot work since the frequency of most edges is zero. In this case, the algorithm must be stopped. In this algorithm, the minimum vertex degree is used as a terminal condition. If $d_{min} = 2$ or 3, the algorithm is terminated and outputs a residual graph. In the paper [17], another terminal condition related to $N$ is used. Compared with $d_{min} = 2$ or 3, the experiments illustrated the terminal condition related to $N$ is more relaxed to compute a residual graph containing an $OHC$ and more edges.

In the algorithm, we assume the three sum distances $d(A, B) + d(C, D)$, $d(A, C) + d(B, D)$ and $d(A, D) + d(B, C)$ are unequal in each quadrilateral $ABCD$. If a $TSP$ instance contains many equal-weight edges, two or three of the sum distances will be equal in a number of quadrilaterals. In this case, it is hard to choose the right optimal four-vertex paths with respect to the $OHC$. As many inappropriate optimal four-vertex paths are used, the frequency of some $OHC$ edges will be lowered. In some iteration, these $OHC$ edges will be cut due to their small frequency.

## 4   The Distances of the Eliminated and Preserved Edges

After edges elimination, we expect to obtain a residual graph containing the short edges. Even though a few $OHC$ edges are lost, we will find a good approximation in the sparse graph. Theorem 1 compare the distances of the eliminated edges with that of the longest preserved edges.

**Theorem 1.** *Given an edge $e_1$ contained in $N$ quadrilaterals in a graph, if the distance of $e_1$ is bigger than two times of that of the longest adjacent edges $e_2$ in the $N$ quadrilaterals, $e_1$ must be eliminated.*

*Proof.* Two edges are adjacent if they have one common vertex. A vertex $v$ is contained in $e_1$ and the other adjacent edges of $e_1$. Among the adjacent edges of $e_1$, there is a longest edge $e_2$. Their distances comply with $d(e_1) > 2d(e_2)$. $e_1$ and $e_2$ will be contained in a quadrilateral $ABCD$. Let $v = A$, $e_1 = (A, C)$ and $e_2 = (A, B)$ shown in Fig. 1 (a). In $ABCD$, there are three sum distances $d(A, B) + d(C, D)$, $d(A, D) + d(B, C)$, $d(A, C) + d(B, D)$. Since $d(C, D) \leq d(A, B)$, $d(A, B) + d(C, D) \leq 2d(A, B) = 2d(e_2) \leq d(e_1) < d(A, C) + d(B, D)$. Since $(A, B)$ is longer than the other adjacent edges except $(A, C)$, $d(A, D) + d(B, C) \leq 2d(A, B) \leq d(e_1) < d(A, C) + d(B, D)$. Thus, $e_1 = (A, C)$ has frequency 1 in the corresponding frequency quadrilateral. In the other $N - 1$ quadrilaterals $AXCY$ containing $(A, C)$, $(A, C)$ still has the frequency 1 since $d(A, B)$ is the longest edge among the adjacent edges of $e_1$. In other words, $d(A, B) \geq d(A, X)$, $d(A, B) \geq d(C, Y)$, $d(A, B) \geq d(A, Y)$ and $d(A, B) \geq d(C, X)$ in any one $AXCY$. Thus, $(A, C)$ has the minimum average frequency 1 based on the $N$ frequency quadrilaterals. Therefore, it must be eliminated until there are less than three edges containing $v$.

In each iteration, an edge whose distance is bigger than two times of that of the longest adjacent edges will be eliminated. The long edges will be cut step by step. Compared with the original graph, an algorithm will find a better solution in the preserved graph in the worst case. Given an algorithm $\mathcal{A}$, it computes an $x$-approximation for $TSP$ in the worst case where $x > 1$. Before edges elimination, $\mathcal{A}$ computes an approximation with some long edges containing a vertex $v$. Given the longest edge has the distance $2d > 0$ and the other edge has a distance $c > 0$, the distance of the two edges is $2d + c$ in the approximation. After edges elimination, the preserved edges containing $v$ is at most $d$. $\mathcal{A}$ will find some edges with the distance $d$ containing $v$. Thus, the distance of the two edges containing $v$ in the approximation will at least decrease by $c$. After several iterations, the preserved edges will become shorter and shorter until the distances of the preserved edges containing each vertex do not have much difference. The approximation found by $\mathcal{A}$ according to these preserved graphs will be improved accordingly. As $\mathcal{A}$ always finds an optimal solution, it will be accelerated due to the large reduction of the search space after edges elimination.

In fact, it is not necessary that the distance of an eliminated edge must be bigger than two times of that of its longest adjacent edge. If an edge $e_1$ has frequency 1 in above $\left\lceil \frac{N}{3} \right\rceil$ frequency quadrilaterals, it will be eliminated with a big probability stated as Theorem 2.

**Theorem 2.** *Given an edge $e_1$ contained in $N$ quadrilaterals in a graph, if the distance of $e_1$ is bigger than two times of that of some longest edges $e_2$ among the adjacent edges of $e_1$ in above $\left\lceil \frac{N}{3} \right\rceil$ quadrilaterals, $e_1$ will be eliminated with an accumulative probability bigger than $\frac{1}{2}$ according to the Formula (1).*

*Proof.* $e_1$ is contained in $N$ quadrilaterals $ABCD$ in a graph. We first assume there are $\left\lceil \frac{N}{3} \right\rceil$ quadrilaterals where the distance of $e_1$ is bigger than two times of that of the longest adjacent edges $e_2$. Let $e_1 = (A, C)$ and $e_2 = (A, B)$ shown in Fig. 1 (a). Since $d(A, B) + d(C, D) \leq 2d(A, B) \leq d(A, C) < d(A, C) + d(B, D)$

and $d(A, D) + d(B, C) \leq 2d(A, B) \leq d(A, C) < d(A, C) + d(B, D)$, $(A, C)$ has
frequency 1 in frequency quadrilateral $ABCD$. Since $e_2 = (A, B)$ is the longest
adjacent edges of $e_1 = (A, C)$ in the $[\frac{N}{3}]$ quadrilaterals, $(A, C)$ has frequency 1
in each of the corresponding frequency quadrilaterals. Thus, the probability that
$e_1$ is cut is $p(e_1) = \frac{1}{3}$ according to 3. As we choose $N$ frequency quadrilaterals
containing $e_1$, the number $X$ of frequency quadrilaterals where $e_1$ has frequency
1 conforms to a binomial distribution $P(X = i) = \binom{N}{i}(\frac{1}{3})^i(\frac{2}{3})^{N-i}$. As $i \geq [\frac{N}{3}]$,
$e_1$ will be cut based on the Formula (1) since $p(e_o) < \frac{1}{3}$. The accumulative
probability of such cases is $Pr(X \geq [\frac{N}{3}]) = \sum_{i=[\frac{N}{3}]}^{N} P(X = i) \geq \frac{1}{2}$. Second, as
the number of frequency quadrilaterals where $e_1$ has frequency 1 is above $[\frac{N}{3}]$,
the probability that it is eliminated is $p(e_1) > \frac{1}{3}$ according to 3. The binomial
distribution is $P(X = i) = \binom{N}{i}(p(e_1))^i(1 - p(e_1))^{N-i}$ and the accumulative
probability $Pr(X \geq [\frac{N}{3}])$ is surely bigger than $\frac{1}{2}$ for $p(e_1) > \frac{1}{3}$.

Given each edge is contained in $N$ quadrilaterals, the number of edges containing a vertex is $\lceil 1 + \sqrt{2N} \rceil$. Based on Theorem 1, none of the edges containing a vertex might be eliminated for some complex $TSP$ instances. As Theorem
2 is applied, we can estimate the number of eliminated edges for a vertex. $e_1$
is contained in above $[\frac{N}{3}]$ frequency quadrilaterals where $d(e_1)$ is bigger than
$2d(e_2)$ of the longest adjacent edges $e_2$. We choose $[\frac{N}{3}]$ of these quadrilaterals which contain at most $K$ adjacent edges of $e_1$ visiting a vertex. $e_1$ and two
of the $K$ edges will compose a quadrilateral $ABCD$. The formula (2) holds
and $K = \lceil 1 + \sqrt{\frac{2N}{3}} \rceil$. If there are just $[\frac{N}{3}]$ frequency quadrilaterals where
$e_1$ has frequency 1, the average frequency of $e_1$ will be 3 on average. Accord-
ing to Formula (1), $e_1$ will be eliminated. As the $K$ short edges are preserved,
$\lceil \sqrt{2N} - \sqrt{\frac{2N}{3}} \rceil \approx \lceil 0.598\sqrt{N} \rceil$ long edges for the vertex will be eliminated. In
the algorithm, only $\frac{1}{3}$ of the edges with the smallest frequencies are eliminated
for each vertex. Each eliminated edge has more number of 1s among the $N$
frequency quadrilaterals containing them.

$$\binom{K}{2} = \frac{N}{3} \tag{2}$$

One can construct a $TSP$ instance where the distances of edges have much
difference. Given the distances of edges are in $(0, M]$ and $M$ is a finite number, the edges whose distance is smaller than $\frac{M}{2}$ will be preserved according to
Theorems 1 and 2 at the first iteration. After $k \geq 1$ iterations, the distance of
the preserved edges will be less than $\frac{M}{2^k}$ unless the ratio of the distances of two
longest adjacent edges is less than 2. Provided that $M$ is $c$ times of the distance
of the longest $OHC$ edge $e_o$, i.e., $M = c \cdot d(e_o)$, the ratio $\frac{c}{2^k}$ between the distance
of the preserved longest edges and $d(e_o)$ decreases exponentially in proportion
to a factor $\frac{1}{a}$ ($a \geq 2$) according to $k$ until the distances of the preserved long
adjacent edges do not have much difference. If a long edge is not cut, there must
exist another long adjacent edges of the nearly equal distance.

In applications, the distance of the eliminated edges $e_1 = (A, C)$ will be smaller since $d(B, D) > 0$ in $ABCD$ and it only requires $d(A, B) + d(C, D) < d(A, C) + d(B, D)$. As a number of inequalities $d(A, B) + d(C, D) < d(A, C) + d(B, D)$ hold, $e_1 = (A, C)$ will have frequency 1 in these frequency quadrilaterals. Although the distance of $e_1 = (A, C)$ is smaller than two times of that of $(A, B)$, it will be cut due to its small frequency. For example, if $d(A, B) + d(C, D) < d(A, C) + d(B, D)$ occurs more than $\left\lceil \frac{N}{2} \right\rceil$ times among $N$ quadrilaterals, the average frequency of $(A, C)$ must be smaller than 3. For the vertices contained in a number of short edges, the relatively longer edges have more chances to have frequency 1 in the frequency quadrilaterals since they are apt to improve the sum distances with their non-adjacent edges in a quadrilateral. After the relatively longer edges are cut, the preserved edges containing each vertex do not have much difference according to their distances. This guarantees that the residual graph contains the $OHC$ or a good approximation. As a residual graph is obtained, one can build an integer programming model with more constraints, or replace the distances of the eliminated edges with a big number [14]. The current algorithms can be applied to resolve the changed $TSP$.

## 5    The Experiments and Analysis

In this section, we shall do experiments to eliminate edges for certain real-world $TSP$ instance in $TSPLIB$ [18]. Given a $TSP$ instance, the iterative algorithm is used to eliminate the edges for each vertex until a sparse graph is computed. According to iteration $k$, one can estimate the maximum vertex degree of the preserved graph. To testify the performance of the iterative algorithm, all the quadrilaterals in $K_n$ and preserved graphs are used to compute the frequencies of edges. The time complexity of the iterative algorithm is $O(n^4)$. It will consume a long time for large size of $TSP$ instances. In this experiment, the small and moderate size of $TSP$ instances are tried.

The iterative algorithm computes the bounded degree graphs for four types of $TSP$ instances which have different distance functions. One will see that the algorithm works well to eliminate many non-$OHC$ edges for different types of $TSP$ instances. At an iteration $k$, the number $M_k$ of the preserved edges and the number $l_k$ of the eliminated known $OHC$ edges are recorded. The known $OHC$ of each $TSP$ instance is computed with the Concorde Solver [19]. As the termination condition is met, the maximum, average and minimum vertex degrees $d_{mx}$, $d_{ag}$ $d_{mn}$ will be output. It is generally believed that a graph containing $O(n \log_2 n)$ edges is sparse [20]. $S = \lceil n \log_2 n \rceil$ is computed for comparisons. The experimental results are shown in Table 1. The first column is the $TSP$ name and the number in each name represents the $TSP$ scale. As the known $OHC$ is preserved in the first several iterations, these datum $M_k/l_k$ are not shown.

The iterative algorithm eliminates $\frac{1}{3}$ of the edges for each vertex according to the frequencies of edges. $M_k$ decreases exponentially in proportion to the factor $\frac{2}{3}$ according to $k$. All $OHC$ edges are preserved in the first several iterations although many non-$OHC$ edges are eliminated. It says the frequencies of the

**Table 1.** The experimental results for certain $TSP$ instances.

| $TSP$ | $M_3/l_3$ | $M_4/l_4$ | $M_5/l_5$ | $M_6/l_6$ | $M_7/l_7$ | $M_8/l_8$ | $S$ | $d_{mx}$ | $d_{ag}$ | $d_{mn}$ |
|---|---|---|---|---|---|---|---|---|---|---|
| att48 | 277 | 189/1 | 135/1 | | | | 268 | 8 | 5 | 3 |
| hk48 | 275 | 178 | 121 | | | | 268 | 7 | 5 | 3 |
| berlin52 | 482 | 342 | 235 | 157 | 114/1 | | 296 | 9 | 4 | 2 |
| gr96 | 1088 | 700 | 460 | 317 | | | 632 | 10 | 6 | 3 |
| kroA100 | 1166 | 751 | 491 | 332/1 | | | 664 | 11 | 6 | 3 |
| gr120 | 1610 | 1048 | 693 | 469 | | | 829 | 13 | 7 | 2 |
| pr136 | 2190 | 1401 | 907 | 601 | 408 | | 963 | 9 | 6 | 2 |
| gr137 | 2172 | 1388 | 901 | 602 | 403/1 | | 972 | 9 | 5 | 2 |
| ch150 | 2666 | 1738 | 1129 | 745 | 508/1 | | 1084 | 10 | 6 | 2 |
| kroA200 | 4542 | 2901 | 1866 | 1220 | 823 | | 1529 | 13 | 8 | 3 |
| gr202 | 4938 | 3201 | 2082 | 1368 | 908 | 616/1 | 1547 | 9 | 6 | 2 |
| gr229 | 6273 | 4046 | 2618 | 1696 | 1126 | 762/2 | 1795 | 11 | 6 | 3 |
| a280 | 9034 | 5828 | 3754 | 2424 | 1602 | 1084 | 2276 | 12 | 7 | 2 |
| pr299 | 10112 | 6442 | 4148 | 2690/1 | 1776/4 | 1180/4 | 2459 | 13 | 7 | 3 |
| lin318 | 11850 | 7592 | 4845 | 3090 | 1997 | 1333/2 | 2643 | 14 | 8 | 3 |
| rd400 | 19263 | 12374 | 7983 | 5137 | 3352 | 2233 | 3458 | 17 | 11 | 3 |

| $TSP$ | $M_5/l_5$ | $M_6/l_6$ | $M_7/l_7$ | $M_8/l_8$ | $M_9/l_9$ | $M_{10}/l_{10}$ | $S$ | $d_{mx}$ | $d_{ag}$ | $d_{mn}$ |
|---|---|---|---|---|---|---|---|---|---|---|
| gr431 | 9166 | 5940 | 3871 | 2560 | 1718/4 | | 3772 | 13 | 7 | 3 |
| pr439 | 9195 | 5887 | 3750 | 2415/1 | | | 3854 | 18 | 11 | 3 |
| pcb442 | 9380 | 6083 | 3982 | 2639 | 1768/4 | | 3884 | 13 | 8 | 3 |
| d493 | 11997 | 7811 | 5049 | 3312 | 2221/3 | | 4410 | 13 | 9 | 3 |
| att532 | 13371 | 8684 | 5635 | 3690 | 2469/2 | | 4817 | 15 | 9 | 3 |
| ail535 | 14370 | 9218 | 5927 | 3801/1 | 2515/5 | | 4849 | 15 | 9 | 3 |
| si535 | 12386 | 7960/2 | 5469/10 | 4205/30 | 2766/32 | | 4849 | 19 | 10 | 2 |
| pa561 | 15122 | 9745 | 6321 | 4139/1 | 2771/2 | 1879/7 | 5123 | 12 | 6 | 2 |
| u574 | 15654 | 10112 | 6553 | 4262/1 | 2825/3 | | 5261 | 15 | 9 | 2 |
| rat575 | 15813 | 10320 | 6782 | 4473 | 2996 | 2031/3 | 5271 | 11 | 7 | 2 |
| p654 | 17357 | 11237 | 7388 | 4885/1 | 3220/4 | | 6117 | 20 | 9 | 2 |
| d657 | 21178 | 13636 | 8829 | 5746 | 3829/1 | | 6149 | 18 | 11 | 2 |
| gr666 | 22831 | 14654 | 9501 | 6183 | 4084 | 2728/2 | 6247 | 13 | 8 | 3 |
| u724 | 24921 | 16106 | 10455 | 6793 | 4508 | 3024/3 | 6878 | 13 | 8 | 3 |
| rat783 | 29068 | 18889 | 12210 | 7888 | 5114 | 3303/1 | 7527 | 12 | 8 | 3 |
| pr1002 | 47608 | 30372 | 19416 | 12443 | 7955 | 5133/1 | 99989 | 18 | 10 | 2 |
| u1060 | 52686 | 33881 | 21806 | 14099 | 9159/1 | 6068/6 | 10653 | 19 | 11 | 3 |
| vm1084 | 53726 | 34071 | 21599 | 13792 | 8908/1 | 5901/4 | 10929 | 18 | 10 | 2 |
| fl1400 | 77851 | 49955 | 32159 | 20562 | 13283/1 | 8649/5 | 14632 | 26 | 12 | 3 |

| $TSP$ | $M_7/l_7$ | $M_8/l_8$ | $M_9/l_9$ | $M_{10}/l_{10}$ | $M_{11}/l_{11}$ | $M_{12}/l_{12}$ | $S$ | $d_{mx}$ | $d_{ag}$ | $d_{mn}$ |
|---|---|---|---|---|---|---|---|---|---|---|
| si1032 | 20677/2 | 15177/2 | 12514/8 | 8113/8 | 5335/9 | | 10332 | 19 | 10 | 3 |
| pcb1173 | 28065 | 18361 | 11999/1 | 7976/1 | 5348/4 | | 11960 | 14 | 9 | 2 |
| d1655 | 54011 | 34967 | 22761 | 14809/3 | 9832/4 | | 17696 | 21 | 11 | 3 |
| rl1889 | 69520 | 44350 | 28283 | 18128/3 | 11874/10 | | 20559 | 22 | 12 | 2 |
| u1817 | 65691 | 42572 | 27727 | 18075 | 11999/1 | 8086/2 | 19673 | 15 | 8 | 2 |
| pr2392 | 113473 | 73501 | 47581 | 30951 | 20418 | 13575/1 | 26848 | 19 | 11 | 3 |

$OHC$ edges are much higher than those of the eliminated edges. Through comparing $M_k$ with $S$, the iterative algorithm computes a sparse graph containing the $OHC$ for most of the $TSP$ instances before the terminal iteration $d_{min} = 2$ or 3 is met. These sparse graphs are surely connected. Moreover, vertices are locally adjacent because the long edges are eliminated. In these bounded degree graphs, the number of $HC$s is less than $\left(\frac{M_k}{n}\right)^n$ owing to Sharir and Welzl [9]. As more and more edges are cut, the number of quadrilaterals becomes smaller and smaller. At the terminal iteration, a very sparse graph is computed for each $TSP$ instance. A few $OHC$ edges are cut since there are no quadrilaterals for these $OHC$ edges in the preserved graphs. The experimental results illustrate the terminal condition $d_{min} = 2$ or 3 is too strict to compute a bounded degree graph containing the known $OHC$ for most $TSP$ instances.

It mentions that some $TSP$ instances, such as si535 and si1032, contain many equal-weight edges. This case causes the difficulty of choosing the right optimal four-vertex paths for the $OHC$ edges in many quadrilaterals. In the program, we choose the optimal four-vertex paths according to the lexicographic orders of vertices. The frequencies of some $OHC$ edges will be lowered due to the usage of the inappropriate optimal four-vertex paths. One sees a few $OHC$ edges of si535 and si1032 are cut at the early stage of the algorithm. For the $TSP$ with few equal-weight edges, the algorithm works well to cut many non-$OHC$ edges and compute a sparse graph containing the known $OHC$. Theorems 1 and 2 guarantee that the long edges are eliminated. Thus, the $OHC$ or a good approximation will be find in the residual graphs. As the distances of the eliminated edges are replaced with a big number, Hougardy and Schroeder [14] have shown that the computation time of Concorde Solver is greatly reduced for some big size of $TSP$ instances.

## 6    Conclusions

An iterative algorithm is presented to convert $K_n$ of a $TSP$ instance into a bounded degree graph based on the frequency quadrilaterals. The theorems guarantees that the relatively longer edges are eliminated and the short edges are preserved for each vertex. As a few $OHC$ edges are cut, a good approximation will be found in the bounded degree graphs. The experiments demonstrate the bounded degree graphs with $O(n \log_2 n)$ edges are computed for the real-life $TSP$ instances. In the future, the terminal conditions will be improved to compute a bounded degree graph containing an $OHC$. In addition, we will explore the algorithms for $TSP$ on a bounded degree graph.

## References

1. Gutin, G., Punnen, A.P.: The Traveling Salesman Problem and its Variations. Springer, New York (2007). https://doi.org/10.1007/b101971
2. Karp, R.M.: On the computational complexity of combinatorial problems. Networks (USA) **5**(1), 45–68 (1975)

3. Held, M., Karp, R.M.: A dynamic programming approach to sequencing problems. J. Soc. Ind. Appl. Math. **10**(1), 196–210 (1962)
4. Bellman, R.E.: Dynamic programming treatment of the travelling salesman problem. J. ACM **9**(1), 61–63 (1962)
5. Carpaneto, G., Dell'Amico, M., Toth, P.: Exact solution of large-scale, asymmetric traveling salesman problems. ACM Trans. Math. Softw. (TOMS) **21**(4), 394–409 (1995)
6. de Klerk, E., Dobre, C.: A comparison of lower bounds for the symmetric circulant traveling salesman problem. Discret. Appl. Math. **159**(16), 1815–1826 (2011)
7. Levine, M.S.: Finding the right cutting planes for the TSP. ACM J. Exp. Algorithmics (JEA) **5**, 1–20 (2000)
8. Applegate, D.L., et al.: Certification of an optimal TSP tour through 85900 cities. Oper. Res. Lett. **37**(1), 11–15 (2009)
9. Sharir, M., Welzl, E.: On the number of crossing-free matchings, cycles, and partitions. SIAM J. Comput. **36**(3), 695–720 (2006)
10. Gebauer, H.: Enumerating all Hamilton cycles and bounding the number of Hamiltonian cycles in 3-regular graphs. Electr. J. Comb. **18**(1), 1–28 (2011)
11. Björklund, A., Husfeldt, T., Kaski, P., Koivisto, M.: The traveling salesman problem in bounded degree graphs. ACM Trans. Algorithms **8**(2), 1–18 (2012)
12. Borradaile, G., Demaine, E.D., Tazari, S.: Polynomial-time approximation schemes for subset-connectivity problems in bounded-genus graphs. Algorithmica **68**(2), 287–311 (2014)
13. Gharan, S.O., Saberi, A.: The asymmetric traveling salesman problem on graphs with bounded genus. In: Proceedings of the Twenty-Second Annual ACM-SIAM Symposium on Discrete Algorithms (SODA 2011), pp. 1–12. ACM, NY, USA (2011)
14. Hougardy, S., Schroeder, R.T.: Edge elimination in TSP instances. In: Kratsch, D., Todinca, I. (eds.) WG 2014. LNCS, vol. 8747, pp. 275–286. Springer, Cham (2014). https://doi.org/10.1007/978-3-319-12340-0_23
15. Wang, Y.: An approximate method to compute a sparse graph for traveling salesman problem. Expert Syst. Appl. **42**(12), 5150–5162 (2015)
16. Wang, Y., Remmel, J.B.: A binomial distribution model for the traveling salesman problem based on frequency quadrilaterals. J. Graph Algorithms Appl. **20**(2), 411–434 (2016)
17. Wang, Y., Remmel, J.: A method to compute the sparse graphs for traveling salesman problem based on frequency quadrilaterals. In: Chen, J., Lu, P. (eds.) FAW 2018. LNCS, vol. 10823, pp. 286–299. Springer, Cham (2018). https://doi.org/10.1007/978-3-319-78455-7_22
18. Reinelt, G.: TSPLIB (1995). http://comopt.ifi.uni-heidelberg.de/software/TSPLIB95/
19. Mittelmann, H.: NEOS Server for Concorde (2018). http://neos-server.org/neos/solvers/co:concorde/TSP.html
20. Nešetřil, J., de Mendez, P.O.: Sparsity Graphs. Structures and Algorithms. Springer, Heidelberg (2012). https://doi.org/10.1007/978-3-642-27875-4

# Prediction Based Reverse Auction Incentive Mechanism for Mobile Crowdsensing System

Zhen Wang[1], Jinghua Zhu[1,2(✉)], and Doudou Li[1]

[1] School of Computer Science and Technology, Heilongjiang University,
Harbin 150001, China
zhujinghua@hlju.edu.cn
[2] Key Laboratory of Database and Parallel Computing of Heilongjiang Province,
Harbin, China

**Abstract.** With the rapid development of the Internet, Mobile Crowd-sensing System (MCS) is widely used in various fields. Because of the insufficient number of users' participation and the insufficient amount of uploaded sensing data, the research of incentive mechanism is particularly important in MCS. Reverse auction mechanism is one of the efficient incentive mechanism in MCS. The platform is responsible for publishing a group of tasks which are bidded by the workers, and only the winning workers are authorized to complete the tasks. One of the critical factors for successful bidding is the distance between the tasks and the workers. However, most workers in MCS always keep moving. So this paper proposes a prediction based reverse auction incentive mechanism—RMMP. We use the Semi-Markov model to predict the position of workers at the next moment. According to the predicated positions, our reverse auction incentive mechanism selects the winner workers with the minimum movement distance and bidding price. Experiment results on real dataset show that our RMMP mechanism has remarkable performance compared with the existing mechanisms.

**Keywords:** Mobile Crowdsensing · Incentive mechanism · Reverse auction · Semi-Markov model

## 1 Introduction

In recent years, various functional sensors have been embedded in various mobile devices in daily life. For example, smartphones are usually equipped with accelerometer, GPS, camera, etc. With the increasingly powerful wireless network, mobile devices are no longer just simple mobile and communication tools, but also a powerful and mature sensor device, which promotes the prosperity and development of Mobile Crowdsensing (MCS).

MCS has infiltrated into various fields, such as noise monitoring system [1], air pollution monitoring system, traffic congestion monitoring system [2]. The

© Springer Nature Switzerland AG 2019
Y. Li et al. (Eds.): COCOA 2019, LNCS 11949, pp. 541–552, 2019.
https://doi.org/10.1007/978-3-030-36412-0_44

MCS relies on the crowd strength of society to solve the technical problems of large-scale monitoring, transportation, news and so on. However, due to the insufficient number of participants and the uneven quality of perceived data, many works have not progressed smoothly. We need to recruit as many participants as possible to accomplish more tasks while reducing the platform's overhead.

The existing incentive mechanism simply calculates the mobile cost and reward through the distance between source and destination. This simple pricing mechanism can not play an obvious incentive role. Therefore, this paper adopts the form of a reverse auction, as shown in Fig. 1. Platform publishes tasks first, and workers choose tasks they are interested in and then bid. Platform selects winners for each task. Winners upload the data of sensing tasks to obtain rewards. Because workers are constantly moving in the real world, we need to predict the location of the workers. According to the scenarios mentioned above, this paper studies a feasible scheme–a Reverse auction Mechanism based on Mobility Prediction in Mobile Crowdsensing (RMMP), which encourages more participants to participate in the sensing task, and designs a reasonable pricing mechanism to maintain the enthusiasm of workers. The main contributions of this paper are as follows:

- We propose a Winner Selection Mechanism, and a Semi-Markov model is introduced to predict the location of mobile workers.
- We propose a Payment Determination Mechanism, which uses other workers competing for the same task to determine the reasonable value of the task and pay the corresponding payment.
- Experiments show that the proposed mechanism can fulfill task requirements more effectively and ensure the minimum total social cost.

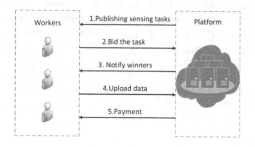

Fig. 1. Reverse auction model

## 2   Related Work

### 2.1   Auction Mechanism

Auction mechanism is widely used in MCS. In 2010, Lee [3] first applied reverse auction in the field of economy to the study of incentive mechanism in Mobile

Crowdsensing. Zhao [4] designed two online mechanisms OMZ and OMG using reverse auction for a more reasonable online real-time mechanism. Li [5] proposed a real stochastic reverse auction mechanism that distributes perceived tasks by weights of probabilities and ensures the approximate authenticity of payments. It can be said that these existing auction mechanisms are clear. Just a little part of register users to complete the task assigned by the server, the scope is not extensive.

In [6], Saadatmand and Kanhere present a Modified Reverse Auction (MRA) to minimize the overhead of the system while strongly motivating the participants. Duan [7] designed a distributed auction framework based on reverse auction. In [8], Wang et al. proposed a multiattribute auction and two-stage auction mechanism to prevent the risk of user privacy leakage.

## 2.2   Mobility

Mobility of workers is an important issue in the real world. In [9], Feng designed an incentive mechanism TRAC. In TRAC, the execution of each mobile user's tasks is location-dependent and can stimulate other users to complete a set of tasks together. Feng [10] proposed a movement-based incentive mechanism for crowdsourcing, in which participants are motivated to move under the instructions of the platform, thereby benefiting participants and the platform. Gao et al. [11] designed an incentive mechanism for vehicle-based uncertain group perception, in which perceptual tasks are executed with different probabilities. In [12], the prediction mechanism is used for mobile workers to efficiently assign tasks to bid workers, which improves the completion of publishing tasks. The reasonable pricing mechanism is the key to the incentive mechanism.

Crowdsourcing also can be extended to other fields. Zhu [13] proposed an innovative hybrid blockchain crowdsourcing platform to secure communications, verify transactions, and preserve privacy. Wang [8] proposed the Discrete Particle Swarm Optimization (DPSO) algorithm, which is part of the incentive mechanism in which the task selection is worker-centric and maximizes its utility.

## 3   System Overview

### 3.1   System Model

We assume that the MCS contains a cloud server platform and several workers. The platform first publishes its own task request scheme, including the location information of the task. First, the platform publishes a task set $T = \{t_1, t_2, \cdots, t_m\}$. The worker chooses the tasks that he or she is interested in and then bid it. After receiving the worker's bid and current location information, the platform predicts the next location when the worker moves, then the platform assigns the task to the worker who wins the bidding, as shown in Fig. 2. Finally, the platform pays the payment. We use the set $W = \{w_1, w_2, \cdots, w_n\}$ denotes these workers, $b_{ij}$ denotes workers $w_i$ bid for the task $t_j$ of interest and defines

$\beta_{ij}$ as a tasks-bid pair, which contains information about workers' bidding tasks, bid, current position, etc. Suppose $B_j$ is a set of task-bid pairs for all bidding task $t_j$. It can be considered that there is not necessarily only one worker to compete for each task.

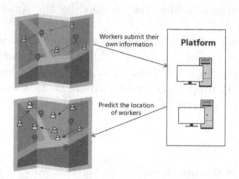

**Fig. 2.** A model for predicting the location of workers

## 3.2   Problem Statement

**Winner Selection Problem.** The optimization goal of Winner Selection Problem to find winners for each task. Let $x(w_i, t_j)$ denote the state of task assignment, when $x(w_i, t_j) = 1$, it means that task $t_j$ has been assigned to the worker $W_i$, otherwise $x(w_i, t_j) = 0$. Therefore, the selection of winners is described as follows.

$$min \quad \sum_{t_j \in T} \sum_{w_i \in W} x(w_i, t_j) C_{ij}$$

$$s.t. \quad x(w_i, t_j) = \{0, 1\}$$

$$\sum_{t_j \in T} x(w_i, t_j) \leq 1, \quad \forall i = 1, 2, \ldots, n \tag{1}$$

$$\sum_{w_i \in W} x(w_i, t_j) = 1, \quad \forall j = 1, 2, \ldots, m$$

The second constraint indicates that each worker can only execute one or zero tasks, and the third constraint means that each task is assigned to a worker.

**Payment Determination Problem.** The Payment Determination Problem is to determine the reward that is ultimately paid to the worker. The benefits available to workers are expressed as follows.

$$U_{ij} = \begin{cases} P_{ij} - RC_{ij}, & if \ x(w_i, t_j) = 1 \\ 0, & otherwise \end{cases} \tag{2}$$

Where $P_{ij}$ represents the payment of worker $W_i$ for completing task $t_j$, and $RC_{ij}$ represents the actual cost of worker $W_i$ for completing task $t_j$. Intuitively speaking, the larger the distance, the greater the cost of $RC_{ij}$.

# 4    Design of the Auction Mechanism

## 4.1    Winner Selection Mechanism

**Main Idea.** In Winner Selection Mechanism, we pick out the workers who make the social cost the smallest. We first calculate the cost of each worker to complete the requested task, and then pick out the minimum cost required for each task, that is, pick out the winner who completes each task.

**Algorithm Details.** We use the Semi-Markov model to solve this problem. In this model, we record the state of user $i$ as Pois $L^i = \{1, 2, 3, \ldots, n\}$, indicating the current Poi id of $i$. $L_n^i$ denotes the $n$th Poi on the user $i$ path. In addition, we use $T_n^i$ to indicate the start time of user i entering the $n$th state. We use the model to calculate the transition probability of user $i$, where the probability of user $i$ enters state $L_{n+1}^i$ from state $L_n^i$ is independent of $L_{n-1}^i$. $S_n^i = T_{n+1}^i - T_n^i$ represents the duration of user $i$ in the nth state. Assuming that $P_{xy}^i$ is the probability matrix of user $i$, where each term is the transition probability of $i$. We calculate the probability that $i$ transits from state $x$ to state $y$ according to the Poi path in $L_n^i$. as shown in Eq. (3):

$$P_{xy}^i = P\left(L_{n+1}^i = y | L_n^i = x\right) = num_{xy}^i / num_x^i \tag{3}$$

Where $num_{xy}^i$ represents the number of transitions from $x$ to $y$ of $i$ in the entire path, and where $num_x^i$ represents the number of transitions from $x$ to any Poi.

In addition, we also need to consider the time series of user $i$ on the Poi path. $G_{xy}^i(t)$ denotes the probability of the $i$ transit from $x$ to $y$ before the arrival of time $t$, as shown in Eq. (4):

$$
\begin{aligned}
G_{xy}^i(t) &= P\left(S_n^i \le t | L_n^i = x, L_{n+1}^i = y\right) \\
&= \sum_{z=1}^{h}\left(S_n^i = z | L_n^i = x, L_{n+1}^i = y\right)
\end{aligned}
\tag{4}
$$

According to the Eqs. (1) and (2), we can find the goal of the Semi-Markov model, that is, the probability matrix $Z_{xy}^i(t)$ of user i transit from state $x$ to state $y$ at $t$, such as Eq. (5):

$$
\begin{aligned}
Z_{xy}^i(t) &= P\left(L_{n+1}^i = y, S_n^i \le t | L_0^i \cdots L_n^i, T_0^i \cdots T_n^i\right) \\
&= P\left(S_n^i \le t \,| L_n^i = x, L_{n+1}^i = y \,| L_n^i = x\right) \\
&= G_{xy}^i(t) \cdot P_{xy}^i
\end{aligned}
\tag{5}
$$

---

**Algorithm 1. Winner Selection Mechanism**

---

**Input:**
    The task set $T$, The worker set $W$, The set $B_j$ of $\beta_{ij}$, threshold $\delta$ ;
**Output:**
    The Winner set $S$, Total Social Cost $Ca$;
1: $S \leftarrow \emptyset, Ca \leftarrow \emptyset, W' \leftarrow W, T' \leftarrow T'$;
2: **for each** $\beta_{ij}$ **in** $B_j$ **do**
3:     **Calculate** $Z$ **based on formula** (3),(4),(5);
4:     $t \leftarrow h$;
5:     $L_i \leftarrow argmax_{y \in A} Z^i_{xy}(h)$;
6:     $\beta_{ij} = \beta_{ij} + \{L_i\}$;
7:     **Calculate** $C_{ij}$ **based on the formula** (6),(7);
8:     $C(T_j) \leftarrow C(T_j) \cup C_{ij}$;
9: **end for**
10: **while** $T' \neq \emptyset$ **and** $W' \neq \emptyset$ **do**
11:     **for all** $C_{ij} \in C(T_j)$ **do**
12:         **if** $(C_{ij} < \delta)$ **then**
13:             $C_j \leftarrow \min C_{ij}$;
14:             **remove** $j$ **from** $T'$;
15:             $S \leftarrow S \cup \{\beta_{ij}\}$;
16:             **remove** $i$ **from** $W'$;
17:         **enf if**
18:     $Ca \leftarrow Ca + C_j$;
19:     **end for**
20: **end while**
21: **return** $S,Ca$

---

So that we can calculate the probability of entering another Poi at a later time according to the present Poi of $i$, then we can predict the next Poi at time $t$. Pick the state with the highest probability and think of it as the next position.

The location information which the platform predicts is added to the user's task-bid pair. The key to choosing a winner is to find the worker who minimizes the social cost. The calculation of the social cost of $i$ completing $j$ is as follows:

$$C_{ij} = C_{ij}^b + C_{ij}^d \tag{6}$$

Because the bid of workers tends to be selfish–put forward a higher price in order to get higher pay. We consider the actual distance between the predicted location and the task as part of the social cost. As shown in formula (6) above, social cost consists of two parts. $C_{ij}^b$ means the cost paid to user $i$ according to user $i$'s bid, $C_{ij}^d$ means the by the platform according to the actual distance, and $\alpha$ represents the weight of the user's bid in social cost.

$$C_{ij}^b = \alpha \cdot b_{ij}$$
$$C_{ij}^d = (1-\alpha) \cdot c \cdot dist_{ij} \tag{7}$$
$$dist_{ij} = \sqrt{(x_1 - x_2)^2 + (y_1 - y_2)^2}$$

Where $b_{ij}$ denotes the user $i$'s bid for task $j$, c represents the unit cost of distance, $dist_{ij}$ represent the actual distance between the $i$'s predicted location and the task $j$'s location. We measure it by Euclidean distance.

This paper proposes the Winner Selection Mechanism as shown in Algorithm 1. h is the time at which the platform informs the worker to complete the task. In order not to make the cost of the task exceed the budget of the platform, we set up a constraint $\delta$ to the social cost of each task. When all tasks are assigned or all workers are assigned to one task, the iteration stops. We pick the winners who pay the least for each task, and finally add them together to get the minimum total social cost value.

**Time Complexity.** We use the number of workers and tasks mentioned earlier as $n$ and $m$, respectively, assuming that the number of workers is sufficient. Lines 2–10 are calculated sequentially for $n$ workers and executed $m$ times, so the time complexity here is $O(mn)$. Lines 11–21 judge $m$ tasks, each task is executed $n$ times and is also executed $mn$ times. In general, the time complexity of Algorithm 1 is $O(mn)$.

### 4.2   Payment Determination Mechanism

**Main Idea.** In Payment Determination Mechanism, we determine the worker's payment from the utility of other workers, and we guarantee that the payment will not be lower than the bid and lowest social cost until the calculated minimum social cost is used up.

**Algorithm Details.** In order to ensure the truthfulness of the mechanism and the reference value of the task, we consider that the payment $p_{ij}$ for a winning task-bid $\beta_{ij}$ calculates according to other workers of the bidding task-bid pairs. Based on the bid $b_{ij}$ for each worker, we assume that the current utilities of all workers can be calculated as follows:

$$U'_{ij} = b_{ij} - RC_{ij} \tag{8}$$

Where $RC_{ij}$ represent the real cost of worker $w_i$ move to task $t_j$:

$$RC_{ij} = c \cdot dist_{ij} \tag{9}$$

Where $c$ represents the cost of unit distance. Suppose $B_{-\beta_{ij}}$ is the set of other task-bid pairs except $\beta_{ij}$. We calculate the average value of utilities of all workers in $B_{-\beta_{ij}}$, then compare it with the utility of $\beta_{ij}$. If the utility of worker $w_i$ is higher, we will pay him the lowest social cost $C_i$ of the task; on the contrary, if the utility of worker $w_i$ is lower than the average value of all other workers, we will pay him a reasonable payment not less than his bid and the social cost of the task. The detailed process is given in Algorithm 2.

**Algorithm 2. Payment Determination Mechanism**

---

**Input:**
     The Winner set $S$, Total Social Cost $Ca$, The set D of $dist_{ij}$ ;
**Output:**
     payment $p_{ij}$;
1: **while** $Ca>0$ **do**
2:     **for all** $\beta_{ij}$ **in** $B_j$ **do**
3:         $p_{ij} \leftarrow 0$;
4:         **if** $\beta_{ij}$ **in** $S$ **then**
5:             remove $\beta_{ij}$ from $B_j$;
6:             **for each** $\beta_{kj}$ **in** $B_{-\beta_{ij}}$ **do**
7:                 $\overline{U'_{-\beta_{ij}}} = \frac{\sum_{k=1}^{n-1} U'_{-\beta_{ij}}}{(n-1)}$
8:                 **if** $U'_{ij} > \overline{U'_{-\beta_{ij}}}$ **then**
9:                     $p_{ij} \leftarrow C_j$;
10:                    **else** $p_{ij} \leftarrow \max \left\{ b_{ij}, \frac{U'_{ij}}{\overline{U'_{-\beta_{ij}}}} \cdot C_j \right\}$
11:                    $\beta_{ij} = \beta_{ij} + \{p_{ij}\}$
12:             **end for**
13:         **end if**
14:         $Ca \leftarrow Ca - p_{ij}$
15:     **end for**
16:     Update $B$
17: **end while**
18: **return** $p_{ij}$

---

**Time Complexity.** There are also $n$ workers and $m$ tasks, line 2 is executed $mn$ times and every worker in each task is still judged. Line 6–10 was executed $n-1$ times. In total, $mn(n-1)$ times should be performed. So the total time complexity of Algorithm 2 is $O\left(mn^2\right)$.

## 5  Experiments and Results Analysis

In this section, we evaluate the performance of the proposed incentive mechanism on a real-world dataset. This paper implements these mechanisms in Python and compares their performance with benchmark mechanisms.

### 5.1  Dataset

In order to ensure the truthfulness of location information, the data set used in this paper is the driving data of 320 taxis in the city of Rome within one month (1th February 2004–2nd March 2014) [14] Through GPS positioning of taxis, the number of different vehicles arriving at different POIs is recorded. We use a part of the data set in this paper, according to the number of visits to different places in these data, remove the data from remote areas. We select the denser urban areas, map the longitude and latitude of the data set to Baidu map, and get the access density of the location as shown in Fig. 3. Task publishing area and poi location are set in longitude (12.443, 12.507) and latitude (41.882, 41.912).

**Fig. 3.** The location of Roma poi

## 5.2   Setup and Metric

The distribution of tasks published by the system on the map is randomly selected. The number of tasks is between 40 and 100. We regard the driving vehicles in the data as workers, in which the number is between 25 and 150. Each worker can be assigned a task at most, but the number of tasks applied for can not exceed 3. The parameters used in the experiment are shown in Table 1.

**Table 1.** Parameter settings.

| Parameter | Value |
|---|---|
| Worker's bid $b_{ij}$ | [1, 5] |
| Number of Pois | 15 |
| Poi radius (m) | 50 |
| TotalTime (s) | 1600 |
| Cost of unit distance c | 1 |
| Weight of bid $\alpha$ | 0.7 |
| Threshold of each task's cost $\delta$ | [3, 5] |
| Threshold of total social cost | 350 |

This paper uses social cost as a metric to evaluate the feasibility, rationality and accuracy of the proposed incentive mechanism. Total Social Cost is the sum of all social costs assigned to each winner's task.

## 5.3    Experimental Comparison and Result Analysis

This paper compares with the following three benchmark methods:

- MCBS [11]: The method employs an approximation algorithm in which the execution of the task is based on the magnitude of the probability. It is a true non-deterministic group perception incentive mechanism.
- No-Bid: Simply calculating social costs by the distance between workers and tasks, without considering workers' bid.
- No-Prediction: A random assignment algorithm that assigns tasks based on information submitted by workers, regardless of whether workers are currently moving or staying in place.

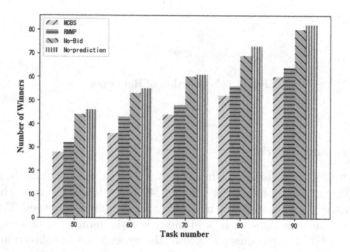

**Fig. 4.** Numbers of winners of different methods

Firstly, this paper compares the number of successful bidders of four different methods under the same number of workers. Assuming that the number of workers is set to 90, the number of tasks increases from 50 to 90, as shown in Fig. 4. It can be seen that the effect of the latter two methods is obviously better than that of the former two methods. Except for some tasks in remote areas, almost every task can be assigned to a worker. Because these two benchmarks do not take into account the uncertainty of workers' movement and the bid price of workers, the number of successful bidders will obviously be much more. Compared with MCBS, our proposed RMMP has advantages.

This experiment compares the total social cost of different algorithms under different numbers of workers and tasks. Experimental data are shown in Figs. 5 and 6. Figure 5 is the value of social cost corresponding to the increase of the number of workers when the number of tasks is 100; Fig. 6 is the value of social cost corresponding to the increase in the number of tasks when the number of workers is 80. It can be seen that the social cost of the proposed method RMMP is the smallest, which is better than the other three benchmarks.

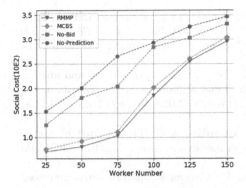
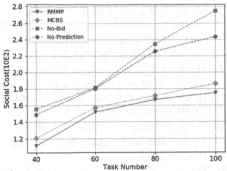

**Fig. 5.** Social costs of the number of different workers

**Fig. 6.** Social cost of the number of different tasks

## 6 Conclusion and Future Work

This paper proposes a prediction based incentive mechanism to recruit workers by reverse auction to optimize the expense of the platform. We utilize the Semi-Markov model to predict the location of workers at the next moment for the workers who are moving in the real world. Then we calculate workers' mobile costs and bidding comprehensively and select winners with a minimum social cost. Finally, the final payment to the winner is determined according to the other workers under the same competitive task. We know from the experimental results that our RMMP mechanism has better effectiveness and performance compared with the existing methods in terms of social costs. In the future, we will explore other more accurate prediction methods and time constraints to select more winners reasonably.

## References

1. Rana, R.K., Chou, C.T., Kanhere, S.S., Bulusu, N., Wen, H.: Ear-Phone assessment of noise pollution with mobile phones. In: ACM Conference on Embedded Networked Sensor Systems (2009)
2. Using Mobile Smartphones, Mohan, P., Venkata, N., Ramjee, R.: Nericell: rich monitoring of road and traffic conditions. In: ACM Conference on Embedded Network Sensor Systems (2008)
3. Lee, J., Hoh, B.: Sell your experiences: a market mechanism based incentive for participatory sensing. In: 2010 IEEE International Conference on Pervasive Computing and Communications (PerCom), pp. 60–68, March 2010. https://doi.org/10.1109/PERCOM.2010.5466993
4. Zhao, D., Ma, H., Liu, L.: Budget-feasible online incentive mechanisms for crowdsourcing tasks truthfully. IEEE/ACM Trans. Netw. 24(2), 1–1 (2016)
5. Li, J., Zhu, Y., Hua, Y., Yu, J.: Crowdsourcing sensing to smartphones: a randomized auction approach. IEEE Trans. Mob. Comput. **PP**(99), 1 (2017)
6. Saadatmand, S., Kanhere, S.S.: MRA: a modified reverse auction based framework for incentive mechanisms in mobile crowdsensing systems. Comput. Commun. **145**, 137–145 (2019). https://doi.org/10.1016/j.comcom.2019.05.020

7. Duan, Z., Li, W., Cai, Z.: Distributed auctions for task assignment and scheduling in mobile crowdsensing systems. In: 2017 IEEE 37th International Conference on Distributed Computing Systems (ICDCS), pp. 635–644, June 2017. https://doi.org/10.1109/ICDCS.2017.121

8. Wang, Y., Cai, Z., Zhan, Z., Gong, Y., Tong, X.: An optimization and auction-based incentive mechanism to maximize social welfare for mobile crowdsourcing. IEEE Trans. Comput. Soc. Syst. **6**(3), 414–429 (2019). https://doi.org/10.1109/TCSS.2019.2907059

9. Feng, Z., Zhu, Y., Qian, Z., Ni, L.M., Vasilakos, A.V.: TRAC: truthful auction for location-aware collaborative sensing in mobile crowdsourcing. In: INFOCOM. IEEE (2014)

10. Feng, T., Bo, L., Xiao, S., Zhang, X., Gui, L.: Movement-based incentive for crowdsourcing. IEEE Trans. Veh. Technol. **66**(8), 7223–7233 (2017)

11. Gao, G., Xiao, M., Jie, W., Huang, L., Chang, H.: Truthful incentive mechanism for nondeterministic crowdsensing with vehicles. IEEE Trans. Mob. Comput. **PP**(99), 1 (2018)

12. Cai, J.L.Z., Yan, M., Li, Y.: Using crowdsourced data in location-based social networks to explore influence maximization. In: IEEE INFOCOM 2016 - The 35th Annual IEEE International Conference on Computer Communications, pp. 1–9, April 2016. https://doi.org/10.1109/INFOCOM.2016.7524471

13. Zhu, S., Cai, Z., Hu, H., Li, Y., Li, W.: zkCrowd: a hybrid blockchain-based crowdsourcing platform. IEEE Trans. Ind. Inform. 1 (2019). https://doi.org/10.1109/TII.2019.2941735

14. Bracciale, L., Bonola, M., Loreti, P., Bianchi, G., Amici, R., Rabuffi, A.: CRAWDAD dataset roma/taxi (v. 2014–07-17) July 2014. https://crawdad.org/roma/taxi/20140717, https://doi.org/10.15783/C7QC7M

# PATRON: A Unified Pioneer-Assisted Task RecommendatiON Framework in Realistic Crowdsourcing System

Yuchen Xia[1], Zhitian Xu[1], Xiaofeng Gao[1(✉)], Mo Chi[2], and Guihai Chen[1]

[1] Shanghai Key Laboratory of Scalable Computing and Systems,
Department of Computer Science and Engineering, Shanghai Jiao Tong University,
Shanghai, China
sjtu1276004354@sjtu.edu.cn, zhitian.xu96@gmail.com,
{gao-xf,gchen}@cs.sjtu.edu.cn
[2] Tencent Inc., Shenzhen, China
harveychi@tencent.com

**Abstract.** Plenty of previous researches focus on crowdsourcing task recommendation to protect data quality and raise task execution efficiency. However, in real life, most crowdsourcing platforms do not allow duplicate task executions due to the budget constraints, and the recommendations are usually made towards new tasks without prior knowledge due to short task lifespan. Therefore, most previous works are noy applicable due to improper assumptions. In this paper, we propose Pioneer-Assisted Task RecommendatiON (PATRON) framework to generate accurate recommendations without analyzing the tasks' contents. We select a set of pioneer workers to collect initial knowledge of the new tasks, and adopt the $k$-medoids clustering algorithm to split the workers into subsets based on the worker similarity. Cluster selection and worker pruning provides accurate and efficient recommendations that satisfies the valid recommendation requirements from requesters. Evaluations conducted based on real datasets collected from Tencent SOHO show the efficiency of our proposed framework on recommendation acceptance rate and recommended worker quality.

**Keywords:** Crowdsourcing · Recommendation system · $k$-Medoids

## 1 Introduction

Crowdsourcing, emerging as an effective service handling human intelligence tasks (HITs), has been a heated research topic in recent years. By utilizing the power of crowd, people can obtain information that is difficult for authorities to disseminate, such as traffic crowdedness [7], real-time air quality [16], even geographic information in disaster areas [4]. However, comparing to traditional expert workers, crowd workers are not controllable and predictable in terms of quality and working will, which will result in relatively low data quality and

© Springer Nature Switzerland AG 2019
Y. Li et al. (Eds.): COCOA 2019, LNCS 11949, pp. 553–564, 2019.
https://doi.org/10.1007/978-3-030-36412-0_45

production rate. Therefore, intended task recommendation for suitable workers should be introduced to improve data quality.

There have been lots of researchers making efforts on improving the stability and efficiency of the crowdsourcing system [3,5,9,12]. However, the realistic crowdsourcing systems, such as Tencent SOHO [1], have many differences with the assumptions in the previous works. First, because of budget limitation and other reasons given by the requester, some crowdsourcing platforms does not allocate same quests (or packages) to multiple workers, which invalidates most traditional truth inference approaches, as there are **no duplicate answers** available to use. Second, the lifespan of each task category on the realistic crowdsourcing platform is usually **relatively short**, thus most task categories requesting recommendations on the platform are **brand new tasks with little prior knowledge**, while most previous works recommend workers to task categories that have ever been executed before, which means that there are plenty of history data of the target to analyze. Last but not least, as the technology advances, more **video- and audio-based quests** are posted, and these materials are much more difficult to analyze its contents comparing to text-based quests, which is the main target of most previous researches. Therefore, a unified task recommendation approach without content analysis is urgent in need.

In this paper, we propose Pioneer-Assisted Task RecommendatiON (PATRON) framework, a novel framework to solve the problems and limitations mentioned above. We design a iterative clustering-based allocation strategy to recommend new task categories to workers, and maximize the recommendation efficiency, while satisfying each new task category's expected valid recommendation requirement given by requesters. This optimization objective means achieving requirements using as few recommendations as possible to save cost, while raising accuracy and keep workers' willingness to accept invitations by giving each worker less recommendations. We select a set of pioneer workers and send them to collect initial knowledges of the new task categories, and use clustering techniques to find groups of similar workers, and approach the expected valid recommendation requirement by selecting clusters and pruning low-will workers. The design of the framework is relatively simple, but it is computational effective, and it can give more efficient recommendations than traditional approaches. In summary, our main contributions are as follows:

- We define the Accurate Recommendation Problem (ARP) to model the task recommendation problem on the crowdsourcing platform under realistic limitations. We model the worker experience by collecting his/her task execution history and apply different confidence on different types of quest packages. Also we model the workers' willingness of accepting task execution invitations by the experience they work on the platform.
- We propose PATRON framework, a clustering-based task recommendation strategy. We select a micro set of pioneer workers to execute new task categorys' test packages to collect their initial knowledge, and use these workers as initial centers to split all workers into clusters with similar features. Worker

pruning is executed to abandon inefficient recommendations, and estimated quality of workers will be the additional information when doing recommendations in the following iterations.

– We conduct several evaluations on the datasets collected from the real crowdsourcing platform, Tencent SOHO. Results show that PATRON returns an overall better result comparing to the other algorithms and strategies, in the terms of recommendation acceptance rate and recommended worker quality.

The rest of this paper is organized as follows. Related work are discussed in Sect. 2. We formally define the Accurate Recommendation Problem (ARP) in Sect. 3. In Sect. 4 we introduce the Pioneer-Assisted Task Recommendation (PATRON) framework in detail, together with some technical discussions and proofs. In Sect. 5, we conduct some evaluations based on real trace data to show the efficiency of our proposed framework. Finally we conclude our work in Sect. 6.

## 2    Related Work

**Task Recommendation.** A proper task recommendation can not only raise the quality of finished tasks by workers, but also accelerate the workers finishing the tasks, and incentivize them to participate in task execution. [8] addresses the problem of forming worker groups for a task in a stream of complex tasks, and compares three strategies on different psychological and quality issues. [9] proposes a quality-bounded task assignment problem with redundancy constraint based on truth inference, and give an approximation algorithm solving it. [11] models worker motivation as the balance between task diversity and payment, test different task allocation strategies and study their overall performance. [5] considers the spontaneously-formed mobile social networks, and introduces two task scheduling problems to minimize operating cost and completion time. [3] works on a new scheme for mobile crowdsourcing in opportunistic mobile social networks, and proposes a heuristic approximation algorithm distributing tasks using minimum forwarders. [13] extends matrix factorization and kNN, and proposes two top-N recommendation algorithms for crowdsourcing systems, giving most suitable tasks to workers and identifying best workers to requesters.

**Worker Quality Discovery.** Learning the crowd workers clearly and efficiently will be a great assistance in avoiding resource loss, while satisfying and incentivizing workers. [6] learns the workers' cognitive ability by conducting a set of cognitive tasks and compare the quality with knowledge of previous crowdtasks, then proposes approaches to predict workers' task performance. [12] allows task requesters and workers to provide feedbacks to task executors or co-workers, and proposes a heuristic task assignment mechanism to distribute skillful workers to all collaborative tasks averagely. [15] determines the unknown quality of IoT devices by peer grading, and crafts a truthful budget feasible task allocation mechanism attaining a threshold quality. [14] jointly estimates the tasks' true answers and workers' quality by filtering, designs an algorithm considering all mappings from tasks to true answers and finds the maximum likelihood.

Y. Xia et al.

## 3   Problem Definition

In this section, we will formally define the Accurate Recommendation Problem (ARP) on the realistic crowdsourcing platform. Firstly we will give the definition of crowdsourcing task category and packages in Definitions 1 and 2.

**Definition 1 (Crowdsourcing Task Category).** *A crowdsourcing task category* $T = (P, v)$ *consists of a number of task packages* $P = \{p_1, p_2, \cdots, p_{|P|}\}$ *under the same execution rule given by it. The workers accept task packages and work on them. When they compelete a task package belonging to the category, the reward* $v$ *will be given to them. The collection of all history crowdsourcing task categories is* $\mathbf{T} = \{T_1, T_2, \cdots, T_n\}$, *while the collection of new task categories is* $\mathbf{T}^N = \{T_1^N, T_2^N, \cdots, T_l^N\}$.

**Definition 2 (Task Packages).** *Each Task package* $p$ *consists of a number of tasks for workers to work on. There are three types of packages, which can be identified by the type function* $type(p)$. **Test** *packages are packages with known truth labels for all tasks. For* **inspector** *packages, a portion* $0 < \Gamma(p) < 1$ *of quests will be inspected manually to estimate the package's accuracy. For* **normal** *packages, no operations are applied, and the package accuracy is not measured.*

We define the workers in the realistic crowdsourcing platform in Definition 3.

**Definition 3 (Crowdsourcing Workers).** *The set of crowdsourcing workers* $W = \{w_1, w_2, \cdots, w_m\}$. *Each worker* $w_j$ *has an execution history set* $H_j = \{h_1^j, h_2^j, \cdots, h_{|H_j|}^j\}$, *listing his/her previous task execution records on the crowdsourcing platform. Each execution record* $h_k^j = \langle q_k^j, a_k^j \rangle$, *where* $q_k^j$ *is the task package the worker worked on, and* $a_k^j$ *is the accuracy when the package is a test or inspector package.*

It is straightforward that, if a worker always produces high accuracy when executing tasks on a certain category, we think that the worker is familiar with it. Also, if a worker executes lots of task packages on a certain task category, we think that the worker will gradually gain experience and familiarity on it. Thus, we can define the worker familiarity to the tasks according to the observation above in Definition 4. Note that, as the inspector packages have only a portion of tasks checked for correctness, the measurement confidence needs to be discounted according to the inspected proportion $\Gamma(p)$. Moreover, as normal packages' accuracy are not measured, they can only be referred by stacking up to show the workers' execution frequency. Therefore, we set accuracy of normal packages to 50%, and confidence to 0.1.

**Definition 4 (Worker Familiarity Vector).** *Give a task category* $T_i$ *and a worker* $w_j$. *The familiarity of* $w_j$ *on* $T_i$, $f(T_i, w_j)$, *is defined as Eq. (1).* $x_i^j$ *is the type indicator, if* $type(q_i) = j$, *then* $x_i^j = 1$, *otherwise* $x_i^j = 0$,

$$f(T_i, w_j) = \sum_{k=1}^{|H_j|} \sum_{q_k^j \in T_i} (a_k^j \cdot x_i^{test} + a_k^j \cdot x_i^{inspector} \cdot \Gamma(q_k^j) + 0.5 \cdot x_i^{normal} \cdot 0.1). \quad (1)$$

We collect the familiarity of worker $w_j$ to each task category in $\mathbf{T}$, and form the worker familiarity vector $\mathbf{f}_j = \langle f_{1j}, f_{2j}, \cdots, f_{nj} \rangle$, where $f_{ij} = f(T_i, w_j)$ $(1 \leq i \leq n)$.

Worker familiarity vector contains the information of quality and quantity of history task execution, and can be regarded as the workers' preference and expertness in each task category. Thus we can define the similarity between different workers as shown in Definition 5.

**Definition 5 (Worker Similarity).** *The similarity $S(w_i, w_j)$ between two workers $w_i$ and $w_j$ is defined as the cosine similarity between their familiarity vector shown in Eq. (2),*

$$S(w_i, w_j) = \frac{\mathbf{f}_i \cdot \mathbf{f}_j}{\|\mathbf{f}_i\| \cdot \|\mathbf{f}_i\|} = \frac{\sum\limits_{k=1}^{n} f_{ki} \cdot f_{kj}}{\sqrt{\sum\limits_{k=1}^{n} f_{ki}^2} \cdot \sqrt{\sum\limits_{k=1}^{n} f_{kj}^2}}. \tag{2}$$

We want to recommend workers in $W$ to new task categories $\mathbf{T}^N$. In order to do make an appropriate recommendation to these new categories with no prior knowledge, we need to get some initial information by selecting a small portion of workers and see how they perform on the new tasks' test packages, and the performance will become the guide on estimating other workers' quality. This small set of workers is called pioneer workers, and the detailed definition to them is shown in Definition 6.

**Definition 6 (Pioneer Worker Set).** *For each new task $T_i^N$, we select a subset of workers with size $z$ $(z \ll m)$ as pioneer workers $P_i = \{w_{i1}^P, w_{i2}^P, \cdots, w_{iz}^P\} \subseteq W$. Each worker in $P_i$ will be sent to complete a test package of task $T_i^N$, and the result collection is $R_i = \{(w_{i1}^P, a_{i1}^P), (w_{i2}^P, a_{i2}^P), \cdots, (w_{iz}^P, a_{iz}^P)\}$.*

Now we are going to define the willingess of the worker accepting the recommendations to execute tasks by the crowdsourcing platform. According to the long tail effect mentioned in [2], in a recommendation system, new users are likely to rate more items than experienced users. Similarly, in a crowdsourcing system, new workers with a short execution history will be more willing to accept recommended tasks, as they are not certain which task categories are suitable for them. Oppositely, experienced workers will be relatively resistant to the recommendations and choose tasks according their own preferences, as they have gained enough experience to judge whether the task categories are efficient to execute. Also it is common sense that the willingness will decrease slower as the experience rises, which is similar to the pattern of long-tail effect. Thus we define the recommendation acceptance willingness as shown in Definition 7.

**Definition 7 (Acceptance Willingness).** *When a worker $w_j$ is recommended with a task category by the crowdsourcing platform, the acceptance willingness $D(w_j)$ of $w_j$ is defined as Eq. (3),*

$$D(w_j) = \frac{1}{\sqrt{\log(|H_j| + 2)}}. \tag{3}$$

Now we will define the Accurate Recommendation Problem (ARP). Our objective is to give recommendations to the non-pioneer workers, and try to minimize the number of recommendations sent, while maximizing the number of workers receving our recommendations at the same time. By doing this, we can make our task recommendations effective to save the potential waste on recommending tasks to bad workers or workers with low willingness to accept them. Also, we want to spread the recommendations accross the worker set, in order not to excessively consume the workers' willingness by giving them too much recommendations at once, correspondingly retaining the workers to work on the platform in the future. Note that we need to meet the expected valid recommendation requirements given by tasks' requesters. Also, there is an upper bound for each worker accepting recommendations at a time, as we want to keep the precision and authority of our framework. Taking all the above into consideration, we define the Accurate Recommendation Problem (ARP) in Definition 8.

**Definition 8 (Accurate Recommendation Problem (ARP)).** *Give the crowdsourcing platform's history task category set* $\mathbf{T} = \{T_1, T_2, \cdots, T_n\}$, *and worker set* $W = \{w_1, w_2, \cdots, w_m\}$. *Give the new task category set as targets* $\mathbf{T}^N = \{T_1^N, T_2^N, \cdots, T_l^N\}$, *and for each new task category* $T_i^N$, *give its valid recommendation requirement* $G_i^N$, *pioneer worker set* $P_i$, *and the corresponding test result* $R_i$. *Give a recommendation count upper bound* $C$. *For each new task category* $T_i^N$, *we need to find a worker recommendation set* $W_i^R \subseteq W - P_i$, *maximizing the recommendation effectiveness, with each worker accepting no more than* $C$, *and each new task* $T_i^N$ *receives at least* $G_i^N$ *valid recommendations.*

With the help of estimated accuracy $a_{ij}^N$ of non-pioneer workers $w_j$ executing tasks in new category $T_i^N$, we can formulate ARP as an Integer Programming shown in (4)–(8). For easier computation and analysis, we integrate the two objectives into one. $d_{ij}$ indicates the recommendation result, if $w_j \in W_i^R$, $d_{ij} = 1$, otherwise $d_{ij} = 0$. $\mathbb{1}(\cdot)$ is the indicator function, which returns 1 if the input statement is true, and returns 0 otherwise.

$$\max \sum_{j=1}^{m} \mathbb{1}\left(\sum_{i=1}^{l} d_{ij} > 0\right) / \sum_{i=1}^{l}\sum_{j=1}^{m} d_{ij} \tag{4}$$

$$\text{s.t.} \sum_{i=1}^{l} d_{ij} \leq C, \forall w_j \in W \tag{5}$$

$$\sum_{j=1}^{m} a_{ij}^N D(w_j) d_{ij} \geq G_i^N, i = 1, 2, \cdots, l \tag{6}$$

$$d_{ij} = 0, \forall w_j \in P_i \tag{7}$$

$$d_{ij} \in \{0, 1\}, \forall i = 1, \cdots, n, j = 1, \cdots, m \tag{8}$$

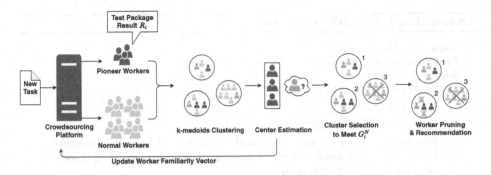

**Fig. 1.** PATRON framework workflow chart.

## 4    PATRON: Pioneer-Assisted Task RecommendatiON

In this section, we will introduce our proposed Pioneer-Assisted Task Recommendation (PATRON) framework to recommend new tasks to workers on the crowdsourcing platform. The workflow is illustrated in Fig. 1. We design an iterative task-by-task recommendation and worker quality estimation strategy based on $k$-medoids [10], the clustering algorithm. When recommending workers to each new category, we estimate the woker quality of each cluster's center using the information collected from the pioneer workers, and use the estimation to represent the quality of the other workers in the corresponding cluster. Primal cluster selection and worker pruning approaches the maximization of recommendation efficiency.

To start with, we need to collect the initial knowledge of the task category to be recommended. we randomly select some workers, and send them invitations to execute the test packages of the new task $T_i^N$. As in realistic situations, the workers may not accept the invitaions, so we need to supplement invitations, until we receive $z$ pieces of results, which is collected in $R_i$. The workers participating in this execution are identified as pioneer workers $P_i$. These information will be an important reference when estimating the quality of normal workers.

Before seeking similarity between different workers, we abstract each worker $w_j$ as a quality-willingness tuple $E_j = (D(w_j), \frac{\sum_{k=1}^n f_{kj}}{n})$. The former of the tuple is the acceptance willingness of $w_j$, while the latter is the average familiarity of $w_j$. We select the pioneer workers $P_i$ as initial cluster centers, choose the Euclidean distance between the abstract tuples $dist(E_i, E_j)$ as the dissimilarities between workers $w_i$ and $w_j$, and execute the $k$-medoids (CLARA [10] as implementation) to split the workers $W$ into $z$ clusters $X = \{X_1, X_2, \cdots, X_z\}$. As $k$-medoids algorithm always selects some data points that exist in the dataset as centers, we can select the cluster center $w_j^X \in X_j$ as the representation of quality and acceptance willingness of all workers in the corresponding cluster $X_j$. In order to avoid the situation that the clustering algorithm does not converge, we set up the round limit $r$ to terminate the process.

---

**Algorithm 1.** PATRON Framework

---

**Input:** $\mathbf{T}, \mathbf{W}, \mathbf{T}^N, \{G_i^N\}, \{P_i\}, \{R_i\}, C$
**Output:** Recommendation $W_i^R$ for each $T_i^N$

1   **for** $i = 1$ *to* $l$ **do**
      // Abstract Tuple Calculation
2     **for** $j = 1$ *to* $m$ **do**
3      $E_j \leftarrow (D(w_j), \frac{\sum_{k=1}^n f_{kj}}{n})$
4     Run $k$-medoids on $W$, with $P_i$ as initial centers, and Euclidean distance $dist(E_i, E_j)$ as dissimilarity between $w_i, w_j$, to get a result of clusters $X = \{X_1, \cdots, X_z\}$ with centers $w_j^X \in X_j$
      // Center Quality Estimation
5     **for** $j = 1$ *to* $z$ **do**
6      $a_{ij}^X = \dfrac{\sum\limits_{k=1}^{z} S(w_j^X, w_{ik}^P) \cdot a_{ik}^P}{\sum\limits_{k=1}^{z} S(w_j^X, w_{ik}^P)};$
7      **foreach** $w_k \in X_j$ **do**
8       $a_{ik}^N \leftarrow a_{ij}^X$
      // Primal Recommendation Set Construction
9    $Valid \leftarrow 0, W_i^R \leftarrow \phi, j \leftarrow 1;$
10   Rank clusters $\{X_j\}$ in descending order of $a_{ij}^X$;
11   **while** $Valid < G_i^N$ **do**
12    $W_i^R \leftarrow W_i^R \cup (X_j - P_i);$
13    $Valid \leftarrow Valid + a_{ij}^X D(w_j^X)|(X_j - P_i)|;$
14    $j \leftarrow j + 1;$
      // Worker Pruning
15   Rank $w_k \in X_j$ decreasingly by $D(w_j) \cdot \frac{C - A(w_j)}{C}$, result in $(w_1^j, w_2^j \cdots, w_{|X_j|}^j);$
16   $k \leftarrow |X_j|;$
17   **while** $Valid > G_i^N$ **do**
18    **if** $Valid - a_{ij}^X D(w_k^j) \geq G_i^N$ **then**
19     **if** $w_k^j \notin P_i$ **then**
20      $Valid \leftarrow Valid - a_{ij}^X D(w_k^j);$
21      $W_i^R \leftarrow W_i^R - \{w_k^j\};$
22     $k \leftarrow k - 1;$
23    **else Break**
      // Familiarity Vector Extension
24   **foreach** $w_j \in W$ **do**
25    $\mathbf{f}_j \leftarrow \langle \mathbf{f}_j; a_{ij}^N \rangle$
26 **Return** $\{W_i^R\};$

---

Now we will estimate the normal worker's quality using the pioneer's test package execution result $R_i$. Note that although we only acquired the test package accuracy of each pioneer worker, it can also be regarded as the possibility of him/her producing the qualified answers, or to say, the quality of the him/her towards the specified task category. Also, according to the discussions about clusters above, we only need to estimate the quality of cluster centers, and use them to represent the quality of workers in their own cluster. To make the estimation more practical, we adopt the weighted average of the test accuracy of all pioneer workers $P_i$ as the estimated quality $a_{ij}^X$ of $w_j^X$: $a_{ij}^X = \dfrac{\sum\limits_{k=1}^{z} S(w_j^X, w_{ik}^P) \cdot a_{ik}^P}{\sum\limits_{k=1}^{z} S(w_j^X, w_{ik}^P)}$.

After the quality estimation of normal workers, we select a primal set of recommendation targets to meet the valid recommendation requirement $G_i^N$ of the new task category. We rank the estimated quality of cluster centers decendingly, and successively select the center $w_j^X$ together with workers in his/her cluster as the primal targets while removing the pioneers in it. Each selected cluster produces $a_{ij}^X D(w_j^X)|X_j - P_i|$ expected valid recommendations. We keep on involving new clusters, until the total expected valid recommendation satisfies the requirement $G_i^N$ given by the requester.

In order to maximize our recommendation efficiency, we need to discard the recommendations that would not be so efficient to reduce the recommendation excess from the primal solution. They might be low in quality with non-relevant experiences and return disqualified answers, or they have a low willingness on accepting platform recommendations. Therefore, for the last cluster we pick up $X_j$ (with center estimated quality $a_{ij}^X$), we rank workers $w_k$ in it in the descending sequence of modified willingness $D(w_k) \cdot \frac{C-A(w_k)}{C}$, where $A(w_k)$ is the recommendations already given to worker $w_k$ in this process. The modified willingness reduces the possibility of recommending multiple task categories to the same worker, while trying to keep workers with high willingness to make the recommendation more likely to convert into a valid one. After the ranking, we prune the worker with the lowest modified will and subtract his/her generated expected valid recommendation, until discarding a worker makes $G_i^N$ unsatisfiable.

After the worker pruning, we will recommend the new category $T_i^N$ to all workers remaining in the primal set, and update the worker profile according to the clustering result. We label the normal worker's estimated quality as the same value of his/her cluster center's $a_{ij}^N$. Furthermore, in the recommendation of the following tasks, we extend the familiarity vector by one dimension using $a_{ij}^N$, and use the new vector to calculate the new worker similarity. We repeat the operations mentioned above, until all new categories are recommended with workers. We collect all the processes and describe it in the pseudo code in Algorithm 1.

## 5   Numerical Evaluation

The PATRON framework is evaluated using real datasets collected from Tencent SOHO [1], a Chinese crowdsourcing platform developed and run by Tencent Inc. We collect the workers' crowdsourcing task execution history on the platform for our evaluation. Each record contains the information of executor ID (worker identicator), task category indicator, package type (test, inspector or normal), number of total quests, checked passed/failed quests, and package status (terminated, abandoned, etc.) for each task package. The raw dataset contains 108480 task execution records. After the filtering and division, we have 5476 available workers, 13 history task categories with 36749 valid records, and 32 new task categories with 6670 valid records remaining. The package type distribution can be seen in Fig. 2. On the horizontal axis, bars on the left of tag $T$ (including $T$) are information of history task categories, while those on the right of tag $V$ (including $V$) are for new task categories.

In order to show the efficiency of PATRON under different valid recommendation requirements, we set the recommendation upper bound $C = 5$, and introduce the following three algorithms as baselines.

- Random Recommendation Algorithm (RAND): Each time we randomly select a worker that has not reached the recommendation upper bound, and a task category that has not reached its requirement. Then we recommend the worker to the task category. We do not estimate the workers' quality, and regard it as 50%.
- Pure Greedy Algorithm (GDY): Rank the workers in decending order of their acceptance willingness, and recommend them with task categories from the top to bottom if they are available on recommendation upper bound, until the task category reaches its requirement. Task categories are processed one by one. We do not estimate the workers' quality, and regard it as 50%.
- Full Estimation Algorithm (FES): We compute the estimated quality of all workers based on the pioneer workers, and keep on selecting the available workers producing highest expected valid recommendation until the requirement is reached. This is the traditional collaborative filtering user-based task recommendation approach.

**Recommendation Success Rate**. We measure the success rate of a recommendation system by finding out how many recommended workers actually accept a task from the category we recommended on the platform, by checking the execution records in the new task categories. The result is presented as the ratio between workers accepting recommended tasks and all recommended workers. We can see from Fig. 3 that PATRON is overall better than the other baseline algorithms, especially when the requirement $G_i^N$ is relatively small, which means that our recommendation is much more precise. This is because our proposed framework discards some workers with low wilingness on accepting recommendations, correspondingly performing bad on producing valid recommendations. Thus this operation can raise the recommendation efficiency. Also, we do recommendations based on experience and feature similarity of workers, and this will attract workers accept their recommended tasks because of their category familiarity, which result in a rise in the acceptance rate.

**Recommended Worker Quality**. We measure the recommended worker quality by observing how well they perform in the recommended new task categories, which is presented as the ratio of the number of all passed quests and tha of all examined quests belonging to the new tasks. We can see from Fig. 4 that

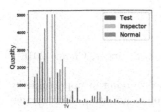

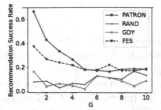

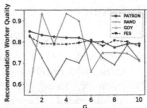

**Fig. 2.** Package type distribution

**Fig. 3.** Acceptance rate

**Fig. 4.** Recommended worker quality

PATRON produces a competitive and more stable worker quality production comparing to other algorithms. Especially, we produce more stability comparing to baselines RAND and GDY with randomization mechanics. This is because we have done sufficient worker quality estimation based on pioneer workers and center workers, and recommend workers with familiar task categories, and the workers are more likely to produce qualified result when receiving the recommendations. The worker pruning also protects the recommended worker quality by discarding some workers with relatively low quality, as the last cluster selected in the primal set building will always be the relatively worst one in quality.

## 6  Conclusion

Real life situations of no duplicate task execution, short task lifespan and non-text tasks makes traditional crowdsourcing task recommendation approaches not appliable due to assumption differences. In this paper, we propose a unified Pioneer-Assisted Task Recommendation (PATRON) framework for efficient crowdsourcing task recommendation. With the help of pioneer workers, we are able to get the initial knowledge of new tasks. Through the combination of cluster selection and worker pruning, we can not only ensure the high quality of recommended workers, but also guarantee their acceptance will when recommended. Experimental results on the data collected from Tencent SOHO demonstrate the high efficiency of our method compared to other baseline methods in recommendation acceptance rate and task execution quality.

**Acknowledgment.** This work was supported by the National Key R&D Program of China [2018YFB1004703]; the National Natural Science Foundation of China [61872238, 61672353]; the Shanghai Science and Technology Fund [17510740200]; the CCF-Huawei Database System Innovation Research Plan [CCF-Huawei DBIR2019002A]; the Huawei Innovation Research Program [HO 2018085286]; the State Key Laboratory of Air Traffic Management System and Technology [SKLATM20180X], and the Tencent Social Ads Rhino-Bird Focused Research Program. Xiaofeng Gao is the corresponding author.

# References

1. Tencent SOHO. https://soho.qq.com/tasks
2. Aggarwal, C.C.: Recommender Systems: The Textbook. Springer, Cham (2016). https://doi.org/10.1007/978-3-319-29659-3
3. Chen, X., Deng, B.: Task allocation schemes for crowdsourcing in opportunistic mobile social networks. In: International Conference on Computing, Networking and Communications (ICNC), pp. 615–619 (2018)
4. Dittus, M., Quattrone, G., Capra, L.: Mass participation during emergency response: event-centric crowdsourcing in humanitarian mapping. In: Conference on Computer Supported Cooperative Work (CSCW), pp. 1290–1303 (2017)
5. Fan, J., Zhou, X., Gao, X., Chen, G.: Crowdsourcing task scheduling in mobile social networks. In: Pahl, C., Vukovic, M., Yin, J., Yu, Q. (eds.) ICSOC 2018. LNCS, vol. 11236, pp. 317–331. Springer, Cham (2018). https://doi.org/10.1007/978-3-030-03596-9_22
6. Feldman, M., Bernstein, A.: Cognition-based task routing: towards highly-effective task-assignments in crowdsourcing settings. In: International Conference on Information Systems (ICIS) (2014)
7. Hu, H., Li, G., Bao, Z., Cui, Y., Feng, J.: Crowdsourcing-based real-time urban traffic speed estimation: from trends to speeds. In: IEEE International Conference on Data Engineering (ICDE), pp. 883–894 (2016)
8. Kumai, K., et al.: Skill-and-stress-aware assignment of crowd-worker groups to task streams. In: AAAI Conference on Human Computation and Crowdsourcing (HCOMP), pp. 88–97 (2018)
9. Liu, C., Gao, X., Wu, F., Chen, G.: QITA: quality inference based task assignment in mobile crowdsensing. In: Pahl, C., Vukovic, M., Yin, J., Yu, Q. (eds.) ICSOC 2018. LNCS, vol. 11236, pp. 363–370. Springer, Cham (2018). https://doi.org/10.1007/978-3-030-03596-9_26
10. Ng, R.T., Han, J.: CLARANS: a method for clustering objects for spatial data mining. IEEE Trans. Knowl. Data Min. (TKDE) **14**(5), 1003–1016 (2002)
11. Pilourdault, J., Amer-Yahia, S., Lee, D., Roy, S.B.: Motivation-aware task assignment in crowdsourcing. In: International Conference on Extending Database Technology (EDBT), pp. 246–257 (2017)
12. Qiao, L., Tang, F., Liu, J.: Feedback based high-quality task assignment in collaborative crowdsourcing. In: IEEE International Conference on Advanced Information Networking and Applications (AINA), pp. 1139–1146 (2018)
13. Safran, M.S., Che, D.: Efficient learning-based recommendation algorithms for top-N tasks and top-N workers in large-scale crowdsourcing systems. ACM Trans. Inf. Syst. (TOIS) **37**(1), 2:1–2:46 (2019)
14. Sarma, A.D., Parameswaran, A.G., Widom, J.: Towards globally optimal crowdsourcing quality management: the uniform worker setting. In: International Conference on Management of Data (SIGMOD), pp. 47–62 (2016)
15. Singh, V.K., Mukhopadhyay, S., Xhafa, F.: A budget feasible peer graded mechanism for IoT-based crowdsourcing. CoRR abs/1809.09315 (2018)
16. Zhang, C., et al.: PMViewer: a crowdsourcing approach to fine-grained urban PM2.5 monitoring in China. In: IEEE International Conference on Mobile Ad Hoc and Sensor Systems (MASS), pp. 323–327 (2017)

# Sequence Submodular Maximization Meets Streaming

Ruiqi Yang[1], Dachuan Xu[1], Longkun Guo[2(✉)], and Dongmei Zhang[3]

[1] Department of Operations Research and Scientific Computing,
Beijing University of Technology, Beijing 100124, People's Republic of China
yangruiqi@emails.bjut.edu.cn, xudc@bjut.edu.cn
[2] College of Mathematics and Computer Science, Fuzhou University,
Fuzhou 350116, People's Republic of China
longkun.guo@gmail.com
[3] School of Computer Science and Technology, Shandong Jianzhu University,
Jinan 250101, People's Republic of China
zhangdongmei@sdjzu.edu.cn

**Abstract.** In this paper, we consider the streaming sequence submodular maximization problem, in which the utility function is defined on sequences of element instead of element sets. We encode the values of different sequences by a weighted directed acyclic graph (W-DAG), where the weight of vertex reveals the utility value in selecting single element and the weight of an edge proclaims the additional benefit according to a certain selected order. In addition, the edges are revealed in a streaming fashion that one edge is known in one time slot. The aim is to output a sequence of vertices of length bounded by a given constant $k$, such that the utility function value is maximized. In this work, we first provide the framework of the sequence submodular maximization under streaming. By utilizing an edge-based thresholding principle, we derive a one pass, $(1 - 2\Delta/(2\Delta + 1 - \varepsilon))$-approximation algorithm with $O(\varepsilon^{-1}k\Delta \log(k\Delta))$ memory and $O(\varepsilon^{-1} \log(k\Delta))$ update time per edge, where $\Delta = \min\{\Delta_{\mathrm{in}}, \Delta_{\mathrm{out}}\}$ and $\Delta_{\mathrm{in}}$ ($\Delta_{\mathrm{out}}$) is the maximum in-degree (out-degree) of the constructed W-DAG. At last, we present a further improved streaming algorithm, which also requires one pass over the stream and attains the same approximation ratio but is with a decreased memory complexity of $O(\varepsilon^{-1}k\Delta)$.

**Keywords:** Submodular maximization · Sequence · Streaming · Approximation algorithm

## 1 Introduction

The traditional submodular functions are usually defined on element sets. Specially, we are given an elements set $\mathcal{V}$, a set function $f : 2^{\mathcal{V}} \to \mathbb{R}$ defined on $\mathcal{V}$ is *submodular* if for any $A, B \subseteq \mathcal{V}$, we have

$$f(A) + f(B) \geq f(A \cup B) + f(A \cap B).$$

© Springer Nature Switzerland AG 2019
Y. Li et al. (Eds.): COCOA 2019, LNCS 11949, pp. 565–575, 2019.
https://doi.org/10.1007/978-3-030-36412-0_46

The above definition of submodular functions ignores the ordering of the elements and hence is called *set submodular*.

However, there are some practical applications which take the appearing order of the elements into consideration, such as recommending system [13], link prediction [11], course sequence design [10], etc. In particular, considering a recommendation task, Tschiatschek et al. [13] introduced the *sequence submodular*. In their paper, the relationship of elements is encoded by a directed graph whose vertices correspond to elements, edges specify the order of selecting elements and the edge weight indicates the additional benefit of the order. We conclude that any set submodular function can be expressed by the sequence submodular function. Tschiatschek et al. [13] investigated a problem of maximizing sequence submodular functions with a cardinality constraint, which can be restated as follows

$$\max_{\sigma \in \Sigma : |\sigma| \le k} F(\sigma), \tag{1}$$

where $\Sigma$ denotes the set of sequences of elements and the utility function $F$ is defined on these sequences. Specifically, let $F(\sigma) = h(\mathcal{E}(\sigma))$, where $h$ is a nonnegative monotone set submodular function over edges of the directed graph and $\mathcal{E}(\sigma)$ denotes the set of edges covered by the sequence $\sigma$.

In recent years, streaming submodular is considered as a popular model for dealing with large scale set submodular optimization problems. In this model, either the elements set can not be stored by the main memory on one computer, or the elements are revealed in a streaming fashion. Any streaming algorithm outputs a series of approximate solutions with a limit memory. Consequently and generally, the performance guarantees of a streaming algorithm can be characterized by the following parameters: (1) the number of passes that the algorithm makes over the elements stream; (2) the size of memory consumed by the algorithm; (3) the update time required by the algorithm, which is the number of oracle queries (evaluations of function); (4) the approximation ratio of the algorithm. In this paper, we first provide the framework of the streaming sequence submodular maximization (SSSM). Using $O(\varepsilon^{-1}k\Delta \log(k\Delta))$ memory complexity and $O(\varepsilon^{-1} \log(k\Delta))$ update time per edge, we obtain a $(1 - 2\Delta/(2\Delta + 1 - \varepsilon))$-approximation algorithm for the SSSM. Furthermore, maintaining the same approximation ratio, we provide an improved streaming algorithm with memory complexity of $O(\varepsilon^{-1}k\Delta)$.

The rest of the paper is organized as follows. First, we review related research works in Sect. 2, and provide some necessary preliminaries in Sect. 3. Then in Sect. 4, we introduce an edge-based thresholding algorithm and show its performance guarantees. In Sect. 5, we present a memory-improved thresholding algorithm. At last, we conclude the paper in Sect. 6.

## 2   Related Work

Under a directed acyclic graph (DAG, excluding self-cycles), Tschiatschek et al. [13] provided a $(1 - e^{1/(2\Delta)})$-approximation algorithm for the problem (1), where

$\Delta_{\text{in}}(\Delta_{\text{out}})$ is the maximum in-degree (out-degree) of the directed graph and $\Delta = \min\{\Delta_{\text{in}}, \Delta_{\text{out}}\}$. Mitrovic et al. [10] considered the sequence submodular maximization under general directed graphs and presented a $(1-e^{-1})/(2\Delta+1)$-approximation algorithm with an improved time complexity of $O(km)$. Further, they extended their algorithm to maximize the sequence submodular function under hypergraphs and derived an generalized $(1-e^{-1})/(r\Delta+1)$-approximation algorithm, where $r$ denotes the maximum size of any edge in the hypergraph. Golovin and Krause [6] provided the framework of the *set adaptive submodular* optimization and presented an adaptive greedy policy which performs nearly optimally for some stochastic set submodular maximization. Recently, Mitrovic et al. [11] introduced the *adaptive sequence submodular* maximization under general directed graphs and provided a policy whose expected value is at least $1/(2\Delta+1)$ times of optimal policy. For the maximization of adaptive sequence submodular under directed hypergraphs, they extended the performance guarantee to $1/(r\Delta+1)$.

For streaming set submodular maximization with a cardinality constraint, Badanidiyuru et al. [2] introduced a one pass, $(1/2-\varepsilon)$-approximation algorithm with $O(\varepsilon^{-1}k\log k)$ memory and $O(\varepsilon^{-1}\log k)$ update time per element. If the elements arrive in random order, Norouzi-Fard et al. [12] provided an improved streaming algorithm with a lower memory for this problem. Recently, Kazemi et al. [9] improved the memory complexity to $O(\varepsilon^{-1}k)$ with the same approximation ratio. Alaluf and Feldman [1] investigated the non-monotone scenario and presented a 0.233-approximation algorithm with $O(k\log k)$ memory. More researchers pay attention to streaming submodular optimization with complex constraints, such as knapsack constraints [7,8], matroid [3], p-matchoid [4,5]. The more work for the streaming set submodular maximization can be found in [14].

## 3 Preliminaries

Let $[n] = \{1, ...., n\}$ for an integer $n$, and $\mathcal{V}$ be a ground set of size $n$ which is not currently known but to be known in advance. Let $\Sigma$ be the set of sequences of $\mathcal{V}$, i.e., $\Sigma := \{(\sigma_1, ..., \sigma_\ell) : \ell \in [n], \sigma_1, ..., \sigma_\ell \in \mathcal{V}, \forall i, j \in [\ell] : i \neq j \Rightarrow \sigma_i \neq \sigma_j\}$. For any $\sigma \in \Sigma$, we denote the length of $\sigma$ by $|\sigma|$.

The benefit of element orders is defined by a weighted directed acyclic graph $\mathcal{G} = (\mathcal{V}, \mathcal{E})$ (excluding self-cycles) in which the vertices correspond to the elements of $\mathcal{V}$ and the weight of an edge reveals the additional utility of selecting the ending point after the starting point. Weight of self-cycle (i.e., the edge which starts and ends at the same vertex) describes the utility in selecting single vertex. The utility function is formally defined as

$$F(\sigma) = h(\mathcal{E}(\sigma)),$$

where $\mathcal{E}(\sigma) = \{(\sigma_i, \sigma_j) : (\sigma_i, \sigma_j) \in \mathcal{E}, i \leq j\}$ denotes the covered edges by $\sigma$ and $h(\cdot) : 2^{\mathcal{E}} \to \mathbb{R}_{\geq 0}$ denotes a non-negative monotone set submodular function over the edges of $\mathcal{E}$. We consider the scenario that the edges arrive in an arbitrary

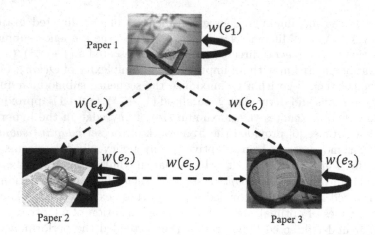

**Fig. 1.** Illustration of the streaming sequence submodular for assembling academic reading lists. Let W-DAG $\mathcal{G} = (\mathcal{V}, \mathcal{E})$, where $\mathcal{V} = \{\text{Paper 1}, \text{Paper 2}, \text{Paper 3}\}$ and $\mathcal{E} = \{e_1, ..., e_6\}$ as the set of edges according to their arriving order. The weights of solid black arrows represent the utilities of the self-cycles of papers and the weights of dotted lines represent additional utilities in reading the papers in the specified order.

order. That is, for each time slot $t$, there is an edge $e_t$ that reveals itself. Any streaming algorithm needs to retain a memory $M_t \subseteq \mathcal{E}$ for edges, and outputs a sequence $\sigma_t$ of length at most $k$ with $\mathcal{E}(\sigma_t) \subseteq M_t$. Whenever a new edge arrives from the stream, we need to decide whether the edge is added into the main memory. We note that the utility function is neither a set function, nor a submodular function on the element sets. Consider an example of assembling academic reading lists from Fig. 1. Let $h(\mathcal{E}(\sigma)) = w(\mathcal{E}(\sigma))$, i.e., the value of a sequence is the sum of weights of edges covered by that sequence. We also assume the weights are non-negative. Consider the sequence $\sigma^1 = (\text{Paper 1})$ where the researcher has learned only Paper 1, the sequence $\sigma^2 = (\text{Paper 2})$ where the researcher has learned only Paper 2, and the sequence of $\sigma^3 = (\text{Paper 1}, \text{Paper 2})$ where the researcher has learned Paper 2 after Paper 1. Then the utilities of these sequences are presented as follows.

- $F(\sigma^1) = w(\mathcal{E}(\sigma^1)) = w(e_1)$,
- $F(\sigma^2) = w(\mathcal{E}(\sigma^2)) = w(e_2)$, and
- $F(\sigma^3) = w(\mathcal{E}(\sigma^3)) = w(e_1) + w(e_2) + w(e_4)$.

It can be concluded above that $F(\sigma^2) \not\geq F(\sigma^3) - F(\sigma^1)$, which violates the diminishing marginal returns property of submodular functions. Let $\sigma^4 = (\text{Paper 2}, \text{Paper 1})$, then we have $F(\sigma^4) = w(e_2) + w(e_1)$ and $F(\sigma^3) \neq F(\sigma^4)$.

In the rest of our paper, we investigate the streaming sequence submodular maximization under weighted directed acyclic graphs (W-DAGs). The restriction is inspired by a good property of DAGs, which can be restated as following: For any set of edges of a DAG, the topological orders of vertices can be efficiently computed in polynomial time. Similarly, for any set $E \subseteq \mathcal{E}$ of a W-DAG, the

**Algorithm 1.** ETA: Edge-based Thresholding Algorithm

**Input:** Stream $\mathcal{E}$, error term $\varepsilon > 0$, parameter $k$, $\alpha = 1 + \frac{1-\varepsilon}{2\Delta}$, and functions $F, h$
1: $m_0 \leftarrow 0$, $t \leftarrow 1$
2: **while** $e_t$ is revealed **do**
3:     compute $m_t \leftarrow \max\{m_{t-1}, h(\{e_t\})\}$
4:     construct $O_t \leftarrow \{(1+\varepsilon)^i | \frac{m_t}{\alpha k \Delta} \leq (1+\varepsilon)^i \leq m_t\}$
5:     **for** each $\tau \in O_t$ **do**
6:         **if** $\tau$ is a new threshold **then**
7:             set $E_\tau \leftarrow \emptyset$ and set $\sigma_\tau \leftarrow ()$
8:         **end if**
9:         **if** $F(A^*(E_\tau \cup \{e_t\})) - F(A^*(E_\tau)) \geq \tau$ and $|\text{nodes}(E_\tau \cup \{e_t\})| \leq k$ **then**
10:             set $E_\tau \leftarrow E_\tau \cup \{e_t\}$
11:             set $\sigma_\tau \leftarrow A^*(E_\tau)$
12:         **end if**
13:     **end for**
14:     set $t \leftarrow t + 1$
15: **end while**
16: **return** $\sigma \leftarrow \arg\max_{\sigma_\tau} F(\sigma_\tau)$

sequence $\sigma_E$ with the maximum utility can also be computed in polynomial time. Let $A^*$ be an efficient algorithm of computing the optimum sequence for clarity. For example, if $A^*$ is fixed as the Depth First Search (DFS), then it will consume a runtime of $O(|V| + |E|)$, where $V$ is the node set covered by the edge subset $E$. Let $\sigma_E = A^*(E)$ be the returned sequence by $A^*$ with respect to $E$. Let $\Delta_{\text{in}}(\Delta_{\text{out}})$ be the maximum in-degree (out-degree) of the W-DAG, and let $\Delta = \min\{\Delta_{\text{in}}, \Delta_{\text{out}}\}$. In the paper, there are three assumptions summarized as following:

1. the value of $\Delta$ is known in advance,
2. there exists an value oracle of $h$, that is, for any subset $E$, we can derive the value of $h(E)$, and
3. there exists an value oracle of $F$, that is, for any sequence $\sigma$, we can acquire the value of $F(\sigma)$.

## 4    Edge-Based Thresholding Algorithm

In this section, we provide an edge-based thresholding algorithm described as in Algorithm 1 for the SSSM. In the model, the edges arrive in a streaming fashion that at each time slot $t$ an edge $e_t$ is revealed. We introduce an auxiliary $m_t$, which denotes the maximum singleton edge value according to function $h$ at time $t$. Let $\tau^* = \frac{F(\sigma^*)}{\alpha k \Delta}$, where $\sigma^*$ and $\alpha$ are respectively an optimum sequence of length at most $k$ and a parameter that will be determined in the following sections. As the optimum value $F(\sigma^*)$ is not known in the current time slot, we lazily construct a candidate set of $\tau^*$ by setting

$$O_t = \{(1+\varepsilon)^i | \frac{m_t}{\alpha k \Delta} \leq (1+\varepsilon)^i \leq m_t\},$$

where $\varepsilon > 0$ is a beforehand constant. For each $\tau \in O_t$, we use $E_\tau$ to denote the set of edges chosen by the algorithm according to $\tau$. Then, we adopt a thresholding strategy as follows. First, we check if the marginal gain of the topology sequence returned by encountering $e_t$ at least $\tau$, i.e., $F(A^*(E_\tau \cup \{e_t\})) - F(A^*(E_\tau)) \geq \tau$, where $A^*(E_\tau)$ denotes an optimal topology order according to $E_\tau$. If the optimal sequences of $E_\tau$ exceed one, we arbitrarily choose a sequence denoted by $A^*(E_\tau)$. Second, we check if the amount of nodes covered by $E_\tau \cup \{e_t\}$ at most $k$, that is, $|\mathrm{nodes}(E_\tau \cup \{e_t\})| \leq k$. We add the encountered edge $e_t$ to $E_\tau$ if the above two conditions are satisfied. After the stream has finished, we output a maximal sequence among all $\sigma_\tau$ with respect to the utility function $F$.

For any $E \subseteq \mathcal{E}$, we restate the alternative set function $g : 2^{\mathcal{E}} \rightarrow \mathbb{R}$ which was introduced by Tschiatschek et al. [13]. Let

$$g(E) = h(\mathcal{E}(\sigma)),$$

where $\sigma = A^*(E)$. It follows that $g(E) = F(\sigma)$ for any subset $E \subseteq \mathcal{E}$ where $\sigma = A^*(E)$. Let $|\mathrm{nodes}(E)|$ be the number of nodes covered by $E$. By defining the alternative maximization problem (P.E) which aims to attain

$$\max_{E \subseteq \mathcal{E}, |\mathrm{nodes}(E)| \leq k} g(E),$$

Tschiatschek et al. [13] proposed three properties which can be restated as in the following lemma.

**Lemma 1** ([13]). *The defined alternative function $g$ given above satisfies the following three properties:*

1. *the alternative function $g$ is non-negative, monotone, and submodular,*
2. *$F(\sigma^*) = g(E^*)$, and*
3. *$|E^*| \leq \Delta \cdot k$,*

*where $E^*$ is an optimal subset of problem (P.E) and $\sigma^*$ is an optimal sequence of problem (1).*

Before proving the main result, we present an invariant property of Algorithm 1 as in the following lemma. For clarity, we call each update of sequence $\sigma$ as one iteration.

**Lemma 2.** *Assuming that edge $e_t$ is revealed at iteration $s$, then for any $\tau \in O_t$, we have*

$$F(\sigma_\tau^s) \geq \frac{\tau}{2} \cdot |\sigma_\tau^s|.$$

*Proof.* We accomplish the proof by induction of iterations. Let () be the empty sequence, then $F(()) = 0$ and the claim holds. We assume the inequality holds at the begin of the iteration $s$, that is, $F(\sigma_\tau^s) \geq \frac{\tau}{2} \cdot |\sigma_\tau^s|$. If the edge $e_t$ does not bring any change of $\sigma_\tau^s$, we immediately have the conclusion that the claim holds in this case. In the rest of this section, we consider the case that $\sigma_\tau^s$ is updated in iteration $s$ by encountering edge $e_t$. By the line 9 of Algorithm 1, for any $\tau \in O_t$, we have

$$F(\sigma_\tau^{s+1}) - F(\sigma_\tau^s) = F(A^*(E_\tau^s \cup \{e_t\})) - F(A^*(E_\tau^s)) \geq \tau,$$

where $E_\tau^s$ denotes the edges set of $E_\tau$ at the begin of iteration $s$. Thus we derive

$$F(\sigma_\tau^{s+1}) \geq F(\sigma_\tau^s) + \tau \geq \frac{\tau}{2} \cdot |\sigma_\tau^s| + \tau \geq \frac{\tau}{2} \cdot |\sigma_\tau^{s+1}|,$$

where the second inequality follows from the induction condition and the third inequality can be obtained by the fact that the amount of the added nodes in each iteration is at most two. This finishes the proof. □

Now we provide our main result by the following theorem, which states that Algorithm 1 deserves a constant approximation for the sequence submodular maximization under streaming.

**Theorem 1.** *Given $\varepsilon > 0$, by setting $\alpha = 1 + \frac{1-\varepsilon}{2\Delta}$, with $O(\varepsilon^{-1} k\Delta \log(k\Delta))$ memory complexity and $O(\varepsilon^{-1} \log(k\Delta))$ update time per edge, Algorithm 1 makes one pass over the stream, keeps at least $(1 - 2\Delta/(2\Delta + 1 - \varepsilon))$-approximation for the SSSM with a cardinality constraint.*

*Proof.* Let $m = \max_{e \in \mathcal{E}} \{h(e)\}$ be the maximum single edge value according to function $h$. Then we can derive $m \leq F(\sigma^*) = h(\mathcal{E}(\sigma^*)) \leq \sum_{e \in \mathcal{E}(\sigma^*)} h(\{e\}) \leq mk\Delta$, where the second inequality follows from submodularity of function $h$ and $\mathcal{E}(\sigma^*)$ denotes the set of edges covered by the optimal sequence $\sigma^*$. Once edge $e_t$ arrives, we construct $O_t = \{(1+\varepsilon)^i | \frac{m_t}{\alpha k\Delta} \leq (1+\varepsilon)^i \leq m_t\}$ to approximate $\tau^*$. Consequently, we derive a threshold value $\tilde{\tau} \in \cup_t O_t$ such that $(1-\varepsilon)\tau^* \leq \tilde{\tau} \leq \tau^*$ for any given $\varepsilon > 0$. Because the value of the sequence $\sigma_{\tilde{\tau}}$ must be a lower bound of the value of the sequence $\sigma$ returned by Algorithm 1, we now provide a lower bound of $F(\sigma_{\tilde{\tau}})$ by the optimum value. The analysis considers two cases: (a). $|\sigma_{\tilde{\tau}}| = k$; (b). $|\sigma_{\tilde{\tau}}| < k$. For the first case, by Lemma 2, we have

$$F(\sigma_{\tilde{\tau}}) \geq \frac{\tilde{\tau}}{2} \cdot |\sigma_{\tilde{\tau}}| \geq \frac{1-\varepsilon}{2\alpha\Delta} \cdot F(\sigma^*). \tag{2}$$

For the second case that $|\sigma_{\tilde{\tau}}| < k$, let $E_{\tilde{\tau}}$ be the set of edges covered by the sequence $\sigma_{\tilde{\tau}}$, we immediately have

$$F(\sigma^*) = g(E^*) \leq g(E^* \cup E_{\tilde{\tau}}) \leq g(E_{\tilde{\tau}}) + \sum_{e \in E^* \setminus E_{\tilde{\tau}}} [g(E_{\tilde{\tau}} \cup \{e\}) - g(E_{\tilde{\tau}})].$$

We estimate the second term of the above inequality by analyzing two cases. Let $E_1$ be the set of the edges of $E^* \setminus E_{\tilde{\tau}}$ encountered before we instantiate the threshold value $\tilde{\tau}$. For any $e_t \in E_1$ and $\tau \in O_t$, we yield

$$F(A^*(E_{\tilde{\tau}} \cup \{e_t\})) - F(A^*(E_{\tilde{\tau}})) = g(E_{\tilde{\tau}} \cup \{e_t\}) - g(E_{\tilde{\tau}}) \leq g(e_t) \leq m_t < \tilde{\tau}.$$

Let $E_2$ be the set of the edges of $E^* \setminus E_{\tilde{\tau}}$ encountered after the threshold value $\tilde{\tau}$ has been instantiated. With the assumption of $|\sigma_{\tilde{\tau}}| < k$, we conclude that

$F(A^*(E_{\tilde{\tau}} \cup \{e\})) - F(A^*(E_{\tilde{\tau}})) \leq \tilde{\tau}$ for any $e \in E_2$. Then we have the following inequality:

$$F(\sigma^*) \leq g(E_{\tilde{\tau}}) + \sum_{e \in E^* \setminus E_{\tilde{\tau}}} [g(E_{\tilde{\tau}} \cup \{e\}) - g(E_{\tilde{\tau}})]$$

$$\leq F(\sigma_{\tilde{\tau}}) + k\Delta\tilde{\tau} \leq F(\sigma_{\tilde{\tau}}) + \frac{F(\sigma^*)}{\alpha}.$$

Thus,

$$F(\sigma_{\tilde{\tau}}) > \left(1 - \frac{1}{\alpha}\right) \cdot F(\sigma^*). \tag{3}$$

Combining inequalities (2) and (3) consequently yields

$$F(\sigma) \geq F(\sigma_{\tilde{\tau}}) \geq \min\left\{\frac{1-\varepsilon}{2\alpha\Delta}, \left(1 - \frac{1}{\alpha}\right)\right\} \cdot F(\sigma^*) = \left(1 - \frac{2\Delta}{2\Delta + 1 - \varepsilon}\right) \cdot F(\sigma^*),$$

where the equality follows by simple calculation when setting $\alpha = 1 + \frac{1-\varepsilon}{2\Delta}$.

The memory complexity can be calculated as follows. For each time $t$, as $O_t = \{(1+\varepsilon)^i | \frac{m_t}{\alpha k\Delta} \leq (1+\varepsilon)^i \leq m_t\}$, the amount of used memory can be upper bounded by $O(\varepsilon^{-1}\log(k\Delta))$. Then the update time per edge is at most $O(\varepsilon^{-1}\log(k\Delta))$. For each threshold $\tau \in O_t$, as $|E_\tau| \leq k\Delta$, the total memory is at most $O(\varepsilon^{-1}k\Delta\log(k\Delta))$. Therefore, we get the desired claim.    □

## 5    Improved Edge-Based Thresholding Algorithm

In this section, we propose a memory improved edge-based thresholding algorithm for the SSSM. The pseudo codes of our improved algorithm are summarized by Algorithm 2. The improved algorithm retains the same approximation ratio, while decreases the size of used memory to $O(\varepsilon^{-1}k\Delta)$. Theorem 2 provides the detailed guarantees of Algorithm 2.

Inspired by the work of [9], we lazily estimate the lower bound of $\tau^* = \frac{F(\sigma^*)}{\alpha k\Delta}$ according to time $t$ in a way that the estimation gets more precise along the iterations. Let $LB = \max_\tau F(\sigma_\tau)$ denote the value of the best sequence output by Algorithm 2. Instead of choosing $m_t$ as the lower bound of $\tau^*$ (as in Sect. 4), we use $\tau_{\min} = \frac{\max\{m_t, LB\}}{\alpha k\Delta}$ as a new lower bound of the marginal gain of encountering the edge $e_t$. So we have $LB \leq F(\sigma^*)$ and discard the thresholds that less than $\frac{\tau_{\min}}{1+\varepsilon}$. As $LB = \max_\tau F(\sigma_\tau)$, the number of edges added to $E_\tau$ is at most $\frac{LB}{\tau}$ for any $\tau$. By the novel changes of the candidate thresholds, we show the memory complexity can be improved by a factor of $O(\log k\Delta)$.

**Theorem 2.** *Given $\varepsilon > 0$, by setting $\alpha = 1 + \frac{1-\varepsilon}{2\Delta}$, with $O(\varepsilon^{-1}k\Delta)$ memory complexity and $O(\varepsilon^{-1}\log k\Delta)$ update time per edge, Algorithm 2 makes one pass over the stream, and retains the approximation ratio $(1 - 2\Delta/(2\Delta + 1 - \varepsilon))$ for the SSSM with a cardinality constraint.*

*Proof.* Let $E_\tau$ be the set of edges maintained for the threshold value $\tau$. As $LB = \max_\tau F(\sigma_\tau) = \max_\tau g(E_\tau) \geq \max_\tau(|E_\tau| \cdot \tau)$, then $|E_\tau| \leq \frac{LB}{\tau}$. We discard

---

**Algorithm 2.** IETA: Improved Edge-based Thresholding Algorithm

---

**Input:** Stream $\mathcal{E}$, error term $\varepsilon > 0$, parameters $k$, $\alpha$, and functions $F, h$
1: $m_0 \leftarrow 0$, $\tau_{\min} \leftarrow 0$, $LB \leftarrow 0$, $t \leftarrow 1$
2: **while** $e_t$ is revealed **do**
3:     compute $m_t \leftarrow \max\{m_{t-1}, h(\{e_t\})\}$
4:     set $\tau_{\min} = \frac{\max\{m_t, LB\}}{\alpha k \Delta}$
5:     discard all $E_\tau$ with $\tau < \tau_{\min}$
6:     construct $O_t \leftarrow \{(1+\varepsilon)^i | \frac{\tau_{\min}}{1+\varepsilon} \leq (1+\varepsilon)^i \leq m_t\}$
7:     **for** each $\tau \in O_t$ **do**
8:         **if** $\tau$ is a new threshold **then**
9:             set $E_\tau \leftarrow \emptyset$ and set $\sigma_\tau \leftarrow ()$
10:        **end if**
11:        **if** $F(A^*(E_\tau \cup \{e_t\})) - F(A^*(E_\tau)) \geq \tau$ and $|\text{nodes}(E_\tau \cup \{e_t\})| \leq k$ **then**
12:            set $E_\tau \leftarrow E_\tau \cup \{e_t\}$
13:            set $\sigma_\tau \leftarrow A^*(E_\tau)$
14:            set $LB \leftarrow \max\{LB, F(\sigma_\tau)\}$
15:        **end if**
16:     **end for**
17:     set $t \leftarrow t + 1$
18: **end while**
19: return $\sigma \leftarrow \arg\max_{\sigma_\tau} F(\sigma_\tau)$

---

the thresholds $\tau < \frac{LB}{\alpha k \Delta}$ and divide the set of the rest thresholds into two groups. In the first group of thresholds, $\frac{LB}{\alpha k \Delta} \leq \tau \leq \frac{LB}{k \Delta}$, we conclude that the number of the thresholds is bounded by $O(\varepsilon^{-1} \log \alpha)$. Since $\alpha$ is no more than two, we can bound this term by $O(\varepsilon^{-1})$. As $|E_\tau| \leq k\Delta$ for each $\tau$, the memory of the first group can be bounded by $O(\varepsilon^{-1} k\Delta)$. For the group of thresholds with $\tau \geq \frac{LB}{k\Delta}$, we sort the sequence of thresholds in non-decreasing order according to their values, that is, $\{\tau_1, \tau_2, ..., \tau_q\}$, s.t. $\tau_{i+1} = (1+\varepsilon)\tau_i$, for any $i \in \{1, ..., q-1\}$. Let $|E_{\tau_1}| = k\Delta$, then $\tau_1 = \frac{LB}{k\Delta}$ and $|E_{\tau_2}| \leq \frac{k\Delta}{1+\varepsilon}$. Apparently, $|E_\tau|$ decreases a coefficient $(1+\varepsilon)$ as threshold value $\tau$ increases a factor $(1+\varepsilon)$. Then the memory of the second group is bounded by

$$\sum_{i=0}^{\log_{1+\varepsilon}(k\Delta)} \frac{k\Delta}{(1+\varepsilon)^i} \leq O(\varepsilon^{-1} k\Delta).$$

This completes the calculation of the memory used by Algorithm 2. We omit the approximation ratio proof, because of its similarly to the proof of Theorem 1. □

## 6 Conclusion

In this paper, we first studied the sequence submodular maximization under streaming and, based on weighted directed acyclic graphs (W-DAGs) modeling, presented a one-pass, $(1 - 2\Delta/(2\Delta + 1 - \varepsilon))$-approximation algorithm with $O(\varepsilon^{-1} k\Delta \log(k\Delta))$ memory and $O(\varepsilon^{-1} \log(k\Delta))$ update time per edge. Further, with better estimation of lower bound of the optimum value, we derived a memory-improved streaming algorithm.

In future, we will consider streaming sequence submodular maximization based on more general graphs rather than weighted directed acyclic graphs (W-DAGs), since there exist applications with more generalized relationships which would be modeled by hypergraphs etc. We will also embark on the task of removing the assumption of $\Delta$ in the approximation ratio. As the edges arrive in streaming fashion, we could learn exactly $\Delta$ as soon as the stream is finished. So it is also interesting to improve the ratio by removing $\Delta$.

**Acknowledgements.** The second author is supported by Natural Science Foundation of China (No. 11531014). The third author is supported by Natural Science Foundation of China (No. 61772005) and Natural Science Foundation of Fujian Province (No. 2017J01753). The fourth author is supported by Natural Science Foundation of China (No. 11871081).

# References

1. Alaluf, N., Feldman, M.: Making a sieve random: improved semi-streaming algorithm for submodular maximization under a cardinality constraint. arXiv preprint arXiv:1906.11237 (2019)
2. Badanidiyuru, A., Mirzasoleiman, B., Karbasi, A., Krause, A.: Streaming submodular maximization: massive data summarization on the fly. In: Proceedings of the 20th ACM SIGKDD International Conference on Knowledge Discovery and Data Mining, pp. 671–680. ACM (2014)
3. Chakrabarti, A., Kale, S.: Submodular maximization meets streaming: matchings, matroids, and more. Math. Program. **154**(1–2), 225–247 (2015)
4. Chekuri, C., Gupta, S., Quanrud, K.: Streaming algorithms for submodular function maximization. In: Halldórsson, M.M., Iwama, K., Kobayashi, N., Speckmann, B. (eds.) ICALP 2015. LNCS, vol. 9134, pp. 318–330. Springer, Heidelberg (2015). https://doi.org/10.1007/978-3-662-47672-7_26
5. Feldman, M., Karbasi, A., Kazemi, E.: Do less, get more: streaming submodular maximization with subsampling. In: Proceedings of the 32nd International Conference on Neural Information Processing Systems, pp. 730–740. Curran Associates Inc. (2018)
6. Golovin, D., Krause, A.: Adaptive submodularity: theory and applications in active learning and stochastic optimization. J. Artif. Intell. Res. **42**(1), 427–486 (2011)
7. Huang, C.-C., Kakimura, N., Yoshida, Y.: Streaming algorithms for maximizing monotone submodular functions under a knapsack constraint. In: Proceedings of the 20th International Workshop on Approximation Algorithms for Combinatorial Optimization Problems and the 21st International Workshop on Randomization and Computation, No. 11. Schloss Dagstuhl-Leibniz-Zentrum fur Informatik GmbH, Dagstuhl Publishing (2017)
8. Jiang, Y., Wang, Y., Xu, D., Yang, R., Zhang, Y.: Streaming algorithm for maximizing a monotone non-submodular function under $d$-knapsack constraint. Optim. Lett. **13**(82), 1–14 (2019)

9. Kazemi, E., Mitrovic, M., Zadimoghaddam, M., Lattanzi, S., Karbasi, A.: Submodular streaming in all its glory: tight approximation, minimum memory and low adaptive complexity. In: Proceedings of the 36th International Conference on Machine Learning, pp. 3311–3320. ACM (2019)

10. Mitrovic, M., Feldman, M., Krause, A., Karbasi, A.: Submodularity on hypergraphs: from sets to sequences. In: Proceedings of the 21st International Conference on Artificial Intelligence and Statistics, pp. 1177–1184. PMLR (2018)

11. Mitrovic, M., Kazemi, E., Feldman, M., Krause, A., Karbasi, A.: Adaptive sequence submodularity. arXiv preprint arXiv:1902.05981 (2019)

12. Norouzi-Fard, A., Tarnawski, J., Mitrović S, Zandieh, A., Mousavifar, A., and Svensson, O.: Beyond 1/2-approximation for submodular maximization on massive data streams. In: Proceedings of the 35th International Conference on Machine Learning, pp. 3826–3835. ACM (2018)

13. Tschiatschek, S., Singla, A., Krause, A.: Selecting sequences of items via submodular maximization. In: Proceedings of the 31st AAAI Conference on Artificial Intelligence, pp. 2667–2673. AAAI Press (2017)

14. Yang, R., Xu, D., Li, M., Xu, Y.: Thresholding methods for streaming submodular maximization with a cardinality constraint and its variants. In: Du, D.-Z., Pardalos, P.M., Zhang, Z. (eds.) Nonlinear Combinatorial Optimization. SOIA, vol. 147, pp. 123–140. Springer, Cham (2019). https://doi.org/10.1007/978-3-030-16194-1_5

# Graph Simplification for Infrastructure Network Design

Sean Yaw[1(✉)], Richard S. Middleton[2], and Brendan Hoover[2]

[1] Montana State University, Bozeman, MT 59717, USA
sean.yaw@montana.edu
[2] Los Alamos National Laboratory, Los Alamos, NM 87545, USA
{rsm,bhoover}@lanl.gov

**Abstract.** Network design problems often involve scenarios where there exist many possible routes between fixed vertex locations (e.g. building new roads between cities, deploying communications and power lines). Many possible routes in a network result in graph representations with many edges, which can lead to difficulty running computationally demanding optimization algorithms on. While producing a subgraph with a reduced number of edges would in itself be useful, it is also important to preserve the ability to interconnect vertices in a cost effective manner. Suppose there is a set of target vertices in a graph that needs to be part of any size-reduced subgraph. Given an edge-weighted, undirected graph with $n$ vertices, set of target vertices $T$, and parameter $k \geq 1$, we introduce an algorithm that produces a subgraph with the number of edges bounded by $\min\{O(n\frac{n^{1/k}}{|T|}), O(n|T|)\}$ times optimal, while guaranteeing that, for any subset of the target vertices, their minimum Steiner tree in the subgraph costs at most $2k$ times the cost of their minimum Steiner tree in the original graph. We evaluate our approach against existing algorithms using data from Carbon Capture and Storage studies and find that in addition to its theoretical guarantees, our approach also performs well in practice.

**Keywords:** Graph spanner · Steiner tree · Network design · Carbon capture and storage

## 1 Introduction

Application-specific networks are commonly represented as generic graph structures, allowing application-agnostic algorithms to be employed for various optimization tasks. Representing a network as a graph requires having information

This research was funded by the US-China Advanced Coal Technology Consortium (under management of West Virginia University) and by the U.S. Department of Energy's (DOE) National Energy Technology Laboratory (NETL) through the Rocky Mountain CarbonSAFE project under Award No. DE-FE0029280.

Y. Li et al. (Eds.): COCOA 2019, LNCS 11949, pp. 576–589, 2019.
https://doi.org/10.1007/978-3-030-36412-0_47

about edge connectivity in the network, which is a challenge in certain geospatial applications (e.g. building road networks, routing fiber) where one knows the vertices that need to be connected, but not how to connect them. For example, if one wanted to connect a set of cities with brand new natural gas transmission pipelines, all that exists is a vertex set without explicit candidate edges from which to select for pipe installation. Edges for these networks can be generated by rasterizing the 2D weighted-cost surface into a grid, placing a vertex in the center of each cell, and determining the cost to visit each of its, up to eight, neighbors. The cost to visit each neighbor cell is based on factors such as the topography, land ownership, and physical edge construction costs of the underlying weighted-cost surface. The result of this process is the discretization of the continuous geospatial surface into a large grid graph with diagonal edges, which achieves the goal of providing a graph representation for these geospatial network design problems [8,16]. However, for even a moderately sized region (e.g. the gulf coast of Texas), this can result in a graph with millions of vertices and edges when the rasterized cell size is approximately a square kilometer. As regions get larger or cell size decreases, graph sizes can quickly become computationally prohibitive to use for complex optimization problems such as Mixed Integer Linear Program (MILP) formulations of optimal infrastructure deployments. Figure 1 shows an example of a rasterized 2D weighted-cost surface with three target vertices, the large grid graph with diagonal edges, and a more manageably sized subgraph.

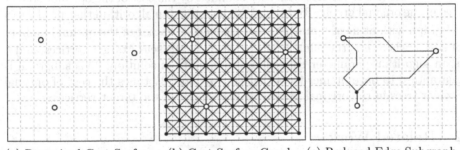

(a) Rasterized Cost Surface    (b) Cost Surface Graph    (c) Reduced-Edge Subgraph

**Fig. 1.** Example of graph construction from weighted-cost surface.

This scenario commonly arises in network design problems embedded in the physical world such as road networks, physical telecommunication networks, and pipeline networks. We are motivated by the design of Carbon Capture and Storage (CCS) networks. CCS is an effort to reduce the dispersal of $CO_2$ into the atmosphere by capturing the $CO_2$ byproduct of industrial processes, and transporting it to geological features where it can be stored [15]. $CO_2$ capture sites are not often collocated with appropriate storage sites. This requires development of a $CO_2$ distribution network to link capture sites with storage sites. Furthermore, like any infrastructure project, CCS networks can be costly undertakings, so care must be taken to select capture sites, storage sites, and intelligent distribution links in a cost effective manner [13,18].

Design of CCS networks, and other geospatial networks, is not as simple as determining edge locations to connect a pre-determined set of capture and storage vertices. Instead, a graph representation of the network needs to be generated that supports deploying an unknown subset of the available capture and storage vertices. This is because the graph feeds a variety of risk analysis studies that determine the best capture and storage locations to open based on cost, capture, and storage objectives, thereby driving final deployment decisions. Contingency plans are also considered in case a storage location's predicted geologic parameters (e.g. capacity, supported $CO_2$ injection rate) do not match reality [17]. Therefore, the graph needs to be able to support various possibilities, while remaining small enough to enable efficient computation and maintaining some guarantee that the solution cost is within some bound of optimal.

The goal of this research is to simplify a graph by reducing the number of edges and vertices in the graph, while maintaining some guarantee that vertices in a target set can all be connected with a Steiner tree of bounded cost. In the case of the CCS problem, the initial graph is the large grid graph output of the rasterizing process shown in Fig. 1b and the target set of vertices is the set of possible capture and storage locations. Formally, we look for a subgraph of a given graph that spans some target vertices, while minimizing the number of edges in the subgraph and ensuring the subgraph contains Steiner trees for any subset of the target vertices with cost bounded by the cost of the optimal Steiner tree in the original graph. To this end, we introduce an algorithm that produces a subgraph with at most a $\min\{O(n\frac{n^{1/k}}{|T|}), O(n|T|)\}$ factor of the optimal number of edges, where $n$ is the number of vertices, $T$ is the set of target vertices, and $k$ is an algorithm parameter controlling the quality of solution. This subgraph is guaranteed to have Steiner trees that cost at most $2k$ times the cost of optimal Steiner trees in the original graph, for any subset of the target vertices.

## 2   Related Work

The problem in the literature that is most closely related to ours is the $k$-spanner problem [5]. A *spanner* (or $k$-spanner) is a subgraph that results from trimming edges from a graph, while preserving some degree of pairwise distance between vertices. Formally, given an edge weighted graph, a $k$-spanner is a subgraph whose distance between each pair of vertices is at most $k$ times the distance between that pair in the original graph. Spanners with a minimal number of edges have been extensively studied [4]. It has been shown that for $k \geq 1$, every graph with $n$ vertices has a $(2k-1)$-spanner with at most $O(n^{1+1/k})$ edges [1]. The problem we consider in this paper differs from finding $k$-spanners in two important ways:

1. Spanners traditionally aim to connect all vertices in the original graph, whereas we look to connect only a designated subset of those vertices. Forcing a solution to include excess vertices unnecessarily enlarges it.
2. Spanners bound the shortest path between each pair of vertices, whereas we aim to bound the minimum Steiner tree cost of any subset of the target subset of vertices.

Prior work has been done on *subset spanners* (or pairwise spanners), which are subgraphs that approximately preserve distances between pairs of a subset of the vertices. However, this prior work differs from ours by considering a spanner variant with additive error instead of multiplicative [6,9]. In other words, distances between each vertex pair is at most $k$ plus, not times, the distance between that pair in the original graph. Work has also been done that seeks spanners that exactly preserve pairwise distance instead of approximately preserving it [2,3]. This could be useful for our application area, but does not enable relaxing the distance requirement in favor of reducing the graph size, which is our primary goal, given the intractability of optimal network design optimizations on large graphs. Furthermore, none of these efforts explicitly consider Steiner trees in the subgraph, motivating a new approach for finding spanners.

Delaunay triangulation has been shown to produce paths between any two points that are most 2.418 times their Euclidean distance [10]. However, this assumes that the graph is embedded in a metric space, which cannot be assumed for general geospatial network where edge weights need not abide by the triangle inequality. Therefore, for non-metric instances, Delaunay triangulation's distance preserving guarantee does not hold. Nonetheless, it is reasonable to assume that edge weights in physical networks are likely distance dependent, so we do evaluate our solution against Delaunay triangulations in Sect. 5.

# 3    Problem Formulation

We consider a graph $G$ consisting of a set of vertices $V$, undirected edges $E$, and a value $k \geq 1$. A subset of vertices is designated as target vertices $T \subseteq V$. A non-negative cost $c(e)$ is associated with each edge $e$ in $E$, reflecting the construction cost for that edge. For any subset of target vertices $T' \subseteq T$, the cost of a minimum Steiner tree on $G$ is denoted $S_{G,T'}$. Our goal is to find a subgraph of $G$ that includes all vertices in $T$, bounds the cost of the minimum Steiner tree for any subset of $T$, and has a minimal number of edges. The problem is formally defined below.

**Definition 1.** *Given $G = (V, E, T, c, k)$, a connected, edge weighted, and undirected graph, where $T \subseteq V$ is a set of target vertices, $c \colon E \to \mathbb{R}_{\geq 0}$ is an edge cost function, and $k \geq 1$, the **Minimal Steiner-Preserving Subset Spanner** problem seeks a subgraph of $G$, $G' = (V', E', T, c, k)$ such that $V' \subseteq V$, $E' \subseteq E$ induced by the set $V'$, $T \subseteq V'$, $|E'|$ is minimized, and for any $T' \subseteq T$, $S_{G',T'} \leq k * S_{G,T'}$.*

### 3.1   Computational Complexity

The minimal Steiner-preserving subset spanner problem is related to the MIN-IMUM EDGE $k$-SPANNER problem: Given $G = (V, E, k)$, a connected graph with $k \geq 1$, find a spanning subgraph $G' = (V, E', k)$ of $G$ with the minimum number of edges such that, for any pair of vertices, the length of the shortest path (i.e. number of edge) between the pair in $G'$ is at most $k$ times the shortest distance between the pair in $G$.

**Theorem 1.** *The MINIMUM EDGE $k$-SPANNER problem is reducible to the minimal Steiner-preserving subset spanner problem via an S-reduction (cost-preserving reduction).*

*Proof.* Deferred to Appendix A.

Theorem 1 means that the following complexity results for the MINIMUM EDGE $k$-SPANNER problem hold for the minimal Steiner-preserving subset spanner problem:

- When $k = 2$, the MINIMUM EDGE $k$-SPANNER problem is NP-Hard to approximate within a bound of $\frac{\alpha \log |V|}{k}$ for some $\alpha > 0$ [11].
- When $k \geq 3$, the MINIMUM EDGE $k$-SPANNER problem is NP-Hard to approximate within a bound of $2^{(\log^{1-\epsilon} n)/k}$, for all $\epsilon > 0$, if $NP \nsubseteq BPTIME(2^{polylog(n)})$ [7].
- When $k \geq 3$, the MINIMUM EDGE $k$-SPANNER problem is NP-Hard to approximate within a bound of $n^{1/(\log \log n)^c}$ if the exponential time hypothesis holds, due to the spanner problem's relationship to the Label Cover and the Densest $k$-Subgraph problems [12].

Because of these complexity results, we pursue suboptimal approaches with performance guarantees for finding solutions to minimal Steiner-preserving subset spanner problem instances in Sect. 4.

## 4   Algorithm

In this section, we detail an approach for finding approximate minimal Steiner-preserving subset spanners within graphs. An initial idea could be to leverage the relationship to the MINIMUM EDGE $k$-SPANNER problem detailed in the proof of Theorem 1. Specifically, we use a variant called the edge-weighted MINIMUM EDGE $k$-SPANNER where the edges in the input graph have weights: For an input graph $G = (V, E, T, c, k)$, run a solution to the MINIMUM EDGE $k$-SPANNER problem on $Q = (V, E, c, k)$. This would result in a solution $Q' = (V, E', c, k)$ embedded with Steiner trees for any subset of $T$ costing at most $k$ times their optimal costs in $G$, since $T \subseteq V$. However, this requires including all vertices in $V$ in the solution, which can dramatically increase the solution size when the size of $V$ is on the order of millions of vertices and the size of $T$ is on

---

**Algorithm 1.** Greedy Subset Spanner, on input $G = (V, E, T, c, k)$

---

Step 1   Find paths for pairs of vertices in $T$.
        Let $c' = c$.
        **forall** vertex pairs in $T$, in order of increasing cheapest path cost in $G$
            Find cheapest path in $F = (V, E, T, c', k)$.
            **forall** Edge $e$ in path
                Mark $e$ and let $c'(e) = \frac{c(e)}{k}$.
            **endforall**
        **endforall**
Step 2   Build $G_{GSS}$.
        Let $E_{GSS} = \emptyset$ and $V_{GSS} = \emptyset$.
        **for** Edge $e$ in $E$
        **if** $e$ is marked
            Add $e$ to $E_{GSS}$ and endpoints of $e$ to $V_{GSS}$.
        **endif**
        **endfor**
        Let $G_{GSS} = (V_{GSS}, E_{GSS}, T, c, k)$.

---

the order of tens of vertices. Instead we pursue an approach that attempts to more rigorously control the size of the solution.

The classic *greedy spanner* $(GS)$ algorithm for the edge-weighted MINIMUM EDGE $k$-SPANNER problem works by iteratively adding individual edges to the solution graph $G_{GS} = (V, E_{GS}, c, k)$ [1]. Given an input of $G = (V, E, c, k)$, $GS$ first sets $E_{GS} = \emptyset$ and sorts $E$ by non-decreasing weight. Then, for each edge $e = (v_1, v_2)$, $GS$ computes the cheapest path from $v_1$ to $v_2$ in $G_{GS}$ and adds $e$ to $E_{GS}$ if the cost of that path is larger than $k$ times the cost of $e$. The algorithm we introduce in this section, $GSS$ for greedy subset spanner, could be seen as a generalization of the $GS$ algorithm. Instead of greedily adding one edge at a time, we greedily add all edges from a path between one pair of vertices in $T$ at a time.

For an input graph $G = (V, E, T, c, k)$, the basic approach we take is to construct an edge set for $G_{GSS}$ by iteratively adding edges from paths between pairs of vertices in $T$. Each pair of vertices from $T$ is first sorted by non-decreasing cost of their cheapest path in $G$. A cheapest path for each pair of vertices from $T$ is generated, one at a time in sorted order. Once a path is generated, its edges are added to $G_{GSS}$ and the cost of each edge is reduced to be $\frac{1}{k}$ times its original cost. Reducing the cost of selected edges encourages their use in the cheapest paths for the remaining pairs in $T$. Reducing the edge cost by $\frac{1}{k}$ ensures that cheapest paths selected will be within $k$ of their actual cheapest paths and is used in the analysis of the algorithm's performance. Note that the cost does not keep lowering if an edge is selected multiple times. Edges costs are either their original costs, or $\frac{1}{k}$ times their original costs. This process builds an edge and vertex set (vertices included in the selected edges) for $G_{GSS}$. Details of the basic $GSS$ algorithm are presented in Algorithm 1.

Step 1 of $GSS$ runs in $O(|V|^4)$ time since it executes a shortest path search $(O(|V|^2))$ for each pair of points in $T$ $(O(|V|^2))$. Step 2 goes through each edge, and thus runs in $O(|V|^2)$ time. Thus the running time of $GSS$ is $O(|V|^4)$.

Since $GSS$ runs in polynomial time and Theorem 1 shows the problem to be NP Hard, we cannot hope solutions produced by $GSS$ to be optimal. In the following theorems, we establish a guarantee on the cost of Steiner trees for any subset of $T$ and provide an approximation ratio for the number of edges in solutions output by $GSS$.

**Theorem 2.** *Given a graph $G = (V, E, T, c, k)$ and a solution provided by $GSS$, $G_{GSS} = (V_{GSS}, E_{GSS}, T, c, k)$, for any subset $T'$ of $T$, the cost of the minimum Steiner tree for $T'$ in $G_{GSS}$ is at most $2k$ times the cost of the minimum Steiner tree for $T'$ in $G$.*

*Proof.* Suppose that $G = (V, E, T, c, k)$ and $G_{GSS} = (V_{GSS}, E_{GSS}, T, c, k)$ are the input and output respectively of $GSS$. The *metric closure* of a set of vertices $S$ in a graph $G$ is the complete (over $S$) subgraph in which each edge is weighted by the cheapest path distance between vertices in $G$. It is well known that the cost of a Steiner tree constructed as a minimum spanning tree on a metric closure is at most twice the cost of a minimum Steiner tree in the original graph [19]. If $\mathrm{MST}_{MC_T,G}$ denotes the minimum spanning tree on the metric closure of $T$ in $G$ and $\mathrm{ST}_{T,G}$ denotes the minimum Steiner tree of $T$ on $G$,

$$\mathrm{cost}(\mathrm{MST}_{MC_T,G}) \leq 2\,\mathrm{cost}(\mathrm{ST}_{T,G}) \tag{1}$$

If $k = 1$, $G_{GSS}$ contains the metric closure of $T$ since $G_{GSS}$ is composed of the cheapest paths in $G$ between each element of $T$. If $k > 1$, $G_{GSS}$ no longer contains the metric closure of $T$ in $G$, but contains a path between each pair of vertices from $T$ that costs at most $k$ times the cost of that path in the metric closure in $G$. Thus, a minimum spanning tree on $T$ in $G_{GSS}$, $\mathrm{MST}_{T,G_{GSS}}$, would be at most $k$ times the cost of the minimum spanning tree of the metric closure of $T$ in $G$:

$$\mathrm{cost}(\mathrm{MST}_{T,G_{GSS}}) \leq k\,\mathrm{cost}(\mathrm{MST}_{MC_T,G}), \forall k \geq 1 \tag{2}$$

Combining Eqs. 1 and 2 shows that $G_{GSS}$ contains a Steiner tree for $T$ of cost at most $2k$ times the optimal cost:

$$\mathrm{cost}(\mathrm{MST}_{T,G_{GSS}}) \leq k\,\mathrm{cost}(\mathrm{MST}_{MC_T,G}) \leq 2k\,\mathrm{cost}(\mathrm{ST}_{T,G}), \forall k \geq 1 \tag{3}$$

Equation 3 holds for all subsets of $T$, since the metric closures of subsets of $T$ are embedded in the metric closure for $T$ itself, and for all $k \geq 1$, $G_{GSS}$ contains, within $k$, cheapest paths between all pairs of vertices from any subset of $T$, since it contains them for all pairs of vertices from $T$ itself. $\square$

**Theorem 3.** *Given a graph $G = (V, E, T, c, k)$ and a solution provided by $GSS$, $G_{GSS} = (V_{GSS}, E_{GSS}, T, c, k)$, $|E_{GSS}| \in O(n\,|T|^2)$, where $n$ is the number of vertices in $V$.*

*Proof.* For each pair of points from $T$, $GSS$ provides a path that is within $k$ of its least cost path. Since edge costs are non-negative, each path can be made loopless, thereby being less than or equal to $n-1$ edges. Since there are $O(|T|^2)$ pairs of vertices, $|E_{GSS}| \in O(n|T|^2)$.     □

The next theorem bounds the size of $E_{GSS}$ in relation to $E_{GS}$. To avoid pathological cases discussed in the proof, the input graph needs to be pre-processed by running $GSS$ on it with $T = V$. This does not affect the conclusions from Theorem 2 or 3, since the pre-processed graph still contains a path between each pair for vertices that costs at most $k$ times the original cost. Since $GSS$ is run twice (first on $T = V$ and then on the actual subset $T \subseteq V$), the running time of this modification is still $O(|V|^4)$.

**Theorem 4.** *Given a graph $G = (V, E, T, c, k)$ and a solution provided by $GSS$, $G_{GSS} = (V_{GSS}, E_{GSS}, T, c, k)$, $|E_{GSS}| \in O(n^{1+1/k})$ when $k \geq 3$, where $n$ is the number of vertices in $V$.*

*Proof.* Deferred to Appendix B.

**Corollary 1.** *$GSS$ is a $\min\{O(n\frac{n^{1/k}}{|T|}), O(n|T|)\}$-approximation algorithm when $k \geq 3$ and a $O(n|T|)$-approximation algorithm when $k < 3$.*

*Proof.* Consider a graph $G = (V, E, T, c, k)$, a solution provided by $GSS$ $G_{GSS} = (V_{GSS}, E_{GSS}, T, c, k)$, and an optimal solution $G_{OPT} = (V_{OPT}, E_{OPT}, T, c, k)$. $|E_{OPT}| \geq |T|-1$, since the optimal edge set must, at minimum, span all vertices in $T$. By Theorem 4, $|E_{GSS}| \in O(n^{1+1/k}) = O(n\frac{n^{1/k}}{|T|}|T|)$ when $k \geq 3$. Thus, $|E_{GSS}| \leq O(n\frac{n^{1/k}}{|T|})|E_{OPT}|$. By Theorem 3, $|E_{GSS}| \in O(n|T|^2)$. Thus, $|E_{GSS}| \leq O(n|T|)|E_{OPT}|$. Therefore, $|E_{GSS}| \leq \min\{O(n\frac{n^{1/k}}{|T|}), O(n|T|)\}|E_{OPT}|$ when $k \geq 3$ and $|E_{GSS}| \leq O(n|T|)|E_{OPT}|$ otherwise.     □

## 5   Evaluation

In this section, we evaluate the performance of the algorithm presented in Sect. 4 using realistic data and scenarios. We label this algorithm the Greedy Subset Spanner ($GSS$) algorithm. In order to characterize $GSS$'s performance, we implement two additional solutions:

1. Greedy Spanner ($GS$) is the simple solution to the edge-weighted MINIMUM EDGE $k$-SPANNER problem introduced in [1], with its implementation outlined in the beginning of Sect. 4.
2. Delaunay triangulation ($DT$) is a simple spanning strategy of calculating a Delaunay triangulation and replacing each edge in the triangulation with the cheapest path in the original graph between those points. Theoretical limitations of this approach are detailed in Sect. 2.

We seek to compare these algorithms based on the quality of solution, as quantified by the size of the output graphs (i.e. number of edges) and cost of minimum Steiner trees in the output graphs, for various subsets of vertices. The dataset we used to evaluate these algorithms encompasses southern Indiana, in the Illinois Basin [14]. Figure 2 shows the geographical area of the dataset as well as the possible capture and storage locations and the edges selected by $GSS$ when $k = 1.5$. The base graph generated by this dataset was constructed as the rasterization of the 2D cost surface as described in Sect. 1. The number of edges in the base graph is 118563 and the number of vertices is 29879, 43 of which are the possible capture and storage locations that compose the set of target vertices.

When considering the number of edges in the output subgraph, we collapse degree 2 vertices that are not in $T$ by making a single edge whose cost is the sum of the costs of the two replaced edges. This is done because those vertices dictate the routing of the edge, but not the overall connectivity or cost of the graph. For each edge in the graph, we keep track of its collapsed vertices for determining actual edge routes. To calculate the minimum Steiner trees, we use the popular minimum spanning tree based approximation algorithm [19].

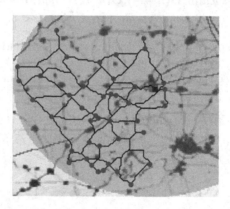

**Fig. 2.** Dataset map indicating capture (square) and storage (circle) target vertex locations as well as the subset of edges selected by the $GSS$ algorithm when $k = 1.5$.

First, we explore the tradeoff between the parameter $k$ and the quality of output for the $GSS$ algorithm by running it with $k$ valued between 1.0 and 2.0. The number of edges in the resulting graph is shown in Fig. 3a. As expected, the size of the output graph decreases as the parameter $k$ increases, since a larger $k$ allows for more expensive Steiner trees and thus more sharing of edges. The number of edges was reduced by 94% going from $k = 1.0$ to $k = 1.25$, 40% going from $k = 1.25$ to $k = 1.5$, and about 20% from $k = 1.5$ to $k = 1.75$ and $k = 1.75$ to $k = 2.0$. The average cost of Steiner trees for 30 random subsets of capture and storage vertices ranging in size from 10 to 30 vertices is shown in Fig. 3b. The increase in Steiner tree cost is much less dramatic than the decrease in graph size for increased $k$ values. The average cost of Steiner trees increased by 4.5% going from $k = 1.0$ to $k = 1.25$ and about 1% for all other jumps in $k$. This suggests that *GSS* is an efficient way to reduce graph size while preserving inexpensive Steiner trees.

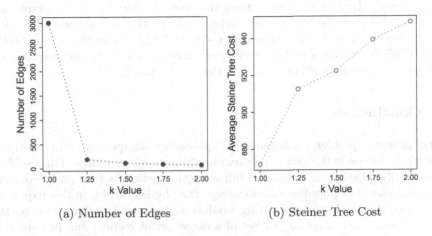

(a) Number of Edges   (b) Steiner Tree Cost

**Fig. 3.** *GSS* performance metrics for various values of $k$.

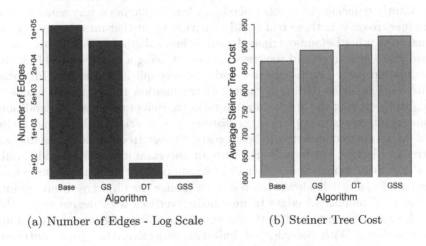

(a) Number of Edges - Log Scale   (b) Steiner Tree Cost

**Fig. 4.** Algorithm performance metrics comparison.

Finally, we seek to compare the *GSS* algorithm with *GS* and *DT*. We also include the base graph, Base, to provide baseline comparison. We parameterized *GSS* and *GS* with $k = 1.5$. The number of edges in the output graph for each algorithm is shown in Fig. 4a with a log scale, since the differences between algorithms is quite drastic. The number of edges in the base graph is 118563 whereas the output from $GS$ has 58895 edges, $DT$ has 203 edges, and $GSS$ has 112 edges. This means that $GGS$ reduced the number of edges in the base graph by over 99.9%. The average cost of Steiner trees for 30 random subsets of capture and storage vertices ranging in size from 10 to 30 vertices is shown in Fig. 4b. As expected, the cost of Steiner trees increases as the size of the graph size decreases, however the increase is markedly small. The average cost of Steiner trees using $GS$ was 3% more than the cost of Steiner trees on the base network, while $GSS$ Steiner trees cost on average 6% more and $DT$ Steiner trees cost on average 4% more than Steiner trees on the base network.

# 6   Conclusions

Network design problems using optimal approaches can quickly become intractable when the size of the graph representing the network gets large. This problem is especially prevalent in geospatial infrastructure networks with many edges to represent the many possible routes through the physical world. In this paper, we formalized the problem of generating a reduced edge subgraph that preserves the cost of Steiner trees over any subset of a target set of vertices and proposed an algorithm for it, $GSS$. $GSS$ provides theoretical approximation ratios for both the cost of Steiner trees and the number of edges in the subgraph it generates. Using cost surface and capture and storage location data from CCS studies, we evaluated $GSS$ against other possible approaches. $GSS$ proved to be effective at significantly reducing the number of edges while displaying a very small increase in Steiner tree cost in these real world scenarios. Of further interest was the performance of using Delaunay triangulation to find subgraphs. Despite the lack of a theoretical guarantee and the cost surface not being a metric space, Delaunay triangulation proved quite effective and was very quick in practice. Important future work includes finding a tighter approximation ratio for the number of edges and removing the approximation ratio on Steiner tree cost. An additional avenue of interest is to explore other distance preservation goals beyond Steiner trees. In networks dominated by fixed costs, Steiner trees are a reasonable connectivity objective, but in networks with an uncertain mix of fixed and variable costs (e.g. utilization costs), a Steiner tree may not be the optimal connectivity objective. Finally, it is also of interest is to change the objective from minimizing the total number of edges to minimizing vertices with degree larger than two. A single source-to-sink path can consist of many edges (a route through a cost surface). With the edge minimization objective, that path contributes many edges, whereas with the degree larger than two minimization objective, that path only counts as a single edge. If further optimization processes that use the subgraph only consider paths between vertices, and not underlying routes, it is more accurate to have that path count as a single edge instead of many.

# A    Proof of Theorem 1

The MINIMUM EDGE $k$-SPANNER problem is a special case of the minimal Steiner-preserving subset spanner problem. Given $Q = (V, E, k)$ with $k \geq 1$ as an input to the MINIMUM EDGE $k$-SPANNER problem, construct an instance to the minimal Steiner-preserving subset spanner problem $G = (V, E, T, c, k)$ by using the same $V$, $E$, and $k$, and letting $T = V$ and $c(e) = 1$ for all $e$ in $E$. We wish to show that a minimal Steiner-preserving subset spanner problem solution to $G$ is a solution to the MINIMUM EDGE $k$-SPANNER problem if and only if a MINIMUM EDGE $k$-SPANNER problem solution to $Q$ is a solution to the minimal Steiner-preserving subset spanner problem.

Suppose $G' = (V', E', T, c, k)$ is a solution to the minimal Steiner-preserving subset spanner problem. This means that every subset $T'$ of $T$ has a Steiner tree in $G'$ whose cost is at most $k$ times the cost of the minimum Steiner tree of $T'$ in $G$. Since $T = V$, $V' = V$ and every pair of points in $V$ is a subset of $T$. Thus, because a Steiner tree on two points is merely the shortest path between them, the shortest paths in $G'$ between any pair of vertices in $V$ is at most $k$ times the distance between the pair in $G$. Therefore, $G'$ is a $k$-spanner of $G$, and thus $Q$.

Suppose $Q' = (V, E', k)$ is a solution to the MINIMUM EDGE $k$-SPANNER problem. This means that every pair of points in $V$ have a path whose cost is at most $k$ times the cost of the cheapest path between those points in $Q$. Given a subset $T'$ of $T$ and a minimal Steiner tree of $T'$ in $Q$, adjacent vertices in the Steiner tree can be connected by the shortest path in $Q'$ between those vertices. Since those paths in $Q'$ are at most $k$ times their cost in $Q$, this yields a Steiner tree whose cost is at most $k$ times the cost of the minimal Steiner tree in $Q$. Therefore, $Q'$ is Steiner-preserving subset spanner of $Q$, and thus $G$.

Since any solution to $Q$ suffices as a solution to $G$ and conversely, the minimum solutions must coincide.                                                             □

# B    Proof of Theorem 4

Suppose that $G_{GSS} = (V_{GSS}, E_{GSS}, T, c, k)$ is the solution provided by $GSS$ on input $G = (V, E, T, c, k)$. Further, suppose that $G_{GS} = (V, E_{GS}, c, k)$ is the solution provided by the $GS$ algorithm described at the beginning of this section. It has been shown that the $GS$ algorithm produces a solution with at most $O(n^{1+1/k})$ edges when $k \geq 3$ [1]. We will show that $|E_{GSS}| \leq |E_{GS}|$ for any set $T$ and $k \geq 1$.

Without loss of generality, assume that edge costs are unique. This can be achieved by small perturbations of the edge costs. We begin by showing that for any $k \geq 1$, $E_{GSS} = E_{GS}$ when $T = V$. This result is not dependent on the pre-processing modification to $GSS$ described above.

Let $e = (v_1, v_2)$ be the least cost edge that meets one of the following conditions:

1. $e \in E_{GSS}$ and $e \notin E_{GS}$
2. $e \in E_{GS}$ and $e \notin E_{GSS}$

<u>Case 1.</u> Suppose that $e$ meets condition 1 ($e$ is in $E_{GSS}$, but not in $E_{GS}$). We will show the contradiction that $e$ cannot be in $E_{GSS}$. Since $e$ is not in $E_{GS}$, there is a path $p$ from $v_1$ to $v_2$ in $E_{GS}$ such that:

1. For each edge $e' \in p$, $\text{cost}(e') < \text{cost}(e)$, since $GS$ schedules edges in order of cost.
2. $\text{cost}(p) \leq k\,\text{cost}(e)$, otherwise $e$ would have been scheduled.

All edges from $p$ must be in $E_{GSS}$, since there does not exist a lower cost edge that is in $E_{GS}$ but not in $E_{GSS}$. When $GSS$ considers $e$, all edges in $p$ must have already been added to $E_{GSS}$, since $GSS$ schedules edges in order of increasing cost. Therefore, when $GSS$ considers $e$, the reduced cost of each edge $e'$ in $p$ is $\frac{\text{cost}(e')}{k}$. This means that the reduced cost of $p$ is,

$$\sum_{e' \in p} \frac{\text{cost}(e')}{k} = \frac{1}{k}\,\text{cost}(p) \leq \frac{1}{k}k\,\text{cost}(e) = \text{cost}(e)$$

Therefore, the cost of $p$ found by $GSS$ is at most the cost of $e$. Increasing $k$ by a small amount in $GSS$ yields a strict inequality, which means that $GSS$ would select path $p$ over edge $e$, thereby excluding $e$ from $E_{GSS}$.

<u>Case 2.</u> Suppose that $e$ meets condition 2 ($e$ is in $E_{GS}$, but not in $E_{GSS}$). Since $e$ is not in $E_{GSS}$, there is a path $p$ from $v_1$ to $v_2$ in $E_{GSS}$ such that:

1. For each edge $e' \in p$, $\text{cost}(e') < \text{cost}(e)$, since $GSS$ schedules edges in order of cost.
2. $\frac{1}{k}\,\text{cost}(p) < \text{cost}(e)$, otherwise $e$ would have been scheduled.

All edges from $p$ must be in $E_{GS}$, since there does not exist a lower cost edge that is in $E_{GSS}$ but not in $E_{GS}$. When $GS$ considers $e$, all edges in $p$ must have already been added to $E_{GS}$, since $GS$ schedules edges in order of increasing cost. Therefore, $p$ would have been scheduled instead of $e$, since $\text{cost}(p) < k\,\text{cost}(e)$.

Thus, there cannot be an edge $e$ such that either $e$ is in $E_{GSS}$ but not in $E_{GS}$, or $e$ is in $E_{GS}$ but not in $E_{GSS}$. Therefore, when $T = V$, every edge in $E_{GSS}$ is in $E_{GS}$, and $E_{GSS} = E_{GS}$.

If $T$ is a strict subset of $V$, there is no guarantee that the size of $E_{GSS}$ decreases compared to $E_{GSS}$ when $T = V$. In certain pathological cases, the size of $E_{GSS}$ can in fact increase with a smaller set $T$. One such scenario occurs when the graph contains a complete component of $m$ vertices paired with a vertex $v$ connected to the other $m$ vertices. With careful edges cost assignments and selection of $k$, $GSS$ can be made to select the $m$ edges incident to $v$, when $T = V$, but select the $O(m^2)$ edges in the complete component when $T$ equals that complete component. As such, we leverage the pre-processing modification to $GSS$ discussed before this theorem, where the input graph is pruned to only include edges from $GSS$ when $T = V$. Using the pruned graph with the input $T$, $GSS$ cannot select edges that are not in the pruned graph, so $E_{GSS}$ cannot grow larger than the pruned graph. Thus, if $E_{GSS-V}$ denotes the edges scheduled by $GSS$ when $T = V$ and $E_{GSS-S}$ denotes the edges scheduled by $GSS$ for some strict subset $S \subset V$, $E_{GSS-S} \subseteq E_{GSS-V}$.

Therefore, for $T \subseteq V$, $|E_{GSS}| \leq |E_{GS}|$ and $|E_{GSS}| \in O(n^{1+1/k})$.     □

# References

1. Althöfer, I., Das, G., Dobkin, D., Joseph, D., Soares, J.: On sparse spanners of weighted graphs. Discret. Comput. Geom. **9**(1), 81–100 (1993)
2. Bodwin, G.: Linear size distance preservers. In: Proceedings of the 2017 ACM-SIAM Symposium on Discrete Algorithms (SODA), pp. 600–615 (2017)
3. Bodwin, G., Williams, V.V.: Better distance preservers and additive spanners. In: Proceedings of the 2016 ACM-SIAM Symposium on Discrete Algorithms (SODA), pp. 855–872 (2016)
4. Bose, P., Smid, M.: On plane geometric spanners: a survey and open problems. Comput. Geom. **46**(7), 818–830 (2013)
5. Chew, P.: There is a planar graph almost as good as the complete graph. In: Proceedings of the Second Annual Symposium on Computational Geometry (SoCG), pp. 169–177 (1986)
6. Cygan, M., Grandoni, F., Kavitha, T.: On pairwise spanners. In: 30th International Symposium on Theoretical Aspects of Computer Science, p. 209 (2013)
7. Dinitz, M., Kortsarz, G., Raz, R.: Label cover instances with large girth and the hardness of approximating basic $k$-spanner. ACM Trans. Algorithms **12**(2), 25:1–25:16 (2015)
8. Hoover, B., Middleton, R.S., Yaw, S.: $CostMAP$: an open-source software package for developing cost surfaces using a multi-scale search kernel. Int. J. Geogr. Inf. Sci. (2019)
9. Kavitha, T., Varma, N.M.: Small stretch pairwise spanners. In: Fomin, F.V., Freivalds, R., Kwiatkowska, M., Peleg, D. (eds.) ICALP 2013. LNCS, vol. 7965, pp. 601–612. Springer, Heidelberg (2013). https://doi.org/10.1007/978-3-642-39206-1_51
10. Keil, J.M., Gutwin, C.A.: Classes of graphs which approximate the complete Euclidean graph. Discret. Comput. Geom. **7**(1), 13–28 (1992)
11. Kortsarz, G.: On the hardness of approximating spanners. Algorithmica **30**(3), 432–450 (2001)
12. Manurangsi, P.: Almost-polynomial ratio ETH-hardness of approximating densest $k$-subgraph. In: Proceedings of the 2017 ACM SIGACT Symposium on Theory of Computing (STOC), pp. 954–961 (2017)
13. Middleton, R.S., Bielicki, J.M.: A scalable infrastructure model for carbon capture and storage: SimCCS. Energy Policy **37**(3), 1052–1060 (2009)
14. Middleton, R.S., Ellett, K., Stauffer, P., Rupp, J.: Making carbon capture, utilization and storage a reality: integrating science and engineering into a business plan framework. In: AAPG—SEG International Conference and Exhibition (2015)
15. Middleton, M.S., et al.: The cross-scale science of $CO_2$ capture and storage: from pore scale to regional scale. Energy Environ. Sci. **5**, 7328–7345 (2012)
16. Middleton, R.S., Kuby, M.J., Bielicki, J.M.: Generating candidate networks for optimization: the $CO_2$ capture and storage optimization problem. Comput. Environ. Urban Syst. **36**(1), 18–29 (2012)
17. Middleton, R.S., Yaw, S.: The cost of getting CCS wrong: uncertainty, infrastructure design, and stranded $CO_2$. Int. J. Greenhouse Gas Control **70**, 1–11 (2018)
18. Middleton, R.S., Yaw, S., Hoover, B., Ellett, K.: $SimCCS$: an open-source tool for optimizing $CO_2$ capture, transport, and storage infrastructure. Environ. Modell. Softw. (2019)
19. Vazirani, V.: Approximation Algorithms, chap. 3, pp. 27–29. Springer, Heidelberg (2001). https://doi.org/10.1007/978-3-662-04565-7

# TNT: An Effective Method for Finding Correlations Between Two Continuous Variables

Dayuan Zheng[1], Zhaogong Zhang[2(✉)], and Yuting Zhang[2]

[1] School of Data Science and Technology,
Heilongjiang University, Harbin, China
[2] School of Computer Science and Technology,
Heilongjiang University, Harbin, China
zhaogong.zhang@qq.com

**Abstract.** Determining whether two continuous variables are relevant, either linearly or non-linearly correlated, is a fundamental problem in data science. To test whether two continuous variables have a linear correlation is simple and has a much complete solution, but to judge whether they are in nonlinear correlation is far more difficult. Here, we propose a novel method, Tight Nearest-neighbor prediction correlation Test (TNT), to determine whether two continuous variables are nonlinearly correlated. TNT first use the values of one variable to construct a tight neighborhood structure to predict the value of the other variable and then use the sum of squared errors to measure how well the prediction is. A permutation test based on the sum of squared errors is employed to determine whether two continuous variables are relevant. To evaluate the performance of TNT, we performed extensive simulations comparing with seven existing methods. The results on both simulation and real data demonstrate that TNT is an efficient method to test nonlinear correlations, particularly for some nonlinear correlation which existing methods cannot solve, such as "ring".

**Keywords:** Independent test · Non-linear correlation · Noise · Normal distribution

## 1 Introduction

Various practical problems can be formulated as whether sets of random variables are mutually independent or not. As a result, testing for independence has spurred much research over the years. Examples include independent component analysis [1], gene selection [2], feature selection [3] and so on. Researchers are often interested in whether the variables are related and, if so, how strong the association is. Measures of association are also frequent topics in methodological research.

This work was supported by the Natural Science Foundation of Heilongjiang Province (No. F2017024, No. F2017025).

Two variables, $X$ and $Y$, are claimed to be associated when the value taken by one variable alter the distribution of the other variable. $X$ and $Y$ are claimed to be independent if changes in one variable do not affect the other variable.

Since the early last century, many measures of an association have been proposed to identify the association between random variables. Commonly used statistical methods to evaluate the independence of two continuous random variables include Pearson's correlation coefficient, Spearman's correlation coefficient, Kendall's tau coefficient, distance correlation, Hoeffding's test of independence, maximal information coefficient (MIC), HSIC (Hilbert-Schmidt Independence Criterion), and BNNPT (Bagging Nearest-Neighbor Prediction Test). Pearson's correlation coefficient is the most popularly used measure of statistical dependence [4, 5]. It generates a complete characterization of dependence in the Gaussian case, and also in some non-Gaussian situations. However, it has some shortcomings, in particular for heavy-tailed distributions and in nonlinear cases, where it may produce misleading, and even disastrous results. The Spearman's correlation coefficient is a nonparametric rank statistic measuring the strength of the association between two variables [6]. It can be used for variables measured at the ordinal level when Pearson's correlation coefficient is undesirable. It assesses the monotonic association between two variables, without the assumptions about the distribution of the variables. Kendall's tau is another nonparametric statistical method which can be used as an alternative to Spearman's correlation coefficient for data in the form of ranks [7]. It is a simple function that quantifies the discrepancy between the number of concordant and discordant pairs by finding the minimum number of neighbor swaps needed to produce one ordering from another [8].

Distance correlation is a measure which can quantify both linear and nonlinear association between two random variables or random vectors of arbitrary, not necessarily equal, dimension [9]. It can be used to perform a statistical test of dependence with a permutation approach. If distance correlation were implemented straightforwardly from its definition, its computational complexity could be as high as $O(n^2)$ for a sample size $n$. Hoeffding's independence test is a measure of group deviation which was derived as an unbiased estimator for testing the independence of the two variables [10]. This test can only be applied to continuous variables. The maximal information coefficient is a measure of the strength of the linear or non-linear association between two variables using binning as a means to apply mutual information on continuous random variables [11]. MIC selects the number of bins and picks a maximum mutual information over many possible grids. HSIC introduced by Gretton et al. is a kernel-based approach using the distance between the kernel embeddings of probability measures in the reproducing kernel Hilbert space for detecting independence [12]. Statistical tests based on the estimation of HSIC are computationally expensive with $O(m^2)$ time and space complexity, where m is the number of observations [13]. BNNPT, proposed by Wang et al. recently, measures the nonlinear dependence of two continuous variables [14]. It first uses one variable to construct bagging neighborhoods, and then gets an out-of-bag estimator of the other variable, based on the bagging neighborhood structure. The squared error is calculated to measure how well one variable is predicted by another. Critical values are calculated by permutations in the resulting statistic. Although BNNPT is statistical powerful to test most nonlinear correlations, it has hard time to detect some correlations, such as "loop".

In this study, we propose a novel non-linear dependence test, Tight Nearest-neighbor prediction correlation Test (TNT), for two continuous variables. TNT is based on a permutation test of the sum of squared errors. In pattern recognition, k-Nearest Neighbors algorithm (k-NN) is a nonparametric approach for classification and regression [15, 16]. When k = 1, the k-Nearest Neighbors algorithm is called the nearest neighbor algorithm. In our TNT framework, a tight neighborhood structure is constructed based on one variable. An estimator of the second variable is generated based on the tight neighborhood structure, which is similar to the nearest neighbor algorithm. The sum of squared errors is calculated to measure how good prediction is based on the tight neighborhood structure. The permutation test of the sum of squared errors is applied to detect the significance of the observed sum of squared errors. To evaluate the performance of TNT, we compared the statistical power [17] and the false positive ratio [18] of TNT with other seven commonly used methods to estimate correlation coefficients and perform independence tests through the simulation studies and real data analysis. The results demonstrate that TNT is an efficient method to test non-linear correlations.

## 2 Method

### 2.1 Definitions and Notation

In this section we give some definitions and introduce some basic notations. Let $X$ and $Y$ denote two continuous random variables and their paired data $\{(x_1, y_1), \ldots, (x_n, y_n)\}$ consisting of n pairs. Without loss of generality, we arrange $\{(x_1, y_1), \ldots, (x_n, y_n)\}$ in ascending order based on $X$. A tight neighborhood structure $M$ for $X$ is an $n \times 2$ index matrix, where each cell is holding an index of $X$. $M[i, 1]$ is the index of the largest $x$ among all $x_s$ which is smaller than $x_i$, and $M[i, 2]$ is the index of the smallest $x$ among all $x_s$ which is larger than $x_i$. For $i = 1$, $M[i, 1] = 1$, and for $i = n$, $M[i, 2] = n$.

Based on the tight neighborhood structure, a tight neighbor estimator $\widehat{Y}$ for $Y$ is defined as:

$$\widehat{y}_i = \frac{1}{2} \sum\nolimits_{j=1}^{2} y_{M[i,j]},$$

where the element $y_{M[i,j]}$ is the counterpart value of $Y$ for $x_i$.

We define the sum of squared errors of the tight nearest neighbor estimator as

$$SSE = \sum\nolimits_{j=1}^{n} \left( \widehat{y}_j - y_j \right)^2$$

which indicates how close $\widehat{y}_j$ is predicted by $x_i$. $SSE$ is used as the test statistic in TNT.

## 2.2    Construction of the Independence Test

The details of TNT are shown in the next page's Algorithm. TNT is based on the permutation test of the sum of square errors of the tight nearest neighbor estimator. A tight neighborhood structure $M$ for $X$ is constructed according to its definition (Line 1–5). The tight neighbor estimator for each $y_i$ is calculated using Eq. 1 (Line 6–8). The sum of squared differences between each tight neighbor estimator and observation is denoted as $SSE_0$ (Line 9). To evaluate the significance level of $SSE_0$, TNT randomly shuffle $y_i$ for $K$ times. For $k$th permutation, $SSE_k$ is computed as in step 1 through step 8. TNT count how many times that $SSE_k$ is less than $SSE_0$. The p-value is the ratio of the count to the number of permutations (Line 19).

---

**Algorithm TNT**

**Input:** $X \in R^n$, $Y \in R^n$, Two continuous variables $X$ and $Y$ and $n$ paired data sample $(x_i, y_i)$. Number of permutations $K$.

**Output:** $p$ – value for $X$ and $Y$ being no correlated.

1. **Init:** Initialize an empty $n \times 2$ index matrix $M$;
2. **For** i ← 1 **to** n **do**
3.      $M[i, 1] = \text{argmax}_j \{x_j \mid x_j < x_i, j..i\}$;
4.      $M[i, 2] = \text{argmin}_j \{x_j \mid x_j > x_i, j..i\}$;
5. **End for**
6. **For** i ← 1 **to** n **do**
7.      $\hat{y}_i \leftarrow \dfrac{1}{2}\sum_{j=1}^{2} y_{M[i,j]}$
8. **End for**
9. $SSE_0 \leftarrow \sum_{j=1}^{n} (\hat{y}_j - y_j)^2$
10. $c \leftarrow 0$
11. **For** k ← 1 **to** K **do**
12.      shuffle($Y$)
13.      repeat Step 1 through Step 8;
14.      $SSE_k \leftarrow \sum_{j=1}^{n} (\hat{y}_j - y_j)^2$
15.      **If** $SSE_k < SSE_0$ **then**
16.          $c \leftarrow c + 1$
17.      **End If**
18. **End for**
19. **Return** $p$-value $\dfrac{c}{K}$;

---

# 3    Experiments and Results

## 3.1    Experimental Design

We used two sets of functions to generate the simulation data. Set 1 contains eleven sample functions as shown in Table 1. In Set 1, $x$ is an independent variable uniformly distributed in $(0, 1.175)$, and $y$ is the dependent variable. Set 1 is used to test the correlation between $x$ and $y$. The curves and scatter plots of eight functions in Set 1 are shown in Fig. 1. In addition, we created another set of functions, Set 2, to test if our TNT and other methods can capture complex non-linear relations, like 'ring'. Set 2 contains four paired functions as shown in Table 2, where $x$ is the independent variable uniformly distributed in (0,1.175), and $y1$ and $y2$ are the dependent variable. Using the data generated according to Set 2, we aimed at testing the correlation between $y1$ and $y2$. The curves and scatter plots of four functions in Set 2 are shown in Fig. 2. The Gaussian noise $N$ (0,1) with mean = 0 and variance = 1 were added to $y$ in all the simulations. Note that $y = 0 + N(0,1)$ was used to evaluate the Type I error rate. 1000 datasets with 50 samples were generated for each function. We set significance level to 0.05. The measurement of the discrimination power and type I error are defined as the fraction of 1000 datasets on which the $p$-value no greater than the significance level. The number of permutations were set to 10,000 to get $p$-values for all the methods unless state otherwise.

There are many criteria for the correlation between two variables. The most commonly used are Pearson correlation coefficients, cosine similarity, etc. All of the above criteria measure the value of continuous values, and mutual information has a better performance for discrete attributes. The proposed algorithm is a feature selection for the classification task and needs to measure the correlation between the feature and the label. Therefore, mutual information is used as the correlation criterion in the experiment.

Seven other methods were utilized in our comparison: Pearson's correlation coefficient, Spearman's correlation coefficient, Kendall's tau coefficient, distance correlation, Hoeffding's test of independence, maximal information coefficient (MIC), HSIC (Hilbert-Schmidt Independence Criterion), and BNNPT (Bagging Nearest-Neighbor Prediction Test). The results of TNT and BNNPT were collected in the Java environment. For the other six methods we used their R packages, i.e. 'energy' [19], 'Hmisc' [20], and 'minerva' [21]. The default parameters of the MIC ($\alpha = 0.6$, c = 15) was used in this work. All experiments were conducted on a desktop computer with Intel i7 4770 CPU and 6 GB memory.

**Table 1.** Type I error and simulation power in ten sample functions

| Methods / Functions | TNT | BNNPT | Pearson | Spearman | Kendall | Hoeffding | Distance | MIC |
|---|---|---|---|---|---|---|---|---|
| Type I Error $y = 0 + N$ (0,1) | 0.039 | 0.051 | 0.056 | 0.053 | 0.054 | 0.068 | 0.059 | 0.046 |
| $y = x + N$ (0,1) | 0.699 | 0.841 | **0.959** | 0.953 | 0.951 | 0.941 | 0.945 | 0.596 |
| $y = 2x^2 + N$ (0,1) | 0.705 | 0.972 | 0.995 | **0.996** | 0.994 | 0.941 | 0.947 | 0.602 |
| $y = 2\sin(\pi x + \pi) + 2 + N$ (0,1) | 0.733 | 0.918 | 0.774 | 0.722 | 0.501 | 0.960 | **0.964** | 0.802 |
| $y = 2\cos(10x + \pi) + 4 + N$ (0,1) | **0.999** | 0.976 | 0.313 | 0.325 | 0.127 | 0.132 | 0.098 | 0.457 |
| $y = 2\sin(\pi x + 3) + N$ (0,1) | **0.995** | 0.930 | 0.203 | 0.216 | 0.077 | 0.128 | 0.103 | 0.401 |
| $y = 2\cos(\pi x + 10.5) + 1 + N$ (0,1) | 0.844 | **0.944** | 0.226 | 0.238 | 0.080 | 0.134 | 0.096 | 0.434 |
| $y = 1.5\sin(15.5x + \pi/2) + 1.5 + N$ (0,1) | **0.888** | 0.872 | 0.188 | 0.205 | 0.051 | 0.107 | 0.094 | 0.398 |
| $y = 1.5\cos(16x - \pi) + 4.5 + N$ (0,1) | **0.972** | 0.788 | 0.179 | 0.175 | 0.055 | 0.097 | 0.089 | 0.332 |
| $y = 2\sin(17x + 0.94) + 2 + N$ (0,1) | **0.997** | 0.969 | 0.177 | 0.196 | 0.049 | 0.089 | 0.076 | 0.107 |
| $y = 2\cos(21x + \pi) + 5.8 + N$ (0,1) | **0.998** | 0.712 | 0.196 | 0.208 | 0.056 | 0.103 | 0.091 | 0.331 |

The bolded means the largest power of all methods compared. 10,000 permutations for all methods.

## 3.2    Experiments on Eleven Sample Functions

Experiments based on Set 1 are shown in Table 1. All methods can well control the type I error, although the type I error of the Hoeffding's independence test is the highest out of eight methods, reaching 0.068. So, Hoeffding's independence method may lead to more false positives compared to the other methods. As shown in Table 1, TNT is a little bit worse than the best method when dealing with linear or quadratic functions, but TNT is superior to other approaches most of the time when dealing with sine and cosine functions.

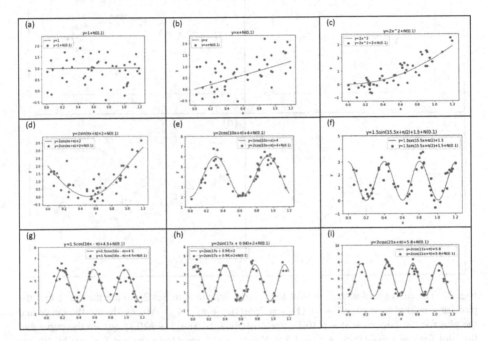

**Fig. 1.** The curves and scatter plots of eight functions in Set 1. (a) are the curve for $y = 1$ and the plot for $y = 1 + N (0,1)$. (b) are the curve for $y = x$ and the plot for $y = x + N (0,1)$. (c) are the curve for $y = 2x^2$ and the plot for $y = 2x^2 + N (0,1)$. (d) are the curve for $y = 2\sin(\pi x + \pi) + 2$ and the plot for $y = 2\sin(\pi x + \pi) + 2 + 2N (0,1)$. (e) are the curve for $y = 2\cos(10x + \pi) + 4$ and the plot for $y = 2\cos(10x + \pi) + 4 + N (0,1)$. (f) are the curve for $y = 1.5\sin(15.5x + \pi/2) + 1.5$ and the plot for $y = 1.5\sin(15.5x + \pi/2) + 1.5 + N (0,1)$. (g) are the curve for $y = 1.5\cos (16x - \pi) + 4.5$ and the plot for $y = 1.5\cos(16x - \pi) + 4.5 + N (0,1)$. (h) are the curve for $y = 2\sin(17x + 0.94) + 2$ and the plot for $y = 2\sin(17x + 0.94) + 2 + N (0,1)$. (i) are the curve for $y = 2\cos(21x + \pi) + 5.8$ and the plot for $y = 2\cos(21x + \pi) + 5.8 + N (0,1)$.

Based on Table 1 and Fig. 1, we have the following observations: (1) Pearson's correlation coefficient performs well when the correlation between two random variables is linear or quadratic. (2) All the methods did well in quadratic functions except TNT and MIC. (3) When the function has a non-linear correlation (such as sine or cosine functions), both TNT and BNNPT achieved high power, and TNT was more powerful than BNNPT for most of the cases. (4) For sine functions with small oscillations, all algorithms are powerful. When the sine functions have large oscillations, only TNT and BNNPT still preform very well. (5) TNT appeared to be more powerful than BNNPT, when the function is highly oscillatory. (6) TNT is always more powerful than MIC in all cases.

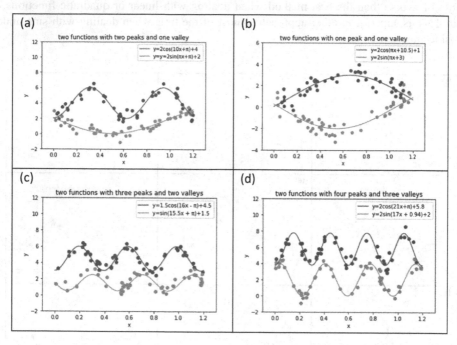

**Fig. 2.** The curves and scatter plots of four paired functions in Set 2. (a) are the curve and the plot for $y = 2\sin(\pi x + \pi) + 2 + N(0,1)$, $y = 2\cos(10x + \pi) + 4 + N(0,1)$. (b) are the curve and the plot for $y = 2\sin(\pi x + 3) + N(0,1)$, $y = 2\cos(\pi x + 10.5) + 1 + N(0,1)$. (c) are the curve and the plot for $y = 1.5\sin(15.5x + \pi) + 1.5 + N(0,1)$, $y = 1.5\cos(16x - \pi) + 4.5 + N(0,1)$. (d) are the curve and the plot for $y = 2\sin(17x + 0.94) + 2 + N(0,1)$, $y = 2\cos(21x - \pi) + 5.8 + N(0,1)$.

### 3.3   Experiments on Four Paired Functions

Table 2 shows the statistic power for eight methods on the four paired functions in Set 2. As shown in Fig. 2 and Table 2, for those complex functions, like 'ring', except TNT and BNNPT, other six methods (Pearson's correlation coefficient, Spearman's correlation coefficient, Kendall's tau coefficient, distance correlation, Hoeffding's independence test, and MIC) are ineffective. BNNPT became ineffective if the oscillations of the 'ring' shape is slightly increased. Only TNT can perform well for all four complex functions with large oscillations. If the trajectories of two random variables do not follow the same distribution similar to Fig. 2, TNT can still work, but other methods have various degrees of invalidity, or even complete invalid. This situation is very common in real life, for example, the stock volume and the unit price of the stock.

Thus, TNT is suitable for testing correlations of two continuous variables with large oscillations.

**Table 2.** Simulation power in four paired functions

| Methods / Functions | TNT | BNNPT | Pearson | Spearman | Kendall | Hoeffding | Distance | MIC |
|---|---|---|---|---|---|---|---|---|
| $y1 = 2\sin(\pi x + \pi) + 2 + N(0,1)$ $y2 = 2\cos(10x + \pi) + 4 + N(0,1)$ | **0.628** | 0.252 | 0.187 | 0.235 | 0.185 | 0.038 | 0.072 | 0.123 |
| $y1 = 2\sin(\pi x +3) + N(0,1)$ $y2 = 2\cos(\pi x + 10.5) + 1+ N(0,1)$ | **0.576** | 0.047 | 0.033 | 0.042 | 0.078 | 0.055 | 0.039 | 0.187 |
| $y1 = 1.5\sin(15.5x +\pi) + 1.5 + N(0,1)$ $y2 = 1.5\cos(16x -\pi) + 4.5 + N(0,1)$ | **0.636** | 0.05 | 0.048 | 0.04 | 0.039 | 0.043 | 0.066 | 0.177 |
| $y1 = 2\sin(17x + 0.94) + 2+N(0,1)$ $y2 = 2\cos(21x - \pi) + 5.8 + N(0,1)$ | **0.746** | 0.181 | 0.011 | 0.039 | 0.028 | 0.039 | 0.036 | 0.089 |

The bolded means the largest power of all methods compared. 10,000 permutations for all methods.

### 3.4 Shanghai Stock Market Data Analysis

To further evaluate our proposed method TNT, we performed a real data analysis based on the Shanghai Stock Market data. We investigated the relationship between two variables: Closing Index and Trading Volume (unit = 10 million shares). The data was collected between July 2, 2018, and December 28, 2018 (total 124 days). This data can be obtained at https://www.juhe.cn/docs/api/id/21?. The means of the Trading Volume and the Closing Index are 14.02 (range: 8.82–24.9) and 2687.1 (range: 24.83.1–2905.6), respectively. Figure 3 shows the scatter plot for the two variables. From Fig. 3, we can see that the relationship between Shanghai Closing Index and Trading Volume is not linear. We analyzed this real data and make a comparison between TNT and the other seven methods. The results are in Table 3. The p-values of BNNPT, Pearson, Spearman, Kendall, Hoeffding, Distance, and MIC are great than 0.05, which indicates that we cannot infer that the two variables are dependent for sure. The p-value of TNT (=0.001) indicates strong dependence between the two variables. This real data analysis demonstrates that our method can uncover the non-linear relationship between two variables, while existing methods failed.

### 3.5 Computation Time

From a practical point of view, a key issue of permutation test is the computational efficiency. In this section, we evaluate the computation time of TNT compared to the rest seven methods. Table 4 summarizes the computation time for 1,000 datasets each with 50 samples with the same setting for each method in Sect. 3.1. Note that computation time for Pearson's correlation coefficient, Spearman's correlation coefficient, or Kendall's tau coefficient is for calculating one correlation coefficient. As the results show, distance correlation took the longest to finish. Pearson's correlation coefficient, Spearman's correlation coefficient, and MIC can finish less than one second. TNT run much faster then BNNPT. It is a reason that the time complexity of TNT is lower than that of BNNPT. The number of neighbors that BNNPT used are $O(n^2)$, where $n$ is the number of sample pairs.

**Table 3.** The $p$-value comparison of benchmarked methods in Shanghai Stock Market data.

| Methods | TNT | BNNPT | Pearson | Spearman | Kendall | Hoeffding | Distance | MIC |
|---------|-----|-------|---------|----------|---------|-----------|----------|-----|
| $p$-value | 0.001 | 0.072 | 0.693 | 0.531 | 0.833 | 0.635 | 0.445 | 0.45 |

Significance level = 0.05.

**Table 4.** Comparison of computing time.

| Methods | TNT | BNNPT | Pearson | Spearman | Kendall | Hoeffding | Distance | MIC |
|---------|-----|-------|---------|----------|---------|-----------|----------|-----|
| Computation time (seconds) | 32 | 169 | 0.005 | 0.0052 | 0.017 | 4.1 | 9746 | 0.06 |

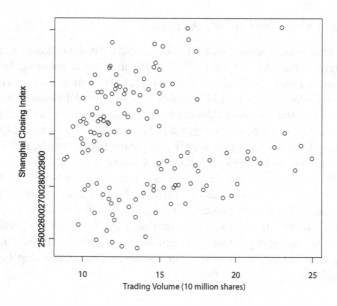

**Fig. 3.** The scatter plot of Shanghai Stock Market data.

## 4 Discussion

TNT is similar to BNNPT which uses bagging to find a set of nearest neighbors, while TNT only uses two closest neighbors. In theory, BNNPT increases robustness. When the data is without any noise or with little noise, to predict $Y$ we only need to consider the value $y_i$ of the two points, right smaller and larger to $x_i$. As shown in this study, when the data noise is weaker than the Gaussian noise disturbance (mean = 0, variance = 1), TNT performed better than BNNPT regarding the statistical power. When the data noise is in the range as Gaussian noise disturbance between (mean = 0, variance = 1) and (mean = 0, variance = 4), the statistical power of TNT and BNNPT is similar. When the data noise is stronger than the Gaussian noise disturbance (mean = 0, variance = 4), BNNPT is more powerful than TNT. The quality of the sample data could be too low, when preprocessing becomes very important.

## 5 Conclusion

As has been shown, this paper provides a novel statistical test, TNT, to detect the correlation between two continuous random variables, especially for non-linear correlations. TNT uses one variable to construct a tight neighborhood structure to predict the value for the second variable and then use the sum of squared errors to evaluate the prediction accuracy. TNT uses the sum of squared error as the test statistic and the permutation test to determine whether two continuous variables are relevant. The experimental and real data results demonstrate that TNT is an efficient method to test non-linear correlations, such as "ring", compared with popular existing methods. In the future, we will test TNT on more real data and expend TNT for multivariate test.

## References

1. Cai, L., Tian, X., Chen, S.: Monitoring nonlinear and non-gaussian processes using gaussian mixture model-based weighted kernel independent component analysis. IEEE Trans. Neural Netw. Learn. Syst. **28**(1), 122–135 (2017)
2. Yamanishi, Y., Vert, J.-P., Kanehisa, M.: Heterogeneous data comparison and gene selection with kernel canonical correlation analysis. In: Kernel Methods in Computational Biology, pp. 209–229 (2004)
3. Fukumizu, K., Bach, F.R., Jordan, M.I.: Dimensionality reduction for supervised learning with reproducing kernel hilbert spaces. J. Mach. Learn. Res. **5**(Jan), 73–99 (2004)
4. Stigler, S.M.: Francis Galton's account of the invention of correlation. Stat. Sci. **4**, 73–79 (1989)
5. Galton, F.: Regression towards mediocrity in hereditary stature. J. Anthropol. Inst. Great Br. Irel. **15**, 246–263 (1886)
6. Hauke, J., Kossowski, T.: Comparison of values of pearson's and spearman's correlation coefficients on the same sets of data. Quaest. Geographicae **30**(2), 87–93 (2011)
7. Kendall, M.G.: A new measure of rank correlation. Biometrika **30**(1/2), 81–93 (1938)
8. Chok, N.S.: Pearson's versus spearman's and kendall's correlation coefficients for continuous data. Ph.D. dissertation, University of Pittsburgh (2010)
9. Huo, X., Szekely, G.J.: Fast computing for distance covariance. Technometrics **58**(4), 435–447 (2016)
10. Hoeffding, W.: A non-parametric test of independence. Ann. Math. Stat. **19**, 546–557 (1948)
11. Reshef, D.N., et al.: Detecting novel associations in large data sets. Science **334**(6062), 1518–1524 (2011)
12. Gretton, A., Bousquet, O., Smola, A., Schölkopf, B.: Measuring statistical dependence with hilbert-schmidt norms. In: Jain, S., Simon, H.U., Tomita, E. (eds.) ALT 2005. LNCS (LNAI), vol. 3734, pp. 63–77. Springer, Heidelberg (2005). https://doi.org/10.1007/11564089_7

13. Zhang, Q., Filippi, S., Gretton, A., Sejdinovic, D.: Large-scale kernel methods for independence testing. Stat. Comput. **28**(1), 113–130 (2018)
14. Wang, Y., et al.: Bagging nearest-neighbor prediction independence test: an efficient method for nonlinear dependence of two continuous variables. Sci. Rep. **7**(1), 12736 (2017)
15. Altman, N.S.: An introduction to kernel and nearest-neighbor nonparametric regression. Am. Stat. **46**(3), 175–185 (1992)
16. Zhou, J., Sander, J., Cai, Z., Wang, L., Lin, G.: Finding the nearest neighbors in biological databases using less distance computations. IEEE/ACM Trans. Comput. Biol. Bioinf. **7**(4), 669–680 (2010)
17. Cohen, J.: Statistical Power Analysis for the Behavioral Sciences. Routledge, Abingdon (2013)
18. Burke, D.S., et al.: Measurement of the false positive rate in a screening program for human immunodeficiency virus infections. N. Engl. J. Med. **319**(15), 961–964 (1988)
19. Szekely, G.J., Rizzo, M.L.: Energy statistics: a class of statistics' based on distances. J. Stat. Plan. Infer. **143**(8), 1249–1272 (2013)
20. Harrell Jr, F.E., Harrell Jr, M.F.E.: Package 'hmisc' (2019)
21. Albanese, D., Filosi, M., Visintainer, R., Riccadonna, S., Jurman, G., Furlanello, C.: Minerva and minepy: a C engine for the MINE suite and its R, Python and MATLAB wrappers. Bioinformatics **29**(3), 407–408 (2012)

# On Conflict-Free Chromatic Guarding
# of Simple Polygons

Onur Çağırıcı[1] [iD], Subir Kumar Ghosh[2], Petr Hliněný[1(✉)] [iD],
and Bodhayan Roy[3]

[1] Faculty of Informatics of Masaryk University, Brno, Czech Republic
`onur@mail.muni.cz, hlineny@fi.muni.cz`
[2] Ramakrishna Mission Vivekananda University, Kolkata, India
`ghosh@tifr.res.in`
[3] Indian Institute of Technology Kharagpur, Kharagpur, India
`broy@maths.iitkgp.ac.in`

**Abstract.** We study the problem of colouring the vertices of a polygon, such that every viewer can see a unique colour. The goal is to minimize the number of colours used. This is also known as the conflict-free chromatic guarding problem with vertex guards, and is motivated, e.g., by the problem of radio frequency assignment to sensors placed at the polygon vertices. We study the scenario in which viewers can be all points of the polygon (such as a mobile robot which moves in the interior of the polygon). We efficiently solve the related problem of minimizing the number of guards and approximate (up to only an additive error) the number of colours required in the special case of polygons called funnels. As a corollary we sketch an upper bound of $O(\log^2 n)$ colours on $n$-vertex weak visibility polygons which generalizes to all simple polygons.

**Keywords:** Computational geometry · Polygon guarding · Visibility graph · Art gallery problem · Conflict-free coloring

## 1 Introduction

The *guarding* of a polygon is placing "guards" into the polygon, in a way that the guards collectively can see the whole polygon. It is usually assumed that a guard can see any point unless there is an obstacle or a wall between the guard and that point. One of the best known problems in computational geometry, the *art gallery problem* is essentially a guarding problem [7,28]. The problem is to find the minimum number of guards to guard an art gallery, which is

O. Çağırıcı—Supported by the Czech Science Foundation, project no. 17-00837S.
S. K. Ghosh—Supported by SERB, Government of India through a grant under MATRICS.
P. Hliněný—Supported by the Czech Science Foundation, project no. 17-00837S.
B. Roy—A significant part of the work was done while the author was affiliated to the Faculty of Informatics of Masaryk University.

© Springer Nature Switzerland AG 2019
Y. Li et al. (Eds.): COCOA 2019, LNCS 11949, pp. 601–612, 2019.
https://doi.org/10.1007/978-3-030-36412-0_49

modeled by an $n$-vertex polygon. This problem was shown to be NP-hard by Lee and Lin [27] and more recently $\exists\mathbb{R}$-complete by Abrahamsen et al. [2]. The Art Gallery Theorem, proved by Chvátal, shows that $\lfloor n/3 \rfloor$ guards are sufficient and sometimes necessary to guard a simple polygon [15].

The guard minimization problem has been studied under many constraints; such as the placement of guards being restricted to the polygonal perimeter or vertices [25], the viewers being restricted to vertices, the polygon being terrains [3,6,16], weakly visible from an edge [8], with holes or orthogonal [10,17,24], with respect to parameterization [11], approximability [23].

For most of these cases the problem remains hard, but interesting approximation algorithms have also been provided [12,19].

In addition to above mentioned versions of art gallery problem (or rather polygon guarding problem), some problems consider not the number of the guards, but the number of colours that are assigned to the guards. The colours, depending on the scope, determine the types of the guards. If any observer in the polygon sees at least one guard with a different type, then that polygon has a *conflict-free chromatic guarding* [4,5,22]. If every guard that any given observer sees is of different type, then that polygon has a *strong chromatic guarding* [18].

*Motivation.* In general, conflict-free colouring of a graph is assigning colours to vertices of that graph such that the neighborhood of each vertex contains at least one unique colour. This problem was first studied by Biggs with the name *perfect code*, which is essentially conflict-free colouring of a graph using only one colour [9,26]. Later on, this topic aroused interest on polygon visibility graphs when the field of robotics became widespread [14,20].

Consider a scenario where a mobile robot traverses a room from one point to another, communicating with the wireless sensors placed on the corners of the room. Even if the robot has full access to the map of the room, it cannot determine its location precisely because of accumulating rounding errors. And thus it needs clear markings in the room to guide itself to the end point in an energy efficient way. To guide a mobile robot with wireless sensors, two properties must be satisfied. First one is, no matter where the robot is in the polygon, it should hear from at least one sensor. That is, the placed sensors must together *guard* the whole room and leave no place uncovered. The second one is, if the robot hears from several sensors, there must be at least one sensor broadcasting with a frequency that is not reused by some other sensor in the range. That is, the sensors must have *conflict-free* frequencies. If these two properties are satisfied, then the robot can guide itself using the deployed wireless sensors as landmarks. This problem is also closely related to frequency assignment problem in wireless networks [1,5]. One can easily solve this problem by placing a sensor at each corner of the room, and assigning a different frequency to each sensor. However, this method becomes very expensive as the number of sensors grow [1, 29]. Therefore, the main goal in this problem is minimize the number of different frequencies assigned to sensors. Since the cost of a sensor is comparatively very low, we do not aim to minimize the number of sensors used.

The above scenario is geometrically modeled as follows. The room is a simple polygon with $n$ vertices. There are $m$ sensors placed in the polygon (usually on some of its vertices), and two different sensors are given two different colours if, and only if they broadcast in different frequencies.

*Basic Definitions.* We consider simple polygons (informally, "without holes"), usually non-convex. Two points $p_1$ and $p_2$ of a polygon $P$ are said to *see* each other, or be *visible* to each other, if the line segment $\overline{p_1 p_2}$ fully belongs to $P$. In this context, we say that a guard $g$ *guards* a point $x$ of $P$ if the line segment $\overline{gx}$ fully belongs to $P$. A polygon $P$ is a *weak visibility polygon* if $P$ has an edge $uv$ such that for every point $p$ of $P$ there is a point $p'$ on $uv$ seeing $p$.

In the paper, we pay close attention to a special type of polygons – *funnels*. A polygon $P$ is a funnel if (Fig. 1) precisely three of the vertices of $P$ are convex, and two of the convex vertices share one common edge – the *base* of the funnel. Funnels attract special interest in the study of visibility graphs, as a very fundamental class of polygons. Other polygons can be decomposed into funnels, giving a good overview of their structure with respect to most geometric problems. Funnels have a simpler structure due to their two concave chains and hence allow for easier handling of visibility problems than other classes of polygons.

A solution of *conflict-free chromatic guarding* of a polygon $P$ consists of a set of *guards* in $P$, and an assignment of colours to the guards (one colour per guard) such that the following holds; every *viewer* $v$ in $P$ (where $v$ can be any point of $P$ in our case) can see a guard of colour $c$ such that no other guard seen by $v$ has the same colour $c$. In the *point-to-point* (P2P) variant the guards can be placed in any points of $P$, while in the *vertex-to-point* (V2P) variant the guards can be placed only at the vertices of $P$. (There also exists a V2V variant in which also viewers are restricted to the vertices.) In all variants the goal is to minimize the number of colours (e.g., frequencies) used.

When writing $\log n$, we mean the binary logarithm $\log_2 n$.

*Related Research.* The aforementioned P2P conflict-free chromatic guarding (art gallery) problem has been studied in several papers. Bärtschi and Suri gave an upper bound of $O(\log^2 n)$ colours on simple $n$-vertex polygons [5]. Later, Bärtschi et al. improved this upper bound to $O(\log n)$ on simple polygons [4], and Hoffmann et al. [22], while studying the orthogonal variant of the problem, have given the first nontrivial lower bound of $\Omega(\log \log n / \log \log \log n)$ colours holding also in the general case of simple polygons.

Our paper deals with the V2P variant in which guards should be placed on polygon vertices and viewers can be any points of the polygon. Note that there are some fundamental differences between point and vertex guards, e.g., funnel polygons (Fig. 1) can always be guarded by one point guard (of one colour) but they may require up to $\Omega(\log n)$ colours in the V2P conflict-free chromatic guarding, as shown in [4]. Hence, extending a general upper bound of $O(\log n)$ colours for point guards on simple polygons by Bärtschi et al. [4] to the more restrictive vertex guards is a challenge, which we can now only approach with

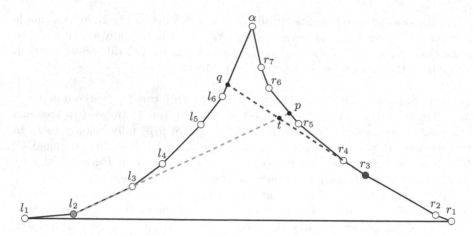

**Fig. 1.** A funnel $F$ with seven vertices in $\mathcal{L}$ labeled $l_1, \ldots, l_7$ from bottom to top, and eight vertices in $\mathcal{R}$ labeled $r_1, \ldots, r_8$, including the apex $\alpha = l_7 = r_8$. The picture also shows the upper tangent of the vertex $l_2$ of $\mathcal{L}$ (drawn in dashed red), the upper tangent of the vertex $r_3$ of $\mathcal{R}$ (drawn in dashed blue), and their intersection $t$. (Color figure online)

an $O(\log^2 n)$ bound (see below). Note that the same bound of $O(\log^2 n)$ colours was attained by [4] when allowing multiple guards at the same vertex.

*Our Results.* We give a polynomial-time algorithm to find the optimum number $m$ of vertex-guards to guard all the points of a funnel, and show that the number of colours in the corresponding conflict-free chromatic guarding problem is $\log m + \Theta(1)$ (Theorem 2). This leads to an approximation algorithm for V2P conflict-free chromatic guarding of a funnel, with only a constant (+4) additive error (Corollary 9). A remarkable feature of this result is that we prove a direct relation (Theorem 8) between the optimal numbers of guards and of colours needed in funnel polygons. Finally, we sketch that a weak visibility polygon on $n$ vertices can be V2P conflict-free chromatic guarded with only $O(\log^2 n)$ guards, and generalize this upper bound to all simple polygons, which is a result incomparable with previous [4].

Note that all our algorithms are simple and suitable for easy implementation. Due to space restrictions, the proofs of a part of the statements are left for the full paper published on arXiv [13], and those statements are marked with (*).

## 2    Minimizing Vertex-to-Point Guards for Funnels

Before turning to the conflict-free chromatic guarding problem, we first resolve the problem of minimizing the total number of vertex guards needed to guard all points of a funnel polygon. We start by describing a simple procedure (Algorithm 1) that provides us with a guard set which may not always be optimal

---

**Algorithm 1.** Simple vertex-to-point guarding of funnels (uncoloured).

---

**Input:** A funnel $F$ with concave chains $\mathcal{L} = (l_1, \ldots, l_k)$ and $\mathcal{R} = (r_1, \ldots, r_m)$.
**Output:** A vertex set guarding all the points of $F$.

1   Initialize an auxiliary digraph $G$ with two dummy vertices $x$ and $y$,
     and declare $ups(x) = \overline{l_1 r_1}$;
2   Initialize $\mathcal{S} \leftarrow \{x\}$;
3   **while** $\mathcal{S}$ *is not empty* **do**
4      Choose an arbitrary $t \in \mathcal{S}$, and remove $t$ from $\mathcal{S}$;
5      Let $s = ups(t)$ ;                    /* s is a segment inside F */
6      Let $q$ and $p$ be the ends of $s$ on $\mathcal{L}$ and $\mathcal{R}$, respectively;
7      Let $i$ and $j$ be the largest indices such that $l_i$ and $r_j$ are not above $q$
       and $p$, resp.;
8      **if** $l_{i+1}$ *can see whole* $s$ **then** $i' \leftarrow i + 1$;
9      **else** $i' \leftarrow i$;     /* the topmost vertex on the left seeing whole s */
10     **if** $r_{j+1}$ *can see whole* $s$ **then** $j' \leftarrow j + 1$;
11     **else** $j' \leftarrow j$;    /* the topmost vertex on the right seeing whole s */
12     Include the vertices $l_{i'}$ and $r_{j'}$ in $G$;
13     **foreach** $z \in \{l_{i'}, r_{j'}\}$ **do**
14        Add the directed edge $(t, z)$ to $G$ ;
15        **if** *segment* $ups(z)$ *includes the apex* $l_k = r_m$ **then**
16          Add the directed edge $(z, y)$ to $G$ ; /* y is the dummy vertex */
17        **else** $\mathcal{S} \leftarrow \mathcal{S} \cup \{z\}$;        /* more guards are needed above z */

18   Enumerate a shortest path from $x$ to $y$ in $G$;
19   Output the shortest path vertices without $x$ and $y$ as the required guard set;

---

(but very close to the optimum, see Corollary 3). This procedure will be helpful for the subsequent colouring results. Then we also refine the simple procedure to compute the optimal number of guards in Algorithm 2.

We use some special notation here. See Fig. 1. Let the given funnel be $F$, oriented in the plane as follows. On the bottom, there is the horizontal *base* of the funnel – the line segment $\overline{l_1 r_1}$ in the picture. The topmost vertex of $F$ is called the *apex*, and it is denoted by $\alpha$. There always exists a point $x$ on the base which can see the apex $\alpha$, and then $x$ sees the whole funnel at once. The vertices on the left side of apex form the *left concave chain*, and analogously, the vertices on the right side of the apex form the *right concave chain* of the funnel. These left and right concave chains are denoted by $\mathcal{L}$ and $\mathcal{R}$ respectively. We denote the vertices of $\mathcal{L}$ as $l_1, l_2, \ldots, l_k$ from bottom to top. We denote the vertices of $\mathcal{R}$ as $r_1, r_2, \ldots, r_m$ from bottom to top. Hence, the apex is $l_k = r_m = \alpha$.

Let $l_i$ be a vertex on $\mathcal{L}$ which is not the apex. We define the *upper tangent* of $l_i$, denoted by $upt(l_i)$, as the ray whose origin is $l_i$ and which passes through $l_{i+1}$. Upper tangents for vertices on $\mathcal{R}$ are defined analogously. Let $p$ be the point of intersection of $\mathcal{R}$ and the upper tangent of $l_i$. Then we define $ups(l_i)$ as the line segment $\overline{l_{i+1} p}$. For the vertices of $\mathcal{R}$, $ups$ is defined analogously: if $q$ is the point of intersection of $\mathcal{L}$ and the upper tangent of $r_j \in \mathcal{R}$, then let $ups(r_j) := \overline{r_{j+1} q}$.

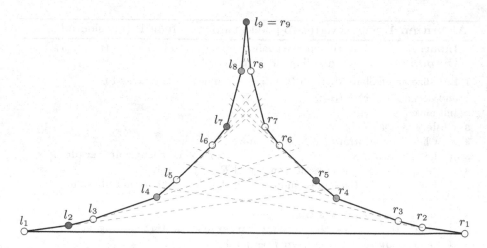

**Fig. 2.** A symmetric funnel with 17 vertices. The gray dashed lines show the upper tangents of the vertices. It is easy to see that Algorithm 1 selects 4 guards, up to symmetry, at $l_2, r_5, l_7, l_9$ (the red vertices). However, the whole funnel can be guarded by three guards at $l_4, r_4, l_8$ (the green vertices). (Color figure online)

The underlying idea of Algorithm 1 is as follows. Imagine we proceed bottom-up when building the guard set of a funnel $F$. Then the next guard is placed at the top-most vertex $z$ of $F$, nondeterministically choosing between $z$ on the left and the right chain of $F$, such that no "unguarded gap" remains below $z$. Note that the unguarded region of $F$ after placing a guard at $z$ is bounded from below by $ups(z)$. The nondeterministic choice of the next guard $z$ is encoded within a digraph, in which we then find the desired guard set as a shortest path.

**Lemma 1 (*).** *Algorithm 1 runs in polynomial time, and it outputs a feasible guard set for all the points of a funnel $F$.*

Unfortunately, the guard set produced by Algorithm 1 may not be optimal under certain circumstances. See the example in Fig. 2; the algorithm picks the four red vertices, but the funnel can be guarded by the three green vertices.

For the sake of completeness, we now refine the simple approach of Algorithm 1 to always produce a minimum size guard set. Our refinement is going to consider also pairs of guards (one from the left and one from the right chain) in the procedure. We correspondingly extend the definition of $ups$ to pairs of vertices as follows. Let $l_i$ and $r_j$ be vertices of $F$ on $\mathcal{L}$ and $\mathcal{R}$, respectively, such that $ups(l_i) = \overline{l_{i+1}p}$ intersects $ups(r_j) = \overline{r_{j+1}q}$ in a point $t$ (see in Fig. 1). Then we set $ups(l_i, r_j)$ as the polygonal line ("V-shape") $\overline{pt} \cup \overline{qt}$.

Algorithm 2, informally saying, enriches the two nondeterministic choices of placing the next guard in Algorithm 1 with a third choice; placing a suitable top-most pair of guards $z = (z_1, z_2)$, $z_1 \in \mathcal{L}$ and $z_2 \in \mathcal{R}$, such that again no "unguarded gap" remains below $(z_1, z_2)$. Figure 2 features a funnel in which placing such a pair of guards ($z_1 = l_4$, $z_2 = r_4$) may be strictly better than using

---

**Algorithm 2. (*)** Optimum vertex-to-point guarding of funnels.

---

**Input:** A funnel $F$ with concave chains $\mathcal{L} = (l_1, \ldots, l_k)$ and $\mathcal{R} = (r_1, \ldots, r_m)$.

**Output:** A minimum vertex set (uncoloured) guarding all the points of $F$.

\* On line 13 of Algorithm 1, consider $z \in \{l_{i'}, r_{j'}, (l_{i''}, r_{j''})\}$, where $i''$ and $j''$ are the largest indices such that $l_{i''}$ lies strictly below $ups(p)$ and $r_{j''}$ strictly below $ups(q)$. (Then $l_{i''}$ and $r_{j''}$ together can see whole $s$.);

\* On line 14 of Algorithm 1, make the edge $(t, z)$ of $G$ weight 2 if $z = (l_{i''}, r_{j''})$.

---

any two consecutive steps of Algorithm 1. On the other hand, we can show that there is no better possibility than one of these three considered steps, giving us:

**Theorem 2 (*).** *Algorithm 2 runs in polynomial time, and it outputs a feasible guard set of minimum size guarding all the points of a funnel $F$.*

Lastly, we establish that the difference between Algorithms 1 and 2 cannot be larger than 1 guard. Let $G^1$ with the source $x^1$ be the auxiliary graph produced by Algorithm 1, and $G^2$ with the source $x^2$ be the one produced by Algorithm 2. We can prove the following detailed statement by induction on $i \geq 0$:

- Let $P^2 = (x^2 = x_0^2, x_1^2, \ldots, x_i^2)$ be any directed path in $G^2$ of weight $k$, let $Q^2$ denote the set of guards listed in the vertices of $P^2$, and $L^2 = \mathcal{L} \cap Q^2$ and $R^2 = \mathcal{R} \cap Q^2$. Then there exists a directed path $(x^1 = x_0^1, x_1^1, \ldots, x_k^1, x_{k+1}^1)$ in $G^1$ (of length $k + 1$), such that the guard of $x_k$ is at least as high as all the guards of $L^2$ (if $x_k \in \mathcal{L}$) or of $R^2$ (if $x_k \in \mathcal{R}$), and the guard of $x_{k+1}$ is strictly higher than all the guards of $Q^2$.

**Corollary 3 (*).** *The guard set produced by Algorithm 1 is always by at most one guard larger than the optimum solution produced by Algorithm 2.*

## 3    V2P Conflict-Free Chromatic Guarding of Funnels

In this section, we continue to study funnels. To obtain a conflict-free coloured solution, we will simply consider the guards chosen by Algorithm 1 in the ascending order of their vertical coordinates, and colour them in the *ruler sequence*, (e.g., [21]) in which the $i^{th}$ term is the exponent of the largest power of 2 that divides $2i$. (The first few terms are $1, 2, 1, 3, 1, 2, 1, 4, 1, 2, 1, 3, 1, 2, 1, 5, 1 \ldots$.) So, if Algorithm 1 gives $m$ guards, then our approach will use about $\log m$ colours.

Our aim is to show that this is always very close to the optimum, by giving a lower bound on the number of necessary colours of order $\log m - O(1)$. To achieve this, we study the following two sets of guards for a given funnel $F$:

- The minimal *guard set $A$* computed by Algorithm 1 on $F$ (which is overall nearly optimal by Corollary 3); if this is not unique, then we fix any such $A$.
- A *guard set $D$* which achieves the minimum number of colours for conflict-free guarding; note that $D$ may be much larger than $A$ since it is the number of colours which matters.

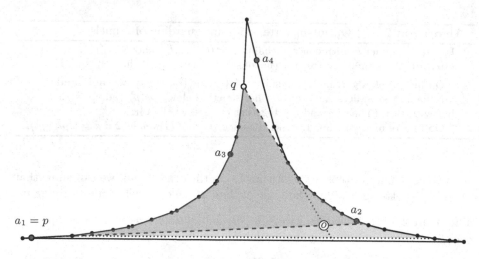

**Fig. 3.** An example of a 2-interval $Q$ of a funnel (green and bounded by $s_1 = ups(p)$ and $s_2 = los(q)$). The red vertices $a_1 = p, a_2, a_3, a_4$ are the guards computed by Algorithm 1, and $a_2, a_3$ belong to the interval $Q$. The shadow of $Q$ (filled light gray) is bounded from below by the bottom dotted line, and the inner point $o$ is the so-called *observer* of $Q$. (Color figure online)

On a high level, we are going to show that the colouring of $D$ must (somehow) copy the ruler sequence on $A$. For that we will recursively bisect our funnel into smaller "layers", gaining one unique colour with each bisection.

Analogously to the notion of an upper tangent from Sect. 2, we define the *lower tangent* of a vertex $l_i \in \mathcal{L}$, denote by $lot(l_i)$, as the ray whose origin is $l_i$ and which passes through $r_j \in \mathcal{R}$ such that $r_j$ is the lowest vertex on $\mathcal{R}$ seeing $l_i$. Note that $lot(l_i)$ may intersect $\mathcal{R}$ in $r_j$ alone or in a segment from $r_j$ up. Let $los(l_i) := \overline{l_i r_j}$. The definition of $lot()$ and $los()$ for vertices of $\mathcal{R}$ is symmetric.

We now give a definition of "layers" of a funnel which is crucial for our proof.

**Definition 4** (*t-interval*).Let $F$ be a funnel with the chains $\mathcal{L} = (l_1, l_2, \ldots, l_k)$ and $\mathcal{R} = (r_1, r_2, \ldots, r_m)$, and $A$ be the fixed guard set $A$ computed by Algorithm 1 on $F$. Let $s_1$ be the base of $F$, or $s_1 = ups(p)$ for some vertex $p$ of $F$ (where $p$ is not the apex or its neighbour). Let $s_2$ be the apex of $F$, or $s_2 = los(q)$ for some vertex $q$ of $F$ (where $q$ is not in the base of $F$). Assume that $s_2$ is above $s_1$ within $F$. Then the region $Q$ of $F$ bounded from below by $s_1$ and from above by $s_2$, excluding $q$ itself, is called an *interval of $F$*. Moreover, $Q$ is called a *t-interval of $F$* if $Q$ contains at least $t$ of the guards of $A$. See Fig. 3.

Having an interval $Q$ of the funnel $F$, bounded from below by $s_1$, we define the *shadow of $Q$* as follows. If $s_1 = ups(l_i)$ ($s_1 = ups(r_j)$), then the shadow consists of the region of $F$ between $s_1$ and $los(l_{i+1})$ (between $s_1$ and $los(r_{j+1})$, respectively). If $s_1$ is the base, then the shadow is empty.

**Lemma 5** (\*). *If $Q$ is a 13-interval of the funnel $F$, then there exists a point in $Q$ which is not visible from any vertex of $F$ outside of $Q$.*

Our second crucial ingredient is the possibility to "almost privately" see the vertices of an interval $Q$ from one point as follows. If $s_2 = los(q)$, then the intersection point of $lot(q)$ with $s_1$ is called the *observer of* $Q$. (Actually, to be precise, we should slightly perturb this position of the observer $o$ so that the visibility between $o$ and $q$ is blocked.) If $s_2$ is the apex, then consider the spine of $F$ instead of $lot(q)$. See again Fig. 3. The following is easy to argue.

**Lemma 6 (*).** *The observer $o$ of an interval $Q$ in a funnel $F$ can see all the vertices of $Q$, but $o$ cannot see any vertex of $F$ which is not in $Q$ and not in the shadow of $Q$.*

The last ingredient before the main proof is the notion of sections of an interval $Q$ of $F$. Let $s_1$ and $s_2$ form the lower and upper boundary of $Q$. Consider a vertex $l_i \in \mathcal{L}$ of $Q$. Then the *lower section of $Q$ at $l_i$* is the interval of $F$ bounded from below by $s_1$ and from above by $los(l_i)$. The *upper section of $Q$ at $l_i$* is the interval of $F$ bounded from below by $ups(l_i)$ and from above by $s_2$. Sections of $r_j \in \mathcal{R}$ are defined analogously. Again, the following is straightforward.

**Lemma 7 (*).** *Let $Q$ be a $t$-interval of the funnel $F$, and let $Q_1$ and $Q_2$ be its lower and upper sections at some vertex $p$. Then $Q_i$, $i = 1, 2$, is a $t_i$-interval such that $t_1 + t_2 \geq t - 3$.*

**Theorem 8.** *Any conflict-free chromatic guarding of a given funnel requires at least $\lfloor \log_2(m+3) \rfloor - 3$ colours, where $m$ is the minimum number of guards needed to guard the whole funnel.*

*Proof.* We will prove the following claim by induction on $c \geq 0$: If $Q$ is a $t$-interval in the funnel $F$ and $t \geq 16 \cdot 2^c - 3$, then any conflict-free colouring of $F$ must use at least $c + 1$ colours on the vertices of $Q$ or of the shadow of $Q$.

In the base $c = 0$ of the induction, we have $t \geq 16 - 3 = 13$. By Lemma 5, some point of $Q$ is not seen from outside, and so there has to be a coloured guard in some vertex of $Q$, thus giving $c + 1 = 1$ colour.

Consider now $c > 0$. The observer $o$ of $Q$ (which sees all the vertices of $Q$) must see a guard $g$ of a unique colour where $g$ is, by Lemma 6, a vertex of $Q$ or of the shadow of $Q$. In the first case, we consider $Q_1$ and $Q_2$, the lower and upper sections of $Q$ at $g$. By Lemma 7, for some $i \in \{1, 2\}$, $Q_i$ is a $t_i$-interval of $F$ such that $t_i \geq (t-3)/2 \geq (16 \cdot 2^c - 6)/2 = 16 \cdot 2^{c-1} - 3$. In the second case ($g$ is in the shadow of $Q$), we choose $g'$ as the lowermost vertex of $Q$ on the same chain as $g$, and take only the upper section $Q_1$ of $Q$ at $g'$. We continue as in the first case with $i = 1$.

By induction assumption for $c - 1$, $Q_i$ together with its shadow carry a set $C$ of at least $c$ colours. The shadow of $Q_2$ is included in $Q$, and the shadow of $Q_1$ coincides with the shadow of $Q$, moreover, the observer of $Q_1$ sees only a subset of the shadow of $Q$ seen by the observer $o$ of $Q$. Since $g$ is not a point of $Q_i$ or its shadow, but our observer $o$ sees the colour $c_g$ of $g$ and all the colours of $C$, we have $c_g \notin C$ and hence $C \cup \{c_g\}$ has at least $c + 1$ colours, as desired.

Finally, we apply the above claim to $Q = F$. We have $t \geq m$, and for $t \geq m \geq 16 \cdot 2^c - 3$ we derive that we need at least $c + 1 \geq \lfloor \log(m + 3) \rfloor - 3$ colours for guarding whole $F$. $\qquad\square$

---

**Algorithm 3.** Approximate conflict-free chromatic guarding of a funnel.

---
**Input:** A funnel $F$ with concave chains $\mathcal{L} = (l_1, \ldots, l_k)$ and $\mathcal{R} = (r_1, \ldots, r_m)$.
**Output:** A conflict-free chromatic guard set of $F$ using $\leq OPT + 4$ colours.

1  Run Algorithm 1 to produce a guard seq. $A = (a_1, a_2, \ldots, a_t)$ (bottom-up);
2  Assign colours to members of $A$ according to the ruler sequence; the vertex $a_i$
   gets colour $c_i$ where $c_i$ is the largest integer such that $2^{c_i}$ divides $2i$;
3  Output coloured guards $A$ as the (approximate) solution;

---

**Corollary 9.** *Algorithm 3, for a given funnel $F$, outputs in polynomial time a conflict-free chromatic guard set $A$, such that the number of colours used by $A$ is by at most four larger than the optimum.*

*Proof.* Note the following simple property of the ruler sequence: if $c_i = c_j$ for some $i \neq j$, then $c_{(i+j)/2} > c_i$. Hence, for any $i, j$, the largest value occurring among colours $c_i, c_{i+1}, \ldots, c_{i+j-1}$ is unique. Since every point of $F$ sees a consecutive subsequence of $A$, this is a feasible conflict-free colouring of the funnel $F$.

Let $m$ be the minimum number of guards in $F$. By Corollary 3, it is $m + 1 \geq t = |A| \geq m$. To prove the approximation guarantee, observe that for $t \leq 2^c - 1$, our sequence $A$ uses $\leq c$ colours. Conversely, if $t \geq 2^{c-1}$, i.e. $m \geq 2^{c-1} - 1$, then the required number of colours for guarding $F$ is at least $c - 1 - 3 = c - 4$, and hence our algorithm uses at most 4 more colours than the optimum.     □

## 4  Concluding Remarks

We have designed an algorithm for producing a V2P guarding of funnels that is optimal in the number of guards. We have also designed an algorithm for a V2P conflict-free chromatic guarding for funnels, which gives only an additive error $(+4)$ with respect to the minimum number of colours required. We believe that the latter can be strengthened to an exact solution by sharpening the arguments involved (though, it would likely not be easy).

Regarding V2P conflict-free chromatic guarding in a more general setting, we provide the following upper bound as a corollary of the previous result.

**Theorem 10 (*).** *There is an algorithm computing in polynomial time a conflict-free chromatic guarding of a weak visibility polygon using $O(\log^2 n)$ colours.*

A rough sketch of the proof is as follows. Each weak visibility polygon can be straightforwardly partitioned into a sequence of maximal (overlapping) funnels, and we can independently guard each of the funnels with $O(\log n)$ colours by applying Algorithm 3. These colourings, unsurprisingly, may conflict with each other, and so we additionally couple the colours in funnels with $O(\log n)$ colours of the ruler sequence assigned to each of the funnels.

Secondly, we can use a polygon decomposition technique introduced by Suri [30] to generalize the upper bound from weak visibility polygons (Theorem 10) to all simple polygons. The technique has already been used by Bärtschi et al. in [4] for the P2P version of our problem. Though, there is still a room for improvement down to $O(\log n)$, which is the worst-case scenario already for funnels and which would match the previous P2P upper bound for simple polygons of [4].

To summarize, we propose the following open problems for future research:

– Improve Corollary 9 to an exact algorithm for guarding a funnel.
– Improve the upper bound in Theorem 10 to $O(\log n)$.

# References

1. Aardal, K.I., van Hoesel, S.P.M., Koster, A.M.C.A., Mannino, C., Sassano, A.: Models and solution techniques for frequency assignment problems. Ann. Oper. Res. **153**, 79–129 (2007)
2. Abrahamsen, M., Adamaszek, A., Miltzow, T.: The art gallery problem is ∃ ℝ-complete. In: Proceedings of the 50th Annual ACM SIGACT Symposium on Theory of Computing, pp. 65–73 (2018)
3. Ashok, P., Fomin, F.V., Kolay, S., Saurabh, S., Zehavi, M.: Exact algorithms for terrain guarding. ACM Trans. Algorithms **14**, 1–20 (2018)
4. Bärtschi, A., Ghosh, S.K., Mihalák, M., Tschager, T., Widmayer, P.: Improved bounds for the conflict-free chromatic art gallery problem. In: Proceedings of the 30th Annual Symposium on Computational Geometry, pp. 144–153 (2014)
5. Bärtschi, A., Suri, S.: Conflict-free chromatic art gallery coverage. Algorithmica **68**, 265–283 (2014)
6. Ben-Moshe, B., Katz, M.J., Mitchell, J.S.B.: A constant-factor approximation algorithm for optimal 1.5D terrain guarding. SIAM J. Comput. **36**, 1631–1647 (2007)
7. de Berg, M., Cheong, O., Kreveld, M., Overmars, M.: Computational Geometry, Algorithms and Applications, 3rd edn. Springer, Heidelberg (2008). https://doi.org/10.1007/978-3-540-77974-2
8. Bhattacharya, P., Ghosh, S.K., Roy, B.: Approximability of guarding weak visibility polygons. Discrete Appl. Math. **228**, 109–129 (2017)
9. Biggs, N.: Perfect codes in graphs. J. Comb. Theory Ser. B **15**, 289–296 (1973)
10. Bonnet, É., Giannopoulos, P.: Orthogonal terrain guarding is NP-complete. In: 34th International Symposium on Computational Geometry, pp. 1–15 (2018)
11. Bonnet, É., Miltzow, T.: Parameterized hardness of art gallery problems. In: 24th Annual European Symposium on Algorithms, pp. 1–17 (2016)
12. Bonnet, É., Miltzow, T.: An approximation algorithm for the art gallery problem. In: 33rd International Symposium on Computational Geometry, pp. 1–15 (2017)
13. Çağırıcı, O., Ghosh, S., Hliněný, P., Roy, B.: On conflict-free chromatic guarding of simple polygons. arXiv:1904.08624, June 2019
14. Chazelle, B.: Approximation and decomposition of shapes. In: Schwartz, J.T. (ed.) Advances in Robotics 1: Algorithmic and Geometric Aspects of Robotics, pp. 145–185 (1987)
15. Chvátal, V.: A combinatorial theorem in plane geometry. J. Comb. Theory **18**, 39–41 (1975)
16. Efrat, A., Har-Peled, S.: Guarding galleries and terrains. Inf. Process. Lett. **100**, 238–245 (2006)
17. Eidenbenz, S., Stamm, C., Widmayer, P.: Inapproximability results for guarding polygons and terrains. Algorithmica **31**, 79–113 (2001)

18. Erickson, L.H., LaValle, S.M.: An art gallery approach to ensuring that landmarks are distinguishable. In: Robotics: Science and Systems VII (2011)
19. Ghosh, S.K.: Approximation algorithms for art gallery problems in polygons. Discrete Appl. Math. **158**, 718–722 (2010)
20. Guibas, L.J., Motwani, R., Raghavan, P.: The robot localization problem. SIAM J. Comput. **26**, 1120–1138 (1997)
21. Guy, R.K.: Unsolved Problems in Number Theory. Springer, New York (1994). https://doi.org/10.1007/978-1-4899-3585-4
22. Hoffmann, F., Kriegel, K., Suri, S., Verbeek, K., Willert, M.: Tight bounds for conflict-free chromatic guarding of orthogonal art galleries. Comput. Geom. **73**, 24–34 (2018)
23. Katz, M.J.: A PTAS for vertex guarding weakly-visible polygons - an extended abstract. CoRR abs/1803.02160 (2018)
24. Katz, M.J., Morgenstern, G.: Guarding orthogonal art galleries with sliding cameras. Int. J. Comput. Geom. Appl. **21**, 241–250 (2011)
25. King, J., Kirkpatrick, D.G.: Improved approximation for guarding simple galleries from the perimeter. Discrete Comput. Geom. **46**, 252–269 (2011)
26. Kratochvíl, J., Křivánek, M.: On the computational complexity of codes in graphs. In: Chytil, M.P., Koubek, V., Janiga, L. (eds.) MFCS 1988. LNCS, vol. 324, pp. 396–404. Springer, Heidelberg (1988). https://doi.org/10.1007/BFb0017162
27. Lee, D.T., Lin, A.K.: Computational complexity of art gallery problems. IEEE Trans. Inf. Theory **32**, 276–282 (1986)
28. O'Rourke, J.: Art Gallery Theorems and Algorithms. Oxford University Press, Oxford (1987)
29. Smorodinsky, S.: Conflict-free coloring and its applications. In: Bárány, I., Böröczky, K.J., Tóth, G.F., Pach, J. (eds.) Geometry — Intuitive, Discrete, and Convex. BSMS, vol. 24, pp. 331–389. Springer, Heidelberg (2013). https://doi.org/10.1007/978-3-642-41498-5_12
30. Suri, S.: On some link distance problems in a simple polygon. IEEE J. Robot. Autom. **6**, 108–113 (1990)

# Author Index

Printed in the United States
By Bookmasters